T0262935

Enzymes in Food and Beverage Processing

Enzymes in Food and Beverage Processing

Edited by
Muthusamy Chandrasekaran

CRC Press
Taylor & Francis Group
Boca Raton London New York

CRC Press is an imprint of the
Taylor & Francis Group, an **informa** business

CRC Press
Taylor & Francis Group
6000 Broken Sound Parkway NW, Suite 300
Boca Raton, FL 33487-2742

First issued in paperback 2018

© 2016 by Taylor & Francis Group, LLC
CRC Press is an imprint of Taylor & Francis Group, an Informa business

No claim to original U.S. Government works

ISBN-13: 978-1-4822-2128-2 (hbk)
ISBN-13: 978-1-138-89417-4 (hbk)

This book contains information obtained from authentic and highly regarded sources. Reasonable efforts have been made to publish reliable data and information, but the author and publisher cannot assume responsibility for the validity of all materials or the consequences of their use. The authors and publishers have attempted to trace the copyright holders of all material reproduced in this publication and apologize to copyright holders if permission to publish in this form has not been obtained. If any copyright material has not been acknowledged please write and let us know so we may rectify in any future reprint.

Except as permitted under U.S. Copyright Law, no part of this book may be reprinted, reproduced, transmitted, or utilized in any form by any electronic, mechanical, or other means, now known or hereafter invented, including photocopying, microfilming, and recording, or in any information storage or retrieval system, without written permission from the publishers.

For permission to photocopy or use material electronically from this work, please access www.copyright.com (http://www.copyright.com/) or contact the Copyright Clearance Center, Inc. (CCC), 222 Rosewood Drive, Danvers, MA 01923, 978-750-8400. CCC is a not-for-profit organization that provides licenses and registration for a variety of users. For organizations that have been granted a photocopy license by the CCC, a separate system of payment has been arranged.

Trademark Notice: Product or corporate names may be trademarks or registered trademarks, and are used only for identification and explanation without intent to infringe.

Visit the Taylor & Francis Web site at
http://www.taylorandfrancis.com

and the CRC Press Web site at
http://www.crcpress.com

Contents

Section I   Enzymes: Basics and Development of Novel Biocatalysts

Section II   Applications of Enzymes in Food and Beverage Industries

Section III Advances in Food Grade Enzyme Biotechnology

Section IV Future Prospects

Preface

Food and beverage production has registered tremendous growth worldwide, concomitant with population growth and demand for processed food and beverages that are nutritive, delicious, and aesthetic in addition to having an extended shelf life. Further advancements in food science and technology have significantly contributed to a better understanding of food and its importance in sustenance of life, to the scope and need for augmentation of a diverse range of quality food and beverages for consumption, and to meeting ever-increasing food demand.

Toward meeting consumer demands, the food and beverage industries have adopted the latest developments in modern food technologies and biotechnologies. Among the biotechnologies, enzyme technology holds the key for implementation of green technologies that are ecofriendly and an ideal alternative technology to conventional chemical technologies, which use chemicals that may cause pollution on disposal into the environment. Thus, recently, there have been large-scale applications of enzymes in selected food processing industries. This is mainly due to the fact that enzymes have become an essential component of food processing as well as part of the foods that are consumed. Enzymes are principally used as additives in food or as processing aids in the production of food and beverages to improve texture, flavor, quality, appeal, and to extend shelf life. Moreover, recent developments in enzyme biotechnology, nanobiotechnology, metagenomics, and proteomics have contributed to the design and development of tailor-made enzymes with novel properties, and catalytic activities have played a significant role in the discovery and development of new enzymes and newer applications in food and beverage processing and valorization of food processing by-products and wastes. Enzymes have also found applications as analytical tools in the food industry.

A huge amount of scientific literature is available in the public domain on rapid developments in enzyme structure, enzyme kinetics, and characteristics; application of a range of food-grade enzymes derived from diverse groups of microorganisms, including extremophiles and plant resources; the scope for effecting desirable modification in the enzyme structures favoring activity under harsh conditions through enzyme engineering and protein modification; enzyme-mediated novel biotransformations of naturally available biologicals; and development of new ranges of functional foods, nutraceuticals, probiotics, and enzyme inhibitors that regulate enzyme activity. These advances and developments in food enzyme research have to be adopted and exploited in food and beverage processing production. This book is an attempt to document such developments for the benefit of and use by the food research fraternity and the food and beverage industries.

Unlike other books that deal with the application of specific enzymes in production of a specific product of interest, this book presents the application of different enzymes in various food and beverage industries in addition to presenting information on basic concepts and fundamental principles of enzymes, emerging enzyme technologies, and enzyme engineering. Further, this book also deals extensively with the latest advances in food science technology with respect to emerging food industries, such as functional foods, nutraceuticals, probiotics, the use of enzyme inhibitors, and enzyme biotransformations in addition to the use of enzymes in valorization of food and beverage processing by-products and waste. In addition, emerging trends and future prospects in the use of enzymes by food industries are also discussed.

The book is presented in 22 chapters grouped under four sections. Section I deals with enzyme basics and development of novel biocatalysts covered in Chapters 1 through 4. Chapter 1 presents comprehensive information on enzymes: their nomenclature, mechanism of action and kinetics, characteristics, and sources of food-grade enzymes. Chapter 2 deals extensively with enzyme technologies: current and emerging technologies for development of novel enzyme catalysts. This chapter includes a discussion on the scope for utilization of new technologies, such as directed evolution for designing new enzymes and use of nanomaterials for applications in food processing. Chapter 3 includes a detailed discussion on the use of enzymes as analytical tools for the assessment of food quality, food safety, and monitoring of food processing with reference to the use of modern developments in techniques such as ELISA and biosensors. Chapter 4 deals with the latest developments and scope for application of enzyme engineering and protein modifications in food processing.

Section II covers various applications of enzymes in food and beverage industries and is presented in Chapters 5 through 15. Chapter 5 includes an overview of market trends in various food and beverage industries in the world and enzyme-producing companies that cater to the needs of the food and beverage industries. Chapter 6 deals with the latest advancements in the application of various enzymes in the starch processing industries and the products derived. Chapter 7 presents an illustrative account of applications of various enzymes in bakery industries that manufacture bakery products. Chapter 8 includes a detailed discussion on the various applications of enzymes in confectionery industries that produce various confections based on sweeteners and different categories of chocolates. Chapter 9 includes an account of the role of enzymes in oil and lipid processing. Chapter 10 describes the importance of enzymes in the processing of fruit juices and vegetables in light of the latest developments in the field. Chapter 11 deals with applications of enzymes in the processing of alcoholic and nonalcoholic beverages. Chapter 12 presents an account of the prospects of enzyme applications in the production of flavors and food additives. Chapter 13 includes advancements in enzyme applications in the production of milk, cheese, and associated dairy products. Chapter 14 includes a discussion on the role of enzymes in meat tenderization and prospects of applications of enzymes for other applications in meat processing industries. Chapter 15 deals with the utilization of current developments in enzyme technologies for the processing of different species of seafood.

Section III exclusively deals with recent advances in food-grade enzyme biotechnology covered under Chapters 16 through 21. Chapter 16 presents advancements in enzyme applications for the synthesis of novel functional food ingredients. Chapter 17 essentially deals with current developments in enzyme-assisted extraction technology for processing of nutraceuticals from plant resources. Chapter 18 presents an in-depth account of the scope of enzyme-mediated novel biotransformations for deriving a new range of biomolecules and products of value in the food and beverage industries. Chapter 19 deals with the prospects of applications of enzymes in the development of probiotics, prebiotics, synbiotics, and cobiotics. Chapter 20 includes a discussion on the recent developments and prospects of using enzyme inhibitors in regulating enzyme processing of food and beverages. Chapter 21 includes a comprehensive account of the use of enzymes in the valorization of food and beverage by-products and waste toward deriving value-added products of commercial importance. Section IV comprises Chapter 22, which presents the various emerging trends and prospects of enzyme applications in food processing industries in the future.

The contents of this book will definitely cater to the needs of food science researchers, enzyme biotechnologists, and food and beverage industries as a reference book for guidance and a roadmap for taking up research toward harnessing the diverse range of enzymes for applications in food and beverage processing, as well as the design and development of novel enzymes utilizing emerging technologies. Moreover, this book holds the potential to serve as a useful text cum reference for advanced courses in food science technology, food biotechnology, food engineering, enzyme biotechnology, enzyme technology, and food waste management. Nevertheless, there may be overlapping content presented under different chapters, by different authors, which could not be avoided. Readers of this book will derive maximal benefits in terms of knowledge on the use of enzymes in food processing.

Acknowledgments

This book could not have been accomplished but for the grace and blessings of the Almighty, who infused me with the physical and mental strength to take up such a challenging task of developing a book on enzymes in the processing of food and beverages, a topic of contemporary interest and significance and for which huge volumes of literature are available in the public domain. I am grateful to the Almighty for the wisdom and intuitive guidance to perform the duties of editor. I am thankful also to all the contributing authors who extended kind cooperation and support in addition to making significant contributions in their own field of specialization. But for their contributions, this book would not have become a reality.

I am grateful to CRC Press LLC for providing me the opportunity to edit this book, and I am very grateful to Stephen Zollo, senior editor at CRC Press, for the kind invitation to edit this book. But for his kind support, encouragement, and warmth, I would not have ventured into this herculean task. I am also thankful to Stephanie Morkert, project coordinator of this project, for her valuable assistance and support.

I record my sincere gratitude to all the authorities of King Saud University, College of Science, Department of Botany and Microbiology, Riyadh, Saudi Arabia, for extending moral and physical support for completing this book. I am very grateful to Prof. Dr. Ali H. A. Bahkali and Prof. Dr. Fahad M. A. Al-Hemaid, former chairmen of the Department of Botany and Microbiology, College of Science, King Saud University, Riyadh, for their constant encouragement and moral support extended during the course of preparation of this book.

The moral and physical support I received from many people who extended valuable assistance directly or indirectly in collecting materials, and extending healthy and useful discussions are also gratefully acknowledged here.

No mission of creativity is complete without the physical and moral support of family. I am very indebted to the untiring support and constant encouragement extended by my wife Prema Chandrasekaran and my other family members, Santhalakshmi, Bibin, Hrishikesh, and Isha.

Editor

Professor Muthusamy Chandrasekaran is a distinguished scientist and a teacher, who has made significant contributions in the fields of marine microbiology and biotechnology. Professor Chandrasekaran earned his BSc degree in zoology from the University of Madras, India; MSc degree in marine biology from Annamalai University India; and PhD degree in microbiology (food microbiology) from Cochin University of Science and Technology, India. He did his postdoctoral research in genetic engineering of bacteria for wastewater treatment at Hiroshima University, Japan. His major areas of research interest are harnessing marine microorganisms for novel enzymes, bioactive molecules, and microbial and enzyme technologies for enzyme production and waste management.

He began his career as a lecturer in the Department of Applied Chemistry in the Cochin University of Science and Technology in 1983 and later in 1991, as founder head, organized the Department of Biotechnology in Cochin University of Science and Technology, India. After more than 31 years of service, he retired from Cochin University of Science and technology in December 2014, and currently serves as a professor of biotechnology in the Department of Botany and Microbiology, College of Science, King Saud University, Riyadh, Saudi Arabia.

Professor Chandrasekaran has made significant contributions to the growth of marine microbiology through publications in peer-reviewed international journals. He demonstrated for the first time that marine bacteria and fungi could be harnessed efficiently for the production of industrial enzymes, such as L-glutaminase, chitinase, alkaline protease, lipase, beta glucosidase, and tannase. He showed for the first time in the world that the L-glutaminase enzyme from marine bacteria is a good antileukemic agent. He also did pioneering work in developing fermentation processes for the large-scale production of these marine microbial enzymes in addition to characterizing the enzymes, finding industrial applications for them, and isolating the full gene coding for an alkaline protease from marine fungi and characterizing it. His studies have significantly advanced the existing knowledge on marine microbial enzymes, which were never studied before for their possible application. He has also worked on value addition, employing microbial enzymes to shrimp processing waste and banana and cabbage waste using solid-state fermentation in addition to environmental solid waste management, among other subjects.

Professor Chandrasekaran has earned recognition from the University Grants Commission, India, as a career awardee for his contribution in microbiology. He was a recipient of the Indian National Science Academy Visiting Fellowship and Overseas Associateship of the Department of Biotechnology, Ministry of Science and Technology, Government of India.

Professor Chandrasekaran edited the book *Valorization of Food Processing By-Products* in 2012, published under the Fermented Foods and Beverages series, CRC Press, Taylor & Francis Group, Boca Raton, Florida, U.S.A.

In 2012 he also coauthored the book *L-Glutaminase Production by Marine Fungi* with Sabu Abdulhameed, Lap Lambert Academic Publishing, Germany.

Professor Chandrasekaran has guided 27 PhD candidates and has several publications in peer-reviewed ISI-listed journals and a number of presentations in international and national symposia, seminars, and conferences. He has completed several sponsored research projects funded by UGC, CSIR, and Department of Biotechnology (DBT), Government of India and has organized many national symposia and popular lecture programs in biotechnology.

Professor Chandrasekaran has served as a member of the editorial board and a reviewer for several international research journals in the fields of food science, microbiology, and biotechnology. He has also served as a subject expert on the boards of studies in microbiology and biotechnology of several universities in India and contributed to the development of the curriculum at the UG and PG levels.

Professor Chandrasekaran founded the Society for Biotechnologists of India in 1995 as founder president. He is also a life member in several professional societies, including the Association of Food Science Technologists of India, the Association of Microbiologists India, the Marine Biological Association of India, the Society of Fishery Scientists and Technologists of India, the Indian Biophysical Society (1996), and the Mycological Society of India.

Professor Chandrasekaran has served as a member of the formulation group for the establishment of the Marine Biotechnology Application Centre, Department of Biotechnology, Ministry of Science and Technology, Government of India, 2000–2001; member, steering committee, National Bioresource Development Board, Department of Biotechnology, Ministry of Science and Technology, Government of India, 2002–2007; nominee of the University Grants Commission (UGC) India, to the Advisory Committee of DRS Programme of Department of Botany, University of North Bengal, India (April, 2002–March 2007); member, Task Force on Biotechnology, ICAR, Government of India (July 2003–2006); member, subcommittee of the Earth Sciences Research Committee on Disaster Preparedness, Council of Scientific & Industrial Research, HRD Group, New Delhi (May 2006 to March 2009); member, Task Force on Aquaculture and Marine Biotechnology, Department of Biotechnology, Ministry of Science and Technology, Government of India, 2006–2009; and member, Research Advisory Committee (RAC) of the Central Institute of Fisheries Education (CIFE) Mumbai, Indian Council of Agricultural Research ICAR, Government of India for the period 2010–2013.

Contributors

Fahad M. A. Al-Hemaid
Department of Botany and Microbiology
College of Science
King Saud University
Riyadh, Saudi Arabia

Caio C. Aragon
Departamento de Biocatálisis
Instituto de Catálisis (CSIC)
Campus UAM
Madrid, Spain

Ali H. A. Bahkali
Department of Botany and Microbiology
College of Science
King Saud University
Riyadh, Saudi Arabia

Soorej M. Basheer
Molecular Biophysics Unit
Indian Institute of Science
Bangalore, India

P. S. Beena
SciGenom Labs Pvt. Ltd.
Cochin, India

Ummalyama Sabeela Beevi
Centre for Biofuels
Biotechnology Division
CSIR-National Institute for Interdisciplinary
    Science and Technology
Thiruvananthapuram, India

Sarita G. Bhat
Department of Biotechnology
Cochin University of Science and Technology
Cochin, India

Cinthia Baú Betim Cazarin
Department of Food and Nutrition
Faculty of Food Engineering
UNICAMP
Campinas, São Paulo, Brazil

Honey Chandran Chundakattumalayil
School of Biosciences
Mahatma Gandhi University
Kottayam, India

Muthusamy Chandrasekaran
Department of Botany and Microbiology
College of Science
King Saud University
Riyadh, Saudi Arabia

Sreeja Chellappan
Molecular Biophysics Unit
Indian Institute of Science
Bangalore, India

Fabiano Jares Contesini
Laboratory of Food Biochemistry
Department of Food Science
Faculty of Food Engineering
UNICAMP
Campinas, São Paulo, Brazil

Juliana Kelly da Silva
Department of Food and Nutrition
Faculty of Food Engineering
UNICAMP
Campinas, São Paulo, Brazil

Clarissa Hamaio Okino Delgado
São Paulo State University "Júlio de Mesquita
    Filho"—UNESP
IBB/DQB
Botucatu, São Paulo, Brazil

Débora Zanoni do Prado
São Paulo State University "Júlio de Mesquita
    Filho"—UNESP
IBB/DQB
Botucatu, São Paulo, Brazil

Angela Dura
Department of Food Science
Institute of Agrochemistry and Food Technology
  (IATA-CSIC)
Spanish Research Council
Paterna, Spain

Pedro Fernandes
Institute for Biotechnology and Bioengineering
Centre for Biological and Chemical Engineering
Department of Bioengineering
Instituto Superior Técnico
Universidade de Lisboa
and
Faculdade de Engenharia
Universidade Lusófona de Humanidades e
  Tecnologias
Lisbon, Portugal

Marco Filice
Departamento de Biocatálisis
Instituto de Catálisis (CSIC)
Campus UAM
Madrid, Spain

Luciana Francisco Fleuri
São Paulo State University "Júlio de Mesquita
  Filho"—UNESP
IBB/DQB
Botucatu, São Paulo, Brazil

Xiao Hua
School of Food Science and Technology
Jiangnan University
Jiangsu, China

Ayman Salih Omar Idris
Centre for Biofuels
Biotechnology Division
CSIR-National Institute for Interdisciplinary
  Science and Technology
Thiruvananthapuram, India

Amit Kumar Jaiswal
School of Food Science and Environmental
  Health
College of Sciences and Health
Dublin Institute of Technology
Dublin, Republic of Ireland

T. R. Keerthi
School of Biosciences
Mahatma Gandhi University
Kottayam, India

Sapna Kesav
Department of Biotechnology
Cochin University of Science and Technology
Cochin, India

Jissa G. Krishna
National Centre for Biological Sciences
University of Agricultural Sciences (UAS)
Gandhi Krishi Vignana Kendra (GKVK)
Bangalore, India

Ravinder Kumar
Animal Biotechnology Centre
ICAR-National Dairy Research Institute
Karnal, India

Sanjeev Kumar
Department of Life Science
Assam University
Silchar, India

Tsz Him Kwan
School of Energy and Environment
City University of Hong Kong
Hong Kong, People's Republic of China

Wan Chi Lam
School of Energy and Environment
City University of Hong Kong
Hong Kong, People's Republic of China

Pengfei Li
School of Food Science and Technology
Jiangnan University
Jiangsu, China

Gláucia Carielo Lima
Department of Food and Nutrition
Faculty of Food Engineering
UNICAMP
Campinas, São Paulo, Brazil

Carol Sze Ki Lin
School of Energy and Environment
City University of Hong Kong
Hong Kong, People's Republic of China

Aravind Madhavan
Centre for Biofuels
Biotechnology Division
CSIR-National Institute for Interdisciplinary
    Science and Technology
Thiruvananthapuram, India

Manzur Ali Pannippara
Department of Biotechnology
MES College
Aluva, India

Mário Roberto Maróstica Jr.
Department of Food and Nutrition
Faculty of Food Engineering
UNICAMP
Campinas, São Paulo, Brazil

Gustavo Molina
Laboratory of Bioflavors
Department of Food Science
Faculty of Food Engineering
UNICAMP
Campinas, São Paulo, Brazil

and

Institute of Science and Technology
Food Engineering
UFVJM
Diamantina, Minas Gerais, Brazil

Renu Nandakumar
Department of Medicine
Columbia University Medical Center
New York, New York

T. Nidheesh
Department of Meat and Marine Sciences
Academy of Scientific and Innovative Research
CSIR-Central Food Technological Research
    Institute
Mysore, India

Paula Kern Novelli
São Paulo State University "Júlio de Mesquita
    Filho"—UNESP
IBB/DQB
Botucatu, São Paulo, Brazil

Gaurav Kumar Pal
Department of Meat and Marine Sciences
Academy of Scientific and Innovative Research
CSIR-Central Food Technological Research
    Institute
Mysore, India

Jose M. Palomo
Departamento de Biocatálisis
Instituto de Catálisis (CSIC)
Campus UAM
Madrid, Spain

Gláucia Maria Pastore
Laboratory of Bioflavors
Department of Food Science
Faculty of Food Engineering
UNICAMP
Campinas, São Paulo, Brazil

Mayara Rodrigues Pivetta
São Paulo State University "Júlio de Mesquita
    Filho"—UNESP
IBB/DQB
Botucatu, São Paulo, Brazil

Anil K. Puniya
College of Dairy Science and Technology
Guru Angad Dev Veterinary and Animal
    Sciences University
Ludhiana, India

Monica Puniya
Dairy Microbiology Division
ICAR-National Dairy Research Institute
Karnal, India

Cristina M. Rosell
Department of Food Science
Institute of Agrochemistry and Food Technology
    (IATA-CSIC)
Spanish Research Council
Paterna, Spain

Utpal Roy
Department of Biological Sciences
Birla Institute of Technology and Science, Pilani
K. K. Birla Goa Campus
Goa, India

Meena Sankar
Centre for Biofuels
Biotechnology Division
CSIR-National Institute for Interdisciplinary
    Science and Technology
Thiruvananthapuram, India

Vani Sankar
Centre for Biofuels
Biotechnology Division
CSIR-National Institute for Interdisciplinary
    Science and Technology
Thiruvananthapuram, India

Vaisakhi Satheesh
Centre for Biofuels
Biotechnology Division
CSIR-National Institute for Interdisciplinary
    Science and Technology
Thiruvananthapuram, India

Hélia Harumi Sato
Laboratory of Food Biochemistry
Department of Food Science
Faculty of Food Engineering
UNICAMP
Campinas, São Paulo, Brazil

Samriti Sharma
School of Biomolecular and Biomedical Science
University College Dublin
Dublin, Republic of Ireland

Juliana Wagner Simon
São Paulo State University "Júlio de Mesquita
    Filho"—UNESP
IBB/DQB
Botucatu, São Paulo, Brazil

S. Raghul Subin
Department of Biotechnology
Cochin University of Science and Technology
Cochin, India

Rajeev K. Sukumaran
Centre for Biofuels
Biotechnology Division
CSIR-National Institute for Interdisciplinary
    Science and Technology
Thiruvananthapuram, India

P. V. Suresh
Department of Meat and Marine Sciences
Academy of Scientific and Innovative Research
CSIR-Central Food Technological Research
    Institute
Mysore, India

Mamoru Wakayama
Department of Biotechnology
College of Life Sciences
Ritsumeikan University
Kusatsu, Japan

Ruijin Yang
School of Food Science and Technology
Jiangnan University
Jiangsu, China

Wenbin Zhang
School of Food Science and Technology
Jiangnan University
Jiangsu, China

Section I

Enzymes: Basics and Development of Novel Biocatalysts

1

Enzymes: Concepts, Nomenclature, Mechanism of Action and Kinetics, Characteristics and Sources of Food-Grade Enzymes

S. Raghul Subin and Sarita G. Bhat

CONTENTS

1.1 Introduction

The use of enzymes in the food processing industry dates back to 6000 BC or earlier with the brewing of beer, bread baking, cheese and wine making, and vinegar production (Poulsen and Buchholz 2003). Enzymatic degradation in meat was first observed by Spallanzani (1729–1799), and a substance in barley was found by Kirchhoff (1815) to be capable of starch paste liquefaction into sugar (Roberts 1995). The term *diastase* was coined by Payen and Persoz for this saccharification process (Payen and Persoz 1833), and this is still in use for amylases in the brewing industry. Now the enzymes find applications in a wide range of food industries including dairy, baking, brewing, food processing (vegetables, fruits, and egg), sweetener production, protein hydrolysis, distilling, and fruit juice and wine production and in lipid modification. In all of these industries, enzymes are meant mainly for the production of fermentation substrates, flavor development and enhancement, or for product production.

The history of the food industry reveals the efficient use of enzymes as they were exploited for their specificity and catalytic activity in food processing. The first immobilized enzyme used in the food industry was invertase in the production of invert sugar syrup. Even as early as the 1960s, the large-scale application of enzymes in the food industry was established with the acid hydrolysis of starch replaced by the application of amylases and amyloglucosidases.

Enzymes are biological catalysts and are natural in origin unlike chemical catalysts. They are polymers catalyzing chemical reactions that are fundamental to life, comprising rapid synthesis of complex compounds, degradation of high-molecular-weight structures and processing of biomolecules into their active state. Enzymes are superior to any other chemical catalysts, with much greater catalytic power, and stereo specific with the ability to convert nonchiral substrates to chiral products, and these properties are exploited in the food industry. Further, they are highly specific, and even absolute specificity is seen in others. With the exception of a few catalytic RNA molecules, the ribozymes, most enzymes are proteins. All the enzymes used in the food industry are protein in nature, and food technologists normally use them in the manufacture, processing, preparation, and treatment of food. In this chapter, the basic enzymology, including the concepts of nomenclature, mechanism of action, enzyme kinetics, their characteristics, and source of food enzymes, is discussed to present the reader with an instant reference and emphasize the significance of enzymes in food and beverage processing.

1.2 Enzyme Nomenclature and Classification

Traditionally, enzyme names end in "-ase." Some exceptions to this rule are the proteolytic enzymes, such as trypsin and chymotrypsin, which end in "-in." Names of some other enzymes involve the use of the substrate names, such as lactase, which hydrolyze the disaccharide lactose into glucose and galactose. Transaminases indicate the nature of the reaction catalyzed, that is, the transfer of amino groups without specifying the names of the substrate, whereas the name catalase, which neither indicates the name of the substrate nor the reaction catalyzed, catalyzes the degradation of hydrogen peroxide into water and oxygen.

This lack of consistency in the nomenclature of the enzymes was apparent when the list of enzymes grew longer, thereby necessitating the requirement of a systematic method of naming and classifying enzymes.

Reports of the commission appointed by the International Union for Biochemistry and Molecular Biology (IUBMB, formerly the International Union for Biochemistry) were published in 1964, and they were updated from time to time (in 1972, 1978, 1984, and 1992). The additions made to the enzyme database are available at the website dedicated to enzyme nomenclature: http://www.enzyme-database.org/.

Enzymes are classified according to the IUBMB report in consultation with the IUPAC-IUBMB Joint Commission on Biochemical Nomenclature (JCBN). Each enzyme is assigned a recommended name and a four-part distinguishing number by the enzyme commission. Nevertheless, it is clear that some alternative names still remain in such common usage that they will be used, where appropriate, in the text.

The specific property that differentiates one enzyme from another is the chemical reaction that is catalyzed, and it is reasonable to utilize this as the basis for the classification and naming of enzymes. The trivial system is another type of naming of enzymes, not recognizing the rules of any formal system of nomenclature. Recommended names or trivial names are original names or named by appending "-ase" to either the name of the substrate or to the type of catalytic reaction. Some common examples of trivial names are lipase (acts on lipids), protease (acts on proteins), and cellulase (acts on cellulose). The IUPAC name is vital for unambiguous communication between biochemists and product chemists. However, the trivial name of the enzyme, derived from the truncated substrate name with "-ase" added, identifies the substrate or substrate range better for food technologists than does its systematic name or its International Union of Biochemistry Enzyme Commission (IUB or EC) number (Dixon and Webb 1999). In certain cases, food technologists use traditional names for enzymes, such as malt, pepsin, and rennet. The nomenclature of major enzymes used in the food industry is presented in Table 1.1.

The enzyme commission (EC) has divided enzymes into six main groups according to the type of reaction catalyzed (http://www.chem.qmul.ac.uk/iubmb/enzyme/rules.html). Each enzyme was assigned a code number, consisting of four digits, separated by dots. The first digit shows the main class to which the enzymes belong, which is as follows.

Class 1: Oxidoreductases catalyze redox reactions in which hydrogen or oxygen atoms or electrons are transferred between molecules. This extensive class includes the dehydrogenases (hydride transfer), oxidases (electron transfer to molecular oxygen), oxygenases (oxygen transfer from molecular oxygen), and peroxidases (electron transfer to peroxide). The second digit in the code indicates the donor of the reducing equivalents involved in the reaction. For example, glucose oxidase (EC. 1.1.3.4, systematic name, β-D-glucose: oxygen 1-oxidoreductase) is used in dough strengthening in the food processing industry. Laccase (EC. 1.14.18.1) and lipoxygenase (EC. 1.13.11.12) are other major enzymes of this class used in the food industry.

Class 2: Transferases catalyze the transfer of an atom or group of atoms between two molecules except those enzymes included in other groups (e.g., oxidoreductases and hydrolases). EC recommends that the names of the transferases should end as X-transferase, with which X is the group transferred. For example, glucanotransferase (EC. 2.4.1.19) is used for the modification of starch into cyclodextrins. The second digit describes the group that is transferred.

Class 3: Hydrolases include enzymes catalyzing the hydrolytic cleavage of bonds such as C–O, C–N, C–C, and some other bonds, including phosphoric anhydride bonds. They are classified according to the type of bond hydrolyzed. This is presently the most commonly encountered class of enzymes in the field of enzyme technology. In the food industry, a majority of the enzymes categorized as such includes α-amylase (EC. 3.2.1.1), β-amylase (EC. 3.2.1.2), lactase (EC. 3.2.1.23), lipase (EC. 3.1.1.3), and proteases, which include aminopeptidase (EC. 3.4.11), trypsin (EC. 3.4.21.4), subtilisin (EC. 3.4.21.62), papain (EC. 3.4.22.2), ficin (EC. 3.4.22.3), pepsin (EC. 3.4.23.1), and chymosin (EC. 3.4.23.4).

Class 4: Lyases are involved in nonhydrolytic removal of groups from substrates. These are elimination reactions in which a group of atoms are removed from the substrate, often leaving double bonds. The second digit in the classification indicates the broken bond. This includes the aldolases, decarboxylases, and dehydratases. For example, acetolactate decarboxylase (EC. 4.1.1.5) is used in the beer industry.

Class 5: Isomerases are those enzymes that can catalyze different molecular isomerization reactions. These enzymes catalyze geometric or structural changes within one molecule. According to the type of isomerism involved, they may be called racemases, epimerases, cis trans isomerases, isomerases, tautomerases, mutases, or cyclo-isomerases, for example, phospho glucose isomerase (EC. 5.3.1.9).

Class 6: Ligases, also known as synthetases, catalyze the synthesis of new bonds between two molecules. They join molecules together with covalent bonds in biosynthetic reactions. These reactions require the input of energy by the hydrolysis of a diphosphate bond in ATP or a similar triphosphate, and this property adds to the difficulty in their commercial application. The second digit indicates the type of bond synthesized.

In the code, the second and third digits describe the kind of reaction being catalyzed. There is no general rule here as the meaning of these digits is specified separately for each class of enzymes.

TABLE 1.1

Nomenclature of Major Enzymes Used in the Food Industry

EC No.	Systematic Name (Other Names)	Application in Food Industry	References
	Oxidoreductase		
1.1.3.4	β-D-glucose: O_2 1-oxidoreductase (notatin; glucose oxidase)	Oxygen removal from food packaging and as food preservative, also has been used in baking, dry egg powder production, and wine production	Wong et al. 2008
1.8.3.2	Thiol: O_2 oxidoreductase (sulphydryl oxidase)	Application in dairy and baking industry; flavor enhancement	Faccio et al. 2011
1.11.1.6	H_2O_2: H_2O_2 oxidoreductase (catalase)	Prevent food spoilage, soft drink manufacturing	Whitehurst and Law 2002
1.11.1.7	Donor: H_2O_2 oxidoreductase (lactoperoxidase)	Cold sterilization of milk, prevents spoilage of food	Whitehurst and Law 2002
1.13.11.12	Linoate: O_2 13-oxidoreductase (lipoxygenase)	Dough strengthening, bread whitening	Kirk et al. 2002
1.14.18.1	*o*-Diphenol: O_2 oxidoreductase (laccases)	The production and treatment of beverages, including wine, fruit juice, and beer; baking industry, flavor enhancer	Gianfreda et al. 1999; Kirk et al. 2002
1.1.3.5	D-hexose: oxygen 1-oxidoreductase (hexose oxidase)	Acts as an oxygen scavenger in food products, bread and wheat flour industry, bakery product preparation, preparation of milk products	Smith and Olempska-Beer 2004
	Transferase		
2.3.2.13	Protein-glutamine: γ-glutamyltransferase (transglutaminase)	Manufacture of cheese and other dairy products, in meat processing, to produce edible films, and to manufacture bakery products	Kieliszek and Misiewicz 2014
2.4.1.19	1,4-α-D-glucan: amine 4-α-D(1,4-α-D-glucano)-transferase (cyclodextrin, glucanotransferase)	Widely used in cyclodextrin production, which is used as food-grade micro-encapsulants for colors, flavors, and vitamins	Van Der Maarel et al. 2002
	Hydrolase		
3.1.1.3	Triacyl glycerol acylhydrolase (lipase, tributyrase)	Widely used in food processing, which includes oil and fat modification, flavor development, and improving quality	Hasan et al. 2006
3.1.1.4	Phosphatidylcholine 2-acylhydrolase (phospholipase A, lechithinase A)	Dairy, baking products, emulsifying agents, manufacture of edible oils	Casado et al. 2012
3.1.1.11	Pectin pectylhydrolase (pectinesterase, pectase)	Fruit and vegetable juice preparations	Jayani et al. 2005
3.1.1.20	Tannin acyl hydrolase (tannase)	Tea and cold drink manufacturing; gallic acid and propylgallate preparation, used as food preservative and antioxidant	Chae et al. 1983
3.2.1.1	1,4-α-D-glucan glucanohydrolase (α-amylase, diastase, ptyalin)	Liquefaction in starch industry	Van Der Maarel et al. 2002
3.2.1.2	1,4-α-D-glucan maltohydrolase (β-amylase)	Production of high malt syrups	Van Der Maarel et al. 2002

(Continued)

TABLE 1.1 (CONTINUED)

Nomenclature of Major Enzymes Used in the Food Industry

EC No.	Systemic Name (Other Names)	Application in Food Industry	References
3.2.1.3	1,4-α-D-glucan glucohydrolase (amyloglucosidase, glucoamylase)	Conversion of dextrins to glucose, corn syrup production, beer and malt liquor preparation, fruit juice preparation	Van Der Maarel et al. 2002
3.2.1.4	1,4-(1,3;1,4)-β-D-glucan 4-glycanohydrolase (cellulase)	Fruit juice production, solubilization of pentosan in baking, natural flavors and color extracts, used in tea industry, additive in detergents	Uhlig 1998
3.2.1.7	2,1-β-D-fructan fructanohydrolase (inulinase)	Inulin hydrolysis	Fernandes and Jiang 2013
3.2.1.8	1,4-β-D-xylan xylanohydrolase (xylanase)	Separation and isolation of starch and gluten from wheat flour	Heldt-Hansen 1997
3.2.1.15	Poly-(1,4-α-D-galacturonide) glycanohydrolase (pectinase, endopolygalacturonase)	Production of high-quality tomato ketchup and fruit pulps, production of cloudy vegetable juice of low viscosity, fruit and vegetable juice preparations, coffee and tea manufacturing industry, retting and degumming of fiber crops	Bhat 2000; Grassin and Auquembergue 1996; Heldt-Hansen 1997; Jayani et al. 2005; Kashyap et al. 2001; Uhlig 1998
3.2.1.17	Peptidoglycan N-acetylmuramoylhydrolase (lysozyme, muramidase)	Prevention of late blowing defects in cheese by spore-forming bacteria	Whitehurst and Law 2002
3.2.1.23	β-D-galactoside galactohydrolase (β-galactosidase, lactase)	Lactase-reducing enzyme, ice cream preparation	Uhlig 1998
3.2.1.26	β-D-fructofuranoside fructohydrolase (invertase, saccharase, glucosucrase)	Hydrolysis of sucrose in confectionery, flavor development in fruit juices	Uhlig 1998
3.2.1.37	4-β-D-xylan xylohydrolase (pentosanase)	Beer manufacturing, degerming, distillery mesh, vineger mesh, and production of nonstandardized bake goods	Uhlig 1998
3.2.1.40	α-L-rhamnoside rhamnohydrolase (naringinase)	Processing of citrus fruit juice	Whitehurst and Law 2002
3.2.1.41	Pullulan 6-α-glucanohydrolase (pullulanase)	Starch saccharification (improves efficiency)	Van Der Maarel et al. 2002
3.4.11	Aminopeptidases	Cheese manufacturing	Whitehurst and Law 2002
3.4.21.4	Trypsin	Production of hydrolysates for food flavoring	Whitehurst and Law 2002
3.4.21.62	Subtilisin (alkaline proteinase)	Milk coagulation for cheese making, hydrolysate production for soups and savory foods, beer manufacturing, bread and flour; meat tenderizer	Uhlig 1998; Whitehurst and Law 2002
3.4.22.2	Papain	Meat tenderization, chill haze prevention in brewing industry	Uhlig 1998; Whitehurst and Law 2002
3.4.22.3	Ficain (ficin)	Hydrolyzation of animal, milk, and plant proteins; meat tenderization	Uhlig 1998; Whitehurst and Law 2002
3.4.22.32	Stem bromelain (bromelain)	Meat tenderization, beer and malt liquor preparation	Uhlig 1998; Whitehurst and Law 2002
3.4.22.33	Fruit bromelain (juice bromelain)	Meat tenderization, beer and malt liquor preparation	Uhlig 1998; Whitehurst and Law 2002

(Continued)

TABLE 1.1 (CONTINUED)

Nomenclature of Major Enzymes Used in the Food Industry

EC No.	Systemic Name (Other Names)	Application in Food Industry	References
3.4.23.1	Pepsin A (pepsin)	Manufacture of fat-free soy flour, precooked instant cereals, beer and cheese production	Uhlig 1998
3.4.23.4	Chymosin (renin)	Cheese production	Uhlig 1998
3.5.1.1	L-asparagine amidohydrolase (L-asparaginase)	Degradation of asparagine in food industry, especially baked and fried foods	Yadav et al. 2014
3.5.1.2	L-glutamine amidohydrolase (L-glutaminase)	Flavor-enhancing enzyme in food and tea industry	Nandakumar et al. 2003
3.5.1.5	Urea amidohydrolase (urease)	Reduction of ethyl carbomate in wine manufacturing industry	Sujoy and Aparna 2013
	Lyases		
4.1.1.5	(2S)-2-hydroxy-2-methyl-3-oxobutanoate carboxy-lyase (acetolactate decarboxylase)	Beer maturation, reduction of wine maturation time	Godfredsen and Ottesen 1982
4.2.2.10	(1→4)-6-O-methyl-α-D-galacturonan lyase (pectin lyase)	Fruit and vegetable juice preparations, coffee and tea manufacturing industry, retting and degumming of fiber crops	Jayani et al. 2005
	Isomerase		
5.3.1.9	D-glucose-6-phosphate ketolisomerase (phospho glucose isomerase)	Conversion of high glucose syrup into fructose syrup	Van Der Maarel et al. 2002

Source: *Enzyme Nomenclature* (recommendations of the Nomenclature Committee of the International Union Biochem. & Mol. Biol. On the Nomenclature and Classification of Enzymes) Academic Press, San Diego, New York, London, 1992.

Apart from this technical classification, enzymes may be differentiated in terms of their location and mode of synthesis:

- Intracellular enzymes: Synthesized and stored inside the cell. They are responsible for cellular metabolism.
- Extracellular enzymes: Synthesized and transported outside the cell into the environment. They may be either released directly into the environment or may be located in the periplasmic space of the cell. They are responsible for the breaking down of complex polymeric substances into monomeric forms before uptake into cells.
- Constitutive enzymes: Enzymes routinely synthesized and stored in the cell for various functions.
- Induced enzymes: Enzymes that are synthesized on induction by respective substrates or biosynthetic signaling molecules.

1.3 Enzyme Structure and Chemistry

Enzymes may be differentiated into *monomeric enzymes* and *oligomeric enzymes* based on their chemical structure.

Monomeric enzymes are those that consist of a single polypeptide chain so that they cannot be dissociated further. Several proteases are monomeric in nature. In order to prevent hydrolytic cleavage and therefore cellular damage due to generalized cleavage of proteins, several of these enzymes are produced

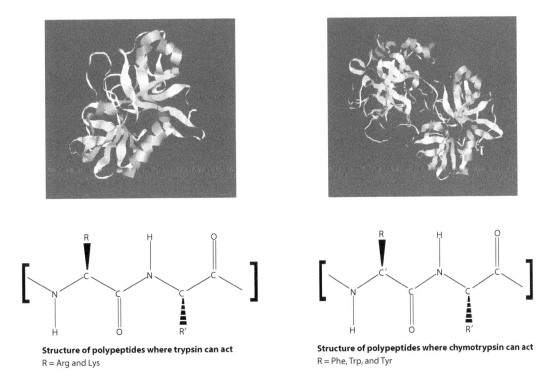

Structure of polypeptides where trypsin can act
R = Arg and Lys

Structure of polypeptides where chymotrypsin can act
R = Phe, Trp, and Tyr

FIGURE 1.1 Tertiary structures of trypsin (PDB ID 1PTN) and chymotrypsin (PDB ID 1YPH).

as zymogens, or proenzymes, which require activation. Serine proteases, so called due to the presence of serine in their active site, include trypsin and chymotrypsin produced by the mammalian pancreas. The tertiary structures of these endopeptidases are similar, and their catalytically important residues match exactly although they show only 40% similarity in their primary structure (Figure 1.1).

Oligomeric enzymes consist of two or more polypeptide chains termed as subunits, usually linked by noncovalent interactions and never by peptide linkages. These have large molecular weights, in excess of 35,000. Lactate dehydrogenase is an oligomeric enzyme, which has five isoenzyme forms. Pyruvate dehydrogenases in bacteria and in eukaryotes are multienzyme complexes that catalyze the conversion of pyruvate to acetyl-CoA.

1.4 Solubility of Enzymes

Enzymes are globular proteins and are soluble in aqueous solvents or dilute salt solutions. Their solubility is enhanced by weak ionic interactions, such as hydrogen bonds between the solute and water. All factors that influence or interfere with this process have an effect on solubility. The four factors that influence solubility of enzymes are salt concentration, pH, temperature, and organic matter of the solvent. Solubility of enzyme proteins depends on the concentration of dissolved salt. It can be increased by the addition of neutral salt in low concentrations. The added ions can interfere with the ionization of the side chains of amino acids, which, in turn, can interfere with interactions within the protein molecule but increase interactions with the solute and solvent. This process of increasing solubility by the addition of salt in low concentrations is known as "salting in." When higher salt concentrations are used, there is greater interaction between the ions and water. This leads to reduced protein–water interactions, often causing precipitation from the solution. This process is known as "salting out." Divalent ions are more effective than monovalent ions. When salts such as ammonium sulfate that have higher solubility are

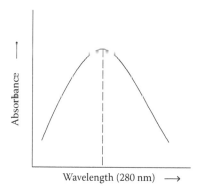

FIGURE 1.2   Absorption maximum of proteins (at 280 nm).

used, some proteins will precipitate at particular salt concentrations, and others will not. Most proteins will precipitate out in more than 80% $(NH_4)_2SO_4$ saturation.

Cations such as Zn^{2+} and Pb^{2+} decrease solubility by forming insoluble complexes with the enzyme protein. Proteins are also precipitated by the addition of acids such as trichloroacetic acid or picric acid due to the formation of acid-insoluble salts, a property used in analytical techniques to separate proteins from the solution before estimation of other substances.

When ethanol, a water-miscible molecule, is added into the solvent, its dielectric point is altered. This causes an increased attraction between the oppositely charged groups within the protein, further reducing its interaction with water molecules and consequently decreasing protein solubility.

Enzymes are charged proteins due to the presence of amino acids with the charge depending on the pH of the solution. At low pH, the amino acids are fully protonated, and there is a positive charge on the protein. As pH is increased, the protein loses a proton to neutralize the OH^- ions and becomes a zwitter ion. As more alkali is added, the NH_3^+ gives its H^+ ions, and the protein becomes positively charged. Under extremes of pH, the ionizable side chains of the protein have charges that are very different from those under normal physiological conditions. This change in the charge pattern of the enzyme protein causes a disruption of the tertiary structure of the protein, a process termed "denaturation." The tertiary structure of the protein has hydrophilic amino acids on the outside, shielding the hydrophobic groups well hidden within the molecule.

The disruption of the tertiary structure brings these hydrophobic groups in close proximity to the aqueous solvent, thereby decreasing the solubility. Restoring the normal conditions may sometimes cause the refolding of the protein into its original tertiary structure required for function and activity. Solubility of enzyme protein will decrease over a narrow pH range called its "isoelectric point" when there is no net charge on the protein; that is, they are electrically neutral. When the temperature is between 40°C and 50°C, solubility of these enzymes increases. At temperatures above this, the tertiary structure is disrupted; the protein is denatured and loses its activity. Rates of enzyme-catalyzed reactions increase with an increase in temperature as the frequency of collisions between molecules increases until the enzyme is denatured and loses its catalytic activity.

Enzymes are proteins and therefore give maximum absorption at 280 nm due to their content of aromatic amino acids (trp, tyr, phe) (Figure 1.2).

1.5 Mechanism of Enzyme Action

The reactant in an enzyme-catalyzed reaction is called "substrate," and the *active site* of an enzyme is responsible for the catalytic action of the enzyme. An enzyme usually contains one or more active sites, which may comprise only a few amino acid residues; the rest of the protein is required for maintaining the three-dimensional structure. The mechanism of enzyme action is well explained by several researchers. The *three point combination* concept by Ogston (1948) pointed out that there were at least three points of interaction between the enzyme and substrate (Figure 1.3), which can explain the stereo-specificity of enzymes. These interactions have either a binding or a catalytic function.

The binding sites link to specific groups in the substrate, ensuring that the enzyme and the substrate molecules are held in a set orientation with respect to each other with the reacting group in the vicinity of the catalytic sites. Regions with binding and catalytic sites are termed "active sites" or the "active center of an enzyme." All enzymes contain an active site (Figure 1.4), which is responsible for the catalytic action of the enzyme, and this is the region where substrates bind. These active sites consist of different regions of the protein, brought together by folding and bending of the protein chain by secondary and tertiary structure formation. This indicates the importance of structural arrangement of enzymes and the destructive effect caused by the denaturation of the protein by heating, which changes the three-dimensional configuration.

There are some amino acid residues in active sites that do not have a binding or catalytic action and therefore may interfere with the binding of chemically similar substances. The three-point interaction theory, for that reason, cannot explain the enzyme action and specificity in detail.

There are several other hypotheses that detail the mechanism of enzyme action. It is believed that enzyme action occurs in two steps. First, the active site of the enzyme combines with the substrate to form an *enzyme–substrate complex*. This enzyme–substrate complex then breaks up to form the products and the free enzyme, which can react again.

According to the hypothesis of the *lock-and-key model* by Fischer (1980), the substrate must fit into the active site of the enzyme; it is very specific, and all structures remain fixed throughout the binding process (Figure 1.5a). The lock-and-key model failed to explain the flexibility of enzymes as it did not successfully distinguish between free and substrate-bound enzymes.

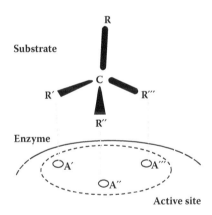

FIGURE 1.3 Three-point interactions between enzyme and substrate. A′, A″, and A‴ are sites on the enzyme that interacts with the groups R′, R″, and R‴, respectively, of the substrate.

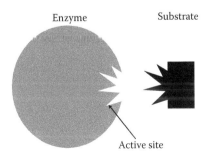

FIGURE 1.4 Active site of enzyme where substrate binds.

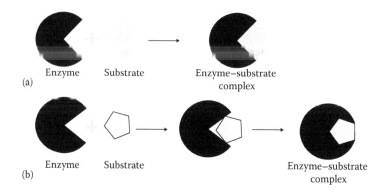

FIGURE 1.5 Models of enzyme action. (a) Lock-and-key hypothesis and (b) induced fit hypothesis.

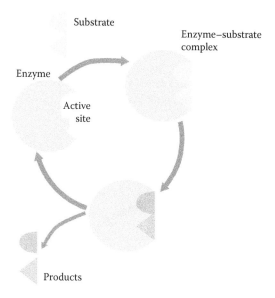

FIGURE 1.6 Mechanism of enzyme action.

The conformational changes of an enzyme, that is, the changes in the three-dimensional structure of an enzyme, resulting from the binding of a substrate to an enzyme, was explained by Koshland (1958) in his *induced fit hypothesis*. He proposed that the structure of a substrate may be complementary to the active site of an enzyme in the enzyme–substrate complex but not in the free enzyme state, and this was due to the conformational change that takes place in the enzyme upon binding with the substrate (Figure 1.5b). This mechanism could explain the flexibility of the enzyme and its high degree of specificity in action.

The mechanism of enzyme action depicted in Figure 1.6 explains the conformational change that happens in the active site of an enzyme that allows substrate to bind to the modified active site and carry out the reaction process.

1.5.1 Mechanism of Catalysis by Enzymes

There are four general mechanisms underlying enzyme catalysis. First, is *catalysis by proximity*; that is, in order to carry out chemical reactions, the substrate and enzymes must come within a bond-forming distance of each other. Thus, as the concentration of substrate increases, the possibility of chemical reactions occurring is greater. When an enzyme binds a substrate at its active site, a zone of higher substrate concentration is developed, causing the substrate molecule to orient spatially in a position ideal for chemical reaction to occur, resulting in enhanced reaction rates. Certain enzymes catalyze reactions by acting as acids and bases due to the presence of ionizable functional groups of aminoacyl side chains and prosthetic groups.

Enzyme-catalyzed lytic reactions are achieved due to straining of the covalent bonds of substrates. When the enzymes bind with substrates, conformational changes unfavorable for the bond that undergoes cleavage is generated, resulting in strain stretches, or distorts the targeted bond, weakening it and making it more susceptible for cleavage.

Covalent catalysis involves the formation of a covalent bond between the enzyme and one or more substrates. The modified enzyme then becomes a reactant and the chemical modification of the enzyme is transient. On completion of the reaction, the enzyme returns to its original state. The proteolysis by trypsin is an example for covalent reactions.

1.6 Specificity of Enzymes

The specificity of enzymes is one of their unique properties. They are specific not only to the reactions they catalyze, but also to the substrates they utilize and can therefore be categorized based on their different exhibited specificities, as *absolute specificity, group specificity, relative specificity, stereospecificity*, and *dual specificity*.

- Enzymes that exhibit *absolute specificity* act only on one substrate, for example, glucokinase, which acts only on glucose, and lactase, which acts only on lactose.
- Enzymes with *group specificity* are specific not only to the type of bond, but also to the structure around it, for example, aminopeptidases and carboxypeptidases are exopeptidases that hydrolyze the peripheral peptide bonds at the amino terminal and carboxy terminal, respectively, and chymotrypsin is an endopeptidase that hydrolyzes the central peptide bonds in which the carboxyl groups belong to aromatic amino acids.
- Enzymes showing *relative specificity* act on substrates that are similar in structure and show the same type of bonds, for example, amylases act on α-1-4 glycosidic linkages of starch, dextrin, and glycogen, and lipases act on ester linkages in triglycerides.
- In *optical specificity*, the enzyme is specific not only to the substrate but also to its optical configuration, for example, L-amino acid oxidases act on L-amino acids, and β-glycosidase acts on β-glycosidic bonds in cellulose.
- Enzymes with *dual specificity* have the ability to act on two substrates, for example, xanthine oxidase converts hypoxanthine to xanthine and xanthine to uric acid.

1.7 Bioenergetics and Enzyme Catalysis

1.7.1 Concepts of Bioenergetics

Bioenergetics deals with changes in energy and in comparable factors, even as a biochemical process takes place, but not with the mechanism of the reactions or the speed thereof.

- The *first law of thermodynamics* states that energy can neither be created nor destroyed but can be converted into many other forms of energy or be used to do work.
- The *second law of thermodynamics* deals with entropy or degree of disorder. It states that entropy in the universe is perpetually increasing. It does not distinguish between different systems,

be it any living cell, a locomotive engine, or any chemical reaction. However, from a state of low entropy maintained by the consumption of chemical energy (in the form of food) in other organisms to light energy by photosynthesis in plants, life ultimately approaches thermodynamic equilibrium via death and decay.

Living systems operating at constant temperatures and pressures cannot use heat energy to perform work. Under these circumstances, the concept of two energy forms has been put forth: one that can be used to perform work, also called *free energy*, and another that cannot.

1.7.2 Enthalpy, Entropy, and Free Energy

Under thermodynamic considerations, a system that allows exchange of energy with its surroundings but not with matter is a *closed system*. In a closed system, if a process that takes place involves transfer of heat to or from the surrounding environment and causes a change in the volume of the system, then, according to the first law of thermodynamics,

$$\Delta E = \Delta H - P\Delta V \tag{1.1}$$

where ΔE is the increase in the intrinsic energy of the system, ΔH is the increase in enthalpy and $P\Delta V$ is the work done on the surrounding environment by increasing the volume of the system by ΔV at constant pressure P and temperature T. Change in enthalpy is defined as the quantity of heat absorbed by the system under the above conditions (at constant pressure P and temperature T) and determined calorimetrically.

The increase in entropy of the surroundings is $-\Delta H/T$ under the above conditions. If the process conditions are thermodynamically reversible (but take place infinitely slowly), then the increase in entropy of the system, ΔS, would be $\Delta H/T$. If the process has to occur spontaneously under thermodynamically irreversible conditions, then ΔS must be greater than $\Delta H/T$. This, as required by the second law of thermodynamics, gives an overall increase in entropy of the system including the surroundings.

Thus,

$$S - \frac{\Delta H}{T} > 0 \tag{1.2}$$

and

$$\Delta H - T\Delta S < 0 \tag{1.3}$$

In 1878, Gibbs defined the increase in free energy of the system ΔG as

$$\Delta G = \Delta H - T\Delta S \tag{1.4}$$

Therefore, at constant pressure and temperature, $\Delta G < 0$ for a spontaneous process.

1.7.3 Standard Free Energy

Standard free energy change $\Delta G°$ can be calculated if the equilibrium constant of a reaction is known.

$$\Delta G° = -hRT \log_e K_{eq} = -2.303 \, RT \log_{10} K_{eq} \tag{1.5}$$

It is also the difference in the standard free energy of formation of the reactants and the standard free energy of formation of the products with each term adjusted according to the stoichiometry of the reaction equation.

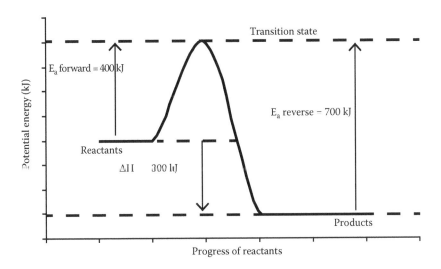

FIGURE 1.7 Reaction profile for hypothetical reaction.

1.7.4 Factors Affecting Rates of Chemical Reactions

Chemical reactions can take place only if the molecules can interact with each other. Molecules can interact only if they are in contact with each other and collide with each other. According to the principles put forth by Arrhenius and van't Hoff (Atkins and DePaula 2006), factors that can increase the rate of collision of molecules will also increase reaction rates, that is, concentration of reactants or increase in temperature. However, it is also important to know that not all colliding molecules will react due to steric hindrance. Further, these colliding molecules may not possess between them sufficient energy for a reaction to take place.

Molecules that are alike may not necessarily possess the same amount of energy even considering the different forms of energy. For example, the energy possessed by individual molecules will depend on the type of collisions they were recently involved in. The energy levels of the colliding molecules must be sufficiently large to overcome a potential barrier known as the "energy of activation."

The *transition state theory* developed by Eyring (1935) explains the requirement for activation energy in a chemical reaction. This postulates that every chemical reaction proceeds via the formation of an unstable intermediate between the reactants and products. The activation energy is required for the formation of the transition state complex from the reactants.

The theory suggests that three major factors determine if a reaction will occur: concentration of transition state complex, rate of breakup of complex, and the way the complex breaks, that is, to reform the reactants or to form the products.

Using a *reaction profile*, the energy necessary to complete a reaction can be determined by plotting energy values on the y-axis of a Cartesian plane. On the x-axis, the reaction progress is plotted.

Figure 1.7 illustrates where the molecules exist as reactants, products, or in the transition state. The relationship between potential energy and reaction rate is clear. Also illustrated is the amount of energy required to initiate a reaction—the activation energy (E_a). A reaction profile can also be used to find the enthalpy (ΔH) of the reaction by subtracting the energy of the products from the energy of the reactants. This example reaction is exothermic. The Gibbs energy change of the reaction (ΔG) is equal to the activation energy of the forward reaction.

1.7.5 Reaction Catalysis

A catalyst increases the rate of a chemical reaction without changing itself and can be separated unchanged or unmodified from the end product of a reaction. This indicates that it has no overall thermodynamic effect, that is, the presence of the catalyst does not change the amount of free energy liberated or utilized (Figure 1.8) when the reaction reaches completion. Catalysts act frequently by reducing the energy of activation for a chemical reaction, wherein a part or the whole of the catalyst interacts with

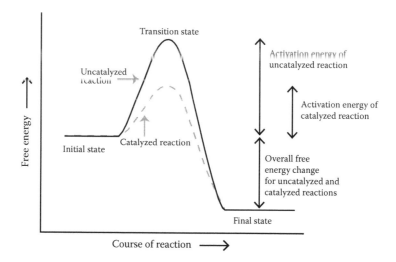

FIGURE 1.8   Effect of catalyst on free energy changes in a chemical reaction.

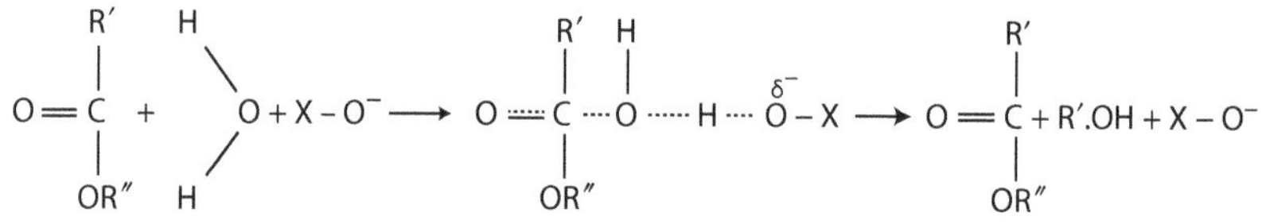

$$O = \overset{\overset{\displaystyle R'}{|}}{\underset{\underset{\displaystyle OR''}{|}}{C}} + \overset{\displaystyle H}{\underset{\displaystyle H}{O}} + X - O^- \longrightarrow O \cdots \overset{\overset{\displaystyle R'}{|}}{\underset{\underset{\displaystyle OR''}{|}}{C}} \cdots \overset{\displaystyle H}{O} \cdots H \cdots \overset{\delta^-}{O} - X \longrightarrow O = \overset{\overset{\displaystyle R'}{|}}{\underset{\underset{\displaystyle OR''}{|}}{C}} + R'.OH + X - O^-$$

FIGURE 1.9   Hydrolysis of ester by a base.

the reactants to form a transition state complex that is not only very different from that produced by an uncatalyzed reaction, but also more stable and hence with lower energy.

Catalysts are often acids or bases. The transition state is stabilized by donation of a proton by acids and acceptance of proton by bases, for example, hydrolysis of an ester by a base (Figure 1.9).

In covalent catalysis, on the other hand, the transition state is stabilized by changes in covalent bonds whereas, in metal ion catalysis, it is stabilized by electrostatic interactions with a metal ion.

1.7.6 Kinetics of Uncatalyzed Reactions

The *law of mass action* proposed by Guldberg and Waage (1867) forms the basis for all kinetic work. This states that the rate of reaction is proportional to the product of the activities of each reactant (each activity) raised to the power of the number of molecules of that reactant participating. For example, for a reaction **xA + yB → products**, the reaction rate is proportional to (activity of A)^x × (activity of b)^y.

However, for all practical purposes, concentration is used instead of activity although this holds good only for ideal gases and in very dilute solutions.

1.7.7 Order of Reactions

The law of mass action helped to develop the concept of the *order of reactions*.

A *first-order reaction* is one in which a reaction rate is dependent on the concentration of a single reactant (Figure 1.10).

Thus, for a first-order reaction **A → P** occurring at constant temperature and pressure in a dilute solution the reaction rate v at any time t is given by

$$v = -\frac{d[P]}{dt} = +\frac{d[P]}{dt} = k[A] \tag{1.6}$$

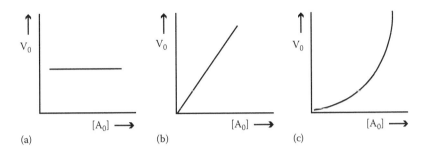

FIGURE 1.10 Graphs representing initial velocity against initial reactant concentration for single-reactant reactions (order of reactions). (a) Zero order, (b) first order, and (c) second order.

where v is the reaction rate at time t, k is the rate constant, [A] is the concentration of reactant A at time t, [P] is the concentration of product P at time t, $-\dfrac{d[P]}{dt}$ is the rate of decrease in [A], and $+\dfrac{d[P]}{dt}$ is the rate of increase in [P].

A *second-order reaction* involves two reactants and proceeds at a rate proportional to the concentration of the two reactants or the second power of a single reactant (Figure 1.10). Thus, for a second-order reaction $A + B \rightarrow P$ occurring at constant temperature and pressure in a dilute solution, the reaction rate v at any time t is given by

$$v = -\frac{d[A]}{dt} = -\frac{d[B]}{dt} = +\frac{d[P]}{dt} = k[A][B] \qquad (1.7)$$

For a second-order reaction $2A \rightarrow P$ the above equation may be rewritten as

$$v = -\frac{d[A]}{dt} = +\frac{d[P]}{dt} = k[A][A] = k[A]^2 \qquad (1.8)$$

However, if one of the reactants is in far greater excess than the other, a two-reactant reaction may be considered a *pseudo single-order reaction*.

Zero-order reactions are also possible. These are reactions in which the reaction rates are independent of the concentrations of any of the reactants involved.

1.7.8 Initial Velocity

The appearance of the product with time can be graphically represented as given in Figure 1.11. The reaction rate at any time is the slope of the curve at that time on the graph, which may be constant for a short while initially; this will later decrease with a decrease in concentration of the reactant(s) as the reaction proceeds, finally reducing to zero. At this point, all reactants have either been converted to products or a reaction equilibrium has been reached, wherein the rate of forward and backward reaction are equal.

Initial velocity (v_0) is the reaction rate when $t = 0$ and can be determined from the graph by drawing a tangent as shown in Figure 1.11. The units for v_0 are those used for product concentration divided by those used for time:

$$v_0 = \frac{[P]_2 - [P]_1}{t_2 - t_1} \qquad (1.9)$$

Initial velocity is a kinetic parameter determined for a reaction, especially when the conditions can be easily specified. Thus, it is evident that initial velocity is dependent on initial concentration of the

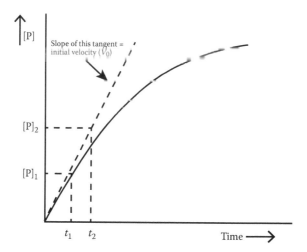

FIGURE 1.11 Graph of product concentration against time for a chemical reaction.

reactants. For a first-order reaction, that is, a single reactant reaction, $v = k$ [A]. Consequently, at time $t = 0$, $v_0 = k$ [A_0], where [A_0] is the initial concentration of A.

Accordingly, for a single-reactant second-order reaction, $v_0 = k$ [A_0]^2, and for a single-reactant zero-order reaction, $v_0 = k$ [A_0]^0.

This approach can be applied to reactions with more than one reactant. The overall order of the reactions is the sum total of the orders of each of the individual reactants in a reaction.

1.7.9 Catalysis Causes Reduction of Energy Barriers

Catalysis causes a change in the stable condition, which can be achieved by supplying energy. Catalysis occurs at active sites, during which process an energy barrier must be surmounted. To surmount the energy barrier to carry out a chemical reaction, enzymes that cause reduction of energy barriers may be used. Lowering of energy can be done with the help of an enzyme–substrate interaction that takes place at the active sites (Uhlig 1998). The active sites of multimeric enzymes are often located at the interface between the subunits and recruit residues from more than one monomer. The three-dimensional active site can shield substrate from solvent and facilitate catalysis.

1.7.9.1 Nature of Enzyme Catalysis

Enzymes behave like other catalysts and combine with reactants to form a transition state with lower free energy than that formed by an uncatalyzed reaction. Formation of enzyme–substrate complexes are not synonymous with transition states. In the case of a single-reactant reaction, the substrate binds with the specific substrate-binding site to form an enzyme–substrate complex; this process is preceded by the formation of the unstable transition state (transition state 1). The groups that react are held in close proximity with each other and with the catalytic site in the enzyme–substrate complex. The reaction proceeds further to form the enzyme–product complex via transition state 2. However, in circumstances in which the product may still be enzyme-bound, there would exist another transition state (3) before the product is released. The free energy profile of this type is depicted in Figure 1.12.

1.8 Kinetics of Enzyme-Catalyzed Reactions

Enzymes can increase the rate of a given reaction. Enzyme kinetics is primarily concerned with the measurement and mathematical description of reaction rates and their associated constants. It is primarily

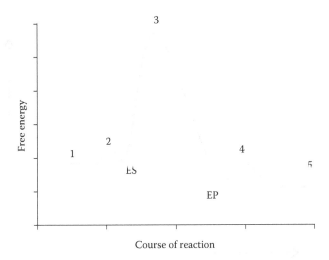

FIGURE 1.12 Free energy profile of an enzyme-catalyzed reaction involving the formation of an enzyme–substrate and enzyme–product complex. (1) Initial state, (2) transition state 1, (3) transition state 2, (4) transition state 3, and (5) final state.

applied to analyze data obtained from an enzymatic reaction and to use the data to optimize the reaction, and these data can be used for detailed characterization of the enzyme.

Both quantitative and qualitative enzyme kinetics are important to determine the minimum level of enzyme needed to carry out a chemical reaction efficiently in particular conditions in order to carry out economical and efficient conversion of food products. The enzyme kinetics of most industrial enzymes have been determined and are available online (www.brenda.uni-koeln.de).

Wilhelmy (1850) demonstrated that the rate of sucrose hydrolysis was proportional to its concentration. Brown (1902) demonstrated that, at low sucrose concentrations, the invertase catalyzed reaction was a first-order reaction, but at higher concentrations, it became zero order. This holds true for all single-substrate reactions. In the case of multisubstrate reactions, this applies only when all but one of the substrate concentrations is kept constant.

A graph of initial velocity, v_0, against initial substrate concentration ($[S_0]$) at constant enzyme concentration ($[E_0]$) was shown to be a rectangular hyperbola (Figure 1.13).

This was first explained by Brown (1902) pertaining to sucrose hydrolysis.

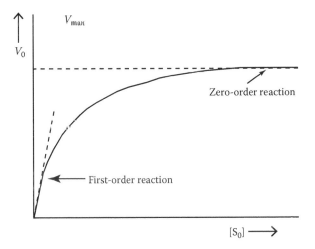

FIGURE 1.13 Graph of initial velocity against initial substrate concentration.

The general equation to explain the graph is

$$v_0 = \frac{V \, [S_0]}{[S_0] + b} \tag{1.10}$$

where V_{max} is maximum value of v_0, and b is the value of $[S_0]$ when $v_0 = 1/2 \, V_{max}$.

Kinetic models to explain this were studied by both Henri (1903) and Michaelis-Menten (1913). Not all enzymes give hyperbolic curves. However, it is characteristic of all enzymes to attain maximum initial velocity with increasing substrate concentration when the enzyme concentration is kept constant. Sufficient increase in substrate concentration will cause enzyme saturation. When this happens, all enzymes will be substrate bound, and there will not be any free enzyme.

Then, $[ES] = [E_0]$, and therefore, the reaction rate will be $k_2 \, [E_0]$.

This is independent of substrate concentration and will therefore not increase with further increase in substrate concentration. Hence, $V_{max} = k_2 \, [E_0]$.

However, in low concentrations of substrate, the enzyme will not be saturated. Therefore, the overall reaction rate will be limited by the rate of formation of the enzyme–substrate complex. When enzyme concentration is constant, the reaction rate is proportional to $[S_0]$ and therefore results in the first-order reaction. Rates of reactions are affected by factors such as enzymes/catalysts, substrates, effectors, temperature, and pH.

1.8.1 Michaelis-Menten Equation

First, the enzyme (E) and the substrate (S) react reversibly and quickly form a noncovalent ES complex:

$$E + S \underset{k_{-1}}{\overset{k_1}{\rightleftarrows}} ES \tag{1.11}$$

Second, the ES complex undergoes a chemical transformation and dissociates to give product (P) and enzyme (E).

$$ES \overset{k_2}{\rightarrow} E + P \tag{1.12}$$

Many enzymatic reactions follow Michaelis-Menten kinetics. The Michaelis-Menten model is a simple model to explain single-substrate enzyme-catalyzed reactions involving the formation of a single intermediate. However, not all enzyme mechanisms are as simple and are always more complicated and can involve more than one substrate or more than one substrate-binding site to name a few variations.

The Michaelis-Menten model is based on several assumptions. It is assumed that the first step is fast and is always at equilibrium.

The catalytic step (ES → E + P, *or* ES → EP) is the rate-limiting step, and therefore, k_1 and k_{-1} are much greater than k_2 (i.e., $k_1, k_{-1} \gg k_2$).

Under these circumstances, Michaelis-Menten equilibrium assumption is valid, and $K_m \cong K_s$.

K_s gives the affinity of the enzyme for the substrate; a low K_s indicates high affinity, and a high K_s indicates a low affinity of the enzyme for the substrate.

Second, the system is assumed to be in a steady state, that is, $d[ES]/dt \approx 0$. It is also assumed that there is only a single reaction or dissociation step (i.e., $k_2 = k_{cat}$). The other assumption of the Michaelis-Menten model is that the initial substrate concentration $[S_0]$ is much greater than the initial enzyme concentration $[E_0]$; hence, the formation of the enzyme–substrate complex will not result in a significant change in the free substrate concentration. Thus, $S_{Tot} = [S_0] + [ES]$, $[S_0] \approx [S]$. Another assumption is that there is no back reaction of P to ES, and initial velocities are measured when $[P] \approx 0$.

1.8.1.1 Briggs-Haldane Modification of Michaelis-Menten Equation

Briggs-Haldane (1925) introduced a more general assumption, such as *steady state*, to the Michaelis-Menten equation. According to them, initially, because the concentration of the enzyme and, therefore, that of the enzyme–substrate complex, was very small compared to the substrate concentration, the rate of change of [ES] would be much less than the rate of change of [P].

Thus, for the reaction,

$$E + S \underset{k_{-1}}{\overset{k_1}{\rightleftharpoons}} ES \overset{k_2}{\rightarrow} E + P \tag{1.13}$$

The rate of formation of ES at time t (initially, product concentration is negligible) $= k_1[E][S]$.

The rate of breakdown of ES at this time $= k_{-1}[ES] + k_2[ES]$ (ES can either form products or reform the reactants).

Using the steady-state assumption,

$$k_1 [E][S] = k_{-1} [ES] + k_2 [ES] = [ES] (k_{-1} + k_2) \tag{1.14}$$

Separating constants from variables, we get

$$\frac{[E][S]}{[ES]} = \frac{k_{-1} + k_2}{k_1} = K_m \tag{1.15}$$

where K_m is another constant.

Substituting $[E] = [E_0] - [ES]$, we get

$$\frac{([E_0] - [ES])[S]}{[ES]} = K_m \tag{1.16}$$

From which we get

$$[ES] = \frac{[E_0][ES]}{[S] + K_m} \tag{1.17}$$

Because $v_0 = k_2[ES]$,

$$v_0 = \frac{k_2[E_0][ES]}{[S] + K_m} \tag{1.18}$$

Because $V_{max} = k_2[E_0]$ and $[S] > [E]$, $[S] \approx [S_0]$, so

$$v_0 = \frac{k_2[E_0][S_0]}{[S_0] + K_m} \text{ at constant } [E_0] \tag{1.19}$$

and hence,

$$v_0 = \frac{V_{max}[S_0]}{[S_0] + K_m} \text{ at constant } [E_0] \tag{1.20}$$

This equation has retained the same form as that of the Michaelis-Menten equation as well as the name, and K_m is called the Michaelis-Menten constant.

A graph of v_0 against $[S_0]$ at constant $[E_0]$ is shown in Figure 1.12. K_m can be obtained from the graph. When $v_0 = 1/2\ V_{max}$, then

$$\frac{V_{max}}{2} = \frac{V_{max}[S_0]}{[S_0] + K_m}$$

(1.21)

Therefore,

$$V_{max}\,([S_0] + K_m) = 2\,(V_{max})\,[S_0]$$

(1.22)

So $K_m = [S_0]$.

1.8.1.2 Significance of Michaelis-Menten Equation

This equation and its Briggs-Haldane modification regard single-substrate enzyme-catalyzed reactions with a single substrate-binding site.

1.8.1.3 Michaelis-Menten Constant (K_m)

K_m is also called the Michaelis-Menten constant. It is the apparent dissociation constant of the enzyme–substrate complex. K_m is the substrate concentration $[S_0]$ required to reach half maximum velocity (i.e., when $v_0 = 1/2\ V_{max}$). Under circumstances in which k_1 and k_{-1} are much greater than k_2 (i.e., k_1, $k_{-1} \gg k_2$), the Michaelis-Menten equilibrium assumption is valid, and $K_m \cong K_s$, is used to describe the enzyme affinity for the substrate.

1.8.1.4 Turnover Number (k_{cat})

The constant k_{cat} is called the turnover number. This is applied to simple single-substrate enzyme-catalyzed reactions. It can be obtained from the expression $V_{max} = k_2[E_0]$. The turnover number indicates the maximum number of substrate molecules that can be converted to product per molecule of enzyme per unit of time. The turnover number ranges from 1 to 10^4 per second for most enzymes.

1.8.1.5 Catalytic Efficiency (k_{cat}/K_m)

The term k_{cat}/K_m is used to denote the catalytic efficiency of an enzyme and is used to rank them. A high value of k_{cat}/K_m (close to that of k_1) indicates that the substrate is bound tightly by an enzyme and the frequency of collisions between the enzyme and substrate is the limiting factor. A low value supports the equilibrium assumption; k_{cat}/K_m is a measure of the enzyme specificity.

1.8.2 Lineweaver-Burk Plot

The Michaelis-Menten graph of v_0 against $[S_0]$ is not able to determine K_m and V_{max} satisfactorily as v_0 becomes V_{max} only at infinite substrate concentrations. Hence, from the Michaelis-Menten plot, the determination of an accurate value for V_{max}, and hence K_m, is difficult. Besides, as the plot is a curve, extrapolation from the values of v_0 at nonsaturating concentrations is inaccurate. This problem was circumvented by Linweaver and Burk (1934) by simply inverting the Michaelis-Menten equation:

$$\frac{1}{v_0} = \frac{[S_0] + K_m}{V_{max}[S_0]} = \frac{[S_0]}{V_{max}[S_0]} + \frac{K_m}{V_{max}[S_0]}$$

(1.23)

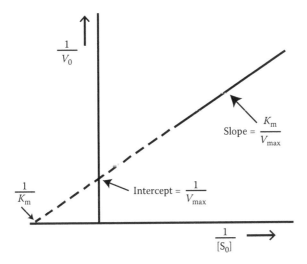

FIGURE 1.14   Lineweaver-Burk plot.

Therefore,

$$\frac{1}{v_0} = \frac{K_m}{V_{max}} \frac{1}{[S_0]} + \frac{1}{V_{max}} \tag{1.24}$$

(Lineweaver-Bulk equation).

This is in the form of $y = mx + c$, which is an equation for a straight-line graph on which the plot of y against x gives a slope m with an intercept c on the y-axis.

Because the graph (Figure 1.14) of $1/v_0$ against $1/[S_0]$ is linear for systems obeying the Michaelis-Menten equation, it can be extrapolated to determine the values of V_{max} and K_m.

1.8.3 Eadie-Hoftree and Hanes Plots

In the Eadie-Hoftree plot (Figure 1.15a), based on the Lineweaver-Burk equation, (which, in turn, is based on the Michaelis-Menten equation), both sides of the equation are multiplied by the factor

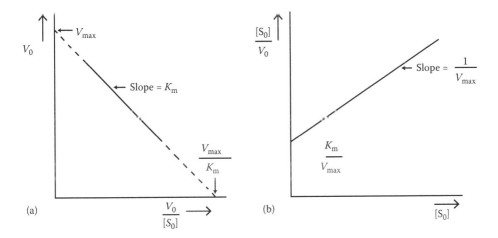

FIGURE 1.15   (a) Eadie-Hoftree plot, (b) Hanes plots.

v_0 V_{max} to give the Eadie-Hoftree equation, which is a straight line from which V_{max} and K_m can be determined.

$$v_0 = -K_m \frac{v_0}{[S_0]} + V_{max} \tag{1.25}$$

The Hanes plot (Figure 1.15b) also starts with the Lineweaver-Burk equation, but here it is multiplied by $[S_0]$ to obtain a linear plot from which both V_{max} and K_m can be determined.

$$\frac{[S_0]}{v_0} = \frac{1}{V_{max}}[S_0] + \frac{K_m}{V_{max}} \tag{1.26}$$

1.9 Enzyme Inhibition

Enzyme activity can be inhibited by several substances, small molecules, or ions that bind with the enzyme. Inhibitors are substances that decrease the rate of enzyme-catalyzed reactions. These can act either on the substrate or the coenzyme or combine directly with the enzyme. The inhibitors that directly interact with the enzyme are of two types: reversible inhibitors that bind enzymes reversibly and can be removed by dialysis to restore enzyme action and irreversible inhibitors that cannot be removed by dialysis.

In reversible inhibition, the dissociation of the enzyme–inhibitor (EI) complex is much faster than in irreversible inhibition as there are no covalent interactions between inhibitor and enzyme. When the inhibitors bind the enzyme, due to the close resemblance to the substrate, and competes for the same binding site as that of the substrate, it is called competitive inhibition. The inhibitor prevents the substrate from binding with the active site. A competitive inhibitor competes with the substrate for the binding site, whereby enzyme activity is reduced due to decreased enzyme-bound substrate. The competitive inhibition can be overcome by increasing the concentration of the substrate. An example of a competitive inhibitor is malonate, which is a structural analogue for succinate and inhibits succinate dehydrogenase activity.

Competitive inhibition is dependent not only upon inhibitor and substrate concentration, but also on their relative affinities for the binding site. Consequently, the degree of inhibition is greater when the inhibitor concentration is greater than that of the substrate. Alternatively, when the substrate concentration is high, the inhibitor competes with the substrate for available binding sites, thereby reducing the extent of inhibition. When substrate concentrations are much higher, the inhibitor molecules are outnumbered by the substrate molecules, leading to insignificant inhibition. Therefore, V_{max} remains unchanged. However, the apparent K_m clearly increases due to the inhibition and is termed K_m' (Figure 1.16).

In the presence of competitive inhibitor I, the Michaelis-Menten equation is

$$v_0 = \frac{V_{max}[S_0]}{[S_0] + K_{m\left(1+\frac{[I_0]}{K_i}\right)}} \tag{1.27}$$

where K_i is the inhibitor constant.

Here, V_{max} remains unaltered, but K_m is altered and becomes K_m'

$$K_m' = K_{m\left(1+\frac{[I_0]}{K_i}\right)} \tag{1.28}$$

where K_m' is the apparent K_m in the presence of initial concentration of competitive inhibitor $[I_0]$. For competitive inhibition, the Lineweaver-Burk equation is given as

$$\frac{1}{v_0} = \frac{K_m'}{V_{max}} \frac{1}{[S_0]} + \frac{1}{V_{max}} \tag{1.29}$$

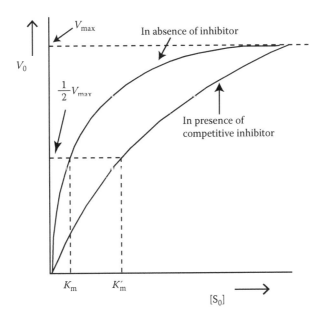

FIGURE 1.16 Michaelis-Menten plot in the presence of a competitive inhibitor.

The Lineweaver-Burk plot showing the effect of competitive inhibitor on enzyme action is given in Figure 1.17.

Although competitive inhibitors bind free enzymes and form the enzyme–inhibitor complex, uncompetitive inhibitors bind only to the enzyme–substrate complex. Here, the inhibitor does not compete for the substrate-binding site but binds at a completely different site. Hence, the inhibition cannot be overcome by increasing the substrate concentration. Both V_{max} and K_m are altered. The Michaelis-Menten form of the equation for uncompetitive inhibition is

$$v_0 = \frac{\dfrac{V_{max}}{\left(1 + \dfrac{[I_0]}{K_i}\right)}[S_0]}{[S_0] + \dfrac{K_m}{\left(1 + \dfrac{[I_0]}{K_i}\right)}} \tag{1.30}$$

where

$$V'_{max} = \frac{V_{max}}{\left(1 + \dfrac{[I_0]}{K_i}\right)} \tag{1.31}$$

and

$$K'_m = \frac{K_m}{\left(1 + \dfrac{[I_0]}{K_i}\right)} \tag{1.32}$$

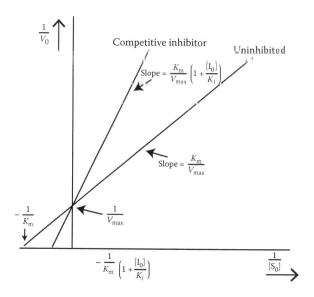

FIGURE 1.17   Lineweaver-Burk plot in the presence of a competitive inhibitor.

where V'_{max} is the value of V_{max} and K'_m is the apparent K_m in the presence of $[I_0]$ concentration of the uncompetitive inhibitor.

The Lineweaver-Burk equation for uncompetitive inhibition is

$$\frac{1}{v_0} = \frac{K'_m}{V'_{max}} \frac{1}{[S_0]} + \frac{1}{V_{max}} \tag{1.33}$$

The Lineweaver-Burk plot showing the effect of the uncompetitive inhibitor on enzyme action is given in Figure 1.18.

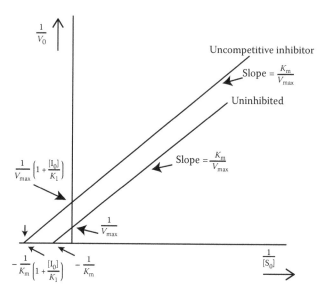

FIGURE 1.18   Lineweaver-Burk plot showing the effect of uncompetitive inhibitor on enzyme action.

In noncompetitive inhibition, there is no competition for the substrate-binding site as the inhibitor molecule binds to the enzyme regardless of the binding of the substrate; that is, the binding sites for the substrate and inhibitor molecules are different. The enzyme turnover number is decreased in this case, and increasing the concentration of substrate molecules can reverse inhibition.

1.10 Factors Affecting Enzyme Activity

Enzyme stability is an important factor to be considered in the application of enzymes in the food industry. The stability and activity of enzymes is influenced by their inherent physical stability, presence of inhibitors/poisons/antagonists in the food/reacting mixture and physicochemical factors such as pH and temperature. Therefore, most enzymes exhibit maximal activity at their optimal conditions that influence enzyme activity. Optimum conditions of enzymes are those favorable conditions that allow them to perform efficiently. Most influencing physicochemical conditions for enzymes are pH, temperature, substrate concentration, and enzyme concentration.

All enzymes exhibit an optimum pH, temperature, and substrate concentration at which they demonstrate maximum activity (Figure 1.19), and they behave according to well-established rules. All enzymes have a specific pH value or pH range for optimal activity (Table 1.2). A change in pH can affect enzymatic activities, and extremely low or high pH values can cause deformity in the structure of enzymes, which, in turn, result in loss of action for most enzymes. The pH value at which an enzyme shows its maximum activity is called its optimum pH, which is one of the important criteria in its application in the food industry. This can be exemplified by the enzymes involved in the brewing industry as a majority of these enzymes have an optimum pH in the range of 3.0–7.0 (Whitehurst and Law 2002).

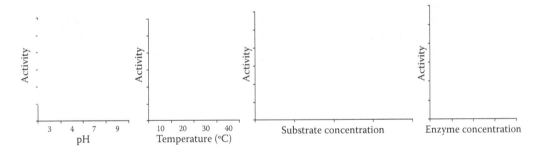

FIGURE 1.19 Effect of temperature, pH, enzyme concentration, and substrate concentration on the initial rate of reactions.

TABLE 1.2

Optimum pH and Temperature Range of Microbial-Derived Enzymes Widely Used in the Food Industry

Sl. No.	Enzyme	Optimum pH Range	Optimum Temperature (°C)	References
1	α-Amylase from *B. licheniformis*	6.0–11.0	76	Saito 1973
2	α-Amylase from *B. stearothermophilus*	4.0–5.2	80	Srivastava and Baruah 1986
3	α-Amylase from *A. oryzae*	5.0	50	Patel et al. 2005
4	Lipase from *Aspergillus nidulans* WG312 strain	6.5	40	Mayordomo et al. 2000
6	*Substilisin* Carlsberg from *B. licheniformis*	8–9		Aunstrup 1980
7	Alkaline protrease from *Aspergillus oryzae*	7.4		Bergkvist 1963
9	Trypsin	5–11	30–60	Kristjansson 1991
10	β-Galactosidase from yeast	7.0	30	Itoh et al. 1982

As with pH, every enzyme has its optimum range of temperature, and enzyme performance usually improves with increasing temperature. Enzymatic reactions increase with every 10°C temperature increase, and this may be applicable up to 10°C–40°C. Some enzymes may be active at low temperatures. Beyond the optimum temperature, enzymes may be denatured, and optimal temperature is therefore another important parameter in the food industry. A wide range of enzymes with various temperature ranges are needed for various purposes in the food industry.

Carrying out enzymatic reactions efficiently requires an optimum level of substrate concentration. The substrate should saturate the enzyme. At a constant enzyme concentration, enzyme activity increases with substrate concentration, and the substrate concentration value at which the activity of the enzyme is found to be maximum is called the optimum substrate concentration, and a further increase in substrate concentration will not cause any elevation in the enzymatic activity. Likewise, each enzyme has its own optimum value for enzyme concentration. In order to carry out reproducible and profitable usage of an enzyme in the food industry, all these physicochemical parameters should be optimized. Performance of enzymes is optimum in aqueous solutions as they must have constant and unimpaired contact for maximal enzymatic activity.

Inhibitors are substances that decrease the rate of enzyme-catalyzed reactions. These can act on either the substrate or the coenzyme or combine directly with the enzyme. Enzyme inhibitors are of two types: (i) reversible inhibitors that bind enzymes reversibly and can be removed by dialysis to restore enzyme action, and (ii) irreversible inhibitors that cannot be removed by dialysis.

1.10.1 Role of Prosthetic Groups, Cofactors, and Coenzymes

Catalysis can be promoted by small nonprotein molecules and metal ions that participate directly in substrate binding, termed "prosthetic groups," "cofactors," and "coenzymes." They are distinguished by their mode of action and binding strength (Murray et al. 2009).

Prosthetic groups are tightly integrated with enzyme structure by covalent or noncovalent forces. Metals, including Co, Cu, Mg, Mn, Se, and Zn, are the most common prosthetic groups, and enzymes containing tightly bound metal ions are called "metalloenzymes." The biotin and flavin dinucleotide are considered to be prosthetic groups other than metal ions. They facilitate the binding and orientation of substrates or the formation of covalent bonds with reaction intermediates, or they interact with substrates to make them more susceptible to chemical reaction. *Cofactors* serve the same function as that of prosthetic groups, but the binding is not as tight as in prosthetic groups. Unlike metalloenzymes, enzymes requiring a metal cofactor are termed "metal-activated enzymes." In the case of cofactors, they should be in the medium surrounding the substrate and enzyme in order to carry out catalytic reactions.

Coenzymes are mainly meant for transfer of substrates from the point of origin to the point of utilization, for example, coenzyme A, tetrahydrofolate.

1.11 Sources of Enzymes in the Food Industry

Enzyme preparations used in food industries are obtained from animal, plant, or microbial sources and may consist of whole cells, parts of cells, or cell-free extracts of the source used. They may contain one or more active components as well as carriers, solvents, preservatives, antioxidants, and other substances consistent with good manufacturing practice. They may be available in liquid, semiliquid, dry, or an immobilized form (www.fao.org). Various enzymes, their sources, and their applications in the food industry are detailed in Table 1.3.

Because enzymes are present in almost all raw food material, their high substrate specificity, rapid reaction rates, and ability to work effectively at relatively low temperatures make them ideal for food processing. Plants, animals, and microorganisms are the primary sources of enzymes in the food industry. For instance, intrinsic enzymes found in plant materials such as amylases and pectinases have a significant role in bread making, beer brewing, and fruit juice production. Plant-based proteases such as papain, bromelain, and ficin find wider application in meat tenderization and hydrolysis of animal, milk, and plant proteins. These proteases are characterized by the sulfhydryl group in the active site, responsible

TABLE 1.3

Sources of Major Food Enzymes

Enzymes	Source	References
Glucose oxidase	**Fungi**	
	Aspergillus niger	Kona et al. 2001
	Penicillium notatum	Keilin and Hartree 1948
Sulphydryl oxidase	**Fungi**	
	Aspergillus niger	De la Motte and Wagner 1987
Catalase	**Animal**	
	Liver of *Bos taurus*	Uhlig 1998
	Fungi	
	Aspergillus niger	Fiedurek and Gromada 2000
Lactoperoxidase	**Animal**	
	Cheese whey: bovine colostrums	Kussendrager and Hooijdonk et al. 2000
Lipoxygenase	**Plant**	
	Soy bean	Brash 1999
Laccases	**Plant**	
	Rhus vernifera	Yoshida 1883
	Fungi	
	Agaricus brunnescens	Fagan and Fergus 1984
	Aspergillus nidulans	Aramayo and Timberlake 1990
	Bacteria	
	Azospirillum lipoferum	Givaudan et al. 1993
Hexose oxidase	**Red algae**	
	Chondrus crispus	Sullivan and Ikawa 1973
Cyclodextrin	**Bacteria**	
	Bacillus sp.	Tonkova 1998
Transglutaminase	**Bacteria**	
	Streptoverticillium sp.	Ando et al. 1989
	Animal	
	Muscles of atka mackerel, botan shrimp, carps, rainbow trout, and scallop	Fernandes 2010
Lipase	**Animal**	
	Pig pancreas, gullet of goat and lamb, calf abomasums	Uhlig 1988
	Fungi	
	Penicillium roqueforti	
	Rhizomucor miehei	Herrgard et al. 2000
	Bacteria	Macedo et al. 2003
	Bacillus subtilis	
		Ruiz et al. 2005
Phospholipase A, lechithinase A	**Animal**	
	Pancreatic glands of porcine	Uhlig 1988
	Bacteria	
	Bacillus mycoides strain 970	Chang et al. 2010
Pectinesterase	**Plants**	
	Carica papaya, *Lycopersicum esculentum*	Fayyaz et al. 1993; Warrilow et al. 1994
	Fungi	
	Aspergillus niger	Maldonaldo and de Saad 1998
Tannase	**Bacteria**	
	Bacillus sphaericus	Raghuwanshi et al. 2011

(Continued)

TABLE 1.3 (CONTINUED)

Sources of Major Food Enzymes

Enzymes	Source	References
	Bacillus cereus	Mondal et al. 2001
	Lactobacillus plantarum	Ayed and Hamdi 2002
	Fungi	
	Aspergillus niger	Barthomeuf et al. 1994
	Aspergillus awamori	Beena et al. 2010, 2011
α-Amylase, diastase	**Animal**	
	Hog or pig pancreas	Uhlig 1998
	Fungi	
	Aspergillus sp.	Kvesitadze et al. 1978
	Bacteria	
	Bacillus stearothemophilus	Srivastava and Baruah 1986
	Bacillus licheniformis	Madsen et al. 1973
	Bacillus amyloliquefaciens	Pazur and Okada 1966
β-Amylase	**Plant**	
	Sweet potato	Laurière et al. 1992
	Cereal seeds	
	Fungi	
	Aspergillus sp.	Abe et al. 1988
Amyloglucosidase	**Fungi**	
	Aspergillus niger	Fogarty and Benson 1983;
	Aspergillus awamori	Bertolin et al. 2003
Cellulase	**Fungi**	
	Aspergillus niger	Kang et al. 2004
	Trichoderma sp.	Cheol et al. 2003
Inulinase	**Fungi**	
	Aspergillur sp.	Gupta et al. 1994; Viswanathan and Kulkarni 1996
	Penicillium sp.	Viswanathan and Kulkarni 1996
	Cladosporium sp.	Viswanathan and Kulkarni 1996
Xylanase	**Fungi**	
	Aspergillus sp.	Gawande and Kamat 1998
	Trichoderma sp.	Cheol et al. 2003; Royer and Nakas 1989
	Bacteria	
	Bacillus thermophilus	Khasin et al. 1993
Pectinase, endopolygalacturonase	**Fungi**	Said et al. 1991
	Penicillium frequentans	
	Saccharomyces cerevesia	Blanco et al. 1994
	Bacteria	
	Bacillus sp.	Kelly and Fogarty 1978
Lysozyme, muramidase	**Animal**	
	Hen egg white	Whitehurst and Law 2002
β-Galactosidase, lactase	**Bacteria**	
	Escherichia coli	Hall and Hartl 1974
	Fungi	
	Aspergillus sp.	Park et al. 1979
Invertase, saccharase, glucosucrase	**Fungi**	
	Neurospora crassa	Braymer et al. 1971

(Continued)

TABLE 1.3 (CONTINUED)

Sources of Major Food Enzymes

Enzymes	Source	References
Pentosanase	**Fungi**	
	Alternaria sp.	Simpson 1954
	Aspergillus sp.	Simpson 1954
	Fusarium sp.	Simpson 1954
	Trichothecium sp.	Simpson 1954
	Trichoderma sp.	Simpson 1954
	Bacteria	
	Bacillus sp.	Simpson 1954
Naringinase	**Fungi**	
	Aspergillus niger	Puri and Kalra 2005
Pullulanase	**Bacteria**	
	Bacillus acidopollulyticus	Martin and Birgitte 1984
Aminopeptidases	**Animal**	
	Pig kidney	Delange and Smith 1971
Trypsin	**Animal**	
	Bovine/porcine pancreas	Keil 1971
Subtilisin	**Bacteria**	
	Bacillus subtilis	Ottesen and Spector 1960
	Bacillus amyloliquefaciens	Peng et al. 2003
Papain	**Plant**	
	Latex of *Carica papaya*	Uhlig 1998
Ficin	**Plant**	
	Latex of *Ficus carica*	Whitaker 1957
	Latex of *Ficus glabrata*	
Bromelain	**Plant**	
	Pineapples such as *Ananas comosus* and *Ananus bracteatus*	Uhlig 1998
Pepsin A	**Animal**	
	Bovine abomasums	Fruton 1971
	Arctic capelin	Fernandes 2010
	Atlantic cod	Fernandes 2010
	Spheciospongia vesperiam	
Chymosin, rennin	**Animal**	
	Calf abomasums	Cheeseman 1981
Asparaginase	**Plants**	
	Sphagnum fallax	Yadav et al. 2014
	Lupine araboreuse	Borek et al. 2004
	Lupin amgustplius	Borek et al. 2004
	Bacteria	
	Pseudomonas flourescence	Mardashev et al. 1975
	Fungi	
	Aspergillus oryzae	Hendriksen et al. 2009
L-Glutaminase	**Bacteria**	
	Bacillus circulans	Kikuchi et al. 1971
	Fungi	
	Saccharomyces cerevisiae	Soberon and Gonzalez 1987

(Continued)

TABLE 1.3 (CONTINUED)

Sources of Major Food Enzymes

Enzymes	Source	References
Urease	**Bacteria**	
	Lactobacillus reureri	Kakimoto et al. 1989
	Klebsiella aerogenes	Mulrooney et al. 2005
	Fungi	
	Rhizopus oryzae	Geweely 2006
Acetolactate	**Bacteria**	
decarboxylase	*Lactobacillus casei DSM 2547*	Godtfredsen et al. 1984
Pectin lyase	**Bacteria**	
	Penicillium italicum	Alaña et al. 1990
	Bacillus sp. *DT 7*	Kashyap et al. 2000
Glucose isomerase	**Bacteria**	
	Pseudomonas sp.	Marshall and Kooi 1957
	Lactobacillus sp.	Kent and Emery 1974
	Actinomycetes	
	Streptomyces sp.	Tsumura and Sato 1965

for their catalytic activity. Similarly, animal-based proteases such as chymosin from calf abomasums, trypsin from bovine and porcine pancreas, pepsin from bovine abomasums, and lysozyme from hen egg white have a role in the food manufacturing industry. Animal-based lipases extracted from the gullet of goat and lamb, calf abomasums, and pig pancreas were mainly used in the past for flavor enhancement in the cheese-making industry. Catalases and lactoperoxidases derived from animal sources are known to prevent food spoilage as they have antimicrobial activity. However, it is important that animal tissues used for the preparation of enzymes comply with meat inspection requirements and good hygienic practice. Similarly, plant materials used in the production of enzyme preparations must not contribute to health problems in the processed finished food under normal conditions of use. However, a majority of animal- and plant-derived enzymes cannot survive the extreme conditions of food processing. In spite of the fact that the animal- and plant-based enzymes are widely acceptable in the food industry, the use of microbial enzymes is also desired and preferred because they have more advantages due to the higher specificity and stability of the enzymes produced. For example, rennet of microbial origin replaced the use of animal rennet to curdle milk in the cheese-manufacturing industry with one third of rennet used in cheese manufacturing being of microbial origin in recent years.

Sources of microbial enzyme preparation may vary from native strains or variants of microorganisms or be derived by genetic modification. The Joint FAO/WHO Expert Committee on Food Additives (JECFA) is an international expert scientific committee that is administered jointly by the Food and Agriculture Organization of the United Nations (FAO) and the World Health Organization (WHO). It gives general specifications and considerations for enzyme preparations used in food processing, which are available at http://www.fao.org/ag/agn/jecfa-additives/docs/enzymes_en.htm. According to them, the production strains for food enzyme preparations should be nonpathogenic and nontoxigenic. Care should be taken in the case of fungal enzyme preparation to avoid mycotoxin contamination. Enzyme preparations by commercial agencies are produced in accordance with good food manufacturing practice. Care must be taken to ensure that the carriers, diluents, supports, and other additives and ingredients (including processing aids) used in the production, distribution, and application of enzyme preparations are acceptable for the relevant food uses of the enzyme preparations concerned, or that they are insoluble in food and removed from the food material after processing (www.fao.org).

Among the commercially available microbial enzymes, fungus-derived enzymes are significant. More than 240 enzymes are listed as commercialized by the Association of Manufacturers and Formulators of Enzyme Products (AMFEP; http://www.amfep.org/). Approximately 55% of the commercial enzymes are of fungal origin. More than 25% of all industrial enzymes are from *Aspergillus*, followed by *Penicillium* and *Rhizopus*. Fungi are heterotrophic organisms, they absorb only smaller molecules through their cell wall, and extracellular digestion of complex organic matters is required. They secrete a complex battery of extracellular enzymes for extracellular digestion, and these exo-enzymes have evolved naturally under harsh conditions, making them ideal candidates for industrial biocatalysts. The solid-state fermentation is more suitable for fungal production of enzymes, making downstream processing easy and more economical. As with fungal enzymes, both natural and recombinant bacterial enzymes are widely used in the food industry. The genus *Bacillus* is dominant among the bacterial enzyme producers in the list of enzyme producers for commercialized enzymes by AMFEP. Nevertheless, scientists worldwide continue to search for novel potential biocatalysts especially from extreme environments. There are reports of enzymes from extremophiles, which are often difficult to grow under typical laboratory conditions. The advancement of recombinant DNA technology aids in raising recombinant microorganisms using mesophiles as hosts with which the genes of interest from extremophiles have been expressed (Fujiwara 2002). There are several strategies available for screening novel genes of potential biocatalysts, which include sequence-based and activity-based metagenomic and metatranscriptomics approaches. There has been remarkable progress in this field of research due to the recent advances of next generation sequencing (NGS) technologies (Schuster 2007).

Plants, animals, and microorganisms from terrestrial environments have been the routine source for potent industrial and food enzymes over the years. However, extreme environments and marine environments hold promise as potential sources of new enzymes with unique and desirable properties for application in food industries. In this context, it must be noted that the marine environment constituting 75% of the earth is the least explored environment for novel biocatalysts. In marine environments, including hydrothermal vents, deep sea sediments, salt marshes, coral reefs, giant kelps, and estuaries, as all life forms are subject to perpetual competition and stress, it is not surprising that the organisms living in these environments produce an enormous range of biocatalysts. Thermo-stable proteases, lipases, esterases, amylases, and xylanases have been actively sought and, in many cases, were found in bacterial and archaeal hyperthermophilic marine microorganisms. The higher marine organisms are also known as potent food enzyme producers (Fernandes 2010). Muscles of atka mackerel, botan shrimp, carp (*Cyprinus carpio*), rainbow trout, and scallop can be exploited for the production of trans-glutaminase. From higher marine organisms, enzymes such as proteases from mud crab (*Scylla serrata*) and sardine orange roughy (*Hoplostethus atlanticus*); amylases from gilt-head (sea) bream, turbot, and deepwater redfish; chymotrypsin from Atlantic cod, crayfish, and white shrimp; and pepsin from Arctic capelin, Atlantic cod, and marine sponges (*Spheciospongia vesperiam*) were reported (Fernandes 2010).

1.12 Conclusion

In the food industry, enzyme substitutes are not as successful as enzymes, which show unique properties such as supreme specificity; the ability to operate under mild conditions of pH, temperature, and pressure while displaying high activity and turnover numbers; and the property of biodegradability, as enzymes are biostatic and more or less biological in origin, which makes them ideal candidates for application in the food industry (Bommarius and Riebel-Bommarius 2004; Illanes 2008). A further significant property of enzymes that makes them suitable for industry is that very small amounts are needed for the bioconversion of a large amount of substrates. They have been used to modify flavor, texture, appearance, and storage stability of foods in the food industry. With the development of enzyme technology, specific enzymes that mediate such effects have been identified, isolated, and characterized. Apart from that, a large number of intrinsic enzymes in plants and the genes coding them, mediating flavor generation, tissue softening, color production, and other qualities that makes food acceptability, have been identified. Thus, enhancement of desirable properties or delay in the development of undesirable properties in foods is available for food processors. The advancement in enzyme protein engineering, recombinant enzyme production, and site-directed mutagenesis enable the introduction of potential biocatalysts in the food industry. Although a search for novel enzymes with unique properties are being pursued through intensive screening of natural sources, biotechnology and enzyme engineering hold immense promise for tailor-making potential enzymes for use in food and beverage processing.

REFERENCES

Abe, J. I., Bergmann, F. W., Obata, K. and Hizukuri, S. 1988. Production of the raw-starch digesting amylase of *Aspergillus* sp. K-27. *Applied Microbiology and Biotechnology* 27(5–6): 447–450.

Alaña, A., Alkorta, I., Dominguez, J. B., Llama, M. J. and Serraet Al, J. L. 1990. Pectin lyase activity in a *Penicillium italicum* strain. *Applied and Environmental Microbiology* 56(12): 3755–3759.

Ando, H., Adachi, M., Umeda, K., Matsuura, A., Nonaka, M. and Uchio, R. 1989. Purification and characteristics of a novel transglutaminase derived from microorganisms. *Agricultural and Biological Chemistry* 53(10): 2613–2617.

Aramayo, R. and Timberlake, W. E. 1990. Sequence and molecular struture of the *Aspergilus nidulans* yA (laccase I) gene. *Nucleic Acids Research* 18(11): 3415.

Atkins, P. and De Paula, J. 2006. *Physical Chemistry, Volume 1.* W. H. Freeman & Co., New York.

Aunstrup, K. 1980. Proteinases. In: *Microbial Enzymes and Bio Conversions Economics Microbiology*, ed. Rose, A. H. 5: 50–114. Academic Press Ltd, New York.

Ayed, L. and Hamdi, M. 2002. Culture conditions of tannase production by *Lactobacillus plantarum*. *Biotechnology Letters* 24(21): 1763–1765.

Barthomeuf, C., Regerat, F. and Pourrat, H. 1994. Production, purification and characterization of a Tannase from *Aspergillus niger* LCF 8. *Journal of Fermentation and Bioengineering* 77(3): 320–323.

Beena, P. S., Soorej, M. B. M., Elyas, K. K., Sarita, G. B. and Chandrasekaran, M. 2010. Acidophilic tannase from marine *Aspergillus awamori* BTMFW032. *Journal of Microbiology and Biotechnology* 20(10): 1403–1414.

Beena, P. S., Basheer, S. M., Bhat, S. G., Bahkali, A. H. and Chandrasekaran, M. 2011. Propyl gallate synthesis using acidophilic tannase and simultaneous production of tannase and gallic acid by marine *Aspergillus awamori* BTMFW032. *Applied Biochemistry and Biotechnology* 164(5): 612–628.

Bergkvist, R. O. L. F. 1963. The proteolytic enzymes of *Aspergillus oryzae*. *Acta Chemica Scandinavica* 17: 8.

Bertolin, T. E., Schmidell, W., Maiorano, A. E., Casara, J. and Costa, J. A. V. 2003. Influence of carbon, nitrogen and phosphorous sources on glucoamylase production by *Aspergillus awamori* in solid state fermentation. *Zeitschrift Fur Naturforschung C* 58(9/10): 708–712.

Bhat, M. K. 2000. Cellulases and related enzymes in biotechnology. *Biotechnology Advances* 18(5): 355–383.

Blanco, P., Sieiro, C., Diaz, A. and Villa, T. G. 1994. Production and partial characterization of an endopolygalacturonase from *Saccharomyces cerevisiae*. *Canadian Journal of Microbiology* 40(11): 974–977.

Bommarius, A. S. and Riebel-Bommarius, B. R. 2004. *Biocatalysis: Fundamentals and Applications.* Wiley-VCH, Weinheim, Germany.

Borek, D., Michalska, K., Brzezinski, K., Kisiel, A., Podkowinski, J., Bonthron, D. T., Krowarsch, D., Otlewski, J. and Jaskolski, M. 2004. Expression, purification and catalytic activity of *Lupinus luteus* aspergines beta. Amidohydrolase and its *Escherichia coli* homolog. *European Journal of Biochemistry* 271(15): 3215–3226.

Brash, A. R. 1999. Lipoxygenases: Occurrence, functions, catalysis, and acquisition of substrate. *Journal of Biological Chemistry* 274(34): 23679–23682.

Braymer, H. D., Meachum, Z. D., Jr. and Colvin, H. J. 1971. Chemical and physical studies of *Neurospora crassa* invertase. Molecular weight, amino acid and carbohydrate composition, and quaternary structure. *Biochemistry* 10(2): 326–332.

Briggs, G. E. and Haldane, J. B. 1925. A note on the kinetics of enzyme action. *Biochemical Journal* 19(2): 338–339.

Brown, A. J. 1902. Enzyme action. *Journal of Chemical Society* 81: 373–386.

Casado, V., Martín, D., Torres, C. and Reglero, G. 2012. Phospholipases in food industry: A review. *Lipases and Phospholipases* 861: 495–523.

Chae, S. K., Yu, T. J. and Kum, B. M. 1983. Experimental manufacture of acorn wine by fungal tannase. *Hanguk Sipkum Kwahakhoechi* 15: 333–334.

Chang, G. W., Ming, K. C. and Tao, C. 2010. Improved purification and some properties of a novel phospholipase C from *Bacillus mycoides* strain 970. *African Journal of Microbiology Research* 4(5): 396–399.

Cheeseman, G. C. 1981. Rennet and cheesemaking. In: *Enzymes and Food Processing*, eds. Birch, G. G., Blakebrough, N. and Parker, K. J. 195–211. Springer, Netherlands.

Cheol, K. K., Yoo, S. S., Oh, Y. A. and Kim, S. J. 2003. Isolation and characteristics of *Trichoderma harzianum* FJ1 producing cellulases and xylanase. *Journal of Microbiology and Biotechnology* 13(1): 1–8.

De la Motte, R. S. and Wagner, F. W. 1987. *Aspergillus niger* sulfhydryl oxidase. *Biochemistry* 26(23): 7363–7371.

Delange, R. J. and Smith, E. L. 1971. *The Enzymes*, ed. Boyer, P. D. 3rd edition, 3: 81–118. Academic Press, New York.

Dixon, M. and Webb, E. C. 1979. *Enzymes*, 3rd ed. Longman, London.

Eyring, H. 1935. The activated complex in chemical reactions. *Journal of Chemical Physics* 3(2): 107–115.

Faccio, G., Nivala, O., Kruus, K., Buchert, J. and Saloheimo, M. 2011. Sulfhydryl oxidases: Sources, properties, production and applications. *Applied Microbiology and Biotechnology* 91(4): 957–966.

Fagan, S. M. and Fergus, C. L. 1984. Extracellular enzymes of some additional fungi associated with mushroom culture. *Mycopathologia* 87(1–2): 67–70.

Fayyaz, A., Asbi, B. A., Ghazali, H. M., Che Man, Y. B. and Jinap, S. 1993. Pectinesterase extraction from papaya. *Food Chemistry* 47(2): 183–185.

Fernandes, P. 2010. Enzymes in food processing: A condensed overview on strategies for better biocatalysts. *Enzyme Research* 1–19.

Fernandes, M. R. V. S. and Jiang, B. 2013. Fungal inulinases as potential enzymes for application in the food industry. *Advance Journal of Food Science and Technology* 5(8): 1031–1042.

Fiedurek, J. and Gromada, A. 2000. Production of catalase and glucose oxidase by *Aspergillus niger* using unconventional oxygenation of culture. *Journal of Applied Microbiology* 89(1): 85–89.

Fischer, E. 1890. Synthese des Traubenzuckers. *Berichte der Deutschen Chemischen Gesellschaft* 23: 799–805.

Fogarty, W. M. and Benson, C. P. 1983. Purification and properties of a thermophilic amyloglucosidase from *Aspergillus niger*. *European Journal of Applied Microbiology and Biotechnology* 18(5): 271–278.

Fruton, J. S. 1971. 4 Pepsin. *The Enzymes* 3: 119–164.

Fujiwara, S. 2002. Extremophiles: Developments of their special functions and potential resources. *Journal of Bioscience and Bioengineering* 94(6): 518–525.

Gawande, P. V. and Kamat, M. Y. 1998. Preparation, characterization and application of *Aspergillus* sp. xylanase immobilized on Eudragit S-100. *Journal of Biotechnology* 66(2): 165–175.

Geweely, S. I. N. 2006. Purification and characterization of intracellular urease enzyme isolated from *Rhizopus oryzae*. *Biotechnology* 5(3): 358–364.

Gianfreda, L., Xu, F. and Bollag, J. M. 1999. Laccases: A useful group of oxidoreductive enzymes. *Bioremediation Journal* 3: 1–25.

Givaudan, A., Effosse, A., Faure, D. and Potier. P. 1993. Polyphenol oxidase in *Azospirillum lipoferum* isolated from rice rhizosphere: Evidence for laccase activity in non-motile strains of *Azospirillum lipoferum*. *FEMS Microbiology Letters* 108: 205–210.

Godtfredsen, S. E. and Ottesen, M. 1982. Maturation of beer with α-acetolactate decarboxylase. *Carlsberg Research Communications* 47: 93–102.

Godtfredsen, S. E., Rasmussen, A. M., Ottesen, M., Rafn, P. and Peitersen, N. 1984. Occurrence of α-acetolactate decarboxylases among lactic acid bacteria and their utilization for maturation of beer. *Applied Microbiology and Biotechnology* 20(1): 23–28.

Grassin, C. and Fauquembergue, P. 1996. Fruit juices. In: *Industrial Enzymology*, eds. Godfrey, T. and West, S. MacMillan Press, London.

Guldberg, C. M. and Waage, P. 1867. E'tudes sur les Affinite's Chimiques. Christiania University Press, Oslo.

Gupta, A., Gill, A., Kaur, N. and Singh, R. 1994. High thermal stability of inulinases from *Aspergillus* species. *Biotechnology Letters* 16: 733–734.

Hall, B. G. and Hartl, D. L. 1974. Regulation of newly evolved enzymes. I. Selection of a novel lactase regulated by lactose in *Escherichia coli*. *Genetics* 76(3): 391–400.

Hasan, F., Shah, A. A. and Hameed, A. 2006. Industrial applications of microbial lipases. *Enzyme and Microbial Technology* 39(2): 235–251.

Heldt-Hansen, H. P. 1997. Development of enzymes for food applications. In: *Biotechnology in the Food Chain–New Tools and Applications for Future Foods*, ed. Poutanen, K. 45–55. Technical Research Center, Espoo, Finland.

Hendriksen, H. V., Kornbrust, B. A., Østergaard, P. R. and Stringer, M. A. 2009. Evaluating the potential for enzymatic acrylamide mitigation in a range of food products using an asparaginase from *Aspergillus oryzae*. *Journal of Agricultural and Food Chemistry* 57(10): 4168–4176.

Henri, V. 1903. *Lois Générales de l'Action des Diastases*. Hermann, Paris.

Herrgard, S., Gibas, C. J. and Subramaniam, S. 2000. Role of electrostatic network of residues in the enzymatic action of *Rhizomucor miehei* lipase family. *Biochemistry* 39: 2921–2930.

Illanes, A. 2008. *Enzyme Biocatalysis Principles and applications.* Springer, New York.

Itoh, T., Suzuki, M. and Adachi, S. 1982. Production and characterization of β-galactosidase from lactose-fermenting yeasts. *Agricultural and Biological Chemistry* 46(4): 899–904.

Jayani, R. S., Saxena, S. and Gupta, R. 2005. Microbial pectinolytic enzymes: A review. *Process Biochemistry* 40(9): 2931–2944.

Kakimoto, S., Sumino, Y., Akiyama, S. and Nakao, Y. 1989. Purification and characterization of acid urease from *Lactobacillus reureri*. *Agricultural and Biological Chemistry* 53: 1119–1125.

Kang, S. W., Park, Y. S., Lee, J. S., Hong, S. I. and Kim, S. W. 2004. Production of cellulases and hemicellulases by *Aspergillus niger* KK2 from lignocellulosic biomass. *Bioresource Technology* 91(2): 153–156.

Kashyap, D. R., Chandra, S., Kaul, A. and Tewari, R. 2000. Production, purification and characterization of pectinase from a *Bacillus* sp. DT7. *World Journal of Microbiology and Biotechnology* 16(3): 277–282.

Kashyap, D. R., Vohra, P. K., Chopra, S. and Tewari, R. 2001. Applications of pectinases in the commercial sector: A review. *Bioresource Technology* 77(3): 215–227.

Keil, B. 1971. Trypsin. *The Enzymes* 3: 249–275.

Keilin, D. and Hartree, E. F. 1948. Properties of glucose oxidase (notatin): Addendum. Sedimentation and diffusion of glucose oxidase (notatin). *Biochemical Journal* 42(2): 221.

Kelly, C. T. and Fogarty, W. M. 1978. Production and properties of polygalacturonate lyase by an alkalophilic microorganism *Bacillus* sp. RK9. *Canadian Journal of Microbiology* 24(10): 1164–1172.

Kent, C. A. and Emery, A. N. 1974. The preparation of an immobilised glucose isomerase II. Immobilisation and properties of the immobilised enzyme. *Journal of Applied Chemistry and Biotechnology* 24(11): 663–676.

Khasin, A., Alchanati, I. and Shoham, Y. 1993. Purification and characterization of a thermostable xylanase from *Bacillus stearothermophilus* T-6. *Applied and Environmental Microbiology* 59(6): 1725–1730.

Kieliszek, M. and Misiewicz, A. 2014. Microbial transglutaminase and its application in the food industry. A review. *Folia Microbiologica* 59(3): 241–250.

Kikuchi, M., Hayashida, H., Nakano, E. and Sakaguchi, K. 1971. Peptidoglutaminase. Enzymes for selective deamidation of gamma-amide of peptide-bound glutamine. *Biochemistry* 10(7): 1222–1229.

Kirchhoff, G. S. C. 1815. Formation of sugar in cereal grains converted to malt, and in flour steeped in boiling water. *Journal of Chemistry and Physics* 14: 389–398.

Kirk, O., Borchert, T. V. and Fuglsang, C. C. 2002. Industrial enzyme applications. *Current Opinion in Biotechnology* 13(4): 345–351.

Kona, R. P., Qureshi, N. and Pai, J. S. 2001. Production of glucose oxidase using *Aspergillus niger* and corn steep liquor. *Bioresource Technology* 78(2): 123–126.

Koshland, D. E., Jr. 1958. Application of a theory of enzyme specificity to protein synthesis. *Proceedings of the National Academy of Sciences of the United States of America* 44: 98–104.

Kristjansson, M. M. 1991. Purification and characterization of trypsin from the pyloric caeca of rainbow trout (*Oncorhynchus mykiss*). *Journal of Agricultural and Food Chemistry* 39(10): 1738–1742.

Kussendrager, K. D. and Hooijdonk, V. 2000. Actoperoxidase: Physico-chemical properties, occurrence, mechanism of action and applications. *British Journal of Nutrition* 84: 19–25.

Kvesitadze, G. I., Svanidze, R. S., Buachidze, T. and Bendianishvili, M. B. 1978. Acid-stable and acid-unstable alpha-amylases of the mold fungi *Aspergillus*. *Biokhimiia* 43(9): 1688–1694.

Laurière, C., Doyen, C., Thévenot, C. and Daussant, J. 1992. β-Amylases in cereals: A study of the maize β-amylase system. *Plant Physiology* 100(2): 887–893.

Lineweaver, H. and Burk, D. 1934. The determination of enzyme dissociation constants. *Journal of the American Chemical Society* 56(3): 658–666.

Macedo, G. A., Lozano, M. M. S. and Pastore, G. M. 2003. Enzymatic synthesis of short chain citronellyl esters by a new lipase from *Rhizopus* sp. *Journal of Biotechnology* 6: 72–75.

Madsen, G. B., Norman, B. E. and Slott, S. 1973. A new, heat stable bacterial amylase and its use in high temperature liquefaction. *Starch-Stärke* 25(9): 304–308.

Maldonaldo, M. C. and de Saad, A. S. 1998. Production of pectinesterase and polygalacturonase by *Aspergillus niger* in submerged and solid state systems. *Journal of Industrial Microbiology and Biotechnology* 20(1): 34–38.

Mardashev, S. R., Nikolaev, A. Y., Sokolov, N. N., Kozlov, E. A. and Kutsman, M. E. 1975. Isolation and properties of an homogeneous L asparagenase preparation from *Pseudomonas flourescens* AG. *Biokhimia* 40(5): 984–989.

Marshall, R. O. and Kooi, E. R. 1957. Enzymatic conversion of D-glucose to D-fructose. *Science* 125(3249): 648–649.

Martin, S. and Birgitte, H. 1984. Characterization of a new class of thermophilic pullulanases from *Bacillus acidopullulyticus*. *Annals of the New York Academy of Sciences* 434(1): 271–274.

Mayordomo, I., Randez-Gil, F. and Prieto, J. A. 2000. Isolation, purification, and characterization of a cold-active lipase from *Aspergillus nidulans*. *Journal of Agricultural and Food Chemistry* 48(1): 105–109.

Michaelis, L. and Menten, M. L. 1913. Kinetik der Invertinwirkung. *Biochemische Zeitschrift* 49: 333–369.

Mondal, K. C., Banerjee, D., Banerjee, R. and Pati, B. R. 2001. Production and characterization of tannase from *Bacillus cereus* KBR9. *The Journal of General and Applied Microbiology* 47(5): 263–267.

Mulrooney, S. B., Ward, S. K. and Hausinger, R. P. 2005. Purification and properties of the *Klebsiella aerogenes* UreE metal-binding domain, a functional metallochaperone of urease. *Journal of Bacteriology* 10: 3581–3585.

Nandakumar, R., Yoshimune, K., Wakayama, M. and Moriguchi, M. 2003. Microbial glutaminase: Biochemistry, molecular approaches and applications in the food industry. *Journal of Molecular Catalysis B: Enzymatic* 23(2): 87–100.

Ogston, A. G. 1948. Interpretation of experiments on metabolic processes, using isotopic tracer elements. *Nature* 162: 963.

Ottesen, M. and Spector, A. 1960. A comparison of two proteinases from *Bacillus subtilis*. *C R Trav Lab Carlsberg* 32: 63–74.

Park, Y. K., Santi, M. S. S. and Pastore, G. M. 1979. Production and characterization of β-galactosidase from *Aspergillus oryzae*. *Journal of Food Science* 44(1): 100–103.

Patel, A. K., Nampoothiri, K. M., Ramachandran, S., Szakacs, G. and Pandey, A. 2005. Partial purification and characterization of alpha-amylase produced by *Aspergillus oryzae* using spent-brewing grains. *Indian Journal of Biotechnology* 4(4): 336–341.

Payen, A. and Persoz, J. 1833. Memoire Sur la Diastase. *Annales De Chimie et de Physique* 53: 73–92.

Pazur, J. H. and Okada, S. 1966. A novel method for the action patterns and the differentiation of α-1, 4-glucan hydrolases. *Journal of Biological Chemistry* 241(18): 4146–4151.

Peng, Y., Huang, Q., Zhang, R. H. and Zhang, Y. Z. 2003. Purification and characterization of a fibrinolytic enzyme produced by *Bacillus amyloliquefaciens* DC-4 screened from *douchi*, a traditional Chinese soybean food. Comparative biochemistry and physiology part b. *Biochemistry and Molecular Biology* 134(1): 45–52.

Poulsen, P. B. and Buchholz, H. K. 2003. History of enzymology with emphasis on food production. In: *Handbook of Food Enzymology*, eds. Whitaker, J. R., Voragen, A. G. J. and Wong, D. W. S. 1–20. Marcel Dekker, New York.

Puri, M. and Kalra, S. 2005. Purification and characterization of naringinase from a newly isolated strain of *Aspergillus niger* 1344 for the transformation of flavonoids. *World Journal of Microbiology and Biotechnology* 21(5): 753–758.

Raghuwanshi, S., Dutt, K., Gupta, P., Misra, S. and Saxena, R. K. 2011. *Bacillus sphaericus*: The highest bacterial tannase producer with potential for gallic acid synthesis. *Journal of Bioscience and Bioengineering* 111(6): 635–640.

Roberts, S. M. 1995. *Introduction to Biocatalysis Using Enzymes and Microorganisms*. Cambridge University Press, New York.

Royer, J. C. and Nakas, J. P. 1989. Xylanase production by *Trichoderma longibrachiatum*. *Enzyme and Microbial Technology* 11(7): 405–410.

Ruiz, C., Pastor, F. I. and Diaz, P. 2005. Isolation of lipid- and polysaccharide degrading micro-organisms from subtropical forest soil, and analysis of lipolytic strain *Bacillus* sp. CR-179. *Letters in Applied Microbiology* 40: 218–227.

Said, S., Fonseca, M. J. V. and Siessere, V. 1991. Pectinase production by *Penicillium frequentans*. *World Journal of Microbiology and Biotechnology* 7(6): 607–608.

Saito, N. 1973. A thermophilic extracellular α-amylase from *Bacillus licheniformis*. *Archives of Biochemistry and Biophysics* 155(2): 290–298.

Schuster, S. C. 2007. Next-generation sequencing transforms today's biology. *Nature Methods* 5(1): 16–18.

Simpson, F. J. 1954. Microbial pentosanases: 1. A survey of microorganisms for the production of enzymes that attack the pentosans of wheat flour. *Canadian Journal of Microbiology* 1(2): 131–139.

Smith, J. and Olempska-Beer, Z. 2004. Hexose oxidase from *Chondrus crispus* expressed in *Hansenula polymorpha*, Chemical and Technical Assessment (CTA) FAO 1-5 prepared at 63rd JECFA (General Specifications for Enzyme Preparations Used in Food Processing).

Soberon, M. and Gonzalez, A. J. 1987. Glutamine degradation through the *w*-amidase pathway in *Saccharomyces cerevisiae*. *General Microbiology* 133: 9–14

Srivastava, R. A. K. and Baruah, J. N. 1986. Culture conditions for production of thermostable amylase by *Bacillus stearothermophilus*. *Applied and Environmental Microbiology* 52(1): 179–184.

Sujoy, B. and Aparna, A. 2013. Enzymology, immobilization and applications of urease enzyme. *International Research Journal of Biological Science* 2(6): 51–56.

Sullivan, J. D., Jr. and Ikawa, M. 1973. Purification and characterization of hexose oxidase from the red alga, *Chondrus crispus*. *Biochimica et Biophysica Acta-Enzymology* 309(1): 11–22.

Tonkova, A. 1998. Bacterial cyclodextrin glucanotransferase. *Enzyme and Microbial Technology* 22(8): 678–686.

Tsumura, N. and Sato, T. 1965. Enzymatic conversion of d-glucose to d-fructose: Part V partial purification and properties of the enzyme from *Aerobacter cloacae* Part VI, Properties of the enzyme from *Streptomyces phaeochromogenus*. *Agricultural and Biological Chemistry* 29(12): 1123–1134.

Uhlig, H. 1998. *Industrial Enzymes and Their Applications*. John Wiley & Sons, New York.

Van Der Maarel, M. J., Van Der Veen, B., Uitdehaag, J., Leemhuis, H. and Dijkhuizen, L. 2002. Properties and applications of starch-converting enzymes of the α-amylase family. *Journal of Biotechnology* 94(2): 137–155.

Viswanathan, P. and Kulkarni, P. R. 1996. Inulinase producing fungi and actinomycetes from dahlia rhizosphere. *Indian Journal of Microbiology* 36: 117–118.

Warrilow, A. G., Turner, R. J. and Jones, M. G. 1994. A novel form of pectinesterase in tomato. *Phytochemistry* 35(4): 863–868.

Whitaker, J. R. 1957. Properties of the proteolytic enzymes of commercial ficin. *Journal of Food Science* 22(5): 483–493.

Whitehurst, R. J. and Law, B. A. 2002. *Enzymes in Food Technology*. Sheffield Academic Press, CRC Press, Sheffield, UK.

Wilhelmy, L. 1850. The law by which the action of acid on cane sugar occurs. *Annales de Chimie et de Physique* 81: 413–433.

Wong, C. M., Wong, K. H. and Chen, X. D. 2008. Glucose oxidase: Natural occurrence, function, properties and industrial applications. *Applied Microbiology and Biotechnology* 78(6): 927–938.

Yadav, S., Verma, S. K., Singh, J. and Kumar, A. 2014. Industrial production and clinical application of L-asparaginase: A chemotherapeutic agent. *International Journal of Medical, Health, Pharmaceutical and Biomedical Engineering* 8(1): 54–60.

Yoshida, H. 1883. Chemistry of lacquer (Urushi). Part 1. *Journal of Chemical Society* 43: 472–486.

FURTHER READINGS

Kraut, J. 1988. How do enzymes work? *Science (New York)* 242(4878): 533–540.

Murray, R. K., Rodwell, V. W, Bender, D. A., Botham, K. M., Kennely, P. J. and Weil, P. A. 2009. *Harper's Illustrated Biochemistry*. McGraw-Hill, New York.

Nelson, D. L., Lehninger, A. L. and Cox, M. M. 2005. *Lehninger Principles of Biochemistry*, 4th ed. W.H. Freeman, New York.

Nicholls, D. G. and Ferguson, S. 2013. *Bioenergetics*. Academic Press, London.

Palmer, T. and Bonner, P. L. 2007. *Enzymes: Biochemistry, Biotechnology, Clinical Chemistry*. Elsevier, Chichester, UK.

Stryer, L. 1995. *Biochemistry*, 4th ed. VH Freeman and Company, New York.

Voet, D. and Voet, J. G. 1995. *Biochemistry*. John Wiley & Sons, New York.

Webb, E. C. 1992. *Enzyme Nomenclature 1992. Recommendations of the Nomenclature Committee of the International Union of Biochemistry and Molecular Biology on the Nomenclature and Classification of Enzymes*. Academic Press, San Diego, California.

2

Enzyme Technologies: Current and Emerging Technologies for Development of Novel Enzyme Catalysts

Rajeev K. Sukumaran, Vani Sankar, Aravind Madhavan, Meena Sankar, Vaisakhi Satheesh, Ayman Salih Omar Idris, and Ummalyama Sabeela Beevi

CONTENTS

2.1 Introduction

Enzymes are biological catalysts ubiquitous in all life forms, allowing chemical reactions to proceed at rates otherwise unachievable. Although the indirect use of enzymes may date back several centuries, use of isolated enzymes with the knowledge of their function is relatively new. Although industrial processes depend largely on chemical catalysts, the scenario is gradually changing toward green technologies with enzymes emerging as preferred catalysts for several such processes. Enzymatic catalysis represents potential alternatives to many of the existing processes, but not every one of the current manufacturing processes is amenable to enzymatic catalysis. New routes for synthesis and novel replacement products are in consideration with a large number of them based on enzymatic catalyses. Also, there is a renewed interest in developing novel enzyme catalysts tailored for the reactions of interest, making use of modern biotechnological tools for achieving this.

The benefits of using enzyme-based catalytic technologies are multitude with the processes being more efficient, requiring milder conditions for operation, having better (chemo-, regio-, and enantio-) selectivity, nontoxic nature, mutual compatibility with other enzymes, and biodegradability in addition to being green and environmentally benign. Enzyme catalyses are no longer restricted to biologically derived substrates or products or to the naturally occurring properties, operating conditions, or diversity of the enzymes themselves.

Enzyme technologies have multiple advantages but, at the same time, are not devoid of drawbacks. Enzymes are highly efficient catalysts that show a very high degree of selectivity, allowing them to act in a mixture of different reactants and accomplish the reaction with their substrates. They are stereo- and regio-selective, providing definitive advantages against chemical catalysts that are nonspecific. Enzyme-based catalyses operate in ambient conditions and are hence energy efficient. They are nontoxic and biodegradable and are mutually compatible, allowing single-pot reactions involving multiple enzymes, thereby eliminating the need to separate the intermediary products for multistep reactions. Because enzyme reactions are specific, there is no generation of unwanted by-products. Although these properties make them highly desirable for industrial catalyses, there are also downsides of enzyme technologies. These include the inability to function at high temperatures, instability at pH extremes, substrate and product inhibitions and feedback controls that regulate the activity, incompatibility with several solvents, inhibition by metal ions and certain chemical inhibitors, degradation by proteases, requirement of cofactors in some cases, and in several cases the relative increase in cost compared to corresponding chemical catalysts.

Modern biotechnological tools as well as computational methods have now made it possible to design enzymes suitable for a target reaction and not just improvising with existing reaction modes/mechanisms or properties of the enzymes. Biocatalysis is likely to be the future of chemical manufacturing, and tailor-made/designer enzymes might be available in the future for most of the industrially important catalytic processes. This has been made possible by advances in knowledge on protein biochemistry and structure biology, which help in interpreting and exploiting the structure function relationships in enzymes. This chapter focuses on the current and emerging trends for enzyme use in industry, strategies for creating advanced enzyme biocatalysts, and, finally, on the emerging and futuristic trends in the development for and use of enzymes in industry.

2.2 Technologies Facilitating Industrial Use of Enzymes

Technologies facilitating enzyme use in industrial catalyses, such as immobilization, blending of enzyme cocktails, formulation of enzyme preparations, stabilization by external chemicals, development of multienzyme cascading reaction systems, and enzyme systems with cofactor regeneration, are rapidly evolving and improving with the possibility of converting a majority of chemical manufacturing processes to enzymatic catalyses.

There is also an enormous advancement in the development of novel enzyme catalysts themselves with the technology slowly moving from rational, semirational, and directed evolution approaches for enzyme modification to de novo synthesis of novel proteins with desired enzyme activities. Powerful algorithms for in silico modeling and design of active sites or substrate-binding pockets are being developed for the ambitious goal of enzyme design from scratches (Hilvert 2013; Kiss et al. 2013).

2.2.1 Enzyme Stabilization

Applications of enzymes in industry often require them to function in conditions that are different from their natural environment, such as higher temperatures, in organic solvents, and the enzymes should ideally be active for a longer duration and be reusable. These can be made possible by enzyme stabilization to improve prolonged performance under the conditions of use, and quite often this involves immobilization of enzymes (Hwang and Gu 2013). Methods ranging from addition of polyol cosolvents to protein engineering for enhancing stability are practiced, and the method of choice often depends on the enzyme as well as the application with which it is being used. Quite often, industrial use of enzymes requires them to work in organic solvents, such as in the case of flavor synthesis by lipases (Kwon et al. 2000; Romero et al. 2007). Such applications require the enzymes to be active and stable in organic solvents. The methods for stabilization of enzymes can be grouped into (i) modification of the biocatalysts and (ii) modification of the solvent environment (Stepankova et al. 2013). Biocatalyst modifications include those that affect the structure of the biocatalysts, such as genetic modifications that mutate the concerned gene to generate novel enzymes that have the required property, or chemical modifications that affect the amino acid residues that make up the enzyme protein. Because the former aspects are addressed in the section on enzyme engineering, the discussions here are restricted to chemical modifications of proteins and the modification of the solvent environment for stabilizing enzymes. Enzyme immobilization, which can also act as a method for stabilization, is discussed separately from this topic.

2.2.1.1 Stabilization by Chemical Modification of the Enzyme

Surface modification by polyethylene glycol (PEG) or its derivatives is one of the most widely used techniques for improving enzyme stability. PEGylation, the covalent attachment of PEG to proteins, has been used to improve the stability and catalytic activity and even to alter the enantioselectivity of enzymes (Stepankova et al. 2013). The monomethoxylated form of PEG is the one generally used in protein surface modification as its monofunctionality yields cleaner chemistries. PEGylation masks the surface of the proteins and can also alter the polydispersivity and solubility of the enzymes. In fact, this can be beneficial in industry because the enzyme may become soluble and active in an organic solvent (Veronese 2001). Although earlier PEGylation methods were rather nonspecific, currently the PEGylation reagents are mostly amino acid-specific, and methods are available to make site-specific modifications (Rodríguez-Martínez et al. 2009). Amino groups, typically the alpha amino or the epsilon amino of the lysine, are the usual sites for PEGylation. The type of PEGs and their conjugation with proteins have been detailed by Veronese (2001). PEGylation of proteins is intensively used in the in vivo stabilization and improving bioavailability of therapeutic proteins (Pasut and Veronese 2009; Veronese and Mero 2008). Applications of PEGylated enzymes and some investigations into the mechanisms of enzyme stabilization by PEGylation have been reported (Inada et al. 1995; Rodríguez-Martínez et al. 2009; Stepankova et al. 2013; Turner et al. 2011; Veronese 2001).

Other strategies for chemical modification of the enzyme proteins to improve stability include propanol rinsing of enzyme preparation, coating with ionic liquids, etc. Repeated rinsing of enzymes with dry n-propanol is a method for stabilizing enzymes for reactions in low water. The method is believed to achieve high activity by preventing the loss of water associated with the protein. The method was demonstrated successfully for immobilized subtilisin Carlsberg and for α-chymotrypsin (Partridge et al. 1998). The steps involved in preparation of propanol rinsed enzyme preparations (PREPs) are immobilization of the enzyme on a support matrix in an aqueous buffered solution; water removal by decanting, taking care to minimize contact with air; and rapid dehydration by rinsing with a suitable water-miscible organic solvent, such as 1-propanol (Moore et al. 2001). The dehydration step is recommended to be carried out using solvent containing low levels (0.5%–10%) of water (may be chosen to match the water activity of the reaction in which it has to be used) to ensure essential water molecules remain bound to the protein. More recently, the technique has been used in stabilization of lipase for enantioselective transacetylation of (R, S)-beta-citronellol (Majumder et al. 2007).

Yet another method for modification is the coating of enzymes with ionic liquids (IL). Zhao (2009) has reviewed some of the aspects of enzyme stabilization by coating with ILs. Coating of lipase with the IL [PPMIM][PF6] (PPMIM =1 (3′ phenylpropyl) 3 methylimidazolium) was effective in enhancing the activity and enantioselectivity of the enzyme (Lee and Kim 2002). Dang et al. (2007) reported an enhancement in activity and stability of ionic liquid pretreated lipase when the IL-coated lipases were more than 1.5 times more active than the controls, and the pretreated lipases were stable for more than 7 days at 60°C with retention of activity whereas the untreated enzymes were completely inactivated after 12 h of incubation. More recently, Lozano et al. (2011) demonstrated the use of 1-butyl-3-methyl imidazolium chloride for stabilizing cellulase for use in saccharification applications. Although there is still ambiguity regarding the mechanisms of stabilization by IL coatings, it is recognized that the IL binds to the enzyme and thus provides a microenvironment favorable for the reaction (Stepankova et al. 2013). The ability of ILs to preserve the secondary structure elements has also been demonstrated (Karimi et al. 2013).

2.2.1.2 Stabilization by Modification of the Solvent Environment

Stabilization of an enzyme/protein is dependent on maintaining its native three-dimensionally folded state (native conformation). Although enzymes maintain their native conformation in their natural environment, which is aqueous, industrial applications often require their use in conditions that are far from natural, including their use in organic solvents. Native conformation of the enzyme is only marginally more stable than its denatured form, and maintaining this more stable state is very much dependent on the interaction of the enzyme protein with the water, cosolvents, and solutes in its vicinity (Canchi and Garcia 2013). Small molecule additives, such as sugars, polyols, and neutral amino acids, have been used traditionally as stabilizers in enzyme preparations because these cosolvents/osmolytes are known to prevent loss of enzyme activity, inhibit reversible aggregation of proteins, and protect the enzymes against thermal and chemical degradation. It is believed that these osmolytes act by inducing preferential hydration of the proteins, which, in turn, is achieved by the preferential exclusion of these cosolvents from near the protein molecules (Fagain 2003; Kumar et al. 2012). The most widely used among these cosolvents are polyols, such as sorbitol, mannitol, etc., and the disaccharides trehalose and sucrose. Although the molecular mechanisms for the observed stabilization by these polyol cosolvents are still unknown to a large extent, the molecular exclusion theory is the one that is widely accepted. Because the protein–solvent interface increases with denaturation, it will increase the degree of thermodynamically unfavorable interaction between the additive and the protein, and hence the enzymes remain in their folded native conformation/more stable form. Most of the possible mechanisms currently known are discussed by Kumar et al. (2012). It is also interesting to note that the mechanisms by which sugars and polyols enhance protein stability are different, and an elaborate study encompassing several of these osmolytes has been published recently in which the role of osmolytes as chemical chaperones is highlighted (Levy-Sakin et al. 2014).

Other stabilizing additives include polymers such as polyetheleneimine (PEI) and surfactants (Fagain 2003). Surfactants are believed to have both a stabilization effect and enhancement of enzyme activities. Spreti et al. (2001a) reported the activation and stabilization effects of cationic surfactants on α-chymotrypsin. The same group also reported that the stabilizing effect of zwitterionic surfactants, such as sulfobetains, were dependent on the nature of the head group of the surfactant and the type of its interaction with the protein (Spreti et al. 2001b). Stabilization effects of surfactants are also studied as microemulsions of them with water organic phase when the enzyme protein is encapsulated in reverse micelles or water in oil microemulsions (Stepancova et al. 2013). The major advantage of such systems are projected to be the possibility of keeping the enzyme in aqueous phase and, at the same time, keeping the hydrophobic substrates in organic phase where the reaction to be catalyzed involve substrates that are soluble in organic solvents and not in water.

Salts, typically alkyl halides (e.g., NaCl), are also widely used additives in enzyme preparations for stabilization. Effects of salts on protein stabilization have been recognized for a long time, and chaotropicity and kosmotropicity of salt additives may explain protein stabilization in aqueous solutions (Eppler 2007). In fact, the very definitions of kosmotropicity and chaotropicity have evolved from their historically defined roles on protein stabilization and destabilization. In general, kosmotropic salts (e.g., $[NH_4]_2 SO_4$) have a net constructive effect on water structuring whereas the chaotropic salts (e.g., urea) have a net destructive effect (Collins 1995; De Xammar Oro 2001; Galinski et al. 1997). The former increases the surface tension whereas the latter decreases it. Similar to the effects of osmolyte cosolvents, there is a preferential exclusion of solute from the protein hydration layer due to the increased surface tension associated with solute water interactions in the case of kosmotropic salt. This stabilizes the enzyme against denaturation because the denatured state is energetically unfavorable (Eppler 2007).

From the forgoing discussions, it is apparent that there are multiple methods for stabilization of an industrial enzyme preparation so as to improve its activity, thermostability, and storage characteristics. Apart from the thermodynamic and kinetic stabilizations, the enzyme protein would require stabilization from microbial or protease attacks, and quite often antimicrobials and protease inhibitors also form part of the enzyme formulations in addition to stabilizers, emulsifiers, dispersants, surfactants, and preservatives. Techniques such as enzyme immobilization also stabilize the industrial enzymes, and the choice of methods often depends on the enzyme to be immobilized and the end applications. A detailed discussion on immobilization as a strategy for enzyme stabilization is not included here although immobilization is indeed an important technique for enzyme stabilization. The following section therefore concentrates on the newer concepts in enzyme immobilization and their applications. Those who are interested in details on conventional strategies for immobilization may refer to books such as *Immobilization of Enzymes and Cells: Methods in Biotechnology* (Guisán 2006).

2.2.2 Enzyme Immobilization: Novel Methods and Applications

Enzymes have multiple advantages as industrial catalysts, which include high selectivity, greater turnovers, ability to work at ambient conditions, higher efficiencies, purer products, and lots more. However, the lack of storage and operational stability, susceptibility to elevated temperatures, presence of organic solvents, etc., hinder their use in several industrial catalyses. Besides, as the enzymes are used in aqueous solutions, their recovery and reuse are challenging (Sheldon 2011). Immobilization of enzymes allows reuse of the enzyme for several reaction cycles, and this probably is the most important application of immobilization (Mateo et al. 2007). In several instances, industrial uses of enzymes require them to be immobilized. This allows easy control of the reaction and avoids contamination of products with the enzyme. Repeated reuse of the biocatalyst requires its immobilization, and this also implies that the enzyme should be stable or should be stabilized enough by the immobilization itself. Stabilization of the enzyme is a welcome consequence of immobilization in several cases because any immobilization process avoids some of the causes of enzyme inactivation, such as aggregation and proteolysis (Barbosa et al. 2013). The structural rigidity offered by the immobilization stabilizes the enzyme although sometimes it is at the cost of activity reduction. Nevertheless, it offers stability in all cases in which the inactivation is due to conformational changes and in several cases in which the decreased operational stability is due to an inactivation agent. The latter is made possible because the inactivation agent is partitioned away from the enzyme environment (Barbosa et al. 2013). Methods for immobilization are wide and varied and have been classically grouped into the techniques that involve adsorption, entrapment, covalent binding, and encapsulation (Chaplin and Buck 1990). It is not important to detail each method here, and the reader may refer to articles authored by Hanefeld et al. (2009) and Chaplin (2014) for detailed descriptions.

Although the classifications of immobilization strategies still hold true, it may be interesting to look at the strategies in light of the techniques that do not require a support matrix. Accordingly, the strategies may be grouped into those that require a carrier/support and those that do not. Carrier-bound immobilization of enzymes may involve enzyme attachment to an inert carrier by ionic or hydrophobic interactions, covalent attachment, or encapsulation in a membrane or entrapment in some matrix creating superior operational performance compared to a free enzyme. However, these systems offer low productivity (kilogram product/kilogram enzyme) due to the large amount of noncatalytic support matrix/carrier compared to enzymes. Also, enzyme immobilization on carriers often leads to significant losses in activity (Cao et al. 2003; Cui et al. 2014; Sheldon 2011). Quite often, the matrices are also expensive because most of the support matrices dedicated for such applications require sophisticated processes and advanced materials for their manufacturing. In contrast, the carrier-free immobilization platform, which is versatile and with different options, such as cross-linked enzyme crystals (CLECs), cross-linked (dissolved) enzymes (CLEs), cross-linked spray-dried enzymes (CLSDs), and cross-linked enzyme aggregates (CLEAs), looks very promising. The advantages on offer include highly concentrated enzyme activity, stability, and lower cost because no carriers are required (Cao et al. 2003). On the other hand, advances in carrier bound enzyme immobilization strategies include the use of magnetic nanoparticles, carbon nanotubes and fibers, hetero-functional materials, etc., as carriers, some of which offer the advantages of both worlds: operational stability of the carrier-bound enzymes and the mobility of carrier-free ones. In the following section, discussion is limited to the carrier-free immobilization platforms because immobilization on/using nanomaterials are dealt with in the section on nanotechnology in enzymes.

2.2.2.1 Carrier-Free Immobilized Enzymes

Carrier-free enzymes are generally synthesized by direct cross-linking of enzyme preparations, and the starting form of enzyme preparation can range from crude enzyme solutions to highly purified enzyme crystals. The nomenclature of such cross-linked enzymes is based on the starting form of enzyme preparation. Thus, when the starting enzyme preparation is an enzyme solution, the cross-linked preparation can be a CLE (cross-linked [dissolved] enzyme), and when the starting preparation is pure crystallized enzymes, they are called CLEC (cross-linked enzyme crystals). Glutaraldehyde, a bifunctional agent, is usually the cross-linking agent. Cao et al. (2003) and Sheldon (2007) have reviewed the carrier-free platform for enzyme immobilization. The following sections provide concise information on the different carrier-free immobilization technologies available today.

Cross-linked enzymes (CLEs) are prepared by chemical cross-linking of dissolved enzymes. Enzyme cross-linking could successfully enhance the thermal stability of enzymes, but these preparations have several drawbacks. These include the loss of activity, lack of mechanical stability, and the difficulties in handling due to gelatinous nature of CLEs (Cao et al. 2003). Lack of mechanical stability was addressed by some researchers by use of carrier-binding strategies for the cross-linked enzyme (Sheldon 2007).

Cross-linked enzyme crystals (CLECs) are produced by cross-linking of crystallized enzymes, which requires the crystallization of enzymes, which, in turn, require the enzyme to be highly pure. Compared to the CLEs, these had better thermo- and mechanical stability and were more tolerant to organic solvents (Cao et al. 2003). It is also known now that the CLEC properties can be modulated by careful selection of crystallization parameters (Margolin 1996). CLECs of varying sizes from 1 to 100 µm have been described for varying applications.

Cross-linked spray-dried enzymes (CLSDs) are synthesized by cross-linking of spray-dried enzymes. Although it is technically possible to control several parameters, such as the size of the enzyme particles, the applications are limited due to the deactivation of enzymes that happens during spray-drying.

Sheldon and coworkers pioneered the synthesis of cross-linked enzyme aggregates (CLEAs) as an alternative to CLECs and demonstrated that instead of using pure enzyme crystals, aggregates of enzyme produced through precipitation of protein from enzyme solutions can be cross-linked effectively to form CLEAs (Cao et al. 2000). Classical precipitating agents, such as ammonium sulfate, or solvents, such as acetone or ethanol, may be used for generating the enzyme aggregates. It was also observed that the properties of CLEAs change with the type of precipitant used, which allows modulation of the properties of CLEAs by careful selection of precipitants and the conditions used for enzyme precipitation. The most important aspect about the strategy was the ability to use crude enzyme preparations and the elimination of the need to crystallize the enzyme protein, which makes it possible to perform the immobilization in any lab. The cross-linking agent is typically glutaraldehyde (Sheldon 2011). Although CLEA is a relatively newer technology for enzyme immobilization, this is a very promising strategy due to the relative ease of preparation and the ability to start with a crude enzyme preparation. It is also possible to coimmobilize different enzymes and to have CLEA–polymer composites by performing the cross-linking along with a second enzyme preparation or in presence of a monomer that polymerizes under the conditions of cross-linking, respectively. CLEAs cross-linked in the presence of magnetic nanoparticles could also yield "smart" magnetic CLEAs (mCLEAs). A review of the synthesis and applications of CLEAs was presented by Sheldon (2011). A schematic diagram representing the different strategies used for carrier-free enzyme immobilization is presented in Figure 2.1.

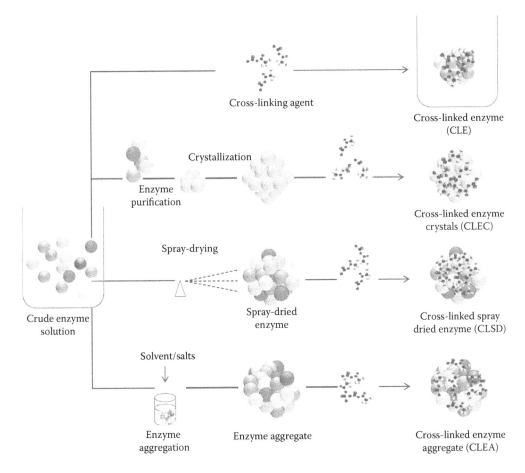

FIGURE 2.1 Schematic diagram representing the different strategies used for carrier free enzyme immobilization.

2.3 Development of Novel Enzyme Catalysts: Current and Emerging Technologies

The demand for new and improved biocatalysts in varied areas of industrial and environmental applications is rapidly increasing due to the increased awareness on the need for green technologies in the future toward sustainable utilization of bioresources in addition to better understanding of the vast range of economic benefits of the use of enzymes in bioprocesses. The industry requires novel and competent catalytic abilities that are as yet undiscovered, nonexistent in nature, or currently unable to meet the challenges posed and therefore require improvement. New enzyme catalysts could be obtained from nature through deliberate modification/assembly of existing enzymes/proteins or through de novo creation of novel enzymes (Adrio and Demain 2014). Each method has its own merits and demerits, and the following sections are meant not to bring these out but would describe these approaches in more detail.

2.3.1 Discovery of Previously Undescribed Enzyme Activities from Nature

Although it is true that a large number of enzymes representing a variety of activities are discovered, it is also true that the currently known enzyme diversity may represent only a minute fraction of the diversity actually available in nature as it is estimated that more than 99% of the microbes are not amenable for laboratory cultivation. Methods to explore these would involve bioprospecting of unique ecological niches, such as extreme geographic locations, and methods such as metagenomics. The discovery of novel activities from such environments would also involve use of unconventional microbial culture strategies, development of novel methods for culturing, and high throughput screening for desired activities/properties.

2.3.1.1 Bioprospecting Extremophiles

Microorganisms that can thrive in extreme environments, such as hydrothermal vents, hyper saline or soda lakes, extreme cold environments, etc., are a relatively underexplored class, particularly so because of the difficulty of culturing them in the laboratory. However, these organisms could be very potent sources of novel enzymes due to the unique environments under which they survive. This is corroborated by the fact that the enzymes that are required to act at conditions uncommon in normal life are already sourced from such microbes. The most illustrious example is Taq DNA polymerase sourced from the extreme thermophile *Thermusaquaticus* (Chien et al. 1976). The sale of this single enzyme was about $500 million in 2009 (Adrio and Demain 2014). Enzymes from extremophiles would be expected to be tolerant to extremes of conditions such as temperature, pH, salinity, radiation, etc., depending on the conditions of growth of their host organisms, and these properties are highly desirable in several industrial applications. For example, several esterases/lipases are expected to work in nonaqueous media, and one of the properties that helps in enzyme stability in organic solvents include the presence of an increased number of negatively charged amino acid residues on their surface. This is also the mechanism by which halophiles cope with high salt concentrations, and hence, enzymes from such microbes may be expected to fare better in organic solvents (Karan et al. 2012; Klibanov 2001). Similarly, low-temperature active enzymes with potential applications in detergents and the textile industry, fruit juice clarification, environmental remediation, etc., could be sourced from psychrophiles, which are naturally adapted to function in extremely low temperatures. Compared to their mesophilic counterparts, enzymes from psychrophiles have a higher flexibility in their structure to counter low water activity, which, in turn, is accomplished by having structural features supporting this destabilization. This is made possible by clustering of glycine residues (providing local mobility), a decreased number of prolines in loops (providing enhanced chain flexibility), a decrease in the number of arginines (providing a reduction in the number of salt bridges and hydrogen bonds, a weakening of charge dipole interactions in α-helices, etc. Also, these enzymes have catalytic sites that are larger in size, allowing better accessibility to ligands in comparison to the mesophilic analogues (Hoyoux et al. 2004). Apparently, bioprospecting of extreme environments can provide hitherto undescribed novel activities, which might even help to perform reactions in unnatural conditions such as in organic solvents. The potential of bioprospecting extremophiles for biocatalysts has been extensively reviewed by several researchers (Antranikian et al. 2005; Demirjian et al. 2001; Lopez-Lopez et al. 2014; van den Burg 2003). Knowledge of how the extremophilic enzymes achieve their unique properties in relation to their structural adaptations is extremely important for developing enzymes with desirable properties, and today, there is a wealth of information available on the structural properties of extremophilic proteins. A detailed discussion on the structural features contributing to the performance of extremophilic enzymes in their habitat is out of scope for this chapter, but this information may be found elsewhere (Liszka et al. 2012; Reed et al. 2013; Turner et al. 2011).

2.3.1.2 Metagenomics for Enzymes from Uncultivable Microbes

Uncultivable microbial biodiversity is considered to be a highly potent resource to look for novel/improved enzyme activities. This can also be a treasure trove for discovery of novel structural features, which may provide the sought-after biocatalytic activity. Metagenomic libraries prepared from the total DNA of a profiled niche/community is the most comprehensive representation of the community's sequence information, and this allows the culture independent screening for novel enzymes (Beloqui et al. 2008). The method for prospecting the metagenome involves isolation of DNA from a sample of the community's immediate habitat (e.g., soil, sea water, etc.), construction of a metagenomic library using an appropriate vector (e.g., phage or plasmid vector for small insert libraries, cosmid or fosmid vectors for libraries with insert sizes up to 40 kb, and BAC vectors for large insert libraries) and screening them for the desired activity.

Two complementary approaches are used for screening of the metagenomic libraries: (i) *function-based screening* and (ii) *sequence-based screening*. In the first approach, libraries are typically constructed in suitable vectors such as lambda phage, plasmid, cosmid, or copy-controlled fosmids that will allow expression of the genes. The libraries are then subjected to a screening based on direct or indirect profiling of activity that will allow the evaluation of a large number of clones in a single screen. The screening strategies commonly used include activity-based screens performed on culture plates, such as the starch iodine test for amylase, the cellulose plate assay for cellulases, etc.; in some cases, appropriately modified to identify desirable properties, such as pH tolerance, salinity tolerance, etc. The major advantage of this type of screening is the possibility to discover previously unknown genes and their encoded enzymes. Prior information about the sequence of the gene is not necessary, and totally novel enzyme activities, which do not share any sequence or structure with the known enzymes, may, in principle, be discovered. The major disadvantage of this technique is the possibility for failure in expression of genes, which can happen due to multiple reasons. This frequently happens when the functional enzyme requires expression of more than one subunit, incompatibility problems with promoter recognition and translation initiation, improper folding or post-translational modifications of expressed proteins, etc. These problems may be overcome with use of vectors that allow larger insert sizes so that entire gene clusters/operons are included, use of broad host range vectors, which will allow expression in hosts other than *Escherichia coli*, use of suitable host strains such as the ROSETTA strains (containing tRNA genes for rare amino acid codons) coexpression of chaperonins, etc. (Perner et al. 2011; Uchiyama and Miyazaki 2009).

The second strategy for screening metagenomic libraries is the sequence-based approach. Here, the libraries may be screened using a nucleotide probe for the enzyme gene sought by using colony hybridization. Alternatively, genes for the activity sought may be amplified by PCR from the metagenomic libraries using specific or degenerate probes and cloned in appropriate vectors. Yet another approach is to search for genes of the sought-after enzyme in a metagenomics sequence database by sequence comparisons followed by the cloning of the identified gene. In any of these conditions, prior knowledge about the sequence of the enzyme protein is required. Also, the technique is limited to discovery of only those enzyme proteins whose sequences and the sequence to function relationship is known and, therefore, eliminates the possibility of discovery of entirely new activities and the activities for which previously undescribed sequences contribute to the function. Nevertheless, this is a powerful tool to discover activities that share the same sequence–function relationships. There are several reports on the discovery of entirely new activities as well as improved activities of enzymes through metagenomic profiling of a variety of ecological niches, such as sea water, soil, insect guts, animal guts, etc. (Brennan et al. 2004; Fu et al. 2013; Pinnell et al. 2014; Vester et al. 2014).

From the foregoing discussions, the suitability of metagenomics as a powerful tool for discovery of novel biocatalytic activities becomes apparent. More information about the technicalities of metagenomic library construction and screening and the cases of discovery of novel activities are presented by several investigators (Guazzaroni et al. 2010; Ilmberger et al. 2012; Li et al. 2009; Rabausch et al. 2013; Streit and Daniel 2010; Thomas et al. 2012).

2.3.2 Development of Novel Biocatalysts: Rational Methods, Directed Evolution and De Novo Design

Enzymes can accelerate the rates of biological reactions several thousandfold without compromising the specificity of the reaction. At the same time, they can be very slow or unstable in non-natural conditions, and often they are inhibited by products of their own reactions. Often the industrial applications of enzymes demand the biocatalysts to be stable under a wide range of conditions (e.g., unnatural temperatures or pH) or in organic solvents. The enzymes are also required to stay active longer and perform with high catalytic rates. Now enzymes are also sought for reactions, which seldom occur in nature. These limitations have led to the engineering of enzymes with the end goal of adapting them to perform a desired catalysis at desired conditions and rates. The properties generally sought include temperature and pH tolerance and stability, solvent tolerance, increased catalytic rates, improving or introducing selectivity, and, in some cases, even to have an entirely new function. Such engineering is performed by changing the amino acid(s), which makes up the structure of the enzyme, which, in turn, is accomplished by modifications to the genes coding the concerned enzyme protein. The engineering of an enzyme protein has been done by either a "rational design," which make use of the existing knowledge on sequence to structure and function relationship to make educated guesses on which amino acids to modify, or a "directed evolution" approach, which works by making random changes in the protein sequence(s) followed by selection for desired functionality (Weissman 2004). Either of these approaches have their own advantages and disadvantages, and "semirational approaches" or even "de novo" enzyme syntheses have also become popular recently. The following sections highlight these contrasting approaches for development of novel biocatalysts.

2.3.2.1 Rational Approaches in Biocatalyst Development

With the increasing availability of protein structures as well as reliable models, biochemical methods, and computational tools, enzyme engineering is evolving from totally random approaches to knowledge-based designing of enzyme function or rational designing. In the rational approach for enzyme engineering, all possible knowledge from the biochemical studies, structure information, modeling data, etc., are utilized to propose site-specific mutations with the objective of imparting a particular property. Although prior information about the molecular structure is a prerequisite, rational design has an improved probability of achieving the desired target, and the number of mutants to be screened (library size) also becomes lesser (Steiner and Schwab 2012). Knowledge of the structure of the enzyme in relation to its function helps to predict the most appropriate amino acid residue(s) to be modified. Often only one or a very few of the residues are modified. The choice of residues to be modified are usually helped by the fact that those residues that will make a difference in activity are often located close to or in the active site or in the substrate binding pocket although it is not a necessary requirement. Quite often, mutations leading to changes in amino acid residues farther from the active site or binding pocket can seriously impact enzyme activity and/or stability of the enzyme and can even impart allosteric regulation capability (Mathieu et al. 2010; Morley and Kazlauskas 2005; Wang et al. 2014).

Rational approaches in engineering enzymes ideally starts with a three-dimensional structure obtained through x-ray crystallography, NMR, or even at times through homology models from related structures of the target enzyme(s). Sometimes, if this information is not available, sequence alignments that identify signature sequences, motifs, or domains can serve as starting points because they provide information on the amino acid sequence conservation or divergence. Target regions to be mutated can be identified by structure/sequence comparisons and often the probable structural change(s) can be predicted in silico using tools such as Deep-view (Guex and Peitsch 1997). Site-directed mutagenesis is the most widely used method for incorporating point mutations to the starting enzyme protein. Detailed descriptions on the technique are out of scope of this chapter, and the reader is referred to Braman (2002) for the protocols. Methods for introducing mutations at multiple points in the sequence of the enzyme protein are also now widely available (Jensen and Weilguny 2005; Kim and Maas 2000; Liu and Naismith 2008; Mikaelian and Sergeant 1992; Wang and Malcolm 1999). Large segments of the protein sequence may also be altered, which allows swapping of domains, which creates chimeric enzymes that might have improved functionality or functionalities that were originally not existing in the starting enzyme. Swapping of domains may be achieved by swapping of regions from the cDNA of two enzymes when there are common restriction sites in their sequence or through peptide ligation approaches (Penning and Jez 2001). The success of rational approaches is largely dependent on knowledge about the structure and catalytic mechanism of the target enzyme. An increase in the number of enzymes with resolved structures and better understanding of their functions can definitely contribute to the better success rates for rational design of an enzyme. Moreover, there are multiple computational tools available to aid in the rational design of enzymes (Barrozo et al. 2012; Li et al. 2012). Selection of amino acid(s) to be mutated can be made through consensus approaches (which assumes that the consensus amino acids of a multiple sequence alignment will contribute more to the structural stability and/

or activity of an enzyme than the nonconsensus amino acids) or a structure-guided approach with which structural information of the related enzyme proteins will be used to decide on which amino acid is to be mutated. The structure-guided approach can be used in conjunction with the consensus approach to have a structure-guided consensus approach, which was successfully used in several instances to improve the enzyme properties. Examples include the creation of a more thermostable penicillin G acylase (Polizzi et al. 2006), thermostable glucose dehydrogenase (Vazquez-Figueroa et al. 2007), engineering of solvent tolerance in *Geobacillus thermophilus* lipase T6 (Dror et al. 2014), etc. There are several recent examples of rational design of enzyme properties, which include engineering of promiscuity (alternate activities, substrate recognition, reaction conditions, sites, etc.); specificities; physical properties, such as pH, temperature, saline, or solvent tolerances and even including or removing allosteric controls. These are achieved by structure-guided recombination, active site or substrate-binding pocket engineering, consensus or structure-guided consensus approaches. Notable examples of rational engineering of enzymes include changing of the activity of pyruvate decarboxylase to an enantioselective carboligase (Meyer et al. 2011), imparting activity toward hydrophobic electrophilic substrates in *E. coli* 2-keto-3-deoxy-6-phosphogluconate (KDPG) aldolase (Cheriyan et al. 2012), improved tolerance of yeast hexokinase toward xylose-induced inactivation (Bergdahl et al. 2013) etc. Details on rational approaches for enzyme engineering may be found in the review by Steiner and Schwab (2012).

2.3.2.2 Directed Evolution

Directed enzyme evolution mimics nature's way of evolving enzymes but at a tremendously faster pace. Where nature takes millions of years to evolve a new activity (as the mutations are accumulated over a large span of time with the selection pressure being the survival of species), directed evolution strategies are extremely fast, taking, in some cases, only a few days. Random mutations are introduced deliberately, and the selection is based on analysis of the predefined target function, which is performed by the experimenter.

Directed evolution does not require prior knowledge of the enzyme structure and sequence structure–function relationships and are often accomplished through iterative accumulation of beneficial mutations (Tracewell and Arnold 2009). Most of the directed-evolution approaches work by introducing relatively small changes to the existing enzyme protein because the strategy works best for modifying existing weak promiscuous activities than creating an entirely new one. An increase in the number of individual mutations has a higher effect on the destabilization of the enzyme protein compared to modifying its activity, and hence, multiple mutations introduced simultaneously can result in an exponential decline in chance of obtaining functional proteins (Bloom et al. 2005).

Any strategy for directed evolution has two steps (i) creation of the variability and (ii) selection of the fittest (beneficial) mutant. Random mutagenesis of the DNA coding for the parent enzyme is the starting step for directed evolution. Mutations are introduced through one of the several strategies available to create sequence diversity, followed by expression using an appropriate vector and host system. The mutant enzyme functionality is tested by a screening procedure, which will be used to select the transformants bearing the enzymes with desired properties. The mutated gene sequence is then amplified, and the cycle of mutagenesis, screening, and amplification may be repeated if further improvements are necessary (Labrou 2010).

Mutagenesis strategies are wide and varied, and these include chemical mutagenesis using agents such as ethyl methyl sulfonate, nitrous acid, mitomycin C, etc., use of mutator strains lacking DNA repair mechanisms (e.g., *E. coli* XL1-red) as hosts to express the enzyme; error-prone PCR (epPCR); DNA shuffling; rolling circle error-prone PCR (RCA); site-saturation mutagenesis (which allows substitution of specific sites against all 20 possible amino acids at once); random chimeragenesis on transient templates (RACHITT); incremental truncation for the creation of hybrid enzymes (ITCHY), etc. More information on the methods used for generating sequence diversity may be obtained from the literature (Labrou 2010; Sen et al. 2007; Tee and Wong 2013).

Although different strategies are available, most of the studies on directed evolution still use error-prone PCR for generation of sequence diversity probably due to the ease with which the experiment can be performed and also because it is easy to increase the error rates by manipulating the conditions of

PCR such as addition of Mn2+ to reduce base pairing specificity or provide an unbalanced dNTP ratio to allow misincorporation of bases. Screening steps are specific for the property to be improved and should allow rapid selection of the mutants with desirable properties.

Directed evolution of enzymes, even using the simplest single mutation libraries, can be quite demanding with respect to the number of mutants to be screened because, with equal possibility of each amino acid site getting mutated, even a 100-residue enzyme can have 20^{100} different possibilities of mutants being generated. Most of these mutants will be nonfunctional, and to screen the limited number of functional mutants from among the huge number of nonfunctional mutants is a Herculean task. Although it is true that beneficial mutations need not necessarily be within the active site or the substrate-binding pocket, it has been observed that a majority of the beneficial mutations were in regions that contribute to substrate binding, catalysis, or the conformation and dynamics of the active site (Dalby 2003). Hence, it is expected that mutations targeted to such regions may result in substantial gains and may contribute to the overall success of directed evolution for enzyme engineering.

The most important advantage of directed evolution approaches for enzyme engineering is that it does not require prior knowledge on enzyme structure. Because the engineering approach is similar to Darwinian evolution, the success rates are also much higher in directed evolution. Nevertheless, knowledge-based selection of regions to be mutated can help in the success of random mutagenesis as well. There have been numerous reports on enzyme engineering using directed evolution approaches. The method was successfully used to engineer pH dependence (Callanan et al. 2007), improving the thermostability (Jochens et al. 2010; Zhao and Arnold 1999), thermolability (Reetz et al. 2009), improvements in specific activity (Giger et al. 2013), modification of substrate specificity (Cheriyan et al. 2011), creation of entirely novel activities for existing enzymes (Chen et al. 2012; Park et al. 2006), and even creating naturally nonexisting enzyme activity out of noncatalytic proteins (Chao et al. 2013). The field of directed evolution is actively developing, and there are multiple excellent reviews available for the methods and approaches used (Acevedo-Rocha et al. 2014; Dalby 2011; Goldsmith and Tawfik 2012; Sen et al. 2007; Tee and Wong 2013) as well as for examples of enzymes developed through these methods (Dalby 2003; Kumar and Singh 2013; Martinez and Schwaneberg 2013; Turner 2009; Wang et al. 2012).

2.3.2.3 *Semirational/Combinatorial Methods*

Rational approaches and directed evolution have their own merits and demerits, and it is obviously advantageous to combine the advantages of both the approaches in creating new engineered enzymes. Thus, semirational approaches help in the creation of smarter and smaller libraries by focusing the directed evolution to target region-based knowledge derived from biochemical and/or structural data (Lutz 2010; Steiner and Schwab 2012). Preselection of the promising target site and the amino acid variability to be engineered is aided by computational methods. The focus brought in through the knowledge of structure–function relationships, evolutionary conservation of residues for a given activity, implications of amino acid substitutions on the folding of protein, its stability and activity results in only certain regions of the template enzyme to be considered for modification, and that too with only a reduced pool of amino acid substitutions. This, in turn, results in dramatically reduced library sizes, eliminating the major limitation of directed evolution approaches. Semirational approaches for enzyme redesign can be based on sequence or structure information of activities related to the one to be engineered, or it can be based purely on theoretical estimation of the energetics of amino acid variations on the enzyme protein structure implemented through computational methods. Computational tools such as HotSpotWizard (Pavelka et al. 2009) and 3DM database (Kuipers et al. 2010) implemented over the Internet provide the experimenter with information on the possible target site for enzyme engineering.

HotSpotWizard (http://loschmidt.chemi.muni.cz/hotspotwizard/) (Pavelka et al. 2009) identifies hot spot sites for engineering of substrate specificity, activity, or enantio-selectivity of enzymes and integrates several bioinformatics databases and computational tools. Structural analysis is performed to select residues contributing to substrate binding, transition state stabilization, or product release, and evolutionary conservation is also analyzed to distinguish highly conserved sites, which help in their elimination as targets for mutagenesis.

3DM (https://3dm.bio-prodict.nl/) (Joosten 2007) is a proprietary database and, similar to HotSpotWizard, can be used to predict the effect of mutations. The 3DM information systems are protein super family platforms that collect, combine, and integrate many different protein-related data. Semirational approaches have been used to improve the enantio-selectivity (Ivarsson et al. 2007; Sandström et al. 2012; Tang et al. 2012), to shift the substrate specificity and/or selectivity (Cheriyan et al. 2007; Panizza et al. 2015), to change the reaction type (Schneider et al. 2008), to improve thermostability (Anbar et al. 2012), etc. Detailed information on semirational approaches for enzyme redesign with examples may be obtained from Chica et al. (2005) and Lutz (2010).

2.3.2.1 De Novo Design of Enzymes

De novo design of the enzyme is the design of a catalytic activity from scratch aided by the knowledge of the configuration of the set of atoms (hence, the specific residues) required for stabilizing the transition state and thereby lowering the activation energy of the reaction to be catalyzed. The design of the enzyme starts with the design of the active site with appropriate functionality. This requires thorough knowledge of the reaction mechanism and the molecular interactions necessary for the desired catalysis. The active site is modeled in silico using quantum-mechanic simulations so that the residues are positioned in the optimized geometry for stabilizing the transition state (Quin and Schmidt-Dannert 2011). This design is called a theozyme for theoretical enzyme (Kiss et al. 2011). Once the structure data for the potential active site (theozyme) is derived, protein scaffolds capable of hosting the new catalytic machinery are selected from the Protein Data Bank, and these are used as templates for grafting the geometry of the active site using molecular modeling tools such as RosettaMatch or Gess (Barker and Thornton 2003; Malisi et al. 2009; Smith et al. 2014; Zanghellini et al. 2006). When the protein backbone structure accommodating the theozyme is generated, additional noncatalytic mutations are introduced to the active site to provide favorable interactions with the transition state as well as to optimize the general configuration of residues. The *RosettaDesign* algorithm is used for performing such 3-D structure optimizations (Kuhlman and Baker 2000; Liu and Kuhlman 2006). The Rosetta modeling software is implemented as web servers at the Rosetta Commons (https://www.rosettacommons.org/software/servers). The designs are then evaluated, and finally, experiments are conducted to synthesize the genes, clone, and express the protein(s). An overview of the de novo design strategy is presented by Kiss et al. (2011, 2013).

The most successful examples of de novo designed enzymes are the biocatalysts developed for the retro-aldol reaction, the Kemp elimination, and, more recently, the Diels-Alder reaction. In the first example (Jiang et al. 2008), carbon–carbon bond breaking in a non-natural substrate was accomplished by the de novo designed retro-aldolases. A computational design strategy was used to produce eight active enzymes that promoted the base catalyzed ring opening of 5-nitrobenzisoxazole, a Kemp elimination reaction (Rothlisberger et al. 2008). The Diels-Alder reaction, which is highly important in organic synthesis can form two C–C bonds and up to four stereogenic centers in one step, and no known natural enzymes are reported to possess the bimolecular Diels-Alder reaction capability (Kiss et al. 2011). Computational design of enzymes catalyzing the reaction with high stereo selectivity and substrate specificity has been described (Siegel et al. 2010). Interestingly, directed evolution methods have been used to further fine-tune the de novo designed enzymes with significant improvements in catalytic activity. The de novo designed Kemp eliminase was optimized by incorporating mutations suggested computationally through rational approaches as well as through random mutagenesis and DNA shuffling (Khersonsky et al. 2011). In nine rounds of directed evolution, rate enhancements of more than 400 fold were achieved compared to the parent enzyme. Although not a de novo design in the strict sense, a very interesting case is the development of novel catalytic function from a noncatalytic protein scaffold achieved through directed evolution. A novel RNA ligase from a noncatalytic human retinoid X-receptor protein with two zinc finger motifs was developed. In this, two adjacent loops of the protein were randomized to generate a combinatorial library of mutants, which were used as templates for selection and the directed evolution. The final protein with RNA ligase activity lost its original structure and adopted an entirely novel structure (Chao et al. 2013).

Enzyme development through de novo design is a challenging field and still under development. Although most of the enzymes developed through the de novo approach may not match the efficiencies or rate accelerations of natural enzymes, improvements in the design and experimental methods and computational tools are happening at a rapid pace, and it may be expected that the development of entirely new activities from scratch would be easier in the future.

2.4 New and Emerging Avenues of Enzyme Technology and Enzyme Applications

Enzyme technology is constantly evolving with rapid developments in protein biochemistry, structural biology, and even material sciences. It is possible today to engineer an enzyme for its efficient multipoint attachment to an immobilization matrix and also to have a novel *nanoscale* matrix with properties tuned to suit the enzyme that is to be attached to it and to the reaction the preparation has to catalyze. Multistep reactions are now possible in a *single pot* by coimmobilization of different enzymes as is the regeneration of expensive cofactors. The following sections address some of these emerging areas in enzyme technology.

2.4.1 Nanotechnology in Enzymes: Nanobiocatalysts

Nanobiocatalysis is a subfield of biocatalysis exploring more advanced materials as enzyme carriers (Kim et al. 2008). Nanotechnology has made it possible to have nanostructured scaffolds as carriers for enzyme immobilization, in several cases eliminating the drawbacks of conventional carriers used for enzyme adsorption, covalent binding, and entrapment. Nanoscale dimensions of such advanced carriers as well as the enzymes themselves allow the enzyme bound nanoscaffold to behave similar to free enzymes. The advantages of classical immobilization are retained while new features/properties are acquired on enzyme immobilization on nanoscale scaffolds. Wang (2006) and Verma et al. (2013a) have detailed these as the following:

- High surface area-to-volume ratio: Nanomaterials in general have large surface area-to-volume ratios. Although this is highest for spherical/near spherical nanoparticles, even the one-dimensional nanofibers offer about two thirds of the surface-to-volume ratio for an identical volume of nanoparticles.
- Higher enzyme loading: Due to the large surface area-to-volume ratios, nanomaterials allow a higher loading of enzymes per unit volume of the material. This can also translate to higher productivities for the nanobiocatalysts as more product is formed per unit volume of catalyst due to minimization of the noncatalytic part of the biocatalyst preparation.
- Decreased mass transfer limitations: Enzymes immobilized on nanoscale materials have lower mass transfer resistance compared to their counterparts on micro-/macroscale matrices.
- Ease of separation: Nanoscaffold in general offers structural rigidity, and problems due to compaction are minimized, allowing separation that is otherwise not possible with conventional immobilized catalysts. Their properties (e.g., shape) can be modulated to suit the separation technique to be used. More importantly, use of magnetic nanomaterials allows the quick and efficient separation of the nanobiocatalysts from the bulk reaction. This will also result in lower product contamination with the catalyst.
- Mobility: Unlike macro-/microsized immobilized catalysts, nanoparticles dispersed in a solution are mobile in the form of Brownian motion. The mobility and diffusivity in this case are between homogenous catalysts (free enzymes) and conventional heterogenous catalysts (enzymes immobilized in/on conventional matrices). The mobility of catalysts is an important factor in determining their activities, and in general, enzymes immobilized on nanoscale particles offered higher activities.
- Adaptability to reactor configurations: Nanomaterials offer greater flexibility in reactor design due to their unique properties. For example, nanofiber membranes have a small pressure drop

and high flow rate compared to traditional enzyme-immobilized membranes and fixed bed reactors. The possibility of confinement to nanospaces also offers the unique possibility of forming molecular machinery capable of catalyzing multiple reactions and with regeneration of cofactors, almost mimicking biological processes. An interesting example is the case of construction of a multienzyme biocatalytic system using spatial confining of enzymes and cofactors in nanopores where the tethered cofactor (NADH) could shuttle between the enzymes allowing its regeneration and reuse (El-Zahab et al. 2004).

- Ability to self-assemble: Enzymes conjugated with hydrophobic polymer groups can self-assemble at oil–water interfaces. Such conjugates can be used effectively for reactions that happen at the interface or aqueous and organic phases. Capabilities of self-assembly offered by nanoscale protein scaffolds such as S-layer proteins open up an entirely new avenue of possibilities. S-layer fusion proteins where catalytic proteins are conjugated to the S-layer proteins offer the unique possibility of a structured self-assembly with which catalysts are arranged on surfaces with a defined order and distance. The possibilities are much more with the capability of self-assembly on solid surfaces providing a method for highly ordered immobilization, formation of structured self-aggregates, etc. Interesting examples of enzymes that have been immobilized using S-layer mediated self-assembly include the works by Tschiggerl et al. (2008) and Shu (2014).

Nanobiocatalysts may be polymer-based, carbon-based, magnetic, or protein-based, and with developments in materials sciences, the possibilities are expanding. Based on the physical form of the nanomaterials, these scaffolds may be grouped as nanoparticles, nanofibers, nanosheets, etc. The following section will elaborate more on the different types of nanoscaffolds currently being utilized for synthesizing the nanostructured biocatalysts.

2.4.1.1 Nanoparticles

Different types of nanoparticles have been tried as carriers for enzymes, and the most common of them include polymer nanoparticles such as that of polylactic acid, polystyrene, grapheme, polyvinyl alcohol, chitosan, magnetic and super paramagnetic nanoparticles, etc. (Verma et al. 2013a). Major advantages of immobilization on nanoparticles include maximum surface area per unit mass, high enzyme loading, reduced mass transfer limitations, and diffusional limitations in the case of nanoporous materials. Effective enzyme loadings up to 10% wt have been achieved on nanoparticles (Chen and Su 2001). Nanoparticle-immobilized enzymes also offer unique solution behaviors that are in between the free enzymes and the enzymes immobilized on conventional matrices. It has been demonstrated that the reduction in size of carrier particles (to nanoscales) could affect the activity of the immobilized enzymes (Jia et al. 2003). Magnetic nanoparticles as carriers offer yet another important feature: easy recovery for reuse. Magnetic iron oxide is used in a large number of studies as the nanoparticle carrier for enzyme immobilization. The material is highly stable, biocompatible, has large surface area, possesses super paramagnetic properties, and is amenable to surface derivatizations in addition to being easy to synthesize. The major limitations with nanoparticle carriers other than the magnetic ones are issues of dispersion in reaction mixtures and the difficulties in recovery.

2.4.1.2 Nanofibers

Nanofibers provide a solution to the major drawback of the nanoparticle carriers in that they are easy to recover and reuse while, at the same time, providing the advantages of nanosized scaffolds (Kim et al. 2006). Similar to nanoparticles, they have a high specific surface area allowing high loading of enzymes and offer the additional advantages of being molded into porous scaffolds minimizing the diffusional resistance. Enzyme loadings up to two thirds of the levels offered by nanoparticles are possible in the nanofibers despite being more or less one-dimensional. The enzymes can be confined within the fiber matrix offering structural stability to the enzyme while, at the same time, allowing exchange of reactants and products with the bulk environment. Nanofibers are typically produced using electrospinning technology, which is driven by electrical force in comparison to the conventional spinning protocols, which use mechanical forces along with some thermosetting phenomenon to create nanofibers. Different polymers can be used for creating nanofibers, and these include nylon, polyurethane (PU), polycarbonate (PC), polyvinyl alcohol (PVA), polylactic acid (PLA), polystyrene (PS), etc. (Huang et al. 2003). Carbon nanofibers, however, can be produced using any material with a carbon backbone and are generally produced from cellulose, pitch, or, more commonly, polyacrylonitrile (PAN) (Zhang et al. 2014). Interesting examples for use of enzymes immobilized on nanofibers include studies in which lipase immobilized on nanofibers were used in transesterification reactions. PVA immobilized lipase showed esterase activities equivalent to commercially available membrane-immobilized enzymes (Nakane et al. 2007). Esterase CLEA immobilized on polymer nanofibers was used for synthesis of cephalosporin-derived antibiotics (Sekhon et al. 2014). Tang et al. (2014) demonstrated the advantages of immobilization in nanofibers with respect to enhancement of thermostability.

2.4.1.3 Nanotubes

Carbon nanotubes (CNTs) are allotropes of carbon that are tubular with diameters less than 100 nm made from graphite (Saifuddin et al. 2012; Tasis et al. 2006). CNTs can be multiwalled with at least two layers, and often many more, which are called multiwalled carbon nanotubes (MWCNTs). Single-walled CNTs (SWCNTs) can also be synthesized following the same route as that of MWCNTs. SWCNTs are narrower (1–2 nm) and tend to be curved. The techniques used for synthesis of CNTs include arc discharge, laser ablation, and chemical vapor deposition (Saifuddin et al. 2012). Among the various nanomaterial scaffolds, carbon nanotubes have unique structural, mechanical, thermal, and biocompatibility properties and have attracted interest in various end applications, especially in the development of biosensors (Verma et al. 2013b). CNTs are highly hydrophobic and can form insoluble aggregates. They are also less reactive, which is an issue for attachment of enzymes on the CNT surfaces. However, there are rapid developments in the functionalization of CNT surfaces aimed at improving the solubility in aqueous solutions. This is usually achieved via attachment or organic moieties by direct bonding to the surface double bonds or to carboxylic acid groups formed through oxidation with strong acids (Tasis et al. 2006). Immobilization on CNTs offers similar advantages as nanofibers. They are more stable, provide higher enzyme density, enhanced catalytic efficiency, mechanical and thermal stability, etc., and reduce diffusional and mass transfer limitations (Jia et al. 2003; Kim et al. 2006). Methods for synthesizing CNTs, their surface modifications, and current strategies for immobilizing enzymes/proteins on them are reviewed by Saifuddin et al. (2012). In a recent study, it was noted that immobilization of inulinase on CNT could enhance its thermo tolerance and as well as storage stability (Garlet et al. 2014). CNT-immobilized enzymes have also been tried for bioremediation applications in the remediation of organophosphates (Mechrez et al. 2014).

2.4.1.4 Nanosheets

Graphene, a monoatomic sheet of honeycomb sp^2-hybridized carbon atoms, has unique mechanical, electrical, and thermal properties and is a subject of intense research owing to this combination of interesting properties (Sun et al. 2014). The graphene sheets, owing to the planar structure with two sides/surfaces, offer a huge area for enzyme immobilization. In addition, there are several functional groups (epoxide, hydroxyl, and carboxylic) with a high amount of water solubility—all of which pose it as an ideal nanoscaffold for enzyme immobilization (Verma et al. 2013b). The most common approach used for production of graphene nanosheets involves the oxidation of graphite followed by exfoliation to yield graphene oxide (GO) (Stankovich et al. 2006). There are also other methods for exfoliation that use ultrasonication or modified acrylate polymers (Nicolosi et al. 2013; Sun et al. 2014). Graphene has also been used extensively for immobilization of enzymes and with applications in various fields. Graphene immobilized beta galactosidase was used for analytical applications based on lactose reduction (Kishore et al. 2012), and thermostability and reusability of lipase immobilized on GO sheets was reported to be enhanced in comparison to the free enzyme (Pavlidis et al. 2012). Another interesting application is the use of enzymes immobilized on graphene for biosensors in detection of glucose (Alwarappan et al. 2010).

2.4.1.5 Other Nanoscale Carriers

Nanoscale carriers are not just limited to nanoparticles, nanofibers, nanotubes, and nanosheets. Other possibilities include mesoporous nanomaterials, such as mesoporous silica, sol-gel approaches, and single enzyme nanoparticles (SENs). Mesoporous materials offer high surface areas, high pore volume, and a well-ordered structure. Mesoporous silica particles, such as SBA15 and MCF, have pore sizes of 5–40 nm and are suitable for enzyme immobilization. Several applications have been reported for enzymes immobilized on mesoporous materials, which include their use in biosensors, peptide synthesis, pulp biobleaching, etc. (Kim et al. 2006). The sol-gel process is a wet-chemical technique used for the fabrication of both glassy and ceramic materials. In this process, the sol (or solution) evolves gradually toward the formation of a gel-like network, containing both a liquid phase and a solid phase. Typical precursors are metal alkoxides and metal chlorides, which undergo hydrolysis and polycondensation reactions to form a colloid. The basic structure or morphology of the solid phase can range anywhere from discrete colloidal particles to continuous chain-like polymer networks (Klein and Garvey 1980). In a typical synthetic protocol, the metal alkoxide precursor is hydrolyzed to form the *sol* into which the enzyme solution is added to initiate the condensation reaction forming the *gel*. During the process, various pores and channels are formed in the silicate matrix ranging from 0.1 to 500 nm in size (Kim et al. 2006). Several applications have been reported for sol-gel immobilized enzymes, which include kinetic resolution of secondary alcohols by a sol-gel entrapped lipase (Ursoiu et al. 2012). A detailed review of sol-gel immobilized enzymes and their applications can be found in Kandimalla et al. (2006).

Another unique concept for developing nanoscale enzyme biocatalysts is the single enzyme nanoparticles (SENs). This strategy has been pioneered by Kim and Grate (2003). Each enzyme molecule is surrounded with a porous composite organic/inorganic network of less than a few nanometers thick. To make SENs, the enzyme surface is covalently modified so as to allow anchoring of the monomer of a polymer and to make them soluble into a solvent (e.g., hexane). Silane monomers containing both vinyl and trimethoxysilane groups are added to initiate polymerization, which will create linear polymers attached to the enzyme surface. In a second polymerization step, which is achieved by extracting these intermediates into aqueous phase when the pendant trimethoxysilane groups are hydrolyzed and condensed to each other, forming a network around the enzyme molecule. SENs are extremely stable and have been tested for applications in bioremediation of organic contaminants, including phenols, polyaromatics, chlorinated compounds, and pesticides (Kim and Grate 2003).

2.4.2 Multienzyme Systems/Reaction Cascades and Cofactor Regeneration

Reactions involving multiple enzymes physically separated in biological compartments (e.g., chloroplast, mitochondria) are nature's strategy to achieve transformations with efficiencies unmatched by any chemical synthetic processes (van Oers et al. 2014). Syntheses of most of the industrially and pharmaceutically relevant compounds are multistep reactions, which are increasingly being studied to be performed efficiently under ambient conditions using enzymatic catalyses. Chemists have been trying multienzymatic reactions either in homogenous systems employing free enzymes or in heterogeneous systems using enzymes compartmentalized through various strategies of immobilization. The former approach has been around for quite some time and generally uses sequential addition of enzymes involved in the catalysis based on the reaction order. The latter approach, which helps to perform the multiple steps of biocatalysis without product separation, is generally referred to as cascade reactions. Although there are multiple definitions to cascade reactions, the generic features of cascade reactions are a sequential array of reactions in which the product of one reaction serves as the substrate for the next, and that the transformations happen in inseparable steps without recovery of product after each step (Findrik and Vasic-Racki 2009 and references therein). Most of the important chemical building blocks and pharmaceutically important compounds today are products of multistep reactions, synthesized by conventional chemical catalyses. Multienzyme cascade reactions offer enormous possibilities in the biosynthesis of compounds. This includes enantioselectivity, stereoselectivity, high yields, and lower expenses for downstream processing, reaction equilibrium shifting by removing one of the products, etc. (Findrik and Vasic-Racki 2009).

Several reactions that use two, three, or four enzymes in cascades have been described in the literature although reports on catalyses involving more enzymes are rather rare. Some of the interesting examples include Degussa's industrial scale synthesis of L-*tert*-leucine using L-leucine dehydrogenase for reductive amination, with which the regeneration of cofactor (NAD) was carried out using formate dehydrogenase. The process was operated continuously in the two-stage cascade of identical membrane bioreactors (Kragl et al. 1996a,b). In another study on production of L-phenylalanine from the racemic mixture of DL-phenyl lactate, a three-enzyme cascade system in a membrane reactor was demonstrated (Schmidt et al. 1987). In the first step, the racemate is dehydrogenated to phenyl pyruvate by two enzymes: D- and L-hydroxyisocaproate dehydrogenase. In the second step, phenyl pyruvate is reductively aminated to L-phenylalanine by L-phenyl alanine dehydrogenase. Because the first step requires NAD and the second NADH, cofactor regeneration is automatically taken care of (Findrik and Vasic-Racki 2009). There are also highly complex cascade systems being demonstrated as in the case of polyketide assembly line reconstruction in vitro for aflatoxin pathway intermediates (Crawford et al. 2008a,b) and even the construction of the complex pathway for synthesis of isotopically labeled purine nucleotides, incorporating a total of 36 enzymes from pentose phosphate and purine synthesis pathways with five different cofactor regenerating systems (Schultheisz et al. 2008).

The kinetics of sequential one-pot reactions may be increased considerably by bringing them in close proximity as would be the case in cellular enzymes functioning in organisms (Lopez-Gallego and Schmidt-Dannert 2010). Coimmobilization of enzymes for cascade reactions is a relatively newer concept, which is rapidly gaining popularity. Nanomaterial-immobilized enzymes are being used increasingly for such kind of reactions. The concept of multienzyme cascade reactions in nanoreactors created by compartmentalization of individual enzymes or enzyme groups taking part in a reaction are also gaining popularity (van Oers et al. 2014). The field may be considered to be in its infancy because lots of issues, such as mutual compatibility of enzymes, recycling of cofactors, sensitivity to side reactions, and positional control, need to be addressed. Nevertheless, there are very promising leads and concepts in areas including cofactor regeneration and in the use of nanoscale compartmentalized enzymes for cascade reactions. One of the very interesting concepts is the use of a tethered NADH cofactor in a multistep biotransformation for production of α-ketoglutarate and lactate using silica nanoparticles attached to glutamate dehydrogenase and lactate dehydrogenase enzymes. NADH attached to the nanoscaffold in the vicinity of the enzymes served the catalysis by a dynamic shuttling between the enzymes (Liu et al. 2009). In another very interesting study, a biomimetic enzyme nanocomplex containing alcohol oxidase and catalase could reduce blood alcohol levels in mice with potential application as an antidote and prophylactic for alcohol intoxication (Liu et al. 2013). Probably one of the most challenging applications for engineered multienzyme cascades are replicating the biological pathways and processes for potent future applications, and an artificial photosynthetic system assembly would be a real breakthrough in storage of solar energy in chemicals. A review of the recent advances in this direction are presented in Kim et al. (2014).

2.4.3　Artificial Metalloenzymes

Artificial metalloenzymes combining the features of both homogeneous/organometallic catalysts and enzymes offer the benefits of both systems (Lu et al. 2009). Artificial metalloenzymes are hybrid catalysts resulting from the incorporation of transition metal species within a biomolecular scaffold. Here, the catalytic scope of transition metal complexes is combined with the high activity and selectivity offered by the enzymes (Onoda et al. 2014). Hybrid enzymes built on this concept has been reported for several reactions, including Diels-Alder reactions (Bos et al. 2012; Talbi et al. 2010), transfer hydrogenation (Creus et al. 2008), imine reduction (Robles et al. 2014), etc. Several different concepts for creating such enzymes by positioning catalytically active metal center(s) in proteins/enzymes are also evolving. These include strategies based on the biotin-avidin technology (Ward 2011). Metal-conjugated affinity labels (Reiner et al. 2013) site-directed mutagenesis (Podtetenieff et al. 2010), cross-linking proteins (Tabe et al. 2014). Protein designs for metalloenzymes, including the redesign of natural metallozymes, are reviewed by Yu et al. (2014). Artificial metalloenzymes combining the advantages of both organo-metallic catalysts and enzymes seems to be poised for rapid developments in the future. Tuning of the properties of both the metal center as well as the protein scaffold is increasingly becoming possible, and innovations are also coming in the use of hybrid catalysts in cascade reactions. Combining natural enzymes and artificial metalloenzymes provide possibilities in synthetic cascades hitherto not possible. Challenges in such reactions involving organometallic catalysts include mutual inactivation of both types of catalysts, which can be addressed through use of the artificial metalloenzymes (Köhler et al. 2013). Possibilities also exist in creating nanofactories with compartmentalized bio- and homogenous catalysts.

2.5 Conclusions and Future Perspectives

Most of the advancements in enzyme technology, especially the ones that poise them as feasible indus-trial catalysts, have happened only in the last few decades. These include the discovery of several unique catalytic abilities useful in organic syntheses and biotransformations; methods for purification and char-acterization; techniques for operational, thermodynamic, and storage stability enhancements; methods to study their structure and function; cloning of enzymes and their overexpression; methods for their reuse, including techniques for immobilization; development of reactors for enzymatic catalyses; and development of artificial enzymes.

Although originally the enzymes were used in industrial catalyses in which no chemical equivalents were available, today industry is looking at enzymatic catalyses as a means to manufacture most of their products in environmentally friendly green processes. Consequently, the increased demand for enzymes and enzymatic processes is driving a new wave in enzyme technology, aided by advances in biotech-nology and chemical sciences. Be it bioprospecting for new activities from nature or de novo design of enzymes, high-end science and technology is contributing significantly to the field. Thus, we have previously undescribed activities being discovered from the environmental genomes through metage-nomics, and artificial enzymes, hybrid catalysts, and enzyme-based nanoreactors being developed with the wealth of knowledge that has been gained in understanding the enzyme structure and function and in nucleic acid and protein engineering. Aiding further are the advances in chemistry and computational sciences, which helps in the design of active centers of novel catalysts that do not exist in nature or for modifying the existing enzymes to suit the requirements. Importantly, most of this know-how and these techniques and tools are coming in the public domain and are accessible to anybody, which will acceler-ate the developments in this field.

Enzyme reusability, which once was a challenge, is no longer an issue with the development of highly efficient methods and matrices for immobilization. Nanomaterials are poised to revolutionize this field in which limitations of traditional matrices are being resolved and new possibilities being explored. Compartmentalization on the nanoscale has made it possible to have cofactor regenerating systems, single enzyme nanocataysts, and even biomimetic multienzyme cascades, which will eventually help in replicating entire biochemical pathways in vitro. It is apparent that the growth in development of enzyme catalysts will be enormous in the coming years, especially in the areas of novel enzyme development, including de novo development of artificial and hybrid enzymes and in the development of nanobiocata-lysts. In the immediate future, it is expected that better and green synthetic processes based on enzyme catalysis would be available to the industry.

REFERENCES

Acevedo-Rocha, C. G., Hoebenreich, S., Reetz, M. T. 2014. Iterative saturation mutagenesis: A powerful approach to engineer proteins by systematically simulating Darwinian evolution. *Methods Mol. Biol.* 1179, 103–128.

Adrio, J. L., Demain, A. L. 2014. Microbial enzymes: Tools for biotechnological processes. *Biomolecules.* 4(1), 117–139.

Almonacid, D. E., Babbitt, P. C. 2011. Towards mechanistic classification of enzyme functions. *Curr. Opin. Chem. Biol.* 15, 435–442.

Alwarappan, S., Liu, C., Kumar, A., Li, C. Z. 2010. Enzyme-doped graphene nanosheets for enhanced glucose. *Biosensing Phys. Chem.* 114, 12920–12924.

Anbar, M., Gul, O., Lamed, R., Sezerman, U. O., Bayer, E. A. 2012. Improved thermostability of clostridium thermocellum endoglucanase Cel8A by using consensus-guided mutagenesis. *Appl. Environ. Microbiol.* 78, 3458–3464.

Antranikian, G., Vorgias, C. E., Bertoldo, C. 2005. Extreme environments as a resource for microorganisms and novel biocatalysts. *Adv. Biochem. Eng. Biotechnol.* 96, 219–262.

Barbosa, O., Torres, R., Ortiz, C., Berenguer-Murcia, A., Rodrigues, R. C., Fernandez-Lafuente, R. 2013. Heterofunctional supports in enzyme immobilization: From traditional immobilization protocols to opportunities in tuning enzyme properties. *Biomacromolecules.* 14(8), 2433–2462.

Barker, J. A., Thornton, J. M. 2003. An algorithm for constraint-based structural template matching: Application to 3D templates with statistical analysis. *Bioinformatics.* 19, 1644–1649.

Barrozo, A., Borstnar, R., Marloie, G., Kamerlin, S. C. L. 2012. Computational protein engineering: Bridging the gap between rational design and laboratory evolution. *Int. J. Mol. Sci.* 13, 12428–12460. doi:10.3390/ijms131012428.

Beloqui, A., de María, P. D., Golyshin, P. N., Ferrer, M. 2008. Recent trends in industrial microbiology. *Curr. Opin. Microbiol.* 11(3), 240–248.

Bergdahl, B., Sandström, A. G., Borgström, C., Boonyawan, T., van Niel, E. W. J., Gorwa-Grauslund, M. F. 2013. Engineering yeast hexokinase 2 for improved tolerance toward xylose-induced inactivation. *PLoS One.* 8(9), e75055. doi:10.1371/journal.pone.0075055.

Bloom, J. D., Meyer, M. M., Meinhold, P., Otey, C. R., MacMillan, D., Arnold, F. H. 2005. Evolving strategies for enzyme engineering. *Curr. Opin. Struct. Biol.* 15(4), 447–452.

Bos, J., Fusetti, F., Driessen, A. J. M., Roelfes, G. 2012. Enantioselective artificial metalloenzymes by creation of a novel active site at the protein dimer interface. *Angew. Chem. Int. Ed.* 51(30), 7472–7475.

Braman, J. (ed.), 2002. *In Vitro Mutagenesis Protocols. Methods in Molecular Biology*, 2nd ed., Vol. 182. Humana Press, Totowa, NJ.

Brennan, Y., Callen, W. N., Christoffersen, L. et al. 2004. Unusual microbial xylanases from insect guts. *Appl. Environ. Microbiol.* 70(6), 3609–3176.

Callanan, M. J., Russel, W. M., Klaenhammer, T. R. 2007. Modification of *Lactobacillus* beta-glucuronidase activity by random mutagenesis. *Gene.* 389, 122–127.

Campbell, R. E., Tour, O., Palmer, A. E., Steinbach, P. A., Baird, G. S., Zacharias, D. A., Tsien, R. Y. 2002. A monomeric red fluorescent protein. *Proc. Natl. Acad. Sci. U.S.A.* 99, 7877–7882.

Canchi, D. R., Garcia, A. E. 2013. Cosolvents effects on protein stability. *Annu. Rev. Phys. Chem.* 64, 273–293.

Cao, L., van Rantwijk, F., Sheldon, R. A. 2000. Cross-linked enzyme aggregates: A simple and effective method for immobilization of penicillin acylase. *Org. Lett.* 2(10), 1361–1364.

Cao, L., van Langen, L., Sheldon, R. A. 2003. Immobilized enzymes: Carrier bound or carrier free? *Curr. Opin. Biotechnol.* 14, 387–394.

Carter, P. 1986. Site directed mutagenesis. *Biochem. J.* 237, 1–7.

Cefic. 2004. A European technology platform for sustainable chemistry. Available at ftp://ftp.cordis.europa.eu/pub/nanotechnology/docs/tp_sust_chem.pdf.

Chao, F. A., Morelli, A., Haugner, J. C., Churchfield, L., Hagmann, L. N., Shi, L., Masterson, L. R., Sarangi, R., Veglia, G., Seelig, B. 2013. Structure and dynamics of a primordial catalytic fold generated by in vitro evolution. *Nat. Chem. Biol.* 9(2), 81–83.

Chaplin, M. 2014. Enzyme technology. Available at http://www1.lsbu.ac.uk/water/enztech/index.html, accessed October 6, 2014.

Chaplin, M., Buck, C. 1990. *Enzyme Technology.* Press Syndicate of the University of Cambridge, Cambridge, NY.

Chen, J. P., Su, D. R. 2001. Latex particles with thermo-flocculation and magnetic properties for immobilization of α-chymotrypsin. *Biotechnol. Progress.* 17, 369–375.

Chen, M. M., Snow, C. D., Vizcarra, C. L., Mayo, S. L., Arnold, F. H. 2012. Comparison of random mutagenesis and semi-rational designed libraries for improved cytochrome P450 BM3-catalyzed hydroxylation of small alkanes. *Protein Eng. Des. Sel.* 25(4), 171–178.

Cheriyan, M., Toone, E. J., Fierke, C. A. 2007. Mutagenesis of the phosphate-binding pocket of KDPG aldolase enhances selectivity for hydrophobic substrates. *Protein Sci.* 16(11), 2368–2377.

Cheriyan, M., Walters, M. J., Kang, B. D., Anzaldi, L. L., Toone, E. J., Fierke, C. A. 2011. Directed evolution of a pyruvate aldolase to recognize a long chain acyl substrate. *Bioorg. Med. Chem.* 19(21), 6447–6453.

Cheriyan, M., Toone, E. J., Fierke, C. A. 2012. Improving upon nature: Active site remodeling produces highly efficient aldolase activity toward hydrophobic electrophilic substrates. *Biochemistry.* 51(8), 1658–1668.

Chica, R. A., Doucet, N., Pelletier, J. N. 2005. Semi-rational approaches to engineering enzyme activity: Combining the benefits of directed evolution and rational design. *Curr. Opin. Biotechnol.* 16(4), 378–384.

Chien, A., Edgar, D. B., Trela, J. M. 1976. Deoxyribonucleic acid polymerase from the extreme thermophile Thermusaquaticus. *J. Bacteriol.* 127(3), 1550–1557.

Choudhary, R. B., Jana, A. K., Jha, M. K. 2004. Enzyme technology applications in leather processing. *Ind. J. Chem. Technol.* 11(5), 659–671.

Collins, K. D. 1995. Sticky ions in biological systems. *Proc. Natl. Acad. Sci. U.S.A.* 92, 5553–5557.

Crawford, J. M., Thomas, P. M., Scheerer, J. R., Vagstad, A. L., Kelleher, N. L., Townsend, C. A. 2008a. Deconstruction of iterative multidomain polyketide synthase function. *Science.* 320(5873), 243–246.

Crawford, J. M., Vagstad, A. L., Whitworth, K. P., Ehrlich, K. C., Townsend, C. A. 2008b. Synthetic strategy of nonreducing iterative polyketide synthases and the origin of the classical "starter-unit effect." *Chembiochem.* 9, 1019–1023.

Creus, M., Pordea, A., Rossel, T. et al. 2008. X-ray structure and designed evolution of an artificial transfer hydrogenase. *Angew. Chem. Int. Ed.* 47, 1400–1404.

Cui, J. D., Li, L. L., Zhao, Y. M. 2014. Simple technique for preparing stable and recyclable cross-linked enzyme aggregates with crude-pored microspherical silica core. *Ind. Eng. Chem. Res.* 53, 16176–16182.

Dalby, P. A. 2003. Optimizing enzyme function by directed evolution. *Curr. Opin. Struct. Biol.* 13, 500–505.

Dalby, P. A. 2011. Strategy and success for the directed evolution of enzymes. *Curr. Opin. Struct. Biol.* 21(4), 473–480.

Dang, D. T., Ha, S. H., Lee, S. M., Chang, W. J., Koo, Y. M. 2007. Enhanced activity and stability of ionic liquid-pretreated lipase. *J. Mol. Catal. B-Enzym.* 45(3), 118–121.

De Xammar Oro, J. R. 2001. Role of co-solute in biomolecular stability: Glucose, urea and water structure. *J. Biol. Phys.* 27, 73–79.

Demirjian, D. C., Morís-Varas, F., Cassidy, C. S. 2001. Enzymes from extremophiles. *Curr. Opin. Chem. Biol.* 5(2), 144–151.

DeRegil, R., Sandoval, G. 2013. Biocatalysis for biobased chemicals. *Biomolecules.* 3, 817–847.

Dror, A., Shemesh, E., Dayan, N., Fishman, A. 2014. Protein engineering by random mutagenesis and structure-guided consensus of Geobacillus stearothermophilus Lipase T6 for enhanced stability in methanol. *Appl. Environ. Microbiol.* 80(4), 1515–1527.

Edwards, I. R., MacLean, K. S., Dow, J. D. 1973. Low dose urokinase in major pulmonary embolism. *Lancet.* 2, 409–413.

El-Zahab, B., Jia, H., Wang, P. 2004. Enabling multienzyme biocatalysis using nanoporous materials. *Biotechnol. Bioeng.* 87, 178–183.

Eppler, R. K. 2007. Salt activation of enzymes in organic solvents: Mechanistic insights revealed through magnetic resonance spectroscopy. PhD Thesis. University of California, Berkeley, CA.

Fagain, C. O. 2003. Enzyme stabilization—Recent experimental progress. *Enzyme Microb. Technol.* 33, 137–149.

Farell, R., Hata, K., Wall, M. B. 1997. Solving pitch problems in pulp and paper processes by the use of enzymes or fungi. In Eriksson, K. E. L. et al. (eds.) *Biotechnology in the Pulp and Paper Industry. Advances in Biochemical Engineering/Biotechnology*, Vol. 57, pp. 197–212. Springer, Berlin.

Fersht, A. 1999. *Structure and Mechanism in Protein Science: A Guide to Enzyme Catalysis and Protein Folding.* W. H. Freeman, New York.

Findrik, Z., Vasic-Racki, D. 2009. Overview on reactions with multi-enzyme systems. *Chem. Biochem. Eng.* 23(4), 545–553.

Folsom, B. R., Schieche, D. R., DiGrazia, P. M., Werner, J., Palmer, S. 1999. Microbial desulfurization of alkylated dibenzothiophenes from a hydrodesulfurized middle distillate by Rhodococcuserythropolis I-19. *Appl. Environ. Microbiol.* 65(11), 4967–4972.

Fu, J., Leiros, H. K., de Pascale, D., Johnson, K. A., Blencke, H. M., Landfald, B. 2013. Functional and structural studies of a novel cold-adapted esterase from an Arctic intertidal metagenomic library. *Appl. Microbiol. Biotechnol.* 97(9), 3965–3978.

Galinski, E. A., Stein, M., Amendt, B., Kinder, M. 1997. The kosmotropic (structure-forming) effect of compensatory solutes. *Comp. Biochem. Physiol.* 117A, 357–365.

Garlet, T. B., Weber, C. T., Klaic, R., Foletto, E. L., Jahn, S. L., Mazutti, M. A., Kuhn, R. C. 2014. Carbon nanotubes as supports for inulinase immobilization. *Molecules.* 19, 14615–14624.

Giger, L., Caner, S., Obexer, R., Kast, P., Baker, D., Ban, N., Hilvert, D. 2013. Evolution of a designed retro-aldolase leads to complete active site remodeling. *Nat. Chem. Biol.* 8, 494–498.

Goldsmith, M., Tawfik, D. S. 2012. Directed enzyme evolution: Beyond the low-hanging fruit. *Curr. Opin. Struct. Biol.* 22(4), 406–412.

Guazzaroni, M. E., Beloqui, A., Vieites, J. M. et al. 2010. Metagenomic mining of enzyme diversity. In Timmis, K. N. (ed.) *Handbook of Hydrocarbon and Lipid Microbiology.* Springer Berlin, Heidelberg, 2911–2927.

Guex, N., Peitsch, M. C. 1997. SWISS-MODEL and the Swiss-PdbViewer: An environment for comparative protein modeling. *Electrophoresis.* 18, 2714–2723.

Guisán, J. M. (ed.) 2006. *Immobilization of Enzymes and Cells; Methods in Biotechnology-Book 22.* Humana Press, Totowa, NJ.

Gutiérrez, A., del Río, J. C., Martínez, A. T. 2009. Microbial and enzymatic control of pitch in the pulp and paper industry. *Appl. Microbiol. Biotechnol.* 82(6), 1005–1018.

Hanefeld, U., Gardossi, L., Magner, E. 2009. Understanding enzyme immobilization. *Chem. Soc. Rev.* 38, 453–468.

Hill, K., Rhode, O. 1999. Sugar based surfactants for consumer products and technical applications. *Fett-Lipid.* 101, 25.

Hilvert, D. 2013. Design of protein catalysts. *Annu. Rev. Biochem.* 82, 447–470.

Hirsh, J., Hale, G. S., McDonald, T. G., McCarthy, R. A., Pitt, A. 1968. Streptokinase therapy in acute major pulmonary embolism: Effectiveness and problems. *Br. Med. J.* 4, 729–734.

Hoyoux, A., Blaise, V., Collins, T. et al. 2004. Extreme catalysts from low-temperature environments. *J. Biosci. Bioeng.* 98(5), 317–330.

Huang, Z. M., Zhang, Y. Z., Kotaki, M., Ramakrishna, S. 2003. A review on polymer nanofibers by electrospinning and their applications in nanocomposites. *Compos. Sci. Technol.* 63, 2223–2253.

Hwang, E. T., Gu, M. B. 2013. Enzyme stabilization by nano/microsized hybrid materials. *Eng. Life Sci.* 13(1), 49–61.

Ilmberger, N., Meske, D., Juergensen, J. et al. 2012. Metagenomic cellulases highly tolerant towards the presence of ionic liquids—Linking thermostability and halotolerance. *Appl. Microbiol. Biotechnol.* 95(1), 135–146.

Inada, Y., Furukawa, M., Sasaki, H., Kodera, Y., Hiroto, M., Nishimura, H., Matsushima, A. 1995. Biomedical and biotechnological applications of PEG- and PM-modified proteins. *Trends Biotechnol.* 13(3), 86–91.

Ivarsson, Y., Norrgård, M. A., Hellman, U., Mannervik, B. 2007. Engineering the enantioselectivity of glutathione transferase by combined active-site mutations and chemical modifications. *Biochim. Biophys. Acta.* 1770, 1374–1381.

Jensen, P. H., Weilguny, D. 2005. Combination primer polymerase chain reaction for multi-site mutagenesis of close proximity sites. *J. Biomol. Tech.* 16, 336–340.

Jia, H., Zhu, G., Wang, P. 2003. Catalytic behaviors of enzymes attached to nanoparticles: The effect of particle mobility. *Biotechnol. Bioeng.* 84(4), 406–414.

Jiang, L., Althoff, E. A., Clemente, F. R. et al. 2008. De novo computational design of retro-aldol enzymes. *Science.* 319, 1387–1391.

Jochens, H., Aerts, D., Bornscheuer, U. T. 2010. Thermostabilization of an esterase by alignment-guided focused directed evolution. *Protein Eng. Des. Sel.* 23(12), 903–909.

Joosten, H. J. 2007. 3DM: From data to medicine. PhD Thesis. Wageningen University, Wageningen, The Netherlands, 109 pp.

Kandimalla, V. B., Tripathi, V. S., Ju, H. 2006. Immobilization of biomolecules in sol–gels: Biological and analytical applications. *Crit. Rev. Anal. Chem.* 36(2), 73–106.

Kang, T. S., Stevens, R. C. 2009. Structural aspects of therapeutic enzymes top treat metabolic disorders. *Hum. Mutat.* 30(12), 1591–1610.

Karan, R., Capes, M. D., DasSarma, S. 2012. Function and biotechnology of extremophilic enzymes in low water activity. *Aquat. Biosyst.* 8, 4, doi:10.1186/2046-9063-8-4.

Karimi, S., Ghourchian, H., Banaei, A. 2013. Protein structure preservation by MWCNTs/RTIL nano-composite. *Int. J. Biol. Macromol.* 56, 169–174.

Khersonsky, O., Roodveldt, C., Tawfik, D. S. 2006. Enzyme promiscuity: Evolutionary and mechanistic aspects. *Curr. Opin. Chem. Biol.* 10(5), 498–508.

Khersonsky, O., Rothlisberger, D., Wollacott, A. M. et al. 2011. Optimization of the in silico designed Kemp eliminase KE70 by computational design and directed evolution. *J. Mol. Biol.* 407, 391–412.

Kim, Y. G., Maas, S. 2000. Multiple site mutagenesis with high targeting efficiency in one cloning step. *Biotechniques.* 28, 196–198.

Kim, J., Grate, J. 2003. Single enzyme nanoparticles armored by a nanometer-scale organic/inorganic network. *Nano Lett.* 3(9), 1219–1222.

Kim, J. B., Grate, J. W., Wang, P. 2006. Nanostructures for enzyme stabilization. *Chem. Eng. Sci.* 61(3), 1017–1026.

Kim, J., Grate, J. W., Wang, P. 2008. Nanobiocatalysts and its potential applications. *Trends Biotechnol.* 26, 639–646.

Kim, J. H., Nam, D. H., Park, C. B. 2014. Nanobiocatalytic assemblies for artificial photosynthesis. *Curr. Opin. Biotechnol.* 28, 1–9.

Kishore, D., Talat, M., Srivastava, O. N., Kayastha, A. M. 2012. Immobilization of β-galactosidase onto functionalized graphene nano-sheets using response surface methodology and its analytical applications. *PLoS One.* 7(7), e40708.

Kiss, G., Johnson, S. A., Nosrati, G., Çelebi-Ölçüm, N., Kim, S., Paton, R., Houk, K. N. 2011. Computational design of new protein catalysts. In Comba, P. (ed.) *Modeling of Molecular Properties*. Wiley-VCH Verlag GmbH & Co. KGaA, Weinheim, Germany. doi:10.1002/9783527636402.ch16.

Kiss, G., Celebi-Olcum, N., Moretti, R., Baker, D., Houk, K. N. 2013. Computational enzyme design. *Angew. Chem. Int. Ed.* 52, 5700–5725.

Klein, L. C., Garvey, G. J. 1980. Kinetics of the sol-gel transition. *J. Non. Cryst. Solids.* 38, 45.

Klibanov, A. M. 2001. Improving enzymes by using them in organic solvents. *Nature.* 409, 241–246.

Köhler, V., Wilson, Y. M., Dürrenberger, M. et al. 2013. Synthetic cascades are enabled by combining biocatalysts with artificial metalloenzymes. *Nat. Chem.* 5, 93–99.

Kragl, U., Kruse, W., Hummel, W., Wandrey, C. 1996a. Enzyme engineering aspects of biocatalysis: Cofactor regeneration as example. *Biotechnol. Bioeng.* 52(2), 309–319.

Kragl, U., Vasic-Rackid, D., Wandrey, C. 1996b. Continuous production of L-tert-leucine in series of two enzyme membrane reactors. *Bioprocess Eng.* 14(6), 291–297.

Kuhlman, B., Baker, D. 2000. Native protein sequences are close to optimal for their structures. *Proc. Natl Acad. Sci. U.S.A.* 97, 10383–10388.

Kuipers, R. K., Joosten, H. J., van Berkel, W. J. et al. 2010. 3DM: Systematic analysis of heterogeneous super-family data to discover protein functionalities. *Proteins.* 78(9), 2101–2113.

Kumar, A., Singh, S. 2013. Directed evolution: Tailoring biocatalysts for industrial applications. *Crit. Rev. Biotechnol.* 33(4), 365–378.

Kumar, V., Singh, S. N., Kalonia, D. S. 2012. Mechanism of stabilization of proteins by poly-hydroxy co-solvents: Concepts and implications in formulation development. *Am. Pharma. Rev.* 112491. Available at http://www.americanpharmaceuticalreview.com/Featured-Articles/112491-Mechanism-of-Stabilization-of-Proteins-by-Poly-hydroxy-Co-solvents-Concepts-and-Implications-in-Formulation-Development/.

Kwon, D. Y., Hong, Y. J., Yoon, S. H. 2000. Enantiomeric synthesis of (S)-2-methylbutanoic acid methyl ester, apple flavor, using lipases in organic solvent. *J. Agric. Food Chem.* 48(2), 524–530.

Labrou, N. E. 2010. Random mutagenesis methods for in vitro directed enzyme evolution. *Curr. Protein Pept. Sci.* 11(1), 91–100.

Lee, J. K., Kim, M. J. 2002. Ionic liquid coated enzyme for biocatalysis in organic solvent. *J. Org. Chem.* 67, 6845–6847.

Levy-Sakin, M., Berger, O., Feibish, N. et al. 2014. The influence of chemical chaperones on enzymatic activity under thermal and chemical stresses: Common features and variation among diverse chemical families. *PLoS One.* 9(2), e88541. doi:10.1371/journal.pone.0088541.

Li, L., Li, Q., Li, F., Shi, Q., Yu, B., Liu, F., Xu, P. 2006. Degradation of carbazole and its derivatives by a Pseudomonas sp. *Appl. Microbiol. Biotechnol.* 73, 941–948.

Li, L. L., McCorkle, S. R., Monchy, S., Taghavi, S., van der Lelie, D. 2009. Bioprospecting metagenomes: Glycosyl hydrolases for converting biomass. *Biotechnol. Biofuels.* 2(10). doi:10.1186/1754-6834-2-10.

Li, X., Zhang, Z., Song, J. 2012. Computational enzyme design approaches with significant biological outcomes: Progress and challenges, *Comput. Struct. Biotechnol. J.* 2(3), e201209007. Available at http://dx.doi.org/10.5936/csbj.201209007.

Liszka, M. J., Clark, M. E., Schneider, E., Clark, D. S. 2012. Nature vs nurture: Developing enzymes that function under extreme conditions. *Annu. Rev. Chem. Biomol. Eng.* 3, 77–102.

Liu, Y., Kuhlman, B. 2006. Rosetta Design server for protein design. *Nucleic Acids Res.* 34(Web Server issue), W235–W238. doi:10.1093/nar/gkl163.

Liu, H., Naismith, J. H. 2008. An efficient one-step site-directed deletion, insertion, single and multiple-site plasmid mutagenesis protocol. *BMC Biotechnol.* 8(91). doi:10.1186/1472-6750-8-91.

Liu, W., Zhang, S., Wang, P. 2009. Nanoparticle-supported multi-enzyme biocatalysis with in situ cofactor regeneration. *J. Biotechnol.* 139(1), 102–107.

Liu, Y., Du, J., Yan, M. et al. 2013. Biomimetic enzyme nanocomplexes and their use as antidotes and preventive measures for alcohol intoxication. *Nat. Nanotechnol.* 3, 187–192.

Lopez-Gallego, F., Schmidt-Dannert, C. 2010. Multi-enzymatic synthesis. *Curr. Opin. Chem. Biol.* 14(2), 174–183.

Lopez-Lopez, O., Cerdan, M. E., Gonzalez Siso, M. I. 2014. New extremophilic lipases and esterases from metagenomics. *Curr. Protein Pept. Sci.* 15(5), 445–455.

Lozano, P., Bernal, B., Bernal, J. M., Pucheaut, M., Vaultier, M. 2011. Stabilizing immobilized cellulase by ionic liquids for saccharification of cellulose solutions in 1-butyl-3-methylimidazolium chloride. *Green Chem.* 13, 1406–1410.

Lu, Y., Yeung, N., Sieraki, N., Marshall, N. M. 2009. Review article design of functional metalloproteins. *Nature.* 460, 855–862.

Lutz, S. 2010. Beyond directed evolution—Semi-rational protein engineering and design. *Curr. Opin. Biotechnol.* 21(6), 734–743.

Majumder, A. B., Shah, S., Gupta, M. N. 2007. Enantioselective transacetylation of (R,S)-beta-citronellol by propanol rinsed immobilized Rhizomucor miehei lipase. *Chem. Cent. J.* 18, 1–10.

Makai, D. K., Harsanyi, G. 2009. Biomedical sensing. In Scwartz, M. (ed.) *Smart Materials.* CRC Press. Boca Raton, FL, 29-21, 29-26.

Malisi, C., Kohlbacher, O., Höcker, B. 2009. Automated scaffold selection for enzyme design. *Proteins.* 77(1), 74–83.

Margolin, A. L. 1996. Novel crystalline catalysts. *Trends Biotechnol.* 14, 223–230.

Martinez, R. Y., Schwaneberg, U. 2013. A roadmap to directed enzyme evolution and screening systems for biotechnological applications. *Biol. Res.* 46(4), 395–405.

Mateo, C., Grazú, V., Pessela, B. C. et al. 2007. Advances in the design of new epoxy supports forenzyme immobilization-stabilization. *Biochem. Soc. Trans.* 35, 1593–1601.

Mathey, D. G., Schofer, J., Sheehan, F. H. 1987. Coronary thrombolysis with intravenous urokinase in patients with acute myocardial infarction. *Am. J. Med.* 83(2A), 26–30.

Mathieu, V., Fastrez, J., Soumillion, P. 2010. Engineering allosteric regulation into the hinge region of a circularly permuted TEM-1 β-lactamase. *Protein Eng. Des. Sel.* 23(9), 699–709.

Mechrez, G., Kepker, M. A., Harel, Y., Lellouche, J. P., Segal, E. 2014. Biocatalytic carbon nanotube paper: A "one-pot" route for fabrication of enzyme-immobilized membranes for organophosphate bioremediation. *J. Mater. Chem. B.* 2, 915–922.

Meyer, D., Walter, L., Kolter, G., Pohl, M., Miller, M., Tittmann, K. 2011. Conversion of pyruvate decarboxylase into an enentioselective carboligase with biosynthetic potential. *J. Am. Chem. Soc.* 133, 3609–3616.

Mikaelian, I., Sergeant, A. 1992. A general and fast method to generate multiple site directed mutations. *Nucleic Acids Res.* 20, 376.

Monticello, D. J., Baker, D. T., Finnerty, W. R. 1985. Plasmid-mediated degradation of dibenzothiophene by Pseudomonas species. *Appl. Environ. Microbiol.* 49, 756–760.

Moore, B. D., Partridge, J., Halling, P. J. 2001. Very high activity biocatalysts for low-water systems: Propanol-rinsed enzyme preparations. In Vulfson, E. N., Halling, P. J., Holland, H. L. (eds.) *Enzymes in Nonaqueous Solvents: Methods and Protocols.* Methods in Biotechnology Series, Vol. 15. Human Press, Totowa, NJ, 97–104.

Morley, K. L., Kazlauskas, R. J. 2005. Improving enzyme properties: When are closer mutations better? *Trends Biotechnol.* 23, 231–237.

Moss, G. P. 2014. Recommendations of the Nomenclature Committee of the International Union of Biochemistry and Molecular Biology on the nomenclature and classification of enzymes by the reactions they catalyse. Available at http://www.chem.qmul.ac.uk/iubmb/enzyme/, accessed November 10, 2014.

Nakane, K., Hotta, T., Ogihara, T., Ogata, N., Yamaguchi, S. 2007. Synthesis of (Z)-3-hexen-1-yl acetate by lipase immobilized in polyvinyl alcohol nanofibers. *J. Appl. Pol. Sci.* 106(2), 863–867.

Nicolosi, V., Chhowalla, M., Kanatzidis, M. G., Strano, M. S., Coleman, J. N. 2013. Liquid exfoliation of layered materials. *Science*. 340(6139), 1226419. doi:10.1126/science.1226419.

Nierstrasz, V. A., Warmoeskerken, M. M. C. G. 2003. Process engineering and industrial enzyme applications. In Cavaco Paulo, A., Gubitz, G. M. (eds.) *Textile Processing with Enzymes*. CRC Press LLC, Boca Raton, FI., 120–157.

Oksanen, T., Pere, J., Paavilainen, L., Buchert, J., Viikari, L. 2000. Treatment of recycled kraft pulps with *Trichoderma reesei* hemicellulases and cellulases. *J. Biotechnol.* 78, 39–48.

Onoda, A., Hayashi, T., Salmain, M. 2014. Artificial metalloenzymes containing an organometallic active site. In Jaouen, G., Salmain, M. (eds.), *Bioorganometallic Chemistry. Applications in Drug Discovery Biocatalysis, and Imaging*. Wiley-VCH Verlag GmBH & Co, KGaA, Weinheim, Germany, 305–338.

Oyama, K. 1992. The industrial production of aspartame. In Collins, A. N., Sheldrake, G. N., Crosby, J. (eds.) *Chirality in Industry*. John Wiley & Sons, Chichester, UK, 237–247.

Paice, M. G., Jurasek, L. 1984. Removing hemicellulose from pulps by specific enzymic hydrolysis. *J. Wood Chem. Technol.* 4(2), 187–198.

Panesar, P. S., Marwaha, S. S., Chopra, H. K. 2010. *Enzymes in Food Processing: Fundamentals and Potential Applications*. IK International Publishing House, New Delhi, India.

Panizza, P., Cesarini, S., Diaz, P., Giordano, S. R. 2015. Saturation mutagenesis in selected amino acids to shift Pseudomonas sp. acidic lipase Lip I.3 substrate specificity and activity. *Chem. Commun.* 51, 1330–1333.

Park, H. S., Nam, S. H., Lee, J. K., Yoon, C. N., Mannervik, B., Benkovic, S. J., Kim, H. S. 2006. Design and evolution of new catalytic activity with an existing protein scaffold. *Science*. 311(5760), 535–538.

Partridge, J., Halling, P. J., Moore, B. D. 1998. Practical route to high activity enzyme preparations for synthesis in organic media. *Chem. Commun.* 7, 841–842.

Pasut, G., Veronese, F. M. 2009. PEGylation for improving the effectiveness of therapeutic biomolecules. *Drugs Today (Barc.)*. 45(9), 687–695.

Pavelka, A., Chovancova, E., Damborsky, J. 2009. HotSpot Wizard: A web server for identification of hot spots in protein engineering. *Nucleic Acids Res.* 37, W376–W383. doi:10.1093/nar/gkp410.

Pavlidis, I. V., Vorhaben, T., Gournis, D., Papadopoulos, G. K., Bornscheuer, U. T., Stamatis, H. 2012. Regulation of catalytic behaviour of hydrolases through interactions with functionalized carbon based nanomaterials. *J. Nanopart. Res.* 14, 842–851.

Peixoto, R. S., Vermelho, A. B., Rosado, A. S. 2011. Petroleum-degrading enzymes: Bioremediation and new prospects. *Enzyme Res.* 2011, 475193. Available at http://dx.doi.org/10.4061/2011/475193.

Penning, T. M., Jez, J. M. 2001. Enzyme redesign. *Chem. Rev.* 101(10), 3027–3046.

Perner, M., Ilmberger, N., Köhler, H. U., Chow, J., Streit, W. R. 2011. Emerging fields in functional metagenomics and its industrial relevance: Overcoming limitations and redirecting the search for novel biocatalysts. In de Bruijn, F. J. (ed.) *Handbook of Molecular Microbial Ecology II: Metagenomics in Different Habitats*. John Wiley & Sons, Inc., Hoboken, NJ, 483–498.

Pinnell, L. J., Dunford, E., Ronan, P., Hausner, M., Neufeld, J. D. 2014. Recovering glycoside hydrolase genes from active tundra cellulolytic bacteria. *Can. J. Microbiol.* 60(7), 469–476.

Podtetenieff, J., Taglieber, A., Bill, E., Reijerse, E. J., Reetz, M. T. 2010. An artificial metalloenzyme: Creation of a designed copper binding site in a thermostable protein. *Angew. Chem. Int. Ed.* 49(30), 5151–5155.

Polizzi, K. M., Chaparro-Riggers, J. F., Vazquez-Figueroa, E., Bommarius, A. S. 2006. Structure-guided consensus approach to create a more thermostable penicillin G acylase. *Biotechnol. J.* 1(5), 531–536.

Pollak, P. 2011. *Fine Chemicals: The Industry and the Business*. John Wiley & Sons, Hoboken, NJ.

Quin, M. B., Schmidt-Dannert, C. 2011. Engineering of biocatalysts: From evolution to creation. *ACS Catal.* 1(9), 1017–1021.

Rabausch, U., Juergensen, J., Ilmberger, N. et al. 2013. Functional screening of metagenome and genome libraries for detection of novel flavonoid-modifying enzymes. *Appl. Environ. Microbiol.* 79(15), 4551–4563.

Rastall, R. 2007. *Novel Enzyme Technology for Food Applications*. Woodhead Publishing Ltd., Abington, Cambridge, UK. Available at http://www.hc-sc.gc.ca/fn-an/securit/addit/list/5-enzymes-eng.php.

Reed, C. J., Lewis, H., Trejo, E., Winston, V., Evilia, C. 2013. Protein adaptations in archaea extremophiles. *Archaea*. 2013, 373275. Available at http://dx.doi.org/10.1155/2013/373275.

Reetz, M. T., Soni, P., Fernández, L. 2009. Knowledge-guided laboratory evolution of protein thermolability. *Biotechnol. Bioeng.* 102(6), 1712–1717.

Reikofski, J., Tao, B. Y. 1992. Polymerase chain reaction (PCR) techniques for site-directed mutagenesis. *Biotechnol. Adv.* 10(4), 535–547.

Reiner, T., Jantke, D., Marziale, A. N., Raba, A., Eppinger, J. 2013. Metal-conjugated affinity labels: A new concept to create enantioselective artificial metalloenzymes. *Chem. Open.* 2, 50–54.

Research & Markets 2013. Food Enzymes Market: Global Trends and Forecasts to 2018, Market research report ID 2569707, 288 pp.

Robles, V. M., Durrenberger, M., Heinisch, T., Lledos, A., Schirmer, T., Ward, T. R., Marechal, J. D. 2014. Structural, kinetic, and docking studies of artificial imine reductases based on biotin–streptavidin technology: An induced lock-and-key hypothesis. *J. Am. Chem. Soc.* 136(44), 15676–15683.

Rodríguez-Martínez, J. A., Rivera-Rivera, I., Solá, R. J., Griebenow, K. 2009. Enzymatic activity and thermal stability of PEG-α-chymotrypsin conjugates. *Biotechnol. Lett.* 31(6), 883–887.

Romero, M. D., Calvo, L., Alba, C., Daneshfar, A. 2007. A kinetic study of isoamyl acetate synthesis by immobilized lipase-catalyzed acetylation in n-hexane. *J. Biotechnol.* 127, 269–277.

Rothlisberger, D., Khersonsky, O., Wollacott, A. M. et al. 2008. Kemp elimination catalysts by computational enzyme design. *Nature.* 453, 190–195.

Saifuddin, N., Raziah, A. Z., Junizah, A. R. 2012. Carbon nanotubes: A review on structure and their interaction with proteins. *J. Chem.* 2013, 676815. Available at http://dx.doi.org/10.1155/2013/676815.

Sandström, A. G., Wikmark, Y., Engström, K., Nyhlén, J., Bäckvall, J. E. 2012. Combinatorial reshaping of the Candida Antarctica lipase A substrate pocket for enantioselectivity using an extremely condensed library. *Proc. Natl. Acad. Sci. U.S.A.* 109(1), 78–83. doi:10.1073/pnas.1111537108.

Schmidt, E., Vasiæ-Raèki, Ð., Wandrey, C. 1987. Enzymatic production of l-phenylalanine from the racemic mixture of d,l-phenyllactate. *Appl. Microb. Biotechnol.* 26(1), 42–48.

Schneider, S., Sandalova, T., Schneider, G., Sprenger, G. A., Samland, A. K. 2008. Replacement of a phenylalanine by a tyrosine in the active site confers fructose-6-phosphate aldolase activity to the transaldolase of Escherichia coli and human origin. *J. Biol. Chem.* 283, 30064–30072.

Schultheisz, H. L., Szymczyna, B. R., Scott, L. G., Williamson, J. R. 2008. Pathway engineered enzymatic de novo purine nucleotide synthesis. *ACS Chem. Biol.* 3(8), 499–511.

Sekhon, S. S., Park, J. M., Ahn, J. Y. et al. 2014. Immobilization of para-nitrobenzyl esterase-CLEA on electrospun polymer nanofibers for potential use in the synthesis of cephalosporin-derived antibiotics. *Mol. Cell. Toxicol.* 10(2), 215–221.

Sen, S., VenkataDasu, V., Mandal, B. 2007. Developments in directed evolution for improving enzyme functions. *Appl. Biochem. Biotechnol.* 143(3), 212–223.

Sheldon, R. A. 2007. Enzyme immobilization—The quest for optimum performance. *Adv. Synth. Catal.* 349, 1289–1307.

Sheldon, R. A. 2011. Characteristic features and biotechnological applications of cross linked enzyme aggregates (CLEAs). *Appl. Microbiol. Biotechnol.* 92, 467–477.

Shu, F. D. 2014. Self-assembled surface layer proteins from corynebacterium glutamicum as a matrix for cutinase immobilization. Master's Thesis, Aalborg University, Aalborg, Denmark. Available at http://projekter.aau.dk/projekter/files/198671519/Self_assembled_surface_layer_proteins_from_Corynebacterium_glutamicum_as_a_matrix_for_cutinase_immobilization.pdf.

Siegel, J. B., Zanghellini, A., Lovick, H. et al. 2010. Computational design of an enzyme catalyst for a stereoselective bimolecular Diels–Alder reaction. *Science.* 329, 309–329.

Sikri, N., Bardia, A. 2007. A history of streptokinase use in acute myocardial infarction. *Tex. Heart Int. J.* 34(3), 318–327.

Smith, W. P., Bishop, M., Gillis, G., Maibach, H. 2007. Topical proteolytic enzymes affect epidermal and dermal properties. *Int. J. Cosmet. Sci.* 29(1), 15–21.

Smith, M. D., Zanghellini, A., Grabs-Röthlisberger, D. 2014. Computational design of novel enzymes without cofactors. *Methods Mol. Biol.* 1216, 197–210.

Spreti, N., Di Profio, P., Marte, L., Bufali, S., Brinchi, L., Savelli, G. 2001a. Activation and stabilization of α-chymotrypsin by cationic additives. *Eur. J. Biochem.* 268, 6491–6497.

Spreti, N., Reale, S., Amicosante, G., Di Profio, P., Germani, R., Savelli, G. 2001b. Influence of sulfobetaines on the stability of the *Citrobacter diversus* ULA-27 -lactamase. *Biotechnol. Prog.* 17, 1008–1013.

Stankovich, S., Dikin, D. A., Dommett, G. H. B. et al. 2006. Graphene-based composite materials. *Nature.* 442, 282–286.

Steiner, K., Schwab, H. 2012. Recent advances in rational approaches for enzyme engineering. *Comput. Struct. Biotechnol. J.* e201209010, doi:10.5936/csbj.201209010.

Stepankova, V., Bidmanova, S., Koudelakova, T., Prokop, Z., Chalaukova, R., Damborsky, J. 2013. Strategies for stabilization of enzymes in organic solvents. *ACS Catal.* 3, 2823–2836.

Streit, W., Daniel, R. 2010. *Methods in molecular biology. Metagenomics: Methods and Protocols-XI*, Vol. 668. Humana Press, Springer, 341.

Sun, Z., Vivekananthan, J., Guschin, D. A. et al. 2014. High-concentration graphene dispersions with minimal stabilizer: A scaffold for enzyme immobilization for glucose oxidation. *Chemistry.* 20(19), 5752–5761.

Tabe, H., Fujita, K., Abe, S. et al. 2014. Preparation of a cross-linked porous protein crystal containing Ru carbonyl complexes as a CO-releasing extracellular scaffold. *Inorg. Chem.* 54, 215–220.

Talbi, B., Haquette, P., Martel, A. et al. 2010. (η6-Arene) ruthenium (II) complexes and metallo-papain hybrid as Lewis acid catalysts of Diels–Alder reaction in water. *Dalton Trans.* 24, 5605–5607.

Tang, L., Zhu, X., Zheng, H., Jiang, R., Majerić Elenkov, M. 2012. Key residues for controlling enantio-selectivity of halohydrin dehalogenase from Arthrobacter sp. Strain AD2, revealed by structure-guided directed evolution. *Appl. Environ. Microbiol.* 78(8), 2631–2637.

Tang, C., Saquing, C. D., Morton, S. W., Glatz, B. N., Kelly, R. M., Khan, S. A. 2014. Cross-linked polymer nanofibers for hyperthermophilic enzyme immobilization: Approaches to improve enzyme performance. *ACS Appl. Mater. Interfaces* 6(15), 11899–11906.

Tasis, D., Tagmatarchis, N., Bianco, A., Prato, M. 2006. Chemistry of carbon nanotubes. *Chem. Rev.* 106(3), 1105–1136.

Tee, K. L., Wong, T. S. 2013. Polishing the craft of genetic diversity creation in directed evolution. *Biotechnol. Adv.* 31(8), 1707–1721.

Thomas, T., Gilbert, J., Meyer, F. 2012. Metagenomics—A guide from sampling to data analysis. *Microb. Inform. Exp.* 2(1), 3.

Tracewell, C. A., Arnold, F. H. 2009. Directed enzyme evolution: Climbing fitness peaks one amino acid at a time. *Curr. Opin. Chem. Biol.* 13, 3–9.

Tschiggerl, H., Breitwieser, A., de Roo, G., Verwoerd, T., Schäffer, C., Sleytr, U. B. 2008. Exploitation of the S-layer self-assembly system for site directed immobilization of enzymes demonstrated for an extremo-philic laminarinase from *Pyrococcus furiosus. J. Biotechnol.* 133(3), 403–411.

Turner, N. J. 2009. Directed evolution drives the next generation of biocatalysts. *Nat. Chem. Biol.* 5(8), 567–573.

Turner, K. M., Pasut, G., Veronese, F. M., Boyce, A., Walsh, G. 2011. Stabilization of a supplemental digestive enzyme by post-translational engineering using chemically activated polyethylene glycol. *Biotechnol. Lett.* 33(3), 617–621.

Uchiyama, T., Miyazaki, K. 2009. Functional metagenomics for enzyme discovery: Challenges to efficient screening. *Curr. Opin. Biotechnol.* 20(6), 616–622.

Underkofler, L. A. 1980. Enzymes. In Furia, T. E. (ed.) *CRC Handbook of Food Additives*, 2nd ed., Vol. 2. CRC Press, Boca Raton, FL, 57–124.

Ursoiu, A., Paul, C., Kurtan, T., Peter, F. 2012. Sol-gel entrapped *Candida antarctica* lipase B—A biocatalyst with excellent stability for kinetic resolution of secondary alcohols. *Molecules.* 17, 13045–13061.

US-FDA 2012. Updated questions and answers for healthcare professionals and the public: Use an approved pancreatic enzyme product (PEP). Available at http://www.fda.gov/drugs/drugsafety/postmarketdrug safetyinformationforpatientsandproviders/ucm204745.htm.

US-FDA 2014. Rennet (animal derived) and chymosin preparation (fermentation derived)—Sec 184.1685, In: Code of Federal Regulations. Title 21. Volume 3. CITE: 21CFR184.1685.

van den Burg, B. 2003. Extremophiles as a source for novel enzymes. *Curr. Opin. Microbiol.* 6(3), 213–218.

van Oers, M. C., Rutjes, F. P., van Hest, J. C. 2014. Cascade reactions in nanoreactors. *Curr. Opin. Biotechnol.* 28, 10–16.

Vazquez-Figueroa, E., Chaparro-Riggers, J., Bommarius, A. S. 2007. Development of a thermostable glucose dehydrogenase by a structure-guided consensus concept. *Chem. Biochem.* 8, 2295–2301.

Verma, M. L., Barrow, C. J., Puri, M. 2013a. Nanobiotechnology as a novel paradigm for enzyme immobilisa-tion and stabilisation with potential applications in biodiesel production. *Appl. Microbiol. Biotechnol.* 97(1), 23–39.

Verma, M. L., Naebe, M., Barrow, C. J., Puri, M. 2013b. Enzyme immobilisation on amino-functionalised multi-walled carbon nanotubes: Structural and biocatalytic characterisation. *PLoS One.* 8(9), e73642. doi:10.1371/journal.pone.0073642.

Veronese, F. M. 2001. Peptide and protein PEGylation: A review of problems and solutions. *Biomaterials.* 22, 405–417.

Veronese, F. M., and Mero, A. 2008. The impact of PEGylation on biological therapies. *BioDrugs.* 22(5), 315–329.

Vester, J. K., Glaring, M. A., Stougaard, P. 2014. An exceptionally cold-adapted alpha-amylase from a metagenomic library of a cold and alkaline environment. *Appl. Microbiol. Biotechnol.* doi:10.1007/s00253-014-5931-0.

Viikari, L., Ranua, M., Kantelinen, A., Sundquist, J., Linko, M. 1986. Bleaching with enzymes, In: Proceedings, International Conference. Biotechnology in Pulp and Paper, Stockholm. 67–69.

Wang, P. 2006. Nanoscale biocatalyst systems. *Curr. Opin. Biotechnol.* 17(6), 574–579.

Wang, W., Malcolm, B. A. 1999. Two-stage PCR protocol allowing introduction of multiple mutations, deletions and insertions using QuikChange Site-Directed Mutagenesis. *Biotechniques.* 26, 680–682.

Wang, M., Si, T., Zhao, H. 2012. Biocatalyst development by directed evolution. *Bioresour. Technol.* 115, 117–125.

Wang, K., Luo, H., Tian, J. et al. 2014. Thermostability improvement of a Streptomyces xylanase by introducing proline and glutamic acid residues. *Appl. Environ. Microbiol.* 80(7), 2158–2165.

Ward, T. R. 2011. Artificial metalloenzymes based on the biotin-avidin technology: Enantioselective catalysis and beyond. *Acc. Chem. Res.* 44(1), 47–57.

Weissman, K. 2004. Rational or random? *RSC Chem. World* July 2004, 1(7). Available at http://www.rsc.org/chemistryworld/Issues/2004/July/rational.asp.

Whitehurst, R. J., van Oort, M. 2010. *Enzymes in Food Technology*, 2nd Edition, Wiley Blackwell, IA.

Yamada, H., Kobayashi, M. 1996. Nitrile hydratase and its application in industrial production of acryl amide. *Biosci. Biotechnol. Biochem.* 60, 1391–1400.

Yu, F., Cangelosi, V. M., Zastrow, M. L. et al. 2014. Protein design: Toward functional metalloenzymes. *Chem. Rev.* 114(7), 3495–3578.

Zanghellini, A., Jiang, L., Wollacott, A. M., Cheng, G., Meiler, J., Althoff, E. A., Baker, D. 2006. New algorithms and an in silico benchmark for computational enzyme design. *Protein Science: A Publication of the Protein Society.* 15(12), 2785–2794. doi:10.1110/ps.062353106.

Zhang, L., Aboagye, A., Kelkar, A., Lai, C., Fong, H. 2014. A review: Carbon nanofibers from electrospun polyacrylonitrile and their applications. *J. Mater. Sci.* 49, 463–480.

Zhao, H. 2009. Methods for stabilizing and activating enzymes in ionic liquids—A review. *J. Chem. Technol. Biotechnol.* 85, 891–907.

Zhao, H., Arnold, F. H. 1999. Directed evolution converts subtilisin E into a functional equivalent of thermitase. *Protein Eng.* 12(1), 47–53.

Zoller, M. J. 1991. New molecular biology methods for protein engineering. *Curr. Opin. Biotechnol.* 2(4), 526–531.

3

Enzymes as Analytical Tools for the Assessment of Food Quality, Food Safety, and Monitoring of Food Processing

Muthusamy Chandrasekaran and Ali H. A. Bahkali

CONTENTS

3.1 Introduction

Foods contain high amounts of macronutrients, such as proteins, carbohydrates, and lipids, and extremely low amounts of flavor components, vitamins, antioxidants, etc. Qualitative and quantitative determination of these different components requires analytical techniques that are not only very specific and sensitive, but simple and affordable. Further, the food and beverage industries need rapid and cost-effective methods to determine compounds that have not previously been monitored and to replace existing ones. Although technological advancements in chromatography, electrophoresis, mass spectrometry, etc., have addressed the challenges in analysis of food ingredients, enzymes, by virtue of their specificity, relative sensitivity, and accurate determination of both macro- and micronutrients in foods, have become ideal and versatile candidates (Ashie 2012) for potential applications as analytical tools in food and beverages. Moreover, recent developments in biosensors with improved robustness and versatility has also led to a wide range of applications of enzymes as an analytical tool in several industries besides their use in clinical diagnoses. In this context, the role of enzymes as analytical tools for the analysis of food and beverage components with respect to food quality control, food safety, and monitoring of food processing is discussed in this chapter.

3.2 Need for Analysis of Food and Beverages

The food industry has not only grown immensely over the years, but also evolved into one of the most significant and dependent industries among all other industries, owing to the rapid change in lifestyle. In fact, recently, there has been a tremendous shift in consumer preferences, which have shifted to healthier and more flavorsome food with higher nutritional value. Further, there is an increased awareness of food safety and quality, which is closely related to compliance with established legal standards concerning human health risks, the environment, animal welfare, protection of natural resources, and ethical requirements. The sensory impact due to flavor, smell, and appearance has also assumed an equal importance among the consumer preferences. As a consequence, raw food matrices and processed food products, including end products as well as semifinished products, require compliance with quality requisites with a suitable degree of confidence and compliance with legal standards.

Agricultural food production involves extensive use of insecticides, fungicides, and other compounds applied in agricultural practice and environmental contaminants, such as polycyclic aromatic hydrocarbons (PAHs), polychlorinated biphenyls (PCBs), and the heterogeneous groups of endocrine disrupting chemicals (EDCs). These undesired chemicals often remain as contaminants and storage residues in raw food materials that are processed for food and beverage production. Mycotoxin contamination is often experienced during postharvest storage and in stored finished foods. These substances pose a threat to food safety, and regulatory agencies have emphasized the need for analytical methods that can determine as many residues and contaminants as possible in food products.

Today, food industries produce a diverse range of food products, including beverages, satisfying consumer preferences while complying with legal stipulations. Food quality control is established through systematic analyses, which ensures not only compliance with legal and labeling requirements, but also enables assessment of food authenticity, geographical origin, species discrimination, product quality, determination of nutritive value, detection of adulterations, research and development, etc. Thus, now food analysis has become imperative due to escalation in the demand for the highest nutritional quality food, food safety concerns of food authorities, and the efforts of producers and the industry, are to meet these demands.

The core of food and beverage analyses mainly includes qualitative and quantitative determination of the following (Table 3.1):

- Volatile compounds from vegetable matrices, essential oils, and various extracts; components that contribute flavor and aroma and off odors

TABLE 3.1

Food Ingredients and Contaminants Routinely Analyzed for Quality Control

Category	Variables Analyzed
Carbohydrates	Starch, reducing sugars, maltose, glucose, polydextrose, dextrin, galacto oligosaccharides, lactulose, lactosucrose, isomalto oligosaccharides, malitol, palatinose, soybean oligosaccharides, fructo oligosaccharides (FOS), xylo oligosaccharides (XOS)
Proteins	Protein, nonprotein nitrogen, gluten, amino acids, peptides, casein phosphor peptide, casein dodeca peptide
Oils and fats	Lipids, glycerol, fatty acids, low chain fatty acids, fatty acid methyl esters (FAMEs), triglycerides, sterols, cholesterol, etc.
Vitamins	All vitamins
Minerals	Phosphorus, calcium as citrate malate, heme iron, metals
Others	Alcohol, organic acids, lactic acid, carotenoids, polyphenols, anthocyanins, volatile compounds from vegetable matrices, essential oils and various extracts, components that contribute flavor and aroma and off odors
Food additives and residues	Antioxidants, veterinary drugs, food packaging migration products, food preservatives, histamine, sodium saccharin, hypoxanthine (Hx), and xanthine
Environmental contaminants	Pesticide residues, polycyclic aromatic hydrocarbons (PAHs), polychlorinated biphenyls (PCBs), polychlorinated dibenzo-p-dioxins and dibenzofurans (PCDD.Fs), polybrominated diphenyl ethers (PBDEs), phthalates, mineral oils, heterogeneous groups of endocrine disrupting chemicals (EDCs)
Toxins	Mycotoxins, endotoxins, phytotoxins
Microbial pathogens	*Salmonella, Listeria, Campylobacter, Escherichia coli* 0157:H7, *Pseudomonas, Staphylococcus aureus*

- Nonvolatiles or semivolatiles, including lipids, proteins, carbohydrates, carotenoids, vitamins, antioxidants, polyphenols, etc.

- Fat analysis and characterization (low chain fatty acids, fatty acid methyl esters [FAMEs], triglycerides, sterols, etc.)

- Food additives, contaminants, and residues that include pesticides, veterinary drugs, microbial and phytotoxins, endocrine disrupting chemicals (EDCs), food-packaging migration products, environmental contaminants, such as polycyclic aromatic hydrocarbons (PAHs), polychlorinated biphenyls (PCBs), polychlorinated dibenzo-p-dioxins and dibenzofurans (PCDD.Fs), polybrominated diphenyl ethers (PBDEs), phthalates, and mineral oils

- Authentication of geographical origin and species discrimination in meat products

- Age/storage duration of food

- Microbial contamination and presence of toxins/fermentation products resulting from microbial growth

Several analytical methods (Table 3.2) are in use for routine analysis of food and beverages to ensure freshness, nutritional quality of food, and food safety. They mainly include gravimetric, spectrophotometric, chromatographic, electrophoretic, molecular techniques, and immunological and microbiological methods. Over the years, the progress made in food science technology through the continued efforts in research and development have driven change of classic analytical methods (titrimetric or gravimetric analysis) to instrumental and biochemical ones (spectroscopy, chromatography, electrophoresis, and biosensors) because of the new quantitative and qualitative information generated. In spite of the wide range of applications of the conventional methods involving instrumentation techniques, the scope for potential application of enzymes as analytical tools has recently been recognized by the food science fraternity.

TABLE 3.2

Analytical Methods Routinely Used for Qualitative and Quantitative Determination of Food Ingredients and Contaminants

Category	Method	Instrumentation Involved	Target Chemical/ Biochemical Analyzed
Chemical	Gravimetric/ titrimetric	Titration	Acids, alkali, amines, toxins, metals, etc.
	Colorimetric	Colorimeter	Organic compounds
	Spectrophotometric	UV visible spectrophotometer, mass spectrophotometer (MS), infrared (IR) spectrophotometer, NMR spectrophotometer	Organic and inorganic compounds, carbohydrates, proteins, lipids, vitamins, minerals, toxins, pesticides, etc.
Physical	Chromatography	Column chromatography, high-performance liquid chromatography (HPLC), liquid chromatography, gas chromatography (GC), GC-MS	Separation and purification of biomolecules, amino acids, peptides, toxins, sugars, organic compounds, etc.
	Electrophoretic-gel electrophoresis	Electrophoresis apparatus	Separation and purification of proteins, peptides, nucleic acids
	Microscopy	Light microscopy, electron microscopy	Identification of microorganisms, protein molecules
Biological	Immunological	Enzyme-linked immunosorbent assay (ELISA)	Proteins, pesticides, allergens, toxins
Microbiological	Bacterial counts	Plate assay	Bacterial load, fungal load, bioburden

3.3 Enzymes as Analytical Tools

Enzymes are now used as part of the analytical methods with applications in pharmaceutical, food, agricultural, environmental, and industrial analysis. They are often considered to be ideal analytical tools for measuring the concentration of compounds that serve as substrates, activators, or inhibitors of specific enzymes. The concentration and/or activity of selected enzymes are useful for the determination of nutritionally or physically induced changes in measuring quality indices of foods. Most of the constituents of food, such as protein, sugar, acids, alcohol, and other substances, can be determined except for additives such as antioxidants or preservatives. Moreover, enzymes are used to probe the primary, tertiary, and quaternary structures of polymeric molecules, such as proteins, nucleic acids, carbohydrates, and triglycerides, making it possible to perform routine analytical determinations that would be laborious or otherwise impossible.

The major advantage of use of enzymes in analysis include high specific activity and sensitivity in a reaction, which enables easy determination of the change in substrate or the reaction products, the concentration of the assayed material, and measurement of the degree of activation of an enzymatic reaction when compared to physical and chemical methods (Giannoccaro et al. 2008). The substrate that is measured is quantitatively transformed, and one of the reaction products is detected photometrically or through an easily readable chemisty. Because enzymes are biological catalysts, the equilibrium constants of the catalyzed reactions are not altered, and their use as analytical reagents is based on their selective effect on the reaction kinetics: decreasing the activation energy and forming a lower energy barrier. Generally, end-point and kinetic analyses are possible. End-point analysis refers to the total conversion of substrates into products in the presence of enzymes in a few minutes, and kinetic analysis involves the rate of reaction and substrate/product concentration. The selectivity of enzymes is based on differences between chemical groups, chemical bonds, or stereoisomers, and enzymes are systematically classified according to the type of chemical reaction that they catalyze (Table 3.3). This wide range of chemical activity makes possible the analysis of many compounds with very different chemical characteristics.

TABLE 3.3

Classification of Enzymes Based on Type of Chemical Reaction Catalyzed

Class	Important Subclass	Type of Chemical Reaction Catalyzed	Examples
Oxidoreductase	Dehydrogenases, oxidases, peroxidases, reductase, monoxygenases, dioxygenases	Oxidation-reduction in which oxygen and hydrogen are gained or lost	Lactate dehydrogenase, glucose oxidase, hydrogen peroxidase, laccases, lipoxygenase
Transferase	C1-transferases, glycosyltransferases, aminotransferases, phosphotransferases	Transfer of functional groups, such as amino group acetyl group, or phosphate group	Transglutaminase, fructosyltransferase, cyclodextrin, glycosyltransferase
Hydrolase	Esterases, glycosidases, peptidases, amidases	Hydrolysis (addition of water)	Lipase, protease, sucrase, amylase, invertase, lactase, galactosidase, glucoamylase, pectinase
Lyase (synthases)	C–C lyases, C–O lyases, C–N lyases, C–S lyases	Removal of groups of atoms without hydrolysis	Oxalate decarboxylase, isocitrate lyase, acetolactate decarboxylase
Isomerase	Epimerases, *cis trans* isomerases, intramolecular transferases	Rearrangement of atoms within a molecule	Glucose isomerase, alanine racemate
Ligase (synthetases)	C–C ligases, C–O ligases, C–N ligases, C–S ligases	Joining of two molecules (using energy usually derived from the breakdown of ATP)	Acetyl-CoA synthetase, DNA ligase

Many enzymes require the presence of cofactors in their catalytic activity, which may be soluble and have to be added (coenzymes), for example, nicotinamide adenine dinucleotides for dehydrogenases, or may be firmly attached (prosthetic group) to the protein, for example, flavine adenine dinucleotides for oxidases. The addition of coenzymes allows monitoring of the reaction in cases in which the substrates or products lack detectable properties. A degree of further selectivity could be achieved by measuring the cofactor and minimizing the risk of interferences in complex samples. However, reliability of the assay depends on the control of reaction conditions, the presence of inhibitors of the enzyme in the sample, and contamination of the system by other enzymes.

Enzymatic tests are widely used as analytical tools for the analysis of food products such as fruit juices, wine or beer, dairy products, eggs, and meat (Figures 3.1 and 3.2). Nowadays, numerous enzyme-catalyzed

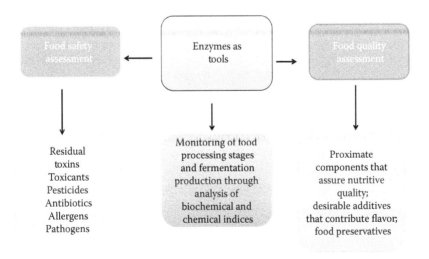

FIGURE 3.1 Enzymes as tools for food and beverage analyses.

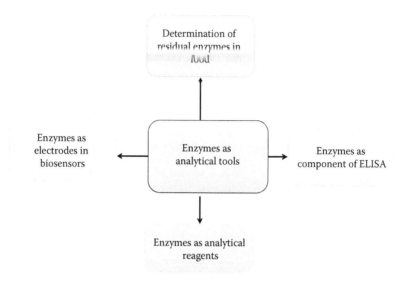

FIGURE 3.2   Types of applications of enzymes in food and beverage analyses.

reactions are applied in bioassays combined with a variety of detection techniques, such as spectroscopic, electrochemical, and mass spectrometry (Hempen and Karst 2006). Enzymatic test kits are used for the determination of sugars, acids, alcohols, and a few other food components. The tests are based on high-quality enzymes, enabling precise and specific measurement of each compound even in complex matrices. Results are measured with a spectrophotometer, and automation is possible.

The first enzyme-based systems for analytical purposes were soluble enzymes in aqueous media, carried out in the batch mode (Bergmeyer 1984). Immobilization of the enzymes results in solid phase reagents, which can be easily recovered and used repetitively (Guilbault 1984). For these reasons, development of the use of immobilized enzymes incorporated into flow systems, flow injection analysis (FIA), and liquid chromatography (LC) has been rapid (Araujo et al. 2010; Carr and Bowers 1980; Frei and Zech 1989; Nenkova et al. 2010). This is demonstrated by the wide range of different compounds analyzed by means of immobilized enzymes, for example, mono-, di-, oligo-, and polysaccharides; alcohols; aldehydes; uric, bile, and amino acids; purines; cholines; steroids; amines; phenols; glucuronides; urea; antibiotics; and vitamins (Marko-Varga and Dominguez 1991).

Several techniques apart from enzyme immunoassays have been developed in recent years for protein or allergen analysis, which utilize enzyme-labeled antibodies for detection. These include lateral flow assays, which are immunochromatographic tests involving movement of immunoreactants along a test strip and dipstick tests, which involve immobilized capture antibodies on a test strip with analyte detection by enzyme-labeled antibodies. Both techniques are generally inexpensive, rapid, and portable, thus making them ideal tools for online monitoring. Details of these techniques and other methods for protein and allergen analysis can be found in Schubert-Ulrich et al. (2009).

The development of immobilized enzyme systems with electrochemical detection methods has increased the applicability of the enzymatic assay in food quality control and also in food processing industries. Commercial kits of tests that include enzyme sensors are well established and have been approved as reference methods by recognized regulatory organizations, such as the International Organization for Standardization (ISO) and Association of Official analytical Chemists (AOAC 1995). The immobilized enzymes are used mainly in biosensors (Bilitewski 2000) and to a lesser extent in contaminant identification test strips (Boer and Beuner 1999).

Among the different enzyme classes, oxidoreductases are most frequently used for chemical analysis owing to their commercial availability in a highly active and purified form. The $NAD^+/NADH$ cofactor couple-dependent dehydrogenases are the largest subclass of oxidoreductases, comprising about 250 different enzymes. Substituting the hydrogen in position 2 in the ribose ring of the NADH-molecule with a phosphate group yields the NADP+/NADPH cofactor couple, which is the soluble cofactor for an additional 200 dehydrogenases (Marko-Varga and Dominguez 1991).

Numerous enzymatic methods have been approved or validated by international organizations, of which the most important are the following:

American Association of Analytical Chemists (AOAC)

European Committee for Standardization (CEN)

International Federation of Fruit Juice Producers (IFU)

International Dairy Federation (IDF)

International Standardization Organization (ISO)

International Organization of Wine (OIV)

(http://www.r-biopharm.com/products/food-feed-analysis/constituents/enzymatic-analysis).

3.4 Enzyme Immunoassays (EIAs)

Enzyme immunoassays (EIAs) encompass all immunoassays that use an enzyme-bound antibody to detect an antigen. The basic principles of immunoassay techniques are based on the strong and highly specific interaction occurring between antigens (Ag) and antibodies (Ab) as the reaction to follow. This recognition reaction is highly specific and possesses a very high constant (up to 10^{11}–10^{12} Lmol^{-1}).

$$Ab + Ag + Ag^+ \rightleftarrows AbAg^+ \tag{3.1}$$

The free antigen (Ag) and enzyme-labeled antigen (Ag+) compete for a fixed and limited number of specific binding sites on the antibody (Ab) molecules. After an incubation period, the free and antibody-bound antigens are separated from each other and the amount of labeled antigen in one of the fractions is determined. At higher concentrations of unlabeled antigens, fewer labeled antigen molecules will be bound by an antibody. Therefore, a calibration graph can be produced, and from this, the concentration of antigen in the samples can be determined (Blake and Gould 1984). Other labeled molecules that act with or upon enzymes that have been used include cofactors, prosthetic group, substrates, and inhibitors. Thus, in food applications, this interaction allows the development of assays for the detection of target compounds, such as pathogens, toxins, hormones, and drugs in low concentrations. Results can be interpreted visually (qualitatively) or using an instrument readout (quantitatively) (Gabaldon et al. 1999).

In comparison to analytical techniques, such as high performance liquid chromatography (HPLC), EIAs offer rapid analysis (a few minutes) with high sensitivity and sensibility without needing expensive equipment and highly trained personnel. Moreover, the desirable feature of such assays is their adaptability for throughput analyses, that is, the analysis of a large number of samples by means of assays in the wells of microtiter plates, in a short time, in an automated fashion. In addition, the portable formats are available for the detection of *Salmonella, Listeria, Campylobacter, E. coli* 0157:H7, *Pseudomonas,* and *Staphylococcus aureus* as commercial kits (Boer and Beuner 1999). The analytical application of the enzyme immunoassays represents a useful tool to detect contaminants in the food field.

Studies have shown that electrochemical transducers using the enzymes horseradish peroxidase (HRP), alkaline phosphatase (ALP), and glucose oxidase (GOx) as labels have good results for combination of sensitivity, selectivity, low cost, and the possibility of portability (Li et al. 2008). In this respect, a miniaturized immunoassay using *on-chip* has been developed to detect diseases such as cancer, arthritis, or heart disease, and can contribute in applications for food quality control and monitoring as screening methods. Another trend is the adaptation of the immunochemical techniques to design immunosuppressors based on biosensor technology, which can be used as rapid testing devices (Marquette and Blum 2006). However, most fluorescent immunoassays are in the enzyme-linked immunosorbent assay (ELISA) format, which uses the fluorescein molecule as the most common label (Kemeny 1991).

3.5 Enzyme-Linked Immunosorbent Assay (ELISA)

ELISA is the most frequently applied type of assay (Riska et al. 2007) for a broad range of food components and other ingredients. General principles underlying the use of sandwich and competitive ELISA (both of which may be direct or indirect) for protein analyses are presented in Figures 3.3 and 3.4.

3.5.1 Sandwich ELISA

In the *direct sandwich ELISA*, a capture antibody specific for the protein being analyzed is first immobilized onto a solid phase. Analytes containing the protein of interest present in the food material bind specifically to the immobilized antibody. Then, a second enzyme-labeled analyte-specific antibody is introduced into the system, which forms a sandwich and facilitates detection of the desired analyte.

In the *indirect sandwich ELISA*, a second analyte-specific antibody binds to a secondary site on the protein of interest (analyte) to form the sandwich before the enzyme-labeled antibody binding. The requirement for a secondary binding site or more than one epitope for specific binding makes this type of ELISA primarily applicable to large molecules such as proteins.

In both types, the binding reaction may be linked to a chromophore with the color intensity being proportional to the protein concentration. Potentiometric, amperometric, electrochemical, and piezoelectric detection methods have since been developed and widely used for these analyses.

3.5.2 Competitive ELISA

Competitive ELISA is more flexible and applicable to analysis of small as well as large molecules. In the *direct form of competitive ELISA*, the analyte of interest is first incubated together with the enzyme-labeled antibody, and the mixture added to the analyte-specific capture antibody with which it reacts competitively and in a concentration-dependent manner for detection (Figure 3.4) whereas the *indirect form of competitive ELISA* involves coincubation of the bound analyte of interest with an analyte-specific antibody. Analyte detection is then achieved by introducing a second enzyme-labeled antibody into the mix, which binds to the analyte-specific antibody. In another form of indirect competitive ELISA, the analyte-specific antibody is first bound to the surface, and the analyte or protein of interest is made to compete with an enzyme-labeled analyte or a tracer for binding to the antibody. In all cases, the binding is concentration-dependent and can be measured by the changes in absorbance or some other signal.

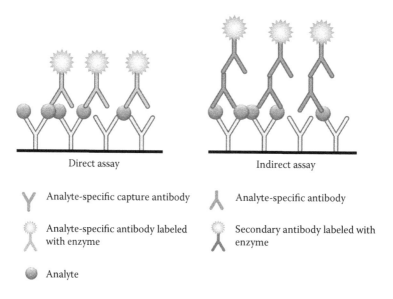

Direct assay   Indirect assay

Y Analyte-specific capture antibody   人 Analyte-specific antibody

Analyte-specific antibody labeled with enzyme   Secondary antibody labeled with enzyme

Analyte

FIGURE 3.3 Sandwich ELISA with direct and indirect enzyme-based antibody detection.

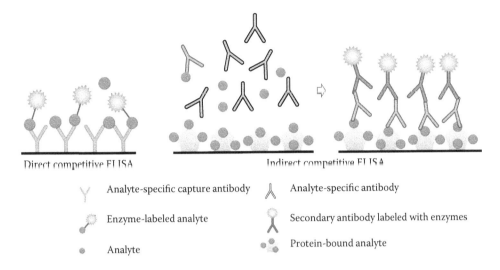

Direct competitive ELISA   Indirect competitive ELISA

Y Analyte-specific capture antibody   Analyte-specific antibody

Enzyme-labeled analyte   Secondary antibody labeled with enzymes

Analyte   Protein-bound analyte

FIGURE 3.4 Competitive ELISA with direct and indirect enzyme-based antibody detection.

ELISA is mostly performed in a 96-well microtiter plate allowing simultaneous analysis of up to 45 samples in duplicate. A chromogenic substance is mostly used as an enzyme substrate with incubation times of 0.5–2 h, and the developed color is usually measured spectrophotometrically (Schneider et al. 2004). Amperometric (Badea et al. 2004) or differential pulse voltammetry (Piermarini et al. 2009) detection is also possible. The most simple method, however, is the visual comparison of color intensity, providing either semiquantitative results (Pal and Dhar 2004) or a yes/no response at a certain concentration level (Prieto-Simón et al. 2007) or concentration range (Schneider et al. 2004). Some other immunoassay types use fluorescent markers, involving a fluorogenic substrate that reacts with the enzyme-linked analyte (de Champdoré et al. 2007; Prieto-Simón et al. 2007), or measure fluorescence polarization induced by increased molecular mass of antibody-bound labeled mycotoxin (Krska et al. 2007; Prieto-Simón et al. 2007).

The most common design is a multiwell microtiter plate used in classic ELISA with spectrophotometric or electrochemical detection (Piermarini et al. 2007). Flow-through or flow-injection immunoassays use immobilized proteins to separate antigen–antibody complexes and detect only the free marker (Zheng et al. 2006). Mycotoxin-tagged liposomes have been prepared to develop a flow-injection liposome immunoanalysis (FILIA) (Ho and Durst 2003). In flow-through tests, antibodies are immobilized on a semipermeable membrane. Such devices are simple to use in field conditions with visual detection but can also be used for quantitative evaluation with instrumental detection in laboratory settings (Pal and Dhar 2004; Schneider et al. 2004). Different immunostrips or immunodipsticks, also named lateral flow devices (tests), have also been developed for mycotoxin testing in the field, using either a membrane with immobilized antibodies or dry reagents on the test strip. In the latter case, the sample would reach and dissolve the reagents by lateral capillary flow, and the test is thus self-developing (Krska et al. 2007; Prieto-Simón et al. 2007; Schneider et al. 2004; Zheng et al. 2006). Some of the interesting applications of ELISA in the food industry include the detection of food allergens, such as gluten (Mermelstein 2011) and proteins from certain nuts (Cuuu ot al. 2011; Gaskin and Taylor 2011) and for detection of microbial pathogens and toxins (AOAC 2011).

3.6 Enzyme-Based Biosensors

Biosensors are analytical tools that measure the electrical signal obtained after conversion of a biological response. They consist of two main components: (i) a bioreceptor (enzyme, antibody, receptor, or microorganism) for selective analyte or food ingredient detection, and (ii) a signal transducer (electrochemical, mass, optical, or thermal) that transmits the resulting effect of the bioreceptor–analyte interaction to a measurable signal. These signals may be amperometric, piezoelectric, potentiometric, fluorometric, or colorimetric (Viswanathan et al. 2009). Based on the working principle, biosensors can be classified into various different types, which include electrochemical, conductometric, calorimetric, potentiometric, amperometric, optical, or even microbial, in which whole microbial cells act as biosensors by responding to the analyte in a monitorable fashion (Thakur and Ragavan 2013).

Biosensors offer advantages as alternatives to conventional methods due to their inherent specificity, simplicity, and quick response; relatively low cost of construction; and storage, potential for miniaturization, facility of automation, and simple and portable equipment construction for a fast analysis and monitoring in platforms of raw material reception, quality control laboratories, or some stage during the food processing (Luong et al. 1991). Biosensors could be applied to the determination of the composition and degree of contamination of raw materials and processed foods and for the online control of the fermentation process. Biosensors or immunosensors can reduce assay time and cost and/or increase the product safety.

Enzymes have most often been used in combination with electrochemical techniques in the development of biosensors. By controlling the potential applied to the electrode, a second degree of selectivity is introduced. The response signal is thus restricted in terms of discrimination toward interfering compounds. An enzyme electrode can be used amperometrically, when the current between the working electrode and the auxiliary electrode is measured, or potentiometrically, when the potential difference generated between the two electrodes is measured. The enzymes can be bound on the electrode surface by simple adsorption or by utilizing other functional groups on the protein for covalent coupling. However, they are most frequently bound in a membrane layer that is in direct contact with the electrode surface. The membranes can be polymerized, incorporating the enzyme(s), and then positioned on the electrode surface (Scheller and Schubert 1989). They can also be electropolymerized, when the monomers with the dissolved enzyme(s) in a batch form the membrane when potential is applied to the electrode. The use of analytical methods, such as HPLC or specific enzymatic methods, may be costly or laborious, and in this context, application of the biosensor can be considered to be an ideal tool for analysis that requires complicated procedures.

Ever since enzymatic electrodes were recognized as a biocomponent in biosensors to determine glucose using amperometric transducers, a large number of applications have been recognized for several other analytes, including other carbohydrates and contaminants and additive compounds in food. Thus, biosensors are being used for detection of pathogens in food (Leonard et al. 2003), for assessment of food quality (Hamada-Sato et al. 2005), or for detection of pesticides in food (Hildebrandt et al. 2008; Schöning et al. 2003). Applications of biosensors in the food industry have been reviewed by Cock et al. (2009) and Thakur and Ragavan (2013).

Some examples of enzymes used as biocomponents in biosensors for analyses of food components in the food and beverage industry are presented in Table 3.4. An immunoaffinity-based optical biosensor is used to determine folic acid concentration levels in milk powder, infant formula, and cereal samples (Malin and Johan 2000). The potential application of biosensor technology in the food and beverage industry was extensively reviewed by Mello and Kubota (2002).

Enzyme sensors fall into various classes, including those that are potentiometric, amperometric, optoelectric, calorimetric, and piezoelectric. Basically, all enzyme sensors work by immobilization of the enzyme system onto a transducer. Among the enzymes commercially available, the oxidases are the most often used. This type of enzyme offers the advantages of being stable and, in some situations, not requiring coenzymes or cofactors (Davis et al. 1995). Some examples of commercial biosensors based on enzymes available for industrial markets are PM-1000 and PM-1000DC (Toyo Joso) and OLGA (online general analyzer; Eppendorf) (Mello and Kubota 2002). The most successfully used biosensors in food applications are the enzyme reaction-based electrochemical types.

Microbial enzymes are commonly used in biosensor design, and glucose oxidase is, by far, the most commonly used enzyme. A glucose electrode in combination with a glucose oxidase reaction and electrochemical determination of oxygen and hydrogen peroxide was recognized as a potential biosensor for the estimation of glucose in various samples. Now, the principle of oxidase reaction has been extended to develop biosensors for other compounds. Oxidase electrodes measuring sugars, cholesterol, acids, amino acids, alcohols, and phenols have been prepared and tested in various foods, such as fruit juices, soft drinks, beer, wines, soy sauce, milk, and yogurt. Dehydrogenases utilize the electron transfer capability

TABLE 3.4

Applications of Some Enzymes in Biosensors Used for Food Analysis

Enzymes	Analyte	Application
Glucose oxidase (GDO)	Glucose	Soft drinks, milk, juices, tomato juice, musts, wine, honey, biscuits, beverages, synthetic samples, Coca-Cola
Glucose oxidase (GDO), galactose oxidase and peroxidase	Glucose and galactose	Yogurt, milk
Glucose oxidase (GDO), D-fructose dehydrogenase, alcohol dehydrogenase, L-lactate dehydrogenase, L-malate dehydrogenase, sulfite oxidase and diaphorase	Glucose, fructose, ethanol, lactate, L-malate, and sulfite (simultan)	Wine
Glucose oxidase (GDO) and glutamate oxidase (GLOX)	Glucose and glutamate	Beverages
Glucose oxidase (GDO), horseradish peroxidase (HRP), and urease	Glucose, ascorbic acids, and citric acids	Fruit drinks
D-fructose dehydrogenase (FDH)	Fructose	Honey, juice, dietetic jelly, sweetener, milk, wine
β-Galactosidase, lactozym, and *Saccharomyces cerevisiae*	Lactose	Milk
D-fructose dehydrogenase (FDH) and β-galactosidase (β-gal)	Lactulose	Milk
α-Amylase, amyloglucosidase (AMG), and glucose oxidase (GOD)	Starch	Wheat flour samples
Alcohol oxidase	Ethanol	Beer, synthetic samples
Alcohol dehydrogenase (ADH) and NaDH oxidase	Ethanol	Alcoholic beverages
Aldehyde dehydrogenase	Acetaldehyde	Alcoholic beverages
Glycerokinase and glycerol-3 phosphate oxidase	Glycerol	Monitoring fermentation
Glycerophosphate oxidase (GPO) and glycerol kinase (GK)	Glycerol	Wines
Tyrosinase	Polyphenols	Olive oil, green tea, grape and olive extracts
Horseradish peroxidase (HRP)	Polyphenols	Wines
Ascorbate oxidase	Ascorbic acid	Juices
Citrate lyase (CL)	Citric acid	Juices and sport drinks
Citrate lyase (CL), pyruvate oxidase (POD), and oxaloacetate descarboxylase (AOCD)	Citric acid, pyruvate acid, and oxaloacetic acid	Synthetic samples (fruits)
Catechol oxidase	Catechol	Synthetic samples
Polyphenol oxidase	Catechol	Beer
L-amino acid oxidase (L-AAO) and horseradish peroxidase (HRP)	L-amino acids	Synthetic samples
D-amino acid oxidase (D-AAO)	L-amino acids	Milk and fruit juices
L-glutamate oxidase (GluOD) and NaDH oxidase (NOD)	L-glutamate	Food seasonings
L-malate dehydrogenase (MDH) and pyruvate oxidase (POP)	L-malate	Wine, juice, and soft drinks
L-malate dehydrogenase, diaphorase and L-lactate dehydrogenase (LDH)	L-malate and L-lactate	Wine
Diamine oxidase (a) and polyamine oxidase (b)	Amines	Fruits and vegetables
Xanthine oxidase (XOD)	Amines	Meat, fish freshness
Diamine oxidase (DOO)	Amines, biogenic amines	Fish
Hypoxanthine oxidase and xanthine oxidase (XOD)	Amines	Fish freshness
Histamine oxidase	Histamine	Seafoods
Nitrate reductase (NR)	Nitrate	Synthetic samples
Oxalate oxidase (OXO)	Oxalate	Spinach, sesame seeds, tea leaves, and strawberry samples

(Continued)

TABLE 3.4 (CONTINUED)

Applications of Some Enzymes in Biosensors Used for Food Analysis

Enzymes	Analyte	Application
Polyphenol oxidase and alkaline phosphatase	Phosphate	Drinking water
Sulfite oxidase	Sulfite	Wine
Acetylcholinesterase (AChE)	Pesticides	Synthetic samples
Acetylcholinesterase (AChE) and butyrylcholinesterase (BChE)	Pesticides	Fruit and vegetable juices
Alcohol oxidase, α-chymotrysin, and catalase	Aspartame	Foods

Source: Mello, L. D. and Kubota, L. T., *Food Chemistry,* 77, 237–256, 2002.

of enzyme cofactors nicotinamide adenine dinucleotide (NAD) and nicotinamide adenine dinucleotide phosphate (NADP) to generate detectable signals, and dehydrogenase-based sensors were utilized to analyze glucose, fructose, lactose, gluconate, lactates, ethanol, and amino acids in foods and fermentation products. The enzyme coupling technique has also been applied in biosensor design. The principle is to link multiple enzyme reactions that convert the analytes into a measurable compound so as to increase the sensitivity (Cheng and Merchant 1995). This technique is widely used in food analysis, and numerous enzyme kits are commercially available.

The coupling of galactosidase with glucose oxidase and catalase in an enzyme electrode is used to determine lactose in milk (Pfeiffer et al. 1990). Detection of multicomponents by enzyme sensors is possible during the analysis of sucrose and glucose in honey (Xu and Guilbault 1990), drinks (Mizutani et al. 1991), and L-malate and L-lactate in wines (Mizutani and Asai 1990; Mulchandani et al. 2001). An alcohol oxidase thermistor has been used to monitor ethanol fermentation (Rank et al. 1992). Fiber optic enzyme sensors have also been used in food analysis. Electrodes made from platinum, graphite, and carbonic composite material have been used in the construction of the enzyme sensor (Ghindilis and Kurochkin 1994; Pandey et al. 1994; Reynolds and Yacynych 1994).

Xanthine oxidase has been used to determine the levels of xanthine and hypoxanthine that are accumulated from purine degradation during muscle aging so as to monitor fish freshness for a long time. Xanthine oxidase enzyme sensor for fish freshness was one of the earlier commercial biosensors. Stability, duration, sensitivity, interference, and availability of substrates to contact enzymes are the criteria for the success of an enzyme in food analysis.

The market for biosensors has followed the steady growth of the food industry. Many advantages are offered by biosensors in food quality control and processing, which include use of these devices for online processes or discrete sampling; it also does not interfere with the process stream because it involves no chemical addition, rapid response, better feedback, and control of the process. Biosensors for food control show rejection of below-standard raw materials on delivery and low cost monitoring of stored products and are expensive in comparison with traditional instruments; hence, they have become increasingly important as commercial portable devices (Song et al. 2006).

3.7 Enzymes in Determination of Food Quality

Food with high nutritional quality and that are fresh in terms of aroma, flavor, and appearance and are devoid of spoilage indices, such as off odors, undesirable amines, and microbial metabolite residues, or the presence of contaminants and toxins are some of the major concerns of consumers and regulatory bodies in addition to the food producers. Quality control analyses mainly focus on the proximate components of food and all other substances that attribute to food quality. Hence, all analytical methods, including chemical and microbiological, focus on determination of specific components and molecules in raw materials that are used in food and beverage production as well as finished and semifinished food products. In this context, enzymes are recently recognized for use as analytical tools in the determination of food quality (Table 3.5).

TABLE 3.5

Applications of Enzymes in Analysis of Food Quality

Category of Application	Variables Analyzed	Enzymes Used	References
Indices of food quality	Residual enzymes in milk and dairy products	Alkaline phosphatase	Murthy et al. 1992
	Residual enzymes in fruits and vegetables	Peroxidase, lipoxygenase, amylase	Powers 1998
	Residual enzyme used as preservative in ripened cheese, hen egg white, wines	Lysozyme (muramidase) (ELISA)	Kerkaert and De Meulenaer 2007; Vidal et al. 2005; Weber et al. 2007
	Metabolites: lactic acid in meat	Xanthine oxidase, diamine oxidase, and polyamide oxidase	Cock et al. 2009
	Metabolites in seafoods: ornithine and amines	Ornithine carbamoyl transferase, nucleoside phosphorylase, xanthine oxidase, and diamine oxidase for ornithine and amines	Cock et al. 2009
	Hypoxanthine	Xanthine oxidase	Cock et al. 2009
Microbial contaminants and pathogens	Salmonella in chicken	Horseradish peroxidase (ELISA)	Salam and Tothill 2009
Major components of food	Protein	Horseradish peroxidase (ELISA)	Schubert-Ulrich et al. 2009
	Gluten	Horseradish peroxidase (ELISA)	Don and Koehler 2014
	Amino acids	Amino acid decarboxylases, tryptophanase, aspartate aminotransferase	Wiseman 1981
	Glucose	Glucose oxidase, glucose-6-phosphate transferase, PQQ-bound dehydrogenases	Wiseman 1981
	Fructose	Pyrroloquinoline quinone (PQQ) enzyme, D-fructose dehydrogenase, D-fructose dehydrogenase	Paredes et al. 1997; Trivedi et al. 2009
	Sucrose	Invertase, mutarotase, glucose oxidase, and peroxidase	Matsumoto et al. 1988; Morkyavichene et al. 1984
	Cholesterol	Cholesterol esterase and cholesterol oxidase	Wisitsoraat et al. 2009
	Antioxidants	Laccase	Prasetyo et al. 2010
	Lactic acid	Malate dehydrogenase and lactate dehydrogenase	Avramescu et al. 2001, 2002; Lima et al. 1998
	Alcohol	NAD-dependent alcohol dehydrogenase, PQQ-bound dehydrogenases	Alpat and Telefoncu 2010
	Glycerol	Glycerophosphate oxidase (GPO) and glycerol kinase (GK)	Kiranas et al. 1997
	Phenolic compounds	Tyrosinase and horseradish peroxidase (HRP)	Campanella et al. 1999; Imabayashi et al. 2001
	Mycotoxins	Horseradish peroxidase (ELISA), alkaline phosphatase	Ammida et al. 2004; Parker and Tothill 2009
	Ochrotoxin	Acetylcholinesterase ELISA based on alkaline phosphatase and horseradish peroxidase	Arduini et al. 2007; Prieto-Simon and Campas 2009
	Trichothecene mycotoxin	Horseradish peroxidase	Lattanzio et al. 2009

(Continued)

TABLE 3.5 (CONTINUED)

Applications of Enzymes in Analysis of Food Quality

Category of Application	Variables Analyzed	Enzymes Used	References
	Histamine	Diamine oxidase, peroxidase	Landete et al. 2004
	Hypoxanthine	Xanthine oxidase	Hernández-Cázares et al. 2010
	Chlroramphenicol	β-Glucuronidase	Bogusz et al. 2004
	Nonsteroidal anti-inflammatory drugs (NSAIDs)	Cyclooxygenase	Campanella et al. 2009
	Acrylamide	Peroxidase-based ELISA	Preston et al. 2008
	Pesticides	Acetylcholinesterase, choline oxidase, cholinesterase, tyrosinase, peroxidases, and cellobiose dehydrogenase	Amine et al. 2006; Mulchandani et al. 2001; Solna et al. 2005

3.7.1 Assay of Residual Enzymes as Indices of Quality

The levels of activities of a number of enzymes as well as the content of certain food components and secondary metabolites resulting from postharvest or postprocessing biochemical and microbial activities have been recognized as quality indices for various foods and consequently monitored.

Measurement of the level of residual peroxidase in fruits and vegetables and residual alkaline phosphatase in milk and dairy products enables determination of the adequacy of heat treatment. Due to their relatively high heat stabilities and ease of determination, these two enzymes are considered to be useful indicators. However, other enzymes may be also useful indicators of the storage stabilities of some foods after heat processing but need validation through research efforts. The adequacy of blanching, an essential step in vegetable processing, is also established by measuring residual lipoxygenase or peroxidase activities (Powers 1998). Amylase activity in malt, also referred to as its diastatic power, is a very essential quality parameter that influences dextrinizing time and, therefore, is closely monitored (Powers 1998).

Milk endogenous alkaline phosphatase is inactivated following heat treatment at 60°C for 5 s. With milk generally subjected to high-temperature short-time (HTST)/flash pasteurization (i.e., 71.5°C–74°C/160°F–165°F for 15–30 s) or ultra-high-temperature treatment (i.e., 135°C/275°F for 1–2 s), residual alkaline phosphatase activity following such treatment provides a good indication of the efficacy of the treatment. The phosphatase test is based on hydrolysis of disodium phenyl phosphate to liberate phenol, which reacts with dichloroquinonechloroimide to form a blue indophenol that is measured colorimetrically at 650 nm (Murthy et al. 1992). An alternative fluorimetric method for alkaline phosphatase analysis has also been developed and commercialized (Rocco 1990).

3.7.2 Enzymes for Analyzing Metabolites

In meats, lactic acid levels tend to increase postmortem, and thus, their levels have been monitored as freshness indicators using xanthine oxidase, diamine oxidase, and polyamide oxidase. In most seafoods, freshness is rapidly compromised postharvest particularly when handled under temperature-abuse conditions that promote endogenous enzyme activity. The relative amounts of the different intermediates, such as ornithine, amines, and hypoxanthine, formed during ATP breakdown in postmortem fish muscle are generally accepted as a fish freshness indicator. These key freshness indicators, which rapidly build up in fresh seafoods, have been analyzed by various enzymes. These include ornithine carbamoyl transferase, nucleoside phosphorylase, xanthine oxidase, and diamine oxidase for ornithine and amines and xanthine oxidase for hypoxanthine (Cock et al. 2009).

3.7.3 Detection of Microbial Contaminants and Pathogens

In addition to the indirect indicators of food quality or safety described, there are other enzymatic methods for direct detection of contaminating microflora. The latter has gained even greater significance for controlling not only food-borne diseases but also the potential for bioterrorism. The United States Department of Agriculture (USDA) Pathogen Reduction Performance Standards set limits for the presence of *Salmonella* for all slaughter facilities and raw ground meat products (Alocilja and Radke 2003). Other well-known pathogens recognized as being responsible for food-borne diseases are *E. coli* 0157:H7, *Campylobacter jejuni, Listeria monocytogenes, Bacillus cereus, Staphylococcus aureus, Streptococci*, etc. Although there are a number of conventional methods for detection of such contaminating microflora or pathogens, they tend to be labor intensive, and results are usually not available until after a couple of days. A number of ELISA methods with high sensitivity have since been developed that address this drawback (Croci et al. 2001; Delibato et al. 2006; Salam and Tothill 2009). Some of these ELISA kits (e.g., LOCATE SALMONELLA produced by R-Biopharm) are also commercially available. In their sandwich ELISA method for *Salmonella* detection in precooked chicken, the capture antibody was immobilized onto a gold electrode surface, and a second antibody conjugated to horseradish peroxidase was used as the detection system for recognition of captured microbial cells (Salam and Tothill 2009). Detection or binding of the enzyme label is then conducted by an electrochemical system using tetramethylbenzidine dihydrochloride as an electron transfer mediator and hydrogen peroxide as the substrate.

3.7.4 Determination of Major Components of Food

Determination of the amount of polymeric materials containing glucose, such as cellulose, glycogen, starch, maltose, and sucrose are rather easy by the use of specific enzymes. The specific hydrolase for the polymer is used to hydrolyze the polymer to glucose, and then the glucose concentration is determined by the appropriate reactions. For example, the amount of starch in a plant material can be determined after liquefaction by hydrolysis with amyloglucosidase (glucoamylase), which hydrolyzes both α-1,4 and β-1,6 glucosidic linkages giving glucose as the only product.

3.7.4.1 Proteins

Proteins, one of the major components of foods, have nutritional and functional impact on foods. Nevertheless, some proteins in food induce allergic reactions in sensitized individuals following consumption. Hence, highly selective and sensitive tools were developed for detection and analysis of such protein components to ensure that food products conform to regulatory standards.

Lysozyme (muramidase), an enzyme derived from hen egg white, is a potential allergen. This is approved for use as an antimicrobial agent for preservation of the quality of ripened cheese. The joint Food and Agricultural Organization/World Health Organization Codex Alimentarius Commission require its presence in foods to be declared on the label as an egg product (Codex Standard I-1985). ELISA has been a major technique for analysis of lysozyme and other proteins with limits of detection ranging from 0.05 to 10 mg/kg (Schubert-Ulrich et al. 2009) due to its specificity, sensitivity, simple sample handling, and high potential for standardization and automation. ELISA is used for analysis of lysozymes in different kinds of cheese, and the method involves the use of a peroxidase-labeled antibody that undergoes a redox reaction with tetramethylbenzidine dihydrochloride to form a product that is measured by the absorbance change at 450 nm (Schneider et al. 2010). ELISA is used for analysis of lysozyme in hen egg white, wines, and other foods (Kerkaert and De Meulenaer 2007; Vidal et al. 2005; Weber et al. 2007) in addition to other proteins in hazelnuts (Holzhauser and Vieths 1999), peanuts (Pomes et al. 2003), walnuts (Doi et al. 2008), soy (You et al. 2008), and sesame (Husain et al. 2010).

Commercial ELISA-based kits are available for analysis of proteins in cereals (gliadins, secalin, hordein, prolamin, gluten), soybeans (soy trypsin inhibitor), mustard (mustard protein), milk (β-lactoglobulin, caseins), nuts (peanut Ara h1 and Ara h2), wine, juice (lupine, for example, γ-conglutin), eggs (ovomucoid, ovalbumin, ovotransferrin, lysozyme), and crustaceans (tropomyosin) from many venders (R, Biopharm, Astor Labs, Crystal Chem Inc., Cosmo Bio USA, ELISA systems, Morinaga Inst., Diagnostic Automation, Neogen, ELISA Technologies US Biologicals, among others).

3.7.4.2 Gluten

Celiac disease (an autoimmune disease) attacks the small intestine due to the presence of gluten, and a gluten-free diet is the only medically accepted treatment. The disease affects an estimated 1% of adults worldwide and appears to be on the increase (Anderson 2011). Gluten is estimated by ELISA using specific antibodies that interact with epitopes of gluten proteins that trigger an immune response in celiac disease patients. Four ELISA methods for gluten determination have been validated by international collaborative studies. The Skerritt ELISA method was validated in 2001. Three proprietary gluten ELISA methods, namely R5 sandwich ELISA, R5 competitive ELISA, and G12 sandwich ELISA kits, have since been developed and were validated between 2010 and 2013 (Don and Koehler 2014). Based on the validation studies, the R5 and G12 antibody-based proprietary test kits have been accepted as AACC International (AACCI) Approved Methods 38-50.01 and 38-52.01, respectively. They are used to verify compliance with the *Codex Alimentarius* threshold for gluten-free food declarations, which has been set at 20 mg of gluten/kg of food (as-is basis) for several cereal-based beverages. The R5 competitive ELISA assay (AACCI Approved Method 38-55.01 and American Society of Brewing Chemists Method Beer-49) are recommended to be used for fermented products and is the only collaboratively validated method currently available that meets the recommended detection limit of ≤10 mg of gluten/kg. However, the gluten peptides formed during fermentation are different than those used in the kit for calibration necessitating further research on preparing more representative calibration materials for hydrolyzed gluten (Don and Koehler 2014).

3.7.4.3 Amino Acids

The amino acid composition and free amino acid content of food influence protein metabolism and food quality. A high content of hydrophobic amino acids tends to cause bitterness in protein hydrolysates, and a high content of free amino acids provides free amino groups that promote Maillard reactions (browning reaction) when these foods are subjected to high-temperature treatments. The products/intermediates of the Maillard reaction may eventually influence the flavor and color of the food product. A Maillard reaction can eventually lead to acrylamide formation (Mottram et al. 2002). Quantitative and qualitative determination of the amino acid content in foods is routinely done using several methods that involve application of enzymes.

Amino acid decarboxylases were employed to decarboxylate amino acids, and the stoichiometric release of carbon dioxide was then analyzed using glass electrodes containing sodium bicarbonate. Tryptophanase was used to catalyze hydrolysis of the tryptophan to produce pyruvate. Coupling of the pyruvate production to lactate dehydrogenase converted NADH to NAD^+ and the corresponding change in absorbance at 340 nm was recorded as the index of tryptophan content. In a similar method, involving coupling to NADH for aspartic acid determination, the aspartate aminotransferase was used for the conversion of aspartic acid to oxaloacetate in the presence of malate dehydrogenase (Wiseman 1981).

3.7.4.4 Sugars

Among the sugars, glucose, fructose, and sucrose are generally analyzed. *Glucose* in foods and beverages is predominantly detected accurately using glucose oxidase. The underlying principle is based on hydrolysis of glucose to form gluconic acid and hydrogen peroxide. The hydrogen peroxide is then coupled to peroxidase catalysis, which oxidizes a colored precursor, resulting in an absorbance change that is proportional to the glucose content. Another method is based on the glucose-6-phosphate transferase catalyzed transfer of an acyl phosphate from a donor molecule to form glucose 6-phosphate. Coupling this reaction to NADP conversion to NADPH resulted in an absorbance change that is measured at 340 nm (Wiseman 1981).

Fructose is of significant interest in the food industry, and the level of fructose in foods provides an indication of the stage of ripening, adulteration of products such as honey, and the sweetness of various foods. Fructose is determined using the highly specific pyrroloquinoline quinone (PQQ) enzyme D-fructose dehydrogenase. The enzyme catalyzes oxidation of D-fructose to 5-keto-D-fructose with a corresponding reduction of the covalently bound cofactor PQQ to $PQQH_2$. Coupling reoxidation of the reduced cofactor to electrodes produces electric current in direct proportion to the amount of fructose (Paredes et al. 1997). In another method, D-fructose dehydrogenase, along with ferricyanide as an electron acceptor, was used for fructose determination. In this reaction, the enzyme oxidizes fructose to keto-fructose with the reduction of ferricyanide and the reduced ferricyanide is subsequently reoxidized, producing electrical current proportional to fructose concentration (Trivedi et al. 2009).

Sucrose (common sugar) is also determined using glucose oxidase. The analysis involves initial conversion of sucrose to D-fructose and α-D-glucose by invertase followed by isomerization of α-D-glucose to β-D-glucose by mutarotase. Later, β-D-glucose is oxidized by glucose oxidase to D-glucono-δ-lactone and hydrogen peroxide. In these catalytic reactions, the amount of peroxide released by the coupled reaction with peroxidase is proportional to the sucrose content (Matsumoto et al. 1988; Morkyavichene et al. 1984).

Glucose and ascorbic acid content of fruit juices have been analyzed using biosensors based on glucose oxidase and metal catalysts (Pt, Pd, and Au–Pd). The metal catalysts improved the sensitivity of the sensor by reducing the oxidation potential of the hydrogen peroxide released as a result of the glucose oxidase activity (Gutes et al. 2006).

Several enzymes are known to be employed in determination of different components in food (Zeravik et al. 2009). PQQ-bound dehydrogenases for the determination of glucose, alcohol, and glycerol; tyrosinase, laccase, and peroxidase for phenolic compounds; and sulfite oxidase for sulfite content were employed. Starch analysis involves basic principles of starch liquefaction and saccharification. Starch liquefaction is carried out with α-amylases releasing maltose units and limiting dextrin. Further treatment with pullulanase and amyloglucosidases (glucoamylases) results in hydrolysis of the maltose units and limit dextrin, ultimately producing glucose units that are determined. Analyses of other polysaccharides have been conducted in similar fashion by initial breakdown to simpler sugars using different carbohydrases depending on the nature of the substrate. Dextranases have been used for dextrans, cellulases for cellulose, hemicellulases for hemicelluloses in dietary fiber, and pectinases and polygalacturonases for polysaccharides in soy sauce and other plant-based foods (Wiseman 1981).

3.7.4.5 Cholesterol

High levels of cholesterol are noticed in cheese, egg yolk, beef, poultry, shrimp, and pork, and determination of cholesterol level in these foods has become mandatory in the food industry. Enzyme-based colorimetric assays of cholesterol include initial de-esterification of cholesterol esters by cholesterol esterase to release free cholesterol, which is further oxidized by cholesterol oxidase to generate hydrogen peroxide. The peroxide produced is subsequently coupled to a colored or fluorescent reagent in a peroxidase-catalyzed reaction (Van Veldhoven et al. 2002). Highly sensitive electrochemical biosensors have been developed with cholesterol esterase and cholesterol oxidase immobilized on electrode surfaces and carbon nanotubes (Wisitsoraat et al. 2009). A commercial kit (Boehringer's Monotest cholesterol or CHOD-PAP (cholesterol oxidase-phenol aminophenazone method) for the colorimetric assay is available, in which 4-aminoantipyrine and phenol are converted to a red cyanogenimine dye in the presence of the peroxide released.

3.7.4.6 Antioxidants

Antioxidants are incorporated in various food product formulations for their ability to maintain food quality during processing and storage. Antioxidants are also considered to protect consumers against some of the potential health risks associated with free radicals.

Antioxidant activity in a wide range of fruits and vegetables (e.g., apples, carrots, garlic, kiwi, lettuce, spinach, tomatoes), beverages (e.g., green tea, coffee), and oils is determined using enzymes. The laccase enzyme used in this antioxidant analysis oxidizes the yellowish compound syringaldazine to produce the purple-colored tetramethoxy azobismethylene quinine (TMAMQ) with maximum absorbance at 530 nm. In the presence of antioxidants, the free-radical intermediates are unavailable or converted, thereby proportionately blocking formation of the purplish TMAMQ (Prasetyo et al. 2010).

3.7.4.7 Lactic Acid

A number of enzymatic methods have been developed to detect and quantify organic acids, such as lactic, acetic, citric, malic, ascorbic, and succinic acids, which are attributed to quality in a number of food products, including wines, milk, vinegar, soy sauce, and fermented products (Wiseman 1981). Among the various organic acids, lactic acid is considered very important due to its impact on a broad range of food products. In yogurt production and cheese maturation, lactic acid is a major intermediate of *Lactobacillus* fermentation and determines the flavor of the final product. In wine production, organic acids impact flavor, protect against bacterial diseases, and may slow down ripening. Excessive lactic acid in wines has a negative effect on wine taste due to formation of acetate, diacetyl, and other intermediates (Avramescu et al. 2001; Shkotova et al. 2008). A method for the simultaneous measurement of both lactate and malate levels in wines using malate dehydrogenase and lactate dehydrogenase has been developed (Lima et al. 1998). These enzymes were injected into a buffer carrier stream flowing to a dialysis unit, where they react with the wine donor containing the two acids. Detection is based on the absorbance change at 340 nm due to reduction of NAD^+ to NADH. Amperometric biosensors were developed using NAD^+-dependent lactate dehydrogenase application for lactate analysis (Avramescu et al. 2001, 2002).

3.7.4.8 Alcohol

Alcohol (ethanol) content in beer, wine, and other alcoholic beverages is a very important quality attribute and is also required for regulatory classification. The NAD-dependent alcohol dehydrogenase is widely used for analysis of alcohol content in foods and alcoholic beverages. The enzyme catalyzes oxidation of ethanol to acetaldehyde in an equilibrium reaction that also produces NADH. The reaction is made to proceed in the forward direction by acetaldehyde reaction with semicarbazine or hydrazine. The absorbance change resulting from the NAD^+-to-NADH conversion is proportional to the alcohol content. Similar NAD-dependent assays have been used for measuring glycerol content in foods (Alpat and Telefoncu 2010).

3.7.5 Contaminants and Residues in Food

3.7.5.1 Toxins

Toxins may be naturally present (endogenously) or process-induced or come from the environment. Food toxins, such as phytotoxins (terpenoids, glycoalkaloids, phytosterols, flavonoids, lignans) and mycotoxins (aflatoxins [AFTs], fumonisins, ochratoxins [OTs], citrinin, mycophenolic acid, etc.) are all naturally occurring, and their toxic effects have been well documented in the literature. These toxins need to be analyzed to ensure food safety and quality.

Mycotoxins are toxic secondary metabolites produced by various fungi (*Aspergillus*, *Fusarium*, *Penicillium*, etc.) and excreted in their substrates. *Aflatoxins* have been shown to occur mainly in the tropics in such food ingredients as peanuts, pistachios, Brazil nuts, walnuts, cottonseed, maize, and other cereal grains, such as rice and wheat. Aflatoxins are considered to induce hepatotoxicity, mutagenicity, teratogenicity, immunosuppressance, and carcinogenic effects, and their presence in foods is highly regulated at a level of 0.05–2 μg/kg in the European Union (EC Report 2006). ELISA is considered to be one of the main methods for aflatoxin analysis. Some of the enzymes that have been used include horseradish peroxidase, which oxidizes tetramethylbenzidene to a product that can be measured by its electrical properties, and alkaline phosphatase for dephosphorylation of naphthyl phosphate to naphthol that is measured by pulse voltametry. Detection limits for these methods were in the range of 20–30 pg/mL (Ammida et al. 2004; Parker and Tothill 2009). A biosensor was developed by immobilizing horseradish peroxidase and aflatoxin antibodies onto microelectrodes (Liu et al. 2006). The presence of aflatoxin in the food results in immunocomplex formation with the antibody, which blocks the electron transfer between the enzyme and the electrode, thereby modifying conductivity in proportion to aflatoxin concentration.

Ochratoxin A (OTA) is considered to be one of the most important mycotoxins because of its high toxicity to both humans and animals and its occurrence in a number of basic foods and agro-products. A rapid and sensitive spectrophotometric method has been developed based on acetylcholinesterase inhibition by aflatoxin (Arduini et al. 2007). ELISA based on alkaline phosphatase and horseradish peroxidase is used for analysis of ochratoxin A in wine (Prieto-Simon and Campas 2009). On comparison of an indirect and a direct competitive enzyme-linked aptamer assay (ELAA) with the classical competitive ELISA for the determination of OTA in spiked red wine samples using two aptamers, designated H8 and H12, that bind with nanomolar affinity with OTA (aptamers produced through in vitro systematic evolution of ligands by exponential enrichment), it was found that the direct competitive ELAA is a useful screening tool for routine use in the control of OTA level in wine because a limit of 1 ng/mL with the midpoint value of 5 ng/mL could be obtained in 125 min for the real sample analysis (Barthelmebs et al. 2011).

A direct competitive chemiluminescent enzyme-linked immunosorbent assay (CL-ELISA) for the determination of OTA was developed using soybean peroxidase (SbP) in combination with 3-(10′-phenothiazinyl), propane-1-sulfonate (SPTZ), and 4-morpholinopyridine (MORPH) as a detection system and compared with colorimetric detection (COL-ELISA). The values of IC10, IC50, and the working range (IC20–IC80) for CL-ELISA and COL-ELISA were 0.01, 0.08, and 0.02−0.3 ng/mL and 0.08, 0.58, and 0.17−2.2 ng/mL, respectively. It was observed that the recovery values of CL-ELISA from three soybean-spiked samples with OTA concentrations of 0.07, 0.1, and 0.15 ng/mL ranged from 72% to 125%. The study reported that COL-ELISA could not detect OTA in eight examined samples among the 21 various agricultural commodities, and in four of these eight samples, the developed CL-ELISA could determine OTA at levels from 0.96 to 4.64 ng/g (Yu et al. 2011).

Trichothecene mycotoxins are commonly found in cereals, such as wheat, barley, maize, oats, and rye, particularly in cold climates. They cause immunosuppressive and cytotoxic effects as well as other disorders due to their inhibition of protein, DNA, and RNA synthesis. AOAC-approved methods (http://www.aoac.org/testkits/testedmethods.html) based on ELISA kits are commercially available for analysis of trichothecenes. These include the RIDASCREEN FAST and AgraQuant manufactured by Biopharm GmbH and Romer Labs, respectively, using horseradish peroxidase (Lattanzio et al. 2009).

3.7.5.2 Biogenic Amines

Biogenic amines are organic bases found in a broad range of foods, such as meat, fish, wine, beer, chocolate, nuts, fruits, dairy products, sauerkraut, and some fermented foods, but they are potentially toxic when consumed. These amines, most of which are produced by microbial decarboxylation of the corresponding amino acids, include putrescine, cadaverine, tyramine, spermine, spermidine, and agmatine. That is, putrescine, histamine, tryptamine, tyramine, agmatine, and cadaverine are formed from ornithine, histidine, tryptophan, tyrosine, arginine, and lysine, respectively (Teti et al. 2002).

Histamine is the most regulated of these amines, and one of the methods for histamine determination is based on the sequential activities of diamine oxidase and horseradish peroxidase. The diamine oxidase catalyzes breakdown of histamine, releasing imidazole acetaldehyde, ammonia, and hydrogen peroxide. The hydrogen peroxide produced is oxidized by the peroxidase, resulting in the color change of a chromogen, which is measured by a colorimeter (Landete et al. 2004). An amperometric biosensor involving use of amine oxidase and horseradish peroxidase has been developed for amine detection (Muresan et al. 2008). In this method, the amines are first separated on a weak acid cation exchange column followed by enzymatic reactions that produce a potential difference, which is measured by a bioelectrochemical detector. An enzyme sensor array for simultaneous detection of putrescine, cadaverine, and histamine has also been reported (Lange and Wittman 2002).

Hypoxanthine (Hx) content in pork meat at different postmortem times as a measure of meat freshness is determined using an oxygen electrode. The amperometric signal obtained due to the oxygen depletion during the Hx oxidation is related to the consumed oxygen at 190 s in the soluble enzyme sensor or the enzymatic rate at 10 s in the immobilized enzyme sensor. In both cases, there is a linear relationship between the signal and the Hx concentration in the range 8.68–26.05 μM (R^2 = 0.999) and 15.63–127 μM (R^2 = 0.995), respectively. Both enzyme sensors exhibit very good working conditions and storage stability. The Hx content measured by both sensors is very reliable as to that measured by HPLC. Both enzyme sensors are reliable and rapid and are an economical alternative for simple or multiple Hx measurements constituting a useful tool as quality control of meat freshness (Hernández-Cázares et al. 2010).

An enzyme sensor was found to be a practical tool to determine hypoxanthine (Hx) and xanthine (X) concentrations in dry-cured ham as an index of minimum curing time (Hernández-Cázares et al. 2011a). Nucleotide degradation during the processing of dry-cured ham is affected when using three types of salting (100% NaCl; 50% NaCl and 50% KCl; and 55% NaCl, 25% KCl, 15% $CaCl_2$, and 5% $MgCl_2$). Divalent salts in the salting mixture depressed the breakdown rate from the beginning of the process (salting and postsalting) up to the ripening stage (7 months) when the inosine (Ino), hypoxanthine (Hx), and xanthine (X) concentrations matched for the three treatments. The evolution of Hx and Hx + X during processing were analyzed by HPLC and an enzyme sensor, respectively. A good correlation between enzyme sensor and HPLC data was observed. The amperometric measurement of hypoxanthine was carried out using an enzyme sensor, based on an oxygen electrode assembly model 20 Dual Digital (Rank Brothers, Bottisham, Cambridge, England) and a membrane with xanthine oxidase (XO) immobilized on an immunodyne ABC membrane (1 cm^2). The XO immobilized membrane was attached to an oxygen-permeable Teflon membrane in the oxygen electrode by an O-ring and clamped into a thermostated reaction cell at 30°C. Forty microliters of an appropriate dilution of Hx standard or tissue extract was injected directly onto the XO membrane, and after each measurement, the enzyme sensor was washed with sodium phosphate buffer and dried with paper tissue. The signal detected in the enzyme sensor, as a decrease of current, corresponds to the consumed oxygen (mg/L) as a result of the catalytic effects of the xanthine oxidase on both Hx and X present in the reaction solution (Hernández-Cázares et al. 2011a).

The presence of biogenic amines in dry-fermented sausages could be detected employing an enzyme sensor based on diamine oxidase (DAO) from porcine kidney in combination with an oxygen electrode (Hernández-Cázares et al. 2011b). The enzyme DAO is immobilized on a preactivated immunodyne membrane using glutaraldehyde as a cross-linking agent, and the enzymatic determination is based on the measurement of the consumed O_2 in a platinum electrode poised at 600 mV versus Ag/AgCl. The reaction is conducted by the direct injection of either the standard or meat extract on the enzymatic membrane. The immobilized enzyme can be used for up to 30 analyses per day without significant loss of sensitivity and can be stored at 4°C showing good stability for at least 6 weeks. A good correlation was observed when comparing data obtained with the enzyme sensor to those obtained with a standard HPLC method, indicating that this sensor may be a reliable screening method for detection of biogenic amines for quality control in the meat industry (Hernández-Cázares et al. 2011b).

3.7.5.3 Sodium Saccharin

Sodium saccharin is a type of artificial sweetener that is sweeter than sucrose but has a bitter aftertaste. This is often used to sweeten soft drinks, candy, biscuits, medicine, and toothpaste. In the 1970s, sodium saccharin was found to be carcinogenic in laboratory rats. On this basis, saccharin was delisted as a food additive worldwide whereas, in May 2000, the U.S. National Toxicology Program (NTP) dropped saccharin from its list of suspected cancer-causing chemicals but has established an acceptable daily intake (ADI) for sodium saccharin. Sodium saccharin residue in food samples is a matter of concern, and its rapid determination is important. A sensitive and specific polyclonal antibody (PcAb)-based indirect competitive enzyme-linked immunosorbent assay (icELISA) was developed as a potential and useful analytical tool for its rapid determination (Wang et al. 2011). In this technique, 6-amino saccharin is coupled to a carrier protein for artificial antigen by diazotization. Antisodium saccharin PcAb was obtained by immunizing New Zealand white rabbits, and then icELISA was developed. The assay was found to be highly sensitive and specific to sodium saccharin with the 50% inhibition value (IC50) of 0.243 μgmL⁻¹, workable range (IC30–IC70) of 0.050–12.8 μgmL⁻¹ and limit of detection (LOD, IC20) of 0.021 μgmL⁻¹. The average recoveries of sodium saccharin in spiked food samples were estimated ranging from 70.7% to 98.8%. A statistically significant correlation of results was obtained between this new ELISA and previously established HPLC approaches with the food-relevant sodium saccharin concentration range 0–320 μmL^{-1} (R^2 = 0.9887–0.9975) (Wang et al. 2011).

3.7.5.4 Dipropyl Phthalate (DPrP)

Phthalates are a class of chemical compounds widely used as plasticizers for polyvinyl chloride resins, adhesives, and cellulose film coating. To date, nearly 20 kinds of phthalates have been used for these purposes. Phthalates are potentially hazardous to human health—especially to children's health due to their classification as endocrine disruptors. This has resulted in regulations regarding the types and levels of phthalates allowable in drinking water and foods. A direct competitive chemiluminescent enzyme-linked immunosorbent assay (CL-ELISA) based on polyclonal antibodies has been developed for enhanced detection of dipropyl phthalate (DPrP) (Zhang et al. 2012). In this method, the immunogen was prepared with bovine serum albumin attached to the hapten with an amino linker introduced on the target molecule by a diazotization method. Optimized assay conditions include the concentration of antibody, dilution ratio of enzyme conjugate, incubation time, effects of buffer concentration, and pH. Under the optimized conditions, the 50% inhibition values of 0.19 μg/L for DPrP were achieved with a limit of detection of 0.03 μg/L. The developed method has been used to quantify DPrP in water and milk samples without purification or preconcentration steps. The results of the study suggested that the developed CL-ELISA would be very suitable for rapid monitoring of DPrP in food samples (Zhang et al. 2012).

3.7.5.5 Pharmaceuticals

The growing use of antibiotics and other drugs for treatment of animals meant for food (e.g., cattle, chicken, pigs) is of great concern to regulatory authorities due to the potential carryover effect on consumers as these drugs enter the food supply. This has led to strict regulations on the prophylactic and therapeutic use of these drugs being introduced in many countries. Therefore, a number of methods have been developed for analysis of these drugs to ensure food safety. A few examples are presented here.

Chloramphenicol is a potent broad-spectrum antibiotic banned in Europe and the United States due to potential toxic effects (aplastic anemia). The high incidence of aplastic anemia in some countries has been attributed to the use of chloramphenicol in the treatment of food-producing animals, particularly aquaculture (Bogusz et al. 2004). Chloramphenicol is generally analyzed using various conventional methods, such as gas chromatography, immunoassay, and mass spectrometry. The drug tends to be conjugated in biological systems. Bogusz et al. (2004) developed an assay to distinguish between the free and conjugated forms by using β-glucuronidase in the extraction process to hydrolyze the conjugation to the free form, enabling selective analysis of the drug in honey, shrimp, and chicken.

Prolonged use of *clenbuterol* is known to cause adverse physiological effects, and incidents of poisoning have been reported in many countries following consumption of foods containing high concentrations of the drug (Pulce et al. 1991). ELISA methods based on monoclonal and polyclonal antibodies are the most widely used of the enzymatic methods for analysis (Matsumoto et al. 2000). He et al. (2009) developed a polyclonal indirect ELISA with high sensitivity for clenbuterol analysis in milk, animal feed, and liver samples.

Another set of drugs subject to European Union Maximum Residual Limits are the *nonsteroidal anti-inflammatory drugs* (NSAIDs) used for treatment of some food-producing animals (porcine and bovine). Consumption of meat products containing NSAIDs has been shown to cause gastrointestinal problems, such as diarrhea, nausea, and vomiting by irritation of the gastric mucosa. A simple enzymatic method involving cyclooxygenase has been developed for analysis of NSAIDs in milk and cheese (Campanella et al. 2009). In absence of NSAIDs, cyclooxygenase catalyzes oxidative conversion of arachidonic acid to prostaglandins. However, the presence of NSAIDs results in competitive inhibition of this activity, thereby reducing prostaglandin synthesis. The NSAID concentration is detected by coupling the catalytic reaction with an amperometric electrode for oxygen.

Fluoroquinolones are antimicrobial agents used for treatment and disease prevention in food-producing animals and sometimes used as feed additives to build up animal body mass. This group of drugs has come under scrutiny due to their misuse, potentially leaving residues in meats, which could lead to drug resistance and allergic hypersensitivity (Martinez et al. 2006). Therefore, maximum residue limits have been instituted in the European Union, and a number of ELISA methods have been developed for their detection in milk, chicken, pork, eggs, and other foods (Coillie et al. 2004; Huet et al. 2006; Lu et al. 2006; Scortichini et al. 2009; Sheng et al. 2009). Euro Diagnostica B.V. (Arnhem, Netherlands) has a commercial kit currently available for analysis of quinolones.

3.7.6 Process-Induced Toxins

Baked and fried products such as potato chips (or crisps), biscuits, crackers, breakfast cereals, and French fries are considered to have high levels of *acrylamide*, which is a process-induced toxin (Swedish National Food Administration 2002) and poses risk to humans from consumption of such foods. The mechanism of acrylamide formation in these foods is attributed to the high levels of asparagine in the raw materials (especially wheat and potatoes) used in their production. Initiation of the Maillard reaction in the presence of reducing sugars with asparagine as an amino group donor channels the complex series of reactions through a pathway that results in the production of acrylamide (Mottram et al. 2002).

Gas chromatography, liquid chromatography, and mass spectrometry are the predominant methods of analysis of this compound. Whereas the potential application of ELISA for analyzing acrylamide content in foods was also reported (Preston et al. 2008). In this method, acrylamide is derivatized with 3-mercaptobenzoic acid (3-MBA) and conjugated to a carrier protein (bovine thyroglobulin) to form an immunogen. Antiserum raised against this immunogen showing high affinity for the 3-MBA-derivatized acrylamide is used in the development of a peroxidase-based ELISA detection system with about 66 µg/kg limit of detection. A commercial ELISA kit has been recently introduced by Abraxis (http://www.abraxis.com) for acrylamide analysis.

3.7.7 Environmental Toxins

Certain toxin-producing phytoplanktons present in the aquatic environment have generally been recognized as the source of marine toxins in the human food chain following consumption of contaminated fish and shellfish. Safety concerns have led to regulatory limits being set for the presence of these toxins in seafoods.

Okadaic acid and its derivatives are considered responsible for diarrhetic shellfish poisoning and are very potent inhibitors of protein phosphatase. A number of methods have been developed around inhibition of the enzyme for its analysis (Campas and Marty 2007; Della Loggia et al. 1999; Tubaro et al. 1996; Volpe et al. 2009). Analysis of this toxin has also been accomplished by ELISA methods (Campas et al. 2008; Kreuzer et al. 1999).

The toxic effects of *yessotoxins* remain to be elucidated even though in vitro studies indicate cardiotoxicity. Analytical methods based on ELISA (Briggs et al. 2004; Garthwaite et al. 2001) and yessotoxin interaction with pectin esterase enzymes are available (Alfonso et al. 2004; Fonfria et al. 2008; Pazos et al. 2005).

Another group of marine toxins of relevance to food are the *brevetoxins* responsible for neurotoxic shellfish poisoning. The inhibition by brevetoxins of the desulfo-yessotoxin interaction with phosphodiesterase provides the basis for an analytical method (Mouri et al. 2009), and various ELISA methods are also reported (Baden et al. 1995; Garthwaite et al. 2001; Naar et al. 2002).

Saxitoxins are responsible for paralytic shellfish poisoning, and their paralytic effect is attributed to their high-affinity binding to specific sites on sodium channels, thereby blocking sodium flux across excitable cells. Their presence in foods is analyzed by ELISA (Chu and Fan 1985; Garthwaite et al. 2001). As with the saxitoxin group of toxins, amnesic shellfish poisoning is associated with consumption of foods contaminated with the domoic acid group of water-soluble neurotoxins produced by the genera *Pseudo-nitzschia*, *Nitzschia*, and *Chondria armata* (Vilarino et al. 2009). Garet et al. (2010) compared relative sensitivities of ELISA and conventional methods (HPLC-UV and mouse bioassay) for detection of amnesic shellfish poisoning and paralytic shellfish poisoning toxins in naturally contaminated fresh, frozen, boiled, and canned fish and shellfish. Their findings showed lower limits of detection (50 µg saxitoxin and 60 μg/kg shellfish meat for saxitoxin and domoic acid, respectively) compared to 350 μg saxitoxin and 1.6 mg domoic acid per kg shellfish meat, respectively, when analyzed by conventional methods. Their results indicated the merit of the ELISA technique for accurate measurement of these toxins.

Palytoxins and their analogues elicit their toxic effects by binding to the Na^+/K^+ ATPase pumps, thereby altering the selectivity of ion flux and membrane potential. Much like the saxitoxins, the tetrodotoxins elicit their neurotoxicity by inhibiting voltage-gated sodium channels. The method for the analysis of tetrodotoxins involves an alkaline phosphatase-labeled antibody, the activity of which produces p-aminophenol that is detected by the amount of current generated. Alkaline phosphatase activity is competitively inhibited by tetrodotoxin, thereby reducing the current produced (Kreuzer et al. 2002). ELISA techniques have been developed for the detection of marine palytoxins and tetrodotoxins (Kreuzer et al. 2002; Neagu et al. 2006).

Ciguatoxin poisoning is characterized by symptoms of nausea, diarrhea, abdominal cramps, memory loss, dizziness, headaches, and death in extreme cases. An ELISA method approved by the AOAC has been adopted in many countries (Kleivdal et al. 2007).

3.7.8 Pesticides

The presence of pesticide residues in foods and their impact on health is an issue of critical concern to the food industry, and therefore, regulatory measures are in place establishing maximum residual limits. Inhibition of acetylcholinesterase by pesticides has formed the basis for a broad range of techniques for pesticide analysis (Amine et al. 2006; Mulchandani et al. 2001). These techniques employed (i) amperometric transducers to measure thiocholine and p-aminophenol produced by acetylcholinesterase hydrolysis of butyrylthiocholine and p-aminophenyl acetate or hydrogen peroxide produced by oxidation of choline following acetylcholine hydrolysis in presence of choline oxidase, (ii) potentiometric transducers for measuring pH changes as a result of acetic acid production from enzyme activity, and (iii) fiber optics for monitoring pH changes using a fluorescein-labeled enzyme. These acetylcholinesterase inhibitory methods, however, are nondiscriminating between the organophosphates, carbamates, and other pesticides. Mulchandani et al. (2001) advocated the use of organophosphorus hydrolase integrated with electrochemical and optical transducers to specifically measure organophosphate pesticides. The enzyme hydrolyzes organophosphates to produce acids or alcohols, which generally tend to be either electroactive or chromophores that can be monitored (Mulchandani et al. 2001).

Bioelectronic tongues have also been proposed for the detection of pesticides, such as organophosphates and carbamates as well as phenols, using an amperometric bioelectronic tongue involving cholinesterase, tyrosinase, peroxidases, and cellobiose dehydrogenase. The corresponding substrates for the enzymes were acetylcholine, phenols, hydrogen peroxide, and cellobiose, respectively, with detection limits in the nanomolar and micromolar range (Solna et al. 2005). A similar bioelectronic tongue with acetylcholinesterase as biosensor has been used to resolve pesticide mixtures of dichlorvos and methylparaoxon (Ramirez et al. 2009). Botrytis, a fungal disease responsible for significant losses of agricultural produce particularly in vineyards (e.g., wine grapes, strawberries), is controlled by preharvest or postharvest treatment with the fungicide fenhexamid, which inhibits growth of the germ tube and mycelia of the fungus. Detection and analysis of fenhexamid levels in must and wines can be accomplished by direct competitive ELISA containing HRP (Mercader and Abad-Fuentes 2009). Atrazine, an herbicide with widespread contamination of waterways and drinking water supplies implicated in certain human health defects (e.g., birth defects, menstrual problems, low birth weight), is tightly regulated in the European Union with a limit of 0.1 g/L in potable water and fruit juices. Conductimetric and amperometric biosensors based on tyrosinase inhibition can be employed for the analysis of atrazine and its metabolites (Anh et al. 2004; Vedrine et al. 2003). This method is based on deactivation of tyrosinase under aqueous conditions. The enzyme normally catalyzes oxidation of monophenols to O-diphenols, which are eventually dehydrogenated to the corresponding O-quinones. Application of enzymatic biosensors for determination of fertilizers (nitrates, nitrites, and phosphates) as well as several pesticides and heavy metals that are potentially toxic to humans when they enter the food chain have been the subject of several investigations. Some of the enzymes applied to these analyses include (i) urease, glucose oxidase, acetylcholinesterase, alkaline phosphatase, ascorbate oxidase, alcohol oxidase, glycerol-3-phosphate oxidase, invertase, and peroxidase for heavy metals and (ii) nitrate reductase, polyphenol oxidase, phosphorylase, glucose-6-phosphate dehydrogenase, and phospho-glucomutase for nitrates, nitrites and phosphates (Amine et al. 2006; Cock et al. 2009).

3.8 Enzymes in Determination of Analytes during Processing of Food and Beverages

During processing of food and beverages, several unit operations are involved (Table 3.6) right from the postharvesting stage to final packaging through processing involving chemical or biological transformations of food raw materials into the final form of food or beverage that is consumed. During these various phases of processing of food and beverages, care is taken to ensure the required quality and safety in the food materials that are processed from the raw material stage to the final stage ready for consumption. This quality assurance is ensured through proper and regular monitoring of various process variables, both physico-chemical and biochemical, that influence the overall quality and safety of food. Physical parameters, such as texture, shape, firmness, and appearance, satisfying consumer necessities are taken care of by several analytical methods and organoleptic assessment, whereas chemical, biochemical, and biological variables listed in Table 3.1 are analyzed, both qualitatively and quantitatively, at various stages employing different analytical techniques mentioned in Table 3.2. In fact, monitoring of all stages of food processing during food and beverage manufacture is warranted in the food industry to comply with food standards across the world.

Proximate components in any food attribute primary nutritive quality to food, and hence, they are routinely analyzed to qualify food for consumption. Whereas there are ample chances for a change in the basic proximate components of raw materials used for food production, which are attributed to food's nutritive quality during the various stages of processing, monitoring of such variables assumes greater significance. Second, during processing, some variables, such as starch, may undergo hydrolysis, either by acid or enzyme, to yield sugars during the pretreatment process. In such instances, their analysis becomes important to ensure optimal hydrolysis and successful application of the hydrolytic process. Several additives are often used during chemical or biological processes in order to improve the quality. The effect of such additives on the final quality of food is ascertained by analysis, and as per requirement, the addition process is optimized. Third, the different environmental contaminants, toxins, and undesirable residues in raw materials of food that pose health concerns need to be assessed for their presence during the production process. Monitoring of the same during processing stages ensures safe removal of the same through appropriate steps and ensures final quality of the food.

Undesirable microbial contamination during the fermentation process is compulsorily monitored in order to avoid loss in productivity. For example, in microbial fermentation production of enzymes, organic acids, amino acids, biomass contamination by undesirable microbes, and their action can lead to overall reduction in the desired product yield as well as contribute to undesirable microbial metabolites in the fermentation broth. Further, production of a desirable product of interest, such as amino acids, lactic acid, citric acid, vitamins, and enzymes utilizing selective and specific precursors during fermentation, requires regular monitoring of residual precursor substrates as well as their end products after transformation is analyzed to monitor the progress of the fermentation process. For example, during wine fermentation, glucose levels in the broth during fermentation determine the level of alcohol formation. Similarly, in glucose syrup production, using starch as a starting material level of glucose determines the efficiency of the enzyme used for conversion of starch to glucose.

Almost all the biochemical variables with respect to polysaccharides, oligosaccharides, sugars, proteins, amino acids, metabolites such as amines, vitamins, and enzymes are monitored in addition to specific chemical species that are qualitatively and quantitatively monitored to ensure success of the food processing methods. All these biochemical and chemical variables are analyzed by one or other methods discussed in the foregoing sections of this chapter, either alone or in combination. In all these methodologies of analyses, enzymes have a significant role as discussed elaborately in foregoing sections with specific examples. Indeed enzymes as analytical tools have a greater role and pivotal importance in the

TABLE 3.6

Unit Operations of the Food Processing Industry by Principal Groups

Group	Unit Operation	Examples of Applications
Cleaning	Washing	Fruits, vegetables, fish
	Peeling	Prawns
	Removal of foreign bodies	Fruits, vegetables, prawns
	Cleaning in place (CIP)	Grains, all food plants
Physical separation	Filtration	Sugar refining
	Screening	Grains
	Sorting	Coffee beans
	Membrane separation	Ultrafiltration of whey
	Centrifugation	Separation of milk
	Pressing, expression	Oilseeds, fruits
Molecular (diffusion based) separation	Adsorption	Bleaching of edible oils
	Distillation	Alcohol production
	Extraction	Vegetable oils
Mechanical transformation	Size reduction	Chocolate refining
	Mixing	Beverages, dough
	Emulsification	Mayonnaise
	Homogenizing	Milk, cream
	Forming	Cookies, pasta
	Agglomeration	Milk powder
	Coating encapsulation	Confectionery
Chemical transformation	Cooking	Meat, biscuits
	Baking	Bread
	Frying	Potato fries
	Fermentation	Wine, beer, yogurt
	Aging, curing	Cheese, wine
	Extrusion cooking	Breakfast cereals
Microbial production	Fermentation	Wine, vinegar, bread, single cell protein, enzymes, amino acids, vitamins, lactic acid, citric acid, acetic acid, L-glutamic acid, L-lysine, sugars
	Biomass	Baker's yeast, *Saccharomyces cervisiae*, mushrooms, lactic starter cultures, probiotics
Preservation	Thermal processing, (blanching, pasteurization, sterilization)	Pasteurized milk, canned vegetables
	Chilling	Fresh meat, fish
	Freezing	Frozen dinners, ice cream, frozen vegetables
	Concentration	Tomato paste, citrus juice concentrate, sugar, salting of fish
	Addition of solutes	Jams, preserves, pickles
	Chemical preservation	Salted fish, smoked fish
	Dehydration	Dried fruit, dehydrated vegetables, milk powder, instant coffee, mashed potato flakes
	Freeze drying	Instant coffee
Packaging	Filling	Bottled beverages
	Sealing	Canned foods
	Wrapping	Fresh salads

Source: Gowthaman, M. K., P. Gowthaman, and M. Chandrasekaran, Principles of Food Technology and Types of Food Waste Processing Technologies, In *Valorization of food processing by-products.* Ed. Chandrasekaran, M. CRC Press, Boca Raton, FL, USA, pp. 109–146, 2012.

monitoring of fermentation, enzymic transformation, or chemical transformation processes employed in food and beverage production. Extensive examples are available in the literature as to monitoring of specific variables during optimization of bioprocesses as well as monitoring the progress of several food production processes. It is out of scope herein to discuss all those examples, and hence, it is suggested that interested readers may refer to specific publications related to their product of interest.

3.9 Future Prospects

Qualitative and quantitative determination of different variables that contribute to the overall quality and safety of food and beverages consumed is of high priority for both food producers as it is of high relevance to food consumers who care for the quality and safety of food due to growing concerns for healthy food. Consequently, there is a need for accurate, appropriate methods for such food analysis that is cost-effective, rapid, and easy to perform. In this context, there is an emerging trend to opt for enzyme-based analytical techniques, such as ELISA and biosensors, which meet such needs and requirements of the food and beverage industries. Although there is tremendous growth in the literature as a result of research and development in the recognition of the right enzymes and their development as suitable enzyme sensors for use in biosensors as well as in ELISA for varied biomolecules and chemical species, there is a need for isolation, identification, and development of new and efficient enzymes for such analytical purposes. The success stories of glucose oxidase in the biosensor industry for food and clinical assays has to be seen with other enzymes for other critical biochemical component analysis that determines food quality. There is a promising future for enzymes as analytical tools for instant assessment of food quality and safety, and such trend may have great implications in the overall growth of the food and beverage industry in addition to contributing to a healthy society in this new millennium.

REFERENCES

Alfonso, A., Vietytes, M. R., Yasumoto, T. and Botana, L. M. 2004. A rapid microplate fluorescence method to detect yessotoxins based on their capacity to activate phosphodiesterases. *Anal Biochem* 326: 93–99.

Alocilja, E. C. and Radke, S. M. 2003. Market analysis of biosensors for food safety. *Biosens Bioelectron* 18: 841–846.

Alpat, S. and Telefoncu, A. 2010. Development of an alcohol dehydrogenase biosensor for ethanol determination with toluidine blue O covalently attached to a cellulose acetate modified electrode. *Sensors (Basel)* 10(1): 748–764. doi:10.3390/s100100748.PMCID:PMC3270867.

Amine, A., Mohammadi, H., Bourais, I. and Palleschi, G. 2006. Enzyme inhibition-based biosensors for food safety and environmental monitoring. *Biosens Bioelectron* 21: 1405–1423.

Ammida, N. H. S., Micheli, L. and Palleschi, G. 2004. Electrochemical immunosensor for determination of aflatoxin B1 in barley. *Anal Chim Acta* 520: 159–164.

Anderson, B. 2011. Coeliac disease: New tests, new genes and rising prevalence. *Med Today* 12(6): 69–71.

Anh, T. M., Dzyadevych, S. V., Van, M. C., Renault, N. J., Duc, C. N. and Chovelon, J. M. 2004. Conductometric tyrosinase biosensor for the detection of diuron, atrazine and its metabolites. *Talanta* 63: 365–370.

AOAC International. 1995. *Official Methods of Analysis*, Association of Official Analytical Chemists, Washington, DC.

AOAC International. 2011. Performance tested methods and validated methods. Available at http://www.aoac.org/testkits/testedmethods.html.

Araujo, A. R. T. S., Lucia, M. F. S., Saraiva, M., and Lima, J. L. F. C. 2010. Flow injection analysis with immobilized enzymes in nonaqueous media. *Curr Anal Chem* 6(3): 193–202. doi:10.2174/157341110791517061.

Arduini, F., Errico, I., Amine, A., Micheli, L., Palleschi, G. and Moscone, D. 2007. Enzymatic spectrophotometric method for aflatoxin B detection based on acetylcholinesterase inhibition. *Anal Chem* 79: 3409–3415.

Ashie, I. N. A. 2012. Enzymes in food analysis. In: *Food Biochemistry and Food Processing*, Second Edition, B. K. Simpson (ed.), Wiley-Blackwell, Oxford, UK. doi: 10.1002/9781118308035.ch3.

Avramescu, A., Noguer, T., Magearu, V. and Marty, J. L. 2001. Chronoamperometric determination of d-lactate using screen-printed enzyme electrodes. *Anal Chim Acta* 433: 81–88.

Avramescu, A., Noguer, T., Avramescu, M. and Marty, J. L. 2002. Screen-printed biosensors for the control of wine quality based on lactate and acetaldehyde determination. *Anal Chim Acta* 458: 203–213.

Badea, M., Micheli, L., Messia, M. C. et al. 2004. Aflatoxin M1 determination in raw milk using a flow-injection immunoassay system. *Anal Chim Acta* 520: 141–148.

Baden, D. G., Melinek, R., Sechet, V. et al. 1995. Modified immunoassays for polyether toxins: Implications of biological matrices, metabolic states, and epitope recognition. *J AOAC Int* 78: 499–508.

Barthelmebs, L., Jonca, J., Hayat, A., Prieto-Simon, B. and Marty, J.-L. 2011. Enzyme-linked aptamer assays (ELAAs), based on a competition format for a rapid and sensitive detection of Ochratoxin A in wine. *Food Control* 22: 737–743.

Bergmeyer, H. U. (ed.). 1984. *Methods of Enzymatic Analysis*, 3rd edn., VCH, Weinheim, Germany.

Bilitewski, U. 2000. Can affinity sensors be used to detect food contaminants? *Anal Chem* 72: 693A–701A.

Blake, C. and Gould, B. J. 1984. Use of enzymes in immunoassays techniques. *Analyst* 109: 533–547.

Boer, E. and Beuner, R. R. 1999. Methodology for detection and typing of foodborne microorganism, *Int J Food Microbiol* 50: 119–130.

Bogusz, M. J., Hassan, H., Al-Enazi, E., Ibrahim, Z. and Al-Tufail, M. 2004. Rapid determination of chloramphenicol and its glucuronide in food products by liquid chromatography—Electrospray negative ionization tandem mass spectrometry. *J Chromatogr B* 807: 343–356.

Briggs, L. R., Miles, C. O., Fitzgerald, J. M., Ross, K. M., Garthwaite, I. and Towers, N. R. 2004. Enzyme-linked immunosorbent assay for the detection of yessotoxin and its analogues. *J Agric Food Chem* 52: 5836–5842.

Campanella, L., Favero, G., Pastorino, M. and Tomassetti, M. 1999. Monitoring the rancification process in olive oils using a biosensor operating in organic solvents. *Biosens Bioelectron* 14(2): 179–186.

Campanella, L., Di Persio, G., Pintore, M., Tonnina, D., Caretto, N., Martini, E. and Lelo, D. 2009. Determination of nonsteroidal drugs in milk and fresh cheese bread on the inhibition of cyclooxygenase. *Food Technol Biotechnol* 47(2): 172–177.

Campas, M. and Marty, J. L. 2007. Enzyme sensor for the electrochemical detection of the marine toxin okadaic acid. *Anal Chim Acta* 605: 87–93.

Campas, M., de la Iglesia, P., Le Berre, M., Kane, M., Diogène, J. and Marty, J.-L. 2008. Enzymatic recycling-based amperometric immunosensor for the ultrasensitive detection of okadaic acid in shellfish. *Biosens Bioelectron* 24: 716–722.

Carr, P. W. and Bowers, L. D. 1980. *Immobilized Enzymes in Analytical and Clinical Chemistry*, Wiley, New York.

Cheng, S. G. G. and Merchant, Z. M. 1995. Biosensors in food analysis. In: *Characterization of Food: Emerging Methods*, A. G. Gaonkar (ed.), Elsevier Science, New York, pp. 329–345.

Chu, F. S. and Fan, T. S. 1985. Indirect enzyme-linked immunosorbent assay for saxitoxin in shellfish. *J Assoc Anal Chem* 68: 13–16.

Cock, L. S., Arenas, A. M. Z. and Aponte, A. A. 2009. Use of enzymatic biosensors as quality indices: A synopsis of present and future trends in the food industry. *Chilean J Agric Res* 69(2): 270–280.

Coillie, E. V., De Block, J. and Reybroeck, W. 2004. Development of an indirect ELISA for flumequine residues in raw milk using chicken egg yolk antibodies. *J Agric Food Chem* 52: 4975–4978.

Croci, L., Delibato, E., Volpe, G. and Palleschi, G. 2001. A rapid electrochemical ELISA for the determination of *Salmonella* in meat samples. *Anal Lett* 34: 2597–2607.

Cucu, T., Platteau, C., Taverniers, I., Devreese, B., de Loose, M. and de Meulenaer, B. 2011. ELISA detection of hazelnut proteins: Effect of protein glycation in the presence or absence of wheat proteins. *Food Addit Contam Part A Chem Anal Control Expo Risk Assess* 28(1): 1–10.

Davis, J., Vaughan, D. H. and Cardosi, M. F. 1995. Elements of biosensors construction. *Enzyme Microb Technol* 17: 1030–1035.

de Champdoré, M., Bazzicalupo, P., De Napoli, L. et al. 2007. A new competitive fluorescence assay for the detection of patulin toxin. *Anal Chem* 79: 751–757.

Delibato, E., Volpe, G., Stangalini, D., De Medici, D., Moscone, D. and Palleschi, G. 2006. Development of SYBR-green real-time PCR and a multichannel electrochemical immunosensor for specific detection of Salmonella enterica. *Anal Lett* 39: 1611–1625.

Della Loggia, R., Sosa, S. and Tubaro, A. 1999. Methodological improvement of the protein phosphatase inhibition assay for the detection of okadaic acid in mussels. *Nat Toxins* 7: 387–391.

Doi, H., Touhata, Y., Shibata, H., Sakai, S., Urisu, A., Akiyama, H. and Teshima, R. 2008. Reliable enzyme-linked immunosorbent assay for the determination of walnut proteins in processed foods. *J Agric Food Chem* 56: 7625 7630.

Doh, C. and Koehler, P. 2014. Enzyme-linked immunosorbent assays for the detection and quantitation of gluten in cereal-based foods. *Cereal Foods World* 59(4): 171–178. Available at http://dx.doi.org/10.1094/CFW-59-4-0171.

European Commission Regulation (EC) No 1881/2006 of 19 December 2006. 2006. *Off J Eur Union* L364/5.

Fonfria, E. S., Vilariño, N., Vieytes, M. R., Yasumoto, T. and Botana, L. M. 2008. Feasibility of using a surface Plasmon resonance-based biosensor to detect and quantify yessotoxin. *Anal Chim Acta* 617: 167–170.

Frei, R. W. and Zech, K. (ed.). 1989. *Selective Sample Handling and Detection in Liquid Chromatography*, part A and part B, Elsevier, Amsterdam.

Gabaldon, J. A., Maquieira, A. and Puchades, R. 1999. Current trends in immunoassays-based kits for pesticide analysis. *Crit Rev Food Sci Nutr* 39: 519–538.

Garet, E., González-Fernández, A., Lago, J., Vieites, J. M. and Cabado, A. G. 2010. Comparative evaluation of enzyme-linked immunoassay and reference methods for the detection of shellfish hydrophilic toxins in several presentations of seafood. *J Agric Food Chem* 58: 1410–1415.

Garthwaite, I., Ross, K. M., Miles, C. O., Briggs, L. R., Towers, N. R., Borrell, T. and Busby, P. 2001. Integrated ELISA screening system for amnesic, neurotoxic, diarrhetic and paralytic shellfish poisoning toxins found in New Zealand. *J Assoc Anal Chem Int* 84: 1643–1648.

Gaskin, F. E. and Taylor, S. L. 2011. Sandwich enzyme-linked immunosorbent assay (ELISA) for detection of cashew nut in foods. *J Food Sci* 76(9): T218–T226.

Ghindilis, A. L. and Kurochkin, I. N. 1994. Glucose potentiometric electrodes based on mediatorless bioelectrocatalysis. A new approach. *Biosens Bioelectron* 9: 353–357.

Giannoccaro, E., Wang, Y. J. and Chen, P. Y. 2008. Comparison of two HPLC systems and an enzymatic method for quantification of soybean sugars. *Food Chem* 106: 324–330.

Gowthaman, M. K., Gowthaman, P. and Chandrasekaran, M. 2012. Principles of food technology and types of food waste processing technologies. In: *Valorization of Food Processing By-Products*, M. Chandrasekaran (ed.), CRC Press, Boca Raton, FL, pp. 109–146.

Guilbault, G. G. (ed.). 1984. *Analytical Uses of Immobilized Enzymes*, Marcel Dekker, New York.

Gutes, A., Ibanez, A. B., Del Valle, M. and Céspedes, F. 2006. Automated SIA e-tongue employing a voltammetric biosensor array for the simultaneous determination of glucose and ascorbic acid. *Electroanalysis* 18: 82–88.

Hamada-Sato, N., Usui, K., Kobayashi, T., Imada, C. and Watanabe, E. 2005. Quality assurance of raw fish based on HACCP concept. *Food Control* 16(4): 301–307.

He, L., Pu, C., Yang, H., Zhao, D. and Deng, A. P. 2009. Development of a polyclonal indirect ELISA with sub-ng g-1 sensitivity for the analysis of clenbuterol in milk, animal feed, and liver samples and a small survey of residues in retail animal products. *Food Add Contamin* 26(8): 1153–1161.

Hempen, C. and Karst, U. 2006. Labeling strategies for bioassays. *Anal Bioanal Chem* 383: 572–583.

Hernández-Cázares, A. S., Aristoy, M. C. and Toldrá, F. 2010. Hypoxanthine-based enzymatic sensor for determination of pork meat freshness. *Food Chem* 123: 949–954.

Hernández-Cázares, A. S., Aristoy, M. C. and Toldrá, F. 2011a. Nucleotides and their degradation products during processing of dry-cured ham, measured by HPLC and an enzyme sensor. *Meat Sci* 87: 125–129.

Hernández-Cázares, A. S., Aristoy, M. C. and Toldrá, F. 2011b. An enzyme sensor for the determination of total amines in dry-fermented sausages. *J Food Eng* 106: 166–169.

Hildebrandt, A., Bragos, R., Lacorte, S. and Marty, J. 2008. Performance of a portable biosensor for the analysis of organophosphorus and carbamate insecticides in water and food. *Sens Actuat B* 133(1): 195–201.

Ho, J. A. and Durst, R. A. 2003. Detection of fumonisin B1: Comparison of flow-injection liposome immunoanalysis with high-performance liquid chromatography. *Anal Biochem* 6: 7–13.

Holzhauser, T. and Vieths, S. 1999. Quantitative sandwich ELISA for determining traces of hazelnut (*Corylus avellana*) protein in complex food matrices. *J Agric Food Chem* 47: 4209–4218.

Huet, A. C., Charlier, C., Tittlemier, S. A., Singh, G., Benrejeb, S. and Delahaut, P. 2006. Simultaneous determination of fluoroquinolone antibiotics in kidney, marine products, eggs and muscle by enzyme-linked immunosorbent assay. *J Agric Food Chem* 54: 2822–2827.

Husain, F. T., Bretbacher, I. E., Nemes, A. and Cichna-Markl, M. 2010. Development and validation of an indirect competitive enzyme-linked immunosorbent assay for the determination of potentially allergenic sesame (*Sesamum indicum*) in food. *J Agric Food Chem* 58: 1434–1441.

Imabayashi, S., Kong, Y. T. and Watanabe, M. 2001. Amperometric biosensor for polyphenols based on horse-radish peroxidase immobilized on gold electrodes. *Electroanalysis* 13(5): 408–412.

Kemeny, D. M. 1991. *A Practical Guide to ELISA*, Pergamon Press, New York, p. 115.

Kerkaert, B. and De Meulenaer, B. 2007. Detection of hen's egg white lysozyme in food: Comparison between a sensitive HPLC and a commercial ELISA method. *Commun Agric Appl Biol Sci* 72: 215–218.

Kiranas, E. R., Karayannis, M. I. and Karayanni, S. M. T. 1997. An enzymatic method for the determination of ATP and glycerol with an automated FIA system. *Anal Lett* 30(3): 537–552.

Kleivdal, H., Kristiansen, S. I., Nilsen, M. V., Goksoyr, A., Briggs, L., Holland, P. and McNabb, P. 2007. Determination of domoic acid toxins in shellfish by biosense ASP ELISA—A direct competitive enzyme linked immunosorbent assay: Collaborative study. *J AOAC Int* 90: 1011–1027.

Kreuzer, M. P., O'Sullivan, C. K. and Guilbault, G. G. 1999. Development of an ultrasensitive immunoassay for rapid measurement of okadaic acid and its isomers. *Anal Chem* 71: 4198–4202.

Kreuzer, M. P., Pravda, M., O'Sullivan, C. K. and Guilbault, G. G. 2002. Novel electrochemical immunosensors for seafood toxin analysis. *Toxicon* 40: 1267–1274.

Krska, R., Welzig, E. and Boudra, H. 2007. Analysis of fusarium toxins in feed. *Animal Feed Sci Technol* 137: 241–264.

Landete, J. M., Ferrer, S. and Pardo, I. 2004. Improved enzymatic method for the rapid determination of histamine in wine. *Food Add Contamin* 21: 1149–1154.

Lange, J. and Wittman, C. 2002. Enzyme sensor array for the determination of biogenic amines in food samples. *Anal Bioanal Chem* 372: 276–283.

Lattanzio, V. M. T., Pascale, M. and Visconti, A. 2009. Current analytical methods for trichothecene mycotoxins in cereals. *Trends Anal Chem* 28(6): 758–768.

Leonard, P., Hearty, S., Brennan, J., Dunne, L., Quinn, J., Chakraborty, T. and O'Kennedy, R. 2003. Advances in biosensors for detection of pathogens in food and water. *Enzyme Microb Technol* 32: 3–13.

Li, X. M., Yang, X. Y. and Zhang, S. S. 2008. Electrochemical enzyme immunoassays using model labels. *Trends Anal Chem* 27: 543–553.

Lima, J. L. F. C., Lopes, T. I. M. S. and Rangel, A. O. S. S. 1998. Enzymatic determination of lactic and malic acids in wines by flow injection spectrophotometry. *Anal Chim Acta* 366: 187–191.

Liu, Y., Qin, Z., Wu, X. and Jiang, H. 2006. Immune biosensor for aflatoxin B1-based bioelectrocatalytic reaction on micro-comb electrode. *Biochem Eng J* 32: 211–217.

Lu, S., Zhang, Y., Liu, J., Zhao, C., Liu, W. and Xi, R. 2006. Preparation of antiperfloxacin antibody and development of an indirect competitive enzyme-linked immunosorbent assay for detection of perfloxacin residue in chicken liver. *J Agric Food Chem* 54: 6995–7000.

Luong, J. H. T., Groom, C. A. and Male, K. B. 1991. The potential role of biosensors in the food and drink industries. *Biosens Bioelectron* 6: 547–554.

Malin, B. C. and Johan, L. 2000. Biosensor-based determination of folic acid in fortified food. *Food Chem* 70: 523–532.

Marko-Varga, G. and Dominguez, E. 1991. Enzymes as analytical tools. *Trends Anal Chem.* 10(9): 290–297.

Marquette, C. A. and Blum, L. J. 2006. State of the art and recent advances in immunoanalytical systems. *Biosens Bioelectron* 21: 1424–1433.

Martinez, M., McDermott, P. and Walker, R. 2006. Pharmacology of the fluoroquinolones: A perspective for the use in domestic animals. *Vet J* 172: 10–28.

Matsumoto, K., Kamikado, H., Matsubara, H. and Osajima, Y. 1988. Simultaneous determination of glucose, fructose, and sucrose in mixtures by amperometric flow injection analysis with immobilized enzyme reactors. *Anal Chem* 60: 147–151.

Matsumoto, M., Maruyama, N., Watanabe, H., Kikuchihara, T., Kido, Y. and Tsuji, A. 2000. Residue analysis of clenbuterol in bovine and horse tissues and cow's milk by ELISA. *J Food Hyg Soc Jpn* 41: 48–53.

Mello, L. D. and Kubota, L. T. 2002. Review of the use of biosensors as analytical tools in the food and drink industries. *Food Chem* 77: 237–256.

Mercader, J. V. and Abad-Fuentes, A. 2009. Monoclonal antibody generation and direct competitive enzyme-linked immunosorbent assay evaluation for the analysis of the fungicide fenhexamid in must and wine. *J Agric Food Chem* 57: 5129–5135.

Mermelstein, N. H. 2011. Testing for gluten in foods. *Food Technol* 65(2). Available at http://www.ift.org/food-technology/past-issues/2011/february/columns/foodsafety-and-quality.aspx?page=viewall.

Mizutani, F. and Asai, M. 1990. Simultaneous determination of glucose and sucrose by a glucose-sensing enzyme electrode combined with an invertase-attached cell. *Anal Chim Acta* 236: 245–250.

Mizutani, F., Yabuki, S. and Asai, M. 1991. L Malate sensing electrode based on malate dehydrogenase and NADH oxidase. *Anal Chim Acta* 245: 145–150.

Morkyavichene, M. B., Dikchyuvene, A. A., Paulyukonis, A. B. and Kazlauskas, D. A. 1984. Choice of initial ratio of enzymes for immobilization in multienzyme systems. *Appl Biochem Microbiol* 20: 60–63.

Mottram, D., Bronislaw, S., Wedzicha, L. and Dodson, A. T. 2002. Acrylamide is formed in the Maillard reaction. *Nature* 419: 448–449.

Mouri, R., Oishi, T., Torikai, K., Ujihara, S., Matsumori, N., Murata, M. and Oshima, Y. 2009. Surface plasmon resonance-based detection of ladder-shaped polyethers by inhibition detection method. *Bioorg Med Chem Lett* 19: 2824–2828.

Mulchandani, A., Chen, W., Mulchandani, P., Wang, J. and Rogers, K. R. 2001. Biosensors for direct determination of organophosphate pesticides. *Biosens Bioelectron* 16: 225–230.

Muresan, L., Valera, R. R., Frébort, I., Popescu, I. C., Csöregi, E. and Nistor, M. 2008. Amine oxidase amperometric biosensor coupled to liquid chromatography for biogenic amines determination. *Microchim Acta* 163: 219–225.

Murthy, G. K., Kleyn, D. H., Richardson, T. and Rocco, R. M. 1992. Phosphatase methods. In: *Standard Methods for the Examination of Dairy Products*, 16th edn., G. H. Richardson (ed.), American Public Health Association, Washington, DC, p. 413.

Naar, J., Bourdelais, A., Tomas, C. et al. 2002. A competitive ELISA to detect brevetoxins from *Karenia brevis* (formerly *Gymnodinium breve*) in seawater, shellfish, and mammalian body fluid. *Environ Health Perspect* 110: 179–185.

Neagu, D., Micheli, L. and Palleschi, G. 2006. Study of a toxin-alkaline phosphatase conjugate for the development of an immunosensor for tetrodotoxin determination. *Anal Bioanal Chem* 385: 1068–1074.

Nenkova, R., Atanasova, R., Ivanova, D. and Godjevargova, T. 2010. Flow injection analysis for amperometric detection of glucose with immobilized enzyme reactor. *Biotechnol Biotechnol Equip* 24(3): 1986–1992.

Pal, A. and Dhar, T. K. 2004. An analytical device for on-site immunoassay. Demonstration of its applicability in semiquantitative detection of aflatoxin B-1 in a batch of samples with ultrahigh sensitivity. *Anal Chem* 76: 98–104.

Pandey, P. C., Pandey, V. and Mehta, S. 1994. An amperometric enzyme electrode for lactate based on graphite paste modified with tetracyanoquinodimethane. *Biosens Bioelectron* 9: 365–372.

Paredes, P. A., Parellada, J., Fernndez, V. M., Katakis, I. and Dominguez, E. 1997. Amperometric mediated carbon paste sensor based on d-fructose dehydrogenase for the determination of fructose in food analysis. *Biosens Bioelectron* 12: 1233–1243.

Parker, C. O. and Tothill, I. E. 2009. Development of an electrochemical immunosensor for aflatoxin M1 in milk with focus on matrix interference. *Biosens Bioelectron* 24: 2452–2457.

Pazos, M. J., Alfonso, A., Vieytes, M. R., Yasumoto, T. and Botana, L. M. 2005. Kinetic analysis of the interaction between yessotoxin and analogues and immobilized phosphodiesterases using a resonant mirror optical biosensor. *Chem Res Toxicol* 18: 1155–1160.

Pfeiffer, D., Ralis, E. V., Makower, A. and Scheller, F. W. 1990. Amperometric bi-enzyme based biosensor for the detection of lactose—Characterization and application. *J Chem Technol Biotechnol* 49: 255–265.

Piermarini, S., Micheli, L., Ammida, N. H. S., Palleschi, G. and Moscone, D. 2007. Electrochemical immunosensor array using a 96-well screen-printed microplate for aflatoxin B-1 detection. *Biosens Bioelectron* 6: 1434–1440.

Piermarini, S., Volpe, G., Micheli, L., Moscone, D. and Palleschi, G. 2009. An ELIME-array for detection of aflatoxin B_1 in corn samples. *Food Control* 20: 371–375.

Pomes, A., Helm, R. M., Bannon, G. A., Burks, A. W., Tsay, A. and Chapman, M. D. 2003. Monitoring peanut allergen in food products by measuring Ara h 1. *J Allergy Clin Immunol* 111: 640–645.

Powers, J. R. 1998. Application of enzymes in food analysis. In: *Food Analysis*, S. S. Nielsen (ed.), Aspen Publication Inc., Gaithersburg, MD, pp. 349–365.

Prasetyo, E. N., Kudanga, T., Steiner, W., Murkovic, M., Nyanhongo, G. S. and Guebitz, G. M. 2010. Laccase-generated tetramethoxy azobismethylene quinone as a tool for antioxidant activity measurement. *Food Chem* 118: 437–444.

Preston, A., Fodey, T. and Elliott, C. 2008. Development of a high throughput enzyme linked immunosorbent assay for the routine detection of the carcinogen in food, via rapid derivatization pre-analysis. *Anal Chim Acta* 608: 178–185.

Prieto-Simon, B. and Campas, M. 2009. Immunochemical tools for mycotoxin detection in food. *Monatsh Chem* 140: 915–920.

Prieto-Simón, B., Noguer, T. and Campàs, M. 2007. Emerging biotools for assessment of mycotoxins in the past decade. *TrAC* 26: 689–702.

Pulce, C., Lamaison, D., Keck, G., Bostvironnois, C., Nicolas, J. and Descotes, J. 1991. Collective human food poisonings by clenbuterol residues in veal liver. *Vet Hum Toxicol* 33: 480–481.

Ramirez, G. V., Gutierrez, M., Valle, M. D., Ramirez-Silva, M. T., Fournier, D. and Marty, J. L. 2009. Automated resolution of dichlorvos and methylparaoxon pesticide mixtures employing a flow injection system with an inhibition electronic tongue. *Biosens Bioelectron* 24: 1103–1108.

Rank, M., Danielsson, B. and Gram, J. 1992. Implementation of a thermal biosensor in a process environment: On-line monitoring of penicillin V in production-scale fermentations. *Biosens Bioelectron* 7: 631–635.

Reynolds, E. R. and Yacynych, A. M. 1994. Direct sensing platinum ultra microbiosensors for glucose. *Biosens Bioelectron* 9: 283–293.

Rocco, R. 1990. Fluorometric determination of alkaline phosphatase in fluid dairy products: Collaborative study. *J Assoc Off Anal Chem* 73: 842–849.

Salam, F. and Tothill, I. E. 2009. Detection of *Salmonella typhimurium* using an electrochemical immunosensor. *Biosens Bioelectron* 24: 2630–2636.

Scheller, F. and Schubert, F. (eds.). 1989. *Biosensoren*, Akademie-Verlag, Berlin.

Schneider, E., Curtui, V., Seidler, C., Dietrich, R., Usleber, E. and Märtlbauer, E. 2004. Rapid methods for deoxynivalenol and other trichothecenes. *Toxicol Lett* 153: 113–121.

Schneider, N., Weigel, I., Werkmeister, K. and Pischetsrieder, M. 2010. Development and validation of an ELISA for quantification of lysozyme in cheese. *J Agric Food Chem* 58: 76–81.

Schöning, M., Arzdorf, M., Mulchandani, P., Chen, W. and Mulchandani, A. 2003. Towards a capacitive enzyme sensor for direct determination of organophosphorus pesticides: Fundamental studies and aspects of development. *Sensors* 3(6): 119–127.

Schubert-Ulrich, P., Rudolf, J., Ansari, P., Galler, B., Führer, M., Molinelli, A. and Baumgartner, S. 2009. Commercialized rapid immunoanalytical tests for determination of allergenic food proteins: An overview. *Anal Bioanal Chem* 395: 69–81.

Scortichini, G., Annunziata, L., Di Girolamo, V., Buratti, R. and Galarini, R. 2009. Validation of an enzyme-linked immunosorbent assay screening for quinolones in egg, poultry muscle and feed samples. *Anal Chim Acta* 637: 273–278.

Sheng, W., Xia, X., Wei, K., Li, J., Li, Q. X. and Xu, T. 2009. Determination of marbofloxacin residues in beef and pork with an enzyme-linked immunosorbent assay. *J Agric Food Chem* 57: 5971–5975.

Shkotova, L. V., Goriushkina, T. B., Tran-Minh, C., Chovelon, J.-M., Soldatkin, A. P. and Dzyadevych, S. V. 2008. Amperometric biosensor for lactate analysis in wine and must during fermentation. *Mat Sci Eng* C28: 943–948.

Solna, R., Dock, E., Christenson, A. et al. 2005. Amperometric screen-printed biosensor arrays with co-immobilized oxidoreductases and cholinesterases. *Anal Chim Acta* 528: 9–19.

Song, S., Xu, H. and Fan, C. 2006. Potential diagnostic application of biosensors; current and future direction. *Int J Nanomed* 1: 433–440.

Swedish National Food Administration (SNFA). 2002. Acrylamide is formed during the preparation of food and occurs in many foodstuffs. Press release from Livsmedelsverket, Swedish National Food Administration, April 24. Available at http://www.slv.se/engdefault.asp.

Teti, D., Visalli, M. and McNair, H. 2002. Analysis of polyamines as markers of pathophysiological conditions. *J Chromatogr B* 781: 107–149.

Thakur, M. S. and Ragavan, K. V. 2013. Biosensors in food processing. *J Food Sci Technol* 50(4): 625–641.

Trivedi, U. B., Lakshminarayana, D., Kothari, I. L., Patel, P. B. and Panchal, C. J. 2009. Amperometric fructose biosensor based on fructose dehydrogenase enzyme. *Sens Actuat B* 136: 45–51.

Tubaro, A., Florio, C., Luxich, E., Sosa, S., Della Loggia, R. and Yasumoto, T. 1996. A protein phosphatase 2A inhibition assay for a fast and sensitive assessment of okadaic acid contamination in mussels. *Toxicon* 34: 743–752.

Van Veldhoven, P. P., Meyhi, E. and Mannaerts, G. P. 2002. Enzymatic quantitation of cholesterol esters in lipid extracts. *Anal Biochem* 258: 152–155.

Vedrine, C., Fabiano, S. and Tran-Minh, C. 2003. Amperometric tyrosinase based biosensor using an electro-generated polythiophene film as an entrapment support. *Talanta* 59: 535–544.

Vidal, M. L., Gautron, J. and Nys, Y. 2005. Development of an ELISA for quantifying lysozyme in hen egg white. *J Agric Food Chem* 53: 2379–2385.

Vilarino, N., Fonfría, E. S., Louzao, M. C. and Botana, L. M. 2009. Use of biosensors as alternatives to current regulatory methods for marine biotoxins. *Sensors* 9: 9414–9443.

Viswanathan, S., Radecka, H. and Radecki, J. 2009. Electrochemical biosensors for food analysis. *Monatsch Chem* 140: 891–899.

Volpe, G., Cotronco, E., Moscone, D., Croci, L., Cozzi, L., Ciccaglioni, G. and Palleschi, G. 2009. A bienzyme electrochemical probe for flow injection analysis of okadaic acid based on protein phosphatase-2A inhibition: An optimization study. *Anal Biochem* 385: 50–56.

Wang, Y., Xu, Z. L., Xie, Y. Y. et al. 2011. Development of polyclonal antibody-based indirect competitive enzyme-linked immunosorbent assay for sodium saccharin residue in food samples. *Food Chem* 126: 815–820.

Weber, P., Steinhart, H. and Paschke, A. 2007. Investigation of the allergenic potential of wines fined with various proteinogenic fining agents by ELISA. *J Agric Food Chem* 55: 3127–3133.

Wiseman, A. 1981. Enzymes in food analysis. In: *Enzymes and Food Processing*, G. G. Birch, N. Blakebrough, and K. J. Parker (eds.), Applied Science Publishers, London, pp. 275–288.

Wisitsoraat, A., Karuwan, C., Wong-Ek, K., Phokharatkul, D., Sritongkham, P. and Tuantranont, A. 2009. High sensitivity electrochemical cholesterol sensor utilizing a vertically aligned carbon nanotube electrode with electropolymerized enzyme immobilization. *Sensors* 9: 8658–8668.

Xu, Y. and Guilbault, G. G. 1990. Fast responding lactose enzyme electrode. *Enzyme Microb Technol* 12: 104–108.

You, J., Li, D., Qiao, S., Wang, Z., He, P., Ou, D. and Dong, B. 2008. Development of a monoclonal antibody-based competitive ELISA for detection of β-conglycinin, an allergen from soybean. *Food Chem* 106: 352–360.

Yu, F. Y., Vdovenko, M. M., Wang, J. J. and Sakharov, I. Y. 2011. Comparison of enzyme-linked immunosorbent assays with chemiluminescent and colorimetric detection for the determination of ochratoxin A in food. *J Agric Food Chem* 59(3): 809–813. doi:10.1021/jf103261u.

Zeravik, J., Hlavacek, A., Lacina, K. and Skládal, P. 2009. State of the art in the field of electronic and bioelectronic tongues towards the analysis of wines. *Electroanalysis* 21(23): 2509–2520.

Zhang, M., Hu, Y., Liu, S., Cong, Y., Liu, B. and Wang, L. 2012. Rapid monitoring of dipropyl phthalate in food samples using a chemiluminescent enzyme immunoassay. *Food Anal Methods* 5: 1105–1113.

Zheng, M. Z., Richard, J. L. and Binder, J. 2006. A review of rapid methods for the analysis of mycotoxins. *Mycopathologia* 161: 261–273.

4

Enzyme Engineering and Protein Modifications

Marco Filice, Caio C. Aragon, and Jose M. Palomo

CONTENTS

4.1 Introduction

Dating from its earliest applications (going back to 6000 BC or earlier with the brewing of beer, bread baking, or the first purposeful microbial oxidation with vinegar production) up to modern day (i.e., the use of rennet, a mixture of chymosin and pepsin, for cheese making or industrial production of bacterial amylases) (Poulsen and Klaus Buchholz 2003; Schäfer et al. 2002; Vasic-Racki 2006), food processing through the use of biological agents has historically been a well-established approach that constantly has undergone process improvement or design of novel strategies.

As a clarifying example, in 1874, Christian D. A. Hansen started the commercial extraction of rennet from the stomach of calves, aiming at the production of cheese. However, with increasing demand and the growth of the cheese industry, the natural production of chymosin was replaced by genetic engineering by cloning the calf prochymosin genes into bacteria, fungi, or yeasts (Vasic-Racki 2006). This classic example illustrates how advances in biotechnology, enzyme engineering, and biocatalyst design have emerged as powerful tools to solve problems in the food and beverage industries, among others. Such food processing is based on biotransformations that are chemical reactions carrying out specific conversion of complex substrates, using plant, animal, or microbial cells or purifying the enzymes directly involved. The biotransformation is conceptually different from biosynthesis (complex products assembled from simple substrates by whole cells, organs, or organisms) or from biodegradation (complex substances broken down to simple ones) (Ramachandra Rao 1996), and it possess great potential to generate novel products or to produce known products more efficiently (Giri et al. 2001). Especially in food processing, the biotransformations involve a number of advantages (Ramachandra Rao 1996):

1. Production of novel compounds
2. Enhancement in the productivity of a desired compound
3. Overcoming the problems associated with chemical analysis
4. Studies on biotransformation can lead to basic information to elucidate the biosynthetic pathway

5. Catalysis can be carried out under mild conditions, such as room temperature, and without the need of high-pressure and extreme conditions, thus reducing undesired by-products, energy needs, and cost
6. Independence of geographical and seasonal variations and various environmental factors
7. Rapidity of production

Generally, the biotransformation can be promoted by using whole cells or directly recurring to purified and engineered enzymes. These are natural, biodegradable, and efficient catalysts that work under mild conditions and catalyze reactions with high specificity and can replace many harsh chemical treatments. However, in addition to the limited supply of a given enzyme (as in the chymosin example), many other factors have forced the adoption of various enzyme engineering and protein modification strategies in order to improve their functions and properties for applications in bioprocessing and production in the food and beverages industries. Some of these factors are the following:

1. Increase of consumer demand and the need to produce more in less time
2. Rise of an exigent consumer market looking for higher-quality products and for those that positively affect health
3. Competitiveness, bringing to the industries the need for innovation at lower prices
4. Concern about the environmental impact during production with the substitution of chemicals whose disposal harms the environment
5. Increasing demand for products that conform to dietary restrictions, such as gluten or lactose intolerance
6. Need of stable enzymes in unusual conditions (certain industrial uses require enzymes stable under very harsh conditions, such as high temperature, pressure, or pH values and the presence of organic solvents)

Consequently, the implementation of native enzyme properties has become an urgent issue to be resolved in order to improve food bioprocessing. In this sense, an interesting quote by Bornscheuer et al. (2012) summarizes this need: "In the past, an enzyme-based process was designed around the limitations of the enzyme; today, the enzyme is engineered to fit the process specifications."

One of the methodologies to obtain improved biocatalysts relies on extensive screening efforts targeting new enzymes with improved characteristics produced by more resistant sources (i.e., extremophile bacteria and, particularly, thermophiles). This strategy has been finely summarized recently (Bertoldo and Antranikian 2002; Kristjansson and Asgeirsson 2003; Sun et al. 2009; Synowiecki et al. 2006; Vieille and Zeikus 2001).

Another general methodology to obtain improved biocatalysts relies on in vitro modifications. In fact, taking advantage of the knowledge gathered on molecular biology, high-throughput processing, and computer-assisted design of proteins, in vitro improvement of biocatalysts have been consistently implemented (Bershtein and Tawfik 2008; Fernandes and Cabral 2010; Vieceli et al. 2006). Such knowledge is also particularly useful for protein engineering of known enzymes (by several substitutions of amino acids in a single mutant) aiming at enhancing stability without compromising catalytic activity (Dalby 2007). The second main strategy used to enhance the catalytic properties of biocatalysts relies on enzyme immobilization.

Hence, in this chapter, some of the most recent and important advances in enzyme engineering and modification techniques in order to obtain improved biocatalysts useful for food bioprocessing are discussed.

4.2 Enzyme Engineering by Molecular Biology

Protein engineering generally aims to modify the protein sequence and hence its tridimensional structure in order to create enzymes with improved functional properties, such as stability, specific activity, inhibition by reaction products, and selectivity toward non-natural substrates (Singh et al. 2013). In particular, enhancing the stability of enzymes is of paramount importance when implementation of industrial processes is expected because it allows for the recyclability and reduction of the enzyme amount used in the process and, consequently, the final overall cost. Given that thermostability is determined by a series of short- and long-range interactions, it can be improved by several substitutions of amino acids in a single mutant in which the combination of each individual effect is usually roughly additive (Lehmann and Wyss 2001). Nevertheless, the targeted improvements have not been restricted to thermostability, but they have also addressed other features, such as broadening the range of pH at which the enzyme is active or lessening the temperature of operation while retaining high activity (Bommarius and Riebel 2004; Wong 2003).

There are two general approaches for protein engineering: rational design and directed evolution. In rational design, the structure, function, and catalytic mechanism of the protein must be well understood in order to make desired changes via site-directed mutagenesis (Scheme 4.1a). However, such understanding is lacking for most proteins of interest. In addition, although computational protein design algorithms were developed to predict optimal mutations at specific residue positions in the protein, only limited success has been demonstrated (Bloom and Arnold 2009). In contrast, the directed evolution approach requires only the knowledge of the protein sequence. This approach involves repeated cycles of random mutagenesis and/or gene recombination followed by screening or selection for positive mutants (Scheme 4.1b) (Tang and Zhao 2009). This methodology, which allows for a high throughput, has been extensively applied, aiming for more efficient biocatalysts (Adamczak and Krishna 2004; Bloom and Arnold 2009; Rubin-Pitel and Zhao 2006; Turner 2009). On continuation, some relevant examples of the successful application of these strategies in the area of food and feed processing are discussed.

The enzyme α-amylase (1,4-α-D-glucan glucanohydrolase, EC. 3.2.1.1) catalyzes the hydrolysis of α-(1,4) glycosidic linkages in starch and related polysaccharides to yield malto-oligosaccharides, such as

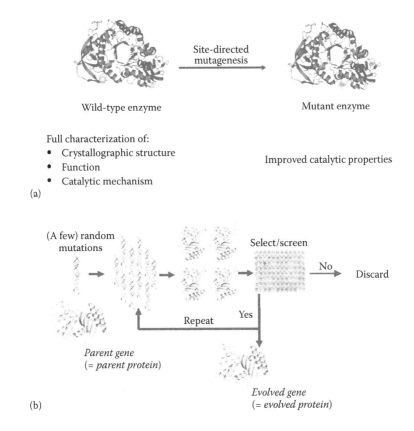

SCHEME 4.1 (a) Rational design of protein mutation. (b) Directed evolution.

maltotriose (G3), maltotetraose (G4), maltopentaose (G5), and maltohexaose (G6). They have attracted attention because they provide useful modifications to flavor and physicochemical characteristics of food in addition to having properties that are beneficial to human health. Major uses of malto-oligosaccharides are in beverages, infant milk powders, confectionery, bakery products, yogurts, and dairy desserts. On the other hand, the price of pure malto-oligosaccharides is extremely high because the chemical manufacture of malto-oligosaccharides larger than maltotriose is very difficult (Subramanian et al. 2012).

The enzyme from *Bacillus stearothermophilus* (AmyUS100) possesses a primary amino acid sequence similar to that (AmyS) of another enzyme produced by a different strain of the same bacterium. Only three of the 516 residues differed in the mature protein, one of which was in the terminal region not involved in catalysis. Both amylases presented approximately the same optimum pH and temperature but differed in their profiles of starch hydrolysis. Thus, it was suspected that the two amino acids (Asn315 and Val450) were related to the catalysis (Ben Ali et al. 2006). Three mutants were then produced from AmyUS100 by site-directed mutagenesis: AmyUS100-D (with mutations N315D), AmyUS100-G (with mutation V450G), and AmyUS100-D/G (with double mutation N315D/V450G). The V450G mutation did not affect the profile of starch hydrolysis; however, the introduction of both substitutions strongly affected the hydrolysis profile, and the main end products shifted from G6/G5 to G3/G2 (Scheme 4.2) (Ben Ali et al. 2006).

In another set of experiments, the same researchers compared the enzyme AmyUS100 with amylases from other bacteria, such as *Bacillus licheniformis* (BLA), and suspected that a small extra loop existing in its structure was involved in the lower thermostability of the enzyme compared with the BLA. Thus, a new mutant was obtained (AmyUS100-ΔIG) with a deletion of the two amino acids that caused the formation of this loop (Ile214 and Gly215). The result was an increase in the enzyme half-life from 15 min to 70 min at 100°C and from 3 min to 13 min at 110°C in the presence of 100 mM $CaCl_2$ (Ben Ali et al. 2006).

As in many other famous historic examples, an apparently bad characteristic of an enzyme can, however, become an important tool in certain industrial applications. For example, in baking, the addition of α-amylase increases dough volume, so the enzyme should be stable enough to partially degrade the starch fraction during the process. Nevertheless, the activity of the enzyme should decrease at a certain stage of the baking process in order to avoid overdosage causing a gummy and inelastic crumb.

To this scope, the α-amylase from *Bacillus amyloliquefaciens* was engineered by site-directed mutagenesis, and the variants exhibited an altered optimum pH-activity profile. The variants improved the volume of the bread as compared to the parent bacterial α-amylase but without the previously described unwanted side effects. In fact, as reported by the authors, such a positive effect may be ascribed to a decrease of pH in the dough during leavening that led to at least partial inactivation of the engineered enzyme (Danielsen and Lundqvist 2008).

Palackal et al. (2004) reported one of the highest stabilizations ever obtained by enzyme engineering by means of directed evolution strategy (increase in Tm of more than 30°C). The starting point was a xylanase discovered by screening 50,000 plaques from a complex environmental DNA library derived from a sample of fresh bovine manure. The use of site-saturation mutagenesis and the screening of

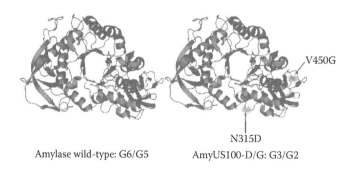

V450G

N315D

Amylase wild-type: G6/G5   AmyUS100-D/G: G3/G2

SCHEME 4.2   Wild-type and mutated amylase structures.

approximately 70,000 clones led to the identification of nine interesting mutations, which, when combined, increased the Tm by 34.2°C (Palackal et al. 2004).

Sorbitol is a sugar alcohol largely used in the food industry as a low-calorie sweetener. De Boeck et al. (2010) reported an elegant and efficient example of rational design of site-directed mutagenesis in order to achieve the large-scale production of desired product using an engineered lactic acid bacterium, such as *Lactobacillus casei*. By means of fine design of gene deletions, the authors produced an engineered strain able to convert lactose, the main sugar from milk, into sorbitol, either using a resting cell system or in growing cells under pH control. A conversion rate of 9.4% of lactose into sorbitol was obtained using an optimized fed-batch system and whey permeate, a waste product of the dairy industry, as substrate (De Boeck et al. 2010).

4.3 Enzyme Engineering by Means of Supporting Heterogeneous Materials

Since the first industrial application of immobilized amino acylase in 1967 for the resolution of amino acids, enzyme immobilization technology has attracted increasing attention, and considerable progress has been made in recent decades. Enzymes are exploited as catalysts in many industrial, biomedical, and analytical processes. There has been considerable interest in the development of enzyme immobilization techniques because immobilized enzymes have enhanced stability compared to soluble enzymes and can easily be separated from the reaction and even permitted to obtain products in continuous reactors. Approaches used for the design of immobilized enzymes have become increasingly more rational and are employed to generate improved catalysts for industrial applications. In general, the main focus of enzyme immobilization is the development of robust enzymes that are not only active but also stable and selective in organic solvents. The ideal immobilization procedure for a given enzyme is one that permits a high turnover rate of the enzyme while retaining high catalytic activity over time. Very recently, in order to further improve their characteristic, the immobilization of proteins either by physical adsorption or by covalent bonding on the surface of previously functionalized nanomaterials was revealed to be a very fast-growing area (Marciello et al. 2012; Palomo et al. 2013). There is a variety of methods used to immobilize enzymes, but generally, they can be divided into four main groups according to the method used: (i) physical adsorption, (ii) entrapment, (iii) covalent coupling, and (iv) cross-linking.

In the food, feed processing, and beverage industries, the main employed enzymes generally are oxidoreductases, transferases, hydrolases, lyases, and isomerases (Binod et al. 2008; Kirk et al. 2002; Schafer et al. 2006). Thousands of scientific papers can be found in the literature describing various methods to immobilize such useful enzymes in order to promote their applicability in industrial processes. However, there is still a huge distance between their development on the laboratory scale and their concrete viability (mainly economic) on an industrial scale. In fact, on one hand, there are several advantages that sustain enzyme immobilization, that is, high-volumetric productivities, simplified downstream or thermal, operational, and storage stabilization (Adlercreutz 2013; Mateo et al. 2007; Sheldon 2007). On the other hand, together with the positive features previously described, the enzyme immobilization strategy presents some intrinsic drawbacks: (i) mass transfer limitations, (ii) loss of activity during immobilization procedures (mainly due to chemical interaction or steric blocking of the active site), (iii) possible operational enzyme leakage, and (iv) deterioration under operational conditions due to mechanical or chemical stress. Furthermore, when scaled-up commercial processes are envisaged, economical issues must to be taken into account (although immobilization has been proven critical for economic viability if costly enzymes are used). In fact, the cost of the support, immobilization procedure, processing the biocatalyst once exhausted, up- and downstream processing of the bioconversion systems, and sanitation requirements could represent hampering elements that have to be taken into consideration.

Therefore, currently, there are few truly immobilized enzymes used in the food and beverage industries. Among them, as representative example, xylose isomerase (glucose isomerase, GI), which converts glucose to its isomer fructose, is required for high fructose corn syrup (HFCS) production. Because GI is an intracellular enzyme and glucose is not its natural substrate (but xylose), its cost is relatively high. Thus, for decades, GI has been industrially used and immobilized on, for example, ion exchangers such as DEAE-cellulose (GI from *Streptomyces rubiginosus*, Genencor/DuPont). By this method, the world production of HFCS is all performed by immobilized biocatalysts, exceeding 107 tons per year (DiCosimo et al. 2013).

4.4 Applications of Engineered Biocatalysts in Food and Beverage

On continuation, some other illustrative examples of some possible uses for the implementation of immobilized enzymes in the food and beverage industries are described (Kourkoutas et al. 2004; Ribeiro et al. 2010; Vanderhaegen et al. 2006).

4.4.1 Naringinase for Debittering Citrus Juice

In the beverage industry, the use of naringinase (an enzymatic heterodimeric complex composed of two subunits, α-L-ramnosidase [EC. 3.2.1.40] and β-D-glucosidase [EC. 3.2.1.21], generally obtained from fungi), represents a valuable alternative for the elimination of excess bitter taste of some citrus fruit juices, such as grapefruit. This enzyme hydrolyzes naringin (a flavonoid that gives such flavor to the juice) to naringenin, glucose, and rhamnose (Scheme 4.3). The high production cost of this enzyme limits its use on an industrial scale, and efforts have been made in order to facilitate its application.

For example, the naringinase from *Aspergillus niger* was entrapped in a polymeric matrix consisting of poly(vinyl alcohol) hydrogel, cryostructured in liquid nitrogen, to obtain biocatalytically active beads. The optimum temperature of the enzyme derivative increased from 60°C up to 70°C thanks to this strategy. This fact proves the stabilization of the tertiary structure of the enzyme as an explanation of its increased operational stability. Nevertheless, the authors warn that this change is not always important because, in the industrial process, the operational temperature is below the optimum in order to prevent changes in the nutritional properties of the juice. The new catalyst was tested on model citrus juice, but only 34% of naringin was hydrolyzed, and after six 24-h cycles of reaction, the efficiency dropped to about 12%, probably due to enzyme inhibition by end products of the reaction (rhamnose and glucose) (Busto et al. 2007).

Following the same research line, Puri and coworkers described a better rate of hydrolysis (but still with inhibition by-products) using, as an unusual immobilization support, wood chips (Puri et al. 2005). The naringinase from *Penicillium* sp. was covalently immobilized on this support previously activated with glutaraldehyde. As a result, a small amount (10 UI) of the obtained derivative with 120% of recovered activity was incubated with freshly prepared kinnow mandarin juice (50 mL), achieving a maximum 76% hydrolysis of naringin in 1 h. The enzyme preparation also showed a more acidic optimum pH (from 4.5 to 3), which is advantageous for this purpose because the juices to be treated with the enzyme are quite acidic (Puri et al. 2005). Another interesting example reporting enzyme activity improvement due to an immobilization protocol has been described by Lei et al. (2011). In fact, the authors describe the immobilization of naringinase from *Penicillum decumbens* on mesoporous silica MCM-41 with glutaraldehyde used to debitter white grapefruit, yielding a naringin conversion on the juice of up to 95%.

4.4.2 β-Glucosidase for Increasing Aroma

Among the huge number of applications of glucosidases, aroma increasing represents one of the most valuable. For example, β-glucosidase can be used to increase the aroma of tea. In fact, among more than 500 kinds of tea aroma constituents, monoterpene alcohols (i.e., linalool and geraniol) and aromatic alcohols (i.e., benzyl alcohol and 2-phenylethanol, which have a floral or fruity smell) are known to largely contribute to the floral aroma of tea (Wan 2003). These aroma compounds are present as monosaccharide or disaccharide flavorless glycoside precursors (β-D-glucopyranosides) in fresh leaves of tea plants. Glycoside precursors can be hydrolyzed by endogenous enzymes, such as β-D-glycosidase to release free aroma constituents (Ogawa et al. 1995; Su et al. 2010; Wang et al. 2001). However, tea plant glycosidases show low activity under natural conditions and are destroyed to a high degree during the tea manufacturing process. Therefore, most of the glycoside precursors will be saved in the final tea products, and there

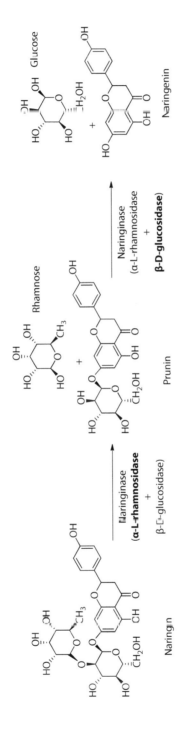

SCHEME 4.3 Naringinase-catalyzed hydrolysis of naringin.

are big potentialities to improve the aroma quality of tea by exogenous glycosidases. For example, after a study conducted with different immobilization techniques, the cross-linking–entrapment–cross-linking on an alginate of a commercial β-glucosidase resulted in a stable enzyme preparation, which was successfully used to treat the total amount of essential oil in green tea, oolong tea, and black tea increasing it by 20.7%, 10.3%, and 6.8%, respectively.

This engineered biocatalyst has been repeatedly used 50 times, maintaining a final residual activity of about 93.6% compared to its initial activity (Su et al. 2010).

In wines, monoterpenes and norisoprenoids from grapes are important aroma compounds, and they are linked to sugars, as nonvolatile compounds, working as a significant reservoir of aromatic precursors. In order to enhance the release of these compounds by means of the hydrolysis of glycosidic moiety, a covalent coimmobilization of β-glucosidase, α-arabinosidase, and α-rhamnosidase on Eupergit C (an epoxide-modified acrylic support) was performed to allow the time control of enzymatic reactions and the achievement of a final product free of catalysts (González-Pombo et al. 2014). In fact, treatment of white Muscat wine with this multienzymatic catalyst derivative caused a significant increase of free monoterpenes (from 1119 to 2132 μg/L) compared with the control wine. Moreover, geraniol (an important monoterpenic alcohol constitutive, which is abundant in many essential oils) was increased 3.4-fold over its flavor thresholds. As a result, sensorial properties were highly improved as certified by nine of 10 judges who considered treated wine more intense in fruit and floral notes (González-Pombo et al. 2014).

4.4.3 Lactose-Free Milk with β-Galactosidases

The enzyme β-galactosidase (EC. 3.2.1.23), most commonly known as lactase, hydrolyses lactose into its monomers glucose and galactose. Many people possess a low amount of this enzyme in the intestine, causing intolerance to the ingestion of the lactose contained in milk and dairy products. In addition to targeting the consumption of these products by intolerant people to this disaccharide, the use of lactase in the food industry can be a solution for other problems related to lactose, such as (Panesar et al. 2010): (i) *easy crystallization*, which limits lactose applications to certain processes in the dairy industry (it is known that cheese manufactured from hydrolyzed milk ripens more quickly than that made from normal milk), (ii) *low relative sweetness*, and (iii) *low relative solubility*.

One of the most crucial drawbacks hampering the industrial use of this enzyme is related to its multimeric structure. In fact, for this class of enzymes, due to the dissociation of some constitutive subunits, the enzyme activity is promptly lost. Consequently, in order to obtain the full stabilization of enzyme activity and its stability improvement, the immobilization of β-galactosidases must necessarily involve all the subunits, for example, in a covalent attachment on the support surface maintaining, at the same time, the correct spatial configuration. To this scope, for example, the stabilization of the quaternary structure of this enzyme was clearly demonstrated by Bernal and coworkers immobilizing the β-gal from *Kluyveromyces lactis* on aldehyde-activated supports by means of a multipoint covalent strategy (Bernal et al. 2013). The resulting highly loaded and stabilized enzyme derivatives (9000 UI/g of support) showed approximately a hundredfold higher stability compared to the unipuntual CNBr-derivatives (a reference enzyme preparation immobilized through just a few points and mimicking the soluble enzyme behavior) after thermal inactivation at pH 7 and 50°C.

When evaluating a method of immobilization for industrial application, increased stability is not always the main goal. For instance, the β-galactosidase from *Thermus* sp. T2 (Htag-BgaA) is competitively inhibited by galactose (3.1 mM) and noncompetitively inhibited by glucose (49.9 mM). Pessela et al. (2003) showed that the immobilization of the enzyme on heterofunctional epoxy-sepabeads promoted slight changes in the conformation of the active center, reducing the inhibitor affinity. As a result, when the authors compared the soluble and immobilized enzyme in the lactose hydrolysis, the former yielded 85% hydrolysis due to the strong inhibition effect of galactose and/or glucose. On the other hand, the immobilized β-galactosidase exhibited more than 99% hydrolysis of lactose, even at 70°C (Pessela et al. 2003).

The same effect was also observed by Mateo and coworkers when sepharose gel activated with aldehyde or glutaraldehyde groups were used as activated supports to immobilize the β-galactosidase from *K. lactis*. The noncompetitive inhibition was greatly reduced, achieving the quantitative hydrolysis of lactose in milk (Mateo et al. 2004).

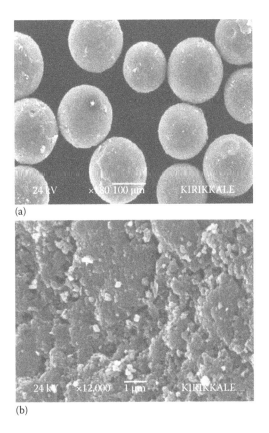

(a)

(b)

FIGURE 4.1 SEM micrograph of the magnetic poly(GMA–MMA) beads, (a) 80× and (b) 12,000× magnification.

As previously introduced, very recently, the use of engineered nanomaterials as novel functional materials to support and improve the enzyme properties is undergoing an impressive growth (Marciello et al. 2012). For example, Bayramoglu et al. (2007) described the preparation of glycidyl methacrylate/ methyl methacrylate magnetic beads in the presence of ethylene dimethyl methacrylate as a cross-linker (Figure 4.1). Finally, the magnetic beads were aminated and used for the covalent immobilization of β-galactosidase via glutaric dialdehyde activation. The immobilized preparation has been employed in a bed reactor at 35°C, and 90% of the enzyme initial activity was retained after 60 h (Bayramoglu et al. 2007).

4.4.4 Prebiotics as Functional Ingredients

In recent years, there has been particular interest in the use of immobilized enzymes in order to produce functional ingredients, especially prebiotics. These compounds are selectively fermentable and, when ingested, allow specific changes in the composition and/or activity of intestinal microbiota, providing benefits to welfare and health (Gibson et al. 2004). In general, prebiotics are composed by short chain carbohydrates, such as galacto-oligosaccharides (GOS), fructo-oligosaccharides (FOS), and xylo-oligosaccharides (XOS), whose production is generally complicated.

It is well known that, in parallel to their hydrolytic activity, β-galactosidases present a transglycosylation ability, allowing the synthesis of galacto-oligosaccharides during the hydrolytic process (Filice and Marciello 2013). The combination of these features, together with the positive effect of enzyme immobilization, result in the development of advanced bioprocesses useful to synthesize these high-value-added prebiotic GOS. For example, a β-galactosidase from *Bifidobacterium bifidum* expressed in *E. coli* was successfully immobilized on Q-sepharose with high yields and recovered activity (Filice). The maximum GOS yield obtained using the Q-sepharose derivative was similar to that obtained using free enzyme (49%–53%), indicating the absence of diffusion limitation and mass transfer issues. However, the supported enzyme showed significantly increased operational stability and GOS synthesis productivity up to 55°C. Moreover, six successive GOS synthesis batches were performed using immobilized β-galactosidase, and all resulted in similar GOS yields, indicating the possibility of developing a robust synthesis process (Osman et al. 2014). Also, for the development of efficient biotransformations useful in feed processing and prebiotic production, magnetic support results were very attractive because of their easy separation from the reaction medium. Consequently, following this research line, Pan and coworkers (2009) reported the application of Fe_3O_4-chitosan nanoparticles as support for β-galactosidase (from *A. oryzae*) immobilization using glutaraldehyde as a linker agent (Pan et al. 2009). The immobilized enzyme retained 92% of its initial activity after 15 cycles, and a good yield of 15.5% (w/v) of GOS was achieved when approximately 50% lactose was hydrolyzed. Another interesting example referring to the use of nanomaterials as support to modify and improve the enzyme properties details the synthesis of GOS by the action of *A. oryzae* β-galactosidase immobilized on magnetic polysiloxane-polyvinyl alcohol particles (Neri et al. 2009). This enzyme nanoderivative allowed the synthesis of a maximum GOS concentration of 26% (w/v) of total sugars at near 55% lactose conversion from 50% (w/v) lactose solution at pH 4.5 and 40°C. Moreover, the magnetic derivative was easily separated from the reaction medium and reutilized 10 times, retaining about 84% of the initial activity (Neri et al. 2009).

Aimed toward the production of FOS (fructo-oligosaccharides) and to overcome some of the limitations of the former gel-based biocatalysts, a modification of the classical calcium alginate enzyme entrapment technique was recently described. This strategy first expects the classical entrapment of a fructosyltransferase from *Aspergillus aculeatus* in calcium alginate gel beads and subsequently their dehydration (DALGEE) (Figure 4.2). Interestingly, the enzymatic dried beads did not reswell upon incubation in highly concentrated (600 g/L) sucrose solutions probably due to the lowered water activity (aw) of such media. The dried beads showed an approximately 30-fold higher volumetric activity (300 U/mL) compared with the calcium alginate gel beads. Moreover, a significant enhancement (40-fold) of the space-time yield of fixed-bed bioreactors of dried beads was observed compared with gel beads (4030 g/day L of FOS vs. 103 g/day L). As claimed by the authors, the operational stability of fixed-bed reactors packed

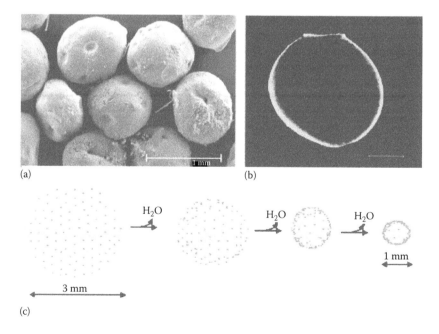

FIGURE 4.2   (a) SEM micrographs of the DALGEE particles. (b) Confocal image of FITC-labeled proteins. (c) Proposed mechanism of protein concentration on the shell of a DALGEE particle upon the drying process.

with these engineered enzymes was extraordinary, providing a nearly constant FOS composition of the outlet during at least 700 h (Fernandez-Arrojo et al. 2013).

As previously described, the multipoint covalent immobilization strategy is generally used to provide better stabilization factors to many enzymes. Based on this argument, an endoxylanase from *Streptomyces halstedii* was stabilized by multipoint covalent immobilization on aldehyde-activated sepharose supports. The obtained derivative was 200 times more thermostable than the reference one-point covalently immobilized derivative with an increment of 10°C in the optimum temperature. After 3 h, 80% of xylan solution (1% w/v) was hydrolyzed, yielding 49% of a XOS mixture ranging from xylobiose up to xylohexose (Aragon et al. 2013).

4.4.5 Lipases

In the food industry and feed processing, lipases are widely employed for a plethora of applications (Fernandes 2010). Among them, the preparation of functional ingredients in general and omega-3 fatty acids in particular are two of the most valuable examples. In fact, recently, there has been a great surge

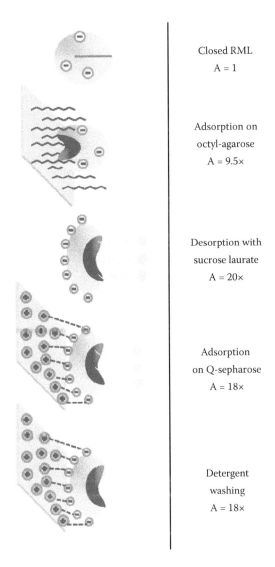

Closed RML
A = 1

Adsorption on
octyl-agarose
A = 9.5×

Desorption with
sucrose laurate
A = 20×

Adsorption
on Q-sepharose
A = 18×

Detergent
washing
A = 18×

SCHEME 4.4 Immobilization of hyperactivated form of RML (A = Activity).

in the interest among health professionals regarding the beneficial effects of omega-3 fatty acids derived from fish oils, mainly consisting of docosahexaenoic acid (DHA) and eicosapentanoic acid (EPA). DHA is required in high levels in the brain and retina as a physiologically essential nutrient to provide optimal neuronal functioning (learning ability and mental development) and visual acuity in the early stages of life (Heird 2001). On the other hand, EPA is considered to have beneficial effects in the prevention of cardiovascular diseases in adults (Demaison and Moreau 2002; Saremi 2009). In this way, the preparation of triglycerides enriched in both of the omega-3 acids (DHA and EPA) or even in only one of them could be very interesting. The first key step for the production of triglycerides of omega-3 is the rapid and selective release of PUFAs from fish oils. To this scope, the hydrolysis of fish oils by lipases immobilized on porous supports has been extensively described (DiCosimo et al. 2013). However, as a main hampering drawback, immobilized lipases cannot undergo interfacial activation via interaction with oil or solvent interfaces because oil drops are unable to penetrate inside the porous structure of the catalyst. This lack of interfacial activation can be compensated by promoting the activation of lipase during its immobilization. To this scope, for example, a novel strategy has been reported, expecting the hyperactivation of *Rhizomucor miehei* lipase (RML) through different concentrations of different detergents and the subsequent immobilization of the most active open form via intense multipoint anion exchange (Scheme 4.4) (Filice et al. 2011). RML immobilized inside porous supports by means of this strategy becomes very active, highly stable, and selective for the mild hydrolysis of fish oils with EPA production also in the absence of interaction with interfaces and of aggregation phenomena. Thus, interfacial activation of soluble lipases on oil drops is not necessary because a similar activation can be achieved via a careful non-natural hyperactivation of the immobilized enzyme.

4.5 Conclusion

The integration of enzymes in food and feed processes is a well-established approach, but evidence clearly shows that dedicated research efforts should be consistently made to make this application of biological agents more effective and/or diversified. This still unmet progress mainly regards the design of new/improved biocatalysts that are more stable (to temperature and pH), less dependent on metal ions, and less susceptible to inhibitory agents and to aggressive environmental conditions while maintaining the targeted activity or evolving to novel activities. This is of particular relevance for application in the food and feed sector because it should allow enhanced performance under operational conditions that minimize the risk of microbial contamination. It also favors process integration by allowing the concerted use of enzymes that naturally have diverse requirements for effective application.

Moreover, despite the achievements obtained in this particular field, there is still the lack of a set of unanimously applicable rules, which furthermore encompass both technical and economic requirements. The latter can be particularly restrictive in the food and feed sector because most products are of relatively low added value. However, it can be foreseen that efforts will be made toward the development of newer immobilized biocatalysts with suitable chemical, physical, and geometric characteristics that can be used in different reactor configurations and that respond to the economic requirements for large-scale application.

The development of newer chemical and genetic strategies aiming to modify the protein structures in order to create semisynthetic and artificial enzymes strengthening the enzyme properties (activity, selectivity, and stability) and minimizing, at the same time, their flaws will undoubtedly be a feasible alternative. It could be foreseen that the application in food and feed processing of all the abovementioned strategies either isolated or, preferably, suitably integrated will undoubtedly be the main road to improving existing processes and implementing new ones.

REFERENCES

Adamczak, M. S., Krishna, H. 2004. Strategies for improving enzymes for efficient biocatalysis. *Food Technol Biotechnol.* 42(4):251–264.

Adlercreutz, P. 2013. Immobilisation and application of lipases in organic media. *Chem Soc Rev.* 42:6406–6436.

Aragon, C. C., Mateo, C., Ruiz-Matute, A. I. et al. 2013. Production of xylo-oligosaccharides by immobilized-stabilized derivatives of endo-xylanase from *Streptomyces halstedii*. *Process Biochem.* 48(3):478–483.

Bayramoglu, G., Tunali, Y., Arica, M. Y. 2007. Immobilization of β-galactosidase onto magnetic poly(GMA–MMA) beads for hydrolysis of lactose in bed reactor. *Catal Commun.* 8(7):1094–1101.

Ben Ali, M., Khemakhem, B., Robert, X., Haser, R., Bejar, S. 2006. Thermostability enhancement and change in starch hydrolysis profile of the maltohexaose forming amylase of Bacillus stearothermophilus US100 strain. *Biochem J.* 394(Pt 1):51–56.

Bernal, C., Marciello, M., Mesa, M. et al. 2013. Immobilisation and stabilisation of β-galactosidase from Kluyveromyces lactis using a glyoxyl support. *Int Dairy J.* 28(2):76–82.

Bershtein, S., Tawfik, D. S. 2008. Advances in laboratory evolution of enzymes. *Curr Opin Chem Biol.* 12(2):151–158.

Bertoldo, C., Antranikian, G. 2002. Starch-hydrolyzing enzymes from thermophilic archaea and bacteria. *Curr Opin Chem Biol.* 6(2):151–160.

Binod, P., Singhania, R. R., Soccol, C. R., Pandey, A. 2008. Industrial enzymes. In *Advances in Fermentation Technology*, A. Pandey, C. Larroche, C. R. Soccol, and C.-G. Dussap, 291–320. Asiatech Publishers, New Delhi, India.

Bloom, J. D., Arnold, F. H. 2009. In the light of directed evolution: Pathways of adaptive protein evolution. *Proc Natl Acad Sci U S A.* 106(1):9995–10000.

Bommarius, A. S., Riebel, B. R. 2004. *Biocatalysis. Fundamentals and Applications*. Wiley-VCH, Weinheim, Germany.

Bornscheuer, U. T., Huisman, G. W., Kazlauskas, R. J., Lutz, S., Moore, J. C., Robins, K. 2012. Engineering the third wave of biocatalysis. *Nature.* 485(7397):185–194.

Busto, M. D., Meza, V., Ortega, N., Perez-Mateos, M. 2007. Immobilization of naringinase from *Aspergillus niger* CECT 2088 in poly(vinyl alcohol) cryogels for the debittering of juices. *Food Chem.* 104(3):1177–1182.

Dalby, P. A. 2007. Engineering enzymes for biocatalysis. *Recent Pat Biotechnol.* 1(1):1–9.

Danielsen, S., Lundqvist, H. 2008. Bacterial alpha-amylase variants. WO Patent 2008/000825.

De Boeck, R., Sarmiento-Rubiano, L. A., Nadal, I., Monedero, V., Pérez-Martínez, G., Yebra, M. J. 2010. Sorbitol production from lactose by engineered Lactobacillus casei deficient in sorbitol transport system and mannitol-1-phosphate dehydrogenase. *Appl Microbiol Biotechnol.* 85(6):1915–1922.

Demaison, L., Moreau, D. 2002. Dietary n-3 polyunsaturated fatty acids and coronary heart disease-related mortality: A possible mechanism of action. *Cell Mol Life Sci.* 59:463–477.

DiCosimo, R., McAuliffe, J., Poulose, A. J., Bohlmann, G. 2013. Industrial use of immobilized enzymes. *Chem Soc Rev.* 42(15):6437–6474.

Fernandes, P. 2010. Enzymes in food processing: A condensed overview on strategies for better biocatalysts. *Enzyme Res.* 2010:862537.

Fernandes, P., Cabral, J. M. S. 2010. Applied biocatalysis: An overview. In *Industrial Biotechnology*, W. Soetaert, and E. J. Vandamme, 227–250. Wiley-VCH, Weinheim, Germany.

Fernandez-Arrojo, L., Rodriguez-Colinas, B., Gutierrez-Alonso, P. et al. 2013. Dried alginate-entrapped enzymes (DALGEEs) and their application to the production of fructooligosaccharides. *Process Biochem.* 48(4):677–682.

Filice, M., Marciello, M. 2013. Enzymatic synthesis of oligosaccharides: A powerful tool for a sweet challenge. *Curr Org Chem.* 17(7):701–718.

Filice, M., Marciello, M., Betancor, L., Carrascosa, A. V., Guisan, J. M., Fernandez-Lorente, G. 2011. Hydrolysis of fish oil by hyperactivated rhizomucor miehei lipase immobilized by multipoint anion exchange. *Biotechnol Prog.* 27(4):961–968.

Gibson, G. R., Probert, H. M., Loo, J. V., Rastall, R. A., Roberfroid, M. B. 2004. Dietary modulation of the human colonic microbiota: Updating the concept of prebiotics. *Nutr Res Rev.* 17(2):259–275.

Giri, A., Dhingra, V., Giri, C. C., Singh, A., Ward, O. P., Narasu, M. L. 2001. Biotransformations using plant cells, organ cultures and enzyme systems, current trends and future prospects. *Biotechnol Adv.* 19:175–199.

González-Pombo, P., Fariña, L., Carrau, F., Batista-Viera, F., Brena, B. M. 2014. Aroma enhancement in wines using co-immobilized *Aspergillus niger* glycosidases. *Food Chem.* 143:185–191.

Heird, W. C. 2001. The role of polyunsaturated fatty acids in term and preterm infants and breastfeeding mothers. *Pediatr Clin North Am.* 48:173–188.

Kirk, O., Borchert, T. V., Fuglsang, C. C. 2002. Industrial enzyme applications. *Curr Opin Biotechnol.* 13(4):345–351.

Kourkoutas, Y., Bekatorou, A., Banat, I. M., Marchant, R., Koutinas, A. A. 2004. Immobilization technologies and support materials suitable in alcohol beverages production: A review. *Food Microbiol.* 21(4):377–397.

Kristjansson, M. M., Asgeirsson, B. 2003. Properties of extremophilic enzymes and their importance in food science and technology. In *Handbook of Food Enzymology*, J. R. Whitaker, A. G. J. Voragen, and D. W. S. Wong, 77–100. Marcel Dekker, New York.

Lehmann, M., Wyss, M. 2001. Engineering proteins for thermostability: The use of sequence alignments versus rational design and directed evolution. *Curr Opin Biotechnol.* 12(4):371–375.

Lei, S., Xu, Y., Fan, G., Xiao, M., Pan, S. 2011. Immobilization of naringinase on mesoporous molecular sieve MCM-41 and its application to debittering of white grapefruit. *Appl Surf Sci.* 257(9):4096–4099.

Marciello, M., Filice, M., Palomo, J. M. 2012. Different strategies to enhance the activity of lipase catalysts. *Catal Sci Technol.* 2:1531–1543.

Mateo, C., Monti, R., Pessela, B. C. C. et al. 2004. Immobilization of lactase from *Kluyveromyces lactis* greatly reduces the inhibition promoted by glucose. Full hydrolysis of lactose in milk. *Biotechnol Prog.* 20(4):1259–1262.

Mateo, C., Palomo, J. M., Fernandez-Lorente, G., Guisan, J. M., Fernandez-Lafuente, R. 2007. Improvement of enzyme activity, stability and selectivity via immobilization techniques. *Enzyme Microb Technol.* 40(6):1451–1463.

Neri, D. F. M., Balcão, V. M., Costa, R. S. et al. 2009. Galacto-oligosaccharides production during lactose hydrolysis by free *Aspergillus oryzae* β-galactosidase and immobilized on magnetic polysiloxane-polyvinyl alcohol. *Food Chem.* 115(1):92–99.

Ogawa, K., Moon, J. H., Guo, W., Yagi, A., Watanabe, N., Sakata, K. 1995. A study on tea aroma formation mechanism: Alcoholic aroma precursor amounts and glycosidase activity in parts of the tea plant. *Z Naturforsch.* 50C:493–498.

Osman, A., Symeou, S., Trisse, V., Watson, K., Tzortzis, G., Charalampopoulos, D. 2014. Synthesis of prebiotic galactooligosaccharides from lactose using bifidobacterial β-galactosidase (BbgIV) immobilised on DEAE-Cellulose, Q-Sepharose and amino-ethyl agarose. *Biochem Eng J.* 82:188–199.

Palackal, N., Brennan, Y., Callen, W. N. et al. 2004. An evolutionary route to xylanase process fitness. *Protein Sci.* 13:494–503.

Palomo, J. M., Filice, M., Romero, O., Guisan, J. M. 2013. Improving lipase activity by immobilization and post-immobilization strategies. In *Immobilization of Enzymes and Cells*, J. M. Guisan, 255–274. Springer Science + Business Media, New York.

Pan, C., Hu, B., Li, W., Sun, Y., Ye, H., Zeng, X. 2009. Novel and efficient method for immobilization and stabilization of β-d-galactosidase by covalent attachment onto magnetic Fe_3O_4–chitosan nanoparticles. *J Mol Catal B Enzyme.* 61(3–4):208–215.

Panesar, P. S., Kumari, S., Panesar, R. 2010. Potential applications of immobilized β-galactosidase in food processing industries. *Enzyme Res.* 2010:473137.

Pessela, B. C. C., Mateo, C., Fuentes, M. et al. 2003. The immobilization of a thermophilic β-galactosidase on Sepabeads supports decreases product inhibition. *Enzyme Microb Technol.* 33(2–3):199–205.

Poulsen, P. B., Klaus Buchholz, H. 2003. History of enzymology with emphasis on food production. In *Handbook of Food Enzymology*, J. R. Whitaker, A. G. J. Voragen, and D. W. S. Wong, 11–20. Marcel Dekker, New York.

Puri, M., Kaur, H., Kennedy, J. F. 2005. Covalent immobilization of naringinase for the transformation of a flavonoid. *J Chem Technol Biotechnol.* 80(10):1160–1165.

Ramachandra Rao, S. 1996. Studies on biotransformation to produce phytochemicals of importance using plant cell cultures, Ph.D. Diss., University of Mysore, Mysore, India.

Ribeiro, D. S., Henrique, S. M. B., Oliveira, L. S., Macedo, G. A., Fleuri, L. F. 2010. Enzymes in juice processing: A review. *Int J Food Sci Technol.* 45(4):635–641.

Rubin-Pitel, S. B., Zhao, H. 2006. Recent advances in biocatalysis by directed enzyme evolution. *Comb Chem High Throughput Screen.* 9(4):247–257.

Saremi, A., R. 2009. The utility of omega-3 fatty acids in cardiovascular disease. *Am J Ther.* 16:421–436.

Schäfer, T., Kirk, O., Borchert, T. V. et al. 2002. Enzymes for technical applications. In *Biopolymers*, S. R. Fahnestock, and S. R. Steinbuchel, 377–437. Wiley-VCH, Weinheim, Germany.

Schafer, T., Borchert, T. W., Nielsen, V. S. et al. 2006. Industrial enzymes. *Adv Biochem Engin/Biotechnol.* 105:59–131.

Sheldon, R. A. 2007. Enzyme immobilization: The quest for optimum performance. *Adv Synth Catal.* 349(8–9):1289–1307.

Singh, R. R. K., Tiwari, M. K., Lee, J.-K. 2013. From protein engineering to immobilization: Promising strategies for the upgrade of industrial enzymes. *Int J Mol Sci.* 14(1):1232–1277.

Su, E., Xia, T., Gao, L., Dai, Q., Zhang, Z. 2010. Immobilization of β-glucosidase and its aroma-increasing effect on tea beverage. *Food Bioprod Process.* 88(2–3):83–89.

Subramanian, G., Ayyadurai, S., Sharma, T., Singh, S. A., Kumar, I. S. 2012. Studies on a maltohexaose (G6) producing alkaline amylase from a novel alkalophilic streptomyces species. *IIOAB.* 3(3):15–30.

Sun, H., Zhao, P., Ge, X., Xia, Y., Hao, Z., Liu, J., Peng, M. 2009. Recent advances in microbial raw starch degrading enzymes. *Appl Biochem Biotechnol.* 160(4):988–1003.

Synowiecki, J., Grzybowska, B., Zdzieblo, A. 2006. Sources, properties and suitability of new thermostable enzymes in food processing. *Crit Rev Food Sci Nutr.* 46(3):197–205.

Tang, W. L., Zhao, H. 2009. Industrial biotechnology: Tools and applications. *Biotechnol J.* 4(12):1725–1739.

Turner, N. J. 2009. Directed evolution drives the next generation of biocatalysts. *Nat Chem Biol.* 5(8):567–573.

Vanderhaegen, B., Neven, H., Verachtert, H., Derdelinckx, G. 2006. The chemistry of beer aging—A critical review. *Food Chem.* 95(3):357–381.

Vasic-Racki, D. 2006. History of industrial biotransformations—Dreams and realities. In *Industrial Biotransformations*, 2nd ed., A. Liese, K. Seelbach, and C. Wandrey, 1–35. Wiley-VCH, Weinheim, Germany.

Vieceli, J., Mullegger, J., Tehrani, A. 2006. Computer-assisted design of industrial enzymes: The resurgence of rational design and in silico mutagenesis. *Ind Biotechnol.* 2(4):303–308.

Vieille, C., Zeikus, G. J. 2001. Hyperthermophilic enzymes: Sources, uses, and molecular mechanisms for thermostability. *Microbiol Mol Biol Rev.* 65(1):1–43.

Wan, X. C. 2003. *Tea Biochemistry*, 3rd ed., 40–43. China Agricultural Press, Beijing.

Wang, D., Kurasawa, E., Yamaguchi, Y., Kubota, K., Kobayashi, A. 2001. Analysis of glycosidically bound aroma precursors in tea leaves. Changes in glycoside contents and glycosidase activities in tea leaves during black tea manufacturing process. *J Agric Food Chem.* 49:1900–1903.

Wong, D. W. S. 2003. Recent advances in enzyme development. In *Handbook of Food Enzymology*, J. R. Whitaker, A. G. J. Voragen, and D. W. S. Wong, 379–387. Marcel Dekker, New York.

Section II

Applications of Enzymes in Food and Beverage Industries

5

Enzymes in Food and Beverage Production: An Overview

Muthusamy Chandrasekaran, Soorej M. Basheer, Sreeja Chellappan, Jissa G. Krishna, and P. S. Beena

CONTENTS

5.1 Introduction

The food and beverage industries are among the major industries that manufacture a wide range of food and drinks to cater to the needs of the growing world population. The food and beverage processing sector, which is a highly fragmented industry, broadly comprises the following subsegments: dairy and milk, grain and cereal, fruits and vegetables, beer and alcoholic beverages, meat and poultry, seafood, packaged or convenience foods, and packaged drinks. The various food and beverage industries grouped in terms of their major food products are presented in Table 5.1. These industries process the raw food and beverage materials and manufacture processed food and beverages according to the consumer demand around the world, and as a consequence, the food and beverage industries are proliferating rapidly. In fact, the principal aims of modern food processing industries are (i) to make safe food, both chemically and microbiologically; (ii) to provide high-quality products, rich in nutrients and organoleptic properties; and (iii) to prepare convenient foods. The benefits of food processing include preservation, an increase in food consistency and verity, and ease in marketing. Further, food processing facilitates the seasonal availability of several foods, safeguards against spoilage, extends shelf life, protects against food-borne pathogens and their poisoning, and facilitates transportation and distribution of perishable foods across long distances. Moreover, large-scale processing of food contributes to cheaper costs of food owing to a reduction in manufacturing costs and enhanced profit potential for the food processing industries.

When foods are processed for commercial use, chemical treatments are generally applied to assist in production. However, enzyme use instead of chemicals in the food processing industry generally creates superior products with improved yields in addition to reducing carbon footprint, energy consumption, and environmental pollution. Further, food enzymes account for low greenhouse gas emissions and less raw material wastage. They offer flexible, high-performing solutions that ensure high-quality, cost-effective products and always meet consumer preferences. Thus, food and beverage manufacturing processes stand to become more environmentally friendly. Food enzymes are cost-effective and provide better food safety, thus increasing their demand in the food industry. Further, enzymes play an essential role in bringing more nutritious and appealing food and beverage products to the modern world, and they offer significant benefits beyond the scope of traditional alternatives. In addition, using enzymes instead of chemicals reduces overall manufacturing costs (http://www.livestrong.com/article /545581-3-advantages-of-enzymes).

In fact, enzymes are known to have played an important part in food production since ancient times. One of the earliest examples of industrial enzyme use was in the production of whiskey. At present, nearly all commercially prepared foods contain at least one ingredient that has been made with enzymes. However, enzymes are known to be used on a large-scale basis only within the latter half of the twentieth century. Now, the food industry spends more than $5 billion on enzyme applications, making them a chief component in the biotechnology field of food processing (http://www.livestrong.com/article /545581-3-advantages-of-enzymes). In this context, an overview of various food and beverage processing industries that utilize enzymes as biocatalysts, their spectra of food and beverages manufactured and marketed, their current status and market potential, and the scope for use of enzymes for introduction of new products by these food industries are discussed here in this chapter.

5.2 Food and Beverage Processing Industries: Global Scenario

5.2.1 Starch Processing

Starch has several roles in the food industry, including as a carbohydrate source, texture modifier, processing aid, thickener, extender, stabilizer, etc. Natural starches are of plant origin and are produced commercially from seeds of plants such as wheat, sorghum, rice, corn, and tuber roots of cassava, potato, or arrowroot. Its abundant availability and low cost are the two factors that make it an important candidate for the food industry. According to the technical market research report from BCC Research (www.bccresearch.com, FOD037B, 2013), sales of starches and derivatives were $51.2 billion in 2012.

TABLE 5.1

Food Industries and Their Manufactured Products

Food Industry		Products Processed
Beverage	Alcoholic drinks	Beer, wine, whiskey, brandy, rum, other distilled spirits
	Nonalcoholic drinks	Fruit juice, soft drinks, mineral water
	Syrups	Syrups
	Other beverages	Coffee, tea
Dairy	Milk	Fresh milk, flavored milk, condensed milk, soy milk
	Milk products	Milk fat, custard, whipped cream, butter, buttermilk, cheese, yogurt, milk powder, cream powder, ice cream, whey, caseinate, lactose from whey
	Starter cultures	*Lactobacillus* sp.
Oils and fats	Oils and fats	Soy oil, seed oil extraction, olive oil, refining oil, fats
Confectioneries	Cocoa	Cocoa butter, cocoa powder, chocolates, candies
	Sugars	Sweets, glucose, fructose
	Others	Chewing gum
Bakery	Cereal, flour	Breakfast cereals, soybeans, couscous
	Starch	Maize starch, wheat starch, potato starch, modified starch
	Biscuits, cookies, toast	Biscuit, puff pastry, shortbread
	Bread	Rusk, rye bread, wheat bread, gingerbread, rice cakes, tortillas, pita bread
	Pastry	Cake, boiled sweets, wet pastry, chocolate teacake, waffles
Fruits and vegetables	Fruit	Fruit salad, strawberries, apple mash, olives, raisins, cranberry jam, marmalade
	Vegetable	Dried vegetables, salad, beans, gherkins, sauerkraut, peas, pearl onions, preserved vegetables, leaf vegetables, tomato, potato chips (French fries), potato crisps, syrup
	Mushrooms	Fresh mushroom cultivation, preserved mushrooms
Meat	Cattle	Beef
	Goat and sheep	Meat
	Pig	Pork meat
Poultry	Chicken	Chicken meat
	Turkey	Meat
	Egg	Raw egg, pasteurized egg, cooked egg, dried egg
Seafood	Finfish	Surimi, fillets, fish meal, fish oil, canned meat, cooked ham, sausages
	Shellfish	Prawns, shrimps, lobsters, crabs, mollusks
Nutraceuticals	Plants	Plant bioactive substances
Probiotics	Probiotics Bacteria	*Lactobacillus* sp., *Brevibacterium* sp., *Bacillus*, *Saccharomyces* sp.
	Prebiotics	Fructo oligosaccharides, inulin
	Digestive aids	Enzymes
Functional foods	Natural foods	Vegtables with functional components such as lycopene, β-carotene, lignans, etc. Fruits with functional components such as caffeic acid, feluric acid, fish oil, ω 3-fatty acids
	Processed foods	Fruit juices, yogurts with probiotics, etc.
Flavor and food additives		Gelatin, artificial sweeteners, alginic acid
Others		Ketchup, salad dressing, margarine, mayonnaise, peanut butter, sauces, mustard

Source: Adapted from M. Chandrasekaran, S. M. Basheer, S. Chellappan, P. Karthikeyan and Elyas K. K. 2012. Food processing industries—An overview. In *Valorization of Food Processing By-Products,* Ed. M. Chandrasekaran, CRC Press. Taylor & Francis Group, Boca Raton, FL.

BCC Research expects the market to reach $77.4 billion by 2018, after increasing at a compound annual growth rate (CAGR) of 7.1% between 2012 and 2018.

Within the global industry, the United States is home to by far the largest starch industry with 51% of world production. In the United States, cereal grains, such as corn and wheat, provide the major source of starch. The average yearly U.S. corn crop contains about 136 million metric tons of starch with only about 15% of the crop being processed to separate the starch or starch-protein (flour) component from the corn kernels. Forty-five percent of starch production is located in the United States, where maize starch converted into fructose and glucose syrups for food industry use is of major importance: It represents 75% of all starch used in the United States. Especially important in the United States and Canada is the use of cornstarch to produce high-fructose syrup for soft drinks. Sweet corn and popcorn are also of considerable economic importance in the United States, but they are seldom used in bakery products; sweet corn kernels are sometimes added to corn muffins or fritters to give texture contrast.

In Europe, maize, wheat, and potato all contribute significant amounts of starch. Tubers, particularly sweet potato and cassava, which are rich in starches, have not yet been exploited for starch when compared with wheat and corn.

In Asia, starch production is about 16 million metric tons (mmt), and demand for starch by both food and nonfood industries is growing by 5%–10% per year. Nearly 98% of starch is produced in Southeast Asia with Thailand as a major player in world markets. More than half of total cassava production is used for starch extraction, including 62% in Indonesia and 52% in Thailand—by far the two largest cassava-producing countries in Asia. China produces a range of starches, including cassava and maize. China is also the major producer of sweet potato starch in the world although most of this occurs on a small scale. In China, Japan, and Korea, sweet potato follows closely behind maize in its overall contribution to starch production. Between 23% and 33% of sweet potato production in this region is used for starch extraction, including 23%–33% in China and as much as 67% in Japan. Japan also produces sweet potato starch and imports maize from the United States.

Interestingly, more than 8% of world starch production is derived from wheat, more than 5% of global starch supplies are derived from potatoes, and the EU is the major source of more than two thirds of the figure. In addition, cassava also contributes about 5% of the world starch output.

Modified starch (corn, tapioca, potato, wheat, and others) has a range of applications in food, feed, and nonfood preparation and is an emerging market. Although the current scenario depicts a wide range of applications, the emerging crops, such as tapioca, are gaining focus among modified starch manufacturers. In food applications, convenience or processed food forms are a major driver due to its increasing consumption in developing as well as developed economies.

Starch derivatives (maltodextrin, cyclodextrin, glucose syrup, hydrolysates, and modified starch) have application in the food and beverage and feed industries. Starch derivative is one of the growing industries of the world and is expected to grow at a rate of 6.2% from 2014 to 2019 to reach $58.2 billion by 2019. The rising demand for convenience foods and beverages in developing countries acts as a driver for the starch derivative industry (http://www.marketsandmarkets.com/Market-Reports/starch-derivatives-market-116279237.html).

5.2.2 Bakery

The baking industry is one of the most stable sectors in the food manufacturing industries (Zhou and Therdthai 2007). Breads have been in existence from ancient days and have been in the human diet from primitive days. In addition to major bakery products such as bread, cake, and biscuits, there are also numerous other types of bakery products, including pastries, pizzas, breakfast cereals, and so on. In 2008, the U.S. bread market recorded $20.5 billion with fresh bread sales at $6.6 billion, compared to $545 million for frozen bread, rolls, biscuits, pastry dough, and $207 million for bread, rolls, bun dough (Toops and Fusaro 2010). Whole-grain breads are expected to remain strong because consumers demand more natural breads with nutritional ingredients, premium breads that feature quality ingredients, and artisanal breads. An April 2009 "Packaged Facts report" estimated the size of the gluten-free market at $1.56 billion and with a compound annual growth rate (CAGR) of 28% from 2004 to 2008. These products are sought by those with medical conditions requiring a lifelong adherence to a gluten-free diet and consumers who believe that a gluten-free diet is more healthful. According to an annual survey of manufacturers in 2001, the total value of the manufacturing of bread and bakery products in the United States was $31.2 billion, including retail bakeries ($2.5 billion), commercial bakeries ($25.7 billion), and frozen products and other pastries ($3.0 billion) (Zhou and Therdthai 2007). Meanwhile, according to the Euromonitor International sources, the total global market value of bread and bakery products stood at $311.1 billion in 2005, representing a 7.32% increase from 2004. The commercial side of the industry is concentrated: The 50 largest companies generate 75% of revenue. But the retail side is highly fragmented: The 50 largest companies account for only about 15% of revenue. The bakery industry has achieved the third position in generating revenue among the processed food sector. The market size for the industry is pegged at $4.7 billion in 2010 and is expected to reach $7.6 billion by 2015 (http://www.researchandmarkets .com/reports/2041431/indian_bakery_industry_2011_2015). The per capita consumption of bakery products in India, as it stands of now, is 1–2 kg per annum, which is comparatively lower than the advanced countries in which consumption is between 10 and 50 kg per annum. The growth rate of bakery products has been tremendous in both urban and rural areas. According to the available data in 2009, the top bread vendors can be listed as follows: Sara Lee Bakery, George Weston Inc., Flowers Foods Bakeries Group, Interstate Bakeries Corp., Bimbo Bakeries, Pepperidge Farm, Stroehmann Bakeries Inc., and La Brea, etc.

5.2.3 Confectioneries

Chocolates and confectionery have influenced mankind in its various tastes and forms ever since it was first made. Confections include sweet foods, digestive aids that are sweet, elaborate creations, and frivolous items. Consumers of all age groups prefer chocolate and confectionery products because of their attractive color, appearance, and flavor. Modern usage may include substances rich in artificial sweeteners as well. Nowadays, varieties of products have gained importance due to their delicious taste and better keeping quality making a direct impact on demand.

The chocolate and confectionery category, the second largest packaged food segment, has been growing steadily in all regions over the last few years. Globally, chocolate confectionery is the largest sector in terms of value, accounting for almost 60% of total sales. The chocolate market is estimated at around 33,000 tons valued at approximately INR 8.0 billion at present. Major players such as Cadbury India, Nestle, Amul, and Campco have captured the heart of the Indian chocolate market driving the industry with an impressive growth rate of almost 18% annually. The Indian confectionery market is valued at approximately INR 50 billion. The Indian confectionery market includes sugar-boiled confectionery, hard-boiled candies, toffees, and other sugar-based candies. Sugar-boiled confectionery has penetrated an estimated 15% of households only, suggesting a large potential for growth (http://www.dofppunjab .info/sector-specific-information/Bakery-Confectionery.php).

5.2.4 Beverages

The beverage industry is the largest food processing industry, which manufactures carbonated beverages and alcoholic drinks. The $110 billion beverage industry is a diverse segment, representing alcoholic and nonalcoholic drinks that include milk, carbonated soft drinks, coffee, and bottled water. It was observed that per capita consumption of selected beverages, as per 2007 statistics, was maximal for carbonated soft drinks (48.8 gallons per person) followed by bottled water (29.1 gallons per person), coffee (24.6 gallons per person), beer (21.8 gallons per person), and milk (20.7 gallons per person).

5.2.5 Dairy and Milk

The dairy industry represents a major and important segment of the food processing industry. According to Fonterra analysis in 2011, about 378 billion liters of milk is processed each year by dairy companies around the world. On the basis of cheese consumption and production details, it is estimated that approximately 9 million tons of cheese per annum is produced within the EU. Raw milk for processing comes mainly from cows and, to a lesser extent, from other mammals, such as goats, sheep, yaks, camels, or horses. Swiss food giant Nestle is the biggest dairy firm in the world with an annual turnover of almost $28.3 billion, followed by Danone, a French company.

5.2.6 Meat and Poultry

The meatpacking industry handles the slaughtering, processing, and distribution of animals, such as cattle, pigs, sheep, and other livestock. In 2012, approximately 301 million tons of meats were produced worldwide. The FAO estimates that by 2050 global meat production will increase to 455 million tons (Alexandratos and Bruinsma 2012). Most of the world's beef is produced in Australia, Brazil, Southern Africa, and the United States. Beef makes up 24% of the world's meat consumption whereas poultry accounts for 34%, and pork accounts for 40% of global meat consumption. The industry, in addition to producing meat for human consumption, also yields a variety of by-products, including hides, feathers, dried blood, and, through the process of rendering, fat, such as tallow, and protein meals, such as meat and bone meal. Some of the major historical and current meatpackers in the United States include Armour and Company, Cargill Meat Solutions, Cudahy Packing Company, Greater Omaha Packing Company, Lomen Company, Hormel Foods, Smithfield Foods, Tyson Foods, Perdue Farms, and Swift Packing. Some outside the United States include Imperial Cold Storage and Supply Company (South Africa), William Davies Company, Maple Leaf Foods (Canada), Schneider Foods (Canada), and JBS S.A. (Brazil).

Poultry is one of the major animal-based food processing industries, and the United States leads the world in poultry production with more than 16.1 million metric tons in 2006. China is the second largest poultry producer, next to the United States, and Brazil is ranked as the third largest poultry producer. These countries are the major players in the world poultry market.

5.2.7 Seafood

The world seafood (finfish and shellfish) industry plays a significant role in providing protein-rich food to a significant part of the world's population. In fact, fishing and fish farming has emerged as one of the major food processing occupations of mankind. Further, fishing and processing activities provide employment to millions of people around the world. The global seafood market is estimated at $100 billion per annum. It is also predicted that the world demand for seafood may increase by 3% each year. The largest seafood consumption in the world is in Japan, followed by the EU. Tokyo's Tsukiji Market is the largest wholesale fish and seafood market in the world, selling more than 400 types of seafood. Currently, only about 30% of seafood produced for human consumption is marketed fresh. The rest is processed and marketed for consumption. In this sector, the frozen fish fillets and fish in the form of ready-to-eat meals and other convenience food products is recording rapid growth in both developed and developing countries. The end products from seafood processing may be fresh, frozen, or marinated fillets; canned fish; fish meal; fish oil; or fish protein products, such as surimi. In the seafood sector, about 75% of world seafood production is used for human consumption, and the rest is used to produce fish meal and oil. Fish meal is used as feed for livestock, such as poultry, pigs, and farmed fish, and fish oil is used as an ingredient in paints and margarine.

5.2.8 Food Stabilizers

Food stabilizers (blends/systems) have applications in bakery, confectionery, dairy, sauces, dressings, meat, poultry, convenience, beverage for stability, texture, and moisture retention. The rising popularity for processed food, such as ready-to-eat and convenience foods, has increased the demand for stabilizer blends from food and beverage manufacturers. The stabilizer blends market is projected to grow at a compound annual growth rate (CAGR) of 4.2% from 2013 to 2018 (http://www.marketsandmarkets.com /Market-Reports/functional-stabilizer-blends-market-41788893.html, accessed January 15, 2015).

5.2.9 Nutraceuticals

The nutraceutical ingredients are an emerging market with increased awareness of people and growing health concerns. People lean toward consuming healthy food that fulfills the need of essential nutrients in the body. Manufacturers are also taking into consideration the convenience factor for consumers and providing them with healthy nutrients in the form of food and beverages instead of supplements. Animal nutrition is developing with the growing meat and milk market. Animal feed manufacturers tend to attain more output from animals, and for this purpose, various nutritious ingredients are used in animal feed. Major applications of nutraceutical ingredients include, for example, functional food, functional beverages, dietary supplements, animal nutrition, etc. In 2010, the U.S. nutraceutical market stood at $50.4 billion and was by far the largest nutraceutical market in the world, whereas the total European nutraceutical industry was valued at $35 billion (Frost and Sullivan 2011).

5.2.10 Digestive Health Enzymes, Prebiotics, and Probiotics

Digestive health is one of the fastest growing markets in the United States, along with the heart health, bone health, weight management, and immune health markets. A large number of gut health or digestive health products are hitting the market, which purely reflects increased consumer interest in a diet that supports gut health. Products aimed at digestive health have been far ahead in popularity in European countries than in the United States.

5.2.11 Sports and Energy Drinks

Global demand for sports and energy drinks is growing at a CAGR of more than 10%. Rising health awareness among consumers is playing a major role in the popularity of this beverage. Companies such as Coca-Cola, Pepsi, Glaxo-SmithKline, and Red Bull are innovating constantly and adding new products to this niche segment.

5.2.12 Natural Colors and Flavors

Natural colors and flavors are types of food additives that are added to food and beverages to make products more appealing and tasty. Their use in the food and beverage industry has been increasing since the last decade. Rising demand for natural foods and consumer avoidance has led to strong progress for the natural colors and flavors market. Hyperactivity and behavioral problems in children due to artificial colors and flavors are other driving factors for this market. Colors and flavors derived for natural products are being exempted from certification. Color and flavor degradation with change in pH, light, temperature, and oxidation with other ingredients is a major restraint for the global natural colors and flavors market.

Food flavors are used by food and beverage manufacturers to enhance the quality and taste of products. Flavors and flavor enhancers are key building blocks that impart taste in processed food and beverage products. The flavors segment is the largest component of overall food ingredients; this industry is highly competitive and concentrated, and the top three companies—Givadudan (Switzerland), IFF (United States), and Firmenich (Switzerland)—hold 44% (estimated) of the global flavors market. Introduction of new flavors and better technology drive the market.

5.2.13 Ready-to-Drink (RTD) Tea and Coffee

RTD tea and coffee is a subgroup of soft drinks, but it differs in the function with proven health benefits. The market for RTD tea and coffee is growing exponentially. Various types of RTD tea and coffee products are introduced in the market to serve the wide consumer base spread across geographies. Various nutraceutical ingredients are being incorporated to make them more functional. Asia-Pacific is the most dominating market whereas North America is the fastest growing RTD tea and coffee market.

5.3 World's Major Players in the Food and Beverage Processing Industries

The world's top ranked 25 major food industries as of 2014 and their marketed food products are presented in Table 5.2. Among them, Nestlé is the world's largest food and beverage company, and Pepsi Co. is the largest U.S.-based food and beverage company. Unilever, an Anglo-Dutch company, owns many

TABLE 5.2

Top-Ranked Food and Beverage Processing Companies in the World (2014)

No.	Name of Company	Major Product Area
1	PepsiCo Inc. www.pepsico.com	Beverages, bakery, cereal, snacks, miscellaneous
2	Tyson Foods Inc. www.tyson.com	Beef, pork, poultry, frozen and prepared foods
3	Nestle (U.S. and Canada) www.nestle.com	Beverages, canned, frozen and preserved foods, confectionery, dairy, pet foods, miscellaneous
4	JBS Swift & Company www.jbsswift.com	Beef and pork
5	The Coca-Cola Co. www.cocacola.com	Beverages
6	Anheuser-Busch Cos. Inc. www.anheuser-busch.com	Beer, malt beverages
7	Kraft Foods Inc. www.kraftfoods.com	Biscuits/crackers, cheese, grain products, meat and poultry, confectionery, packaged meals, miscellaneous
8	Smithfield Foods Inc. www.smithfieldfoods.com	Pork, beef and turkey products, esp. ham
9	General Mills Inc. www.generalmills.com	Cereals, refrigerated and prepared foods, dough products, baking products, snacks, yogurt
10	ConAgra Inc. www.conagra.com	Snacks, grocery, dairy, frozen foods, specialty potato products
11	Mars Inc. www.mars.com	Confectionery, snacks, packaged foods, pet foods, ice cream/frozen novelties
12	Kellogg Co. www.kelloggs.com	Cereal, cookies and crackers, fruit leather, vegetarian/soy products
13	Dean Foods Co. www.deanfoodscom	Dairy
14	Hormel Foods Corp. www.hormel.com	Meat and poultry, canned, frozen and preserved foods, fats/oils, miscellaneous
15	Cargill Inc. www.cargill.com	Meat and poultry, fats/oils, grain mill products
16	MillerCoors www.millercoors.com	Beer and malt beverages
17	Saputo Inc. Saint-Leonard, Quebec www.saputo.com	Dairy products (mostly cheese), bakery
18	Pilgrim's Pride Corp. www.pilgrimspride.com	Poultry, meat, eggs
19	Hershey Co. www.hersheys.com	Confectionery
20	Mondelez International Inc. www.mondelez.com	Biscuits (cookies and crackers), chocolate, gum and candy, coffee and beverages
21	Unilever www.unilever.com	Misc. grocery products, ice cream
22	Bimbo Bakeries USA www.grupobimbo.com; www.bimbobakeriesusa.com	Bakeries
23	Dr Pepper Snapple Group www.drpeppersnapplegroup.com	Beverages
24	J. M. Smucker Co. www.smucker.com	Canned, frozen and preserved foods, beverages
25	Campbell Soup Co. www.campbellsoup.com	Meat and poultry, canned, frozen and preserved food, bakery, sugar/confectionery, beverages, miscellaneous

Source: http://www.foodprocessing.com/top100/top-100-2014/ accessed on January 14, 2015.

of the world's consumer product brands in foods and beverages whereas Kraft is apparently the world's second largest food company following its acquisition of Cadbury in 2010. Dole Food Company is the world's largest fruit company. Chiquita Brands International, another U.S.-based fruit company, is the leading distributor of bananas in the United States. Sunkist Growers Inc. is a U.S.-based grower cooperative. JBS S.A. is the world's largest processor and marketer of chicken, beef, and pork. Grupo Bimbo is one of the most important baking companies.

5.4 Applications of Enzymes in the Food and Beverage Industry

Major classes of enzymes and different enzyme groups used in food and feed processing are presented in Table 5.3. Food enzymes primarily include carbohydrase, protease, and lipase with carbohydrase

TABLE 5.3

Enzyme Groups Used in Food and Feed Processing

Enzyme Class	Enzymes	Functions
Oxidoreductases	Glucose oxidase	Dough strengthening
	Laccases	Clarification of juices, flavor enhancer (beer)
	Lipoxygenase	Dough strengthening, bread whitening
Transferases	Cyclodextrin-glycosyl transferase	Cyclodextrin production
	Fructosyltransferase	Synthesis of fructose oligomers
	Transglutaminase	Modification of viscoelastic properties, dough processing, meat processing
Hydrolases	Amylases	Starch liquefaction and saccharification, increasing shelf life and improving quality by retaining moist, elastic, and soft nature, bread softness and volume, flour adjustment, ensuring uniform yeast fermentation, juice treatment, low calorie beer
	Galactosidase	Viscosity reduction in lupins and grain legumes used in animal feed, enhanced digestibility
	Glucanase	Viscosity reduction in barley and oats used in animal feed, enhanced digestibility
	Glucoamylase	Saccharification
	Invertase	Sucrose hydrolysis, production of invert sugar syrup
	Lactase	Lactose hydrolysis, whey hydrolysis
	Lipase	Cheese flavor, in situ emulsification for dough conditioning, support for lipid digestion in young animals, synthesis of aromatic molecules
	Proteases (chymosin, papain)	Protein hydrolysis, milk clotting, low-allergenic infant-food formulation, enhanced digestibility and utilization, flavor improvement in milk and cheese, meat tenderizer, prevention of chill haze formation in brewing
	Pectinase	Mash treatment, juice clarification
	Peptidase	Hydrolysis of proteins (namely, soy, gluten) for savory flavors, cheese ripening
	Phospholipase	In situ emulsification for dough conditioning
	Phytases	Release of phosphate from pyruvate, enhanced digestibility
	Pullulanase	Saccharification
	Xylanases	Viscosity reduction, enhanced digestibility, dough conditioning
Lyases	Acetolactate decarboxylase	Beer maturation
Isomerases	Xylose (glucose) isomerase	Glucose isomerization to fructose

Source: P. Fernandes, *Enzyme Research*, Article ID 862537, 2010. doi:10.4061/2010/862537.

encompassing prevalent types, such as amylase, cellulase, pectinase, lactase, and others. Food enzymes are mainly used in the baking industry and fruit juice and cheese manufacturing as well as wine making and brewing to improve flavor, texture, digestibility, and nutritional value. Carbohydrases dominate the food enzyme market followed by protease and lipase. Carbohydrases such as α-amylases and glucoamylases, pectinases, glucose isomerases, cellulases, and hemicellulases, etc., have various applications in the food and beverage industry, making them the most important class of food enzyme. Protease and lipase have their main application in the dairy sector. Some of the typical applications include enzyme use in the production of bakery products, dough conditioning, sweeteners, chocolate syrups, candy, cheese, dairy products, alcoholic beverages, chill proofing of beer, precooked cereals, fruit juice, soft drinks, vegetable oil, infant foods, meat tenderizing, egg products, seafood, flavor extracts and flavor development, and liquid coffee, among others. The details with some examples are presented in Table 5.4.

5.4.1 Types of Enzyme Applications

Enzymes are used either as food additives or in the manufacture of foods, the latter being the more common use of enzymes in this industry (Figure 5.1). From a regulatory point of view, the distinction between processing aids and additives is very important because the national regulatory context of enzymes differs significantly in different countries even among the EU member states. In terms of enzyme regulations, enzymes used in food can be distinguished into food additives and processing aids.

TABLE 5.4

Application of Enzymes in Different Food, Beverage, and Feed Industries

Industry	Product/Process	Application	Enzyme
Dairy	Cheese	Cheese manufacturing: breaks down and destabilizes casein molecules in milk	Rennin (chymosin), lysozymes
		Cheese ripening	Lipase
	Enzyme modified cheese (EMC)	Degrades protein and enables aging of cheese	Protease
	Dairy products	Used to degrade hydrogen peroxide (H_2O_2), which is added to dairy products to aid in pasteurization and subsequently removed	Catalase
	Milk	Breaks down lactose to glucose and galactose in milk processing to avoid lactose intolerance	β-Galactosidase, lactase
	Infant formula	Predigest casein	Trypsin
Cereals	Baby foods	Precooked baby foods	Amylases
	Breakfast foods	Breakfast foods condiments	Protease-papain, bromelain, pepsin
Baking	Bread	Degrading starch in flours and controlling the volume and crumb structure of bread	α-Amylases
		Giving increased gluten strength	Oxidoreductases
	Dough	Starch saccharification to maltose	α-Amylases
		Improving dough handling and dough stability	β-Xylanases
		Improving stability of the gas cells in dough	Lipases
	Dough conditioner	Reducing the protein in flour	Proteases
Juice industry	Juice	Breaks down starch into glucose, clarifying cloudy juice, especially for apple juice	Amylases, glucoamylases
		Degrading pectins, which are structured polysaccharides present in the cell wall	Pectinases
		Acting on soluble pectin hydrolysis and on cell wall composition with pectinases; lowering viscosity and maintenance of texture	Cellulases, hemicellulases

(Continued)

TABLE 5.4 (CONTINUED)

Application of Enzymes in Different Food, Beverage, and Feed Industries

Industry	Product/Process	Application	Enzyme
		Increasing the susceptibility of browning during storage	Laccase
		Acting on compounds that cause bitterness in citrus juices	Naringinase and limoninase
	Color removal	Decolorization of fruit juices	Anthocyanase
Confectioneries	Soft-centered confections and chocolates	Extends shelf life by prevention of formation of sugar crystals by conversion of sucrose to glucose and fructose	Invertase
	Chocolate	Addition of flavor	Protease, polyphenol oxidase
		Cocoa butter equivalents/substitutes	Lipase
	High-fructose corn syrup	Conversion of glucose to fructose	Glucose isomerase
	Glucose syrup	Manufacture of glucose from corn starch	α-Amylase, glucoamylase
	Chewing gum	Activates mouths own protective system. Effective protection against cariogenic bacteria	Lactoperoxidase
Starch processing	Liquefaction of starch	Cleaving α-1,4-glycosidic bonds in the inner region of the starch, causing a rapid decrease in substrate molecular weight and viscosity	α-Amylases
		Attacking α-1,6-linkages, liberating straight-chain oligosaccharides of glucose residues linked by α-1,4-bonds	Pullulanases
		Acting on both α-1,6- and α-1,4-linkages	Neopullulanases, amylopullulanases
	Maltose	Cleaving α-1,4-linkages from nonreducing ends of amylose, amylopectin, and glycogen molecules; producing low-molecular weight carbohydrates, such as maltose and β-limit dextrin	α-Amylases
	Glucose	Attacking α-1,4-linkages and α-1,6-linkages from the nonreducing ends to release β-d-glucose	Glucoamylases
		Hydrolysing α-1,6-linkagess in glycogen and amylopectin	Isoamylases
	Fructose	Catalyzing isomerization of glucose to fructose	Glucose isomerases
	Modified starch	Transferring a segment of a 1,4-α-D-glucan chain to a primary hydroxyl group in a similar glucan chain to create 1,6-linkages, increasing the number of branched points to obtain modified starch with improved functional properties, such as higher solubility, lower viscosity, and reduced retrogradation	Glycosyltransferases
	Cyclodextrin	Cyclodextrin from starch	Cyclomaltodextrin glucanotransferase
Brewing	Mashing	Hydrolyzing starch to reduce viscosity, liquefying adjunct, increasing maltose and glucose content	α-Amylases
	Mashing	Starch saccharification to maltose	β-Amylases
	Wort	Hydrolysing glucans into soluble oligomers and leading to lower viscosity and better filterability, improving wort separation	β-Glucanases
		Hydrolyzing a-1,6 branch points of starch, securing maximum fermentability of the wort	Pullulanases

(Continued)

TABLE 5.4 (CONTINUED)

Application of Enzymes in Different Food, Beverage, and Feed Industries

Industry	Product/Process	Application	Enzyme
	Beer	Increasing glucose content, increasing 1% fermentable sugar in "light" beer	Amyloglucosidases
		Increasing soluble protein and free amino-nitrogen (FAN), malt improvement, improving yeast growth	Proteases
		Hydrolyzing pentosans of malt, barley, and wheat; improving extraction and beer filtration	Pentosanases, xylanases
		Converting α-acetolactate to acetoin directly, decreasing fermentation time by avoiding formation of diacetyl, making beer taste right	α-Acetolactate-decarboxylases (ALDC)
	Chill-proof beer	Removes protein haze in cooled beer	Protease, tannase
	Wine	Decolorization of wine	Anthocyanase
	Green tea	Develops flavor by creating aldehydes and terpenes	Polyphenoloxidases, peroxidases
	Instant tea	Improved solubility	Tannase
	Coffee	Coffee bean fermentation, coffee concentrates	Pectinase, hemicellulase
Carbonated beverages	Nonalcoholic beverages and soft drinks	Oxygen removal	Glucose oxidase
Flavor		Removal of starch, oxygen removal	Amylase, glucose oxidase
Edible oils	Vegetable oils	Purification of vegetable oil	Phospholipase
Meat	Soft meat	Tenderization of meat	Protease, papain, bromelain, transglutaminase
Egg	Dried egg white	Used to degrade sugars, such as in dried egg whites because the sugars remain in the dried egg white and caramelize during heat treatment	Glucose oxidase
Seafood	Finfish and shellfish	Removal of scales and skin from fish, production of fish sauce, recovery of protein and protein hydrolysate from seafood by-products	Protease
	Fish oils	Enrichment of ω-3 PUFA, isolation of oil and fats from seafood products	Lipase
	Surimi and fish meat	Surimi processing and restructured fishery products, meat mince formation	Transglutaminase
	Extending shelf life and preservation	Preservation of seafood	Glucose oxidase lysozyme, catalase
	Finfish products	Removal of off-odor and fishy taste	Urease
Animals feeds industry		Degrading fiber in viscous diets	Xylanases
		Degrading phytic acid to release phosphorus, and liberating calcium, magnesium cations	Phytases
		Degrading proteins into its constituent peptides and amino acids to overcome antinutritional factors	Proteases (substilisin)
		Digesting starch	α-Amylases
		Feed additive	Tannase

Source: Adapted from S. Li, X. Yang, S. Yang, M. Zhu, X. Wang, *Computational and Structural Biotechnology Journal*, 2, 3, 2012, e201209017, http://dx.doi.org/10.5936/csbj.201209017.

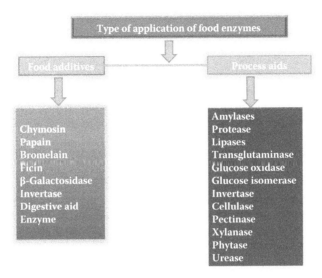

FIGURE 5.1   Types of enzyme applications.

Most food enzymes are considered to be processing aids, and only a few are used as additives, such as lysozyme and invertase (European Commission Report 2002). But in general, the processing aids are used during the manufacturing process of foodstuff and do not have a technological function demand in the final food. Enzymes used in food processing are typically sold as enzyme preparations, which contain not only the desired enzyme, but also metabolites from the production strain and several added substances, such as stabilizers. All these materials are expected to be safe under the guidance of good manufacturing practice (GMP).

Examples of enzymes directly used as food additives include papain, bromelain, and ficin, and these enzymes are even available as chewable tablets to aid in digestion. Another important enzyme used as an additive in foods is invertase, which is typically used in the manufacture of invert sugar, which is much more soluble than sucrose. Due to this property, it is employed in preventing sugar crystallization, a property essential in the manufacture of chocolates and candies, especially of the soft-centered variety. This enzyme is sourced from *Saccharomyces cereviseae*. β-Galactosidase (lactase) sourced from *Saccharomyces lactis* or *Aspergillus niger* is another important enzyme additive in food. This enzyme is used to hydrolyze lactose in whey allowing its up-gradation. β-Galactosidase is also used in production of sweeter milk, milk concentrates, and ice creams to prevent lactose crystallization (Underkofler 1980).

Apart from direct addition to food, enzymes are also widely employed in the processing/manufacture of food, in which they are considered to be processing aids (Adrio and Demain 2014). Classical examples of such applications include the use of calf rennin (chymosin) in the manufacture of cheese. Although the enzyme was originally sourced from calf stomach, recombinant chymosin expressed from *Escherichia coli*, *Kluyveromyces marxianus*, and *A. niger var. awamori* are now approved by the Food and Drug Administration (FDA) as food ingredients and are commonly employed in cheese manufacturing (US FDA 2014).

5.4.2 Starch Processing and Production of Sugar Syrups

Starch is the main reserve polysaccharide of many higher plants. It forms about 75% of the dry mass in cereal grains, 65% in potato tubers, and about 71% in corn (Qureshi and Blaschek 2006). Starch is made of amylose and amylopectin. It is a polymer containing D-anhydroglucose, monomeric units, which on hydrolysis yield D-glucose. The abundance of hydroxyl groups in starch attributes to its absorbing moisture, and it has the property of gelatinization. The corn sugar and syrup industry derive glucose, maltose, and higher oligosaccharides from uncooked starch. Two types of reactions are utilized in the conversion of starch to syrup and sugars: acid hydrolysis by hydrochloric acid or enzyme hydrolysis employing amylases or glucosidic enzyme. During acid hydrolysis, the starch is converted to dextrose. Amylolytic enzyme contains α- and β-amylase, and maltose units are formed as a result of enzymatic action on starch. Glucosidase is used to supplement acid or amylolytic enzyme conversion, and that yields a higher proportion of dextrose or glucose in the product mix. Pullulanase, an important debranching enzyme, has been widely utilized to hydrolyze the α-1,6 glucosidic linkages in starch, amylopectin, pullulan, and related oligosaccharides, which enables a complete and efficient conversion of the branched polysaccharides into small fermentable sugars during the saccharification process (Hii et al. 2012).

Sugar is the expensive element of most sweet products, such as candy or cola. Enzymes enable corn, cassava, or wheat/potato starch to be transformed into sugar syrup. The enzymes work by rearranging and cutting up the starch molecules, turning them into liquid sugar. When the process is complete, the syrups and modified starches, which have different compositions and physical properties, can be used in a wide variety of foodstuffs, including soft drinks, confectionery, meats, baked products, ice cream, sauces, baby food, tinned fruit, preserves, and much more.

During the latter half of the twentieth century, enzymes replaced acid hydrolysis in starch processing and in the manufacture of sugar syrups because they contributed to higher-quality products, energy efficiency, and a safer working environment. In the 1970s, high fructose corn syrup (HFCS) was developed that closely mimicked the sweetness of sucrose (table sugar). Today, the production of HFCS is a major industry, which converts large quantities of corn (maize) and other natural starches to this and other useful sweeteners. These sweeteners are used in soft drinks, candies, baking, jams and jellies, and many other foods. HFCS from starch is an alternative to sugar cane and sugar beets.

5.4.3 Bakery

Modern bread production is often reliant upon oxidative compounds that can help in forming the right consistency of the dough. Enzymes are usually added to modify dough rheology, gas retention, and crumb softness in bread manufacture; to modify dough rheology in the manufacture of pastry and biscuits; to change product softness in cake making; and to reduce acrylamide formation in bakery products (Cauvain and Young 2006). Enzymes such as glucose oxidase have been used to replace the unique effect of chemical oxidants, such as bromates, azodicarbonamide, and ascorbic acid, which have been widely used to strengthen gluten when making bread with quality standards that satisfy consumer demand. Further consumers enjoy soft bread. Use of enzymes in the baking industry started more than 100 years ago. The three main categories of enzymes that are commonly used in baking are enzymes that hydrolyze carbohydrates, mainly amylases, cellulases, and pentosanases, enzymes that hydrolyze proteins called proteases and enzymes hydrolyzing fats and oils called lipases and lipoxygenases. Enzymes are often used to modify starch, which, in turn, keeps bread softer for a longer period of time. By choosing the right enzyme, starch can be modified during baking to retard staling. The bread stays soft and flavorful for a longer time (3–6 days).

5.4.4 Confectioneries

Sweetening agents are the most important of the ingredients used in the manufacture of a large variety of confectionery products. Among them, starch syrups (glucose) and dextrose occupy a position second only to sucrose. Both cereal and tuber starches are utilized in glucose and glucose syrup manufacture (Sreekantiah 1979). Glucose syrups containing more than 50% maltose or even more than 70% maltose (extra-high-maltose syrup) can be produced by using β-amylase or fungal α-amylase. High-maltose glucose syrup has a great advantage in the production of hard candy. At given moisture levels and temperatures, a maltose solution has a lower viscosity than a glucose solution but will still set to a hard product. Maltose is also less humectant than glucose, so candy produced with high-maltose syrup will not become sticky as easily as candy produced with standard glucose syrup (Hull 2010). Soft candy and other treats made with sugar, especially soft-center candy, such as chocolate-covered cherries, often have a short shelf life because the sugar sucrose contained in the product begins to crystallize soon after the confection is produced. A similar change occurs in soft cookies and other specialty bakery items. An enzyme, invertase, converts the sucrose to two simple sugars, glucose and fructose and thus prevents the formation of sugar crystals that otherwise would severely shorten the shelf life of the product or make some products virtually unavailable at reasonable prices.

5.4.5 Fruit Juices

Enzymes are ideal for fruit juices due to their efficiency of specific action, high purification, and standardization. Enzymes are either an integrated part of the juice or added to it, providing a number of advantages in the process, such as high substrate specificity, high reaction rate under mild operational costs and investment, and fast and continuous readily controlled reaction with generally low operational costs and investment. Enzymes have many applications in fruit processing, including prepeeling, pulp washing, peel juice extraction, juice clarification, and also in other beverage industries. The main reasons for the use of enzymes are the following: (i) to improve the yield of juice production, (ii) to liquefy the entire fruit for maximal utilization of the raw material, (iii) to improve color and aroma, (iv) to clarify juice, and (v) to break down all insoluble carbohydrates, such as pectins, hemicelluloses, and starch (Olsen 1995).

The shelf life of citrus fruit juices can be prolonged with the combination of the enzymes glucose oxidase and catalase, because the oxidative reactions leading to aroma deterioration are retarded. Other applications of catalase are in removal of O_2 from the air present in the head space of bottled and canned drinks and reduction of nonenzymatic browning in wines (Rolle 1998). Hydrolases are the major enzymes used in the fruit juice industries (Kashyap et al. 2001). Production of fruit juice with enzymes is an essential practice in the juice industry throughout the world, and apple juice is the most popular juice overall. Other issues, such as increasing yield, controlling clarity with enzymes, sweetness, and shelf life, are all important factors for juice processors. Enzymes in fruit juice also affect the color and flavor of the juice. Many of the naturally occurring enzymes help to form esters, aldehydes, and alcohols— all important flavor volatiles in fruit (http://www.foodproductdesign.com/articles/1996/09/enzymes-that -aid-beverages.aspx).

Juices extracted from ripe fruit contain a significant amount of pectin that imparts a cloudy appearance to the juice that many consumers do not find appealing. Pectinase enzymes act on pectin, yielding a crystal-clear juice with the appearance, stability, mouth-feel, taste, and texture characteristics preferred by consumers. The use of enzymes in juice processing helps ensure that the maximum amount of juice is removed from the fruit, thereby reducing waste and controlling costs. The sweet taste of the soft drink comes from syrup, which is made using enzymes. Enzymes are able to make fruit juices completely clear. By exploiting enzymes in the breakdown process, the pressing process becomes much easier. The juice becomes crystal clear and contains all the nutrients from the apple due to the enzymatic breakdown of the fibers.

5.4.6 Beverages

Enzymes play a crucial role in the production of beer and other types of malted liquor, such as whiskey. In these products, enzymes provide three major functions: the formation of sugars to be used during fermentation; viscosity control; and, in beer, "chill-proofing." Enzymes are also helpful in wine production, helping in the preservation of wine quality, sometimes over many years in storage. They also reduce fermentation time and promote clarification, filtration, and stabilization. Winemakers need to produce high-quality wines year after year, regardless of annual variations in the weather, and enzymes can help as they are the perfect tool for transferring valuable components, such as aroma, color, and tannins from the grape to the wine (Kashyap et al. 2001). The main benefits of using these enzymes during winemaking include better maceration, improved color extraction, easy clarification, easy filtration, improved wine quality, and improved stability (Galante et al. 1998). Beer brewing is based on the action of enzymes activated during malting and fermentation.

Calorie-conscious consumers can enjoy *reduced calorie beer* prepared using special enzymes in the brewing process. Major ingredients used in the production of beer include barley, rice, and other grains.

Enzymes transform the complex carbohydrates in the grains to simpler sugars, and hence, the desired alcohol content can be achieved with a smaller amount of added grain. This results in a beer with fewer carbohydrate calories and, ultimately, a lower calorie beer with good taste.

Green tea leaves must undergo a fermentation step to create the colors, flavors, and astringency associated with black tea. This fermentation is actually an enzymatic degradation of various compounds. Naturally occurring polyphenoloxidase and peroxidase cause most of the desired chemical changes, but other enzymatic reactions probably help develop the flavor by creating aldehydes and terpenes.

5.4.7 Dairy

Rennet, an enzyme mixture from the stomach of calves and other ruminant mammals, is a critical element in *cheese making*. Rennet has been the principle ingredient facilitating the separation of the curd (cheese) from the whey for thousands of years. A purified form of the major enzyme in rennet, chymosin, is produced from genetically modified microorganisms and is commercially available now. This helps assure availability of excellent cheeses at a reasonable cost. Lipases contribute to the distinctive flavor development during the ripening stage of cheese production. They act on the butter fat in cheese to produce flavors that are characteristic of different types of cheese.

Lactose-free dairy products are important as a significant portion of the adult population is unable to consume normal portions of dairy products as they cause gastrointestinal (GI) upset in the form of bloating, gas, or diarrhea, or a combination of GI symptoms. Lactase, an enzyme that occurs naturally in the intestinal tract of children and many adults, is either absent or not present in sufficient quantity in lactose-intolerant adults. Lactase converts the milk sugar found in dairy products, such as milk, ice cream, yogurt, and cheese, to two readily digestible sugars, glucose and galactose. Without adequate lactase, the lactose in the food ferments in the intestine, producing undesirable side effects. People who historically could not consume dairy products can now enjoy these nutritious foods, thanks to the commercial availability of the digestive enzyme, lactase. Many products present in the dairy case today are labeled "lactose-free" as the result of pretreatment of the milk or final product with the enzyme lactase. Additionally, lactase is available at retail for use in treating lactose-containing dairy products in the home.

The other minor enzymes having limited applications in dairy processing include glucose oxidase, catalase, superoxide dismutase, sulphydryl oxidase, lactoperoxidase, and lysozymes. Glucose oxidase and catalase are often used together in selected foods for preservation. Superoxide dismutase is an antioxidant for foods and generates H_2O_2 but is more effective when catalase is present. Thermally induced generation of volatile sulphydryl groups is thought to be responsible for the cooked off-flavor in ultrahigh-temperature (UHT) processed milk. Use of sulphydryl oxidase under aseptic conditions can eliminate this defect. The natural inhibitory mechanism in raw milk is due to the presence of low levels of lactoperoxidase (LP), which can be activated by the external addition of traces of H_2O_2 and thiocyanate. Cow milk can be provided with protective factors by the addition of lysozyme, making it suitable as an infant milk. Lysozyme acts as a preservative by reducing bacterial counts in milk (Neelakantan et al. 1999).

5.4.8 Meat

The world market for meat and poultry is predicted to grow at almost 4% a year to reach nearly $640 billion in 2015. Chilled meat products are the leading product segments in the industry, accounting for sales around $236 billion, which corresponds to 45% of the global market. The world's fresh meat market is expected to hit the 300 million ton mark by 2015 reports "Global Industry Analysts." Asia-Pacific is a leading region in meat consumption, where demand is rising due to strong economic growth and higher incomes (http://www.reportlinker.com/ci02026/Meat-and-Poultry.html).

Fermented meat and sausages were manufactured by Romans and Greeks even in ancient times and was expanded throughout Europe in the middle ages. Today, a wide variety of fermented sausages are available in the market, depending on the large number of variables related to the raw materials, microbial population, and processing conditions. But today, they are mostly produced using the starter cultures of lactic acid bacteria alone (LAB) or a combination with coagulase negative *Staphylococci* (CNS) and yeasts or molds. These microorganisms provide a good number of enzymes involved in relevant biochemical changes such as the enzymatic breakdown of carbohydrates, proteins, and lipids that affect color, flavor, and texture.

With *meat tenderizing*, the major meat proteins responsible for tenderness are the myofibrillar proteins and the connective tissue proteins. In principle, enzymes may be used in two different ways to alter the structure of meat and meat products. First, enzymes may catalyze breakdown of covalent bonds in proteins, thereby generating smaller peptide fragments or amino acids. This structure breakdown may increase the tenderness of the meat. Second, enzymes may promote the formation of new covalent bonds between meat proteins. In meat gels, such enzymes may enhance the firmness and water-holding capacity of gels. Protease enzymes such as papain and bromelain are used to tenderize tougher cuts of meat for many years. To improve this process, more specific proteases have also been introduced to make the tenderizing process more robust.

5.4.9 Seafood

In recent times, products and processes have emerged in the seafood (finfish and shellfish) processing industry, which utilizes enzymes in a deliberate and controlled fashion. In the seafood industry, enzymes have been used for deskinning, specialty product development, and quality assurance.

Quality assurance is of the utmost importance in the seafood business, perhaps even more so than in most other areas of food processing because of the fragility and ease of spoilage of seafood flesh. Thus, it is no surprise that this is the area of seafood processing that has seen the most rapid development in recent years. Recent advances in this area include the development of rapid detection methods for seafood spoilage bacteria and typical seafood-borne pathogens, such as *Vibrio parahaemolyticus* (Venkateswaran et al. 1996), *Vibrio cholerae* (Karunasagar et al. 1995), and others (Fung 1994), and also for toxic compounds such as histamine, one of the causative agents of scombroid poisoning (Lopez-Sabater et al. 1993; Male et al. 1996).

Determination of trimethylamine (TMA) or hypoxanthine levels is the commonest in freshness testing. A rapid assay was devised to monitor the levels of inosine monophosphate (IMP), inosine, and hypoxanthine in seafood flesh (Luong et al. 1992) and an even simpler method that utilizes a xanthine oxidase enzyme electrode have been devised (Shen et al. 1996). A rapid enzymatic method for measuring TMA levels in a perchloric acid extract of fish flesh has also been devised.

5.4.10 Digestive Aids

Enzymes can be used to improve the nutritional quality of food for humans and animals. The full utilization of the potential nutritive value in legumes and soy-based foods is limited by the presence of nondigestible sugars, such as raffinose and stachyose. These sugars contain chemical linkages that cannot be broken by the natural enzymes produced by the body. Consequently, the sugars proceed through the digestive tract until reaching the large intestine in which they are hydrolyzed by the natural microflora in the intestine. These organisms utilize the sugars that are converted to gas during this metabolism, causing discomfort and flatulence. The enzyme, α-galactosidase, is used to convert stachyose and raffinose to simple sugars that are adsorbed by the human digestive tract, thereby preventing the flatulence often caused by legumes, such as beans and soy based foods. This enzyme can be used to hydrolyze raffinose and stachyose during soy processing, during the food preparation process or by addition to the food itself immediately before ingestion.

The animal feed industry is another major consumer of industrial enzymes with several of the feeds supplemented with enzymes such as amylases and proteases to aid in digestion of the feed. Enzymes such as β-glucanases and arabinoxylanases added to feed cereals can help in breaking down nonstarch polysaccharide antinutritional factors. Poultry and hog feed grains contain phosphorus, which is bound to phytic acid. In this form, the phosphorus is not available to the animals and is excreted in the animals' waste. Further phytic acid is an antinutritional component in plant-derived food that interferes with dietary absorption of essential minerals such as iron and zinc (Gavrilescua and Chisti 2005). The digestion of phytic acid by phytase releases the bound phosphorus, making it available for the animal (Kirk et al. 2002). Several hydrolases are also employed in preparation of feed supplements. For example keratinase is employed for digestion of chicken feathers to obtain amino acids and peptides, which can be supplemented in animal and poultry feed.

5.5 Food Enzyme Market Trends

The global food and beverage enzymes market covers various types of food and beverage enzymes, which can be broadly categorized as carbohydrases, proteases, lipases, and others. Among them, carbohydrases, which have diverse applications within the food and beverage industry, dominate the market, accounting for almost 70% of this market followed by proteases and lipases. According to a research report from the Austrian Federal Environment Agency (European Commission Report 2002), approximately 158 enzymes were used in the food industry, 64 enzymes in technical applications, and 57 enzymes in feedstuff, of which 24 enzymes were used in three industrial sectors. Almost 75% of all industrial enzymes are hydrolytic enzymes. The global food and beverage enzymes market valued at $1355.8 million in 2012 is estimated to reach $2306.4 million by 2018, growing at a CAGR of 8% from 2013 to 2018 (Food Enzymes Market: Global Trends and Forecasts to 2018). The food enzymes market is divided into four regions: North America, Europe, Asia-Pacific, and the rest of the world (ROW). In 2012, North America led the global food enzymes market with a share of 35% in value terms, followed by Europe (29%), and Asia-Pacific (24%) (http://www.marketsandmarkets.com/Market-Reports/food-enzymes-market-800.html). The North American region constitutes the largest market with the United States commanding the leading share in this industry on grounds of a strong market for healthy foods rich in nutritional content. Europe is at the second position and is expected to witness a below-average growth in the market. Asia-Pacific has the third largest share and is expected to be the fastest growing market, followed by the rest of the world. Brazil, South Africa, and India are some of the emerging markets for food enzymes. The United States is one of the largest players in the food and beverage enzymes business due to excessive demand for processed food products, followed by China (http://www.marketsandmarkets.com/Market-Reports/food-enzymes-market-800.html).

The global market is set to be driven by the Asia-Pacific region, which is witnessing a growing demand for processed foods due to the increasing disposable income of the middle class population. Their use in various industries, such as cheese processing, brewing, baking, meat processing, and fruit juice processing, is on the rise as the yield, quality, and shelf life of end products are improved. The market in Asia-Pacific is expected to grow at the fastest rate of 10.1%, reaching $608.9 million by 2018. The ROW market is anticipated to be the second fastest growing market after Asia-Pacific as the emerging economies drive the need for processed foods. Globally, bakery is the largest application market as enzyme use improves dough stability, crumb structure, and shelf life of end products. Wide use of enzymes in cheese processing drives the use of enzymes in the dairy industry, which is the second largest application market followed by the beverages industry (http://www.marketsandmarkets.com/Market-Reports/food-enzymes-market-800.html).

Growing demand for processed foods coupled with increasing awareness about health-conscious products is significantly propelling growth in the food industry, which, in turn, is driving growth of the enzymes market. The manufacturers are keenly aware of the emerging Asian and Latin American markets as potential drivers of the food and beverage enzymes business. Reduction in the use of emulsifiers in the baking segment is a major market driver for baking enzymes, and recognition of several new applications for food enzymes, such as egg processing and protein fortification, are also expected to drive the market forward in the coming years.

In the case of the animal feed industry, the global market for feed enzymes is definitely one promising segment in the enzyme industry. The market for feed enzymes, which was estimated at approximately $344 million in 2007, is expected to reach $727 million in 2015 (Frost and Sullivan 2007). Use of enzymes as feed additives is restricted in most countries by local regulatory authorities (Pariza and Cook 2010), and their applications may vary from country to country. Globally, the market for feed enzymes is growing in regions such as the United States, China, and Southeast Asia.

TABLE 5.5

Leading Food Enzyme Manufacturers

Company	Location	Established Year	Major Products	Market Share (%)
Novozymes	Bagsvaerd, Denmark	1921	Food and beverage, feed and other technique enzymes	47
Genencor	Copenhagen, Denmark	1982	Food, animal nutrition	21
DSM	Delft, the Netherlands	1952	Food ingredients, animal nutrition	6
Amano	Nagoya, Japan	1899	Food processing, diagnostics	20
AB Enzymes	Feldbergstrasse, Germany	1907	Food, food additives	
BASF	Ludwgshafen, Germany	1865	Feed additives	
Chr Hansen	Horsholm, Denmark	1870	Enzyme for cheese	
Shin-Nihon	Aichi, Japan		Food animal nutrition	
ADM	Illinois, USA	1923	Food feed	
Verenium	San Diego, CA, USA		Animal nutrition, grain processing	5
Iogen	Ontario, Canada	1970	Grain processing and brewing, animal feed	
Dyadic	Florida, USA	1979	Food brewing animal feed	
Meiji	Tokyo, Japan	1916	Food	
Nagase	Osaka, Japan	1832	Food	

Source: Adapted from S. Li, X. Yang, S. Yang, M. Zhu, X. Wang, *Computational and Structural Biotechnology Journal*, 2, 3, 2012, e201209017, http://dx.doi.org/10.5936/csbj.201209017.

5.6 Major Enzyme Manufacturing Companies

Commercial enzyme production has grown in volume and number of products over the years in response to increasing demand for novel biocatalysts, especially for the heavy environmental burden and expanding markets. Currently, food and beverage enzymes are mainly produced from microbial sources because of their lower production cost and higher productivity when compared to plants and animals. In fact, food enzymes sourced from microorganisms dominate the global market share in 2012. With the advent of new production strains using genetically modified microorganisms, the large-scale manufacturing of food enzymes is possible with functions aimed at specific requirements without the risk of undesired side effects. Further, the enzymes thus produced can function in adverse operational conditions and improve process efficiency.

At present, almost 4000 enzymes are known, and of these, approximately 200 microbial original types are used commercially. However, only about 20 enzymes are produced on truly industrial scale (Li et al. 2012). To date, the production of enzymes has been relatively concentrated in a few developed nations: Denmark, Switzerland, Germany, the Netherlands, and the United States although many small- to medium-sized enzyme-producing companies are located in other developing countries. The world enzyme demand is satisfied by about 12 major producers and 400 minor suppliers. The market is highly competitive, has small profit margins, and is technologically intensive (Li et al. 2012). A list of some of the leading food enzyme manufacturers around the world is presented in Table 5.5. Among the enzyme manufacturers, Denmark based Novozymes, U.S. based DuPont (which acquired Denmark-based Danisco/Genencor in 2011), and Netherlands-based DSM, command more than 70% of the market followed by Chr. Hansen (Denmark), AB Enzymes (Germany), and Amano Enzymes Inc. (Japan) (http://www.marketsandmarkets.com/Market-Reports/food-enzymes-market-800.html).

5.7 Conclusion

Today's food and beverage manufacturers face an increasing number of demands, such as the need to produce more healthy and tasty foods, to make them at competitive costs, and to meet high ethical standards. The worldwide enzyme market for the food and beverage industry is expected to reach $1.69 billion by 2018 with a moderate growth over the next five years. The use of enzymes is growing in the food and beverage processing industries as they are an essential ingredient in food items. Increasing health concerns and declining use of emulsifiers in food and beverage products provide growth opportunities to food and beverage processing enzyme manufacturers to produce more innovative products. The food industries are also developing new foods and appropriate strategies for production of food that meets food hygiene standards and that can maintain freshness during storage and have extended shelf life on delivery. From the discussion presented in this chapter, it is obvious that food and beverage processing industries have witnessed rapid growth and proliferation across the globe, and they adopted modern biotechnologies and particularly enzyme technologies for production of consumer-oriented processed food and beverages, meeting various consumer demands in addition to coping with the stringent environmental regulations that call for environmentally friendly technologies toward sustainable growth and development for food processing industries and economies around the world. There is definitely an emerging trend in proliferation of enzyme manufacturers solely utilizing microbial resources around the world, meeting demands of food industries, and exploring new horizons in developing enzyme-based new applications and new products. The global food enzyme market is evolving, and in the coming years, it may take on many challenges posed by food and beverage processing.

REFERENCES

Adrio, J. L., and Demain, A. L. 2014. Microbial enzymes: Tools for biotechnological processes. *Biomolecules* 4(1): 117–139.

Alexandratos, N., and Bruinsma, J. 2012. World agriculture towards 2030/2050: The 2012 revision. ESA Working paper No. 12-03. Rome, FAO.

Cauvain, S., and Young, L. 2006. Ingredients and their influences. In *Baked Products. Science, Technology and Practice*, S. Cauvain and L. Young (eds.), pp. 72–98. Oxford: Blackwell Publishing.

European Commission Report. 2002. Collection of information on enzymes. Final Report by Austrian Federal Environment Agency.

Food Enzymes Market: Global Trends and Forecasts to 2018. Available at http://www.researchandmarkets .com/research/3xgvqh/food_enzymes.

Frost and Sullivan. 2007. Feed enzymes: The global scenario. Available at http://www.frost.com/sublib/display -market-insight.do?id=115387658; accessed on January 13, 2015.

Frost and Sullivan. 2011. Global nutraceutical industry: Investing in healthy living. FICCI White paper 2.

Fung, D. Y. C. 1994. Rapid methods and automation in food microbiology: A review. *Food Reviews International* 10: 357–375.

Galante, Y. M., DeConti, A., and Monteverdi, R. 1998. Application of *Trichoderma* enzymes in food and feed industries. In *Trichoderma and Gliocladium-Enzymes, Vol. 2, Biological Control and Commercial Applications*, G. F. Harman and C. P. Kubicek (eds.), pp. 311–326. London: Taylor and Francis.

Gavrilescua, M., and Chisti, Y. 2005. Biotechnology—A sustainable alternative for chemical industry. *Biotechnology Advances* 23: 471–499.

Hii, S. L., Tan, J. S., Ling, T. C., and Ariff, A. B. 2012. Pullulanase: role in starch hydrolysis and potential industrial applications. *Enzyme Research* 2012: 921362. Available at http://dx.doi.org/10.1155/2012/921362.

Hull, P. 2010. *Glucose Syrups: Technology and Applications*. Wiley-Blackwell, Iowa.

Karunasagar, I., Sugumar, G., Karunasagar, I., and Reilley, A. 1995. Rapid detection of *Vibrio cholerae* contamination of seafood by polymerase chain reaction. *Molecular Marine Biology and Biotechnology* 4: 365–368.

Kashyap, D. R., Vohra, P. K., Chopra, S., and Tewari, R. 2001. Applications of pectinases in the commercial sector: A review. *Bioresource Technology* 77(3): 215–227.

Kirk, O., Borchert, T. V., and Fugslang, C. C. 2002. Industrial enzyme applications. *Current Opinion in Biotechnology* 13: 345–351.

Li, S., Yang, X., Yang, S., Zhu, M., and Wang, X. 2012. Technology prospecting on enzymes: Application, marketing and engineering. *Computational and Structural Biotechnology Journal* 2: e2012209017. doi:10.5936/csbj.201209017.

Lopez-Sabater, E. I., Rodriguez-Jerez, J. J., Roig-Sagues, A. X., and Mora-Ventura, M. T. 1993. Determination of histamine in fish using an enzymic method. *Food Additives and Contaminants* 10: 593–602.

Luong, J. H. T., Male, K. B., Masson, C., and Nguyen, A. L. 1992. Hypoxanthine ratio determination in fish extract using capillary electrophoresis and immobilized enzymes. *Journal of Food Science* 57: 77–81.

Male, K. B., Bouvrette, P., Luong, J. H. T., and Gibbs, B. F. 1996. Amperometric biosensor for total histamine, putrescine and cadaverine using diamine oxidase. *Journal of Food Science* 61: 1012–1016.

Neelakantan, S., Mohanty, A. K., and Kaushi, J. K. 1999. Production and use of microbial enzymes for diary processing. *Current Science* 77(1): 143–148.

Olsen, H. S. 1995. Use of enzymes in food processing. In *Biotechnology, 2nd Ed., Vol. 9, Enzymes, Biomass, Food and Feed*, H.-J. Rehm, G. Reed, A. Puhler and P. Stadler (eds.), pp. 663–736, Weinheim: VCH Verlagsgesellschaft mbH.

Pariza, M. W., and Cook, M. 2010. Determining the safety of enzymes used in animal feed. *Regulatory Toxicology and Pharmacology* 56: 332–342.

Qureshi, N., and Blaschek, H. P. 2006. Butanol production from agricultural biomass. In *Food Biotechnology*, K. Shetty, G. Paliyath, A. Pometto and R. E. Levin (Eds.), pp. 525–549. Boca Raton, FL: Taylor & Francis.

Rolle, R. S. 1998. Review: Enzyme applications for agro-processing in developing countries: An inventory of current and potential applications. *World Journal of Microbiology and Biotechnology* 14: 611–619.

Shen, L. Q., Yang, L. J., and Peng, T. Z. 1996. Amperometric determination of fish freshness by a hypoxanthine biosensor. *Journal of the Science of Food and Agriculture* 70: 298–302.

Sreekantiah, K. R. 1979. Utilisation of microbial enzymes for the manufacture of sweeteners required in the confectionery industries. *Proceedings of the Symposium on the Status and Prospects of the Confectionery Industry in India*, pp. 66–72.

Toops, D., and Fusaro, D. 2010. Bakery and bread: Rise to the occasion. Available at http:/storage.globalcitizen .net/data/topic/knowledge/uploads/2010101993431705.pdf; accessed on January 14, 2015.

Underkofler, L. A. 1980. Enzymes. In *Handbook of Food Additives*, 2nd ed., T. E. Furia (ed.), pp. 57–124. Boca Raton, FL: CRC Press.

US-FDA. 2014. Rennet (animal derived) and chymosin preparation (fermentation derived)—Sec 184-1685. In: Code of Federal Regulations, 21:3, CITE: 21CFR184.1685.

Venkateswaran, K., Kurusu, T., Satake, M., and Shinoda, S. 1996. Comparison of a fluorogenic assay with a conventional method for rapid detection of *Vibrio parahaemolyticus* in seafoods. *Applied and Environmental Microbiology* 62: 3516–3520.

Zhou, W., and Therdthai, N. 2007. Manufacturing of bread and bakery products. In *Handbook of Food Products Manufacturing*, Y. H. Hui (ed.), pp. 261–278. Hoboken, NJ: Wiley & Sons, Inc.

6

Enzymes in Starch Processing

Xiao Hua and Ruijin Yang

CONTENTS

6.1 Introduction

6.1.1 Starch

Starch is a white, granular, organic chemical produced by all green plants as the energy store. It is the main carbohydrate in the human diet and is contained in many staple foods. Cereals (maize, wheat, and rice) and root vegetables (potatoes, cassava, tapioca, sorghum, etc.) are the major sources of starch (Table 6.1).

Starch consists of a large number of glucose units jointed by glycosidic bonds. It contains two types of molecules: amylose and amylopectin. Amylose is a linear and helical polysaccharide consisting of 300~3000 glucose units via α-(1,4) glycosidic bonds. It commonly makes up approximately 20%~30% of the starch structure (Table 6.1). Amylose is thought to be completely unbranched, and it employs a tightly packed structure, resulting in insolubility in water or alcohol and resistance to digestion. Amylose chains can be organized in disordered amorphous conformation or in two types of helical forms: the double helix of the amylose chains (A or B form) or a single-helix structure (V-form), which is induced in the presence of a hydrophobic guest, such as iodine, fatty acids, and aromatic compounds (Cohen et al. 2008).

Amylopectin is a highly branched polysaccharide making up approximately 70%~80% of starch by weight, depending on the source (Table 6.1). It is usually formed by 2,000~200,000 glucose units. Its backbone consists of glucose units linearly linked by α-(1,4) glycosidic bonds, and the branching takes place with α-(1,6) bonds occurring every 24 to 30 glucose units. Branch links, accounting for about 5% of the amylopectin structure (Ball et al. 1996), have brought many end points for enzyme action; therefore, amylopectin is highly soluble in water and can be quickly degraded in the body. Current models for amylopectin suggest that it consists of A-, B-, and C-chains (Oates 1997). A-chains are unbranched and attached to the molecule by a single linkage whereas B-chains are branched and connected to two or more other chains (Figure 6.1). The C-chain is single in amylopectin and has the sole reducing terminal.

Starch molecules arrange themselves in the plant in semicrystalline granules, and each plant species has a unique starch granular size (Table 6.2). According to the granule model, which was first proposed in the 1920s (Sponsler 1923) and subsequently extended, the starch granule employs an alternating region structure (Figure 6.1): crystalline lamella formed by tightly parallel packing of side chains of amylopectin and the amorphous lamella comprised by branch points. In addition, amylose molecules are assumed to be the mobile entities largely separated from the amylopectin fraction. Because the starch granule is inert and structurally stable, the destruction of the granule structure has been the prerequisite during starch processing for degradation of starch macromolecules.

In the presence of water and heating, starch granules swell and burst, the semicrystalline structure is lost due to breaking of intermolecular bonds, and the smaller amylose molecules start leaching out of the granule, forming a network that holds water and increasing the mixture's viscosity. This irreversible

TABLE 6.1

Starch Content and Composition in Different Sources

	Starch %	Amylose %	Amylopectin %
Maize	70~75	26	74
Wheat	57~75	25	75
Rice	62~86	18	82
Potato	12~14	20	80
Cassava	14~40	17	83
Sorghum	60~65	23	77
Waxy corn	60~70	1	99

Source: Summarized based on Sikorski, Z. E., *Chemical and Functional Properties of Food Components* (3rd Edition, p. 100). Boca Raton: Taylor & Francis, 2007; Fennema, O. R., *Food Chemistry* (3rd Edition, p. 193). New York: Marcel Dekker, Inc., 1996.

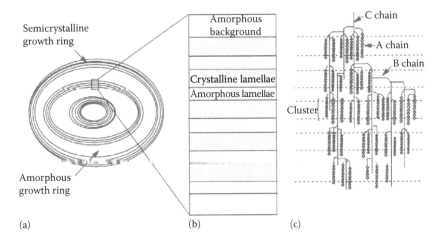

FIGURE 6.1 Densely packed semicrystalline starch granule structure. (a) A single granule comprising concentric rings of alternating amorphous and semicrystalline composition. (b) Expanded view of the internal structure. The semicrystalline growth ring contains stacks of amorphous and crystalline lamellae. (c) Currently accepted cluster structure for amylopectin within the semicrystalline growth ring. (Reproduced from Jenkins, P. J., and Donald, A. M., *International Journal of Biological Macromolecules*, 17, 6, 315–321, 1995.)

TABLE 6.2

Granular Size of Starch in Different Plant Species

Source	Size (μm)
Bimodal	
Barley	2~3 and 12~32
Immature sweet corn	1~5 and 10~20
Rye	2~3 and 22~36
Triticale	5 and 22~36
Wheat	<10 and 10~35
Small Granule Starch	
Buckwheat	2~14
Cattail	2~15
Dropwort	0.5~10
Durian	3~14
Grain teff	2~6
Oat	2~14
Parsnip	1~6
Rice	2~10
Small millet	0.8~10
Wild rice	2~8
Very Small Granule Starch	
Amaranth	1~2
Canary grass	1.5~3.5
Cow cockle	0.3~1.5
Dasheen	0.5~3
Pigweed	1.5~4
Quinoa	0.5~3
Taro	2~3

Source: Lindeboom, N., Chang, P. R., Tyler, R. T., *Starch*, 56, 89–99, 2004.

processing is called starch gelatinization. The gelatinization temperature of starch depends upon the plant type; the amount of water present; and the pH, types and concentration of salts, sugar, fat, and protein in the recipe as well as the derivatization technology used. During cooling, the semicrystalline structure is partially recovered because of amylose retrogradation, resulting in starch paste thickening and expelling of water.

6.1.2 Starch Industry

The starch supply chains can be divided into three stages (Agrosynergie 2010): the agricultural sector, which is cultivating crops storing starch; the starch-processing sector, which includes starch production (native starch) and starch processing for different applications (hydrolyzed starch and modified starch); and the end-user sector, which includes food and nonfood industries consuming starch and starch products.

Worldwide starch production has continuously increased in recent years due to a steady demand from starch-consuming industries. In 2012, global starch production has reached 75 million tons, of which the United States produces more than half, and the EU production figures are approximately 14%. Approximately 60% of the starch was used for food uses, and the rest was for nonfood industrial applications, such as papermaking, cardboard, medicine, cosmetics, textiles, and renewable packing.

6.1.2.1 Native Starch

Native starch production is the isolation of starch from plants. At first, the raw material for starch preparation was wheat, but currently, the starch source is maize in America and potato in Europe. Starch from other sources, such as cassava, rice, barley, etc., is also produced for specific applications. In its native form, starch has a relatively limited range of uses, such as for a thickener, binder, and stabilizer in food processing and papermaking because of the property of retrogradation. In 2008, the annual production of native starch reached 11 million tons and increased by 14% in comparison with that in 2000 (Agrosynergie 2010).

6.1.2.2 Hydrolyzed Starch

Hydrolyzed starch, or starch sugars, is produced from native starch using an acid or enzymatic treatment or a combination of the two. With a specific catalyst and controlling the reaction, fragments of a certain size with a specific dextrose equivalent (DE) can be obtained. In different countries, hydrolyzed starch can be categorized according to different standards.

In Europe, hydrolyzed starch can be divided into three categories: maltodextrins, which can be used as bulking materials and as an ingredient in foods and an excipient in medicine; glucose syrup, which is a sweetener in the food and beverage industries; and other hydrolysates, which are raw materials for the fermentation industry, such as alcohol, citric acid, and amino acids, and production of other starch derivatives, such as isoglucose, high fructose syrup, etc.

In China, starch sugars are classified according to the National Standard of China GB/T28720-2012 (Figure 6.2). Products containing a high content of glucose/maltose are defined based on their glucose/maltose content. Maltodextrins with various DE are finely classified as MD 10, MD 15, MD 20, and MD 30 for guidance of their use. Both high-fructose syrups (F42, F55, and F90) and crystalline fructose (>99%) are included in the fructose sector. Fructose syrups are made by treating a glucose solution with glucose isomerase, and crystalline fructose is produced by further separation (industrial chromatography) and purification (recrystallization) of fructose syrup. In addition, the generalized "starch sugar" also includes functional oligosaccharides that are synthesized by enzymatic transglycosylation using starch hydrolysates both as glycosyl donor and acceptor and sugar alcohol manufactured by reduction of glucose, fructose, and other monosaccharides (xylose, sorbose, and erythrose).

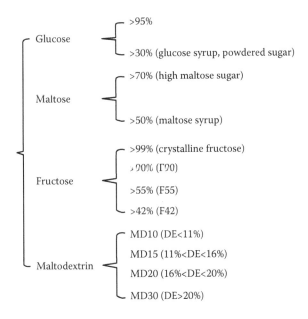

FIGURE 6.2 Overview of starch sugars classified in National Standard of China GB/T28720-2012.

6.1.2.3 *Modified Starch*

A modified starch is a starch that has been chemically or physically modified to allow the starch to function properly under conditions frequently encountered during processing or storage, such as high heat, high shear, varied pH, freeze/thaw, and cooling. Modified starch has become an attractive material in many food and nonfood sectors. Current worldwide demand for modified starch has reached nearly 10 million tons. According to modification of the molecular structure, modified starch can be classified as substituted starches: substitution of hydroxyl groups in glucose units in starch, resulting in a significant change in the physicochemical properties. Typical substituted starches include starch ethers and starch esters with a degree of substitution in the range of 0.002 to 0.2.

1. Degraded/converted starch: Acid-treated or hydrolase-treated starch with a partially degraded structure for easier and quicker dissolving in water. Sometimes dextrin is considered to be one kind of degraded starch. Particularly, oxidized starches with a low viscosity at high solid concentrations, in which hydroxyl groups are transformed to carbonyl and/or carboxyl groups, are important raw materials for paper and textile industries.
2. Cross-linked starch: Starch molecules are cross-linked by chemical linking reagents for acquiring high resistance to processing conditions.

Actually, most modified starches are both cross-linked and substituted/converted, depending on the end use. However, most modifications are achieved by chemical methods, especially for substitution and cross-linking. Enzymatic-modified starch is mainly restricted in hydrolysis products and partially in transglycosylation products, determined by the commercial enzymes currently available for industrial production. Nevertheless, enzymatic modification is still a promising method in future due to its safety and friendliness to environment.

Modified food starches have been E-coded by the International Numbering System for Food Additives (INS). For example, E1400 is the dextrin roasted starch with hydrochloric acid; E1401 is acid-treated starch (thin boiling starch), obtained by treating starch with inorganic acids with reduced viscosity; and E1405 is the enzyme-treated starch.

6.2 Enzymes Involved in Starch Conversion

6.2.1 History of Amylolytic Enzymes

The use of enzymes in the starch industry has been known for a long time. As early as the late seventeenth and early eighteenth centuries, the conversion of starch to sugars by plant extracts and saliva were known. In 1833, Anselme Payen and Jean-François Persoz discovered the first enzyme, diastase, which is now defined as any one of a group of enzymes catalyzing the breakdown of starch into maltose. In 1894, Jokichi Takamine isolated the enzyme takadiastase, an enzyme that catalyzes the breakdown of starch from koji (*Aspergillus oryzae*). Takadiastase is the first enzyme produced industrially. In 1919, a French scientist successively discovered *Bacillus subtilis* that secretes heat-stable and more reactive α-amylase. In 1949, the Japanese successfully conducted the production of α-amylase by submerged fermentation. In the 1960s, major usage of enzymes in the starch industry has been considered to be of great progress in the industrial enzyme history (Binod et al. 2013). Traditional acid hydrolysis was completely replaced by α-amylases and glucoamylases, which could convert starch with more than 95% yield to glucose. The most significant event came in 1973 with the development of immobilized glucose isomerase, which made the industrial production of high-fructose syrup feasible. Now the starch industry is the second largest (after the detergent industry) and longest-standing market for enzymes. Many important products are manufactured by enzymatic starch conversion (Figure 6.3).

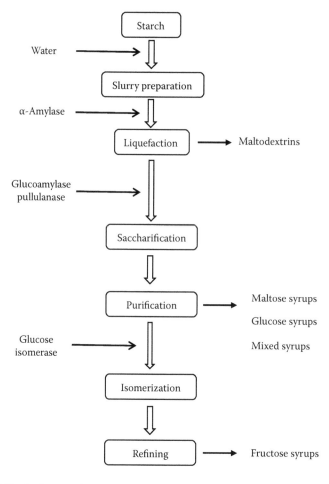

FIGURE 6.3   Enzymatic starch conversion.

Enzymatic processing of starch has many advantages. First of all, enzymatic processing acts on specific glycosidic bonds, therefore avoiding undesirable side effects and waste disposal and further simplifying post-treatments and reducing cost. Second, the level of enzymatic reaction can be easily controlled by temperature, time, and ratio of substrates; meanwhile, the enzyme amount needed for accomplishing the reaction is small. Additionally, special saccharides and syrups that cannot be produced using conventional chemical hydrolysis were the first compounds made entirely using enzymatic processes.

6.2.2 Reactions Catalyzed by Amylolytic Enzymes

Many hydrolases that use α-1,4-glucan as a substrate can hydrolyze starch, and they are categorized in EC. 3.2.1.X (Table 6.3). Most of them are exo-hydrolase; namely, they break α-1,4-glucosidic bonds from the nonreducing end of starch whereas they release products from glucose to gluco-oligosaccharides due to different hydrolysis mechanism. α-Amylase is the only endo-hydrolase randomly breaking α-1,4-glucosidic bonds in starch, resulting in fragments of different sizes.

For rapid and complete degradation of amylopectin, an enzyme being able to hydrolyze α-1,6-glucosidic bonds is required. Oligo-1,6-glucosidase (EC. 3.2.1.10), pullulanase (EC. 3.2.1.41), and isoamylase (EC. 3.2.1.68) are specially employed in starch processing for this purpose. Because α-1,6-glucosidic bonds take place in the position in which branches connect to the glucan backbone, these enzymes have the alternative name "debranching enzyme." Glucan 1,4-α-glucoamylase (EC. 3.1.2.3), which has another name, "gamma-amylase," can hydrolyze both α-1,4- and α-1,6-glucosidic bonds in starch; however, the most important feature of it is to release β-D-glucose.

Isomerization of glucose to fructose is the key process for production of high-fructose syrup, a sweetener used worldwide in foods. Xylose isomerase (EC. 5.3.1.5), which catalyzes the interconversion of aldose and ketose sugars with broad substrate specificity, has been commercially produced for glucose isomerization. Recently, the isomerization of glucose to fructose has also been considered to be a key intermediate step in the conversion of biomass to fuels and chemicals.

Transglycosylation has become a promising approach for production of biofunctional oligosaccharides and cyclodextins with special physicochemical properties. Starch, as a cheap and abundant glucose source, acts as the substrate for glucosyl tranferases, such as 1,4-α-glucan 6-α-glucosyltransferase (EC. 2.4.1.24), 4-α-glucanotransferase (EC. 2.4.1.25), and oligosaccharide 4-α-D-glucosyltransferase (EC. 2.4.1.161).

Amylolytic enzymes are primarily produced by fermentation using certain microorganism strains. Both fungi and bacteria yield an invaluable array of enzymes when fermented on appropriate substrates. More than 75% of the industrial enzymes are produced using submerged fermentation (SmF), one of the major reasons being that SmF supports the utilization of genetically modified organisms to a greater extent than solid-state fermentation (Subramaniyam and Vimala 2012). Solid-state fermentation (SSF) utilizes solid substrates, such as bran, rice, vegetable waste, and paper pulp as the substrates. Fungi and microorganisms that require less moisture content can employ SSF whereas an organism living with high water activity cannot be used in SSF. Recent investigations of production of amylolytic enzymes are listed in Table 6.4.

6.2.3 Reaction Mechanism

6.2.3.1 Hydrolysis

Mechanisms of *O*-glycosyl hydrolases (EC. 3.1.2.X) have been deeply investigated and explained by two pathways (Davies and Henrissat 1995; Rye and Withers 2000). One is the retaining mechanism (Figure 6.4a), in which the glycosidic oxygen is protonated by an acid catalyst, and nucleophilic assistance to aglycon departure is provided by a base. The resulting glycosyl enzyme is then hydrolyzed by one molecule of water. During the process, the anomeric carbon in the product employs the direction that is the same as that in the substrate. The second pathway is the inverting mechanism (Figure 6.4b), in which protonation of the glycosidic oxygen and aglycon departure are accompanied by a concomitant attack of a water molecule that is activated by the base residue, leading to the opposite direction of anomeric carbon in the product.

TABLE 6.3

Enzymes Involved in Starch Processing

Enzyme	EC No.	Category	Reaction
α-Amylase	3.2.1.1	Endohydrolase	Endohydrolysis of 1,4-α-D-glucosidic linkages in polysaccharides containing three or more 1,4-α-linked D-glucose units
β-Amylases	3.2.1.2	Exohydrolase	Hydrolysis of 1,4-α-D-glucosidic linkages in polysaccharides so as to remove successive β-maltose units from the nonreducing ends of the chains
Glucan 1,4-α-glucoamylase	3.2.1.3	Exohydrolase	Hydrolysis of terminal 1,4-α-D-glucose residues successively from nonreducing ends of the chains with release of β-D-glucose
Oligo-1,6-glucosidase	3.2.1.10	Debranching	Hydrolysis of 1,6-α-D-glucosidic linkages in some oligosaccharides produced from starch and glycogen by α-amylase and in isomaltose
α-Glucosidase	3.2.1.20	Exohydrolase	Hydrolysis of terminal, nonreducing 1,4-α-D-glucose residues with release of α-D-glucose
Amylo-α-1,6-glucosidase	3.2.1.33	Debranching	Hydrolysis of 1,6-α-D-glucosidic branch linkages in glycogen phosphorylase limit dextrin
Pullulanase	3.2.1.41	Debranching	Hydrolysis of 1,6-α-D-glucosidic linkages in pullulan, amylopectin, and glycogen and in the α- and β-limit dextrins of amylopectin and glycogen
Glucan 1,4-α-maltotetraohydrolase	3.2.1.60	Exohydrolase	Hydrolysis of 1,4-α-D-glucosidic linkages in amylaceous polysaccharides to remove successive maltotetraose residues from the nonreducing chain ends
Isoamylase	3.2.1.68	Debranching	Hydrolysis of 1,6-α-D-glucosidic branch linkages in glycogen, amylopectin, and their β-limit dextrins
Glucan 1,6-α-glucosidase	3.2.1.70	Debranching	Hydrolysis of 1,6-α-D-glucosidic linkages in 1,6-α-D-glucans and derived oligosaccharides
Glucan 1,4-α-maltohexaosidase	3.2.1.98	Exohydrolase	Hydrolysis of 1,4-α-D-glucosidic linkages in amylaceous polysaccharides to remove successive maltohexaose residues from the nonreducing chain ends
Glucan 1,4-α-maltotriohydrolase	3.2.1.116	Exohydrolase	Hydrolysis of 1,4-α-D-glucosidic linkages in amylaceous polysaccharides to remove successive maltotriose residues from the nonreducing chain ends
Glucan 1,4-α-maltohydrolase	3.2.1.133	Exohydrolase	Hydrolysis of 1,4-α-D-glucosidic linkages in polysaccharides so as to remove successive α-maltose residues from the nonreducing ends of the chains
1,4-α-glucan branching enzyme	2.4.1.18	Transferase	Transfers a segment of a α-1,4-D-glucan chain to a primary hydroxyl group in a similar glucan chain
Cyclomaltodextrin glucanotransferase	2.4.1.19	Transferase	Cyclizes part of a α-(1,4)-D-glucan chain by formation of a α-(1,4)-D-glucosidic bond
1,4-α-Glucan 6-α-glucosyltransferase	2.4.1.24	Transferase	Transfers an α-D-glucosyl residue in a α(1,4)-D-glucan to the primary hydroxy group of glucose, free or combined in an α(1,4)-D-glucan
4-α-Glucanotransferase	2.4.1.25	Transferase	Transfers a segment of a α(1,4)-D-glucan to a new position in a glucose or a α(1,4)-D-glucan
Oligosaccharide 4-α-D-glucosyltransferase	2.4.1.161	Transferase	Transfers the nonreducing terminal α-D-glucose residue from a α(1,4)-D-glucan to the 4-position of a free glucose or of a glucosyl residue at the nonreducing terminus of a α(1,4)-D-glucan rearrangement of oligosaccharides
Xylose isomerase	5.3.1.5	Isomerase	Isomerizes D-xylose to D-xylulose

Source: Summarized based on http://enzyme.expasy.org/.

TABLE 6.4

Recent Investigation on Production of Amylolytic Enzymes

Enzymes	Microorganism	Fermentation	Substrate	Culture Conditions	Productivity	References
α-Amylase	*Streptomyces* sp.	SmF	Sisal waste or sugarcane bagasse	30°C/6 days	10.1 U/mL, best activity at 50°C/pH 7.0	dos Santos et al. 2012
	Bacillus sp.	SmF	Unhydrolyzed corn starch and wheat bran	30°C/72 h	40 U/mL/min, active in pH 4–9, 40°C–50°C	Shamala et al. 2012
	Bacillus licheniformis	SmF	Starch, rice, wheat, ragi powder	37°C/72 h	2.36–3.64 IU/m/min, best activity at 37°C, pH 7.0	Divakaran et al. 2011
	Bacillus sp. SMIA-2	SmF	Starch/corn steep liquor	50°C/18 h	401 U/mg protein	Corrêa et al. 2011
	Streptomyces erumpens	SSF	Cassava fibrous residue	50°C/60 h	3457.67 U/gds	Kar et al. 2010
	Bacillus subtilis subsp. *spizizenii*	SSF	Wheat bran	37°C	521 391 U/g dry solids	Soni et al. 2012
β-Amylases	*Bacillus subtilis*	SmF	Starch	37°C/initial pH 7	18.32 U/mg	Poddar et al. 2012
	Aspergillus niger	SSF	Plantain peels	29 ± 1°C/6 days	33.2 EU	Adeniran et al. 2010
	Clostridium thermosulfurogenes	SmF	Potato, sorghum, sweet potato, ragee, barley, rice, corn, wheat, and tapioca flours	60°C/24 h	0.87U/mL culture broth	Ramesh et al. 2001
Glucoamylase	Mutants of *Aspergillus awamori* gene	SmF	Starches	30°C/7 days	6.6 U/ml, best active at pH 4.5, 65°C	Pavezzi et al. 2011
	Aspergillus phoenicis	SmF	Maltose	30°C/4 days	306 U/mg, best active at pH 4.5, 60°C	Benassi et al. 2014

(Continued)

TABLE 6.4 (CONTINUED)

Recent Investigation on Production of Amylolytic Enzymes

Enzymes	Microorganism	Fermentation	Substrate	Culture Conditions	Productivity	References
	Aspergillus awamori	SSF	Pastry waste	30°C/10 days	76.1 ± 6.1 U/mL, best activity at pH 5.5, 55°C	Lam et al. 2013
	Aspergillus kawachii	SmF	Glucose, dextrin, and indigestible dextrin	30°C/66 h	281.2 mU/mL	Sugimoto et al. 2011
Cyclodextrin glycosyltransferase	Gene form *Paenibacillus macerans* and expressed in *Escherichia coli*	SmF	Modified TB medium	25°C/pH 7.0/12 h	47.55 U/mL	Liu et al. 2012
	Gene from *Bacillus megaterium* and expressed in *Escherichia coli*	SmF	Xylose	32°C/15 h	48.9 U/mL	Zhou et al. 2012
	Gene form *Paenibacillus macerans* and expressed in *Escherichia coli*	SmF	Lactose	30°C/pH 7.0	275.3 U/ml	Cheng et al 2011
	Alkalophilic *Bacillus* sp.	SmF	Soluble starch	33°C	65 U/mL	Kuo et al. 2009
	Paenibacillus campinasensis	SSF and SmF	Wheat bran, soybean bran, soybean extract, cassava solid residue, cassava starch, corn starch, and other combinations	45°C/96 h for SmF; 45°C/72 h for SSF	50–100 u/mL in SSD; 70–90 U/mL in SmF	Alves-Prado et al. 2006
Pullulanase	*Streptomyces erumpens*	SmF	Soluble starch	50°C/pH 7.0	222.5 U of α-amylase; 69.5 U of pullulanase	Kar et al. 2012
	Raoutella planticola	SmF	Sago starch	30°C/initial pH 7.23	1.94 U/mL	Hii et al. 2012

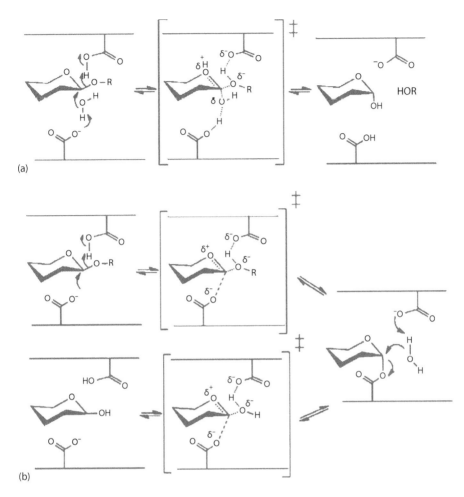

(a)

(b)

FIGURE 6.4 Mechanisms of *O*-glycosyl hydrolases: (a) retaining mechanism; (b) inverting mechanism. (Reproduced from Rye C. S., and Withers, S. G., *Current Opinion in Chemical Biology*, 4, 573–580, 2000.)

6.2.3.2 *Isomerization*

Aldose-ketose isomerization catalyzed by xylose isomerase proceeds via a hydride-shift mechanism (sigmatropic reaction). It is a type of rearrangement reaction wherein the net result is one σ-bond that is changed to another σ-bond in an uncatalyzed intramolecular process.

In the case of isomerization from glucose to fructose, the mechanism involves hydrogen transfer from C-2 to C-1 and from O-2 to O-1 of an α-hydroxyl aldehyde to create the related α-hydroxyl ketone (Nagorski and Richard 2001). The xylose isomerase catalyzed process is thought to be mediated by enolate intermediates generated by histidine-directed base catalysis (Rose et al. 1969; Schray and Rose 1971).

Recent studies have indicated that metal centers in the enzyme are responsible for stabilizing the sugar's open-chain form, and the subsequent aldose-ketose isomerization proceeds by an intramolecular hydride shift (Kovalevsky et al. 2010). The interconversion has been assumed to be a three-stage mechanism, including ring opening, isomerization, and ring closure (Figure 6.5).

6.2.3.3 *Cyclization*

Cyclization catalyzed by cyclodextrin glucanotransferase (CGTase) is the method for production of cyclodextrins (CD) using starch as the substrate. CGTases are unique members in the family of 13 glycosyl

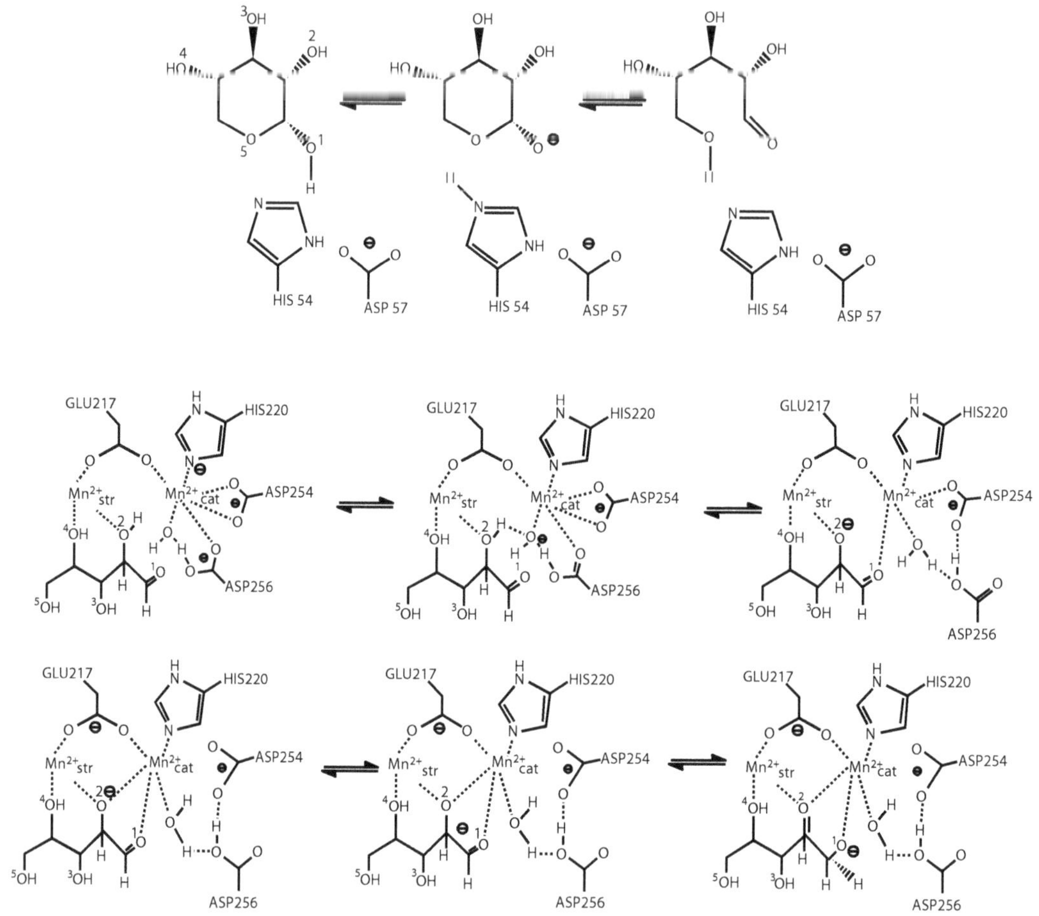

FIGURE 6.5 Mechanism of isomerization of glucose to fructose by xylose isomerase. Top: ring opening, middle: substrate deprotonation, bottom: hydride shift. (Reproduced from Náray-Szabó, G., Oláh, J., Krámos, B., *Biomolecules*, 3, 662–702, 2013.)

hydrolases (GH13). They employ the nonreducing end of a bound oligosaccharide as an acceptor for circular α-(1,4)-linked oligosaccharide formation. The cyclization reaction starts with the binding of a linear saccharide chain across the nine sugar-binding subsites labeled −7 to +2 (Figure 6.6) (Strokopytov et al. 1996) with glycosidic bond breaking occurring between the −1 and +1 subsites to yield a stable covalent glycosyl enzyme intermediate (Uitdehaag et al. 1999). The sugar at the acceptor subsites labeled +1/+2 is then exchanged for a new acceptor, which is the nonreducing end of the covalently bound oligosaccharide, producing a cyclodextrin.

6.2.3.4 Transglycosylation

In addition to the cyclization reaction, the CGTase can catalyze a coupling reaction (opening of the cyclodextrin rings and transfer of the resulting linear malto-oligosaccharides to acceptors) and a disproportionation reaction (transfer of linear malto-oligosaccharides to acceptors) through intermolecular transglycosylation reactions (Figure 6.6). CGTase also transfers the covalently bound oligosaccharide to a water molecule exhibiting weak hydrolyzing activity. The disproportionation of CGTase has been proposed to operate by a ping-pong (substituted-enzyme) mechanism (Nakamura et al. 1994) while the coupling was considered to belong to a random ternary complex mechanism (van der Veen et al. 2000).

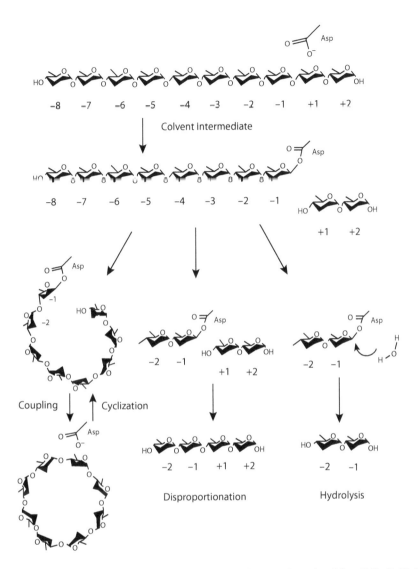

FIGURE 6.6 Schematic representation of the reactions catalyzed by CGTase. (Reproduced from Kelly, R. M., Dijkhuizen, L., Leemhuis, H., *Applied Microbiology Biotechnology*, 84, 119–133, 2009.)

6.2.4 Enzymes Used in Starch Conversion

6.2.4.1 α-Amylase

6.2.4.1.1 Catalytic Activity

α-Amylase (glycogenase, endoamylase, Taka-amylase A, 1,4-α-D-glucan glucanohydrolase) can endo-hydrolyze α-(1,4)-D-glucosidic linkages in polysaccharides containing three or more α-(1,4)-linked D-glucose units, liberating dextrin and ultimately maltose in the α-configuration. α-Amylases are calcium metalloenzymes, completely unable to function in the absence of calcium.

6.2.4.1.2 Source

α-Amylase has been widely found in humans and other mammals, plant seeds, and microorganisms. In humans, α-amylase exists in many tissues, but it is most prominent in pancreatic juice and saliva (ptyalin). Each human amylase has its own isoform with a different isoelectric point; however, all of them link to chromosome 1p21.

α-Amylase used in industry sectors is mainly secreted from bacteria and fungi. α-Amylases produced by the genus *Bacillus*, such as *Bacillus licheniformis*, *Bacillus stearothermophilus*, and *Bacillus amyloliquefaciens* (Konsoula and Liakopoulou-Kyriakides 2007; Pandey et al. 2000), and some thermostable *Actinomycetes*, including *Thermomonospora* and *Thermoactinomyces*, have good thermostability, which can meet the requirement of enzymatic liquefaction (80°C–90°C) and gelatinization (100°C–110°C) in the starch industry (Sindhu et al. 1997).

Fungi that produce α-amylase have been mainly confined to a few species of mesophilic fungi, mostly *Aspergillus* and *Penicillium* (Kathiresan and Manivannan 2006). Production of α-amylase by *Aspergillus* has attracted increasing attention as the strain can secrete extracellular α-amylase with considerable quantities and high hydrolytic activity (Jin et al. 1998). Additionally, *Aspergillus* can tolerate a low fermentation pH (<3) avoiding contamination by bacteria.

6.2.4.1.3 Structure and Catalytic Domains

Most of the microbial α-amylases belong to the GH13 family based on amino acid sequence homology according to the classification of Henrissat (1991). The catalytic domain possesses an $(\beta/\alpha)_8$ barrel structure (or TIM structure), that is, a barrel of eight parallel β-strands surrounded by eight α-helices (domain A) with the active site being at the C-terminal end of the barrel β-strands, an ~70–amino acid calcium-binding domain protruding between β strand 3 and α helix 3 (Abe et al. 2005). In most α-amylases, domain A occurs at the N-terminal end of the protein. Most enzymes have domain C, which is made up of β-strands following the catalytic β/α barrel. Domain C is assumed to stabilize the catalytic domain by shielding the hydrophobic residues in domain A. In addition, several α-amylases contain a β-sheet domain organized as a five-stranded antiparallel β-sheet, usually at the C terminus (Kadziola et al. 1994, 1998).

6.2.4.1.4 Mechanism

Enzymes classified in the α-amylase family (GH13) are believed to have a similar mechanism of action, and so the catalytic amino acid residues are thought to be common to all the enzymes (Svensson 1994).

The double displacement mechanism of α-amylase is illustrated in Figure 6.4. First, the carboxylic group of enzyme protonates the glycosidic oxygen in substrate, resulting in the scission of the C1–O bond and transient formation of an oxocarbenium ion-like transition state. Meanwhile, a dissociated carboxylic group nucleophilicly attacks the anomeric carbon in sugar to give a β-glycosyl enzyme intermediate while the aglycone of the substrate leaves the active site. In the second stage, the process just described is essentially reversed by attack at the anomeric center by a water molecule activated by the carboxylate form of the former proton donor (MacGregor et al. 2001).

6.2.4.1.5 Applications in Starch Processing

In starch processing, α-amylase has been used in ethanol production to break starches in grains into fermentable sugars. The first step in production of high-fructose syrup was hydrolyzing starch by α-amylase, yielding more reducing terminals for further hydrolysis by glucoamylase.

6.2.4.1.6 Commercial α-Amylase

Commercial thermostable α-amylases commonly produced by *Bacillus* are widely used in starch liquefaction and saccharification. Liquozyme® (X, Supra, Supra 2.2X) from Novozyme can be used in primary liquefaction at pH 5.2–5.6, 105°C–108°C for low-temperature pasting and 130°C–160°C for high-temperature pasting, 95°C–98°C for secondary liquefaction. The calcium requirement is 5–15 ppm. Termamyl® (120L Type L, 2X, SC, DS) can act at pH 6.0–6.4 in preliquefaction at 80°C–85°C and continuous liquefaction at 105°C–110°C with a calcium requirement of 40–70 ppm. Spezyme® (XTRA, FRED, HPA) and GC 262 S P (Genencor) produced by *Bacillus licheniformis* are acid-thermostable liquefying enzymes, which can act in the presence of low calcium content (or even calcium free), that bring great viscosity reduction and low panose formation.

Kleistase (E5S, SD80, T10S) commercialized by the Amano Enzyme Group are widely used to produce maltodextrin, glucose syrup, high-maltose syrup, and high-fructose syrup. AMT 1.2L is specially evolved for production of maltotriose-rich syrup.

α-Amylases secreted by fungi usually act at natural to weak basic pH (7~8) and lower temperatures (50°C~80°C); therefore, they can be used as flour correction enzymes in the baking industry. Typical commercialized fungi α-amylases are Fungamyl® 2500SG, 4000BG, 800L, UltraBG, and UltraSG. They help correct a common deficiency of flour by standardizing α-amylase activity. This ensures desired end product characteristics, such as volume, crust color, and crumb structure. Additionally, Ban® 480L (from *Bacillus amyloliquefaciens*, Novozyme) is also developed for the baking industry. It contains 1,4-β-glucanase activity, making it an ideal liquefaction enzyme for raw materials containing high amounts of 1,4-β-glucans.

6.2.4.2 β-Amylases

6.2.4.2.1 Catalytic Activity

β-Amylases (EC. 3.2.1.2), also known as 1,4-α-D-glucan maltohydrolase, can catalyze the α-(1,4) glucosidic bond in polysaccharides, releasing successive maltose units from the nonreducing ends of the chains. The reducing glucose of maltose is in the β-form (Kossmann and Lloyd 2000). However, it is unable to bypass branch linkages in branched polysaccharides, such as glycogen or amylopectin. Very recently, the first transglycosylation activity of β-amylase from sweet potato with malto-oligomer as a substrate has been observed (Fazekas et al. 2013).

6.2.4.2.2 Source

Early purifications of β-amylases are mainly from plants, such as barley, wheat, soybean, pea, potato, and cereal flours. Particularly, barley β-amylases have attracted much attention and have been widely used in industry. An investigation on β-amylases from 6752 worldwide barley genetic resources has suggested that there is a clear geographical differentiation in β-amylase thermostability (Kaneko et al. 2001). The main microorganism source producing thermostable β-amylase is *Clostridium thermosulfurogenes*. Mutation and chemical modification have been applied to enhance the thermostability of β-amylase (Daba et al. 2013).

6.2.4.2.3 Structure and Catalytic Domain

Most β-amylases are monomeric enzymes (Thoma et al. 1971). β-Amylase is the only known activity in family GH14. Three highly conserved sequence regions are found in all known β-amylases. The first region is located in the N-terminal section of the enzymes and contains an aspartate that is involved in the catalytic mechanism (Sakiyama et al. 1989). The second, located in a more central location, is centered around a glutamate that is also involved in the catalytic mechanism (Totsuka et al. 1994). The $(\beta/\alpha)_8$ super-secondary structural core has also been found in β-amylase.

6.2.4.2.4 Mechanism

Hydrolysis catalyzed by β-amylase obeys the invertion mechanism. The Glu residues have been proposed as a catalytic residue, both as the nucleophile or the proton donor. Investigation of the roles of Glu186 and Glu380 in the catalytic reaction of soybean β-amylase has confirmed the critical roles played by Glu186 in the donation of a proton to the glycosidic oxygen of the substrate, and by Glu380 in the activation of an attacking water molecule (Kang et al. 2004). Loss of hydrolytic activity of β-amylase from *Bacillus cereus var. mycoides* also has been observed when either of the catalytic residues (Glu172 or Glu367) was replaced with an alanyl residue (Miyake et al. 2002).

6.2.4.2.5 Applications in Starch Processing

β-Amylase is now used for production of high-maltose syrup from starch in combination with pullulanase. Maltose finds a wide range of applications in the food and pharmaceutical industries because its properties are represented by mild sweetness, good thermal stability, low viscosity in solution, and lack of color formation.

6.2.4.2.6 Commercial β-Amylases

Commercial β-amylases are commonly derived from plants, especially from barley. Optimax® BBA (Genencor) has an optimal condition of pH 4.6, 58°C for high-maltose syrup production (maltose content 55%~60%). The very high-maltose syrup (maltose content >80%) is prepared by using Optimax BBA and pullulanase Optimax®-L 1000 (Genencor) cooperatively.

A novel β-amylase from *Bacillus* has been successfully commercialized by Amano Enzyme Group. It is the world's first microbial β-amylase produced on an industrial scale. It presents its highest activity at pH 5.8, 62°C and its thermostability is about 10°C higher than that of barley β-amylases. This microbial β-amylase is superior to plant β-amylases in maltose production because glucose formation is very slow around pH 6.0. Additionally, this enzyme also possesses the unique ability to work on raw starch, and plant β-amylases cannot act on raw starch.

6.2.4.3 Glucoamylase

6.2.4.3.1 Catalytic Activity

Glucoamylase, with the accepted name glucan 1,4-α-glucosidase, also known as amyloglucosidase or γ-amylase, can decompose the terminal α-1,4 glycosidic linkages in glucans into glucose by tearing off glucose units from the nonreduced end with the inversion of anomeric carbon to produce β-glucose. Fungal glucoamylases present some substrate flexibility and are able to degrade not only α-1,4-glycosidic bonds, but also α-1,6-, α-1,3-, and α-1,2- bonds to a lower degree (Pazur and Ando 1960; Weil et al. 1954). Hydrolysis of starch by glucoamylase results in the high-glucose yield beyond 96% during saccharification (Sauer et al. 2000). The activity toward the β-1,6 linkage is only 0.2% of that for the β-1,4 linkage (Hiromi et al. 1966).

6.2.4.3.2 Source

Glucoamylase can be produced by both fungi and bacteria. The various fungi synthesizing glucoamylase that is active at higher temperatures include *Aspergillus awamori*, *Aspergillus foetidus*, *Aspergillus niger*, *Aspergillus oryzae*, *Aspergillus terreus*, *Mucor rouxians*, *Mucor javanicus*, *Neurospora crassa*, *Rhizopus delmar*, *Rhizopus oryzae* (Pandey et al. 2000), and *Arthrobotrys amerospora* (Jaffar et al. 1993; Norouzian and Jaffar 1993). The molecular weight of glucoamylases from various fungal sources is usually in the range of 48 to 90 kD excepting a few 125-kD GAs produced by *Aspergillus niger* (Suresh et al. 1999).

6.2.4.3.3 Structure and Catalytic Domain

Glucoamylases constitute glycoside hydrolase family 15 (GH15). A majority of glucoamylases are multidomain enzymes consisting of a catalytic domain connected to a starch-binding domain by an *O*-glycosylated linker region (Sauer et al. 2000). The catalytic domain folds as a twisted $(\alpha/\alpha)_6$-barrel with a central funnel-shaped active site in which two key catalytic glutamic acid residues are approximately 200 residues apart in sequence. The starch-binding domain (SBD) is attached to the catalytic domain by a heavily glycosylated linker, folds as an antiparallel β-barrel, and has two binding sites for starch or β-cyclodextrin.

6.2.4.3.4 Mechanism

Family GH15 glycosidases are inverting enzymes, as first shown by Weil et al. (1954), and follow a classical Koshland single-step displacement mechanism, which involves proton transfer from the catalyst to the glycosidic oxygen of the scissile bond. The oxocarbenium ion intermediate mechanism is considered to be suitable for glucoamylase (Chiba 1997), and this mechanism is substantially a general acid–base catalytic mechanism.

6.2.4.3.5 Applications in Starch Processing

Glucoamylase has been used in production of either crystalline glucose or glucose syrup in the starch industry (Kovalenko et al. 2002). The purpose of glucoamylase in commercial food activities is centered on the brewing of beer and the production of bread products and fruit juices.

6.2.4.3.6 Commercial Glucoamylases

Spirizyme® (Excel, Ultra, Fuel HS; Novozyme) are developed for full starch saccharification, giving high-glucose syrup for ethanol production. STARGEN® (001, 002; Genencor) is an α-amylase and glucoamylase blend for the processing of uncooked starch. G-ZYME® 480 and FERMENZYME® L-400 (Genencor) are glucoamylases and blends for hydrolysis of starch to glucose during saccharifying.

Commercial glucoamylases are widely used in baking and brewing. AMG® 300 L (Novozyme) and GoldCrust® (Novozyme) enhance the crust color by generating glucose to enable the Maillard reaction responsible for crust color formation and overcome the challenge of crust separation and moisture loss in pre- and par-baked goods. BakeZyme® AG 800 BG (DSM) can shorten the fermentation time and improve browning, crumb structure, and softness of breads. DIAZYME® X4 (Genencor) is active at normal wort and beer pH, therefore producing highly attenuated low-carbohydrate beers and improved attenuation. Attenuzyme® Flex (Novozyme) is a high-performing, fast-acting combination of glucoamylase, specific α-amylase, and a debranching enzyme that produces low-calorie beer with shorter mashing times.

6.2.4.4 α-Glucosidase

6.2.4.4.1 Catalytic Activity

α-Glucosidases (EC. 3.2.1.20), also named maltase, glucosidosucrase, and glucoinvertase, are enzymes involved in hydrolyzing terminal nonreducing 1-4-linked α-glucose residues in their substrates, such as starch, glycogen, and maltose, releasing a single α-glucose molecule. α-Glucosidases also catalyze transglycosylation of the α-D-glucose moiety; therefore, the enzymes have potential use in industrial biosynthesis of oligosaccharides or glycoconjugates (Kurimoto et al. 1997).

6.2.4.4.2 Source

Thermostable α-glucosidases are produced intracellularly by many microorganisms, such as *Thermoplasma acidophilum*, *Thermoascus aurantiacus*, *Bacillus subtilis*, etc. A thermostable α-glucosidase isolated from *Thermus thermophilus* has presented unchanged activity at 75°C for 5 h and the half-life of 2 h incubated at 85°C (Zdziebło and Synowiecki 2002). The thermostability of α-glucosidase from *Thermoanaerobacter tengcongensis* MB4 can be considerably increased by introducing prolines at suitable sites (Zhou et al. 2010).

6.2.4.4.3 Structure and Catalytic Domain

α-Glucosidase is a calcium-containing enzyme. This class of enzyme displays significant diversity in its amino acid sequence and aglycon specificity. The α-glucosidases characterized to date consist of two major families, GH13 (36) and GH31 (44), and two minor families, GH4 (3) and GH97 (3) (Henrissat and Davies 1999). The structure of enzymes in the GH13 family has been well characterized whereas α-glucosidases in GH31 have been less reported. An α-glucosidase from *Sulfolobus solfataricus* in the GH31 family presents a central catalytic domain (A) with an (β/α) eight-barrel topology flanked by three β-sandwich domains (N, C, and D) (Ernst et al. 2006).

6.2.4.4.4 Mechanism

Two mechanisms of α-glucosidases have been proposed, and they include a nucleophilic displacement and an oxocarbenium ion intermediate (Chiba 1997). In the former, a carboxyl group and a carboxylate group cleave the glucosidic linkage cooperatively by direct electrophilic and nucleophilic attacks against the glucosyl oxygen and anomeric carbon atoms, respectively, resulting in a covalent β-glucosyl-enzyme complex by a single displacement involving anomer inversion. In oxocarbenium mechanism, the carboxylate group is regarded to promote oxocarbenium ion formation and to stabilize the intermediate while the carboxyl group attacks the oxygen of the glucosidic linkage. The subsequent nucleophilic displacement of the hydroxyl ion of water to the oxocarbenium ion is finished obeying Sn1 mechanism.

6.2.4.4.5 Applications in Starch Processing

During starch processing, α-glucosidase is used in saccharification after liquefaction. Screening thermostable α-glucosidase is required because liquefaction is conducted at 90°C~100°C, pH 6, which is not suitable for α-glucosidase action. Some research also proposed a potential single-step method for glucose syrup production using thermostable α-glucosidase incorporation with α-amylase and pullulanase, possessing similar working temperature and pH values (Zdzieblo and Synowiecki 2002). Another important industrial application of α-glucosidase is the manufacture of isomalto-oligosaccharides, which have physiological functions, such as improving intestinal microflora based on the selective proliferation of bifidobacteria stimulation.

6.2.4.4.6 Commercial α-Glucosidase

During starch saccharification, high-glucose syrup was usually produced by the addition of glucoamylase rather than α-glucosidase. Because both glucoamylase and α-glucosidase can hydrolyze starch from the nonreducing end and release glucose, the development of commercial α-glucosidase is relatively slow and few in comparison to commercial glucoamylase.

6.2.4.5 Pullulanase

6.2.4.5.1 Catalytic Activity

Pullulanase, also named a debranching enzyme or α-dextrin endo-1,6-α-glucosidase, is an amylolytic exoenzyme that can degrade pullulan, a polysaccharide polymer consisting of maltotriose units. In detail, three glucose units in maltotriose are connected by an α-1,4 glycosidic bond whereas consecutive maltotriose units are connected to each other by an α-1,6 bond.

To date, five groups of pullulanase enzymes have been reported (Table 6.5), that is, (i) pullulanase type I, (ii) pullulanase type II (amylopullulanase), (iii) neopullulanase, (iv) isopullulanase, and (v) pullulan hydrolase type III (Hii, S. L. et al. 2012). Both pullulanases type I and type II attach α-1,6 glucosidic linkages in pullulan but are unable to degrade cyclodextrin (Duffner et al. 2000). Neopullulanase and isopullulanase are only able to cleave α-1,4 glucosidic linkages in pullulan. They are highly active on cyclodextrins (Matzke et al. 2000) yet show no action on starch.

6.2.4.5.2 Source

Pullulanase type I has been produced from mesophilic bacteria, such as *Aerobacter aerogenes* (Ohba and Ueda 1973), *Bacillus acidopullulyticus* (Jensen and Norman 1984), *Klebsiella pneumonia* (Kornacker and Pugsley 1990), and *Streptomyces* sp. (Takasaki et al. 1993). Moderate thermophilic bacteria, such

TABLE 6.5

Reaction Specificities of Pullulan-Degrading Enzymes

Enzymes	EC Number	Bonds Processed	Preferred Substrate	End Products
Pullulanase type I	3.2.1.41	α-(1,6)	Oligo- and polysaccharides, Pullulan	Trimer (maltotriose)
Pullulanase type II (amylopullulanase)	3.2.1.41	α-(1,6) α-(1,4)	Pullulan Poly- and oligosaccharide (starch)	Trimer (maltotriose) Mixture of glucose, maltose, and maltotriose
Pullulan hydrolase type I (neopullulanase)	3.2.1.135	α-(1,4)	Pullulan	Panose
Pullulan hydrolase type II (isopullulanase)	3.2.1.57	α-(1,4)	Pullulan	Isopanose
Pullulan hydrolase type III	3.2.1.--	α-(1,4) and α-(1,6)	Pullulan Starch, amylose, and amylopectin	Mixture of panose, maltose, and maltotriose Maltotriose and maltose

Source: Hii, S. L., Tan, J. S., Ling, T. C., Ariff, A. B., *Enzyme Research*, 2012, 1–14, 2012.

as *Bacillus flavocaldarius* (Suzuki et al. 1991), *Bacillus thermoleovorans* (Messaoud et al. 2002), and *Fervidobacterium pennavorans* (Koch et al. 1997) also secrete pullulanase type I.

Pullulanase type II is widely distributed among extreme thermophilic bacteria and hyperthermophilic archaea. The most thermostable and thermoactive pullulanase type II reported to date was from the hyperthermophilic archaeon *Pyrococcus woesei* (Rudiger and Jorgensen 1995) and *Pyrococcus furiosus* (Brown and Kelly 1993).

6.2.4.5.3 Structure and Catalytic Domain

The first pullulanase isolated from *Klebsiella pneumoniae* (Wallenfels et al. 1966) is composed of 1150 amino acid residues and has at least four domains containing catalytic and C-terminal domains that have similarities to α-amylase (Katsuragi et al. 1987). The catalytic domain includes four conserved regions that are common to the enzymes of the α-amylase family.

The substitution of His607, Asp677, and His682 in the conserved region of *Klebsiella aerogenes* pullulanase led to the complete loss of enzyme activity, and the substitution of His607 and Asp677 resulted in the loss of the ability to bind cyclodextrins (Yamashita et al. 1997).

A recent crystal structure refining of *Klebsiella pneumoniae* pullulanase revealed that it contains 920–1052 amino acid residues and is composed of five domains (N1, N2, N3, A, and C). The N1 domain was found to be a new type of carbohydrate-binding domain with one calcium site. The structure features determine the different substrate specificities between pullulanase and isoamylase (Mikami et al. 2006).

6.2.4.5.4 Catalytic Mechanism

Pullulanase preferentially hydrolyses pullulan; therefore, it belongs to the glycoside hydrolase family GH13. Kuriki et al. (1989) have pointed out that in cyclomaltodextrin glucanotransferase, pullulanase, isoamylase, and neopullulanase, the existence of the four highly conserved regions that are well known in amylase and contain all the catalytic residues that bind glucosyl residues adjacent to the scissile linkage in the substrate. This means these enzymes have a similar catalytic mechanism.

6.2.4.5.5 Applications

Pullulanase has been widely utilized to hydrolyze the α-1,6 glucosidic linkages in starch, amylopectin, pullulan, and related oligosaccharides, which enables a complete and efficient conversion of the branched polysaccharides into small fermentable sugars during the saccharification process. Particularly, the reversion reaction that involves resynthesis of saccharides from glucose molecules is prevented in the presence of pullulanase.

6.2.4.5.6 Commercial Pullulanase

Commercial pullulanses widely applied in starch processing include Promozyme (Novozyme), Optimax-L 1000 (Genencor), and Pullulanase "Amano" 3, Klelstase PL45 (Amano Enzyme Group). Promozyme from *Bacillus acidopullulyticus* is appropriate to further hydrolysis of liquefied starch at pH 5.0, 40°C. During saccharification, Promozyme can be applied cooperatively with glucoamylase preventing the formation of oligosaccharides and improving glucose yield. During the production of maltose syrup, Promozyme is used with Fungamyl (Novozyme) for improving maltose yield. A typical application of Pullulanase "Amano" 3 (from *Bacillus brevis*) is to produce branched oligosaccharides (isomalto oligosaccharide) containing a series of α-1,6 bonds in its structure. It is produced by liquefaction of starch followed by treatment with β-amylase, pullulanase, and transglucosidase. Its sweetness is about 40% that of sugar, and it is easily utilized by intestinal bacteria but not by bacteria in the mouth, so it does not promote dental caries.

6.2.4.6 Xylose Isomerase

6.2.4.6.1 Catalytic Activity

Xylose isomerase, with the alternative name D-xylose ketoisomerase, can catalyze the interconversion of aldose and ketose sugars with broad substrate specificity. It has been widely used industrially to convert glucose to fructose in the manufacture of high-fructose corn syrup; therefore, it is also referred to as "glucose isomerase."

6.2.4.6.2 Source

The first reported xylose isomerase was from *Pseudomonas hydrophila* (Hochster and Watson 1953), and most commercial glucose isomerases come from *Streptomyces* or *Actinoplanes* species. Recently, xylose isomerase from *Orpinomyces* (Madhavan et al. 2009), *Thermus thermophilus* (Lönn et al. 2002), and *Bacillus licheniformis* (Vieille et al. 2001) has also been reported.

6.2.4.6.3 Structure and Catalytic Domain

From the viewpoint of structure, xylose isomerases have been categorized as Class 1 and Class 2; however, the primary and tertiary structures of the enzymes in the two classes are almost identical, and their sequence differences are few. Xylose isomerase is a tetramer of four identical subunits. Each subunit has two domains: a β-barrel domain with a core of eight β-strands separated by α-helices ($[\alpha/\beta]_8$ barrel) and a C-terminal helical domain that embraces the adjacent subunit (Banner et al. 1975). The active site was tentatively identified in the center of the barrel.

6.2.4.6.4 Mechanism

Catalytic mechanism of xylose isomerase has been summarized to eight steps in detail based on the difference in Fourier analyses of the binding to *Arthrobacter* xylose isomerase by various substrates, inhibitors, and metal ions (Collyer et al. 1990). The reaction pathway can include binding to substrate, ring-opening, chain extension, isomerization, chain contraction, ring-closure, and product dissociation.

McKay and Tavlarides (1979) proposed an enzymatic isomerization mechanism of D-glucose to D-fructose with three component equilibria: (i) α-D-glucose ⇔ β-D-glucose, (ii) α-D-glucose ⇔ β-D-fructose, and (iii) β-D-fructose ⇔ α-D-fructose equilibria, with which equilibrium (iii) was further delineated by the anomerization of α-, β-D-fructopyranose and α-, β-D-fructofuranose. However, experiments also indicate that β-anomer as well as α-anomer is a substrate of the glucose isomerase (Lee and Hong 2000).

6.2.4.6.5 Applications in Starch Processing

Xylose isomerases have been widely used in manufacturing high-fructose corn syrup and fructose crystalline. Recently, glucose isomerization is being considered as an intermediate step in the possible route of biomass to fuels and chemicals.

6.2.4.6.6 Commercial Glucose Isomerase

Sweetzyme® IT/IT Extra (Novozyme) from *Streptomyces murinus* can be operated in the pH range 7.0–8.0 and temperature scope of 55°C–60°C for batch/reuse glucose isomerization. The optimal condition is pH 7.2–7.5, 60°C. It has a long half-life of more than 200 days. During its lifetime (which, depending upon the conditions of use, can be up to a year), 1 kg of immobilized Sweetzyme can convert at least 18,000 kg of fructose syrup (dry mass). Gensweet® IGI (Genencor) is a granular, immobilized glucose isomerase with the conversion of 44% fructose based on standard conditions of 40% w/w glucose at pH 7.5, 60°C.

6.2.4.7 Cyclodextrin Glycosyltransferase

6.2.4.7.1 Catalytic Activity

CGTase (EC. 2.4.1.19), also named cyclodextrin glucanotransferase or cyclodextrin glycosyltransferase, can cyclize part of a (1,4)-α-D-glucan chain by formation of a (1,4)-α-D-glucosidic bond. CGTase also has weak hydrolyzing activity, which consists of cleaving the longer polysaccharidic chains into shorter fragments.

6.2.4.7.2 Source

CGTase is an enzyme common to many bacterial species, in particular of the *Bacillus* genus (e.g., *B. circulans*, *B. macerans*, and *B. stearothermophilus*) as well as to some archaea.

6.2.4.7.3 Structure and Catalytic Domain

CGTase is a bacterial enzyme belonging to GH13, and all the members in this family share a similar α-retaining catalytic mechanism but can have different reaction and product specificities. Particularly, CGTase has uncommonly high transglycosylation activity and is able to form cyclodextrins. Structures of CGTase from different sources have been characterized (Klein and Schulz 1991; Lawson et al. 1994). CGTases have five domains, A–E (Uitdehaag et al. 2002). Domain A is the catalytic $(\alpha/\beta)_8$ domain, and domain B, contributing to substrate binding, is an extended loop region inserted after strand 3 of domain A. Domains C and E have a sheet structure and are specialized in binding to raw starch granules. Domain D also has a sheet structure, and its function remains to be elucidated.

6.2.4.7.4 Catalytic Mechanism

All of the CGTases can catalyze up to four reactions: cyclization, coupling, disproportionation, and hydrolysis. All these activities share the same catalytic mechanism, which is common to all glycosyl-hydrolases. Cyclization is the process through which a linear polysaccharidic chain is cleaved, and the two ends of the cleaved fragment are joined to produce a CD. α-CD (six residues), β-CD (seven residues), and γ-CD (eight residues) are the most common cyclization products. The coupling reaction is the reverse process of cyclization: The enzyme cleaves a cyclodextrin to produce a linear dextrin, which is subsequently joined to a linear oligosaccharide (Figure 6.6). Disproportionation is very similar to coupling, but the cleaved dextrin is not a cyclodextrin, but a linear oligosaccharide that is then joined to a second oligosaccharide.

6.2.4.7.5 Applications

CGTase has important industrial applications. The major interest in the use of CGTase is for the large-scale production of CDs. CDs are cyclic oligomers of glucose that can form water-soluble inclusion complexes with small molecules and portions of large compounds. Cyclodextrins have important applications in food, cosmetics, and pharmaceuticals. CGTase can be used for its coupling and disproportionation reactions for the transfer of oligosaccharides from donor substrates, such as CDs or starch, to various acceptor molecules.

Recently, CGTases have been used for synthesis and modification of oligosaccharides, such as stevioside (Pedersen et al. 1995). Stevioside is a potential high-intensity sweetener isolated from *Stevia rebaudisna*, and it has a bitter aftertaste and low solubility. CGTase has been used to connect glucose to stevioside for decreasing its bitterness and increasing its solubility.

6.2.4.7.6 Commercial CGTase

Relatively few commercial CGTases have been reported. Cyclodextrin glucanotransferase "Amano" produced from *Bacillus macerans* has the optimal condition of pH 6.0, 60°C–65°C.

6.3 Starch Processing with Amylolytic Enzymes

6.3.1 Liquefaction

Many manufacturing processes using starch as the raw material, such as brewing, papermaking, and confections, usually starts from the starch liquefaction (or thinning), which is commonly accomplished with α-amylase as the catalyst.

In liquefaction, the mashes usually contain around 15% starch (Goyal et al. 2005). The natural pH of starch slurry is usually around 4.5; therefore, a high cost will be paid for adjusting slurry pH to the optimal pH (around 6.5) of the most bacterial α-amylase (Sivaramakrishnan et al. 2006). The high liquefaction temperature (90°C–110°C) also puts extreme demands on the thermostability of α-amylases. New α-amylases with improved properties, such as enhanced thermostability, acid tolerance, and ability to function without the addition of calcium, have recently been developed (Bisgaard-Frantzen et al. 1999; Declerck et al. 2000).

A worldwide interest has recently been focused on the raw starch-digesting amylases (Goyal et al. 2005), which suggests enzymatic degradation of raw starch at a much lower temperature of 60°C–70°C, or even without heating. Microorganisms that produce amylases capable of digesting raw starch are mostly fungi (Kelly et al. 1995) and relatively few molds and bacteria (Chiou and Jeang 1995). Moreover, the source of starch also affects the efficiency of raw digestion, for example, potato starch has been reported to present high resistance to most raw digestion amylolytic enzymes (Qates 1997; Tanakai et al. 1997). In addition, the phenolic compounds have recently been reported to interfere with starch amylolysis during liquefaction (Kandil et al. 2012).

6.3.2 Saccharification

Saccharification, followed by liquefaction, is the process for near complete hydrolysis (95%–97% glucose) of starch accomplished by glucoamylase at a much lower temperature of 50°C–60°C. Excess glucoamylase is required to obtain rapid rates because it hydrolyzes α-1,6 linkages very slowly. However, excess glucoamylase also leads to formation of "reversion products," that is oligosaccharides linked by the resistant α-1,6 linkages, which lower the final DE value (Crabb and Shetty 1999). For this reason, the debranching enzyme pullulanase is used along with glucoamylase. In particular, a variety of sweeteners having dextrose equivalents in the ranges 40–45 (maltose), 50–55 (high maltose), and 55–70 (high conversion syrup) can be produced by applying a series of enzymes, including glucoamylase, pullulanase, and β-amylase.

In conventional enzymatic starch saccharification, a variation of parameters in different steps caused many handicaps, such as time consumption, reverse reactions, and low yield; therefore, amylolytic enzymes from thermophiles are highly desired (Satyanarayana et al. 2004).

Immobilization of glucoamylase can improve its stability to heat denaturation and pH extremes and therefore meet the requirement of industrial applications. Novel matrices, including polyaniline polymer (Silva et al. 2005), macrostructured carbonized ceramics (Kovalenko et al. 2002), and magnetic poly(methylmethacrylate) microspheres (Arica et al. 2000), have been investigated. Furthermore, Roy and Gupta (2004) reported the coimmobilization of glucoamylase and pullulanase in calcium alginate beads. The coimmobilized enzymes showed a broader pH optima and increased stability at 80°C and were able to hydrolyze insoluble starch to a relatively greater extent than the free enzyme system.

6.3.3 Glucose Isomerization

Glucose isomerization is the key step for production of fructose syrup, which is now popular in the food and soft drink industries. In industry, continuous production of fructose syrup was performed by flowing glucose syrup through a column filled with commercial immobilized glucose isomerase. Current commercial xylose (glucose) isomerase operates at only 60°C to produce 45% fructose syrups. The thermodynamic equilibrium of the glucose/fructose ratio increases with the isomerization temperature; therefore, thermostable isomerases are desirable for production of 55% fructose syrups at >90°C. The performance of this process is strongly affected by thermal inactivation of the enzyme (Illanes et al. 1996), which is responsible for a progressive decrease in the isomerization rate.

Immobilization of the glucose isomerase can offer several advantages for industrial applications, including repeated use, ease of recycling, improved enzyme stability, continuous operation in a packed-bed reactor, and alteration of the properties of the enzyme. Various matrices, such as DEAE-cellulose (Chen and Anderson 1979; Huitron and Limon-Lason 1978), polyacrylamide gel (Demirel et al. 2006; Strandberg and Smiley 1971) and alginate beads (Rhimi et al. 2007), and copolymer Eupergit (Tükel and Alagöz 2008), have been investigated.

Reaction conditions are another important factor influencing isomerization yield. Although a high glucose conversion rate was obtained in aqueous conditions, successful isomerization has been achieved in aqueous ethanol, indicating that GI can maintain activity at reduced water concentrations (Visuri and Klibanov 1987). The first investigation on glucose isomerization using glucose isomerase in ionic liquid DBAO was reported by Ståhlberg et al. (2012). The fructose yield is about 60% at 80°C within 5 h, and some mannose was synthesized. In addition, it was reported that microwave irradiation significantly enhanced the activity of immobilized glucose isomerase, and the effects of acceleration might be nonthermal microwave effect (Yu et al. 2011).

On the basis of experimental data, kinetic modeling on glucose isomerization has been frequently researched in recent years. Palazzia and Converti (1999) have proposed a linearization technique to calculate the isomerization rate with respect to substrate concentration within a wide range of conditions of actual interest in industrial processes. Later, Palazzia and Converti (2001) evaluated the diffusional resistances of an industrial column for glucose isomerization to fructose by immobilized glucose isomerase and concluded that outer resistance is quite weak. Dehkordi et al. (2009) modeled the isomerization conducted by Sweetzyme IT over the reaction temperature range of 50°C–65°C and the glucose concentration range of 0.1–1.25 M, and they found slight deviations from the Michaelis-Menten kinetic model at low fractional conversions.

In addition, inorganic catalysts have attracted increasing concern in recent years. Inorganic catalysts have a wider operating temperature range, a longer lifetime, and a higher resistance to impurities. Saravanamurugan et al. (2013) investigated the isomerization of glucose catalyzed by commercial zeolites in two steps, and a remarkable 55% yield of fructose was achieved after reacting for 1 h at 120°C. This result is equivalent to that of bioisomerization. Large-pore zeolites containing tin (Sn-Beta) can also convert glucose to fructose in aqueous conditions with fructose yield less than 50% (Moliner et al. 2010).

6.3.4 Cyclodextrin Production

Industrial production of CD is accomplished by enzymatic degradation of starch by the CGTases. The reaction takes place in a 30% starch suspension. The high viscous starch suspension can reduce CD yield; therefore, the suspension is initially partially hydrolyzed with α-amylase. CGTases are subsequently used to break the starch molecules down into segments. Due to the helical structure of the starch, the segments formed consist mostly of seven (β-) and, less commonly, six (α-) or eight (γ-) glucose units, which then react further to form cyclic oligosaccharides in a CGTase-typical reaction.

Different types of starch can be used as substrate although potato starch is the most commonly used for CD production. Maize starch and wheat starch are used, but they contain a higher percentage of amylose, which gives lower yields of CD than amylopectin (Szerman et al. 2007).

All CGTases convert starch to a mixture of CDs, the relative composition of which depends on the reaction conditions and the bacterial source of the enzyme. The product selectivity of industrially used CGTases is not very high; therefore, new techniques are required for production of certain CDs. It has been reported that addition of polar organic solvents can improve the product selectivity of CGTase (Blackwood and Bucke 2000; Lee and Kim 1991).

Thermostable CGTases isolated from thermophilic anaerobic bacteria (Norman and Jorgensen 1992; Wind et al. 1995) are active and stable at high temperatures and low pH values, and are able to solubilize raw starch granules, thereby eliminating the need for α-amylase pretreatment without any traces of oligosaccharides produced in the initial stages of the reaction.

6.4 Starch Derivatives Manufactured with Amylolytic Enzymes

6.4.1 Glucose

Glucose syrup, manufactured by hydrolyzing starch by acids or enzymes (Figure 6.3), is offered in a multitude of varieties, which vary with respect to sweetening power and technological properties (viscosity, consistence, susceptibility to crystallization).

Dextrorotatory glucose crystals (DGCs) are a fast energy source that requires no digestion prior to absorption in contrast to any other carbohydrate source. DGCs are commonly added into functional foods, nutritional supplements, and sports drinks to create a high-energy source.

Glucose syrup containing more than 90% glucose is used in industrial fermentation (Dziedzic and Kearsley 1995), but syrups used in confectionery and bakery manufacturing contain varying amounts of glucose, maltose, and higher oligosaccharides, depending on the grade, and can typically contain 10% to 43% of glucose (Jackson 1995) as well as maltose and maltotriose. In some areas of the world outside of Germany, glucose syrup can be used as a raw material in the manufacture of alcoholic beverages, such as beer and cider. Glucose syrup also can be used in coating frozen foods for retarding oxidation. Addition of glucose syrup can prevent crystallization of sugar in processed foods such as canned fruit, jams, and jellies. It helps in preserving food and saves food from getting spoiled and, at the same time, does not increase the level of sweetness. It is also an instant energy source for baby foods.

6.4.2 High-Fructose Corn Syrup

High-fructose corn syrup (HFCS) is made by conversion of glucose to fructose by xylose isomerase (Figure 6.3). Because of its low price compared to sugar, HFCS now is the predominant sweetener used in processed foods and beverages in many countries. HFCS 55 (55% fructose and 42% glucose) is commonly used in soft drinks, and HFCS 42 (42% fructose and approximately 53% glucose) is used in beverages, processed foods, cereals, and baked goods. HFCS 90 (90% fructose and 10% glucose) is mainly used to blend with HFCS 42 and HFCS 55.

In the United States, HFCS has become an attractive substitute for cane sugar. Many soft drink makers, such as Coca-Cola and Pepsi, use sugar, and other nations have switched to HFCS since 1984. In 2012, the average American consumed approximately 12.3 kg of HFCS. However, health concerns have been raised about HFCS, which allege contribution to obesity, cardiovascular disease, diabetes, and nonalcoholic fatty liver disease (Bocarsly et al. 2010). Now the question "is HFCS healthy?" is still under debate.

6.4.3 High-Maltose Corn Syrup

High-maltose corn syrup (HMCS) produced by β-amylase or fungal α-amylase (Figure 6.3) typically contains 40%~50% maltose (some have as high as 70%) and little to no fructose. Cargill® classified HMCS to different categories: high-conversion syrup (DE 62~63, maltose 30%~45%, glucose 35%), high-maltose concentration syrup (DE 48~52, maltose 48%~52%, glucose 5%), and very high maltose concentration syrup (DE 50~60, maltose 70%~85%, glucose 5%).

HMCS is often used to prolong the shelf life of products, and its sweetness can also meet the requirement as an alternative to glucose syrup. At a given moisture level and temperature, HMCS has a lower viscosity than a glucose solution but will still set to a hard product; therefore, it is currently used in candies for replacing glucose syrup. Additionally, HMCS also can be used in frozen desserts and ice creams for the low freezing point of maltose. In particular, HMCS can be used to substitute malt in the manufacture of beer.

Recently, as fructose has been a source of health concerns, and people are worried about the health effects of high-fructose corn syrup, the use of HMCS as a food additive has increased to a considerable extent. Being completely gluten-free, it cannot have severe high-maltose corn syrup side effects and therefore has not attracted much nutritional criticism.

6.4.4 Isomalto-Oligosaccharide

Isomalto-oligosaccharide (IMO) is a starch-related oligosaccharide consisting of glucose subunits (typically 4~7 units) that contain a series of α (1, 6) bonds in this structure. For manufacturing IMO on a commercial scale, starch is first converted by simple enzymatic hydrolysis into high-maltose syrup with di-, tri-, and oligosaccharides, which are further enzymatically converted into more digestion-resistant α(1,6) glycosidic linkages by α-glucosidase. Nevertheless, new research has suggested that IMO also can be synthesized by dextransucrase (EC. 2.4.1.5) secreted by *Leuconostoc* (Cho et al. 2014).

IMO exerts positive effects on human digestive health: It is a prebiotic, promotes less flatulence, and has a low glycemic index and anticaries activities (Hesta et al. 2003; Rycroft et al. 2001). Therefore, IMO has been used in a broad spectrum of applications in food product categories, including as a functional ingredient in beverage and bakery and as a dietary fiber and low-calorie sweetener in confectioneries, cereal products, and frozen desserts.

6.4.5 Maltodextrin

Maltodextrin consisting of D-glucose units connected in chains of variable length (typically 2–7 units) is produced from starch by partial hydrolysis (Figure 6.3). It is easily digestible, being absorbed as rapidly as glucose, and might be either moderately sweet or almost flavorless.

In frozen novelties, maltodextrin provides full-fat texture to reduced-fat formulations, builds solids for a superior mouthfeel, and replaces dairy solids for lower-cost formulations. Ice creams also benefit from its freezing point control and the inhibition of lactose and large ice-crystal growth. Maltodextrin is sometimes used in beer brewing to increase the specific gravity of the final product. This improves the mouthfeel of the beer, increases head retention, and reduces the dryness of the drink. In the United States, maltodextrin is usually produced from corn by enzymatic hydrolysis. In Europe, the starch source is commonly wheat, and wheat-derived maltodextrin may cause concern for individuals suffering from gluten intolerance.

6.4.6 Cyclodextrin

Because it is able to form complexes with hydrophobic compounds, CDs are currently used in pharmaceutical (drug delivery), cosmetics, and chemical industries as well as agriculture and environmental engineering. Especially in the food industry, CDs are employed in cholesterol-free products, dietary supplement, multifunctional dietary fiber, microparticles for volatile or unstable compounds. β-CD complexes with certain carotenoid food colorants have been shown to intensify color, increase water solubility, and improve light stability (Marcolino et al. 2011). In 2008, α-CD was authorized for use as a dietary fiber in the EU. In 2013, the EU commission has verified a health claim for α-CD, confirming the reducing effect of α-CD on blood sugar peaks following a high-starch meal.

REFERENCES

Abe, A., Yoshida, H., Tonozuka, T., Sakano, Y., Kamitori, S. 2005. Complexes of *Thermoactinomyces vulgaris* R-47 alpha-amylase 1 and pullulan model oligossacharides provide new insight into the mechanism for recognizing substrates with alpha-(1,6) glycosidic linkages. *FEBS Journal*, 272(23): 6145–6153.

Adeniran, H. A., Abiose, S. H., Ogunsua, A. O. 2010. Production of fungal β-amylase and amyloglucosidase on some nigerian agricultural residues. *Food and Bioprocess Technology*, 3(5): 693–698.

Agrosynergie. 2010. Evaluation of common agricultural policy measures applied to the starch sector-Final report. Available at http://ec.europa.eu/agriculture/eval/reports/starch/index_en.htm.

Alves-Prado, H. F., Gomes, E., da Silva, R. 2006. Evaluation of solid and submerged fermentations for the production of cyclodextrin glycosyltransferase by *Paenibacillus campinasensis* H69-3 and characterization of crude enzyme. *Applied Biochemistry and Biotechnology*, 129–132: 234–246.

Arica, M. Y., Yavuz, H., Patir, S., Denizli, A. 2000. Immobilization of glucoamylase onto spacer-arm attached magnetic poly (methylmethacrylate) microspheres: Characterization and application to a continuous flow reactor. *Journal of Molecular Catalysis B: Enzymatic*, 11(2–3): 127–138.

Ball, S., Guan, H. P., James, M. et al. 1996. From glycogen to amylopectin: A model for the biogenesis of the plant starch granule. *Cell*, 86: 349–352.

Banner, D. W., Bloomer, A. C., Petsko, G. A. et al. 1975. Structure of chicken muscle triose phosphate isomerase determined crystallographically at 2.5 Å resolution: Using amino acid sequence data. *Nature (London)*, 255: 609–614.

Benassi, V. M., Pasin, T. M., Facchini, F. D., Jorge, J. A., de Lourdes Teixeira de Moraes Polizeli, M. 2014. A novel glucoamylase activated by manganese and calcium produced in submerged fermentation by *Aspergillus phoenicis*. *Journal of Basic Microbiology*, 54(5): 333–339.

Binod, P., Palkhiwala, P., Gaikaiwari, R., Nampoothiri, K. M., Duggal, A., Dey, K., Pandey, A. 2013. Industrial enzymes-present status and future perspectives for India. *Journal of Scientific & Industrial Research*, 72: 271–286.

Bisgaard-Frantzen, H., Svendsen, A., Norman, B. et al. 1999. Development of industrially important α-amylases. *Journal of Applied Glycoscience*, 46: 199–206.

Blackwood, A. D., Bucke, C. 2000. Addition of polar organic solvents can improve the product selectivity of cyclodextrin glycosyltransferase: Solvent effects on CGTase. *Enzyme and Microbial Technology*, 27: 704–708.

Bocarsly, M. E., Powell, E. S., Avena, N. M., Hoebel, B. G. 2010. High-fructose corn syrup causes characteristics of obesity in rats: Increased body weight, body fat and triglyceride levels. *Pharmacology Biochemistry & Behavior*, 97(1): 101–106.

Brown, S. H., Kelly, R. M. 1993. Characterization of amylolytic enzymes, having bothα-1,4 and α-1,6 hydrolytic activity, from the thermophilic archaea *Pyrococcus furiosus* and *Thermococcus litoralis*. *Applied and Environmental Microbiology*, 59(8): 2614–2621.

Chen, W. P., Anderson, A. W. 1979. Purification, immobilization and some properties of glucose isomerase from *Streptomyces flavogriseus*. *Applied and Environmental Microbiology*, 38: 1111–1119.

Cheng, J., Wu, D., Chen, S., Chen, J., Wu, J. 2011. High-level extracellular production of alpha-cyclodextrin glycosyltransferase with recombinant *Escherichia coli* BL21 (DE3). *Journal of Agriculture and Food Science*, 59(8): 3797–3802.

Chiba, S. 1997. Molecular mechanism in alpha-glucosidase and glucoamylase. *Bioscience, Biotechnology, and Biochemistry*, 61(8): 1233–1239.

Chiou, S. Y., Jeang, C. L. 1995. Factors affecting production of raw starch digesting amylase by the soil bacterium *Cytophagasp*. *Biotechnology Applied Biochemistry*, 22: 377–384.

Cho, S. K., Eom, H.-J., Moon, J. S. et al. 2014. An improved process of isomaltooligosaccharide production in kimchi involving the addition of a *Leuconostoc* starter and sugars. *International Journal of Food Microbiology*, 170: 61–64.

Cohen, R., Orlova, Y., Kovalev, M., Ungar, Y., Shimoni, E. 2008. Structural and functional properties of amylose complexes with genistein. *Journal of Agriculture and Food Chemistry*, 56: 4212–4218.

Collyer, C. A., Henrick, K., Blow, D. M. 1990. Mechanism for aldose-ketose interconversion by d-xylose isomerase involving ring opening followed by a 1,2-hydride shift. *Journal of Molecular Biology*, 212: 211–235.

Corrêa, T. L. R., Moutinho, S. K. dos S., Martins, M. L. L., Martins, M. A. 2011. Simultaneous α-amylase and protease production by the soil bacterium *Bacillus* sp. SMIA-2 under submerged culture using whey protein concentrate and corn steep liquor: Compatibility of enzymes with commercial detergents. *Food Science and Technology (Campinas)*, 31(4): 843–848.

Crabb, W. S., Shetty, J. K. 1999. Commodity scale production of sugars from starches. *Current Opinion in Microbiology*, 2: 252–256.

Daba, T., Kojima, K., Inouye, K. 2013. Chemical modification of wheat β-amylase by trinitrobenzenesulfonic acid, methoxypolyethylene glycol, and glutaraldehyde to improve its thermal stability and activity. *Enzyme and Microbial Technology*, 53(6–7): 420–426.

Davies, G., Henrissat, B. 1995. Structures and mechanisms of glycosyl hydrolases. *Structure*, 3: 853–859.

Declerck, N., Machius, M., Wiegand, G., Huber, R., Gaillardin, C. 2000. Probing structural determinants specifying high thermostability in *Bacillus licheniformis* α-amylase. *Journal of Molecular Biology*, 301: 1041–1057.

Dehkordi, A. M., Tehrany, M. S., Safari, I. 2009. Kinetics of glucose isomerization to fructose by immobilized glucose isomerase (Sweetzyme IT). *Industrial & Engineering Chemistry Research*, 48: 3271–3278.

Demirel, G., Özçetin, G., Şahin, F., Tümtürk, H., Aksoy, S., Hasırcı, N. 2006. Semi-interpenetrating polymer networks (IPNs) for entrapment of glucose isomerase. *Reactive & Functional Polymers*, 66: 389–394.

Divakaran, D., Chandran, A., Chandran, R. P. 2011. Comparative study on production of α-amylase from *Bacillus licheniformis* strains. *Brazilian Journal of Microbiology*, 42(4): 1397–1404.

dos Santos, E. R., Teles, Z. N. S., Campos, N. M. et al. 2012. Production of alpha-amylase from *Streptomyces* sp. SLBA-08 strain using agro-industrial by-products. *Brazilian Archives of Biology and Technology*, 55(5): 793.

Duffner, F., Bertoldo, C., Andersen, J. T., Wagner, K., Antranikian, G. 2000. A new thermoactive pullulanase from *Desulfurococcus mucosus*: Cloning, sequencing, purification, and characterization of the recombinant enzyme after expression in Bacillus subtilis. *Journal of Bacteriology*, 182(22): 6331–6338.

Dziedzic, S. Z., Kearsley, M. W. 1995. *Handbook of Starch Hydrolysis Products and Their Derivatives* (p. 230). London: Blackie Academic & Professional.

Ernst, H. A., Lo, L. L., Willemoës, M., Blum, P., Larsen, S. 2006. Structure of the *Sulfolobus solfataricus* α-glucosidase: Implications for domain conservation and substrate recognition in GH31. *Journal of Molecular Biology*, 358: 1106–1124.

Fazekas, E., Szabó, K., Kandra, L., Gyémánt, G. 2013. Unexpected mode of action of sweet potato β-amylase on maltooligomer substrates. *Biochimica et Biophysica Acta (BBA)—Proteins and Proteomics*, 1834(10): 1976–1981.

Fennema, O. R. 1996. *Food Chemistry* (3rd Edition, p. 193). New York: Marcel Dekker, Inc.

Goyal, N., Gupta, J. K., Soni, S. K. 2005. A novel raw starch digesting thermostable α-amylase from *Bacillus* sp. I-3 and its use in the direct hydrolysis of raw potato starch. *Enzyme Microbiology Technology*, 37: 723–734.

Henrissat, B. 1991. A classification of glycosyl hydrolases based on amino acid sequence similarities. *Biochemical Journal*, 280: 309–316.

Henrissat, B., Davies, G. 1999. Structural and sequence-based classification of glycoside hydrolases. *Current Opinion in Structural Biology*, 271: 619–628.

Hesta, M., Roosen, W., Janssens, G. P., Millet, S., De Wilde, R. 2003. Prebiotics affect nutrient digestibility but not fecal ammonia in dogs fed increased dietary protein levels. *British Journal of Nutrition*, 90: 1007–1014.

Hii, S. L., Roslanizah, M., Ling, T. C., Ariff, A. B. 2012. Statistical optimization of pullulanase production by *Raoultella planticola* DSMZ 4617 using sago starch as carbon and peptone as nitrogen sources. *Food Bioprocess Technology*, 5: 729–737.

Hii, S. L., Tan, J. S., Ling, T. C., Ariff, A. B. 2012. Pullulanase: Role in starch hydrolysis and potential industrial applications. *Enzyme Research*, 2012: 1–14.

Hiromi, K., Hamauzu, Z. I., Takahashi, K., Ono, S. 1966. Kinetic studies on gluc-amylase. II. Competition between two types of substrate having α-1,4 and α-1,6 glucosidic linkage. *Journal of Biochemistry (Tokyo)*, 59: 411–418.

Hochster, R. M., Watson, R. M. 1953. Xylose isomerase. *Journal of the American Chemical Society*, 75: 3284–3285.

Huitron, C., Limon-Lason, J. 1978. Immobilization of glucose isomerase to ion-change materials. *Biotechnology and Bioengineering*, 20: 1377–1391.

Illanes, A., Altamirano, C., Zúñiga, M. E. 1996. Thermal inactivation of immobilized penicillin acylase in the presence of substrate and products. *Biotechnology Bioengineering*, 50: 609–616.

Jackson, E. B. 1995. *Sugar Confectionery Manufacture* (p. 132). Berlin: Springer.

Jaffar, M. B., Bharat, R. P., Norouzian, D., Irani, S. D., Shetty, P. 1993. Production of glucoamylase by nematiphagus fungi *Arthrobotrys* species. *Indian Journal of Experimental Biology*, 31: 87–89.

Jenkins, P. J., Donald, A. M. 1995. The influence of amylose on starch granule structure. *International Journal of Biological Macromolecules*, 17(6): 315–321.

Jensen, B. D., Norman, B. E. 1984. Bacillus acidopullyticus pullulanase: Applications and regulatory aspects for use in food industry. *Process Biochemistry*, 1: 397–400.

Jin, B., van Leeuwen, H. J., Patel, B., Yu, Q. 1998. Utilisation of starch processing wastewater for production of microbial biomass protein and fungal α-amylase by *Aspergillus oryzae*. *Bioresource Technology*, 66: 201–206.

Kadziola, A., Abe, J., Svensson, B., Haser, R. 1994. Crystal and molecular structure of barley alpha-amylase. *Journal of Molecular Biology*, 239(1): 104–121.

Kadziola, A., Søgaard, M., Svensson, B., Haser, R. 1998. Molecular structure of a barley alpha-amylase-inhibitor complex: Implications for starch binding and catalysis. *Journal of Molecular Biology*, 278(1): 205–217.

Kandil, A., Li, J., Vasanthan, T., Bressler, D. C. 2012. Phenolic acids in some cereal grains and their inhibitory effect on starch liquefaction and saccharification. *Journal of Agriculture and Food Chemistry*, 60(34): 8444–8449.

Kaneko, T., Zhang, W. S., Ito, K., Takeda, K. 2001. Worldwide distribution of β-amylase thermostability in barley. *Euphytica*, 121(3): 225–228.

Kang, Y.-N., Adachi, M., Utsumi, S., Mikami, B. 2004. The roles of Glu186 and Glu380 in the catalytic reaction of soybean β-Amylase. *Journal of Molecular Biology*, 339(5): 1129–1140.

Kar, S., Datta, T. K., Ray, R. C. 2010. Optimization of thermostable α-amylase production by *Streptomyces erumpens* MTCC 7317 in solid-state fermentation using cassava fibrous residue. *Brazilian Archives of Biology and Technology*, 53(2): 301–309.

Kar, S., Ray, R. C., Mohapatra, U. B. 2012. Purification, characterization and application of thermostable amylopullulanase from *Streptomyces erumpens* MTCC 7317 under submerged fermentation. *Annals of Microbiology*, 62(3): 931–937.

Kathiresan, K., Manivannan, S. 2006. α-Amylase production by *Penicillium fellutanum* isolated from mangrove rhizosphere soil. *African Journal of Biotechnology*, 5: 829–832.

Katsuragi, N., Takizawa, N., Murooka, Y. 1987. Entire nucleotide sequence of the pullulanase gene of *Klebsiella aerogenes* W70. *Journal of Bacteriology*, 169: 2301–2306.

Kelly, C. T., Tiguo, M. M., Doyle, E. M., Fogarty, W. M. 1995. The raw starch degrading alkaline amylase of *Bacillus* sp. IMD 370. *Journal of Indian Microbiology*, 15: 446–448.

Kelly, R. M., Dijkhuizen, L., Leemhuis, H. 2009. The evolution of cyclodextrin glucanotransferase product specificity. *Applied Microbiology Biotechnology*, 84: 119–133.

Klein, C., Schulz, G. E. 1991. Structure of cyclodextrin glycosyltransferase refined at 2.0 Å resolution. *Journal of Molecular Biology*, 217(4): 737–750.

Koch, R., Canganella, F., Hippe, H., Jahnke, K. D., Antranikian, G. 1997. Purification and properties of a thermostable pullulanase from a newly isolated thermophilic anaerobic bacterium, *Fervidobacterium pennavorans* Ven5. *Applied and Environmental Microbiology*, 63(3): 1088–1094.

Konsoula, Z., Liakopoulou-Kyriakides, M. 2007. Co-production of alpha-amylase and beta-galactosidase by Bacillus subtilis in complex organic substrates. *Bioresource Technology*, 98: 150–157.

Kornacker, M. G., Pugsley, A. P. 1990. Molecular characterization of pulA and its product, pullulanase, a secreted enzyme of *Klebsiella pneumonia* UNF5023. *Molecular Microbiology*, 4(1): 73–85.

Kossmann, J., Lloyd, J. 2000. Understanding and influencing starch biochemistry. *Critical Reviews in Plant Science*, 35(3): 141–196.

Kovalenko, G. A., Komova, O. V., Simakov, A. V., Khomov, V. V., Rudina, N. A. 2002. Macrostructured carbonized ceramics as adsorbents for immobilization of glucoamylase. *Journal of Molecular Catalyst A: Chemical*, 182–183: 73–80.

Kovalevsky, A. Y., Hanson, L., Fisher, S. Z. et al. 2010. Metal ion roles and the movement of hydrogen during the reaction catalyzed by d-xylose isomerase: A joint x-ray and neutron diffraction study. *Structure*, 18: 688–699.

Kuo, C. C., Lin, C. A., Chen, J. Y., Lin, M. T., Duan, K. J. 2009. Production of cyclodextrin glucanotransferase from an alkalophilic *Bacillus* sp. by pH-stat fed-batch fermentation. *Biotechnoloy Letters*, 31: 1723–1727.

Kuriki, T., Imanaka, T. 1989. Nucleotide sequence of neopullulanse gene from *Bacillus stearothermophilus*. *Journal of General Microbiology*, 135: 1521–1528.

Kurimoto, M., Nishimoto, T., Nakada, T., Chaen, H., Fukada, S., Tsujisaka, Y. 1997. Synthesis by an α-glucosidase of glycosyl-trehaloses with an isomaltosyl residue. *Bioscience, Biotechnology, and Biochemistry*, 61: 699–703.

Lam, W. C., Pleissner, D., Lin, C. S. K. 2013. Production of fungal glucoamylase for glucose production from food waste. *Biomolecules*, 3(3): 651–661.

Lawson, C. L., van Montfort, R., Strokopytov, B. et al. 1994. Nucleotide sequence and x-ray structure of cyclodextrin glycosyltransferase from *Bacillus circulans* Strain 251 in a maltose-dependent crystal form. *Journal of Molecular Biology*, 236(2): 590–600.

Lee, Y. D., Kim, H. S. 1991. Enhancement of enzymatic production of cyclodextrins by organic solvents. *Enzyme and Microbial Technology*, 13: 499–503.

Lee, H. S., Hong, J. 2000. Kinetics of glucose isomerization to fructose by immobilized glucose isomerase: Anomeric reactivity of D-glucose in kinetic model. *Journal of Biotechnology*, 84(2): 145–153.

Lindeboom, N., Chang, P. R., Tyler, R. T. 2004. Analytical, biochemical and physicochemical aspects of starch granule size, with emphasis on small granule starches: A review. *Starch*, 56: 89–99.

Liu, H., Li, J., Du, G., Zhou, J., Chen, J. 2012. Enhanced production of α-cyclodextrin glycosyltransferase in *Escherichia coli* by systematic codon usage optimization. *Journal of Industrial Microbiology & Biotechnology*, 39: 1841–1849.

Lönn, A., Gárdonyi, M., van Zyl, W., Hahn-Hägerdal, B., Otero, R. C. 2002. Cold adaptation of xylose isomerase from *Thermus thermophilus* through random PCR mutagenesis. *European Journal of Biochemistry*, 269(1): 157–163.

MacGregor, E. A., Janecek, S., Svensson, B. 2001. Relationship of sequence and structure to specificity in the alpha-amylase family of enzymes. *Biochimica et Biophysica Acta (BBA)*, 1546(1): 1–20.

Madhavan, A., Tamalampudi, S., Ushida, K. et al. 2009. Xylose isomerase from polycentric fungus *Orpinomyces*: Gene sequencing, cloning, and expression in *Saccharomyces cerevisiae* for bioconversion of xylose to ethanol. *Applied Microbiology and Biotechnology*, 82(6): 1067–1078.

Marcolino, V. A., Zanin, G. M., Durrant, L. R., Benassi, M. De T., Matioli, G. 2011. Interaction of curcumin and bixin with β-cyclodextrin: Complexation methods, stability, and applications in food. *Journal of Agricultural and Food Chemistry*, 59(7): 3348–3357.

Matzke, J., Herrmann, A., Schneider, E., Bakker, E. P. 2000. Gene cloning, nucleotide sequence and biochemical properties of a cytoplasmic cyclomaltodextrinase (neopullulanase) from *Alicyclobacillus acidocaldarius*, reclassification of a group of enzymes. *FEMS Microbiology Letters*, 183(1): 55–61.

McKay, G. A., Tavlarides, L. L. 1979. Enzymatic isomerization kinetics of D-glucose to D-fructose. *Journal of Molecular Catalyst*, 6: 57–69.

Messaoud, E. B., Ammar, Y. B., Mellouli, L., Bejar, S. 2002. Thermostable pullulanase type I from new isolated *Bacillus thermoleovorans* US105: Cloning, sequencing and expression of the gene in *E. Coli*. *Enzyme and Microbial Technology*, 31(6): 827–832.

Mikami, B., Iwamoto, H., Malle, D. et al. 2006. Crystal structure of pullulanase: Evidence for parallel binding of oligosaccharides in the active site. *Journal of Molecular Biology*, 359: 690–707.

Miyake, H., Otsuka, C., Nishimura, S., Nitta, Y. 2002. Catalytic Mechanism of β amylase from *Bacillus cereus* var. mycoides: Chemical rescue of hydrolytic activity for a catalytic site mutant (Glu367→Ala) by azide. *The Journal of Biochemistry*, 131(4): 587–591.

Moliner, M., Román-Leshkov, Y., Davis, M. E. 2010. Tin-containing zeolites are highly active catalysts for the isomerization of glucose in water. *Proceedings of the National Academy of Sciences of the United States of America*, 107(14): 6164–6168.

Nagorski, R. W., Richard, J. P. 2001. Mechanistic imperatives for aldose-ketose isomerization in water: Specific, general base- and metal ion-catalyzed isomerization of glyceraldehyde with proton and hydride transfer. *Journal of Agriculture and Food Chemistry*, 123: 794–802.

Nakamura, A., Haga, K., Yamane, K. 1994. The transglycosylation reaction of cyclodextrin glucanotransferase is operated by a Ping-Pong mechanism. *FEBS Letters*, 337: 66–70.

Náray-Szabó, G., Oláh, J., Krámos, B. 2013. Quantum mechanical modeling: A tool for the understanding of enzyme reactions. *Biomolecules*, 3: 662–702.

Norman, B. E., Jorgensen, S. T. 1992. *Thermoanaerobacter* sp. CGTase: Its properties and application. *Denpun Kagaku*, 39: 101–108.

Norouzian, D., Jaffar, M. B. 1993. Immobilization of glucoamylase produced by fungus *Arthrobotrys amerospor*. *Indian Journal of Experimental Biology*, 31: 680–681.

Oates, C. G. 1997. Towards an understanding of starch granule structure and hydrolysis. *Trends in Food Science & Technology*, 8: 375–382.

Ohba, R., Ueda, S. 1973. Purification, crystallization and some properties of intracellular pullulanase from *Aerobacter aerogenes*. *Agricultural and Biological Chemistry*, 37(12): 2821–2826.

Palazzia, E., Converti, A. 1999. Generalized linearization of kinetics of glucose isomerization to fructose by immobilized glucose isomerase. *Biotechnology Bioengineering*, 63: 273–284.

Palazzia, E., Converti, A. 2001. Evaluation of diffusional resistances in the process of glucose isomerization to fructose by immobilized glucose isomerase. *Enzyme and Microbial Technology*, 28(2–3): 246–252.

Pandey, A., Nigam, P., Soccol, C. R., Soccol, V. T., Singh, D., Mohan, R. 2000. Advances in microbial amylases. *Biotechnology and Applied Biochemistry*, 31: 135–152.

Pavezzi, F. C., Carneiro, A. A., Bocchini-Martins, D. A. et al. 2011. Influence of different substrates on the production of a mutant thermostable glucoamylase in submerged fermentation. *Applied Biochemistry Biotechnology*, 163(1): 14–24.

Pazur, J. H., Ando, T. 1960. The hydrolysis of glucosyl oligosaccharides with α-D-(1,4) and α-D-(1,6) bonds by fungal amyloglucosidase. *Journal of Biology Chemistry*, 235: 297–302.

Pedersen, S., Dijkhuizen, L., Dijkstra, B. W., Jensen, B. F., Jørgensen, S. T. 1995. A better enzyme for cyclodextrins. *Chemtech*, 25: 19–25.

Poddar, A., Gachhui, R., Jana, S. C. 2012. Optimization of physico-chemical condition for improved production of hyperthermostable β-amylase from Bacillus subtilis DJ5. *Journal of Biochemical Technology*, 3(4): 370–374.

Ramesh, B., Reddy, P. R. M., Seenayya, G., Reddy, G. 2001. Effect of various flours on the production of thermostable β-amylase and pullulanase by *Clostridium thermosulfurogenes* SV2. *Bioresource Technology*, 76(2): 169–171.

Rhimi, M., Messaud, E. B., Borgi, M. A., Khadra, K. B., Bejar, S. 2007. Co-expression of l-arabinose isomerase and d-glucose isomerase in *E. coli* and development of an efficient process producing simultaneously D-tagatose and D-fructose. *Enzyme and Microbial Technology*, 40: 1531–1537.

Rose, I. A., O'Connell, E. L., Mortlock, R. P. 1969. Stereochemical evidence for a cis-enediol intermediate in Mn-dependent aldose isomerases. *Biochimica et Biophysica Acta-Enzymology*, 178(2): 376–379.

Roy, I., Gupta, M. N. 2004. Hydrolysis of starch by a mixture of glucoamylase and pullulanase entrapped individually in calcium alginate beads. *Enzyme and Microbial Technology*, 34(1): 26–32.

Rudiger, A., Jorgensen, P. L., Antranikian, G. 1995. Isolation and characterization of a heat-stable pullulanase from the hyperthermophilic archaeon *Pyrococcus woesei* after cloning and expression of its gene in *Escherichia coli. Applied and Environmental Microbiology*, 61(2): 567–575.

Rycroft, C. E., Jones, M. R., Gibson, G. R., Rastall, R. A. 2001. A comparative in vitro evaluation of the fermentation properties of prebiotic oligosaccharides. *Journal of Applied Microbiology*, 91(5): 878–887.

Rye, C. S., Withers, S. G. 2000. Glycosidase mechanisms. *Current Opinion in Chemical Biology*, 4: 573–580.

Sakiyama, F., Nitta, Y., Isoda, Y., Toda, H. 1989. Identification of glutamic acid 186 affinity-labeled by 2,3-epoxypropyl alpha-D-glucopyranoside in soybean beta-amylase. *Journal of Biochemistry*, 105(4): 573–576.

Saravanamurugan, S., Paniagua, M., Melero, J. A., Riisager, A. 2013. Efficient isomerization of glucose to fructose over zeolites in consecutive reactions in alcohol and aqueous media. *Journal of the American Chemical Society*, 135(14): 5246–5249.

Satyanarayana, T., Noorwez, S. M., Kumar, S., Rao, J. L. U. M., Ezhilvannan, M., Kaur, P. 2004. Development of an ideal starch saccharification process using amylolytic enzymes from thermophiles. *Biochemical Society Transactions*, 32: 276–278.

Sauer, J., Siguroskjold, B. W., Christensen, U. et al. 2000. Glucoamylase: Structure/function relationship and protein engineering. *Biochimica et Biophysica Acta (BBA)—Protein Structure and Molecular Enzymology*, 1543: 275–293.

Schray, K. J., Rose, I. A. 1971. Anomeric specificity and mechanism of two pentose isomerases. *Biochemistry*, 10(6): 1058–1062.

Shamala, T. R., Vijayendra, S. V. N., Joshi, G. J. 2012. Agro-industrial residues and starch for growth and co-production of polyhydroxyalkanoate copolymer and α-amylase by *Bacillus* sp. CFR-67. *Brazilian Journal of Microbiology*, 43(3): 1094–1102.

Sikorski, Z. E. 2007. *Chemical and Functional Properties of Food Components* (3rd Edition, p. 100). Boca Raton, FL: Taylor & Francis.

Silva, R. N., Asquieri, E. R., Fernandes, K. F. 2005. Immobilization of *Aspergillus niger* glucoamylase onto a polyaniline polymer. *Process Biochemistry*, 40(3–4): 1155–1159.

Sindhu, G. S., Sharma, P., Chakrabarti, T., Gupta, J. K. 1997. Strain improvement for the production of a thermostable α-amylase. *Enzyme and Microbial Technology*, 24: 584–589.

Sivaramakrishnan, S., Gangadharan, D., Nampoothiri, K. M., Soccol, C. R., Pandey, A. 2006. Alpha-amylases from microbial sources-an overview on recent developments. *Food Technology Biotechnology*, 44(2): 173–184.

Soni, S. K., Goyal, N., Gupta, J. K., Soni, R. 2012. Enhanced production of α-amylase from *Bacillus subtilis* subsp. *spizizenii* in solid state fermentation by response surface methodology and its evaluation in the hydrolysis of raw potato starch. *Starch*, 64(1): 64–77.

Sponsler, O. L. 1923. Structural units of starch determined by x-ray crystal structure method. *The Journal of General Physiology*, 5(6): 757–776.

Ståhlberg, T., Woodley, J. M., Riisager, A. 2012. Enzymatic isomerization of glucose and xylose in ionic liquids. *Catalysis Science & Technology*, 2: 291–295.

Strandberg, G. W., Smiley, K. L. 1971. Free and immobilized glucose isomerase from *Streptomyces phaechromogenes. Applied Microbiology*, 21: 588–593.

Strokopytov, B., Knegtel, R. M., Penninga, D. et al. 1996. Structure of cyclodextrin glycosyltransferase complexed with a maltononaose inhibitor at 2.6 angstrom resolution. Implications for product specificity. *Biochemistry*, 35: 4241–4249.

Subramaniyam, R., Vimala, R. 2012. Solid state and submerged fermentation for the production of bioactive substances: A comparative study. *International Journal of Science & Nature*, 3(3): 480–486.

Sugimoto, T., Horaguchi, K., Shoji, H. 2011. Indigestible dextrin stimulates glucoamylase production in submerged culture of *Aspergillus kawachii. Journal of Indian Microbiology Biotechnology*, 38(12): 1985–1991.

Suresh, C., Dubey, A. K., Srikanta, S., Kumar, U. S. 1999. Characterization of starch hydrolyzing enzyme of *Aspergillus niger. Applied Microbiology Biotechnology*, 51: 673–675.

Suzuki, Y., Hatagaki, K., Oda, H. A. 1991. Hyperthermostable pullulanase produced by an extreme thermophile, *Bacillus flavocaldarius* KP 1228, and evidence for the proline theory of increasing protein thermostability. *Applied Microbiology and Biotechnology*, 34(6): 707–714.

Svensson, B. 1994. Protein engineering in the α-amylase family: Catalytic mechanism, substrate septicity, and stability. *Plant Molecular Biology*, 25: 141–157.

Szerman, N., Schroh, I., Rossi, A. L., Rosso, A. M., Krymkiewicz, N., Ferrarotti, S. A. 2007. Cyclodextrin production by cyclodextrin glycosyltransferase from *Bacillus circulans* DF 9R. *Bioresource Technology*, 98: 2886–2891.

Takasaki, Y., Hayashida, A., Ino, Y., Ogawa, T., Hayashi, S., Imada. K. 1993. Cell-bound pullulanase from *Streptomyces* sp. No. 27. *Bioscience, Biotechnology, and Biochemistry*, 57(3): 477–478.

Tanakai, S., Ternishi, K., Yamada, T. 1997. Inner structure of potato starch granule. *Starch*, 49: 387–390.

Thoma, J. A., Spradlin, J. E., Dygert, S. 1971. Plant and animal amylases. In: *The Enzymes*, Boyer, P. D. ed. Academic Press, New York, (Vol. 5, pp. 115–189).

Totsuka, A., Nong, V. H., Kadokawa, H., Itoh, Y., Fukazawa, C., Kim, C. S. 1994. Residues essential for cata lytic activity of soybean beta-amylase. *Europe Journal of Biochemistry*, 221(2): 649–654.

Tükel, S. S., Alagöz, D. 2008. Catalytic efficiency of immobilized glucose isomerase in isomerization of glu cose to fructose. *Food Chemistry*, 111(3): 658–662.

Uitdehaag, J. C. M., Mosi, R., Kalk, K. H. et al. 1999. X-ray structures along the reaction pathway of cyclo dextrin glycosyltransferase elucidate catalysis in the alpha-amylase family. *Nature Structural Biology*, 6: 432–436.

Uitdehaag, J. C. M., van der Veen, B. A., Dijkhuizen, L., Dijkstra, B. W. 2002. Catalytic mechanism and prod uct specificity of cyclodextrin glycosyltransferase, a prototypical transglycosylase from the α-amylase family. *Enzyme and Microbial Technology*, 30: 295–304.

van der Veen, B. A., van Alebeek, G.-J. W. M., Uitdehaag, J. C. M., Dijkstra, B. W., Dijkhuizen, L. 2000. The three transglycosylation reactions catalyzed by cyclodextrin glycosyltransferase from *Bacillus circulans* (strain 251) proceed via different kinetic mechanisms. *European Journal of Biochemistry*, 267: 658–665.

Vieille, C., Epting, K. L., Kelly, R. M., Zeikus, J. G. 2001. Bivalent cations and amino-acid composition contrib ute to the thermostability of *Bacillus licheniformis* xylose isomerase. *European Journal of Biochemistry*, 268(23): 6291–6301.

Visuri, K., Klibanov, A. M. 1987. Enzymatic production of high fructose corn syrup containing 55% fructose in aqueous ethanol. *Biotechnology Bioengineering*, 30: 917–920.

Wallenfels, K., Bender, H., Rached, J. R. 1996. Pullulanase from *Aerobacter aerogenes*; production in a cell-bound state. Purification and properties of the enzyme. *Biochemical and Biophysical Research Communications*, 3: 254–261.

Weil, C. E., Burch, R. J., Van Dyk, J. W. 1954. An α-amyloglucosidase that produces β-glucose. *Cereal Chemistry*, 131: 150–158.

Wind, R. D., Liebl, W., Buitelaar, R. M. et al. 1995. Cyclodextrin formation by the thermostable α-amylase of *Thermoanaerobacterium thermosulfurigenes* EM1 and reclassification of the enzyme as acyclodextrin glycosyltransferase. *Applied Environmental Microbiology*, 61: 1257–1265.

Yamashita, M., Matsumoto, D., Murooka, Y. 1997. Amino acid residues specific for the catalytic action towards α-1,6-glucosidic linkages in *Klebsiella* pullulanase. *Journal of Fermentation and Bioengineering*, 84: 283–290.

Yu, D., Wu, H., Zhang, A. et al. 2011. Microwave irradiation-assisted isomerization of glucose to fructose by immobilized glucose isomerase. *Process Biochemistry*, 46(2): 599–603.

Zdziebło, A., Synowiecki, J. 2002. New source of the thermostable α-glucosidase suitable for single step starch processing. *Food Chemistry*, 79(4): 485–491.

Zhou, C., Xue, Y., Ma, Y. 2010. Enhancing the thermostability of α-glucosidase from *Thermoanaerobacter tengcongensis* MB4 by single proline substitution. *Journal of Bioscience and Bioengineering*, 110(1): 12–17.

Zhou, J., Liu, H., Du, G., Li, J., Chen, J. 2012. Production of α-cyclodextrin glycosyltransferase in *Bacillus megaterium* MS941 by systematic codon usage optimization. *Journal of Agriculture and Food Science*, 60: 10285–10292.

7

Enzymes in Bakeries

Cristina M. Rosell and Angela Dura

CONTENTS

7.1 Introduction

Baking industries provide a whole range of bakery products, of which a common characteristic is that they are mainly fermented, cereal-based products subjected to high-temperature processes for increasing their organoleptic properties and shelf life. The main ingredients in baked goods are flours from cereals such as wheat, corn, and sorghum. Other major cereal crops produced include rice and barley, oats, millet, and rye. In general, all flours contain valuable amounts of energy, protein, iron, and vitamins, but the degree of milling will influence the final nutritional content. Cereals still remain important to human nutrition because they lead to nutrient-dense baked goods, which have a worldwide important contribution in the daily intake of macro- and micronutrients. In addition, the consumption of cereal baked foods produces feelings of satiety, and their regular consumption with main meals appears to be a key driver of healthier dietary patterns (Aisbitt et al. 2008).

The almost ubiquitous consumption of baked goods all over the world confers those products a prominent position in international nutrition. Baked products account for a great part of our daily diet through the consumption of bread, breakfast cereals, cookies, snacks, cereal bars, cakes, and so on (Figure 7.1).

Many different types of bread formulations and many types of bakery goods have been developed so far due to increased awareness on the part of consumers as to the value of eating complex carbohydrates. In addition, bread is handy and very convenient for on-the-go consumers, available in any place and all year around, and affordable, and from a nutritional point of view, it is a source of energy in the form of starch in addition to the supply of dietary fiber and a range of vitamins and minerals.

Bread has changed in many ways since the days of our ancestors, going from grainy flat bread to having an aerated texture. Nevertheless, many types of so-called bread are detected around the world (reviewed by Rosell and Garzon 2014). Sliced pan bread is normally eaten in the United States, but

FIGURE 7.1   Different baked products.

Europe's consumers prefer crispy breads with crusty surface, such as a French baguette. The opposite sensory characteristics can be found in the steamed bread that is consumed in Asian countries. In India, flat bread called chapatti is consumed with the main meals whereas flat bread in Mexico is made of corn, and is named tortilla. In Finland and Germany, a very dark rye bread made of 100% rye flour is the most common bread. In Venezuela, arepas are considered a staple part of the diet, and they are eaten for breakfast, as a snack, or together with a meal. In Brazil, pão de queijo, small, round, cheese stuffed bread balls, are traditionally served for breakfast. It seems that the term "bread" comprises thousands of different types of breads around the world, and there is a large amount of diversity within each country.

Worldwide bread consumption accounts for one of the largest consumed foodstuff with average consumption ranging from 41 to 303 kg/year per capita, and it is an essential part of the human diet, enjoyed at various times of the day. The most popular kind of bread is white bread made from white wheat flour, which attained great prominence up to the 1960s, but its consumption underwent a steady decrease. On the other hand, a variety of bread products have picked up a market share because they have a superior taste and desirable sensory and nutritional qualities; they include multigrain bread, high fiber bread, cracked wheat bread, sourdough bread, milk bread, composite flour bread, high protein bread, wheat germ bread, and gluten-free bread. Increasing consumer interest in health has had an effect on the types and varieties of breads available, such as whole-grain and multigrain breads. Particularly in Europe, more than 250 types of variety of breads are available, and more than 1000 types of biscuits compete for space on grocer's shelves in Western countries.

In the past, the general market image of baked products and some media negativity impacted the prospects for any growth in the consumption of cereal-based foods related to the prevalence of diets that stress the avoidance of carbohydrate-rich foods. Currently, the global baked goods market has shown rapid recovery following the economic recession, recording strong growth in recent years. Factors fuelling market expansion include convenience, affordability, and the health benefits of baked goods products. Interestingly, the rise in cereal-based product consumption that started was also triggered by health concerns. Changing consumer habits have had an important impact on the bread industry. Nowadays, the commercial bakery industry has experienced some recent consolidation and is currently undergoing significant changes in how products are marketed and distributed.

One of the most extensive areas of ongoing research within the industry involves investigating methods of extending shelf life and preserving product freshness.

Bakeries are very active industries in launching innovative foods regarding processing, recipes, shapes, and so on. In this picture, enzymes have played an important role as processing or technological aids. Baking enzymes have become an essential part of the industry throughout the ages. Enzyme supplements have been largely used in baking to make consistently high-quality products by enabling better dough handling, providing antistaling properties, and allowing control over crumb texture and color, taste, moisture, and volume. Lately, a growing interest has been focused on enzymes owing to their generally recognized as safe (GRAS) notation because industries are facing changes due to the greater demand of chemical replacement. Enzymes with some contribution in bakery products can proceed from enzymes naturally present in flour, those associated with the metabolic activity of the dominant microorganism, or those intentionally added during the mixing step as technological or processing aids. The supplementation of flour and dough with enzyme improvers is a usual practice for flour standardization and also as baking aids. Over the years, enzymes gained importance in the bakery industry for their efficiency to modify dough rheology, gas retention, and crumb softness in bread manufacturing; to modify dough rheology in the manufacture of pastry and biscuits; to change product softness in cake making; and to reduce acrylamide formation in bakery products (Rosell and Collar 2008). The enzymes can be added individually or in complex mixtures, which may act in a synergistic, additive, or antagonistic way in the production of baked goods, and their levels are usually very low. There are many enzymes used to alter major and minor biomolecule structures and to achieve desired functionality (Rosell and Collar 2008). In baking, the most commonly used enzymes classes are hydrolases, such as amylases, proteases, hemicellulases, and lipases, and oxidoreductases, which comprise, among others, glucose oxidase and lipoxygenase. The enzymes most frequently used in bread making can be classified depending on the substrate on which they are acting. Particularly in bakeries, cereals are the raw commodity, and thus, when deciding on the inclusion of enzymes as processing aids, it is fundamental to know the cereal constituents that will be present and, consequently, ready to be used as enzyme substrates.

7.2 Enzymes in Baking Industries

There are many enzymes used as processing aids to alter major (starch, nonstarch, and proteins) and minor (lipids) cereal structures and to achieve desired functionality (Rosell and Collar 2008). In baking, the most commonly used enzyme classes fall within the category of oxidoreductases (i.e., glucose oxidase and lipoxygenase) and hydrolases, such as amylases, proteases, hemicellulases, and lipases. Nevertheless, due to the great number of enzymes used in bread making, the next sections will refer to enzymes grouped according to the substrate on which they are acting. Cereal grains contain 66%–76% carbohydrates; thus, this is by far the most abundant group of constituents. The major carbohydrate is starch (55%–70%) followed by minor constituents, such as arabinoxylans (1.5%–8%), β-glucans (0.5%–7%), sugars (~3%), cellulose (~2.5%), and glucofructans (~1%). The second important group of constituents is proteins, which fall within an average range of approximately 8%–11%. With the exception of oats (~7%), cereal lipids belong to the minor constituents (2%–4%) along with minerals (1%–3%) (Rosell 2012).

Considering the importance of the cereal constituents to understanding the role of enzymes in baking, a description of the chemical structure and main characteristics of carbohydrates, proteins, and lipids is included in the following sections in addition to the description of the enzyme actions (Table 7.1).

7.2.1 Starch-Degrading Enzymes

These enzymes are the most extended and commonly used as processing aids in the bakery industry (Table 7.1). The starch-converting enzymes include α-amylases (EC. 3.2.1.1), β-amylases (EC. 3.2.1.2), and glucoamylases (EC. 3.2.1.3). Some other enzymes involved in starch degrading but not commonly used in bread making are the debranching enzymes isoamylase (EC. 3.2.1.68) and pullanase (EC. 3.2.1.41) and the transferases amylomaltase (EC. 2.4.1.25) and cyclodextrin glycosyltransferase (CGTase) (EC. 2.4.1.19).

7.2.1.1 Starch

Starch is polymers of glucose linked to one another through the C1 oxygen, known as the glycosidic bond. It is the most important reserve polysaccharide and the most abundant constituent of many plants, including cereals. Two types of glucose polymers are present in starch: amylose and amylopectin. These two glucan polymers are organized into a complex, semicrystalline granular structure with a particle size ranging from 1 to 100 μm in diameter. Amylose and amylopectin have different structures and properties. Amylose is an essentially linear α-1,4-linked molecule containing very few branch points. There are usually between 1500 and 6000 glucose units in a single amylose molecule. Linear glucose chains, much like protein chains, tend to form helices. An important characteristic of amylose is its ability to form helical inclusion complexes with a number of substances—in particular polar lipids— and this can occur in native as well as in gelatinized starch. In contrast, amylopectin is a very large, highly branched polysaccharide of up to 3 million glucose units, consisting of short α-1,4-linked linear chains of 10–60 glucose units and α,1-6-linked side chains with 15–45 glucose units. The average number of branching points in amylopectin is 5%; the ratio of the two polysaccharides varies according to the botanical origin of the starch with typical levels of amylose and amylopectin of 20%–30% and 70%–80%, respectively (van der Maarel et al. 2002). Although an amylopectin molecule has many nonreducing ends, it has only one reducing end. The "waxy" starches contain less than 15% amylose, "normal" contains 20%–35%, and "high" contains (amylo-) amylose starches greater than approximately 40% (Tester et al. 2004).

The amylose and amylopectin in starch form a semicrystalline aggregate organized in granules. The size, shape, and morphology of these granules are dependent on the botanical origin. These granules are

TABLE 7.1

Enzymes Used in Bakery Industries (EC Numbers and Recommended Names of Baking Enzymes According to IUBMB)

EC 1	Oxidoreductases	(EC. 1.1) Acting on the CH–OH group of donors	(EC. 1.1.3) With oxygen as acceptor	Glucose oxidase (EC. 1.1.3.4)
		(EC. 1.10) Acting on diphenols and related substances as donors	(EC. 1.10.3) With oxygen as acceptor	Laccase (EC. 1.10.3.1)
		(EC. 1.13) Acting on single donors with incorporation of molecular oxygen (oxygenases)	Lipoxygenase (EC. 1.13.11)	
EC 2	Transferases	(EC. 2.3) Acyltransferases	(EC. 2.3.2) Aminoacyltransferases	Transglutaminase (EC. 2.3.2.13)
		(EC. 2.4) Glycosyltransferases	(EC. 2.4.1) Hexosyltransferases	CGTase (EC. 2.4.1.19)
				Amylomaltase (EC. 2.4.1.25)
EC 3	Hydrolases	(EC. 3.1) Acting on ester bonds	(EC. 3.1.1) Carboxylic ester Hydrolases	Lipases (EC. 3.1.1.3)
				Phospholipase A (EC. 3.1.1.4)
				Hemicellulase (EC. 3.1.1.73)
			(EC. 3.1.3) Phosphoric monoester hydrolases	Phytase (EC. 3.1.3.26)
		(EC. 3.2) Glycosidases	(EC. 3.2.1) Enzymes hydrolyzing O- and S-glycosyl compounds	α-Amylase (EC. 3.2.1.1)
				β-Amylase (EC. 3.2.1.2)
				Glucoamylase (EC. 3.2.1.3)
				Xynalase (EC. 3.2.1.8)
				Pullulanase (EC. 3.2.1.41)
				Isoamylase (EC. 3.2.1.68)
		(EC. 3.4) Proteases	Acting on peptide bonds: endopeptidases/exopeptidases	
		(EC. 3.5) Acting on carbon-nitrogen bonds, other than peptide bonds	(EC. 3.5.1) In linear amides	Asparaginase (EC. 3.5.1.1)

Source: http://www.chem.qmul.ac.uk/iubmb/.

approximately 30% crystalline. The amylopectin molecules within the granule provide the crystallinity whereas the amylose is present in an amorphous form. The degree of crystallinity ranges from 20% to 40% and is primarily caused by the structural features of amylopectin. It is thought that the macromolecules are oriented perpendicularly to the granule surface (Buleon et al. 1998) with the nonreducing ends of the molecules pointing to the surface. It has long been recognized that the functional properties of starch depend on a number of integrated factors, which include polymer composition, molecular structure, interchain organization, and minor constituents, such as lipids, phosphate ester groups (typical of potato amylopectin), and proteins. Starch is a major component of bread making and plays an important role in the texture and quality of the dough and bread. Heat treatment during the baking process causes structural and often molecular changes in the granular and polymeric structures of starch. Functional proportion of starch are directly influenced by hydrothermal treatment or processing conditions. During the gelatinization process, the semicrystalline nature of their structure is reduced or eliminated, and the granules break down; starch granules swell, and the molecular order of the granule is gradually and irreversibly destroyed (and thus the birefringence); a (limited) leaching of the polymer molecules, mainly amylose, and a (partial) granule disruption causes the amylose chains to solubilize, and a starch gel is formed, forming a viscous solution. At this point, starch is easily digestible. Gelatinization of granules in excess water causes large changes in the rheological properties of the system and has a major influence on the behavior and functionality of starch-containing systems.

7.2.1.2 Amylases

α- and β-Amylases are naturally present in cereal flours, but although β-amylases are always in a sufficient amount, very often flours require α-amylase supplementation. The α-amylases (α-1,4-glucanohydrolase) are endo-enzymes that randomly catalyze the cleavage of α-1,4-glycosidic bonds in the inner part of amylose or amylopectin chains into a series of branched and linear fragments, called dextrins. Only damage or gelatinized starches are susceptible to enzyme hydrolysis, but after long incubation, even intact starch granules can be attacked, leading to perforated structures (Figure 7.2). These enzymes can be obtained from cereal, fungal, bacterial, and biotechnologically altered bacterial sources (Rosell et al. 2001). Depending upon their origin, they have significant differences in pH, optimum temperature, thermostability, and other chemical stability. In a comparison study, Rosell et al. (2001) showed that α-amylases from cereals (wheat and malted barley) were less sensitive to the presence of ingredients, additives, and metabolites. Fungal amylase (*Aspergillius oryzae*) is the least temperature stable. Bacterial amylase (*Bacillus subtilis*) has high thermal stability; in consequence, some activity might remain in the bread even after baking. Because of that, an α-amylase of intermediate thermostability was isolated and recommended as an antistaling agent in baked goods. Overall, structural changes induced by α-amylase in dough and the bread microstructure are dependent on the α-amylase origin (Blaszczak et al. 2004).

β-Amylases are exo-enzymes that generate maltose by breaking every second α-1,4 linkage from the nonreducing ends of amylose, amylopectin, or dextrins. Therefore, β-amylase needs the previous action of α-amylase that breaks starch into smaller pieces or dextrins for β-amylase efficiency.

Fungal α-amylases are routinely supplemented in bread making to produce small dextrins for the yeast to use during dough fermentation and the early stage of baking. This results in improved bread volume and crumb texture. In addition, the small oligosaccharides and sugars, such as glucose and maltose, produced by these enzymes enhance the reactions for the browning of the crust and baked flavor. Through starch modification, amylases improve moisture retention, have a crumb-softening effect, and decrease

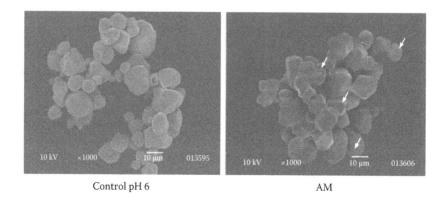

Control pH 6   AM

FIGURE 7.2 Scanning electron micrographs of cornstarch (control pH 6.0) and after treatment with fungal α-amylase (AM). Arrows indicate the enzymatic action.

staling. Amylase supplementation can occur at the flour mill or at the bakery in dough and sponges. Addition of amylases mainly aims at optimizing the amylase activity of the flour (i.e., flour standardization) and at retarding bread staling.

7.2.1.3 Glucoamylase

Glucoamylase or amyloglucosidase acts on starch, dextrins, and oligosaccharides by cleaving α-(1,6)-linkages from the nonreducing ends. It is widely used in the manufacture of glucose and for conversion of carbohydrates to fermentable sugars. β-Amylases and glucoamylases have been found in a large variety of microorganisms (Pandey et al. 2000).

Although damaged starch and gelatinized starch are more susceptible to amyloglucosidase attack, Dura et al. (2014) reported the effect of amyloglucosidase on starch at the subgelatinization temperature, revealing that this mechanism might be an alternative to obtaining porous starch. Amyloglucosidase can be more effective than α-amylase for degrading intact or gelatinized starch (Figure 7.3).

Isoamylase is a debranching amylase that hydrolyzes 1,6-α-D-glycosidic linkages of glycogen, amylopectin, and α- and β-limit dextrins, producing linear malto-oligosaccharides (Rani Ray 2011). For the same purpose, pullulanases are used to exclusively hydrolyze α,1-6 glycosidic bonds in pullulan (a polysaccharide with a repeating unit of maltotriose), amylopectin, and related oligosaccharides (van der Maarel et al. 2002).

Amylomaltase belongs to the 4-α-glucanotransferase group of the α-amylase family. The enzyme can produce cycloamylose or large-ring cyclodextrin through intramolecular transglycosylation or cyclization reactions of α-1,4-glucan. Amylomaltase and CGTase form a new α,1-4 glycosidic bond while the branching enzyme (EC. 2.4.1.18) forms a new α,1-6 glycosidic bond. CGTase is an enzyme that cleaves α-1,4 glycosidic linkages degrading the starch and produces intramolecular transglycosylation or cyclization reaction. In the process, cyclic and acyclic dextrins are originated, which are oligosaccharides of intermediate size. The cyclic products are cyclodextrins (CDs), namely α-, β-, and γ-CDs consisting of six, seven, and eight glucose monomers in cycles, respectively (Alves-Prado et al. 2008). Cyclodextrin glycosyltransferase catalyzes the conversion of starch and related α,1-4 glucans to cyclodextrins. These molecules have a hydrophilic exterior, which can dissolve in water, and a hydrophobic cavity can form inclusion complexes with a width variety of hydrophobic guest molecules (Astray et al. 2009). In addition to cyclization, the enzyme also catalyzes several intermolecular transglycosylations: coupling (opening of CD rings and transfer of resulting linear malto-oligosaccharides to acceptors), disproportionation (transfer of linear malto-oligosaccharides to acceptors), and saccharification (hydrolysis of starch). CGTases are found in a wide array of bacteria and archaea living under various environmental conditions, including high temperature (Biwer et al. 2002; Kelly et al. 2009).

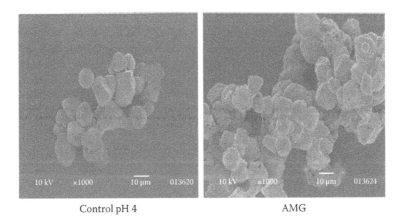

Control pH 4   AMG

FIGURE 7.3 Scanning electron micrographs of cornstarch (control pH 4.0) and after treatment with amyloglucosidase (AMG).

The effect of CGTase on wheat starch functionality was assessed by Gujral and Rosell (2004c). When the ability of starch to form a complex with lipids was tested in the presence of CGTase, α-cyclodextrin occurred in the highest concentration followed by β- and γ-cyclodextrins, and the resulting starch exhibited better emulsifying properties owing to the presence of cyclodextrins. In fact, a bakery improver was proposed containing CGTase and, optionally, a starch debranching enzyme (Rosell and Solis Nadal 2004).

A bacterial strain designated US132, isolated from Tunisian soil, has shown an ability to produce potent CGTase activity (Jemli et al. 2007) that can be used in bread making because it improves significantly the loaf volume and decreases the firmness of bread during storage.

7.2.2 NSP-Degrading Enzymes

7.2.2.1 Nonstarch Polysaccharides (NSP) in Cereal Flours

Polysaccharides other than starch are primarily constituents of the cell walls and are much more abundant in the outer than in the inner layers of the grains. Therefore, a higher extraction rate is associated with a higher content of NSP. From a nutritional point of view, NSPs are dietary fiber, which has been associated with positive health effects (Goesaert et al. 2008). There are two types of NSP, insoluble and soluble. The insoluble NSP content of most cereals is similar, and the composition of the water-soluble NSP varies. Pentosan refers to a polymer of pentose sugars, comprising xylan, xylobiose, arabinoxylan, and arabinogalactan. The insoluble pentosans are made up of the five-carbon sugars arabinose and xylose. Pentosans present in flour have an important role in bread quality due to their water absorption capability and interaction with gluten, which is vital for the formation of the loaf structure. The arabinoxylans (AX) are the major fraction from nonstarch polysaccharide (85%–90%), which consists of a backbone of β-(1,4)-linked xylose residues, which are substituted with arabinose residues on the C(O)-2 and/or C(O)-3 position (Dornez et al. 2009). They are present in water-extractable (WE-AX) and water-unextractable (WU-AX) forms. The former makes up 25%–30% of total AX in wheat and 15%–25% in rye (Izydorczyk and Biliaderis 1995). Both WU-AX and WE-AX are polydisperse polysaccharides with one general structure. Endosperm arabinoxylans contain a backbone of 3-1,4-linked o-xylopyranosyl residues, either unsubstituted or substituted at the C(O)-3 and/or the C(O)-2 position with monomeric OL-L arabinofuranose residues. The xylose "backbone" is very insoluble but becomes more soluble as more arabinose is bound. The molecule becomes less soluble as it increases in size (Goesaert et al. 2008). Their structure and aspect result in unique physicochemical properties that strongly determine their functionality in bread making. In particular, WE-AX form highly viscous aqueous solutions because AX are able to absorb 15–20 times more water than their own weight and, thus, form highly viscous solutions, which may increase the gas-holding capacity of wheat dough via stabilization of the gas bubbles. Numerous studies on the functional role of pentosans in dough development have been performed, studying their effect on bread properties (Ahmad et al. 2014; Butt et al. 2008; Denli and Ercan 2001; Maeda and Morita 2003; Oort et al. 1995).

β-Glucans are another type of nonstarchy polysaccharide and are present in a water-extractable and water-unextractable form with one general structure. β-Glucans are composed solely of glucose with β-1,3 and β-1,4 linkages without any branch points. The β-(1,3) linkages interrupt the extended, ribbon-like shape of β-(1,4)-linked glucose molecules, inducing kinks in the chain and making the β-D-glucan chains more flexible, more soluble, and less inert than cellulose (Goesaert et al. 2008). They are also called lichenins and are present particularly in barley (3%–7%) and oats (3.5%–5%). Although present in low concentrations, they can affect the quality of the flour in significant ways because of their ability to bind large amounts of water.

7.2.2.2 Nonstarch Hydrolyses

A number of other carbohydrate-degrading enzymes comprise those acting on nonstarch cereal carbohydrates. The use of some enzymes, such as pentosanases (hemicelluloses and xylanases), makes the dough easier to handle, and the resulting bread has a bigger loaf volume and an improved crumb structure (Oort et al. 1995). Pentosanases increase the WE-AX and the ratio of xylose to arabinose due to the enzyme-debranching action on arabinoxylans. The extent of their activity is dependent on the enzyme source; some act during fermentation and others increase their activity during baking (Jimenez and Martinez-Anaya 2000). For instance, *Bacillus subtilis* endoxylanase has selectivity for WU-AX whereas *Aspergillus aculeatus* endoxylanase has selectivity for WE-AX, and they mainly act during mixing and fermentation, respectively (Courtin et al. 2001). In addition, breads obtained in the presence of those xylanases confirm that WU-AX is detrimental for bread making whereas WE-AX has a positive impact on bread volume.

Hemicellulases (EC. 3.1.1.73) are a diverse class of enzymes that hydrolyze hemicellulose. Hemicellulose comes in several chemical forms and requires a significantly large number of enzymes and processes to break it down to glucose. In bread making, the hemicellulose reduces the processing time for dough and increases loaf volume without changing the size of the gas cells (Morita et al. 1997). It seems that the presence of hemicellulose slightly increases both gelatinization temperature and enthalpy of starch, but the viscosity coefficient and dough modulus of elasticity decreases.

Xylanase (EC. 3.2.1.8) is a glycosidase that catalyzes the endohydrolysis of 1,4-β-D-xylosidic linkages in xylan and arabinoxylan, yielding arabinoxylo-oligosaccharides. B-D-xylosidases (EC. 3.2.1.37) cleave xylose monomers from the nonreducing end of arabinoxylo-oligosaccharides. The arabinose residues are removed by α-L-arabinofuranosidases (EC. 3.2.1.55), and ferulic acid esterases (EC. 3.1.1.73) cleave ester linkages between arabinose residues and ferulic acid (Collins et al. 2006; Goesaert et al. 2008). Xylanases from families eight and 11 have been found to be effective as baking processing aids (Collins et al. 2006), and within the fungal xylanases, *Aspergillus oryzae* xylanase was the best bread improver and *Trichoderma reesei* xylanase the best antistaling improver (Baskinskiene et al. 2007). In bread making, xylanase increases gluten strength, improves handling and fermentative dough characteristics, and avoids excessive dough hardness and stickiness by interfering with arabinoxylan action on the gluten network (Courtin and Delcour 2002; Rosell and Collar 2008), and that effect is even greater in whole wheat bread (Shah et al. 2006). They also play a significant role in increasing the shelf life of bread and reducing bread staling although the extent of this effect depends on their origin (Jiang et al. 2005, 2010). Nevertheless, we must consider the existence of naturally present endoxylanases inhibitors in flours that are active in the bread making process and can affect the xylanase activity (Trogh et al. 2004).

In rye breads, xylanases make the dough soft and slack and require shorter fermentation times; they induce the fragmentation of cell walls and the release of amylose from the starch granule, and even the release of proteins from aleurone cells is promoted at a high dosage (Autio et al. 1996). These enzymes are also useful for obtaining fiber-enriched breads. Particularly the preliminary treatment of rye bran with hemicelullases or xylanases improves its performance in bread making, reducing the content of total dietary fiber and increasing the amount of soluble pentosans (Laurikainen et al. 1998). Namely, doughs with treated bran are softer and have lower stability, and in general, enzyme mixtures are more effective than individual incorporation of enzymes (Laurikainen et al. 1998).

Endo-glucanases have been also tested in rye breads, resulting in lower proofing times due to their action on the beta-glucans, but they do not influence the content or solubility of the arabinoxylans (Autio et al. 1996).

7.2.3 Enzymes Acting on Cereal Proteins

The different aspects of gluten functionality can be impacted by different enzymes, such as depolymerizing enzymes (proteases) and enzymes enhancing cross-linking reactions (transglutaminases).

7.2.3.1 Proteins

The average protein content varies among cereals (8%–11%) although it depends on the genotype (cereal, species, variety), the growing conditions (soil, climate, fertilization), and the time of nitrogen fertilization. A common means of classifying the different proteins of wheat and other grains was devised by Thomas Osborne in the early 1900s. The Osborne classification system (Osborne 1924) is based on solubility. Albumins, which are water soluble, make up about 15% of the proteins; globulins, which are relatively minor, make up only about 3% of the total protein and are soluble in salt solutions but insoluble in water; and finally, prolamins and glutenins are soluble in alcohol or acids, respectively. The Osborne fractionation does not provide a clear separation of wheat proteins that differ biochemically or genetically or in functionality during bread making. Nowadays, proteins are preferentially classified from a functional point of view in the nongluten and the gluten proteins (Hui et al. 2006). Gluten proteins (ca. 80%–85% of total wheat protein), the main storage proteins of wheat, are insoluble in water and can be divided into gliadins and glutenins. Gliadin, one of the two major components of the wheat gluten complex, is a prolamin that constitutes about 33% of all the proteins. Those have molecular weights between 30,000 and 80,000 (Veraverbeke and Delcour 2002) and are single chained and extremely sticky when hydrated. Gliadins are rich in proline and glutamine and have a low level of charged amino acids. Intrachain cystine disulphide bridges are present. The other major component of gluten, glutenins, is insoluble in neutral aqueous solutions, saline solutions, or alcohol. Glutelins make up approximately 55%–70% of the gluten complex and are larger (from 100,000 to several million) than gliadin molecules because of the high number of disulfide bonds connecting the subunits.

Sulfhydryl–disulfide interchanges are the major reactions responsible for the formation of wheat dough. It appears that a specific pattern of interaction between low molecular weight glutenin (<90 KDa) and high molecular weight glutenin (>90 KDa) is important for the development of a viscoelastic gluten (MacRitchie 1992). The disulfide bonding occurs toward the end of the chains, so in effect, the glutenin molecule is linear. The tertiary structure is thought to be one containing repetitive β-turns, which form a β-spiral structure. This type of structure is stabilized by hydrogen bonding and may explain the elastic nature of glutenin. When stress is applied, this stable conformation is disrupted, but it returns when the stress is absent.

Concerning the amino acid composition of the proteins, glutamine that contains an amide side group constitutes more than 40% of the amino acids comprising these proteins on a molar basis. Another amino acid comprising about 15% of gliadin and 10%–12% of glutenin is proline, which has a cyclic R group structure that puts a bend in a chain of amino acids. Another amino acid of significance is cysteine, which has the ability to form bonds connecting protein chains with their sulfur-containing R group, the so-called disulfide bonds. These three amino acids (glutamine, proline, and cysteine) play a major role in explaining the characteristics of gluten proteins.

7.2.3.2 Proteases

The reaction catalyzed by proteinases, proteases, or peptidases is not reversible because the peptide bonds and the original rheological condition cannot be restored. Protease from bacteria or fungi can be added to dough to reduce the size of the gluten polymers, making dough easier to mix, but it may compromise the dough's gas-holding ability during fermentation and baking (Goesaert et al. 2005). There are two main types of proteases (EC. 3.4) according to their site of action: exopeptidases and endopeptidases. Exopeptidases cleave the peptide bond proximal to the amino- or carboxy-termini of the substrate whereas endopeptidases cleave peptide bonds distant from the termini of the substrate (Rao et al. 1998). Based on the chemistry of their catalytic mechanism, proteases can be classified as serine, thiol, or cysteine, metallo, and aspartic proteases, which require a hydroxyl group (serine residue), a sulfhydryl group (cysteine residue), a metal ion (e.g., zinc), and a carboxylic function (aspartic acid residue), respectively, at the active site to function properly (Mathewson 1998). Proteases shorten the flow process time and increase the relaxation rate of the flow processes occurring at long time (Wikstrom and Eliasson 1998). In bakery products, tyrosinase makes the dough harder and less extendable, which results in breadcrumbs with uneven and large pore size but with a soft texture and increased volume (Selinheimo et al. 2007).

7.2.3.3 Transglutaminase

Transglutaminase (TGase) is a protein-glutamine γ-glutamyl-transferase (EC. 2.3.2.13), which catalyzes an acyl-transfer reaction between the γ-carboxyamide group of peptide-bound glutamine residues and a variety of primary amines (Motoki and Seguro 1998). When the ϵ-amino group of a peptide-bound lysine residue acts as a substrate, the two peptide chains are covalently linked through an ϵ-(γ-glutamyl)-lysine bond (Folk and Finlayson 1977). Thus, the enzyme is capable of introducing covalent cross-links between proteins (Nonaka et al. 1989). In the absence of primary amines in the reaction system, water becomes the acyl-acceptor, and the γ-carboxy-amide groups of glutamine residues are deamidated, becoming glutamic acid residues. The enzyme builds up new inter- and intramolecular bonds with the former comprising new covalent nondisulfide cross-links between peptide chains. Transglutaminases for baking applications are usually obtained from microbial origin. Depending on the accessibility of glutamine and lysine residues in the proteins, transglutaminase shows different activity (Gerrard 2002, Houben et al. 2012).

With regards to cereal proteins, TGase application has shown positive effects on wheat-based baked goods: It reduces the required work input, decreases water absorption of the dough (Gerrard et al. 1998), increases dough stability (Gottmann and Sproessler 1992), increases volume, improves the structure of breads, and strengthens breadcrumbs (Gerrard et al. 1998) and the baking quality of weak wheat flours (Basman et al. 2002).

TGase action on bread making is readily evident even after 15 min mixing, leading to a decrease in extensibility and an increase in the resistance to stretching due to the stronger protein network (Autio et al. 2005). With TGase supplementation, differences are also observed during proofing when the dough contains more small air bubbles than reference dough, and resulting breads have higher specific volume (Autio et al. 2005).

7.2.3.4 Asparaginase

Asparaginase (EC. 3.5.1.1) is an enzyme that catalyzes the hydrolysis of the amino acid L-asparagine to L-aspartic acid and ammonia. Free L-asparagine present in food is the main precursor of acrylamide, which is considered to be a probable human carcinogen (World Health Organization 2006). Acrylamide is formed from L-asparagine and reducing sugars primarily in starchy foods that are baked or fried at temperatures above 120°C. In a fried dough pastry model, Kukurova et al. (2009) confirmed that asparaginase almost completely converted asparagine to aspartic acid, and in consequence, acrylamide formation was reduced up to 90%; this enzyme also converted glutamine in glutamic acid, but neither action affected the browning and Maillard reactions. With that in mind, a enzymatic composition comprising asparaginase and one hydrolyzing enzyme has been reported for decreasing the levels of acrylamide in bakery products such as bread, pastry, cake, pretzels, bagels, Dutch honey cake, cookies, gingerbread, ginger cake, and crispy bread (De Boer and Boer 2007).

7.2.4 Lipid Enzymes

7.2.4.1 Lipids

In spite of the fact that lipids are minor constituents of cereals (1%–3% in barley, rice, rye, and wheat; 5%–9% in corn; and 5%–10% in oats on a dry-matter basis), they have a great impact on the baking performance. Cereal lipids consist of different chemical structures that vary in their content, depending on the cereal species. Cereal lipids have similar fatty acid compositions in which linoleic acid reaches contents of 39%–69%, and oleic acid and palmitic acid make up 11%–36% and 18%–28%, respectively (Delcour and Hoseney 2010).

Based on solubility in selective extraction conditions, they are classified as starch lipids and free and bound nonstarch lipids (Hoseney 1994). Nonstarch lipids (NSL) comprise about 75% of the total flour lipids and consist predominantly of triglycerides as well as of other nonpolar lipids and digalactosyl diglycerides. The lipids bound to starch (25%) are generally polar. Lysophospholipids, in particular lysophosphatidylcholine (lysolecithin), are the major constituents of the starch lipids. Lipids that are strongly bound in starch granules are essentially unavailable until starch is gelatinized (Van Der Borght et al. 2005). Complexes present in native starch (starch lipids) increase the temperature of gelatinization and, thus, prolong the oven spring. Inclusion complexes between amylose helices and polar lipids are responsible for the antistaling effect. The nonpolar lipids are mainly present in the free NSL fraction, and the glyco and phospholipids are mainly associated with proteins and present in the bound NSL fraction (Hoseney 1994). In fact, most of the free nonstarch lipids "bind" to gluten during dough mixing (Hoseney 1994).

Protein–lipid interactions in wheat flour dough also play an important role because both lipids and proteins govern the bread-making quality of flour. Lipids are important components in bread making as they provide a variety of beneficial properties during processing and storage, which reflects their overall diversity. Lipids in bread come from different sources, such as wheat flour lipids, shortenings, and surfactants (Pareyt et al. 2011). Lipids have a positive effect on dough formation and bread volume, namely polar lipids or the free fatty acid component of the nonstarch lipids, whereas nonpolar lipids have been found to have a detrimental effect on bread volume (MacRitchie 1983).

7.2.4.2 Lipases

The use of lipases in bread making is quite recent when compared to that of other enzymes. Lipases (EC. 3.1.1.3) possess the unique feature of acting at the interface between an aqueous and a nonaqueous phase, yielding mono- and diglycerides and free fatty acids. In particular, 1,3-specific lipases acting on 1- and 3-positions improve dough rheological properties and the quality of the baked product when added at 5000 LUS/g, specifically a strengthening effect on the gluten that leads to an increase in dough stability (Jensen and Drube 1998). Moreover, lipases induce an improvement in breadcrumb structure and freshness retention (Poulson et al. 2006, 2010) in addition to an increase of specific volume when added in combination with vegetable oil (Jensen and Drube 1998).

Furthermore, a lipase was found to increase expansion of the gluten network, to increase the wall thickness, and to reduce cell density, enhancing volume and crumb structure of high-fiber white bread (Stojceska and Ainsworth 2008).

Lipase was primarily used to enhance the flavor content of bakery products by liberating short-chain fatty acids through esterification. Along with flavor enhancement, it also prolongs the shelf life of most bakery products. Texture and softness could be improved by lipase as a dough and bread improver, which reduces crumb pore diameter, increases crumb homogeneity, and improves the gluten index in dough (Aravindan et al. 2007; Poulson et al. 2010), and these effects could be enhanced when full-fat soy meal (0.5%) is supplemented (Ertas et al. 2006). More particularly, lipases can be used as a dough strengthener to increase bread oven spring and specific volume (Moayedallaie et al. 2010) or to alter bread crust fracture behavior (Primo-Martin et al. 2008).

Other lipolytic enzymes may also improve bread making, such as phospholipase A (EC. 3.1.1.4), which liberates one fatty acid from phospholipids. This enzyme improves dough-handling properties, suppresses dough stickiness, and increases loaf volume (Inoue and Ota 1986; Sirbu and Paslaru 2007).

7.2.5 Other Enzymes That Can Be Involved in Bakery Performance

7.2.5.1 Phytase

Phytase (EC. 3.1.3.26) is naturally present in cereal flour, and lately there is a growing interest in its action, owing to its nutritional significance. Much of the phosphate contained in cereals is bound in a molecule called phytate that consists of phosphate groups bound to the six-carbon ring molecule inositol. Inositol is a vitamin, but neither the inositol nor the bound phosphate in phytate is available upon digestion in the human gut. Additionally, phytate chelates (binds) minerals such as calcium and magnesium, rendering them inaccessible to the consumer, and hence it is considered to be an antinutritional compound. Phytase liberates the phosphate and increases the bioavailability of all these nutrients (Afinah et al. 2010). In bread making, the supplementation of phytase results in a shorter fermentation process, an increase of the specific bread volume, and softer crumbs, in addition to the nutritional benefits associated to the reduction of phytate content (Haros et al. 2001a,b).

7.2.5.2 Oxidases

Different oxidases (lipoxygenase, sulphhydryl oxidase, glucose oxidase, polyphenoloxidase, and peroxidase) have been used for their beneficial effects on dough development and dough quality, mainly ascribed to dough strengthening and stabilization (Oort 1996), but also as dough-bleaching agents (Gelinas et al. 1998). Glucose oxidase (EC. 1.1.3.4) catalyzes the conversion of β-D-glucose to δ-D-1,5 gluconolactone, which converts spontaneously into gluconic acid and hydrogen peroxide. The hard oxidizing agent hydrogen peroxide interacts with the very reactive thiol groups of the proteins by forming disulphide bonds and promotes the gelation of water-soluble pentosans, changing the rheological properties of wheat dough (Hoseney and Faubion 1981; Primo-Martin et al. 2003). Nonetheless, side activities present in glucose oxidase commercial preparations might have a substantial effect on those reactions (Hanft and Koehler 2006).

The addition of glucose oxidase produces strengthening of wheat dough and an improvement of fresh bread quality. Studies on gluten proteins at the molecular level by high-performance capillary electrophoresis and at the supramolecular level by cryo-scanning electron microscopy revealed that glucose oxidase modifies gluten proteins (gliadins and glutenins) through the formation of disulfide and nondisulfide cross-links, mainly affecting high molecular weight glutenin subunits (Bonet et al. 2006). Treatment of dough with glucose oxidase increases the gluten macropolymer content due to disulphide and nondisulphide cross-linking (Steffolani et al. 2010) and leads to protein aggregates and broken segments (Indrani et al. 2003). Overdosage of glucose oxidase produces excessive cross-linking in the gluten network with a dramatic effect on the bread-making properties. Glucose oxidase decreases the relaxation time of the flow processes occurring in a short time, whereas it increases the relaxation time of the flow processes occurring over a long time (Wikstrom and Eliasson 1998).

A similar strengthening effect to the one obtained with glucose oxidase can be reached with hexose oxidase, which causes a dose-responsive reduction of thiol groups (Poulsen and Hostrup 1998). At the same dosage, hexose oxidase increases dough strength and bread volume more efficiently than glucose oxidase (Gul et al. 2009; Poulsen and Hostrup 1998).

Lipoxygenase (EC. 1.13.11) is present in high levels in cereal germ and catalyzes the peroxidation of certain polyunsaturated fatty acids containing *cis*, *cis*-1,4-pentadiene systems to produce peroxy-free radicals in the presence of molecular oxygen. Its typical substrate is linoleic acid containing a methylene-interrupted, doubly unsaturated carbon chain with double bonds in the *cis*-configuration. Many of the carotenoid pigments of flour are also subject to the action of lipoxygenase. In most flour-based products, the destruction of these pigments and the resulting bleaching effect are positive. Lipoxygenase from soy flour has been used as a bleaching agent as well as dough conditioner to improve the viscoelastic property of the dough (Junqueira et al. 2007).

Recent research (Zhang et al. 2013) evaluated the effects of a purified recombinant lipoxygenase produced extracellularly in *Bacillus subtilis* on flour, dough, and bread property variations. The textural and structural quality parameters of the enzyme-treated bread showed improved properties, including crumb color, specific volume, resilience, chewiness, and hardness.

Polyphenoloxidases that catalyze the polymerization of the phenolic compounds, such as catechol, pyrogallol, and gallic acid, to quinones by molecular oxygen are naturally present in the outer layers of the grains. When oxygen is present, phenolic compounds polymerize to form very dark pigments that lead to a problem known as "gray dough." Free radicals generated in these reactions are mainly responsible for the protein–protein cross-linking, ferulic acid–mediated, protein–arabinoxylan interactions and diferulated oxidation of arabinoxylans. Based on their substrate specificity, polyphenoloxidases are designated as tyrosinase (EC. 1.14.18.1), catechol oxidase (EC. 1.10. 3.2), and laccase (EC. 1.10.3.1).

Laccase is able to stabilize the dough structure by cross-linking proteins and proteins with arabinoxylans, resulting in a strong arabinoxylan network by oxidative dimerization of feruloyl esters through ferulic acid, and this property is mainly responsible for the improvement of wheat flour dough properties (Houben et al. 2012; Labat et al. 2000). Laccase decreases arabinoxylan extractability due to the cross-linking and increases the oxidation of sulfhydryl groups and the rate of protein depolymerization during mixing (Labat et al. 2000). As a result of laccase action, dough increases in strength and stability as well as reducing stickiness, which improves machinability and leads to a softer crumb in baked products (Caballero et al. 2007b; Selinheimo et al. 2006).

Laccase catalyzes the oxidation of various aromatic compounds, particularly o-diphenols, producing semiquinones with the concomitant reduction of molecular oxygen to water. The free radical may lead to polymerization of the semiquinones. In wheat flour, laccase catalyzes the polymerization of feruloylated arabinoxylans by dimerization of their ferulic esters (Figueroa-Espinoza and Rouau 1998). Protein cross-linking may also result from oxidation of sulfhydryl groups, resulting in disulphide bonds (Labat et al. 2000). In wheat flour-based bread-making applications, GO and LAC treatments increase dough strength and stability, increase loaf bread volume, and improve crumb structure and softness (Goesaert et al. 2005; Labat et al. 2000; Si 1994; Vemulapalli and Hoseney 1998). Therefore, glucose oxidase and laccase may well promote the formation of a protein and/or nonstarch polysaccharide network in oat batters, resulting in improved bread-making performances.

7.3 Effect of Enzymes in the Bread-Making Process

Bread is considered a staple food worldwide although it must be considered that this term encompasses a wide variety of products that differ in the making process, recipes, shapes, and so on. The common features to all these types of breads are first that they use cereals as a raw commodity and second that mixing, proofing, and baking are the main bread-making stages. In these stages, enzymes play a fundamental role, which might become evident at the dough level and/or in bread quality. However, it must be stressed that in addition to the role of the enzymes intentionally added to dough recipes, enzymes naturally present in the flour (endogenous enzymes) and the yeast enzymes also act on the bread-making process (Rosell and Collar 2008). Nevertheless, this section is focused on the enzymes used as processing aids in the bakery industry and the effects of those enzymes either in dough or bread (Table 7.2). The inclusion of enzymes in bakery processes does not require any changes in the processing line operation. Enzymes are commercialized in powder or liquid forms, they are usually blended with the rest of the ingredients during the mixing stage, and their action occurs during all bread-making stages.

For years, enzymes such as malt and fungal alpha-amylase have been used in bread making. Due to the changes in the baking industry and the awareness of consumers of more natural products, enzymes have gained real importance in bread making.

Bread making is one of the most common food processing techniques throughout the world. Bread processing can be divided into the following basic operations: mixing of wheat flour and water together with yeast and salt and other specified ingredients in appropriate ratios; fermentation and expansion of the bread dough; and after resting, the dough is divided into loaf-sized pieces, rounded, molded, placed on a baking tray, proofed, and baked. Through the process, a series of physical, chemical (baking time and temperature effect), and biochemical changes (including enzyme-catalyzed reactions) occur to obtain the final baked product. These changes include the development of a gluten network, which traps gas from the yeast fermentation and increases the volume; creation and modification of particular flavor compounds in the dough; evaporation of water; formation of a porous structure; denaturation of protein; gelatinization of starch; crust formation; and browning reactions. In bread making, mixing is one of the key steps that determines the mechanical properties of the dough, which has direct consequences on the quality of the end product. Bread dough is a viscoelastic material that exhibits an intermediate rheological behavior between a viscous liquid and an elastic solid. It is generally known that the formation of gluten, the combination of proteins, which form a large network during dough formation, plays a predominant role in dough development and textural characteristics of the finished bread (Rosell 2011). The number and type of sulfuric bonds between gluten proteins have a major effect on the properties of the three-dimensional glutenin network and the dough rheological properties. At optimal mixing, the dough develops because gluten proteins form a continuous network, and within gluten, gliadin and glutenin associate with different free polar lipid types, which contributes to increasing the gas cell stability throughout the bread-making process. During proofing or fermentation, yeast metabolism results in carbon dioxide release and growth of air bubbles previously incorporated during mixing, leading to expansion of the dough during fermentation and baking, and, in consequence, the bread structure and the volume and texture of the baked product. The yeast uses fermentable carbohydrate to produce carbon dioxide and alcohol during alcoholic fermentation, and the enzymes present in yeast and flour also help to speed up this reaction (Rosell 2011). Baking is the last stage of the bread-making process, in which dough is transformed into an edible final product with excellent organoleptic and nutritive characteristics. The different temperatures reached inside and outside the dough cause the formation of the crust and crumb of bread. The network-like structure of breadcrumb formation is mainly due to starch gelatinization and protein denaturation. The increase

TABLE 7.2

Brief Description of Enzymes Used in Bread Making and Their Effects

Enzymes	Dough	Bread	Staling
α-Amylases	Improve fermentation	Improve volume and crumb texture	Antistaling effect
CGT-ases		Improve volume Decrease firmness	Slower staling kinetics
Pentosanases	Improve handling	Improve volume and crumb texture	
Hemicellulases	Reduce processing time	Increase volume	
Xylanases	Increase gluten strength Improve handling and fermentative characteristics	Increase shelf life	Reduce bread-staling Increase shelf life
Proteases	Easy mixing May compromise gas-holding ability		
Tyrosinase	Increase hardness Decrease extensibility	Uneven crumbs and large pore size Soft texture and increase volume	
Transglutaminase	Decrease water absorption Increase stability Favors tiny gas cells	Increase volume and improve structure	
Lipases	Improve rheological properties Dough strengthener	Improve crumb structure Flavor enhancement Increase specific volume	Antistaling effect when combined with α-amylase
Phospholipase A	Improve handling and suppress stickiness	Increase volume	
Phytase	Shorten fermentation	Increase specific volume and soften crumbs Improve nutritional values	
Glucose oxidase	Dough strengthener	Improve quality	
Lipoxygenase	Improve viscoelasticity Bleaching agent		
Laccase	Improve stabilization and machinability Dough strengthener	Soften crumb	

in temperature and lower moisture content induce a nonenzymatic browning reaction, which results in the crust formation at the surface of the bread.

Many authors have described the effect of various enzymes and combinations of them that directly or indirectly improve the strength of the gluten network and so improve the quality of the bread (Caballero et al. 2007b, Rosell and Collar 2008; Selinheimo et al. 2006; Steffolani et al. 2010). Enzymes, such as xylanases, transglutaminases, glucose oxidase, laccase, or a combination of them all, have been used to increase gluten strength, and consequently, to improve dough functionality for bread making. Apart from the application of individual enzymes, such as amylases, proteases, or lipases, the use of enzymes in combination offers a wide range of alternatives to improve bread quality and antistaling or to provide nutritional benefits. The interactions between enzymes on the various components of bread dough are a very complex subject, and the wrong enzyme combination or even the wrong application rate can result in unfavorable effects on either the dough or the finished baked product (Collar et al. 1998; Rosell et al. 2001c).

7.3.1 Enzymes Added for Improving Processing

Enzymes have been used extensively for improving dough performance in bread making and bread features. Usually enzymes are blended with the rest of the ingredients during the mixing stage, and their action occurs during all bread-making stages. Additionally, the supplementation of enzymes during grain tempering has been proposed for increasing the milling extraction rate or for extending the period of enzyme action and improving the bread-making wheat flour performance. Rosell et al. (2003) tested the effect of transglutaminase and glucose oxidase supplemented during wheat tempering and investigated the performance of the treated flour. Enzymatically treated flour showed the same behavior as the dough supplemented with those enzymes, confirming the cross-linking effect by capillary electrophoresis and also by empirical rheological analysis. Later on, Yoo et al. (2009) reported the effect of an enzyme cocktail consisting of cellulase, xylanase, and pectinase added at different levels during tempering at the laboratory scale. Some of those enzymatic treatment combinations affected flour yield from the break rolls more than that from the reduction rolls; nevertheless, results did not allow the confirmation of a significant effect on milling yield. However, the effects observed on the dough and bread characteristics confirmed that enzymes diffuse through the outer layers and remain in the milled flour.

Changes in bread processing have prompted modifications in recipes and processing lines for answering consumer demands. Frozen dough and bake-off technology of frozen dough or frozen partially baked breads have been important changes in bread making, and processing aids have been developed with this change in mind. The dough matrix is subjected to much stress during freezing, frozen storage, and thawing due to ice-crystal formation, and because of this, an extra reinforcement of the gluten matrix is generally required. Glucose oxidase reduces the dough damage caused by frozen storage due to its strengthening action, which counteracts the depolymeration effect of gluten induced by ice-crystal formation and released reducing substances (Steffolani et al. 2012). A frozen dough additive containing protease, glucose oxidase, gluten, and stearoyl lactate has been recommended for producing frozen dough with improved extensibility and workability and also higher dough stability over a long period of time (Anon. 2000).

The combination of oxidases, such as glucose oxidase and peroxidase, has been proposed as an improver to form a stable gluten network structure with resistance to damage of ice crystals in frozen dough (Huang and Jia 2006). For other enzymes with cross-linking action, the TGase has been combined with a recombinant lipase (*Rhizopus chinensis* lipase) for improving the rheological properties and structure of frozen dough (Li et al. 2011). That enzyme combination increases the water-holding capacity, leading to dough in which starch granules are embedded within the gluten network after 35 days of frozen storage, and after baking, breads have a high specific volume, open network, and uniform crumb structure. Presumably, the new isopeptide bonds within the proteins catalyzed by TGase give sufficient gluten stability to mitigate the damage caused by dough freezing (Steffolani et al. 2012). For the same application, Hsing et al. (2010) recommended a combination of hemicellulose and/or endoxylanase with ascorbic acid, which delayed the staling and also improved the quality of the freshly baked bread.

7.3.2 Enzyme Combinations for Improving Bread Quality

Numerous enzyme combinations have been reported focused on improving bread technological or instrumental quality regarding crumb and crust properties.

For bread made with durum and bread wheat flour, α-amylases, proteases, and a mixture of α-amylase-protease was suggested by Pena-Valdivia and Salazar-Zazueta (1997) confirming the dose-dependent effect of the enzymes on the loaf volumes and the best performance of fungal over bacterial α-amylase in durum wheat flour. Those authors recommended a mixture of α-amylase-protease for producing breads from durum wheat flour with more acceptable quality characteristics.

Martinez-Anaya and Jimenez (1997a) studied nine commercial enzyme preparations and two laboratory-designed mixes with amylase and/or pentosanase activity for obtaining bread with improved volume, a soft crumb, and high aroma intensity. The individual or combined addition of commercial amylases, xylanases, lipases, and glucose-oxidases modifies the rheological behavior of dough; in general, enzyme supplementation results in softening and weakening of dough immediately after mixing although it could be counteracted with glucose oxidase (Martinez-Anaya and Jimenez 1997b). Among the enzymes mentioned previously, pentosan-degrading enzymes cause the main changes, reducing consistency and increasing stickiness (Martinez-Anaya and Jimenez 1998), and pure lipase preparations are the least significant when compared with unsupplemented dough (Martinez-Anaya and Jimenez 1997b). Enzymes act quickly and change the textural properties of dough immediately after mixing and continue during resting (Martinez-Anaya and Jimenez 1998). Similar effects are observed in sourdough wheat bread making (Martinez-Anaya and Devesa 2000).

Due to the complexity of finding the right and optimum enzyme combinations, a central composite design was initially proposed by Collar et al. (2000), comprising bacterial α-amylase, intermediate thermostability α-amylase, xylanase, lipase, and glucose-oxidase. Results showed that xylanase resulted in softer, stickier, and less adhesive dough with higher gas retention capability but with weakened gluten, and when combined with intermediate thermostability, amylase decreased fermentation time and increased dough extensibility. According to these authors, the incorporation of xylanase with glucose oxidase is not recommended because glucose oxidase counteracts the softening effect of xylanases (Collar et al. 2000). This antagonistic effect has been elucidated on the basis of the glucose oxidase activity by Primo-Martin et al. (2005), who explained the glucose action through the formation of protein disulfide bonds and cross-links between arabinoxylans; the latter negatively affected the bread quality. The supplementation of xylanases combined with glucose oxidase counteracts that effect, breaking arabinoxylan complexes and releasing small arabinoxylan fragments that interfere with the cross-linking (Primo-Martin et al. 2005). This action could render a redistribution of the water from the nonstarch carbohydrates to gluten, leading to more extensible gluten (Dagdelen and Gocmen 2007). Conversely, it has been reported that glucose oxidase acts synergistically with phospholipase, ensuring roll shape uniformity and tolerance to process changes in addition to greater bread volume (Novozymes 2001).

Similar to the effects observed with glucose oxidase, laccase leads to dough hardening, but this effect decreases over the dough resting period due to the laccase-mediated depolymerization of the cross-linked arabinoxylans network, which could be hindered by the addition of xylanase (Selinheimo et al. 2006).

Gluten cross-linking enzymes can actively contribute to confer functional properties to dough. Caballero et al. (2007b) used a number of cross-linking enzymes (TGase, glucose oxidase, and laccase) along with polysaccharide and gluten-degrading enzymes in bread making (α-amylase, xylanase, and protease), systematically analyzing the individual and synergistic effects in bread-making systems. Moreover, they observed a significant synergistic effect within the pairs of glucose oxidase–laccase, glucose oxidase–pentosanase, amylase–laccase, amylase–protease, and pentosanase–protease (Caballero et al. 2007a). Further on, Steffolani et al. (2010) investigated the effect of glucose oxidase, TGase, and pentosanase on wheat protein and bread quality.

Bread-making ingredients greatly affect enzyme activities. For instance, TGase in combination with pectin and diacetyl tartaric acid ester improves water absorption and makes highly cohesive dough and gluten strength in addition to suitable pasting performance during cooking (Collar and Bollain 2004).

Xylanase in combination with α-amylase and glucose oxidase, along with ascorbic acid, was reported to exhibit an excellent dough-strengthening effect. The effect of this combination of enzymes to weaken wheat flour was evaluated by Shafisoltani et al. (2014). The authors concluded that a combination of optimum levels of the three enzymes resulted in dough with low stickiness and bread with a higher specific volume, higher quality texture, better shape, and higher total score in a sensory evaluation test.

Less attention has been paid to bread crust although lately different enzymes have been tested for extending the crispiness of baked bread. Hamer and Primo-Martin (2006) proposed the application of an enzyme material with proteolytic activity onto the outside surface of the dough or partially baked bread to retain the crispy perception of the crust for an extended period of time. Lately, amyloglucosidase sprayed onto the partially baked surface has also been reported for modulating the properties of the bread crust and increasing its crispness (Altamirano-Fortoul et al. 2013). The amyloglucosidase treatment affects the color of the crust and decreases the moisture content and water activity of the crust and the force required for crust rupture. It seems that amyloglucosidase induces the disruption of the crust structure by removing the starchy layer that covers the granules and increasing the number of voids, leading to texture fragility.

7.3.3 Enzymes Acting as Antistaling Agents

Bread staling is a complex phenomenon that happens during storage, and it is largely caused by water migration and transformations that occur in the starch leading to crumb hardening. Staling implies a relatively short shelf life for fresh bakery products, leading to loss of consumer acceptance. Alterations related to this phenomenon include an increase in moisture in the crust (loss of crispiness), an increase in crystallinity in the starch granule, an increase in crumb firmness, a loss of organoleptic properties in the loaves, and the crumb's loss of water-holding capacity (Gray and BeMiller 2003; Ribotta and Le Bail 2007). Specific amylases that produce maltotriose and maltotetraose (Min et al. 1998) or maltopentaose (Hyuck et al. 2005) have been reported as effective antistaling agents for retarding amylopectin retrogradation. It seems that the presence of a low degree of polymerization maltodextrins (DP 3-7) (Rojas et al. 2001) or even with a polymerization degree up to 66 (Defloor and Delcour 1999) might be responsible for the antistaling effect of α-amylases. However, Hug-Iten et al. (2001) explained the antistaling action of α-amylase is due to its capacity to partially degrade amylopectin, hindering its recrystallization, and to slightly hydrolyze amylose by an endo-mechanism, which induces a part crystallization of amylose that contributes to the kinetic stabilization of the starch network during aging (Hug-Iten et al. 2003). Later on, Palacios et al. (2004) stated that the antistaling action of α-amylase might be a consequence of the decrease in the rate and extent of starch retrogradation that hampers the formation of double helices, but also the antistaling efficiency of amylases have been related to water immobilization (Goesaert et al. 2009). Additionally, Goesaert et al. (2009) determined the structure of starch in amylase-supplemented breads and correlated these data with crumb firming and amylopectin recrystallization. Results showed that the use of bacterial endo-α-amylase from *B. subtilis* and maltogenic α-amylase from *B. stearothermophilus* reduced crumb firming during storage.

Apart from α-amylase, it has been suggested that CGTase from *B. stearothermophilus* ET1 has potential application in bread making through modulation of the cyclizing and hydrolytic activities (Sung-Ho et al. 2002) because, by protein engineering, it was possible to replace residues Phe191 and Phe255 of CGTase by glycine (Phe191Gly-CGTase) and isoleucine (Phe255Ile-CGTase), respectively, yielding reduced cyclization activity.

α-Amylase antistaling effect can be more effective when combined with lipase, owing to the formation of thermostable amylose–lipid complexes (Leon et al. 2002), which can be enhanced with the addition of distilled monoglyceride, although motivating the formation of amylose–lipid complexes favors the water migration from crumb to crust (Purhagen et al. 2011). In addition, a synergistic effect between intermediate thermostability α-amylase and TGase has been evident in white and whole wheat breads stored up to 20 days (Collar and Bollain 2005). Supplemented breads with both enzymes had slower staling kinetics, particularly for hardness, cohesiveness, chewiness, resilience, and also sensory deterioration (Collar and Bollain 2005).

Different combinations of antistaling agents have been suggested for extending the freshness of breads. Armero and Collar (1996) proposed the combination of emulsifiers, hydrocolloids, and α-amylase for retarding staling in white and whole meal breads, but different synergistic and antagonistic effects were encountered (DATEM*SSL, α-amylase*SSL, α-amylase*HPMC).

The effect of pentosanase, bacterial α-amylase, and lipase added individually, or in their blends on the shelf life of white pan bread during storage, was reported by Gil et al. (1999), who showed that bacterial α-amylase, specially blended with pentosanase and lipase, increased bread crumb elasticity and reduced firmness during storage up to 72 h. That enzyme mixture has been also effective in high-fiber wheat bread, in which enzymes and sourdough lower the changes in firmness, amylopectin crystallinity, and rigidity of polymers (Katina et al. 2006).

The influence of three different starch-degrading enzymes, a conventional α-amylase, a maltogenic α-amylase, and β-amylase were investigated by Hug-Iten et al. (2003). Results suggested that the enzyme-induced fast formation of a starch network contributes to a kinetic texture stabilization, which prevents structure collapse and hinders rearrangements in the starch (amylose) phase, thus contributing to the antifirming effect by preventing further firming over time. Rosell and Collar (2008) described the effect of different enzymes on bread staling, highlighting the antistaling effect of amylases, the combination of bacterial α-amylase with TGase or xynalases, which led to breads with softer and less chewy fresh crumbs, retarding the starch retrogradation and consequent staling effect on bread samples. Barrett et al. (2005) also confirmed the antistaling effect of two amylases reducing the amount of starch recrystallization, and that effect was even intensified in the presence of 6% glycerol.

Nonstarch carbohydrases (cellulase, xylanase, and glucanase) decrease the starch retrogradation, but xylanases have the greatest antistaling effect (Haros et al. 2002). A kinetic study of the firmness during storage using the Avrami equation showed that carbohydrases reduced the rate of bread firming, but the simultaneous analysis of the hardening and starch retrogradation indicated that the antistaling effect of xylanase might be due to the retardation in the starch retrogradation although, in the case of cellulase and beta-glucanase, some other mechanism should be also involved (Haros et al. 2002).

Xylanases used at optimum levels play a significant role in increasing the shelf life of bread and reduce bread staling (Butt et al. 2008). When xylanase is added in a binary combination with high ester pectin and/or hydroxypropyl methylcellulose, a shortened effect on dough extensional parameters is observed (Rosell and Collar 2008). Caballero et al. (2007b) studied the changes in dough rheology, quality, and shelf life of breads caused by the use of cross-linking enzymes (TGase, glucose oxidase, laccase) and hydrolytic enzymes (α-amylase, xylanase, protease) and found that bread staling during storage was increased with TGase but was inhibited by amylase, xylanase, and protease.

In a recent study, Oliveira et al. (2014) investigated the influence of the enzyme cocktail with xylanolytic activity from *T. aurantiacus* fungus CBMAI 756 on bread quality and the staling process as well as to identify the specific products released through its activity on wheat flour arabinoxylans. The main product released through enzyme activity after prolonged incubation was xylose, indicating the presence of xylanase; however, a small amount of xylobiose and arabinose also confirmed the presence of xylosidase and α-L-arabinofuranosidase, respectively. The enzyme mixture in vitro mainly attacked water-unextractable arabinoxylan, contributing to a beneficial effect in bread making. The use of an optimal enzyme concentration improved bread quality by increasing specific volume and reducing crumb firmness.

7.3.4 Enzyme Combinations for Different Bread Specialties

Some bread specialties require specific enzyme combinations adapted to the bread-making processes. Wen et al. (2003) reported the effect of α-amylase on steamed bread, showing an improvement in the elongation of the dough and the elasticity of the bread. They did find a minor effect when flour was supplemented with xylanase. Bacterial maltogenic α-amylase also gave adequate properties to rice-steamed bread, which prevented the staling and maintained the flavor characteristics and the chewy taste (Ahn 2008). Another improver defined for making this type of bread comprised fungal α-amylase, xylanase, ascorbic acid, monoglycoside, and soybean powder (Jia et al. 2006). Specific thermophilic xylanase (131 U/ml) from *Chaetomium* sp. added in the range of 2.5–5.0 ppm produced up to a 25% increase in the specific volume of steamed bread and almost the same decrease in the firmness (Jiang et al. 2010).

Turkish heart bread has been also supplemented with α-amylases obtaining differences depending on the enzyme source (from cereal and fungal sources) (Dogan 2003).

Unleavened Indian flat bread, namely South Indian parotta, improves the overall quality score when flour was supplemented with proteinase. It is likely this enzyme breaks large protein fibrils, leading to small ones that improve the continuous gluten formation (Prabhasankar et al. 2004). α-Amylase and xylanase have also been tested as an antistaling agent in other unleavened Indian flat bread made from whole wheat flour, chapatti, and both enzymes inhibited staling to different extents (Shaikh et al. 2008). Moreover, these enzymes can change the nutritional, nutraceutical, and antioxidant properties of chapatti. α-Amylase and xylanase improve the antioxidant properties of this bread, particularly the amylase that increases antioxidant phenolic acids, such as caffeic, gentisic, and syringic acids, in addition to the phenolic acids as well as the soluble dietary fiber (Hemalatha et al. 2012). Nevertheless, at least one starch-degrading enzyme is advisable to be added for retarding flat bread staling and preferably a mixture of an α-amylase and a glucan 1,4-α-glucosidase (Forman and Evanson 2011).

α-Amylase has been tested as an improver in teftoon, a flat bread prepared from whole wheat flour by hand sheeting and baking (Koocheki et al. 2009). The addition of amylase and emulsifiers decreased the force required to tear the fresh bread, and the best antistaling effect was observed with diacetylated tartaric acid esters of mono- and diglyceride of fatty acids. The optimization of the wheat α-amylase activity determines the softness and staling of the teftoon, and the right α-amylase level can be obtained by germinating wheat flour for a period of 72 h (Moharrami and Shahedi Baghkhandan 2011).

7.3.5 Enzyme Supplementation for Producing Nutritional Benefits

Other enzymes that can be used in bread making, leading to nutritional improvement are the phytases (Haros et al. 2001a,b). Fungal phytases improve the firmness and chewiness of whole meal wheat bread over a period of 3 days of storage, and the extent of the effect was dose dependent (Zyla et al. 2005). Moreover, the supplementation of phytase induces a decrease of phytate content and an increase in the in vitro digestibility of phosphorus, calcium, carbohydrates, and proteins (Zyla et al. 2005).

The impact of the exogenous fungal phytase on the phytate content of whole meal wheat breads has been reported by comparing different bread-making processes (Rosell et al. 2009). There was significant interaction between the bread-making process and enzyme addition concerning specific volume, crumb lightness, and crumb texture. Freezing and frozen storage of whole meal bread in the presence of fungal phytase decreased significantly the phytate content, independently of the bread-making process that was followed. Furthermore, Penella et al. (2008) studied the effect of a combination of fungal phytase and fungal α-amylase, bran content, and particle size distribution of bran on the technological performance of dough and on the amount of phytic acid remaining in the bread. The authors concluded that the combination of bran with amylolytic and phytate-degrading enzymes could be recommended for overcoming the detrimental effect of bran on the mineral availability (phytase) and on the performance of the dough (α-amylase). Fungal phytase can improve nutritional and bread-making performance of whole wheat bread; however, not all the phytates are hydrolyzed (Haros et al. 2001a,b).

7.4 Enzymes in "Gluten-Free" Products

A variety of cereal foods are produced from wheat flour, such as bread, noodles, bulgur, and couscous, because gluten, which is formed by combining wheat proteins, has unique viscoelastic properties. However, many people suffer from gluten-related disorders, which prompted the development of gluten-free products. One of the main challenging aspects in gluten-free bread making is the production of high-quality bread with good structural properties using gluten-free cereals, such as rice or corn, or some pseudocereals, such as quinoa or amaranth. However, the proteins of those flours are incapable of retaining carbon dioxide during fermentation due to the lack of viscoelastic properties, which forced researchers to find technological alternatives to mimic gluten functionality (Arendt et al. 2008). In this respect, the addition of different groups of enzymes in gluten-free bread production plays an important role in achieving desirable properties to improve the quality of the final product, mainly by cross-linking proteins (Rosell 2009).

7.4.1 Transglutaminase

In this respect, transglutaminase has been the enzyme most extensively proposed for creating protein cross-links in rice flour (Gujral and Rosell 2004a). Nevertheless, the protein network created by the TGase is greatly dependent on the protein origin, its thermal compatibility, and the dosage of the enzyme (Marco and Rosell 2008c). Gujral and Rosell (2004a) initially proposed the hypothesis that the enzymatic creation of a protein network in gluten-free dough might mimic gluten functionality. With that aim, the authors studied the addition of increasing amounts of TGase (0.5%, 1.0%, or 1.5% w/w) to rice flour obtaining a progressive increase of the viscous and elastic moduli, but a higher bread volume and softer crumb was obtained with 1.0% TGase and further improvement was obtained with the addition of 2% HPMC.

Nonetheless, the flour source has great influence on the extent of activity of TGase as reported by Renzetti et al. (2008) when compared with the effect of that enzyme on six different gluten-free cereals (brown rice, buckwheat, corn, oat, sorghum, and teff). Three-dimensional confocal laser scanning micrographs confirmed the formation of protein complexes by TGase. Batter fundamental rheological analysis and bread quality indicated an improving effect of 10 U TGase on buckwheat and brown rice batters and breads and a detrimental effect on corn batters but with increased specific volume and decreased crumb hardness and chewiness on corn breads. Conversely, transglutaminase was not effective in making breads from oat, sorghum, or teff (Renzetti et al. 2008). Nevertheless, Onyango et al. (2010b) obtained changes in the rheological properties of gluten-free batters of sorghum blended with pregelatinized cassava starch when adding TGase, specifically, the enzyme decreased the resistance to deformation and compliances and augmented zero shear viscosity and elastic recovery.

Surely, the amount and nature of the proteins present in those flours, considering that, for TGase action, lysine and glutamic acid are needed, must explain the differences encountered among flours. In answer to that possible protein deficiency for TGase activity, protein supplementation to gluten-free flours was proposed to increase the amount of substrate for the enzyme (Marco and Rosell 2008b,c; Marco et al. 2007, 2008).

Marco and Rosell (2008b) reported the effect of transglutaminase on rice flour functionality when it was blended with protein isolates from different sources (pea, soybean, egg albumen, and whey proteins). The cross-linking action of TGase was confirmed by the decrease in the amount of free amino acids. Pea, soybean, and whey proteins decreased the final viscosity, thus affecting amylose recrystallization during cooling. Viscoelastic moduli of the dough were significantly modified, but whereas pea and soybean increased this parameter, egg albumen and whey protein dramatically decreased it. Even a combination of soybean and pea protein optimized with an experimental design was recommended for obtaining a better-structured protein network (Marco and Rosell 2008c). In fact, electrophoretic studies confirmed that TGase action resulted in the formation of isopeptide and disulfide bonds mainly within albumins and globulins and, to a lesser extent, within glutelins, and in consequence, large aggregates between pea and rice proteins were present (Marco et al. 2007). Similarly, soybean proteins were cross-linked with rice proteins through the formation of new intermolecular covalent bonds catalyzed by transglutaminase and the indirect formation of disulfide bonds among proteins, mainly involving β-conglycinin and glycinin of soybean and the glutelins of the rice flour although albumins and globulin also participated (Marco et al. 2008). The combination of gluten-free flours with legume flours has the additional benefit of improving the amino acid balance because the action of TGase does not have any nutritional repercussion.

Concerning the effect of that strategy on bread making, Marco and Rosell (2008a) proposed the protein enrichment of gluten-free breads with the additional benefit of creating a protein network through the use of transglutaminase. For doing that, those authors optimized the amount of water needed to hydrate the system, the level of protein and enzyme, and also included hydroxypropyl methylcellulose for additional structural strength and a more open aerated structure. Although soybean proteins reduced the specific volume of the bread, scanning electron micrographs confirmed the participation of those proteins in the network created by the TGase.

Houben et al. (2012) proposed the addition of TGase in the presence of skim milk powder and egg protein powder to form protein networks that yielded stiff crumbs. An experimental design carried out to obtain protein-enriched rice-based gluten-free breads showed that the combination of TGase (1.35 U of enzyme/g of rice flour protein), egg albumen (0.67 g/100 g of flour), and casein (0.67 g/100 g of flour) yielded the highest specific volume of rice-based bread with the lowest crumb hardness (Storck et al. 2013). With the same aim, different extruded gluten-free flours (rice, potato, corn, buckwheat), proteins (egg-white powder, soybean isolate, caseinate), and TGase were included in an experimental design, showing the best quality was in the bread with extruded buckwheat extrudate, egg-white powder, and 10 IU TGase per gram of protein (Smerdel et al. 2012).

Gluten-free jasmine rice bread improved its specific volume with a simultaneous reduction in the crumb hardness when up to 10% rice flour was replaced by pregelatinized tapioca starch and TGase was added up to 1% (Pongjaruvat et al. 2014).

Nevertheless, some concern about the use of microbial TGase arose due to its homology to tissue TGase that is mediated in celiac disease. A study carried out by Cabrera-Chavez et al. (2008) evaluated the reactivity of the IgA of celiac patients to proteins of wheat and gluten-free breads treated with microbial TGase, showing that the reactivity was higher against prolamins of TGase-treated breads due to two individual patients' sera.

7.4.2 Protease

Even though the creation of a protein network was primarily thought to be the best alternative for improving gluten free bread quality, proteases also have been proposed as processing aids. Proteases induce the release of low molecular weight proteins from macromolecular protein complexes, which results in lower complex modulus and initial viscosity in addition to a decrease in paste viscosity and breakdown but an increase in the paste stability (Renzetti and Arendt 2009). Batters show lower resistance to deformation during proofing and in the early stages of baking while preserving its elasticity. When added to brown rice flour, breads significantly increase the specific volume with a parallel decrease in crumb hardness and chewiness (Renzetti and Arendt 2009).

Kawamura-Konishi et al. (2013) reported improved crumb appearance, high volume, soft texture, and a low staling rate when a commercial protease from *Bacillus stearothermophilus* (thermoase) was added to rice flour. Presumably, this protease majorly hydrolyzed albumins and globulins, leading to many cellular structures in the breadcrumb.

7.4.3 Oxidases

Oxidases such as glucose oxidase and laccase could also yield the formation of networks.

Glucose oxidase has been supplemented to rice flour, yielding an increase in specific volume with a simultaneous reduction of the crumb hardness (Gujral and Rosell 2004b), owing that effect to the protein cross-linking as it reveals the decrease in the amino and thiol groups. In addition, electrophoresis analysis confirmed changes on glutelins (Gujral and Rosell 2004b).

Renzetti et al. (2010) reported the increased specific volume and softening crumb effect of commercial preparations of laccase (0.01%) and protease (0.001% and 0.01%) containing endo-β-glucanase side activity for making gluten-free oat flour. The authors attributed the improvement due to the increase in batter softness, deformability, and elasticity in part due to the β-glucan depolymerization. In oat-based breads, the extensive protein hydrolysis during baking may have improved the functionality of the soluble protein fraction. Flander et al. (2011) also observed an improvement in specific volume of oat bread with the combination of *Trametes hirsute* laccase and xylanase, although crumb softness remained unaltered.

7.4.4 Other Enzymes

A different alternative for improving the quality of gluten-free breads has been the addition of different α-amylases. Gujral et al. (2003a) investigated the use of CGTase as a processing aid in rice-based breads, obtaining a reduction of dough consistency and the elastic modulus and also acted as antistaling additive (Gujral et al. 2003b). At the bread level, using an experimental design, the authors optimized the combination of CGTase, HPMC, and oil amount for producing better specific volume and crumb texture. In addition, CGTase decreased the amylopectin retrogradation during storage (Singh et al. 2003), owing these effects to the hydrolyzing and cyclizing activity of the CGTase. α-Amylase of intermediate stability has been effectively used for antistaling in rice breads but resulted in sticky textures (Singh et al. 2003). Similarly, Onyango et al. (2010a) tested the effect of α-amylase in sorghum-based bread, observing that by increasing enzyme concentration (up to 0.3 U/g) decreased crumb firmness, cohesiveness, springiness, resilience, and chewiness but increased adhesiveness.

Trichoderma reesei tyrosinase has been applied to oat breads, inducing the formation of protein aggregates of high molecular weight, which involved globulins (Flander et al. 2011). That cross-linking resulted in dough hardening and better bread characteristics, and further improvement was obtained when tyrosinase was combined with xylanase.

A complex mixture comprising rice flour, dried albumin, cooking oil, sugar, yeast, emulsifier, salt, and an enzyme improver containing hemicellulase, glucose oxidase, xylanase, and α-amylase has been proposed as a gluten-free bread premix, which requires agitation for degrading the cell wall of rice starch and for improving expansion property (Kim et al. 2009).

TABLE 7.3

Enzymes Used in Sweet Baked Goods

Enzyme	Cakes	Biscuits or Cookies	Pastries and Croissants
Endoxynalases	+	−	−
Thermostable 4-alpha-glucanotransferase	+	−	−
α-Amylase	+	−	+
β-Amylase	+	−	−
Lipase	+	−	−
Xylanase	+	+	+
Proteases	−	+	+
Transglutaminase	−	−	+
Glucose oxidase	−	−	+
Phospholipase	−	−	+
Oxygenases	−	−	+
Oxidases	−	−	+
Peroxidases	−	−	+

7.5 Use of Enzymes in Making Cakes, Cookies, and Pastries

Another application of enzymes in baking is in the production of breakfast cereals, biscuits, buns, cakes, and pastries (Table 7.3). This kind of product, in addition to pasta, forms part of the group of "foods containing fat" and "foods containing sugar," of which the nutritional composition in fat and sugar contents greatly differs from bread and is unique for each elaborated product. Even the process of making these products requires different operations. However, enzymes are also very useful as processing aids in those products although the aim of their application maybe different than the one described for bread making. Mostly, individual enzymes or combinations are added to retard staling and extend freshness perception over time.

7.5.1 Cakes

In sponge cake systems, protein-foaming properties are fundamental in determining the overall textural quality of the product (Celik et al. 2007). Limited enzymatic hydrolysis of wheat gluten significantly increases its foaming and emulsifying properties (Drago and González 2000; Kong et al. 2007), improving cake volume and moistness (Bombara et al. 1997).

Other authors proposed to treat oat or rice bran with endoxylanase (70 and 700 ppm) for improving the properties of the bran when used as an ingredient in cakes (Lebesi and Tzia 2012). This treatment increased the amount of soluble dietary fiber and reduced the water-holding capacity of bran resulting in a softer cake crumb with high specific volume and better porosity and sensory characteristics. In addition, cakes containing treated bran showed better performance during storage with slower deterioration of the sensory characteristics (Lebesi and Tzia 2011).

Kim et al. (2012) recommended the use of rice flours previously treated for 18 h with a thermostable 4-α-glucanotransferase from *Thermus aquaticus* to obtain rice cakes with better textural properties and retarded starch retrogradation. The authors explained the extended shelf life due to the action of the enzyme, reducing the chain length of the starch.

The combination of a bacterial α-amylase with gums and pregelatinized starch have been proposed for making devil's fudge cakes (Sozer et al. 2011). By adding these improves, cakes showed 25% lower toughness and hardness and slower starch retrogradation.

There are some sweet specialties in which enzyme supplementation has been tested leading to improved quality. Specifically, panettone made with the addition of lipase or amylase increases the height, whereas xylanase improves the crumb porosity and dough-handling properties, and the extent of the effect is dose-dependent (Benejam et al. 2009). A steamed, cooked Chinese cake, Mi Gao, stales more slowly when β-amylase, α-amylase, or hydrocolloids were added although considering sensory acceptability the β-amylase was preferred (Ji et al. 2010).

7.5.2 Biscuits or Cookies

Proteases have been very useful for modifying the extensional properties of gluten and improving cookie performance of the wheat flours (Bruno and Oliveira 1995; Kara et al. 2005; HadiNezhad and Butler 2009). The action of proteases degrading gliadin and glutenin subunits of flour gluten led to an increase in the spread ratio values of cookies (Kara et al. 2005).

The addition of xylanases is advisable when high fiber content flours are used. In fact, Jia et al. (2011) could make almond cookies adding up to 20%–23% California almond skin flour in the presence of xylanase that acted on the noncarbohydrate polymers without affecting the spread ratio and the breaking force of the cookies.

7.5.3 Pastries and Croissants

Gerrard et al. (2000) first tested the effect of microbial TGase on puff pastries and croissants, observing a similar cross-linking action as mentioned in the case of wheat breads. Particularly, TGase improved the expansion of puff pastry and the volume of yeasted croissants, and this enzyme also conferred stability to the dough during freezing and 90 days of frozen storage. The stabilization effect of the TGase was explained in terms of cross-linking of albumins and globulins, leading to high molecular weight aggregates and also cross-linking of gluten subunits (Gerrard et al. 2001). A similar cross-linking effect on the albumins and globulins was obtained by using glucose oxidase due to disulfide and nondisulfide linkages formation, but this enzyme was less effective in increasing croissant volume due to its minor action on cross-linking glutenins (Rasiah et al. 2005). In the production of pastries, the combination of TGase (1.5 mg/100 g flour) with fermizyme (5 mg/100 g flour) has been recommended for improving sensory quality (Hozova et al. 2003).

Conversely, proteolytic enzyme material has also been advisable for obtaining laminated dough, such as croissants (Hargreaves et al. 2005). The enzyme material is applied on the outside surface of laminated dough or the partially baked laminated dough; in this way, the baked product retains a crispy crust for an extended period after baking and even after reheating.

In bakery products, such as croissants, pastries, pies, and so on, the water in an oil emulsion is fundamental for obtaining laminated dough that is a layered structure. Phospholipase has been recommended for totally or partially replacing emulsifiers such as lecithin, keeping the water-binding capacity of the spread, obtaining a baked product with thin and uniform layers (Nielsen 2009).

Another improver proposed for increasing the stability during frozen storage of Viennese pastries and brioche dough is comprised of gluten, natural flavoring, bean flours, and a blend of hemicellulose and α-amylase (Muchembled and Julien 2005). Even more complex enzyme combinations have been reported for pastry production that comprises xylanases, α-amylases, oxygenases, oxidases, peroxidases, phospholipases, and proteases (Anon. 2006). This preparation incorporates xylanases to enhance extensibility and improve hydration; α-amylases to produce fermentable sugars; oxidases, oxygenases, laccases, and peroxidases to improve consistency; glucose oxidases to reinforce dough; phospholipases to increase the volume of the bread and to improve crumb texture; and proteases to improve the effectiveness of kneading.

7.6 Future Trends

Enzymes have been extensively applied in bakery industries, initially for modulating dough performance and improving fresh bread quality, and later on for extending shelf life of baked products. Over the years, fundamental insights about the enzyme activity have been gathered, which help to comprehensively apply individual enzymes or enzymatic combinations for further applications. Owing to the intrinsic innovation of the bakery industry, future application of enzymes is foresight in different trends. First, is the search for new enzymes or new sources of enzymes with tailored properties for making baked products using mechanized processing and/or different baked specialties; second, a strong tendency is focused on the replacement of chemicals with enzymes moving to green labels in bakery products for answering consumers' demands for more natural products free of chemical additives, and finally, the interest in nutritional aspects is exponentially growing, which is prompting the development of enzymatic applications for improving nutritional and healthy features of baked products.

ACKNOWLEDGMENTS

The authors acknowledge the financial support of the Spanish Ministry of Economy and Competitiveness (Project AGL2011-23802), the European Regional Development Fund (FEDER), and Generalitat Valenciana (Project Prometeo 2012/064).

REFERENCES

Afinah, S., A. M. Yazid, M. H. Anis Shobirin, and M. Shuhaimi. "Phytase: Application in Food Industry." *International Food Research Journal* 17, no. 1 (2010): 13–21.

Ahmad, Z., M. S. Butt, A. Ahmed et al. "Effect of *Aspergillus Niger* Xylanase on Dough Characteristics and Bread Quality Attributes." *Journal of Food Science and Technology* 51 (2014): 2445–2453..

Ahn, J. B. "Method of Making Rice Steamed Bread Using Wheat Flour and Enzymes to Retard Shaping and Staling Together with Rice Flour." Shany Co Ltd., 2008.

Aisbitt, B., H. Caswell, and J. Lunn. "Cereals—Current and Emerging Nutritional Issues." *Nutrition Bulletin* 33, no. 3 (2008): 169–185.

Altamirano-Fortoul, R., I. Hernando, and C. M. Rosell. "Influence of Amyloglucosidase in Bread Crust Properties." *Food and Bioprocess Technology* 44 (2013): 85–94.

Alves-Prado, H. F., A. A. J. Carneiro, F. C. Pavezzi et al. "Production of Cyclodextrins by CGTase from *Bacillus Clausii* Using Different Starches as Substrates." [In English]. *Applied Biochemistry and Biotechnology* 146, no. 1–3 (2008): 3–13.

Anon. "Frozen Dough Additive for Manufacture of Bread, Comprises Protease Enzyme, Gluten, Stearoyl Lactate and Glucose Oxidase." Showa Sangyo Co., 2000.

Anon. "Enzymatic Preparation Comprises Incorporating Ascorbic Acid, Xylanases, Alpha-Amylases, Oxydases, Oxygenases, Laccases, Peroxidases, Phospholipases and Proteases in the Pastry." 2006.

Aravindan, R., P. Anbumathi, and T. Viruthagiri. "Lipase Applications in Food Industry." [In English]. *Indian Journal of Biotechnology* 6, no. 2 (2007): 141–158.

Arendt, E. K., A. Morrissey, M. M. Moore, and F. Dal Bello. "13 Gluten-Free Breads." In *Gluten-Free Cereal Products and Beverages*, edited by E. K. Arendt and F. Dal Bello, 289-VII. San Diego, CA: Academic Press, 2008.

Armero, E., and C. Collar. "Antistaling Additive Effects on Fresh Wheat Bread Quality." *Food Science and Technology International* 2, no. 5 (1996): 323–333.

Astray, G., C. Gonzalez-Barreiro, J. C. Mejuto, R. Rial-Otero, and J. Simal-Gandara. "A Review on the Use of Cyclodextrins in Foods." [In English]. *Food Hydrocolloids* 23, no. 7 (2009): 1631–1640.

Autio, K., H. Harkonen, T. Parkkonen et al. "Effects of Purified Endo-Beta-Xylanase and Endo-Beta-Glucanase on the Structural and Baking Characteristics of Rye Doughs." *Food Science and Technology-Lebensmittel-Wissenschaft and Technologie* 29, no. 1–2 (1996): 18–27.

Autio, K., K. Kruus, A. Knaapila, N. Gerber, L. Flander, and J. Buchert. "Kinetics of Transglutaminase—Induced Cross-Linking of Wheat Proteins in Dough." *Journal of Agricultural and Food Chemistry* 53, no. 4 (2005): 1039–1045.

Barrett, A. H., G. Marando, H. Leung, and G. Kaletunc. "Effect of Different Enzymes on the Textural Stability of Shelf-Stable Bread." *Cereal Chemistry* 82, no. 2 (2005): 152–157.

Baskinskiene, L., S. Garmuviene, G. Juodeikiene, and D. Haltrich. "Comparison of Different Fungal Xylanases for Wheat Bread Making." *Getreidetechnologie* 61, no. 4 (2007): 228–235.

Basman, A., H. Koksel, and P. K. W. Ng. "Effects of Increasing Levels of Transglutaminase on the Rheological Properties and Bread Quality Characteristics of Two Wheat Flours." *European Food Research and Technology* 215, no. 5 (2002): 419–424.

Benejam, W., M. E. Steffolani, and A. E. Leon. "Use of Enzyme to Improve the Technological Quality of a Panettone Like Baked Product." *International Journal of Food Science and Technology* 44, no. 12 (2009): 2431–2437.

Biwer, A., G. Antranikian, and E. Heinzle. "Enzymatic Production of Cyclodextrins." *Applied Microbiology and Biotechnology* 59, no. 6 (2002): 609–617.

Blaszczak, W., J. Sadowska, C. M. Rosell, and J. Fornal. "Structural Changes in the Wheat Dough and Bread with the Addition of Alpha-Amylases." *European Food Research and Technology* 219, no. 4 (2004): 348–354.

Bombara, N., M. C. Anon, and A. M. R. Pilosof. "Functional Properties of Protease Modified Wheat Flours." *Food Science and Technology-Lebensmittel-Wissenschaft and Technologie* no. 30 (1997): 441–447.

Bonet, A., C. M. Rosell, P. A. Caballero, M. Gomez, I. Perez-Munuera, and M. A. Lluch. "Glucose Oxidase Effect on Dough Rheology and Bread Quality: A Study from Macroscopic to Molecular Level." *Food Chemistry* 99, no. 2 (2006): 408–415.

Bruno, M. E. C., and C. R. de Oliveira Camargo. "Proteolytic Enzymes in the Processing of Cookies and Bread." *Boletim da Sociedade Brasileira de Ciencia e Tecnologia de Alimentos* 29, no. 2 (1995): 170–178.

Buleon, A., P. Colonna, V. Planchot, and S. Ball. "Starch Granules: Structure and Biosynthesis." [In English]. *International Journal of Biological Macromolecules* 23, no. 2 (1998): 85–112.

Butt, M. S., M. Tahir-Nadeem, Z. Ahmad, and M. T. Sultan. "Xylanases and Their Applications in Baking Industry." *Food Technology and Biotechnology* 46, no. 1 (2008): 22–31.

Caballero, P. A., M. Gomez, and C. M. Rosell. "Bread Quality and Dough Rheology of Enzyme-Supplemented Wheat Flour." [In English]. *European Food Research and Technology* 224, no. 5 (2007a): 525–534.

Caballero, P. A., M. Gómez, and C. M. Rosell. "Improvement of Dough Rheology, Bread Quality and Bread Shelf-Life by Enzymes Combination." *Journal of Food Engineering* 81, no. 1 (2007b): 42–53.

Cabrera-Chavez, F., O. Rouzaud-Sandez, N. Sotelo-Cruz, and A. M. Calderon de la Barca. "Transglutaminase Treatment of Wheat and Maize Prolamins of Bread Increases the Serum IgA Reactivity of Celiac Disease Patients." *Journal of Agricultural and Food Chemistry* 56, no. 4 (2008): 1387–1391.

Celik, I., Y. Yilmaz, F. Isik, and O. Ustun. "Effect of Soapwort Extract on Physical and Sensory Properties of Sponge Cakes and Rheological Properties of Sponge Cake Batters." *Food Chemistry* no. 101 (2007): 907–911.

Collar, C., and C. Bollain. "Impact of Microbial Transglutaminase on the Viscoelastic Profile of Formulated Bread Doughs." *European Food Research and Technology* 218, no. 2 (2004): 139–146.

Collar, C., and C. Bollain. "Impact of Microbial Transglutaminase on the Staling Behaviour of Enzyme-Supplemented Pan Breads." *European Food Research and Technology* 221, no. 3–4 (2005): 298–304.

Collar, C., P. Andreu, and M. A. Martinez-Anaya. "Interactive Effects of Flour, Starter and Enzyme on Bread Dough Machinability." *Zeitschrift Fur Lebensmittel-Untersuchung Und-Forschung a-Food Research and Technology* 207, no. 2 (1998): 133–139.

Collar, C., J. C. Martinez, P. Andreu, and E. Armero. "Effects of Enzyme Associations on Bread Dough Performance. A Response Surface Analysis." *Food Science and Technology International* 6, no. 3 (2000): 217–226.

Collins, T., A. Hoyoux, A. Dutron et al. "Use of Glycoside Hydrolase Family 8 Xylanases in Baking." *Journal of Cereal Science* 43, no. 1 (2006): 79–84.

Courtin, C. M., and J. A. Delcour. "Arabinoxylans and Endoxylanases in Wheat Flour Bread-Making." [In English]. *Journal of Cereal Science* 35, no. 3 (2002): 225–243.

Courtin, C. M., G. G. Gelders, and J. A. Delcour. "Use of Two Endoxylanases with Different Substrate Selectivity for Understanding Arabinoxylan Functionality in Wheat Flour Breadmaking." *Cereal Chemistry* 78, no. 5 (2001): 564–571.

Dagdelen, A. F., and D. Gocmen. "Effects of Glucose Oxidase, Hemicellulase and Ascorbic Acid on Dough and Bread Quality." *Journal of Food Quality* 30, no. 6 (2007): 1009–1022.

De Boer, L., and L. D. Boer. "Enzymatic Composition for Decreasing Acrylamide Levels in the Food Product, e.g., Bread and Pastry, Comprises Asparaginase and at Least One Hydrolyzing Enzyme." *Dsm Ip Assets Bv*, Boer L. D., 2007.

Defloor, I., and J. A. Delcour. "Impact of Maltodextrins and Antistaling Enzymes on the Differential Scanning Calorimetry Staling Endotherm of Baked Bread Doughs." *Journal of Agricultural and Food Chemistry* 47, no. 2 (1999): 737–741.

Delcour, J. A., and R. C. Hoseney. "Principles of Cereal Science and Technology Authors Provide Insight into the Current State of Cereal Processing." [In English]. *Cereal Foods World* 55, no. 1 (2010): 21–22.

Denli, E., and R. Ercan. "Effect of Added Pentosans Isolated from Wheat and Rye Grain on Some Properties of Bread." *European Food Research and Technology* 212, no. 3 (2001): 374–376.

Dogan, I. S. "Effect of Alpha-Amylases on Dough Properties During Turkish Hearth Bread Production." *International Journal of Food Science and Technology* 38, no. 2 (2003): 209–216.

Dornez, E., K. Gebruers, J. A. Delcour, and C. M. Courtin. "Grain-Associated Xylanases: Occurrence, Variability, and Implications for Cereal Processing." *Trends in Food Science and Technology* 20, no. 11–12 (2009): 495–510.

Drago, S. R., and R. J. González. "Foaming Properties of Enzymatically Hydrolysed Wheat Gluten." *Innovative Food Science & Emerging Technologies* no. 1 (2000): 269–273.

Dura, A., W. Błaszczak, and C. M. Rosell. "Functionality of Porous Starch Obtained by Amylase or Amyloglucosidase Treatments." *Carbohydrate Polymers* 101, no. 1 (2014): 837–845.

Ertas, N., N. Bilgicli, and S. Turker. "The Effects of Soy Flour, Glucose Oxidase and Lipase Enzymes Addition on Flour, Dough and Bread Properties." *Gida* 31, no. 3 (2006): 143–149.

Figueroa-Espinoza, M. C., and X. Rouau. "Oxidative Cross-Linking of Pentosans by a Fungal Laccase and Horseradish Peroxidase: Mechanism of Linkage between Feruloylated Arabinoxylans." *Cereal Chemistry* no. 75 (1998): 259–265.

Flander, L., U. Holopainen, K. Kruus, and J. Buchert. "Effects of Tyrosinase and Laccase on Oat Proteins and Quality Parameters of Gluten-Free Oat Breads." *Journal of Agricultural and Food Chemistry* 59, no. 15 (2011): 8385–8390.

Folk, J. E., and J. S. Finlayson. "The Epsilon-(Gamma-Glutamyl)Lysine Crosslink and the Catalytic Role of Transglutaminases." [In English]. *Advances in Protein Chemistry* 31 (1977): 1–133.

Forman, T. M., and D. N. Evanson. "Antistaling Process for Flat Bread." Novozymes North America Inc., 2011.

Gelinas, P., E. Poitras, C. M. McKinnon, and A. Morin. "Oxido-Reductases and Lipases as Dough-Bleaching Agents." *Cereal Chemistry* 75, no. 6 (1998): 810–814.

Gerrard, J. A. "Protein–Protein Crosslinking in Food: Methods, Consequences, Applications." [In English]. *Trends in Food Science and Technology* 13, no. 12 (2002): 391–399.

Gerrard, J. A., S. E. Fayle, A. J. Wilson, M. P. Newberry, M. Ross, and S. Kavale. "Dough Properties and Crumb Strength of White Pan Bread as Affected by Microbial Transglutaminase." *Journal of Food Science* 63, no. 3 (1998): 472–475.

Gerrard, J. A., M. P. Newberry, M. Ross, A. J. Wilson, S. E. Fayle, and S. Kavale. "Pastry Lift and Croissant Volume as Affected by Microbial Transglutaminase." [In English]. *Journal of Food Science* 65, no. 2 (2000): 312–314.

Gerrard, A. J., S. E. Fayle, P. A. Brown, K. H. Sutton, L. Simmons, and I. Rasiah. "Effects of Microbial Transglutaminase on the Wheat Proteins of Bread and Croissant Dough." *Journal of Food Science* 66, no. 6 (2001): 782–786.

Gil, M. J., M. J. Callejo, G. Rodriguez, and M. V. Ruiz. "Keeping Qualities of White Pan Bread Upon Storage: Effect of Selected Enzymes on Bread Firmness and Elasticity." *Zeitschrift Fur Lebensmittel-Untersuchung Und-Forschung a-Food Research and Technology* 208, no. 5–6 (1999): 394–399.

Goesaert, H., K. Brijs, W. S. Veraverbeke, C. M. Courtin, K. Gebruers, and J. A. Delcour. "Wheat Flour Constituents: How They Impact Bread Quality, and How to Impact Their Functionality." *Trends in Food Science and Technology* 16, no. 1–3 (2005): 12–30.

Goesaert, H., C. M. Courtin, and J. A. Delcour. "Use of Enzymes in the Production of Cereal-Based Functional Foods and Food Ingredients," 237–265. Elsevier Inc., 2008.

Goesaert, H., L. Slade, H. Levine, and J. A. Delcour. "Amylases and Bread Firming—An Integrated View." [In English]. *Journal of Cereal Science* 50, no. 3 (2009): 345–352.

Gottmann, K., and B. Sproessler. "Baking Agent and Baking Flour for Bread, Rolls, Etc.—Contg. Transglutaminase as Enzyme with Emulsifiers, Flour and Sugar," 492406-B1: Roehm Gmbh (Rohg), 1992.

Gray, J. A., and J. N. BeMiller. "Bread Staling: Molecular Basis and Control." [In English]. *Comprehensive Reviews in Food Science and Food Safety* 2, no. 1 (2003): 1–21.

Gujral, H., and C. M. Rosell. "Functionality of Rice Flour Modified with a Microbial Transglutaminase." [In English]. *Journal of Cereal Science* 39, no. 2 (2004a): 225–230.

Gujral, H. S., and C. M. Rosell. "Improvement of the Breadmaking Quality of Rice Flour by Glucose Oxidase." *Food Research International* 37, no. 1 (2004b): 75–81.

Gujral, H. S., and C. M. Rosell. "Modification of Pasting Properties of Wheat Starch by Cyclodextrin Glycosyltransferase." [In English]. *Journal of the Science of Food and Agriculture* 84, no. 13 (2004c): 1685–1690.

Gujral, H. S., I. Guardiola, J. V. Carbonell, and C. A. Rosell. "Effect of Cyclodextrinase on Dough Rheology and Bread Quality from Rice Flour." *Journal of Agricultural and Food Chemistry* 51, no. 13 (2003a): 3814–3818.

Gujral, H. S., M. Haros, and C. M. Rosell. "Starch Hydrolyzing Enzymes for Retarding the Staling of Rice Bread." *Cereal Chemistry* 80, no. 6 (2003b): 750–754.

Gul, H., M. Sertac Ozer, and H. Dizlek. "Improvement of the Wheat and Corn Bran Bread Quality by Using Glucose Oxidase and Hexose Oxidase." *Journal of Food Quality* 32, no. 2 (2009): 209–223.

HadiNezhad, M., and F. Butler. "Association of Glutenin Subunit Composition and Dough Rheological Characteristics with Cookie Baking Properties of Soft Wheat Cultivars." [In English]. *Cereal Chemistry* 86, no. 3 (2009): 339–349.

Hamer, R. J., and C. Primo-Martin. "Preparing Bread Dough/Part Baked Bread for Baked Bread Involves Mixing Flour, Water and Bakery Ingredients to Obtain Bread Dough, Followed by Part Baking to Obtain Part Baked Bread and Applying Enzyme Material." Wageningen Cent Food Sci; Wageningen Ct for Food Sci, 2006.

Hanft, F., and P. Koehler. "Studies on the Effect of Glucose Oxidase in Bread Making." *Journal of the Science of Food and Agriculture* 86, no. 11 (2006): 1699–1704.

Hargreaves, N. G., E. Brinker, U. Scharf et al. "Preparation of Laminated Dough or Part-Baked Laminated Dough Product for Preparing, e.g., Croissant, by Mixing Flour and Water, Laminating and Part-Baking Resultant Dough, and Applying Enzyme Material with Proteolytic Activity." *Csm Nederland Bv*, 2005.

Haros, M., C. M. Rosell, and C. Benedito. "Fungal Phytase as a Potential Breadmaking Additive." [In English]. *European Food Research and Technology* 213, no. 4/5 (2001a): 317–322.

Haros, M., C. M. Rosell, and C. Benedito. "Use of Fungal Phytase to Improve Breadmaking Performance of Whole Wheat Bread." [In English]. *Journal of Agricultural and Food Chemistry* 49, no. 11 (2001b): 5450–5454.

Haros, M., C. M. Rosell, and C. Benedito. "Effect of Different Carbohydrases on Fresh Bread Texture and Bread Staling." *European Food Research and Technology* 215, no. 5 (2002): 425–430.

Hemalatha, M. S., S. G. Bhagwat, P. V. Salimath, and U. J. S. Prasada Rao. "Enhancement of Soluble Dietary Fibre, Polyphenols and Antioxidant Properties of Chapatis Prepared from Whole Wheat Flour Dough Treated with Amylases and Xylanase." *Journal of the Science of Food and Agriculture* 92, no. 4 (2012): 764–771.

Hoseney, R. C. *Principles of Cereal Science and Technology* [In English]. Principles of Cereal Science and Technology, St. Paul, MN: American Association of Cereal Chemists, 1994.

Hoseney, R. C., and J. M. Faubion. "A Mechanism for the Oxidative Gelation of Wheat-Flour Water-Soluble Pentosans." [In English]. *Cereal Chemistry* 58, no. 5 (1981): 421–424.

Houben, A., A. Hochstotter, and T. Becker. "Possibilities to Increase the Quality in Gluten-Free Bread Production: An Overview." [In English]. *European Food Research and Technology* 235, no. 2 (2012): 195–208.

Hozova, B., I. Kukurova, and L. Dodok. "Application of Transglutaminase and Fermizyme for Sensory Quality Improvement of Pastry." *Nahrung-Food* 47, no. 3 (2003): 171–175.

Hsing, I. L., J. M. Faubion, and C. E. Walker. "Using Enzyme-Oxidant Combinations to Improve Frozen-Dough Bread Quality." *Getreidetechnologie* 64, no. 2 (2010): 124–134.

Huang, W., and C. Jia. "Anti-Freezing Fermented Dough Containing Glucose Oxidase and Peroxidase and Its Producing Method." Univ. Jiangnan, 2006.

Hug-Iten, S., F. Escher, and B. Conde-Petit. "Structural Properties of Starch in Bread and Bread Model Systems: Influence of an Antistaling Alpha-Amylase." *Cereal Chemistry* 78, no. 4 (2001): 421–428.

Hug-Iten, S., F. Escher, and B. Conde-Petit. "Staling of Bread: Role of Amylose and Amylopectin and Influence of Starch-Degrading Enzymes." *Cereal Chemistry* 80, no. 6 (2003): 654–661.

Hui, Y. H., H. Corke, I. de Leyn, W.-K. Nip, and N. Cross. *Bakery Products. Science and Technology* [In English], edited by Y. H. Hui, H. Corke, I. de Leyn, W.-K. Nip and N. Cross, Oxford, UK: Blackwell Publishing, 2006.

Indrani, D., P. Prabhasankar, J. Rajiv, and G. V. Rao. "Scanning Electron Microscopy, Rheological Characteristics, and Bread-Baking Performance of Wheat-Flour Dough as Affected by Enzymes." *Journal of Food Science* 68, no. 9 (2003): 2804–2809.

Inoue, S., and S. Ota. "Bread or Other Cereal-Based Food Improver Composition Involving the Addition of Phospholipase a to the Flour." Kyowa Hakko Kogyo Co. Ltd., 1986.

Izydorczyk, M. S., and C. G. Biliaderis. "Cereal Arabinoxylans: Advances in Structure and Physicochemical Properties." *Carbohydrate Polymers* 28, no. 1 (1995): 33–48.

Jemli, S., E. Ben Messaoud, D. Ayadl-Zouail, B. Nalli, B. Khemakhem, and S. Bejar. "A Beta-Cyclodextrin Glycosyltransferase from a Newly Isolated Paenibacillus Pabuli Us 132 Strain: Purification, Properties and Potential Use in Bread-Making." [In English]. *Biochemical Engineering Journal* 34, no. 1 (2007): 44–50.

Jensen, B., and P. Drube. "Effect and Functionality of Lipases in Dough and Bread." *Getreide, Mehl und Brot* 52, no. 3 (1998): 153–157.

Ji, Y., K. Zhu, Z.-C. Chen, H. Zhou, J. Ma, and H. Qian. "Effects of Different Additives on Rice Cake Texture and Cake Staling." *Journal of Texture Studies* 41, no. 5 (2010): 703–713.

Jia, C. L., W. N. Huang, M. A. S. Abdel-Samie, G. X. Huang, and G. W. Huang. "Dough Rheological, Mixolab Mixing, and Nutritional Characteristics of Almond Cookies with and without Xylanase." *Journal of Food Engineering* 105, no. 2 (2011): 227–232.

Jia, Y., H. Li, and Y. Tian. "Enzyme Composition as Modifier Dedicated to Flour for Steamed Bread." Univ. Hebei Agric, 2006.

Jiang, Z. Q., S. Q. Yang, S. S. Tan, L. T. Li, and X. T. Li. "Characterization of a Xylanase from the Newly Isolated Thermophilic Thermomyces Lanuginosus Cau44 and Its Application in Bread Making." *Letters in Applied Microbiology* 41, no. 1 (2005): 69–76.

Jiang, Z., Q. Cong, Q. Yan, N. Kumar, and X. Du. "Characterisation of a Thermostable Xylanase from Chaetomium Sp and Its Application in Chinese Steamed Bread." *Food Chemistry* 120, no. 2 (2010): 457–462.

Jimenez, T., and M. A. Martinez-Anaya. "Characterization of Water Soluble Pentosans of Enzyme Supplemented Dough and Breads." *Food Science and Technology International* 6, no. 3 (2000): 197–205.

Hyuck Joong, A., L. S. Yong, Y. S. Seok et al. "A Novel Maltopentaose-Producing Amylase as a Bread Antistaling Agent." *Food Science and Biotechnology* 14, no. 5 (2005): 681–684.

Junqueira, R. M., F. Rocha, M. A. Moreira, and I. A. Castro. "Effect of Proofing Time and Wheat Flour Strength on Bleaching, Sensory Characteristics, and Volume of French Breads with Added Soybean Lipoxygenase." [In English]. *Cereal Chemistry* 84, no. 5 (2007): 443–449.

Kara, M., D. Sivri, and H. Koksel. "Effects of High Protease-Activity Flours and Commercial Proteases on Cookie Quality." [In English]. *Food Research International* 38, no. 5 (2005): 479–486.

Katina, K., M. Salmenkallio-Marttila, R. Partanen, P. Forssell, and K. Autio. "Effects of Sourdough and Enzymes on Staling of High-Fibre Wheat Bread." *LWT-Food Science and Technology* 39, no. 5 (2006): 479–491.

Kawamura-Konishi, Y., K. Shoda, H. Koga, and Y. Honda. "Improvement in Gluten-Free Rice Bread Quality by Protease Treatment." *Journal of Cereal Science* 58, no. 1 (2013): 45–50.

Kelly, R. M., L. Dijkhuizen, and H. Leemhuis. "Starch and A-Glucan Acting Enzymes, Modulating Their Properties by Directed Evolution." *Journal of Biotechnology* 140, no. 3 4 (2009): 184–193.

Kim, D. C., H. Kim, O. W. Kim, S. S. Kim, S. E. Lee, and J. H. Park. "Preparing Gluten-Free Bread Premix, Involves Blending Rice Flour, Dried Albumin, Cooking Oil, Sugar, Yeast, Emulsifier, Salt, Enzyme Conjugate Comprising Hemicellulase, Glucose Oxidase, Xylanase and Amylase, and Agitating Mixture." Korea Food Res Inst., 2009.

Kim, Y., Y. L. Kim, K. S. Trinh, Y. R. Kim, and T. W. Moon. "Texture Properties of Rice Cakes Made of Rice Flours Treated with 4-Alpha-Glucanotransferase and Their Relationship with Structural Characteristics." [In English]. *Food Science and Biotechnology* 21, no. 6 (2012): 1707–1714.

Kong, X. Z., H. M. Zhou, and H. F. Qian. "Enzymatic Preparation and Functional Properties of Wheat Gluten Hydrolysates." *Food Chemistry* 101 (2007): 615–620.

Koocheki, A., S. A. Mortazavi, M. N. Mahalati, and M. Karimi. "Effect of Emulsifiers and Fungal Alpha-Amylase on Rheological Characteristics of Wheat Dough and Quality of Flat Bread." *Journal of Food Process Engineering* 32, no. 2 (2009): 187–205.

Kukurova, K., H. J. Morales, A. Bednarikova, and Z. Ciesarova. "Effect of L-Asparaginase on Acrylamide Mitigation in a Fried-Dough Pastry Model." *Molecular Nutrition & Food Research* 53, no. 12 (2009): 1532–1539.

Labat, E., M. H. Morel, and X. Rouau. "Effects of Laccase and Ferulic Acid on Wheat Flour Doughs." *Cereal Chemistry* 77, no. 6 (2000): 823–828.

Laurikainen, T., H. Harkonen, K. Autio, and K. Poutanen. "Effects of Enzymes in Fibre-Enriched Baking." *Journal of the Science of Food and Agriculture* 76, no. 2 (1998): 239–249.

Lebesi, D. M., and C. Tzia. "Staling of Cereal Bran Enriched Cakes and the Effect of an Endoxylanase Enzyme on the Physicochemical and Sensorial Characteristics." [In English]. *Journal of Food Science* 76, no. 6 (2011): S380–S387.

Lebesi, D. M., and C. Tzia. "Use of Endoxylanase Treated Cereal Brans for Development of Dietary Fiber Enriched Cakes." [In English]. *Innovative Food Science & Emerging Technologies* 13 (2012): 207–214.

Leon, A. E., E. Duran, and C. B. de Barber. "Utilization of Enzyme Mixtures to Retard Bread Crumb Firming." *Journal of Agricultural and Food Chemistry* 50, no. 6 (2002): 1416–1419.

Li, Z., X. Tang, W. Huang, J. G. Liu, M. Tilley, and Y. Yao. "Rheology, Microstructure, and Baking Characteristics of Frozen Dough Containing Rhizopus Chinensis Lipase and Transglutaminase." *Cereal Chemistry* 88, no. 6 (2011): 596–601.

MacRitchie, F. "Role of Lipids in Baking." In *Lipids in Cereal Technology*, edited by P. J. Barnes, 165–188. London: Academic Press, 1983.

MacRitchie, F. "Physicochemical Properties of Wheat Proteins in Relation to Functionality." *Advances in Food and Nutrition Research* 36 (1992): 1–87.

Maeda, T., and N. Morita. "Flour Quality and Pentosan Prepared by Polishing Wheat Grain on Breadmaking." *Food Research International* 36, no. 6 (2003): 603–610.

Marco, C., and C. M. Rosell. "Breadmaking Performance of Protein Enriched, Gluten-Free Breads." [In English]. *European Food Research and Technology* 227, no. 4 (2008a): 1205–1213.

Marco, C., and C. M. Rosell. "Effect of Different Protein Isolates and Transglutaminase on Rice Flour Properties." [In English]. *Journal of Food Engineering* 84, no. 1 (2008b): 132–139.

Marco, C., and C. M. Rosell. "Functional and Rheological Properties of Protein Enriched Gluten Free Composite Flours." [In English]. *Journal of Food Engineering* 88, no. 1 (2008c): 94–103.

Marco, C., G. Perez, P. Ribotta, and C. M. Rosell. "Effect of Microbial Transglutaminase on the Protein Fractions of Rice, Pea and Their Blends." [In English]. *Journal of the Science of Food and Agriculture* 87, no. 14 (2007): 2576–2582.

Marco, C., G. Perez, A. E. Leon, and C. M. Rosell. "Effect of Transglutaminase on Protein Electrophoretic Pattern of Rice, Soybean, and Rice-Soybean Blends." [In English]. *Cereal Chemistry* 85, no. 1 (2008): 59–64.

Martinez-Anaya, M. A., and T. Jimenez. "Functionality of Enzymes That Hydrolyse Starch and Non-Starch Polysaccharide in Breadmaking." *Zeitschrift Fur Lebensmittel-Untersuchung Und-Forschung a-Food Research and Technology* 205, no. 3 (1997a): 209–214.

Martinez-Anaya, M. A., and T. Jimenez. "Rheological Properties of Enzyme Supplemented Doughs." *Journal of Texture Studies* 28, no. 5 (1997b): 569–583.

Martinez-Anaya, M. A., and T. Jimenez. "Physical Properties of Enzyme-Supplemented Doughs and Relationship with Bread Quality Parameters." *Zeitschrift Fur Lebensmittel-Untersuchung Und-Forschung a-Food Research and Technology* 206, no. 2 (1998): 134–142.

Martinez-Anaya, M. A., and A. Devesa. "Influence of Enzymes in Sourdough Wheat Breadmaking. Changes in Pentosans." *Food Science and Technology International* 6, no. 2 (2000): 109–116.

Mathewson, P. R. "Common Enzyme Reactions." [In English]. *Cereal Foods World* 43, no. 11 (1998): 798–803.

Min, B. C., S. H. Yoon, J. W. Kim, Y. W. Lee, Y. B. Kim, and K. H. Park. "Cloning of Novel Maltooligosaccharide-Producing Amylases as Antistaling Agents for Bread." *Journal of Agricultural and Food Chemistry* 46, no. 2 (1998): 779–782.

Moayedallaie, S., M. Mirzaei, and J. Paterson. "Bread Improvers: Comparison of a Range of Lipases with a Traditional Emulsifier." [In English]. *Food Chemistry* 122, no. 3 (2010): 495–499.

Moharrami, E., and M. Shahedi Baghkhandan. "Optimization of Flour Alpha-Amylase Activity with Germinated Wheat Flour and Its Effect on Staling of Taftoon Bread." *Iranian Journal of Food Science and Technology*, Fall 8 (2011): 23–33.

Morita, N., Y. Arishima, N. Tanaka, and S. Shiotsubo. "Utilization of Hemicellulase as Bread Improver in a Home Baker." *Journal of Applied Glycoscience* 44, no. 2 (1997): 143–152.

Motoki, M., and K. Seguro. "Transglutaminase and Its Use for Food Processing." [In English]. *Trends in Food Science & Technology* 9, no. 5 (1998): 204–210.

Muchembled, J. J., and P. Julien. "Improver for Frozen Dough for Bread, Viennese Pastries and Brioches Comprises Gluten, Acerola and/or Cynorrhodon and/or Camu-Camu Powder and Enzymatic Preparations Providing Hemicellulase and Alpha-Amylase." Lesaffre & Cie Sa, 2005.

Nielsen, P. M. "Producing a Water-In-Oil Emulsion Useful for Making Bakery Product, e.g., Croissant, Danish Pastry, Pie, Pastry Involves Mixing a Composition Comprising Oil, Water, Phospholipase, and Phosphatide." Hansen as Chr; Novozymes As, 2009.

Nonaka, M., H. Tanaka, A. Okiyama et al. "Polymerization of Several Proteins by Ca-2+-Independent Transglutaminase Derived from Microorganisms." [In English]. *Agricultural and Biological Chemistry* 53, no. 10 (1989): 2619–2623.

Novozymes, A. S. "Combined Use of Glucose Oxidase and Phospholipase for Baking." *Research Disclosure* 442 (2001): 231.

Oliveira, D. S., J. Telis-Romero, R. Da-Silva, and C. M. L. Franco. "Effect of a Thermoascus Aurantiacus Thermostable Enzyme Cocktail on Wheat Bread Quality." *Food Chemistry* 143 (2014): 139–146.

Onyango, C., C. Mutungi, G. Unbehend, and M. G. Lindhauer. "Batter Rheology and Bread Texture of Sorghum-Based Gluten-Free Formulations Modified with Native or Pregelatinised Cassava Starch and Alpha-Amylase." *International Journal of Food Science and Technology* 45, no. 6 (2010a): 1228–1235.

Onyango, C., C. Mutungi, G. Unbehend, and M. G. Lindhauer. "Rheological and Baking Characteristics of Batter and Bread Prepared from Pregelatinised Cassava Starch and Sorghum and Modified Using Microbial Transglutaminase." *Journal of Food Engineering* 97, no. 4 (2010b): 465–470.

Oort, M. van. "Oxidases in Baking. A Review of the Uses of Oxidases in Bread Making." *International Food Ingredients* 4 (1996): 42–47.

Oort, M. van, F. van Straaten, and C. Laane. "Pentosans and Pentosanases in Bread Making." *International Food Ingredients* 2 (1995): 23–27.

Osborne, T. B. *The Vegetable Proteins*, 2nd edition. Longmans Green and Co, London, 1924.

Palacios, H. R., P. B. Schwarz, and B. L. D'Appolonia. "Effect of Alpha-Amylases from Different Sources on the Retrogradation and Recrystallization of Concentrated Wheat Starch Gels: Relationship to Bread Staling." *Journal of Agricultural and Food Chemistry* 52, no. 19 (2004): 5978–5986.

Pandey, A., P. Nigam, C. R. Soccol, V. T. Soccol, D. Singh, and R. Mohan. "Advances in Microbial Amylases." *Biotechnology and Applied Biochemistry* 31, no. 2 (2000): 135–152.

Pareyt, B., S. M. Finnie, J. A. Putseys, and J. A. Delcour. "Lipids in Bread Making: Sources, Interactions, and Impact on Bread Quality." [In English]. *Journal of Cereal Science* 54, no. 3 (2011): 266–279.

Pena-Valdivia, C. B., and A. Salazar-Zazueta. "Effect of Amylases and Proteases on Bread Making Quality of Durum Wheat Flour." *Journal of Food Science and Technology-Mysore* 34, no. 6 (1997): 516–518.

Penella, J. M. S., C. Collar, and M. Haros. "Effect of Wheat Bran and Enzyme Addition on Dough Functional Performance and Phytic Acid Levels in Bread." [In English]. *Journal of Cereal Science* 48, no. 3 (2008): 715–721.

Pongjaruvat, W., P. Methacanon, N. Seetapan, A. Fuongfuchat, and C. Gamonpilas. "Influence of Pregelatinised Tapioca Starch and Transglutaminase on Dough Rheology and Quality of Gluten-Free Jasmine Rice Breads." *Food Hydrocolloids* 36 (2014): 143–150.

Poulsen, C., and P. B. Hostrup. "Purification and Characterization of a Hexose Oxidase with Excellent Strengthening Effects in Bread." *Cereal Chemistry* 75, no. 1 (1998): 51–57.

Poulson, C. H., J. B. Soe, P. Rasmussen, S. M. Madrid, and M. R. Zargahi. "Lipase and Use of Same for Improving Doughs and Baked Products." Danisco A/S, 2006.

Poulson, C. H., J. B. Soe, P. Rasmussen, S. M. Madrid, and M. R. Zargahi. "Lipase and Use of Same for Improving Doughs and Baked Products." Danisco A/S, 2010.

Prabhasankar, P., D. Indrani, R. Jyotsna, and G. V. Rao. "Influence of Enzymes on Rheological, Microstructure and Quality Characteristics of Parotta—An Unleavened Indian Flat Bread." *Journal of the Science of Food and Agriculture* 84, no. 15 (2004): 2128–2134.

Primo-Martin, C., R. Valera, and M. A. Martinez-Anaya. "Effect of Pentosanase and Oxidases on the Characteristics of Doughs and the Glutenin Macropolymer (Gmp)." [In English]. *Journal of Agricultural and Food Chemistry* 51, no. 16 (2003): 4673–4679.

Primo-Martin, C., M. W. Wang, W. J. Lichtendonk, J. J. Plijter, and R. J. Hamer. "An Explanation for the Combined Effect of Xylanase-Glucose Oxidase in Dough Systems." *Journal of the Science of Food and Agriculture* 85, no. 7 (2005): 1186–1196.

Primo-Martin, C., H. de Beukelaer, R. J. Hamer, and T. Van Vliet. "Fracture Behaviour of Bread Crust: Effect of Ingredient Modification." [In English]. *Journal of Cereal Science* 48, no. 3 (2008): 604–612.

Purhagen, J. K., M. E. Sjoo, and A.-C. Eliasson. "Starch Affecting Anti-Staling Agents and Their Function in Freestanding and Pan-Baked Bread." *Food Hydrocolloids* 25, no. 7 (2011): 1656–1666.

Rani Ray, R. "Microbial Isoamylases: An Overview." [In English]. *American Journal of Food Technology* 6, no. 1 (2011): 1–18.

Rao, M. B., A. M. Tanksale, M. S. Ghatge, and V. V. Deshpande. "Molecular and Biotechnological Aspects of Microbial Proteases." [In English]. *Microbiology and Molecular Biology Reviews* 62, no. 3 (1998): 597+.

Rasiah, I. A., K. H. Sutton, F. L. Low, H. M. Lin, and J. A. Gerrard. "Crosslinking of Wheat Dough Proteins by Glucose Oxidase and the Resulting Effects on Bread and Croissants." *Food Chemistry* 89, no. 3 (2005): 325–332.

Renzetti, S., and E. K. Arendt. "Effect of Protease Treatment on the Baking Quality of Brown Rice Bread: From Textural and Rheological Properties to Biochemistry and Microstructure." *Journal of Cereal Science* 50, no. 1 (2009): 22–28.

Renzetti, S., F. Dal Bello, and E. K. Arendt. "Microstructure, Fundamental Rheology and Baking Characteristics of Batters and Breads from Different Gluten-Free Flours Treated with a Microbial Transglutaminase." *Journal of Cereal Science* 48, no. 1 (2008): 33–45.

Renzetti, S., C. M. Courtin, J. A. Delcour, and E. K. Arendt. "Oxidative and Proteolytic Enzyme Preparations as Promising Improvers for Oat Bread Formulations: Rheological, Biochemical and Microstructural Background." [In English]. *Food Chemistry* 119, no. 4 (2010): 1465–1473.

Ribotta, P. D., and A. Le Bail. "Thermo-Physical Assessment of Bread During Staling." [In English]. *LWT-Food Science and Technology* 40, no. 5 (2007): 879–884.

Rojas, J. A., C. M. Rosell, and C. Benedito de Barber. "Role of Maltodextrins in the Staling of Starch Gels." [In English]. *European Food Research and Technology* 212, no. 3 (2001): 364–368.

Rosell, C. M. "Enzymatic Manipulation of Gluten-Free Breads," 83–98. Wiley-Blackwell, 2009.

Rosell, C. M. "The Science of Doughs and Bread Quality," 3–14. Elsevier Inc., 2011.

Rosell, C. M. "The Nutritional Enhancement of Wheat Flour." In *Breadmaking: Improving Quality*, 2nd edition, edited by S. Cauvain, 687–710. Cambridge, UK: Woodhead Publishing, 2012.

Rosell, C. M., and D. R. Solis Nadal. "Composition for Improving Baking Doughs and Baked Products, Comprises a Cyclodextrin Glucanotransferase and Optionally Starch Debranching Enzyme, and Particularly Has an Antistaling Effect." Consejo Superior Investigaciones Cientif; Aplic Enzimaticas Alimentarias Sa; Consejo Sup Invest Cientificas, 2004.

Rosell, C. M., and C. Collar. "Effect of Various Enzymes on Dough Rheology and Bread Quality." In *Recent Research Developments in Food Biotechnology. Enzymes as Additives or Processing Aids*, 165–183. Kerala, India: Research Signpost, 2008.

Rosell, C. M., and R. Garzon. "Chemical Composition of Bakery Products." In *Handbook of Food Chemistry*. Berlin: Springer-Verlag, 2014.

Rosell, C. M., M. Haros, C. Escriva, and C. Benedito de Barber. "Experimental Approach to Optimize the Use of Alpha-Amylases in Breadmaking." [In English]. *Journal of Agricultural and Food Chemistry* 49, no. 6 (2001): 2973–2977.

Rosell, C. M., J. A. Rojas, and C. Benedito de Barber. "Combined Effect of Different Antistaling Agents on the Pasting Properties of Wheat Flour." [In English]. *European Food Research and Technology* 212, no. 4 (2001c): 473–476.

Rosell, C. M., J. Wang, S. Aja, S. Bean, and G. Lookhart. "Wheat Flour Proteins as Affected by Transglutaminase and Glucose Oxidase." [In English]. *Cereal Chemistry* 80, no. 1 (2003): 52–55.

Rosell, C. M., E. Santos, J. M. Sanz Penella, and M. Haros. "Wholemeal Wheat Bread: A Comparison of Different Breadmaking Processes and Fungal Phytase Addition." *Journal of Cereal Science* 50, no. 2 (2009): 272–277.

Selinheimo, E., K. Kruus, J. Buchert, A. Hopia, and K. Autio. "Effects of Laccase, Xylanase and Their Combination on the Rheological Properties of Wheat Doughs." [In English]. *Journal of Cereal Science* 43, no. 2 (2006): 152–159.

Selinheimo, E., K. Autio, K. Krijus, and J. Buchert. "Elucidating the Mechanism of Laccase and Tyrosinase in Wheat Bread Making." *Journal of Agricultural and Food Chemistry* 55, no. 15 (2007): 6357–6365.

Shafisoltani, M., M. Salchifar, and M. Hashemi. "Effects of Enzymatic Treatment Using Response Surface Methodology on the Quality of Bread Flour." *Food Chemistry* 148 (2014): 176–183.

Shah, A. R., R. K. Shah, and D. Madamwar. "Improvement of the Quality of Whole Wheat Bread by Supplementation of Xylanase from *Aspergillus foetidus*." *Bioresource Technology* 97, no. 16 (2006): 2047–2053.

Shalkh, I. M., S. K. Ghodke, and L. Ananthanarayan. "Inhibition of Staling in Chapati (Indian Unleavened Flat Bread)." *Journal of Food Processing and Preservation* 32, no. 3 (2008): 378–403.

Si, J. Q. "Use of Laccase in Baking." Report Novo Nordisk A/S. (1994).

Sirbu, A., and V. Paslaru. "Phospholipase Using as Bread-Making Improver." *Journal of Environmental Protection and Ecology* 8, no. 1 (2007): 1–10.

Smerdel, B., L. Pollak, D. Novotni et al. "Improvement of Gluten-Free Bread Quality Using Transglutaminase, Various Extruded Flours and Protein Isolates." *Journal of Food and Nutrition Research* 51, no. 4 (2012): 242–253.

Sozer, N., R. Bruins, C. Dietzel, W. Franke, and J. L. Kokini. "Improvement of Shelf Life Stability of Cakes." [In English]. *Journal of Food Quality* 34, no. 3 (2011): 151–162.

Steffolani, M. E., P. D. Ribotta, G. T. Pérez, and A. E. León. "Effect of Glucose Oxidase, Transglutaminase, and Pentosanase on Wheat Proteins: Relationship with Dough Properties and Bread-Making Quality." *Journal of Cereal Science* 51, no. 3 (2010): 366–373.

Steffolani, M. E., P. D. Ribotta, G. T. Perez, M. C. Puppo, and A. E. Leon. "Use of Enzymes to Minimize Dough Freezing Damage." *Food and Bioprocess Technology* 5, no. 6 (2012): 2242–2255.

Stojceska, V., and P. Ainsworth. "The Effect of Different Enzymes on the Quality of High-Fibre Enriched Brewer's Spent Grain Breads." *Food Chemistry* 110, no. 4 (2008): 865–872.

Storck, C. R., E. D. Zavareze, M. A. Gularte, M. C. Elias, C. M. Rosell, and A. R. G. Dias. "Protein Enrichment and Its Effects on Gluten-Free Bread Characteristics." [In English]. *LWT-Food Science and Technology* 53, no. 1 (2013): 346–354.

Sung-Ho, L., K. Young-Wan, L. Suyong et al. "Modulation of Cyclizing Activity and Thermostability of Cyclodextrin Glucanotransferase and Its Application as an Antistaling Enzyme." *Journal of Agricultural and Food Chemistry* 50, no. 6 (2002): 1411–1415.

Tester, R. F., J. Karkalas, and X. Qi. "Starch—Composition, Fine Structure and Architecture." [In English]. *Journal of Cereal Science* 39, no. 2 (2004): 151–165.

Trogh, I., J. F. Sorensen, C. M. Courtin, and J. A. Delcour. "Impact of Inhibition Sensitivity on Endoxylanase Functionality in Wheat Flour Breadmaking." *Journal of Agricultural and Food Chemistry* 52, no. 13 (2004): 4296–4302.

Van Der Borght, A., H. Goesaert, W. S. Veraverbeke, and J. A. Delcour. "Fractionation of Wheat and Wheat Flour into Starch and Gluten: Overview of the Main Processes and the Factors Involved." [In English]. *Journal of Cereal Science* 41, no. 3 (2005): 221–237.

van der Maarel, M. J. E. C., B. van der Veen, J. C. M. Uitdehaag, H. Leemhuis, and L. Dijkhuizen. "Properties and Applications of Starch-Converting Enzymes of the A-Amylase Family." *Journal of Biotechnology* 94, no. 2 (2002): 137–155.

Vemulapalli, V., and R. C. Hoseney. "Glucose Oxidase Effects on Gluten and Water Solubles." *Cereal Chemistry* 75 (1998): 859–862.

Veraverbeke, W. S., and J. A. Delcour. "Wheat Protein Composition and Properties of Wheat Glutenin in Relation to Breadmaking Functionality." [In English]. *Critical Reviews in Food Science and Nutrition* 42, no. 3 (2002): 179–208.

Wen, J., J. Lin, X. Wang, and J. Zhao. "Effects of Enzymes on the Rheologic Peculiarities and Food Making Properties of Flour Dough." *Food Science and Technology* no. 4 (2003): 44–46.

Wikstrom, K., and A. C. Eliasson. "Effects of Enzymes and Oxidizing Agents on Shear Stress Relaxation of Wheat Flour Dough: Additions of Protease, Glucose Oxidase, Ascorbic Acid, and Potassium Bromate." *Cereal Chemistry* 75, no. 3 (1998): 331–337.

World Health Organization, Geneva. "Safety Evaluation of Certain Contaminants in Food. Prepared by the Sixty-Fourth Meeting of the Joint FAO/WHO Expert Committee on Food Additives (Jecfa)." [In English]. *FAO Food and Nutrition Paper* 82 (2006). 1–778.

Yoo, J., B. P. Lamsal, E. Haque, and J. M. Faubion. "Effect of Enzymatic Tempering of Wheat Kernels on Milling and Baking Performance." *Cereal Chemistry* 86, no. 2 (2009): 122–126.

Zhang, C., S. Zhang, Z. Lu et al. "Effects of Recombinant Lipoxygenase on Wheat Flour, Dough and Bread Properties." *Food Research International* 54, no. 1 (2013): 26–32.

Zyla, K., M. Mika, H. Gambus, A. Nowotny, and B. Szymczyk. "Fungal Phytases in Wholemeal Breadmaking. I: 3-Phytase a Improves Storage Stability and in Vitro Digestibility of Nutrients in Wheat Breads." *Electronic Journal of Polish Agricultural Universities* 8, no. 4 (2005).

8

Enzymes in Confectioneries

Muthusamy Chandrasekaran and P. S. Beena

CONTENTS

8.1 Introduction

Chocolates and confectioneries have influenced mankind in its various tastes and forms ever since they were first made. Consumers of all age groups prefer chocolate and confectionery products because of their attractive color, appearance, and flavor. Nowadays, varieties of products have gained importance due to their delicious taste and shelf life, making a direct impact on demand. All of these delicious products range from colorful wrapped chocolates to chocolate bars, baking chocolate to cocoa powder. The chocolate and confectionery category, the second largest packaged food segment, has been growing steadily in all regions over the last few years. Globally, chocolate confectionery is the largest sector in terms of value, accounting for almost 60% of total sales.

8.2 Types of Confectionery

"Confectionery" is a term that refers to food items that are rich in sugar content and the art of creating sugar-based dessert forms. Nowadays, this term is also used to include substances rich in artificial sweeteners as well. In general, confections include sweet foods, sweetmeats, digestive aids that are sweet, elaborate creations, and some amusing and frivolous things (http://innutri.ch/confectionery/). Confectionery is broadly classified as two types, namely "bakers confections" and "sugar confections" (International Food Information Service 2009). "Bakers" confectionery, also called "flour confections," include principally sweet pastries, cakes, and similar baked goods whereas "sugar confectionery" includes sweets, candied nuts, chocolates, chewing gum, sweetmeats, pastillage, and other confections that are made primarily of sugar. The words "candy" (United States and Canada), "sweets" (UK and Ireland), and "lollies" (Australia and New Zealand) are also used for the extensive variety of sugar confectionery. In certain cases, "chocolate confections" (confections made of chocolate) are treated as a separate category as are sugar-free versions of sugar confections (Edwards 2000; http://lexbook.net/en/confection).

Based on sugar content, the confectionery can be of many types, particularly whether the products are made by crystallized sugar or not: boiled sweets; caramel, toffee, and fudge; gums and jellies; liquorice paste, cream paste, and aerated confectionery; tablets, lozenges, and sugar panning; medicated confectionery and chewing gum; centers, fondants, marzipan, and crystallized confectionery. Various types of confectionery products are summarized in Table 8.1.

8.3 Ingredients of Confectionery

Confections are low in micronutrients but rich in calories. The compositions comprise a sweetening ingredient, a proteinaceous material, and a thickening agent and can further comprise various additives, such as colorants, flavoring agents, and preservatives. The compositions can be provided in the form of a dry mix, which can be stored for long periods of time, or in the form of a ready-to-use composition. Ingredients used in confectioneries are listed in Table 8.2.

8.3.1 Sweetening Ingredients

Sweetening ingredients form the main inclusive component in confectionery products and natural sweeteners, and sweetening substances derived from plant materials are commonly used. The most important commercially available sweetener is sucrose, which is generally referred to as sugar. Since the fourteenth century AD, when sugar was first refined, sucrose has been used both at home and throughout the industry as a sweetening agent. Until recently, it was the only natural sweetener so used. Sucrose is the sugar generally used in confectionery factories in different forms and in different grades, such as granulated, icing, coarse sugar, medium sugar, powdered sugar, caster sugar, ultrafine sugar, fine sugar etc. Commercially, sucrose is processed from cane (60%) and beet (40%). Molasses, the material left when no more sugar can be extracted from sugar beet or cane, and treacle are used in the preparation of several products. In particular, clarified molasses are blended with sugar syrups in some confectioneries. Invert sugar is also used in confectioneries as this improves the flavor of the product. But nowadays, invert sugar has been replaced with glucose syrup in the industry.

The phenomenal increase in the price of this commodity coupled with advances in starch technology has led to the production of natural sweeteners from unconventional sources other than the sugar cane and beet root (Ahure and Ariahu 2013). A review of recent developments in this area of research indicates that starch, the normal form of stored carbohydrate in plants, is the most promising source of sweeteners (Inglet 1981). The principle sources of starch for the industry are corn, cassava, wheat, potato, sorghum, and similar plants. The contribution from these sources depends on their availability and cost (Ahure and Ariahu 2013).

TABLE 8.1

Various Types of Confectionery Products Commercially Available in the Market

Type	Product	Characteristics, Raw Materials, and Additives
Candies	Boiled sweets and hard candy	Hard candies and boiled sweets with sucrose/glucose syrup as major ingredients
	Pulled candy	Hard candy similar to fondant with low moisture content
	Fondant	Partially crystallized sugar glucose syrup
Caramels	Caramels	Low-cooked confections; has equal glucose-to-sugar ratio. The syrup–milk mixture is cooked to varying degrees to give hard caramels or soft caramels
	Toffees	High-boiled caramels with less milk
	Fudge	Like a fondant made using sugar
	Nougats and marshmallows	Made in grained or ungrained form, usually covered with chocolate to prevent drying out
Jellies and gums	Jellies and gums	Single-phase low-boiled, high-moisture confection with equal ratio of sugar to corn syrup. Starch, high methoxyl, and low methoxyl pastries, gelatin, and agar-agar are used in confectionery gels
	High-fructose corn syrup (HFCS)	Liquid sweetener alternative to sucrose (table sugar) used in many foods and beverages. The glucose-to-fructose ratio in HFCS is nearly 1:1, similar to the ratio in sucrose, invert sugar, and honey
	Butterscotch	Primary ingredients are brown sugar and butter although other ingredients, such as corn syrup, cream, vanilla, and salt, are part of some recipes
Chocolates	Chocolates, milk chocolates, milk chocolates with biscuits	Cocoa powder, cocoa butter, milk
Other confections	Dragee, (Jordal almonds, panned chocolates, medicinal dragees, and metallic decorative balls)	Colorful form of confectionery with a hard outer shell. Used for decorative, symbolic, medicinal purposes in addition to consumption purely for enjoyment
	Dodol	A sweet toffee-like confection made with coconut milk, jaggery, and rice flour
	Halvah	Dense flour-based sweet confections, made from grain flour, clarified butter, and sugar; nut butter-based made from sesame paste or other nut butters, such as sunflower seed butter and sugar
	Ice cream	Frozen food usually made from dairy products, such as milk and cream, and often combined with fruits or other ingredients and flavors
	Lollies	Sugar is the principal ingredient

8.3.1.1 Function of Sugars in Confectionery

The main reasons for the use of sugars in confectionery include (i) their ability to preserve food; (ii) their contribution to body, flavor, and texture; and (iii) their ability to enhance the flavor of other ingredients (http://www.niir.org/books/book/confectionery-products-handbook-chocolate-toffees-chewing-gum-sugar-free-onfectionery/). However, in the technology of confectionery manufacturing, the combination of these sugars is the important factor. The presence of glucose syrup, for example, will retard sucrose crystallization, increase viscosity, and provide a syrup concentration that resists microbiological action during its shelf life.

8.3.2 Other Raw Materials

Although the sugars described are the primary raw material, other ingredients also play an important part in the formulation of the variety of confections available today.

TABLE 8.2

Ingredients Used in Confectioneries

Ingredient	Function and Applications
Sweetening agents are major ingredients. Sugar in the form of sucrose, used as brown sugar or refined sugar derived from cane sugar or molasses; invert sugar; corn syrup; liquid glucose; glucose; dextrose; high-fructose corn syrup, maltodextrin, cyclodextrin	Major component; preserve food; contribute to body, flavor, and texture; enhance the flavor of other ingredients. Used in all types of confectionery products
Milk: condensed milk, sweetened milk; modern confectioneries use evaporated milk, that is, unsweetened condensed milk	Used in the manufacture of caramel, fudge, and toffee
Oils and fats: vegetable oils, natural oils, cocoa butter, lactic acid butter, sweet cream butter, and whey butter	Major component used in many recipes
Egg albumen	Used as whipping agent
Gelatin	Depression of surface tension in concentrated sugar syrup compared to in water. Used as a gelling agent as well as whipping agent particularly for mallow products
Agar, pectin	Gelling agent. Used in jellies, whipped and aerated products
Starch, modified starch	Major energy storage of confectioneries; stabilizer; preparation of jellies
Soy flour, minor ingredient	Nutritional filler
Fruit jams; dried fruit; nuts	Flavor, texture, and appearance
Vegetable lecithin, glyceryl monostearate	Emulsifiers stabilize the fat dispersion when fats are included in syrup-based formulations

8.3.2.1 Oils and Fats

Oils and fats are another major component of confectioneries. Fats that are used in confectioneries conventionally are milk fats, such as buttercream, whole milk powder, or condensed milk. Butter is used principally as an ingredient in toffees and butterscotch. Lactic acid butter, sweet cream butter, and whey butter are the generally used butters in confectioneries. Whey is used in confectionery as whey powder, but it is not widely used as it is not pleasant. Vegetable fats are used alternatively, and they are cheaper than milk fat. Vegetable fats can be more blended, hydrogenated, or interesterified; they are used in many recipes and are usually refined and hardened natural oils. The one exception is cocoa butter, the natural fat of the cocoa bean.

8.3.2.2 Dairy Ingredients

Milk is used in the manufacture of caramel, fudge, and toffee. Dairy ingredients in confectioneries are not made directly from liquid milk. Concentrated milk and sweetened milk are considered best. Solid milks are used either as milk solids or sweetened condensed milk. But most modern confectioneries use evaporated milk, that is, unsweetened condensed milk. Skim milk solids are an essential part of toffees and fudge. Toffees made from sweetened condensed milk are normally smoother than those made from milk powder.

8.3.2.3 Major Energy Storage

Different types of starch form the major energy storage in confectioneries, and polysaccharides derived from cereal crops serve the purpose. Starches in confectionery are classified based on their uses as "gelling" and "nongelling." Starch with gelling property is utilized to make jellies, pastilles, and wine gums as thin boiling starches, and starch with nongelling property is utilized as a substitute for gum, such as oxidized waxy maize starch. Different starches that have been used in confectioneries are pregelatinized starches and oxidized starches that reduce the micelle formation and make paste more stable. Gum tragacanth that comes from the shrub *Astragalus* is a polysaccharide polymer that has glucuronic acid and arabinose as its monomers are used in confectioneries, such as lozenges (Edwards 2000). Locust bean or carob bean gum obtained from the carob tree *Cevatonia silique* is a galactomannan compound of β-D-mannose and α-D-galactose units, and it is used as a thickener (Panda et al. 2012). Xanthan gum is the first microbial gum used in confectionery produced by aerobic fermentation of *Xanthomonas campestris*.

0.3.3 Minor Ingredients

Gelatin, agar, pectin, and egg albumen are used as "gelatinizing" and "aerating" agents for making jellies and whipped or aerated products. Starches and modified starches are used as stabilizers and also in the preparation of a class of jellies. Soy flour may be regarded as a nutritional filler. Preserved fruit jams and dried fruit are often included in confectionery for their flavor, texture, and appearance. Nuts may be added in a comminuted form as in marzipan or in chopped pieces. In the latter case, the confection must be one of low moisture, such as hard caramel toffee or nougat. In higher-moisture materials, nuts lose their crispness because of the differences in water activity between the nuts and the matrix. Emulsifiers are needed to stabilize the fat dispersion when fats are included in syrup-based formulations. The main emulsifiers are vegetable lecithin and glyceryl monostearate.

8.4 Enzymes in Confectionery Production

Various hydrolytic enzymes are used to facilitate the manufacture of candies and confectioneries. Invertase, for example, is used to prepare invert sugar and to bring about the partial liquefaction of fondant centers. Amylases and invertases are used to make chocolate-converted soft cream candies, and they are used with proteases to assist in the recovery of sugars from candy scraps (Cowan 1983). An enzyme mix of pectinases, cellulases, invertase, and proteases are used to make candied fruits, and amylases are also employed for the production of high-maltose and high-glucose syrups as sweeteners for use in hard candies, soft drinks, and caramels (Mochizuki et al. 1971). Lipolytic enzymes are used to modify butterfat to increase buttery flavors in candies and caramels and reduce sweetness in these products (Burgess and Shaw 1983). Other enzymes are used to reclaim defective products in syrup by the destruction of pectin and egg albumen residues, which hinder filtration. Phospholipase (NzyCake PLC, a product of Nature Biochem, Bangalore, India) is used for improvement in cake structure and shelf life. It contributes to softness that lasts and improves the overall quality of pound cakes, sponge cakes, and muffins (http://naturebiochem.com/downloads/Baking NATBIO.pdf). SternEnzym, Germany, produces "Sweetase" (used in chocolate) and "Bactozym" (used in chewing gum) for application in confectioneries. The principle enzyme in Sweetase is invertase, and that in Bactozym is lactoperoxidase. Novozymes, Denmark, is a world leader in enzyme manufacturing and provides a variety of enzymes to be used in food and confectioneries. A summary of enzymes used in confectioneries and chocolates is presented in Figure 8.1 and in Table 8.3.

8.4.1 Invertase

Invertase is used in the production of confectionery with liquid or soft centers. A well-known application is in the production of soft-centered sweets, such as after-dinner mints. The centers are formulated using crystalline sucrose and tiny (about 100 U kg^{-1}, 0.3 ppm, w/w) amounts of invertase. The small amount of enzyme facilitates very slow inversion of sucrose so that the center remains solid long enough for enrobing with chocolate to be completed. Then, over a period of days or weeks, sucrose hydrolysis occurs, and the increase in solubility causes the centers to become soft or liquid, depending on the water content of the center preparation (http://www1.lsbu.ac.uk/water/enztech/sucrose.html). During the process, invertase splits the disaccharide sucrose into the monosaccharides glucose and fructose. A little water

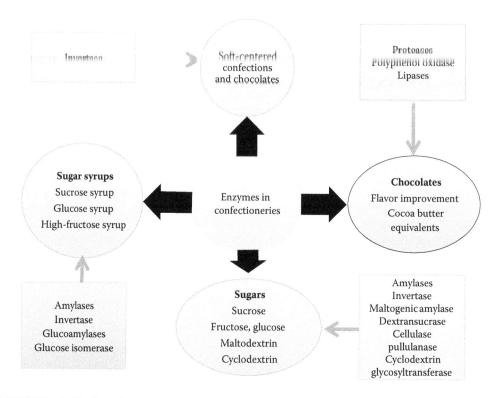

FIGURE 8.1   Applications of enzymes in confectionery production.

is consumed in this process: exactly one molecule per molecule of sucrose. The mixture of sucrose, glucose, and fructose has a lower viscosity and shows fewer tendencies to crystallize than sucrose alone. So the product stays softer. Moreover, fructose is hygroscopic: it attracts and binds water. This reduces the tendency of the water to evaporate. Invertase can therefore be used to prolong the shelf life of cream fillings and marzipan, etc. (http://www.ncbe.reading.ac.uk/ncbe/materials/enzymes/invertase.html). A solid paste with the consistency of fudge is made using sucrose (table sugar). A small amount of invertase is added to this fondant before it is enrobed in chocolate. During storage for a couple of weeks at 18°C, the enzyme partially liquefies the sucrose within the chocolate shell (http://www.ncbe.reading.ac.uk/ncbe/materials/enzymes/invertase.html).

The quality of many confectionery products depends on the softness and smoothness of their texture. But the moisture, which is the crucial component of this texture, is easily lost because water evaporates and, when this happens, sugar crystallizes out, leaving the product dry and hard. There are two ways of avoiding or retarding this process. Sorbitol syrup has the property of binding water and can be substituted for part of the sucrose or glucose content. The other method relies on the action of invertase, an enzyme that gradually converts sucrose into invert sugar (equal parts of glucose and fructose)—a form of sugar that does not readily crystallize and is also hygroscopic and moreover also tends to impart this property to its environment.

Invertin 104738 (Merck®, Germany) is a sweetish, nearly clear, light yellow, viscous liquid with a slightly yeasty smell. Its active component is the enzyme β-fructosidase (derived from yeast cells) carried in glycerol, and its enzymic activity (as measured by the Weidenhagen method) is always constant. Because the enzyme (unlike Sorbitol) is destroyed by temperatures above 65°C, it can only be added to products requiring boiling after they have cooled. It is similarly deactivated by an excess of alcohol and can therefore not be used in mixes containing more than 20% (vol) of this. The rate of inversion can be accelerated by adjusting the pH to 4.5–5.5, for example, by adding to a neutral (pH 7) mix of approximately 2 ml of 50% citric acid solution per 10 kg of composition, provided that this is not objectionable on flavor grounds. When flavor considerations prevent the use of citric acid, the amount of Invertin can be

TABLE 8.3

Enzymes Used in the Confectionery Industry

Enzyme	Applications	References
Invertase	Conversion of sucrose into glucose and fructose and sugar syrup; inverted syrup production; production of confectionery, including chocolates with liquid or soft centers; prolong the shelf life of cream fillings and marzipan	http://www1.lsbu.ac.uk/water/enz tech/sucrose.html http://www.ncbe.reading.ac.uk
Pullulanase	Saccharification	Brienzo et al. 2008
Glucoamylase	Starch saccharification	Kovalenko et al. 2010
Amylases	Production of high-maltose and high-glucose syrups as sweetener for use in hard candies, soft drinks, and caramels	Burgess and Shaw 1983
Amylases and invertases	Chocolate-converted soft cream candies	Cowan 1983
Pectinases, cellulases, invertase, and proteases	Make candied fruits	Mochizuki et al. 1971
Dextransucrase	Dextran production	http://www1.lsbu.ac.uk/water/enz tech/sucrose.html
Fungal raffinase	Raffinose	http://www1.lsbu.ac.uk/water/enz tech/sucrose.html
α-Amylase and glucoamylase	High-fructose syrup (HFS) from corn starch	Johnson et al. 2009
α-Amylases and maltogenic enzymes (fungal alpha-amylases, vegetable beta-amylases)	Maltose syrups	Pontoh and Low 1995
Cellulase, α-amylase, and glucoamylase	Glucose and high-fructose syrup from tapioca starch	Johnson and Padmaja 2012
α-Amylase, glucoamylase, glucose isomerase	High-fructose corn syrup	Parker et al. 2010
Malto-oligosaccharide-forming amylase	Malto-oligosaccharide syrup	http://www.amano-enzyme.co .jp
α-Amylases, glucoamylases	Maltodextrins (MD)	Fontana et al. 2001
α-Amylase, glucoamylase, and pullulanase	Maltodextrins (MD) and glucose syrup (GS) from cassava	Lambri et al. 2014
Cyclodextrin glycosyltransferase (CGTase)	Cyclodextrin production	Li et al. 2014; van der Veen et al. 2000a,b
Pectinase and cellulase	Date syrup	Abbès et al. 2011
β-fructosidase (Invertin)	Marzipan products	
Polyphenoloxidase	Reduction in bitter taste and astringency of cocoa products; improvement in flavor	Brito et al. 2004; Hammer 2013
Proteases	Recovery of sugars from candy scraps	Cowan 1983
Aspartic endoprotease and a carboxipeptidase	Cocoa flavor development	Oliveria et al. 2011; Voigt et al. 1994a,b
Lipase	Produce cocoa butter equivalents	Nazaruddin et al. 2011; Kim et al. 2014
Lipases	Modify butterfat to increase buttery flavors in candies and caramels and reduce sweetness in these products	Burgess and Shaw 1983
Phospholipase	Improvement in cake structure and shelf life	http://naturebiochem.com/

increased instead. Invertin is particularly beneficial for (and widely used) in marzipan products. To protect these from fermentation, yeasts, and molds when not stored under ideal conditions, the addition of a small amount of sorbic acid (1.5 g per kg), uniformly distributed through the mix, are strongly recommended. Sometimes, a viscous, opaque fondant cream (or near-solid fondant) has to be used to fill chocolate hollows so that these can be closed with couverture without difficulty, but it is usually desirable that the center should subsequently become a clear liquid. This can be achieved with the aid of Invertin added to the composition shortly before filling and after the temperature has dropped below 65°C, but preferably before the addition of alcohol and at a rate of between 150 and 250 ml of Invertin per 100 kg of composition (http:// www.keylink.org/knowledgebank/invertin-an-enzyme-confectionery-humectant.php).

8.4.2 Sucrose Syrup

The hydrolysis ("inversion") of sucrose, completely or partially, to glucose and fructose provides sweet syrups that are more stable (i.e., less likely crystallize) than pure sucrose syrups. Yeast (*Saccharomyces cerevisiae*) invertase is used to produce other types of syrup from a cane sugar refinery. In spite of the fact that this enzyme suffers from substrate inhibition at high sucrose levels (>20%, w/w), this enzyme is commercially used at even higher concentrations (http://www1.lsbu.ac.uk/water/enztech/sucrose.html). Traditionally, invertase was produced on site by autolyzing yeast cells. The autolysate was added to the syrup (70% sucrose, w/w) to be inverted together with small amounts of xylene to prevent microbial growth. Inversion was complete between 48 and 72 h at 50°C and a pH 4.5. The enzyme and xylene were removed during the subsequent refining and evaporation. Partially inverted syrups were (and still are) produced by blending totally inverted syrups with sucrose syrups. Now, commercially produced invertase concentrates are employed (http://www1.lsbu.ac.uk/water/ enztech/sucrose.html).

During sugar syrup production from cane sugar refining, enzymes are also used as aids to remove materials that inhibit crystallization or cause high viscosity. In some parts of the world, sugar cane contains significant amounts of starch, which becomes viscous, and retards filtration processes in addition to making the solution hazy when the sucrose is dissolved. This problem is solved with the most thermostable α-amylases (e.g., Termamyl-Novozymes at about 5 U kg⁻¹), which are entirely compatible with high temperatures and pH values that prevail during the initial vacuum evaporation stage of sugar production (http://www1.lsbu.ac.uk/water/ enztech/sucrose.html).

A dextran is produced by the action of dextransucrase (EC. 2.4.1.5) from *Leuconostoc mesenteroides* on sucrose and found to be slime on damaged cane and beet tissue, especially when processing has been delayed in hot and humid climates. Raffinose, which consists of sucrose with α-galactose attached through its C-1 atom to the six-position on the glucose residue, is produced at low temperatures in sugar beets. Both dextran and raffinose have the sucrose molecule as part of their structure, and both inhibit sucrose crystal growth. Dextran can produce extreme viscosity in process streams and even bring plants to a stop. Extreme dextran problems are frequently solved by the use of fungal dextranases produced from *Penicillium* species (Novo Nordisk A/S 1977). These are used (e.g., 10 U kg⁻¹ raw juice, 55°C, pH 5.5, 1 h) only in times of crisis as they are not sufficiently resistant to thermal denaturation for long-term use and are inactive at high sucrose concentrations. Raffinose may be hydrolyzed to galactose and sucrose by a fungal raffinase (Davies 1953). The enzyme is also used to remove raffinose and stachyose from soybean milk, which causes flatulence. It is an active ingredient in the production of cocoa (chocolate). Due to all of these industrial uses, raffinase is gaining a high level of importance in industry.

8.4.2.1 Starch-Derived Syrups

Root and tuber crops, along with corn (maize), supply most of the starch in Asian markets. Globally, the major commodities from which starch is derived are corn, cassava, sweet potato, potato, and wheat, with corn accounting for 37% of the starch production. In developing countries, root crops are relatively more important as sources of starch than cereal crops. Root crops account for 60% of Asian starch production, especially cassava (29%), sweet potato (26%), and potato (5%) (Johnson and Padmaja 2013). The transformation of starch into sugar is an important branch of the starch industry and is one of the most important applications of biotechnology. Many industrial products, such as modified starches, maltodextrins, high-fructose syrup (HFS), glucose syrup, gums, adhesives, bioethanol, etc., are made from starch (Bindumole and Balagopalan 2001; Shetty et al. 2007). The biggest user of starch worldwide is the sweetener industry. It can easily be hydrolyzed to form syrup containing dextrose, maltose, and other oligosaccharides. Enzymes are the key to these chemical reactions; enzymes that are predominantly produced with the help of genetically modified microorganisms. The production of sugar syrups by an enzymatic method is among the most advanced food technologies, characterized by higher yields, a wide range of products, higher product quality, and energy economy (Blanchard and Katz 1995). HFSs are widely employed as nutritional sweeteners because of characteristics such as a noncrystalline nature, high sweetness, better solubility, etc. (Hanover and White 1993; White 1992). Cornstarch is the raw material for high-fructose syrup production in the United States and in many other parts of the world (Cabello 1999). However, alternative starch sources have been reported as potential raw materials in Europe, South America, and Asia (Schenck and Hebeda 1992). Production of HFS, glucose syrups, and maltodextrins from nonconventional sources of starch, such as amaranth, cassava, potato, and sorghum, has been reported (Abraham et al. 1988; Aschengreen et al. 1979; Gorinstein and Lii 1992). There are several reports on the studies conducted to decrease production costs and also to find a new and economically attractive starch source for glucose and fructose syrup production (Bandlish et al. 2002; van der Veen et al. 2005; Voragen 1998). Studies were done by several workers to economize the production of HFS through synergistic action of α-amylase and glucoamylase on raw corn (Johnson et al. 2009).

8.4.3 Glucose Syrup

Glucose syrup (GS) is a food ingredient obtained from the hydrolysis of starch, and it consists of glucose monomer and varying quantities of dimer, oligosaccharides, and polysaccharides, dependent on the GS in question and its process of manufacture (Dziedzic and Kearsley 1984). Several industries in West Africa, especially in Ghana and Nigeria, use glucose syrup in their manufacturing operations, and the bulk of the syrups are made from cassava and rice malt. The aim is to provide local supply sources for glucose syrup to substitute for the imported product (Dziedzoave et al. 2003).

High-glucose syrup (HGS) is mainly used as a source of crystalline dextrose or as a substrate for HFS production. HGS finds application in the beverage and confectionery industries, and HFS is predominantly used in ice cream, yogurt, processed foods, and as feed for honeybees (Johnson et al. 2009).

Total enzyme conversion using thermostable α-amylases and maltogenic enzymes (fungal α-amylases, plant-derived β-amylases) allows maltose syrups of higher maltose levels (approximately 55%) to be produced. These have led to significant improvements in confectionery, particularly in high-boiled products, in which better control of the rheology of the sugar mass has resulted in the ability to use higher sugar replacement levels and thus to obtain better texture, better shelf life, low color formation, and other properties.

Enzymatic production of glucose syrup from starch is a multistage process, involving liquefaction, saccharification, purification, and concentration. Generally, the key features of the liquefaction enzymes are (i) high-dextrose yields with minimal by-product formation; (ii) fast viscosity reduction, enabling high dry substance levels; and (iii) low color formation and reducing the refinery costs (Pontoh and Low 1995). The saccharification is done by using an exo-acting glucoamylase, which specializes in cleaving α-1,4 glucosidic bonds and slowly hydrolyzes α-1,6 glucosidic bonds present in maltodextrins. This will result in accumulation of isomaltose. Therefore, recently, pullulanase, which efficiently hydrolyzes α-1,6 glucosidic bonds, is used (Brienzo et al. 2008; Ezeji and Bahl 2006; van der Maarel et al. 2002). It is therefore well suited for "debranching" starch after liquefaction. Novozym® 26062 is a commercial pullulanase preparation supplied by Novozymes, Denmark. The operating conditions of some commercial amylolytic enzyme preparations for performance of starch hydrolysis and production of corn syrup were studied in addition to identification of the carbohydrate profile of the produced corn syrup (Eshra et al. 2014).

Five commercial amylolytic enzyme preparations, two liquefying (Termamyl Supra and Clarase L40, 000) and three saccharifying (AMG E, Dextrozyme DX, and Optimax 4060 VIIP), were used for the performance of starch hydrolysis to produce corn syrup (Eshra et al. 2014). The operating conditions of these preparations showed that AMG E was the least effective enzyme within the saccharifying enzymes tested. Four enzyme combinations from the other four liquefying and saccharifying enzymes were tested for starch hydrolysis. The results indicated that the liquefaction period by Termamyl S and Clarase must not exceed more than 90 min whereas the best starch concentrations were 30 gdl^{-1} for Termamyl S and 40 gdl^{-1} for Clarase. It was found that the combinations of Clarase followed by Optimax or Dextrozyme were more effective than those of Termamyl S, followed by the same two enzymes. The carbohydrate profile of the produced corn syrup showed that glucose is the main component (86.92%). The values of dextrose equivalent (DE) and the true dextrose equivalent (DX) of corn syrup were 79.587 and 85.334, respectively (Eshra et al. 2014).

8.4.4 Glucose-Fructose Syrups

Starch-derivative products contribute to a considerable increase in the shelf life of confectionery prod ucts (candies, waffles, and cakes). Further, consumption of "sweet" products containing no sucrose and having dietary properties is recommended for diabetes mellitus patients and people suffering from overweight. Consequently, starch and sweeteners, such as starch treacle, glucose, and glucose-fructose syrups, are not only used on a large scale in the food industry, but they are also in high demand. These sweeteners derived from starch-containing renewable resources (potatoes, grains of corn, rye, wheat, barley, ambercane, and millet) completely or partially replace cane or beet sugar, for instance, by 90%–100%, 50%, 35%, and 25% in nonalcoholic beverages, confectionery products, dairy products, and bakery products, respectively.

Since the 1980s, glucose-fructose syrups (GFSs) have witnessed a substantial growth of annual production volume and consumption such that the annual world volume of syrup reached ~15 million tons by 2005 (Bucholz et al. 2005). In the United States, the GFS production records an annual increase by 4%–5%, and according to the specialists' estimations, in 2015, the annual consumption of GFSs will exceed 17 million tons, which will comprise 35% of the sweetener market. It is speculated that processing of vegetable starch-yielding plants, such as potatoes, grains of corn, rye, wheat, barley, ambercane, and millet (by means of partial or complete hydrolysis of starch), could contribute to the increase in the production of sweeteners.

Further use of relatively inexpensive starch-containing products (corn and wheat) as the initial products for production of GFSs could facilitate economic production of the syrups produced from them by reducing the cost by 10%–50% compared to that of natural sugar produced from cane (beet).

Along with GFS, in the food industry, and particularly on confectionery plants and nonalcoholic beverage plants, the "liquid sugar" in form of the solution of high-purity sucrose, inverted syrup (invert), or inverted sugars with various additives are in great demand. In the United States and England, the production of liquid sugar, mainly in the form of partially or completely inverted sugar, comprises more than 30% of the total sugar production. Today inverts (inverted sugar), which comprise the formulation of a wide range of confectionery products (pastry, cakes, and biscuits), are produced at confectionery plants by continuous acid hydrolysis of sugar syrups at elevated (70°C–80°C) temperatures according to the primary needs of a particular production. The product produced using acid technology has many undesirable side impurities, which are sometimes toxic (in particular, oxymethyl-furfurol, which makes the final product brown). The completely inverted sugar syrup composition is identical to GFS-50 (with a 50% fructose content) by its carbohydrates.

The biocatalyst with glucoamylase activity prepared by adsorption of the enzyme GlucoLux (Sibbiopharm Ltd., Russia) on the nanoporous carbon support Sibunit was observed to possess an ability for significant starch saccharification (Kovalenko et al. 2010). The biocatalyst possessed remarkably high activity (ca. 700 U/g) and stability (the half-inactivation time t1/2 > 350 h) at 60°C. For glucose isomerization, the biocatalyst with glucose isomerase activity was prepared by inclusion of biomass of bacterial strains producing intracellular enzyme and insoluble hydroxo compounds of Co(II), which are required for enzyme stabilization, inside the silica xerogel. The stability of the biocatalyst in continuous process of glucose isomerization was sufficiently high (t1/2 > 500 h) at 65°C. For sucrose inversion the biocatalyst with invertase activity was prepared by inclusion of biomass of baker's yeast inside the silica xerogel. At an optimal ratio of biomass to SiO_2, the invertase activity of the biocatalyst reached up to ca. 600 U/g, and t1/2 was more than 200 h at 50°C (Kovalenko et al. 2010).

Cassava (*Manihot esculenta* Crantz) is one of the largest sources of starch and sago. The cassava fibrous residue (CFR) contains 61% to 63% starch and other polysaccharides and causes major disposal problems. An investigation was conducted on the production of glucose syrup from cassava fibrous residue by hydrolyzing starch and cellulose with various enzyme treatments, followed by conversion to high-fructose syrup (Johnson and Padmaja 2012). The investigators standardized various conditions for the enzyme hydrolysis and found that a combination of α-amylase and glucoamylase resulted in 52.88%–54.24% conversion of CFR to glucose. They also noted that the yield could be enhanced to 58.70%–60.00% by adding the cellulase enzyme complex Accellerase 1000 in combination with α-amylase and glucoamylase. Cellulase activity helped to reduce the viscosity of the CFR slurry and also improved the starch hydrolysis by α-amylase and glucoamylase. According to Johnson and Padmaja (2012), the most suitable enzyme treatment was found to be simultaneous hydrolysis of CFR with glucoamylase, Dextrozyme GA, and cellulase enzyme Accellerase 1000 at pH 4.5 and 60°C for 48 h, followed by 1 h liquefaction with α-amylase, Liquezyme X. Hydrolysis of a higher substrate concentration, that is, 25%, was possible with the enzyme treatment, and production of high-fructose syrup from higher slurry concentrations reduces the cost of concentrating glucose syrup. Further, saccharification and isomerization of the two different samples of cassava fibrous residue also indicated that the yield of glucose and fructose depended on the starch content in the initial raw material (Johnson and Padmaja 2012).

Glucose and high-fructose syrup (HFS) are made extensively from cornstarch, and the high cost of production, demands the lookout for alternative starches as raw material. Johnson and Padmaja (2013) compared the potential of tuber starches, such as arrowroot, cassava, *Curcuma*, *Dioscorea* (air potato), sweet potato, and *Xanthosoma*, with cornstarch for HFS production, employing liquefaction followed by saccharification and isomerization using three enzymes, such as Liquezyme (thermostable α-amylase from Novozymes, Denmark), Dextrozyme (glucoamylase from Novozymes, Denmark), and Sweetzyme (glucose isomerase from NovoNordisk Biochem, USA), respectively. A high-performance liquid chromatographic (HPLC) profile showed that the starch conversion to glucose for starches was equivalent or superior to that for cornstarch. The sugar profile of the saccharified slurry had a composition of 98.28% to 98.84% glucose, maltose (1.03% to 1.69%), and maltotriose (0.03% to 0.10%) for arrowroot, Curcuma, and cassava, and a lower range of glucose (94.76%–97.28%) and higher range of maltose and maltotriose (2.0%–4.3% and 0.49%–0.75%, respectively) for the other starches. The percentage of conversion to fructose as well as fructose yield (g/100 g starch) was the highest for arrowroot and *Curcuma* starches. The study showed that the fructose yield and percentage of conversion to fructose from arrowroot, cassava, *Curcuma*, *Dioscorea*, and sweet potato were superior to corn and *Xanthosoma*; the five tuber starches could be considered as potential feedstocks for glucose and HFS production (Johnson and Padmaja 2013).

8.4.5 High-Fructose Corn Syrup Production

High-fructose corn syrup (HFCS) is a liquid alternative sweetener to sucrose that is made from corn, the "king of crops," using chemicals (caustic soda, hydrochloric acid) and enzymes (α-amylase and glucoamylase) to hydrolyze corn starch to corn syrup containing mostly glucose and a third enzyme (glucose isomerase) to isomerize glucose in corn syrup to fructose to yield HFCS. HFCS products are classified according to their fructose content: HFCS-90, HFCS-42, and HFCS-55. HFCS-90 is the major product of these chemical reactions and is blended with glucose syrup to obtain HFCS-42 and HFCS-55 (Figure 8.2).

HFCS has become a major sweetener and additive used extensively in a wide variety of processed foods and beverages, ranging from soft and fruit drinks, to yogurts and breads. HFCS has many advantages compared to sucrose that make it attractive to food manufacturers. These include its sweetness, solubility, acidity, and its relative cheapness in the United States. The use of HFCS in the food and beverage industry has increased over the years in the United States. The increase in its consumption in the United States has coincided with the increase in incidence of obesity, diabetes, and other cardiovascular diseases and metabolic syndromes (Parker et al. 2010). Grocery foods items found to contain HFCS are numerous. These include baked goods, such as pastries; biscuits, breads, cookies, and shortcakes; soft drinks; juice drinks; carbonated drinks; jams and jellies; dairy products, including ice creams, flavored milks, eggnog, yogurts, and frozen desserts; canned ready-to-eat foods, including sauces and condiments; cereals and cereal bars; and many other processed foods. A majority of processed foods in the United States, HFCS is used to attribute nutritional functionality in a majority of processed foods (Parker et al. 2010). In Europe, the product is not so widely used due to the imposition of a production quota by the EEC.

8.4.6 Malto-Oligosaccharide Syrup

Malto-oligosaccharide syrup is used in different food products, such as Japanese-style confectionery, western confectionery, candies, tsukudani (food boiled down in sweetened soy sauce), and various types

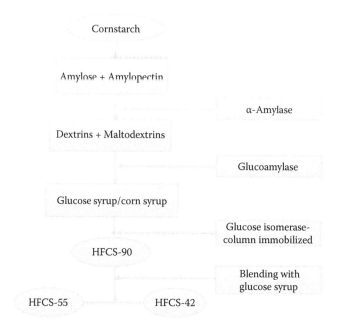

FIGURE 8.2 High-fructose corn syrup production. (Adapted from Parker, K., Salas, M. and Nwosu, V. C., *Biotechnology and Molecular Biology Review*, 5, 5, 71–78, 2010.)

of sauces and beverages. Malto-oligosaccharides carry a mellow sweetness and have a high moisture-retaining capacity and have an ability to suppress aging of starchy food products. As far as commercial products are concerned, starch syrup containing elevated levels of specific malto-oligosaccharides is being traded (http://www.amano-enzyme.co.jp/eng/enzyme/22.html).

Malto-oligosaccharide is derived from starch by the action of a Malto-oligosaccharide-forming amylase (EC. 3.2.1.l), which breaks down starch into oligosaccharides that are composed of glucose molecules joined by α-1,4 linkages, such as maltotriose, maltotetraose, maltopentaose, and maltohexaose. However, different enzymes are responsible for the production of each of the oligosaccharides listed above. Malto-oligosaccharide-forming amylases from different sources are used to produce malto-oligosaccharides with different degrees of polymerization of glucose molecules. Maltotriose is produced by malto-oligosaccharide-forming amylases obtained from *Microbacterium imperiale*, *Streptomyces griseus*, *Bacillus subtilis*, and other strains. Maltotetraose is produced by malto-oligosaccharide-forming amylases obtained from *Bacillus circulans*, *Pseudomonas stutzeri*, and other strains. Maltopentaose is produced by *Bacillus licheniformis*, *Bacillus subtilis*, *Bacillus cereus*, and other strains. Maltohexaose is produced by malto-oligosaccharide-forming amylases obtained from *Klebsiella pneumoniae*, *Bacillus subtilis*, *Bacillus circulans*, and other strains (http://www.amano-enzyme.co.jp/ eng/ enzyme/22.html).

8.4.7 Maltodextrins

Maltodextrins (MDs) are enzymatic and/or acid hydrolysis products of starch, consisting of α-(1,4) linked D-glucose oligomers and/or polymers, which are normally defined as having a dextrose equivalent (DE) value < 20. They are commonly used as spray-drying aids for flavors and seasonings, carriers for synthetic sweeteners, texture providers, fat replacers, film formers, and bulking agents in the food industry (Chronakis 1998). In starch processing for maltodextrin and glucose syrup production, thermoresistant α-amylases (maltogenic for amylose, malto-, and dextrinogenic for amylopectin; EC. 3.2.1.1) and amylo-glucosidases or glucoamylases (glucogenic for both starch fractions; EC. 3.2.1.3) are most intensively employed. The combination of both types of enzyme involves two steps, such as liquefaction in which the enzyme α-amylase partially hydrolyzes starch to maltodextrins and saccharification in which the low DE syrup is completely converted to glucose by glucoamylase (Fontana et al. 2001).

Although cassava is primarily used as a food crop, it is also widely used for the production of starch, and its role has been increasingly recognized for the production of polylactic acid, bioethanol, GS, and HFS (Chinnawornrungsee et al. 2013; Lauven et al. 2013; Moore et al. 2005). Further, the low content of lipids (0.1%) in cassava starch ensures that amylose–lipid complex formation is negligible, and consequently, good liquefaction can be achieved at a lower temperature, and retrogradation problems are less severe. An additional advantage of cassava starch is its low protein content. Hence, less color is developed during hydrolysis, and refining requirements are reduced (Carpio et al. 2011).

Lambri et al. (2014) produced maltodextrins (MDs) and glucose syrup (GS) from the direct conversion of cassava roots collected in Burundi. The detoxified cassava slices were previously gelatinized and then liquefied aimed at obtaining MD with a dextrose equivalent (DE) value < 20 using (V_{enzyme}/$W_{fresh\ mash}$) thermostable α-amylase (Liquezyme-X) at pH 6.5 before and after 10 min—90°C step at atmospheric pressure (P_{atm}) or 143.27 kPa (110°C) allowing starch gelatinization. The saccharification step followed the liquefaction in order to obtain a GS with DE close to 99. The hydrolyzed cassava mash from liquefaction experiments was added at pH 5.4 and 60°C with (V_{enzyme}/$W_{fresh\ mash}$) glucoamylase (Dextrozyme-GA) and pullulanase (Dextrozyme-GX). Results showed that the 10 min, 143.27 kPa (lab scale) and the 12 min 145/152 kPa (small scale) burst of starch granules in 1:1.6 cassava: water mash with 0.013% (V_{enzyme}/$W_{fresh\ mash}$) thermostable α-amylase at pH 6.5 followed by a 15 min, 90°C liquefaction phase at P_{atm} yielded MD with DE value < 20. According to the investigators, in order to gain a GS having a DE value close to 99, a 4 h, 60°C saccharification phase at pH 5.4 with 0.019% (V_{enzyme}/$W_{fresh\ mash}$) glucoamylase and pullulanase should be carried out. Finally, highly significant correlations were found between the water amount in the cassava mash, the concentration of the α-amylase enzyme, and the liquefaction time. The study inferred that this type of process is advantagesous because it is simple and practical with reduced working times and enzyme doses and is popular especially in developing countries (Lambri et al. 2014).

8.4.8 Cyclodextrins

α-Cyclodextrin is known to have use as a carrier and stabilizer for flavors, colors, and sweeteners; as a water-solubilizer for fatty acids and certain vitamins; as a flavor modifier in soy milk; and as an absorbent in confectionery products under good manufacturing practices (Li et al. 2014; WHO 2002). Cyclodextrin production is performed by two different processes based on the use of solvent: (i) The nonsolvent process does not require complexing agents and produces a cyclodextrin mixture that is further separated by chromatographic procedures or selective precipitation, and (ii) the solvent process involves the use of an organic complexing agent to selectively extract one type of cyclodextrin and thus directs the enzymatic reaction to produce the cyclodextrin of interest (Li et al. 2007). The solvent process is commonly used to produce α-cyclodextrin on an industrial scale. Further, it was found that addition of acetonitrile, ethanol, and tetrahydrofuran favored the production of α-cyclodextrin by the α-cyclodextrin glycosyltransferase (α-CGTase) from *Thermoanaerobacter* sp. (Blackwood and Bucke 2000).

CGTase is a multifunctional enzyme that can catalyze three transglycosylation reactions (disproportionation, cyclization, and coupling) and a hydrolysis reaction (van der Veen et al. 2000a,b). A variety of bacteria and archaea produce CGTase as an extracellular enzyme. The most extensively studied CGTases are from *Bacillus* species although species of *Paenibacillus, Klebsiella, Thermoanaerobacterium, Thermoanaerobacter,* and *Actinomycetes* (Qi and Zimmermann 2005; Tonkova 1998) also produce the CGTase. CGTases have been further classified, according to their major cyclodextrin products, into α-, β-, and γ-CGTases (Li et al. 2007; Penninga et al. 1995). The α-CGTases from *Bacillus macerans* or *Paenibacillus macerans* have been most commonly used in the commercial production of α-cyclodextrin (Li et al. 2014).

During the production of α-cyclodextrin from starch, a 20%–30% solution of starch is gelatinized first and liquefied using a thermostable α-amylase, α-CGTase, or acid to make the starch suitable for incubation with an α-CGTase at lower temperatures. The liquefied starch is treated with α-CGTase under conditions of controlled pH and temperature. 1-Decanol is added to form an insoluble 1:1 α-cyclodextrin/1-decanol inclusion complex. After the enzymatic reaction, the complex of α-cyclodextrin/1-decanol is separated from the reaction mixture by centrifugation. The supernatant contains unused starch, maltodextrin, glucose, oligosaccharides, CGTase, excess 1-decanol, and water. The recovered complex is resuspended in water and dissolved by heating. Subsequent cooling leads to reprecipitation of the complex. The precipitate is recovered by centrifugation, and the 1-decanol is removed by steam distillation. Upon concentration and cooling, α-cyclodextrin crystallizes from solution. The crystals are removed by filtration and drying, yielding a white powder with a water content <11%. The purity on a dried basis is at least 98% (Li et al. 2014; WHO 2002). In the presence of pullulanase (EC. 3.2.1.41), the α-CGTase from *B. macerans* could convert starch, maltodextrin, and glycogen into α-cyclodextrin in yields higher than those obtained in the absence of pullulanase. Using 1-decanol as a complexant, 10% amylopectin was converted into α-cyclodextrin in 84% yield at 15°C, although the process took place over 5 days (Rendleman 1997). Duan et al. (2013) achieved 84.6% (w/w) total yield of cyclodextrins from 10% (w/v) potato starch by synchronous utilization of isoamylase (EC. 3.2.1.68) and α-CGTase from *P. macerans* JFB05-01. This yield was 31.2% higher than that obtained with α-CGTase alone. Furthermore, the α-cyclodextrin content of the total cyclodextrins reached >94%.

8.4.9 Hydrogenated Starch Hydrolysate

Hydrogenated starch hydrolysates (HSHs), also known as hydrogenated glucose syrup, are mixtures of sorbitol, maltitol, and hydrogenated oligosaccharides. HSHs are used increasingly in so-called "sugar-free" confectionery; sorbitol is also used in diabetic products. HSHs serve a number of functional roles, including use as bulk sweeteners, viscosity, or bodying agents, humectants (moisture-retaining ingredient), crystallization modifiers, cryoprotectants, and rehydration aids (Eberhardt 2001). The use of all enzyme hydrolysis has allowed starch hydrolysates of high dextrose (glucose) content to be made (>97%). This has allowed the achievement of higher yields and glucose/dextrose production, whether anhydrous or monohydrate, and the manufacture of hydrolysates of specific use in the fermentation industry. Dextrose can be used to produce confectionery tablets with a smooth cooling sensation on the palate.

8.5 Date Syrup

Date syrup is directly consumed or used as an ingredient in some food formulations, such as ice cream products, beverages, confectionery, bakery products, sesame paste/date syrup blends, jam, and butter (Barreveld 1993). The processing conditions for the extraction of date juice and yield of soluble solids and turbidity were optimized by investigating the effect of enzyme treatment (pectinase and cellulase) at different times and enzyme concentrations. The studies indicated enzymatic treatment could give high-value addition to dates of low quality (with hard texture) and consequent production of high-quality date syrups (Abbès et al. 2011).

8.6 Chocolates

Chocolate is a fast-growing product within the confectionery industry. Chocolate's unique texture and flavor make it very pleasing to eat and are the main reasons for its expanding consumption around the world. Sucrose appears in chocolate in the range of 35%–50%, conferring multiple functional properties on the chocolate, including sweetness, particle size distribution, and mouthfeel. However, the sugar content in chocolate varies between 35% and 50%, an issue for those with health and nutrition concerns. In fact, growing health and nutrition problems—including diabetes, weight control, and tooth decay—along with public demands have motivated the food industry to produce low-calorie and reduced-sugar chocolate.

Cream-centered confections are not only of high quality, but are also a highly desirable treat due to their rich taste and creamy mouthfeel. Consumers are usually willing to pay a high price for these products because of their taste and quality. The center can consist of many different components, such as dairy cream, milk products, chocolate, sugar, corn syrup, starches, egg albumen, gelatin, fruit pieces, nuts, flavors, acids, colors, and invertase. Confectionery products, including truffles and other chocolates with cream centers, are usually manufactured through starch molding, which is then enrobed in chocolate. The center must be firm enough to withstand handling and processing, yet will soften to a desirable creamy consistency by the time the consumer eats the product. Invertase causes this transformation by converting the crystalline sucrose present in the center to the syrup phase, making it possible to handle a cream consistency directly from the starch molds. In addition to the softening that occurs, the fructose generated, which is hygroscopic, helps to minimize the drying of the confection. This, in combination with the prevention of sugar crystallization, will increase the shelf life of the product by maintaining a desirable consistency for an extended period of time. Further, the process also reduces the water activity and helps to minimize microbial growth and contamination.

8.6.1 Chocolate Flavor

The popularity of cocoa and cocoa-derived products, in particular chocolate, can be ascribed to the unique and complex flavors of these delicious foods (Jinap et al. 1995; Ziegleder 1991). The flavors and, in particular, the flavor precursors of cocoa are developed during primary processing of the cocoa beans, that is, fermentation and drying. This development of flavor precursors involves the action of various micro-organisms in the cocoa pulp and the action of enzymes on carbohydrates, proteins, and polyphenols in the cocoa beans. Unlike many other fermented raw materials, endogenous enzymes play a crucial role in cocoa flavor development. However, there is no flavor in cocoa beans without fermentation (Lehrian and Patterson 1983). Fermentation and roasting are key to developing cocoa and chocolate flavor, respectively. Chocolate flavor is an extremely complex mixture of more than 550 compounds, and as analytical methods improve, this number increases (Nijssen et al. 1996). Whereas undesirable bitter and astringent tastes decrease during fermentation, desirable fruity, floral, and cocoa flavors develop during fermentation and drying. Essential precursors of the cocoa-specific flavor components are generated during microbial fermentation of the pulp surrounding the beans as well as by enzymatic processes in the beans (Camu et al. 2007; de Brito et al. 2000; Hansen et al. 1998).

Fermented and dried cocoa beans constitute the raw material used by the chocolate industry to produce various different products, which are largely appreciated for their characteristic flavor. During fermentation, the pulp surrounding the seeds is metabolized by the micro-organisms, resulting in a temperature rise and a drop in pH, which are responsible for cotyledon death. During this, some other substances, such as ethanol, acetic, and lactic acids, and the formation of flavor precursors, basically peptides, free amino acids, and reducing sugars, are also produced (Mohr et al. 1976). These precursors participate in the Maillard reaction during cocoa roasting, leading to the characteristic chocolate flavor.

Proteolysis is very important for cocoa flavor development (Voigt et al. 1994a,b). Amino acids and peptides are produced during fermentation by the combined action of an aspartic endoprotease and a carboxipeptidase. The main substrate for these enzymes is a globulin described as a cocoa vicilin. A specific cocoa aroma was obtained when this globulin was degraded by exogenous proteases, and the resulting products were roasted in the presence of reducing sugars (Voigt et al. 1994a,b). However, other cocoa proteins should be considered in flavor precursor formation (Lerceteau et al. 1999). During subsequent roasting, amino acid levels were reduced, but this reduction was not complete, and there were different rates of reaction for the distinct amino acid groups (Abeygunasekera and Jansz 1989).

Enzymes were found to be useful in reducing polyphenol contents by nearly 25% in cocoa nibs, and in autoclaved nibs, a greater polyphenol reduction was noted probably due to facilitated enzyme penetration into tissues and action because of autoclaving (Barbery 1999; Yoshiyama and Ito 1996). Another study also demonstrated that bitter taste and astringency were reduced by polyphenoloxidase action, which contributed to a better acceptance of the cocoa products (Barbery 1999). The formation of precursors and the effect on cocoa flavor was further confirmed by protein hydrolysis on cocoa nibs, using an exogenous protease (Brito et al. 2004). Influence of temperature (30°C to 70°C) and enzyme:substrate ratio [E/S] (97.5 to 1267.5 U g^{-1} of protein) was evaluated and was found that the percentage degree of hydrolysis (% DH) was affected mainly by [E/S] leading to a fourfold increase (from 5% to 20%) after 6 h of treatment. During cocoa nib roasting, there was a greater consumption of hydrolysis compounds in the sample treated with protease as compared to the control, indicating their participation in the Maillard reaction. An increased perception of chocolate flavor and bitter taste was observed in a product formulated with protease-treated cocoa. The study concluded that proteolytic enzyme was useful in improving cocoa flavor precursors and affected the flavor perception in their products. The difficulty in establishing a model could be suppressed by changing the process in such a way as to enhance the access of enzyme to substrate and a better heat transfer (Brito et al. 2004).

8.6.1.1 Polyphenol Oxidase in Cocoa Processing

In chocolate production, polyphenol oxidase is one component responsible for the formation of flavor precursors, beginning in the oxidative phase of the fermentation and continuing into the drying phase. Among the changes affecting flavor are reduction in the bitterness and astringency that result from polymerized polyphenol–protein interactions. An increase in oxygen penetration into the cocoa bean mass during drying induces maximum oxidation of epicatechin and pro-cyanidins, causing the production of melanin and melanoproteins that are responsible for the color in brown chocolate (Hammer 2013).

Cocoa almonds, the fermented and dried seeds of cocoa fruits, are the raw materials used in the production of chocolate. The commercial value of cocoa is based not only on the melting characteristics of its fat, which melts in a very narrow range that is close to body temperature and provides a unique mouthfeel, but also on the chocolate flavor that is developed in properly processed seeds (Oliveria et al. 2011). Fresh beans, extracted from the ripe cocoa fruit have no chocolate flavor and are, in fact, extremely bitter and astringent. For the desired flavor to develop, the seeds must go through a curing process that involves a stage of fermentation and drying, which leads to the formation of flavor precursors. The recurring problem in the chocolate industry is poor quality of some almonds, which leads to reduced fermentation and poor flavor formation. Because the fermentation and drying processes still take place on farms without any controlled conditions, a significant percentage of cocoa almonds from each batch do not undergo the necessary changes (acidification and temperature increase) for the necessary enzymatic reactions to occur. As a result, a significant portion of the roasted almonds do not develop the characteristic chocolate flavor, which reduces the quality of the chocolate produced. These enzymes hydrolyze proteins present in the almonds releasing the flavour precursors that were not produced during the fermentation period. Thus, it would be possible to standardize the quality of cocoa produced, ensuring the quality of the chocolate (Gray 2011).

The production of flavor precursors improves the quality of these almonds, which are usually responsible for the low quality of the liquor produced. An enzymatic treatment with proteases and carboxypeptidases was developed by Oliveria et al. (2011) to ensure the formation of chocolate flavor precursors and improve the quality of the chocolate produced using poor-quality cocoa almonds (known as "slate"). They tested three commercial enzymes for their ability to improve the flavor attributes of cocoa slate, which included swine pepsin, carboxypeptidase A (purchased from Sigma-Aldrich), and Flavorzyme (Novozymes), against enzymes extracted from unfermented coca beans (vegetable enzyme). The enzymatic treatments were evaluated by chemical analysis (hydrolysis efficiency) and sensory analysis of the treated material compared to good-quality cocoa almonds. They recorded that almonds treated with microbial enzymes (Flavorzyme) developed better chocolate flavor, and Flavorzyme, which contains aspartic proteases and carboxypeptidases of microbial origin, provided an improvement of 50% in relation to the chocolate flavor. Although the hydrolysis achieved was similar to all tested enzymes, microbial enzymes were able to produce more of the desired precursors than after roasting, leading to the formation of the chocolate flavor. Their results indicate that it is possible to use microbial enzymes to improve the quality of cocoa almonds, which is advantageous because microbial enzymes are low in cost and can be supplied in significant quantities, making them more viable for industrial applications. The results showed that it is possible, through the use of microbial enzymes, to generate the mixture of compounds that will release, after roasting, the characteristic chocolate flavor in poor-quality almonds. However, it is necessary to optimize the conditions of enzymatic treatment to obtain better results and thus establish a process that can be used for industrial purposes for manufacturing cocoa and chocolate (Oliveria et al. 2011).

8.6.2 Cocoa Butter Substitutes and Cocoa Butter Equivalents

Cocoa butter, the fat component of chocolate, is the most expensive constituent in chocolate formulations and is composed mainly of symmetric monounsaturated triacylglycerols (SMUT) containing an oleoyl (O) residue at the *sn*-2 position and palmitoyl (P) or stearoyl (S) residues at the *sn*-1,3 positions of the glycerol backbone, represented as POP (15%–19%), POS (36%–41%), and SOS (25%–31%) (Shukla 2005; Yamada et al. 2005). The large amounts of total SMUT (79%–87%) in cocoa butter are responsible for its unique physical attributes, such as the dominating existence of stable b crystals and rapid melting at body temperature, thereby providing chocolate with snap, gloss, and a smooth mouthfeel (Lipp et al. 2001). Cocoa butter equivalents (CBE), which are produced from cheaper fat/oil resources than cocoa butter and have similar total SMUT content to cocoa butter, have been used to replace cocoa butter in the manufacturing of chocolate products because they reduce production costs. Typical commercial CBEs contain more SOS and POP and less POS than cocoa butter because they are produced by blending fats rich in POP, such as palm midfraction with SOS-rich fats, which are obtained by fractional crystallization (e.g., shea stearin, sal stearin) or enzymatic interesterification (e.g., high oleic sunflower oil esterified with stearic acid) (Yamada et al. 2005).

In recent years, CBE manufacturers and researchers have attempted to prepare POS-rich fats for formulating CBE with weight ratios between the three SMUT species that are similar to those of cocoa butter. The POS-rich fats are produced by esterifying POP-rich fats with stearic acid (or its alkyl esters) or tristearin (SSS) using *sn*-1,3-specific lipases as biocatalysts (Pinyaphong and Phutrakul 2009; Tchobo et al. 2009). A possible alternative for producing POS-rich fats is to esterify triolein (OOO)-rich vegetable oils, such as high oleic sunflower oil (Carrin and Crapiste 2008), olive–pomace oil (Ciftci et al. 2009), and tea seed oil (Wang et al. 2006) with mixtures of palmitic and stearic acids (or their methyl esters) with the same enzymes. These published studies have employed batch reaction systems for the production of POS. However, despite using *sn*-1,3-specific lipases, use of a stirred batch reactor for performing lipase-catalyzed esterification often leads to significant migration of acyl groups from one site on the glycerol backbone to another site, thereby forming positional isomers of POS, such as PSO and SPO.

Kim et al. (2014) optimized the lipase-catalyzed transesterification of high oleic sunflower oil (A) with a mixture of ethyl palmitate and ethyl stearate (B) to produce cocoa butter equivalents with a weight ratio of 1-palmitoyl-2-oleoyl-3-stearoyl-rac-glycerol (POS) to total symmetric monounsaturated triacylglycerols (SMUT) that is similar to that of cocoa butter in a continuous packed bed reactor, using 0.45 g of Lipozyme RM IM (*Rhizomucor miehei* lipase immobilized on macroporous anion exchange resins, supplied by Novozymes, Denmark) as the biocatalyst by response surface methodology. The effects of temperature (Te), residence time (RT), substrate molar ratio (SR, B/A), and water content (WC) of the substrates on the composition of reaction products were elucidated using the models established. Optimal reaction conditions for maximizing total SMUT and POS contents while minimizing the levels of diacylglycerol formation and acyl migration were Te, 60°C; RT, 28.5 min; SR, 8.5; WC, 300 mg/kg. The contents of total SMUT, POS, and diacylglycerol in the reaction products and the content of palmitoyl and stearoyl residues at the *sn*-2 position of triacylglycerols in the products were 52.0%, 25.1%, 9.4%, and 4.8%, respectively, under these conditions. Successful scale-up of the reaction was achieved under optimal conditions, using 5 g of the lipase. A silver-ion HPLC analysis showed that the products obtained by the larger-scale reaction contained 49.1% total SMUT and 6.1% of their positional isomers (Kim et al. 2014).

The need for low-calorie fats has increased because of recent changes in consumers' concerns about leading healthier lifestyles. Three types of commercial cocoa butter substitutes (CBSs) were studied for use as substrates in the production of low-calorie structured lipids (Nazaruddin et al. 2011). Mixtures of commercial CBSs, ethanol, and medium-chain fatty acids in a molar ratio of 1:2:1 were incubated in a water bath at 50°C for 24 h at 250 rpm, using 10% immobilized lipase as a catalyst, and the solid fat content (SFC), the slip melting point, the iodine value, the free fatty acid content, the thermal behavior, and the fatty acid composition were determined before and after enzymatic alcoholysis. Hisomel CBSs (supplied by KL-Kepong Cocoa Products SDN BHD, Malaysia) was chosen as a substrate for further analysis because of its sharp melting profile between room temperature (25°C) and body temperature (37.5°C). At room temperature, the SFC of Hisomel was 60.56%, and after the reaction with either caprylic or capric acid, it became 12.50% and 14.26%, respectively, which was the highest value among other CBS samples. Furthermore, Hisomel also had a similar fatty acid profile to cocoa butter before the enzymatic alcoholysis, which was 31.13% palmitic acid, 50.05% stearic acid, and 18.80% oleic acid (Nazaruddin et al. 2011).

8.7 Conclusion and Future Prospects

Chocolates and confectioneries are among the major food industries that witness rapid growth owing to the demand from consumers of all age groups because of their attractive color, appearance, and flavor. As discussed in the foregoing sections, confectioneries and chocolates largely depend on sweeteners, both natural (sucrose) and artificial, as an important constituent of confections. Thus, production of sugar syrups, sweeteners, and sugars from conventional sources, such as corn, and renewable resources, such as tapioca, forms the major activity of the food industry to meet the demands of the confectioney industries. Further, it has been convincingly proven by several investigators that hydrolytic enzymes such as invertase and amylases have a key role in the production of syrups and sugar for the confectioneries. Today, enzymes have a larger role in confectioneries as well as in the improvement of flavor in chocolates and production of cocoa butter equivalents. There is no doubt that newer applications of enzymes in confectioneries are possible, provided more intense research studies are undertaken in isolation of new enzymes with novel properties from unexplored microorganisms, in addition to enzyme/protein engineering of potential enzymes that have the potential for application in sugar and sugar syrup production from underutilized starch sources.

REFERENCES

Abbès, F., Bouaziz, M. A., Blecker, C., Masmoudi, M., Attia, H. and Besbes, S. 2011. Date syrup: Effect of hydrolytic enzymes (pectinase/cellulase) on physico-chemcial characteristics, sensory and functional properties. *LWT-Food Science and Technology*, 44(8), 1827–1834. Available at http://hdl.handle.net/2268/119978.

Abeygunasekera, D. D. and Jansz, E. R. 1989. Effect of the maturation process on fermented cocoa beans I: Free amino acids and volatile carbonyls. *Journal National Science Council Sri Lanka*, 17, 23–33.

Abraham, E. T., Krishnaswamy, C. and Ramakrishna, S. V. 1988. Hydrolytic depolymerization of starch raw materials. *Starch/Stärke*, 40, 387–392.

Ahure, D. and Ariahu, C. C. 2013. Quality evaluation of glucose syrup from sweet cassava hydrolyzed by rice malt crude enzymes extract. *Journal of Food Technology*, 11(1), 1–3.

Aschengreen, N. H., Nielsen, B. H., Rosendal, P. and Ostergaard, J. 1979. Liquefaction, saccharifaction and isomerization of starches from sources other than corn. *Starch/Stärke*, 31(2), 64–66.

Bandlish, R. K., Hess, J. M., Epting, K. L., Vieille, C. and Kelly, R. M. 2002. Glucose-to-fructose conversion at high temperatures with xylose (glucose) isomerases from *Streptomyces murinus* and two hyperthermophilic *Thermotoga* species. *Biotechnology and Bioengineering*, 80(2), 185–194.

Barbery, S. D. F. 1999. *Estudo do melhoramento do sabor de cacau* (Theobroma cacao L.) *utilizando polifenoloxidase extraida da pinha* (Annona squamosa L.) *etratamento térmico não convencional*. Tese (Doutorado), UNICAMP, Campinas, Brasil.

Barreveld, W. H. 1993. Date palm products. *Nume'roAgricole 101 de buletin de services de la FAO*.

Bindumole, V. R. and Balagopalan, C. 2001. Saccharification of sweet potato flour for ethanol production. *Journal of Root Crops*, 27(1), 89–93.

Blackwood, A. D. and Bucke, C. 2000. Addition of polar organic solvents can improve the product selectivity of cyclodextrin glycosyltransferase. Solvent effects on cgtase. *Enzyme and Microbial Technology*, 27, 704–708.

Blanchard, P. H. and Katz, F. R. 1995. Starch hydrolyzates. In: Stephan, A. M. (ed.), *Food Polysaccharides and Their Applications*. Marcel Dekker, Inc., New York, pp. 99–122.

Brienzo, M., Arantes, V. and Milagres, A. M. F. 2008. Enzymology of the thermophilic ascomycetous fungus *Thermoascus aurantiacus*. *Fungal Biology Reviews*, 22(3–4), 120–130.

Brito, E. S., Nelson, H. P. G. and Allan, C. A. 2004. Use of a proteolytic enzyme in cocoa (Theobroma cacao L.) processing. *Brazilian Archives of Biology and Technology*, 47(4), 553–558.

Bucholz, K., Kasche, V. and Bornscheuer, U. T. 2005. *Biocatalysts and Enzyme Technology*. Wiley VCH, Weinheim, Germany.

Burgess, K. and Shaw, M. 1983. Dairy. In: Godfrey, T. and Reichelt, J. (eds.), *Industrial Enzymology: The Application of Enzymes in Industry*. The Nature Press, New York.

Cabello, C. 1999. Amylases using in glucose syrup production. In: *Seminário Brasileiro de Tecnologia Enzimática—Enzitec*, 4, Rio de Janeiro, Anais, 1, V1–V3.

Camu, N., De Winter, T., Verbrugghe, K. et al. 2007. Dynamics and biodiversity of populations of lactic acid bacteria and acetic acid bacteria involved in spontaneous heap fermentation of cocoa beans in Ghana. *Applied and Environmental Microbiology*, 73, 1809–1824.

Carpio, C., Escobar, F., Batista Viera, F. and Ruales, G. 2011. Bone-bound glucoamylase as a biocatalyst in bench-scale production of glucose syrups from liquefied cassava starch. *Food and Bioprocess Technology*, 4, 566–577.

Carrin, M. E. and Crapiste, G. H. 2008. Enzymatic acidolysis of sunflower oil with a palmitic-stearic acid mixture. *Journal of Food Engineering*, 84, 243–249.

Chinnawornrungsee, R., Malakul, P. and Mungcharoen, T. 2013. Life cycle energy and environmental analysis study of a model biorefinery in Thailand. *CET*, 32, 439–444.

Chronakis, I. S. 1998. On the molecular characteristics, compositional properties, and structural-functional mechanisms of maltodextrins: A review. *Critical Reviews in Food Science and Nutrition*, 38, 599–637.

Ciftci, O. N., Fadiloglu, S. and Gogus, F. 2009. Utilization of olive pomace oil for enzymatic production of cocoa butter-like fat. *Journal of the American Oil Chemists' Society*, 86, 119–125.

Cowan, D. 1983. "Proteins." In industrial enzymology. In: Godfrey, T. and Reichelt, J. (eds.), *The Application of Enzymes in Industry*. The Nature Press, New York, Chapter 4.14, pp. 352–374.

Davies, R. 1953. Enzyme formation in *Saccharomyces fragilis*. 1. Invertase and raffinase. *Biochemical Journal*, 55, 484–497.

de Brito, E. S., Pezoa García, N. H., Gallão, M. I., Cortelazzo, A. L., Fevereiro, P. S. and Braga, M. R. 2000. Structural and chemical changes in cocoa (Theobroma cacao L) during fermentation, drying and roasting. *Journal of the Science of Food and Agriculture*, 81, 281–288.

Duan, X., Chen, S., Chen, J. and Wu, J. 2013. Enhancing the cyclodextrin production by synchronous utilization of isoamylase and alpha-CGTase. *Applied Microbiology and Biotechnology*, 97, 3467–3474.

Dziedzic, S. Z. and Kearsley, M. W. 1984. *Glucose Syrups: Science and Technology*. Elsevier Applied Sciences, London, UK, p. 276.

Dziedzoave, N., Graffham, A. and Boateng, B. O. 2003. Travelling manual for the production of glucose syrup. DFID-Funded Project on New Markets for Cassava Food Research Institute Technical Report. FRI, Africa.

Eberhardt, L. 2001. Hydrogenated starch hydrolysates and maltitol syrups. In: O'Brien Nabors, L. (ed.), *Alternative Sweeteners*, 3rd edition. Marcel Dekker, Inc., New York.

Edwards, W. P. 2000. *The Science of Sugar Confectionery*. Royal Society of Chemistry, Cambridge, UK, p. 1.

Eshra, D. H., El-Iraki, S. M. and Abo Bakr, T. M. 2014. Performance of starch hydrolysis and production of corn syrup using some commercial enzymes. *International Food Research Journal*, 21(2), 815–821.

Ezeji, T. C. and Bahl, H. 2006. Purification, characterization, and synergistic action of phytate-resistant α-amylase and α-glucosidase from *Geobacillus thermodenitrificans* HRO 10. *Journal of Biotechnology*, 125(1), 27–38.

Fontana, J. D., Passos, M., Baron, M., Mendes, S. V. and Ramos, L. P. 2001. Cassava starch maltodextrinization/monomerization through thermopressurized aqueous phosphoric acid hydrolysis. *Applied Biochemistry and Biotechnology*, 91–93, 468–478.

Gorinstein, S. and Lii, C. 1992. The effects of enzyme hydrolysis on the properties of potato, cassava and amaranth starches. *Starch/Stärke*, 44(2), 461–466.

Gray, N. 2011. Enzymes may boost chocolate flavour: Study. Available at http://www.foodnavigator.com.

Hammer, F. E. 2013. Oxidases. In: *Enzymes in Food Processing*, Nagodawithana, T. and Reed, G. (eds.), Academic Press, CA, 221–278.

Hanover, L. M. and White, J. S. 1993. Manufacturing, composition and applications of fructose. *The American Journal of Clinical Nutrition*, 58, 724S–732S.

Hansen, C. E., del Olmo, M. and Burri, C. 1998. Enzyme activities in cocoa beans during fermentation. *Journal of the Science of Food and Agriculture*, 77, 273–281.

Inglet, G. E. 1981. Sweeteners—A review. *Food Technology*, 35, 37–41.

International Food Information Service (ed.). 2009. *Dictionary of Food Science and Technology*, 2nd edition. Wiley–Blackwell, Chichester, UK, p. 106.

Invertin-An Enzyme Confectionery Humectant. 2015. Available at http://www.keylink.org/knowledge bank/invertin-an-enzyme-confectionery-humectant.php (accessed on January 19, 2015).

Jinap, S., Dimick, P. S. and Hollender, R. 1995. Flavour evaluation of chocolate formulated from cocoa beans from different countries. *Food Control*, 6, 105–110.

Johnson, R. and Padmaja, G. 2012. Utilization of cassava fibrous residue for the production of glucose and high fructose syrup. *Industrial Biotechnology*. December 2011, 7(6), 448–455. doi:10.1089/ind.2011.0015.

Johnson, R. and Padmaja, G. 2013. Comparative studies on the production of glucose and high fructose syrup from tuber starches. *International Research Journal of Biological Sciences*, 2(10), 68–75.

Johnson, R., Moorthy, S. N. and Padmaja, G. 2009. Comparative production of glucose and high fructose syrup from cassava and sweet potato roots by direct conversion techniques. *Innovative Food Science & Emerging Technologies*, 10, 616–620.

Kim, S., Kim, I.-H., Akoh, C. C. and Kim, B. H. 2014. Enzymatic production of cocoa butter equivalents high in 1-Palmitoyl-2-oleoyl-3-stearin in continuous packed bed reactors. *Journal of the American Oil Chemists' Society*, 91, 747–757. doi:10.1007/s11746-014-2412-7.

Kovalenko, G. A., Perminova, L. V. and Sapunova, L. I. 2010. Heterogeneous biocatalysts for production of sweeteners—Starch treacle and syrups of different carbohydrate composition catalysis in industry. *Catalysis in Industry*, 2(2), 180–185.

Lambri, M., Dordoni, R., Roda, A. and De Faveri, D. M. 2014. Process development for maltodextrins and glucose syrup from cassava. *Chemical Engineering Transactions*, 38, 469–474. doi:10.3303/CET1438079.

Lauven, L. P., Liu, B. and Geldermann, J. 2013. Investigation of the influence of plant capacity on the economic and ecological performance of cassava-based bioethanol. *CET*, 35, 595–600.

Lehrian, D. W. and Patterson, G. R. 1983. Cocoa fermentation. In: Reed, G. (ed.), *Biotechnology, a Comprehensive Treatise*, vol. 5. Verlag Chemie, Basel, pp. 529–575.

Lerceteau, E., Rogers, J., Pétiard, V. and Crouzillat, D. 1999. Evolution of cacao bean proteins during fermentation: A study by two-dimensional electrophoresis. *Journal Science Food Agriculture*, 79, 619–625.

Li, Z., Wang, M., Wang, F. et al. 2007. Gamma-Cyclodextrin: A review on enzymatic production and applications. *Applied Microbiology and Biotechnology*, 77, 245–255.

Li, Z., Chen, S., Gu, Z., Chen, J. and Wu, J. 2014. Alpha-cyclodextrin: Enzymatic production and food applications. *Trends in Food Science & Technology*, 35, 151–160.

Lipp, M., Simoneau, C., Ulberth, F. et al. 2001. Composition of genuine cocoa butter and cocoa butter equivalents. *Journal of Food Composition and Analysis*, 14, 399–408.

Mochizuki, K., Isobe, K. and Sawada, Y. 1971. Method for producing candied fruit. US Patent No. 3615687.

Mohr, W., Landschreiber, E. and Severin, T. 1976. Zur spezifitat des kakaoaromas. *Fette Seifen Anstrichmittel*, 78, 88–95.

Moore, G. R. P., do Canto, L. R. and Amante, E. R. 2005. Cassava and corn starch in maltodextrin production. *Quimica Nova*, 28, 596–600.

Nazaruddin, R., Nurul Zakiyani, S. and Mamot, S. 2011. The effect of enzymatic alcoholysis on the physicochemical properties of commercial cocoa butter substitutes. *Pakistan Journal of Nutrition*, 10(8), 718–723.

Nijssen, L. M., Visscher, C. A., Maarse, H., Willemsens, L. C. and Boelens, M. H. 1996. *Volatile Compounds in Foods. Qualitative and Quantitative Data*, 7th edition. TNO Nutrition and Food Research Institute, Zeist, The Netherlands.

Novo Nordisk A/S. 1977. Dextranase Novo 25L. A dextran decomposing enzyme for the sugar industry. Product data information 112-GB. Novo Enzyme Division, Dagsvaerd, Denmark.

Oliveria, H. S. S., Mamede, M. E. O., Goes-Neto, A. and Koblitz, M. G. B. 2011. Improving chocolate flavor in poor quality cocoa almonds by enzymatic treatment. *Journal of Food Science*, 76(5), C755–C759. doi:10.1111/j.1750-3841.2011.02168.x.

Panda, S., Botta, G. B., Pattnaik, S. and Maharana, L. 2012. A complete review on various natural biodegradable polymers in pharmaceutical use. *Journal of Pharmacy Research*, 5(12), 5390.

Parker, K., Salas, M. and Nwosu, V. C. 2010. High fructose corn syrup: Production, uses and public health concerns. *Biotechnology and Molecular Biology Review*, 5(5), 71–78.

Penninga, D., Strokopytov, B., Rozeboom, H. J. et al. 1995. Site-directed mutations in tyrosine 195 of cyclodextrin glycosyltransferase from *Bacillus circulans* strain 251 affect activity and product specificity. *Biochemistry*, 34, 3368e3376.

Pinyaphong, P. and Phutrakul, S. 2009. Synthesis of cocoa butter equivalent from palm oil by Carica papaya lipase-catalyzed interesterification. *Chiang Mai Journal of Science*, 36, 359–368.

Pontoh, J. and Low, N. H. 1995. Glucose syrup production from Indonesian palm and cassava starch. *Food Research International*, 28(4), 379–385.

Qi, Q. and Zimmermann, W. 2005. Cyclodextrin glucanotransferase: From gene to applications. *Applied Microbiology and Biotechnology*, 66, 475–485.

Rendleman, J. A. J. 1997. Enhancement of cyclodextrin production through use of debranching enzymes. *Applied Biochemistry and Biotechnology*, 26, 51–61.

Schenck, F. W. and Hebeda, R. E. (eds.). 1992. *Starch Hydrolysis Products—Worldwide Technology, Production and Applications*. VCH Publishers, New York.

Shetty, J., Chotani, G., Gang, D. and Bates, D. 2007. Cassava as an alternative feedstock in the production of renewable transportation fuel. *International Sugar Journal*, 109(1307), 3–11.

Shukla, V. K. S. 2005. Cocoa butter, cocoa butter equivalents, and cocoa butter substitutes. In: Akoh, C. C. (ed.), *Handbook of Functional Lipids*. CRC Press, New York, pp. 279–307.

Tchobo, F. P., Piombo, G., Pina, M., Soumanou, M. M., Villeneuve, P. and Sohounhloue, D. C. K. 2009. Enzymatic synthesis of cocoa butter equivalent through transesterification of *Pentadesma butyracea* butter. *Journal of Food Lipids*, 16, 605–617.

Tonkova, A. 1998. Bacterial cyclodextrin glucanotransferase. *Enzyme and Microbial Technology*, 22, 678–686.

van der Maarel, M. J. E. C., van der Veen, B., Uitdehaag, J. C. M., Leemhuis, H. and Dijkhuizen, L. 2002. Properties and applications of starch converting enzymes of the α-amylase family. *Journal of Biotechnology*, 94(19), 137–155.

van der Veen, B. A., Uitdehaag, J. C., Dijkstra, B. W. and Dijkhuizen, L. 2000a. The role of arginine 47 in the cyclization and coupling reactions of cyclodextrin glycosyltransferase from *Bacillus circulans* strain 251 implications for product inhibition and product specificity. *European Journal of Biochemistry*, 267, 3432–3441.

van der Veen, B. A., van Alebeek, G. J., Uitdehaag, J. C., Dijkstra, B. W. and Dijkhuizen, L. 2000b. The three transglycosylation reactions catalyzed by cyclodextrin glycosyltransferase from Bacillus circulans (strain 251) proceed via different kinetic mechanisms. *European Journal of Biochemistry*, 267, 658–665.

van der Veen, M. E., van der Goot, A. J. and Boom, R. M. 2005. Production of glucose syrups in highly concentrated systems. *Biotechnology Progress*, 21(2), 598–602.

Voigt, J., Biehl, B., Heinrichs, H., Kamaruddin, S., Marsoner, G. C. and Hugi, A. 1994a. In vitro formation of cocoa-specific aroma precursors: Aromarelated peptides generated from cocoa-seed protein by cooperation of an aspartic endoprotease and a carboxipeptidase. *Food Chemistry*, 49, 173–180.

Voigt, J., Heinrichs, H., Voigt, G. and Biehl, B. 1994b. Cocoa-specific aroma precursors are generated by proteolytic digestion of the vicilin-like globulin of cocoa seeds. *Food Chemistry*, 50, 177–184.

Voragen, A. G. J. 1998. Technological aspects of functional food-related carbohydrates. Review. *Trends in Food Science and Technology*, 9(6), 328–335.

Wang, H. X., Wu, H., Ho, C. T. and Weng, C. X. 2006. Cocoa butter equivalent from enzymatic interesterification of tea seed oil and fatty acid methyl esters. *Food Chemistry*, 97, 661–665.

White, J. S. 1992. Fructose syrup: Production, properties and applications. In: Schenck, F. W. and Hebeda, R. E. (eds.), *Starch Hydrolysis Products-Worldwide Technology, Production and Applications*. VCH Publishers Inc, New York, pp. 177–200.

WHO. 2002. Safety evaluation of certain food additives and contaminants, a-cyclodextrin. In *WHO Food Additives Series*, vol. 48. Available at http://www.inchem.org/documents/jecfa/jecmono/v48je10.htm (accessed on July 10, 2013).

Yamada, K., Ibuki, M. and McBrayer, T. 2005 Cocoa butter, cocoa butter equivalents, and cocoa butter replacers. In: Lai, O. M. and Akoh, C. C. (eds), *Healthful Lipids*. AOCS Press, Champaign, IL, pp. 642–664.

Yoshiyama, M. and Ito, Y. 1996. Decrease of astingency of cacao beans by an enzymatic treatment. *Nippon Shokuhin Kagaku Kaishi*, 43, 124–129.

Ziegleder, G. 1991. Composition of flavor extracts of raw and roasted beans. *Zeitschrift für Lebensmittel-Untersuchung und-Forschung*, 192, 521–525.

9

Enzymes in Oil- and Lipid-Based Industries

Wenbin Zhang, Pengfei Li, and Ruijin Yang

CONTENTS

9.1 Introduction

Oils and fats are generally considered to be a chemically heterogeneous group of substances, having in common the property of insolubility in water but solubility in nonpolar solvents, such as chloroform, hydrocarbons, or alcohols (Keefe 2007). Lipids supply energy, support structural aspects of the body, and provide substances that regulate physiological processes. In addition, they contain such essential fatty acids as linolenic acid. These are metabolized eventually to provide eicosanoids, substances that possess hormone-like activity, and thus may regulate many body functions. Lipids are also the transport vehicle for vitamins A, D, E, and K. Other contributions from oils and fats include, but are not limited to, providing desirable appearance, particular flavor, color, and nutritional value in baking and cooking goods (Ruan et al. 2014).

The oil- and lipid-based industry generally covers four aspects of the food industry, namely oil recovery, refining, conversion/modification, and stabilization (Johnson 2007). Oil recovery is often referred to as extraction or crushing when processing plant sources and rendering in the case of processing animal tissues. Oil extraction involves pressing the oil-bearing material to separate crude oil from the solids high in protein or washing flaked or modestly pressed material with solvent, almost always hexane. The defatted solids after pressing are known as cake and after solvent extraction as meal. Oil refinery aims at removing contaminants and producing high-quality edible oils necessary for crude oil, which contains undesirable components, such as pigments, phosphatides, free fatty acids, and off-flavors and off-odors (Jiang et al. 2011; Jiang et al. 2014). Conversion of oil, including selective hydrolysis and interesterification, provides a series of derived products, including triacylglycerol (TAG), diacylglycerol (DAG), monoacylglycerol (MAG), and free fatty acid (FFA) for industries. Stabilization with plasticizing, tempering, and Stehling could generate shortenings and margarines for the food industry.

Conventionally, edible oil is produced by extraction with an organic solvent alone or by screw pressing prior to solvent extraction. The most prevalent solvent used is n-hexane, which has many uses in the chemical and food industries (de Moura et al. 2008; Rosenthal et al. 1996). Although n-hexane achieves a crude oil yield over 96%, the acute inhalation exposure of humans to high levels of hexane causes mild central nervous system effects, and its attributes of being highly flammable and explosive may jeopardize the safety of plants and human beings. Therefore, n-hexane is classified as a neurotoxin and a volatile organic compound by the National Institute for Occupational Safety and Health (NIOSH) and the United States Environmental Protection Agency (USEPA), respectively (ATSDR 2011; USEPA 2000; Wu et al. 2009). Meanwhile, a conventional chemical refinery also expects new technologies owing to the desire for an increasingly refined oil yield and economic attractiveness as well as environmental friendliness. With the expansion of the world population and dietary changes in both developed and developing countries, consumption of edible oils requires both quantitative and qualitative improvements. Industries are welcoming desirable changes for traditional processing of oilseeds and lipids. Therefore, enzymes, are gaining importance as processing aids in the technology of oilseeds, oils, and fats.

In the past few decades, enzyme-assisted aqueous extraction processing (EAEP) has experienced a quick development stage, and use of enzymes to facilitate the recovery of oils from oilseeds and other oil-bearing materials has become known quickly. Interests in enzymatic degumming has also been observed mainly due to the commercial availability of several new, cost-effective, and stable phospholipases with sufficiently high enzyme activity. Meanwhile, the development of low-cost enzyme immobilization methods, coupled with the ability to greatly increase fermentation yields, resulted in the development of an enzymatic interesterification process that could compete with the conventional chemical process.

9.2 Enzyme Application in Oil Extraction

Application of enzymatic technology in industry has been restricted to the academic level for many years due to the cost of enzyme preparations and consequent economic disadvantage. In the case of enzyme-assisted aqueous processing, the cost could also be very high if an economic way for utilization of the protein-rich liquid phase has not been found. Another reason is the lack of consideration for environmental protection as a consequence of predominant use of solvent extraction in the oil extraction industry, which contributes to much environmental concern. However, with the change in the human dietary model, increasing interest in plant protein consumption, cost decrease of enzyme preparations, and more concern for environmental protection, enzyme application in oil extraction, especially those processes bringing quality protein products simultaneously would have a bright future.

Present knowledge on enzyme application in oil extraction mainly covers three aspects (de Moura et al. 2008; Jiang et al. 2010; Rosenthal et al. 1996; Zhang et al. 2007a). First, oil-bearing material pretreatment technology based on an understanding of its microstructure. Second, various enzyme-assisted processes, especially those in an aqueous system for oil extraction. Third, downstream processes include oil recovery from an emulsion system. Hereafter, major enzymes influencing oil extraction from the most popular oilseeds, such as soybean, peanut, and sunflower, will be discussed. In addition, a compensatory strategy combining enzyme pretreatment and conventional processing would also be discussed due to its outstanding economic advantage.

9.2.1 Principle of Enzyme-Assisted Aqueous Extraction from Oilseeds

Extraction processing of oil-bearing material mainly consists of dispersing the ground oilseeds into water and providing a motive force to free the oil from cellular confines. In addition, pH, duration, temperature, solid-to-liquid ratio, and particle size of the ground seeds affect the oil and protein extraction rate. Optimization of these parameters is necessary to achieve high efficiency of oil and protein extraction. The EAEP process scheme is shown in Figure 9.1. Centrifugation is indispensable in subsequent steps to separate the free oil from cream, skim, and solids. However, the extraction technology still has problems, such as the formation of a stable emulsion during the process, the high cost of enzymes, and high effluent generation. Therefore, the development of a demulsification method, improvement and innovation of water recycling, membrane technology, and processing technology are needed for an industry applicable process.

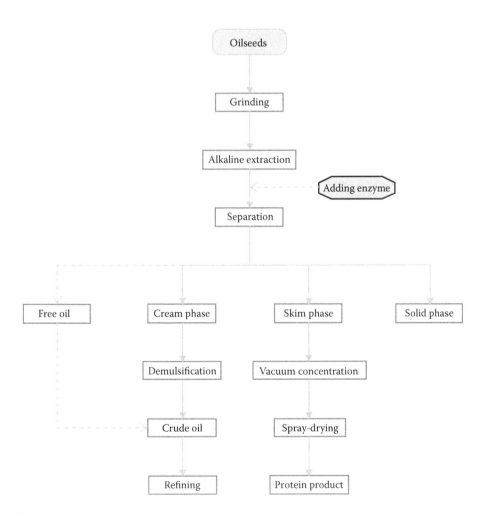

FIGURE 9.1 Process scheme for enzyme-aided aqueous extraction.

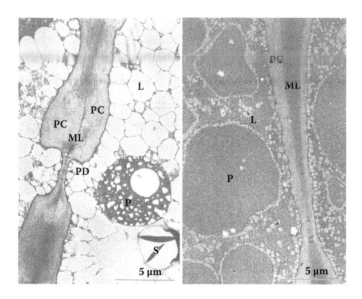

FIGURE 9.2 Electron micrograph of the cotyledon cells of peanuts and soybeans. (From Yamamoto, N., and S. Tamura, *Journal of Home Economics of Japan*, 50 (4), 313–321, 1999.)

TABLE 9.1

Average Diameters of Oil Bodies Isolated from the Seeds of Various Species

Oil Seed Species	Rape	Mustard	Cotton	Flax	Maize	Peanut	Sesame
Average diameter (μm)	0.65	0.73	0.97	1.34	1.45	1.95	2.00

Source: Tzen, J. T. C., Y. Z. Cao, P. Laurent, C. Ratnayake, and A. H. C. Huang, *Plant Physiology*, 101, 1, 267–276, 1993.

9.2.1.1 Microstructure of Oilseeds

To better understand the influence of enzymes on oil extraction, it is necessary to outline the microstructure of oilseeds and the oil extraction processes. Diverse plant seeds store lipids and proteins in the cotyledon tissue in intracellular organelles called oil bodies (or oleosomes) and protein bodies, respectively. The oil bodies, which consist of oil droplets, are covered by a monolayer of phospholipids and proteins called oleosins. The lipids will be used during germination and postgerminative growth. The monolayer coat makes the oil bodies remarkably stable and prevents oil droplets from aggregating or coalescing. The rest of the space between the oil bodies is occupied by protein bodies, which account for a large amount of total protein content (Beisson et al. 2001; Frandsen et al. 2001; Murphy et al. 2001; Tzen et al. 1993). Oilseeds generally have significant variances on their microstructure. As shown in Figure 9.2, the cotyledon cells of peanut and soybean are quite different from each other on a subcellular level.

The average diameter of oil bodies from the seeds of various species ranges from 0.5 to 2.0 μm (Tzen et al. 1993) as shown in Table 9.1. The sizes of oil bodies and protein bodies are affected by the variety of plant, maturity of the seeds, the environment, the rainfall and sunshine of cultivation year, soil fertility, and moisture. Meanwhile, high oil content means high ratio of oils to oleosins, and the oil bodies are larger and spherical; while low oil content seeds generally have smaller oil bodies (Huang 1996). Because the oil bodies have a complex constitution and structure, limiting the oil droplet in oilseeds, the destruction of the seed is generally very important as a privilege step for oil extraction.

9.2.1.2 Mechanical Comminution

One common view on enzyme-assisted aqueous extraction processing is that the cell wall must be disrupted prior to aqueous or enzyme-assisted aqueous extraction processing. The effect of mechanical and enzymatic treatment on oilseed particle disruption is shown in Figure 9.3. There are many mechanical

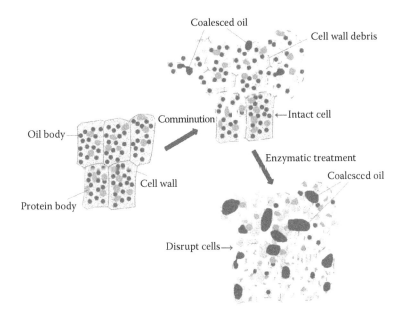

FIGURE 9.3 Effect of mechanical comminution and enzymatic treatment on oilseed cell structure.

methods that can disrupt the cell wall of various species, such as milling, flaking, extrusion, crack, ultrasonication, and microwave treatment (Gai et al. 2013; Lamsal et al. 2006; Li et al. 2014). Rhee et al. (1973) compared the effect of different grinding degrees on the extraction of peanut protein and oil and found that pregrinding of the peanut kernels with 0.01 inches prior to extraction is necessary to insure maximum recovery, particularly of the oil. However, the effect of high extent disruption contributed less to protein extraction compared with oil. The particle size of the soybean meal was reported to have a significant effect on protein and oil extraction (Rosenthal et al. 1998). The yields of both components were inversely proportional with particle sizes. Therefore, taking a longer time for grinding would obtain a smaller particle size and enable better oil and protein release. When flaking and extruding, the most common methods of comminution used for commercial hexane extraction nowadays, were applied, oil yield of 56% and 60% only was obtained with flaked soybean and ground flours as starting material, respectively (Lamsal et al. 2006). Meanwhile, when extruded flakes were employed, the oil extraction yield was up to 88%, indicating that different particle size reduction or cell disruption methods would bring about different results for specific oil-bearing material.

The study (Lamsal et al. 2006) also showed that yields of oil and protein were higher when the extruding temperature was 100°C than when it was 120°C. The reason could be the disadvantageous effect of the higher extrusion temperature, because it might cause protein denaturation and increase oil binding with the denatured protein at the high temperature. The combination of flaking and extruding achieved complete cellular disruption. Meanwhile, they are industrially practical to extrude full-fat soy flakes for hexane extraction.

In a study on the effect of ultrasound on aqueous enzymatic extraction of oil from perilla (*Perillafrutescens* L.) seeds, it was observed that when ultrasound power of 250 W was employed, the increase of ultrasound treatment time from 5 to 30 min would bring the oil yield from 41.37% to 50.32% (Li et al. 2014).

In the case of high protein content oilseeds, wet grinding usually caused the formation of a stable emulsion, which is quite often observed in soybean, peanut, and sunflower seeds (de Moura et al. 2008; Dominguez et al. 1995; Rhee et al. 1973; Wang 2005). Although this would increase the cost of demulsification, it does not mean the dry grinding methods are the better choice. Researchers also found that dry grinding could generate serious heat during the comminution process and thus limit the milling to batch treatment processes. A systematically designed process can only be obtained when the science, technology, and engineering of the process is fully understood.

9.2.1.3 Enzymatic Treatment

The main components of the cell wall are cellulose, hemicellulose, and pectin. With the help of the cell wall, compartmentalization between the cells is possible, which aids in protection of the cell and mainte nance of the cell shape. However, the cotyledon cell, which contains a large amount of oils and proteins as energy reserves, was wrapped in the cell wall, which may obstruct the release of oils and proteins. The enzymes involved in the hydrolysis of fruit and oilseed cell walls are mainly pectinases, cellulases, and hemicellulases, all of which act in concert. Cellulases, hemicellulases, and pectinases affected oil and protein extraction yields as well as dissolution of nonlipid material, but the degradation of cellulose was generally very limited.

Researchers attempted different combinations of cell wall-degrading enzymes (carbohydrases and proteases) for individual seeds or fruits. Celluclast 1.5 L and Multifect were applied to assist hexane to extract soybean oil (Dominguez et al. 1995). There were 5% and 8%–10% increases after adding enzymes, which were simultaneously carried out with the oil extraction or treated prior to the solvent extraction, respectively. The oil yields with both processes were almost independent of enzyme concentrations in the range they investigated. In addition, the addition of enzymes rendered 20–30 min shorter than control.

The effect of pectinase on the oil solvent extraction was evaluated (Perez et al. 2013) by addition of 2% (w/w) pectinase to a fine meal of sunflower seeds prior to solvent extraction. Due to the effect of the enzyme on the fiber structure, a higher oil release and a quick extraction was achieved. At the same time, it was found that pectinase treatment was highly effective in the tocopherol extraction from sunflower seeds, obtaining a 32.3% increase on average.

9.2.1.4 Separation

The amount of emulsion that formed during the AEP or EAEP process needs to be further separated into free oil or cream, skim, and residual solid. Therefore, a longer time is needed to render the emulsion naturally settled. Actually, a centrifugal apparatus was preferred for most processes. A pilot-scale test was conducted while studying enzymatic aqueous extraction of peanut oil (Jiang et al. 2009). A three-phase decanter centrifuge was employed to separate the emulsion into free oil/cream, skim, and residual solid. Meanwhile, a disc decanter centrifuge was used to separate the free oil from stabilized cream, which was treated by enzymes to improve free oil yield. However, the processes of centrifuge and pipeline could create problems, such as stable emulsion formation. Therefore, ways to reduce emulsion formation during the centrifuge and pipeline processes are necessary to scale up laboratory EAEP or AEP processes.

9.2.1.5 Cream Demulsification

Because extraction of oil from oil-bearing material and obtaining free oil are almost equally important for EAEP/AEP processes, the contribution of these enzyme preparations to destabilize the emulsion cannot be neglected for its evaluation. More details about the emulsion that is formed in the AEP or EAEP were discussed (Chabrand et al. 2008; de Moura et al. 2008; Lamsal and Johnson 2007). Emulsion formed during aqueous or enzyme-assisted extraction of vegetable oil was the major obstacle to releasing free oil. So far, the common methods, such as ultrasound treatment, heating treatment, ethanol treatment, the phase inversion method, freezing–thawing, and enzymatic treatment, were widely used in cream demulsification. The freezing–thawing method can lower the stability of the emulsion and achieve the highest oil recovery in many oilseeds. Fat crystals may form and pierce the water phase during the freezing–thawing process. However, the disadvantages, including quite high energy consumption and difficulty in scale-up, were obvious.

Ethanol is regarded as a "green" solvent for the food industry, but manufacturers and customers still worry about residual solvent in the product. And this limits the application of ethanol for demulsification. In the case of enzymatic demulsification, after the addition of proteases to the emulsion, the interface protein was hydrolyzed, and the major substance to stabilize the emulsion was destroyed. It could be very specific due to the difference of starting material as well as EAEP processes (Fang and Moreau 2014;

Latif et al. 2008; Wu et al. 2009; Zhang et al. 2013). Its industrial application in EAEP processes would be promising with advantages such as mild reaction temperature, less by-product, and good quality of oil along with the decreased cost of enzyme, the recycling and utilization of water, and development of new ingredients/foods.

9.2.2 AEP/EAEP Applications in Major Oilseeds

Oilseeds have complex cell structures and components. According to the components of oilseeds, cellulase and hemicellulase are necessary for oil release when the oilseeds and fruits contain high cellulose and hemicellulose content. In the case of oilseeds and oil-bearing material with high protein contents, proteases are generally required. The expense of the enzymes should also be considered, and commercial enzyme preparation is preferred due to its comparatively lower cost. If a crude enzyme was employed in the process, freedom from lipase activity is necessary (Rosenthal et al. 1996). So far, EAEP has been widely researched on both laboratory scale and pilot scale for different material as described in Table 9.2.

9.2.2.1 Rice Bran

Rice bran is a by-product of the rice milling process, which typically contains 19.97% ± 0.07% crude oil, 14.12% ± 0.03% protein, 18.22% ± 0.88% total dietary fiber and 22.04% ± 0.39% starch. In an investigation on the effect of enzyme concentration, incubation time, and temperature on oil and protein extraction yields, it was found that the maximal extraction yields of oil and protein were 79% and 68%, respectively, and the quality of oil recovered from the process in terms of free fatty acid, iodine value, and saponification value was comparable with solvent-extracted oil and commercial rice bran oil although the peroxide value was higher (Hanmoungjai et al. 2001). A mixture of amylase (80 U), protease (368 U), and cellulase (380 U) recorded the highest oil yield (77%) under pH 7.0, 65°C, extraction time 18 h with constant shaking at 80 rpm (Sharma et al. 2001). Further, during demulsification of cream formed in EAEP of rice bran, the oil recovery from emulsion reached 87.8% with Protex 6 L (Fang and Moreau 2014).

9.2.2.2 Corn Germ

Extraction of oil from wet milling corn germ by EAEP, which contained 41% oil, 14.9% protein, 6.2% starch, and 0.7% pectin on a dry solids basis employing hydrothermal pretreatment before germ milling, indicated that hydrothermal pretreatment has positive effects on the quality of oil (Bocevska et al. 1993). Time and pH of the hydrothermal pretreatment had an obvious effect on free fatty acid. Meanwhile, the peroxide value is not significantly increased. Various enzymes (Gamanase 1.5 L, Pextinex Ultra SP-L, Celluclast 15 and SP-348) were used either individually or in mixtures, and their effects on oil yield have been studied. The oil yield reached 84.7% of the available oil with Celluclast 1.5 L.

Corn oil yield was reported to be higher than 90% by EAEP when the most effective cellulase enzymes, Multifect GC and Celluclast, were used in the study, and oil yields of about 93% and 91% were obtained, respectively (Moreau et al. 2004). Proteases were tested in the study, and there was almost no positive effect observed on oil yield. Researchers are in unanimous agreement that heating treatment is needed to destroy the endogenous enzymes. With equipment that could be reasonably scaled to full scale for a typical dry-grind plant, the oil yield could reach 72% with aqueous enzymatic process (Dickey et al. 2008). In the case of the effect of a blender and colloid mill on improving the oil yields, it was found that the colloid mill was better to release oil droplets.

9.2.2.3 Soybeans

Use of protease in heated soybean flours was found to enhance the extraction yield of protein and oil significantly compared with normal processing, from 27.8% to 66.2%, and 41.8% to 58.7%, respectively (Rosenthal et al. 2001). Protease was also useful when nonheated flour with large particle sizes was treated in the extraction. However, the yields of protein and oil had a slight decrease when nonheat-treated

TABLE 9.2

Enzyme-Aided AEP for Extracting Oil from Different Oilseeds and Oil-Bearing Materials

Material			Process Conditions	Extraction Yield (%)	References
Rice bran	1	Control	Water/bran = 5, hexane/bran = 1.5, 4 h, 50°C	54.1	Sengupta and
	2		Water/bran = 5, hexane/bran = 1.5, 4 h, 50°C, 2% cellulase + 2% pectinase	82.1	Bhattacharyya 1996
	3	Control	S/L = 1:5, preheated at 90°C, stirred at 1000 rpm, pH 9, 3 h, 50°C	73.3	Hanmoungjai et al. 2001
	4		Preheated at 90°C, 1% protease (alcalase), 3 h, 60°C, stirred at 1000 rpm, pH 9	79	
	5		Rice bran preheated at 90°C, 15 min, 1% alcalase 0.6 L, 1 h, pH 9, 50°C, stirred at 1000 rpm, S/L = 1:5	75	Hanmoungjai et al. 2002
	6		Rice bran preheated at 90°C, 15 min, 1% papain, 1 h, pH 6, 50°C, followed by pH shift to 8, stirred at 1000 rpm, S/L = 1:5	70	
	7		2.6% amylase + 3.8% crude cellulose (380 U) + 0.1% Protizyme™, pH 7 stirred at 20 rpm, 30 min, 65°C, 18 h	77	Sharma et al. 2001
	8		0.1% Protizyme (368 U), pH 5, stirred at 20 rpm, 30 min, 40°C, 18 h	6	
	9		3.8% cellulase (380 U), pH 5, stirred at 20 rpm, 30 min, 40°C, 18 h	4	
	10		0.1% Protizyme™ + 3.8% cellulase (368 U + 380 U), pH 5, stirred at 20 rpm, 30 min, 40°C, 18 h	11	
	11		3.8% cellulase + 2.6% amylase (380 U + 80 U), pH 5, stirred at 20 rpm, 30 min, 40°C, 18 h	10	
	12		2.6% amylase + 0.1% Protizyme™ (80 U + 368 U), pH 5, stirred at 20 rpm, 30 min, 40°C, 18 h	15	
	13		Pretreatment: steam + 0.5% amylase + 0.2% protease, extraction conditions: 6 h, 60°C, 1.2% cellulase, pH 5.0, S/L = 1:5	85.76	Yang et al. 2004
	14		Pretreatment at 100°C, 1 h, 1.2% cellulase and neutral protease (2:1), S/L = 1:7, 50°C, pH 5, 5 h	87.23	Liu et al. 2011
	15		Extrusion conditions: material moisture 14%, die-hole diameter 18 mm, temperature 110°C, screw speed 105 r/min. Extraction conditions: protex 6L was used, but no details about the parameters	89.61	Zhao et al. 2014
Corn germ	1	Control	Wet-milled corn germ, no enzyme, pH 4, boiling 20 min, 20 h at 65°C, shaking at 160 rpm	36.6	Moreau et al. 2004
	2		0.5% Multifect GC (82 GCU/g), pH 4.0, 50°C (4 h), 65°C (16 h), shaking at 160 rpm	82.1	
	3		Same as 2 except using 0.5% celluclast 1.5L (790 EGU/g) at pH 6.0	83.5	
	4	Control	Precooked at pH 4, 122°C, 20 min, wet milled germ, S/L = 1:10, 50°C, 24 h	0	Dickey et al. 2009
	5		Precooked at pH 4, 122°C, 20 min, wet-milled germ to water or buffer solution, S/L = 1:10, 1.25% accellerase TM 1000 cellulase complex, 50°C, 24 h	79.7	
	6		Thawed–frozen corn germs to buffer (0.05 mol/L citrate solution) 1:8, heat treatment at 112°C, 65 min, 2.0% proteinase and 1.5% cellulase, S/L = 1:6, at the optimal temperature and pH of the enzyme	91	Qian et al. 2004
	7		Same as 6 except the pretreatment thaw–freeze was not employed	61	
	8		Preheated germ, 2.5% cellulose, and α-amylase (a mass ratio of 4:3), S/L = 1:5, pH 6, 7 h	89	Zhao et al. 2010

(Continued)

TABLE 9.2 (CONTINUED)

Enzyme-Aided AEP for Extracting Oil from Different Oilseeds and Oil-Bearing Materials

Material			Process Conditions	Extraction Yield (%)	References
Peanut	1	Control	Grinding soaked peanut, pH 4.0, 40°C, 18 h, S/L − 1:5, shaking at 80 rpm	44	Sharma et al. 2002
	2		Grinding soaked peanut, 2.5% Protizyme™, pH 4.0, 40°C, 18 h, S/L = 1:5, shaking at 80 rpm	92	
	3	Control	Grinding blanched peanut, pH 8.5, 60°C, 8 h, S/L = 1:5, stir at 200 rpm	–	Wang et al. 2008
	4		Grinding blanched peanut, 1.5% alcalase 2.4L, pH 8.5, 60°C, 8 h, S/L = 1:5, stir at 200 rpm	92.2	
	5	Control	Grinding blanched peanut, pH 8.5, 60°C, 5 h, S/L = 1:5, slowly stir	30.59	Jiang et al. 2010
	6		Grinding blanched peanut, 1.5% alcalase 2.4L, pH 8.5, 60°C, 5 h, S/L = 1:5, slowly stir	79.32	
	7		Grinding blanched peanut, 1.5% alcalase 2.4L + 1.0% As1398, 55°C, pH 8.5, 2 + 3 h, S/L = 1:5, slowly stir	73.59	
	8		Grinding blanched peanut, 5% Protizyme, pH 4, 40°C, 18 h, S/L = 1:5, slowly stir	55.02	
	9		Grinding blanched peanut, 2% As1398, pH 7, 45°C, 5 h, S/L = 1:5, slowly stir	66.35	
	10		Grinding blanched peanut, 2% Nutrase, pH 7.5, 55°C, 5 h, S/L = 1:5, slowly stir	60.08	
	11		Grinding blanched peanut, 1.5% Protamex, pH 6, 40°C, 5 h, S/L = 1:5, slowly stir	48.89	
	12		Grinding blanched peanut, 1.0% complex cellulose, pH 4.5, 2 h + 1.5% Alcalase, pH 8.5, 3 h, 55°C, S/L = 1:5, slowly stir	74.14	
	13		Grinding roasted peanut (190°C, 20 min), 2% Alcalase 2.4L, pH 9.5, 3 h, S/L = 1:5, stir at 120 rpm	78.6	Zhang et al. 2011
Soybean	1	Control	No enzyme, S/L = 0.10, 50°C, pH 7 (1 h), pH 8 (15 min)	68	Lamsal et al. 2006
	2		0.5% protease (MultifectNeutra), S/L = 0.10, 50°C, pH 7 (1 h), pH 8 (15 min)	88	
	3		Extruded flakes, 1% Protex 6L, S/L = 0.10, 50°C, pH 9, 1 h	97	de Moura et al. 2008
	4		Extruded flakes, 1% Protex 7L, S/L = 0.10, 50°C, pH 7, 1 h and pH 8, 15 min	93	
	5		Countercurrent 2-stage extraction, 0.5% Protex 6L, S/L = 0.20, 50°C, pH 8 (1 h), pH 9 (15 min)	98	de Moura and Johnson 2009
	6	Control	No enzyme, S/L = 0.10, 50°C, pH 8, 2 h	72	Campbell and Glatz 2009
	7		0.5% Protex 7L, S/L = 0.10, 50°C, pH 8, 2 h	81	
	8		No enzyme, 3% SDS, S/L = 0.10, 50°C, pH 8, 2 h	85	
	9		Extrude condition: die length 20 mm, material moisture 14.5%, screw speed 105 r/min, 90°C, 2% Alcalase 2.4L, 57°C, pH 9.5, 3.75 h, S/L = 1:6.5	93.02	Li et al. 2010
Rapeseed/ canola	1	Control	–	–	
	2		Seed/water = 1:3, boiled 5 min, wet-grinding 3 min, addition 2.5% (pectinase + cellulase + β-glucanase) (4:1:1) for 4 h at pH 5, 48°C stirring at 200 rpm, then shift pH to 10, 1.5% Alcalase 2.4L, 60°C, 3 h	75.7	
	3	Control	S/L = 1:6, boiled 5 min, 45°C, 2 h, stirring at 120 rpm	38.28	Latif et al. 2008
	4		Same as 3 except Protex 7L was used, but the optimal pH and enzyme concentration were not mentioned	54.29	
	5		Same as 4 except Multifect Pectinase FE was employed	51.50	
	6		Same as 4 except Multifect CX 13L was employed	60.32	
	7		Same as 4 except Natuzyme was employed	52.67	

(Continued)

TABLE 9.2 (CONTINUED)

Enzyme-Aided AEP for Extracting Oil from Different Oilseeds and Oil-Bearing Materials

Material			Process Conditions	Extraction Yield (%)	References
Sunflower seed	1	Control	–	–	Leng et al. 2006
	2		Dry-grinding, addition citric buffer solution at pH 4.8, S/L = 1:5, 2.5% Viscozyme L, stirring at 250 r/min, 50°C, 7 h	85.73	
	3	Control	–	–	Ren 2008
	4		Preheat at 110°C, 60 min in pH 4.8 citric buffer, 2% cellulase and pectinase (2:1), pH 4.7, 50°C	89.8	
	5	Control	S/L = 1:6, boiled 5 min, 45°C, 2 h, stirring at 120 rpm	40.22	Latif and Anwar 2009
	6		Protex 7L, the optimal pH and enzyme concentration were not mentioned, S/L = 1:6, boiled 5 min, 45°C, 2 h, stirring at 120 rpm	62.20	
	7		Same as 6 except Kemzyme was employed	70.77	
	8		Same as 6 except Alcalase 2.4L was employed	58.46	
	9		Same as 6 except Natuzyme was employed	78.02	
	10		Same as 6 except Viscozyme L was employed	87.25	

material was used. Extrusion-facilitated protease (0.5% wt/wt) action resulted in oil recovery increasing from 71% to 88% (extrusion temperature at 100°C; AEP was carried out at 50°C, pH 8, 15 min for extruded flakes) (Lamsal et al. 2006). The oil yield was similar to the result obtained while applying extrusion to disrupt the soybean cell walls without flaking (Freitas et al. 1997). However, cellulase (1%, wt/wt) was ineffective with extruded flakes in increasing oil and protein extraction over the free enzyme control. The reason may be that the compression and high shear generated during flaking and extrusion could have crushed the cell wall materials to the extent that further cellulase treatment was invalid. With the addition of 0.5% Protex 51FP, Protex 6 L, or Protex 7 L to the extruded flakes during extraction, 90%–93% extraction yield could be achieved, which was significantly higher than 73% of the control (Wu et al. 2009).

Heating was found to be ineffective in breaking the emulsion during an investigation of the effect of some physical methods to demulsify and recover free oil from the soybean cream phase, including heating (at 95°C, 3–4 h in a water bath) and freezing–thawing (freezing at −20°C, thawing at room temperature) (Lamsal and Johnson 2007). Only 22% ± 6% of the free oil was obtained, and 36% ± 6% was still left in the cream. The freezing–thawing method resulted in the highest free oil yield (86% ± 4%) compared to other demulsification methods (32% ± 6% of control, 68% ± 9% of Lysomax/G-zymc cocktail, and 73% ± 5% of phospholipase C). The protease used in cream demulsification had a significant effect on the emulsion stability, implying that protein was an important component to maintain system stability.

The emulsion formed during the enzyme-assisted aqueous extraction of oil from soybean flour was not stable, and the stability of cream was enhanced by the presence of proteins and phospholipids, either of which could serve as emulsifiers (Chabrand and Glatz 2009). Phospholipase facilitated the highest oil recovery (103% ± 1%) when it was applied to destabilize the emulsion at 2% concentration. After two-stage demulsification (protex 6 L), recovery of 95% free oil from the cream was obtained. However, by one-stage demulsification with Protex 6 L, only 72% recovery of free oil was obtained. Protease was found to release more free oil than phospholipase at concentrations less than 2%. At 0.2% concentration, 88% and 48% of the free oil were obtained by using protease and phospholipase, respectively (Wu et al. 2009). These observations indicated that phospholipase could be beneficial for cream destabilization.

9.2.2.4 Peanuts

In the 1950s, the processing of peanuts or other oil-bearing seeds for the simultaneous extraction of oil and protein received increasing attention (Subrahmanyan et al. 1959), but the grinding methods such as wet milling, which may cause vigorous emulsification, limited the processing used in high-protein materials. In an investigation when a commercial mixture of three proteases and Protizyme™ were used in enzyme-assisted aqueous extraction of peanut oil, it was found that the oil yield with the assistance of Protizyme™ (86%–92%) was remarkably higher than that of papain (76%), chymotrypsin (61%), and Trypsin (67%) (Sharma et al. 2002). However, a high concentration of enzyme (2.5% w/w) was used in the study, and the cost of the enzyme is a major concern for industrial application of this technology. The investigation also adopted high-speed centrifugation (18,000 × g, 20 min) and 18 h incubation to achieve a higher oil yield. In another study on the effect of different enzymes on oil and protein extraction from peanuts, it was found that Alcalase 2.4 L exhibited an excellent capacity on oil extraction, and 92.2% of the total oil was extracted. In addition, when Neutral As1398 (85.9% ± 0.97%), papain (84.5% ± 2.0%), and Protease-N (86.9% ± 1.9%) were used for EAEP, the oil extraction yields were slightly reduced (Wang et al. 2008). The rich perfume obtained with roasted peanut seeds during aqueous enzymatic extraction was observed to be better than that extracted from unroasted peanuts (Zhang et al. 2011). But roast processing lowered the free peanut oil yield by about 10%. Under optimized roast time and temperature, the free oil yield reached 86%–90% after the cream demulsification. In another study on the demulsification of the oil-rich emulsion from the peanut oil extraction process, the optimal demulsification rate (94%) was achieved by adding Mifong®2709 (alkaline endopeptidase from *Bacillus licheniformis*, optimal pH 9.0, optimal temperature 50°C). Compared with soybeans, peanuts showed significant differences on cream demulsification with phospholipase. It was found that phospholipase A2 had a marginal effect on releasing free oil from peanut cream. The possible reason could be that peanuts contained a much lower content of phospholipids (0.18% ± 0.03%) (Zhang et al. 2013). Therefore, phospholipids might not be a major contributor to the stability of peanut emulsion.

Freeze–thaw and ethanol addition could remarkably aggregate the oil droplets in stubborn emulsion formed during aqueous extraction from peanuts, especially after 50% ethanol addition, most oil droplets were combined, and 90% of the oil in stubborn emulsion could be recovered. Under these conditions, the total free oil yield could be increased to 93% from 88% in the overall process (Chi et al. 2014).

9.2.2.5 Sunflower Seeds

During aqueous extraction of sunflower seeds without using enzymes, 41.6% of the total oil was released after treatment. With addition of 2% cellulase, hemicellulase, pectinase, pectinex, or acid protease to ground sunflower seeds during extraction, protease showed superiority on oil extraction to its counterparts. The oil yields were 55.77%, 54.07%, 44.47%, 46.8%, and 57.08%, respectively (Badr and Sitohy 1992). It was observed that the oil inside the cells was connected with starch, cellulose, hemicellulose, protein, and pectin, and the links between oil and other molecules were broken by proteases, cellulases, and pectinases, resulting in the release of oil. Another study on the effect of five enzyme preparations on the extraction of oil and protein from sunflower seeds reported oil yield improvement in the following order: Viscozyme L (87.25%) > Natuzyme (78.02%) > Kemzyme (70.77%) > Protex 7 L (62.20) > Alcalase 2.4 L (58.46%). Further, with Protex 7 L in the extraction, the maximum protein extraction yield in the skim was obtained (Latif and Anwar 2009).

In an investigation on optimization of the extraction parameters of sunflower oil during enzyme-assisted aqueous extraction processing, it was found that particle size has a substantial influence on the oil extraction yield, and compared with traditional solvent extraction methods, the meal obtained by EAEP had a lighter color. This could be attributed to the removal of sugar and phenolic compounds in the skim, thus reducing the reaction between phenolics and protein (Sineiro et al. 1998).

9.2.2.6 Rapeseed

The enzymes produced by *Aspergillus niger* and *Bacillus* were employed in rapeseed oil extraction, and it was found that the enzyme with hemicellulase activity (3% *Aspergillus niger* enzyme) effected the highest (72.2%) oil yield (Fullbrook 1983). In a study on the effect of enzyme content, seed moisture content, incubation time, and temperature on rapeseed oil extraction, it was noted that an oil extraction yield increase of 6% could be obtained compared with the control (Sarker et al. 1998).

Various carbohydrases were used to achieve hydrolysis of the wet-milled slurry during optimization of the extraction process of rapeseed oil and protein, using emulsified oil yield to evaluate different conditions with a single enzyme or an enzyme cocktail. The mixture of pectinase, cellulase, and β-glucanase (4:1:1, 2.5% v/w, 4 h, pH 5, 48°C) supported the highest emulsified oil yield. In the subsequent steps, demulsification by protease may be viable option, and Alcalase 2.4 L was employed for demulsification at 60°C for 1 h. Under the optimized conditions, the free oil yield of 73%–76% was obtained (Zhang et al. 2007b).

9.2.3 Summary

To summarize, enzyme-assisted aqueous extraction processing is an effective technology for obtaining a higher oil recovery from oilseeds. This process may achieve the extraction of oil and protein at the same time. Meanwhile, it is an environmentally friendly process and possesses the advantages of less equipment investment, relatively simple operation, and less hazardous production. But it still has some disadvantages, such as the cost of enzymes, which is the main limiting factor of EAEP. In addition, approximately 10% of the oil still existed in the skim and residual fraction, which is hard to recover. With the development of new technologies for both oil extraction and emulsion breakdown and an increasing requirement for desirable protein-rich ingredients from oilseeds, as well as environment concerns, commercial bulk-scale application of EAEP for oilseeds and oil-bearing material at competitive prices would facilitate the oil industry significantly.

9.3 Enzyme Application in Oil Refining

Oil without refining is termed "crude" oil because it contains a number of impurities including phospholipids, FFAs, pigments, sterols, carbohydrates, proteins, and their degradation products. Degumming is an indispensable part of the oil refining process. The stand or fall of the degree of degumming has a direct influence on the refined oil product quality. Phospholipids that generally exist in crude oil may form insoluble and gummy precipitate. Meanwhile, the presence of large amounts of phospholipids can cause oil discoloration and serve as a precursor of off-flavors (Jiang et al. 2014). Therefore, the degumming process is necessary for vegetable oil production.

Phospholipids are a kind of lipid containing triglyceride with two fatty acid radicals and one side chain formed by a phosphate ester (Figure 9.4). According to their hydrophilic characteristics, phospholipids can be divided into two categories: hydratable and nonhydratable phospholipids (NHP). Most

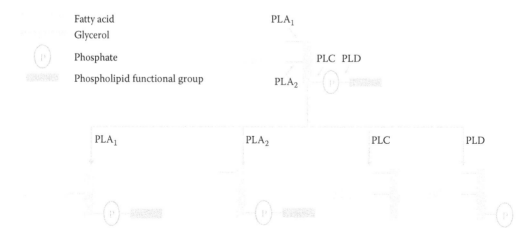

FIGURE 9.4 Structure of phospholipids and reactions catalyzed by phospholipase enzymes. Phospholipids comprise diacylglycerol coupled to a phosphate ester. All phospholipase enzymes act to break phospholipids into water-soluble and oil-soluble fragments.

TABLE 9.3

Relative Rates of Phospholipid Hydration

Phospholipid	Relative Rate of Hydration
Phosphatidylcholine	100
Phosphatidylinositol	44
Phosphatidylinositol (calcium salt)	24
Phosphatidylethanolamine	16
Phosphatidic acid	8.5
Phosphatidylethanolamine (calcium salt)	0.9
Phosphatidic acid (calcium salt)	0.6

Source: http://lipidlibrary.aocs.org/processing/degum-enz/index.htm.

phospholipids have amphiphilic characteristics. An absence of water renders them able to dissolve in the oil. However, the presence of water makes them insoluble in oil. The highest hydration rate of phospholipids is phosphatidylcholine (PC), followed by phosphatidylinositol (PI), phosphatidylinositol calcium salt, phosphatidylethanolamine (PE), phosphatidic acid (PA), phosphatidylethanolamine calcium salt, and phosphatidic acetate (calcium salt) (Table 9.3). The hydratable phospholipids are effectively removed by water degumming. But NHPs are hard to remove without special treatments. They consist mainly of calcium and magnesium salts of phosphatidic acid (PA) and free phosphatidylethanolamine (PE) (Dijkstra 2013). Therefore, the major purposes of degumming are removing the NHPs or transforming them into hydratable phospholipids and then eliminating them with water.

9.3.1 Principle of Enzyme Degumming

As one of the first industrial applications of enzymes in edible oil refining, enzymatic degumming has been extensively studied. Generally, phospholipases are used to increase the hydratability of the phospholipids in the oil (Dijkstra 2013). From a biochemical perspective, there are four main types of phospholipase. Their common mode of action is depicted in Figure 9.4. Phospholipases (A1 and A2) that remove one of the fatty acids from the glycerol backbone to produce a lysophospholipid are the most commonly used in edible oil refining. Thus, nonhydratable phospholipids are converted into a more hydratable form, making it easy to remove by water as part of the physical degumming process. Commercial Lecitase®Ultra was applied to optimize the enzymatic degumming process of rice bran and obtained residual phosphorus concentrations less than 10 mg/kg. Optimum operating conditions were found to be a reaction time of 4.07 h, enzyme dosage of 50 mg/kg, added water of 1.5 ml/100 g, and temperature of 49.2°C (Jahani et al. 2008).

Phospholipase D removes the phospholipid head group to produce a glycerophosphate, and it was found to play an important role in the formation of nonhydratable PA even at a temperature of 65°C in the presence of water-saturated hexane. However, in aqueous media, this enzyme is readily deactivated by heat (De Maria et al. 2007). Recently, phospholipase C has been introduced to oil degumming. The hydrolysis of the bond between the glycerol backbone and the phosphate group results in the formation of a diglyceride and a water-soluble portion containing the phosphorus and head group portions of the phospholipid (Dijkstra 2011). These enzymes can be applied when total hydrolysis of the phospholipid is required to maximize release of free fatty acids and reduce gum volumes as much as possible. Phospholipases C and D could be used together with A1 and A2, but A1 phospholipase generally requires an A2 lysophospholipase and vice versa if complete hydrolysis is required.

Phospholipase B (PLB), also known as lysophospholipase, can remove the remaining fatty acid from a lysophospholipid. PLB is an enzyme with a combination of both PLA1 and PLA2 activities. This enzyme was successfully applied in the degumming of soybean oil and peanut oil (Huang et al. 2014). The PLB was heterologously overexpressed in *Pichia pastoris*, and it has higher enzyme productivity than the native one. Under the optimal conditions, the phosphorus contents of soybean oil and peanut oil decreased from 125.1 mg kg^{-1} to 4.96 mg kg^{-1} and 96 mg kg^{-1} to 3.54 mg kg^{-1}, respectively.

Mei et al. (2013) compared acid degumming and enzymatic degumming processes on silybum marianum seed oil. Lecitase®Ultra (with phospholipase A1 activity) was employed in their study, and when an enzyme reaction time of 6 h and dosage of 100 mg kg⁻¹ was used, the phospholipid content of the degummed seed oil was reduced to 17.95 mg kg⁻¹, which showed remarkable superiority to the acid degumming method (128.1 mg kg⁻¹). Compared with traditional degumming methods, enzymatic degumming could happen at mild reaction temperatures (40°C–50°C), much lower than that of traditional methods (85°C–90°C), thus helping to benefit oil quality (Huang et al. 2014).

9.3.2 Enzymatic Degumming Process

A general flowchart of an enzymatic degumming process is shown in Figure 9.5. As the first step, acid conditioning/pH adjustment of the crude or water degummed oil is to make the nonhydratable phospholipids more accessible for enzyme degradation at the oil–water interface and to bring the pH closer to the optimal pH of the enzyme. Afterward, the enzyme is added. High shear mixing is applied to ensure optimal distribution in the oil. Enzyme dosing depends on the type of enzyme and on the phospholipid content of the oil, which usually varies between 50~200 ppm. The optimal reaction temperature is 50°C–60°C, and the required reaction time mainly depends on the enzyme dosage. Finally, the heavy phase consisting of water and gums or phosphate esters is separated by centrifugation from the degummed oil.

Enzymatic degumming processes can be classified into two different types, that is, enzymatic water degumming and deep enzymatic degumming. Enzymatic water degumming is typically applied in (soybean) crushing plants. The main concern is to increase oil yield. The expected yield increase depends on the type of oil and the type of enzyme used. The highest increase (up to 2%) can be expected when crude soybean oil is enzymatically degummed with PLC (Hitchman 2009). A lower yield increase (1.0%–1.5%) will be obtained from PLC degumming of crude rapeseed oil or when PLA1 or PLA2 is used on crude soybean oil (Kellens and De Greyt 2010).

Typically, PLC degummed soybean oil has 100–150 ppm residual P (mainly present in PA and PC). A significantly better degumming efficiency (P < 10 ppm) can be obtained when crude or water-degummed vegetable oils are enzymatically degummed with commercial PL-A1 or PL-A2. This so-called "deep enzymatic degumming" is already applied in several industrial plants. In addition to the increased oil yield, the very efficient phospholipid removal—making the degummed oil suitable for physical refining—is of great interest to refiners. As an alternative, a combination of PL-C and PL-A1/PL-A2 can be used for deep enzymatic degumming (Dayton and Galhardo 2008). The two enzymes can be added either separately or as a cocktail, depending on the plant design. Although the potential advantages of the latter process are well described in the literature, it is still rarely applied on an industrial scale.

Direct enzymatic de-oiling of the lecithin fraction resulting from the water degumming of crude oils is a potential alternative to enzymatic degumming. Kellens and De Greyt (2010) proposed a process with phospholipase added to the wet lecithin to degrade phospholipids into much less hydrophobic lysophospholipids. As a result, 80%–90% of the entrapped neutral oil can be recovered by simple static decantation or centrifugation (Kellens and De Greyt 2010). Enzymatic lecithin de-oiling processes showed significant superiority to enzymatic degumming processes due to its lower enzyme consumption (~50% less) and independence on the oil degumming/refining process.

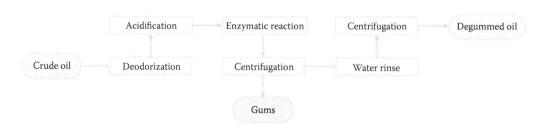

FIGURE 9.5 Typical flow chart of an enzymatic degumming process.

Process optimization pays attention to not only oil yield, but oil quality as well. The enzymatic degumming process influences the composition of crude sunflower oil in addition to the quality and oxidative stability of the crude oil. Thus, the combination of two commercial phospholipases, Lecitase®Ultra and LysoMax®Oil, led to the phosphorus content in crude sunflower seed oil below 3 mgkg^{-1} (Lamas et al. 2014). However, the degummed oil was found suitable for physical refining (Lamas et al. 2014).

9.3.3 Summary

Degumming has been part of edible oil processing for decades, and the technology has been developed during its history. Nowadays, the quality of degummed oils has been much improved in terms of residual phosphorus and iron levels, which, in turn, facilitates the subsequent refining steps and ensures the high quality of the final products. However, most degumming processes currently applied still rely on the use of added reagents. Therefore, the operating cost has been increased due to the large amount of chemical use. Meanwhile, environmental concern has also increased as most of them must be discharged into wastewater.

Degumming with enzymatic methods became attractive because it eliminated or reduced chemical use and reduced the cost and environmental impact of degumming. The most widely used enzyme preparations include phospholipases A1 (such as Lecitase Novo and Ultra), and phospholipase C (such as Purifine), and lipid acyl transferase (such as LysoMax) with PLA2 activity. The Lecitases and the LysoMax enzymes catalyze the hydrolysis of all common phosphatides and differ from the Purifine enzyme, which is specific for phosphatidyl choline and phosphatidyl ethanolamine. These phosphatides are hydrolyzed to oil-soluble diacylglycerol and water-soluble phosphate esters. Because these diacylglycerols remain in the oil during refining, they contribute to the oil yield. In addition, all enzymes cause less oil to be retained by the gums by decreasing the amount of gums and/or their oil retention, improving oil yields in another way.

As to nonhydratable phosphatides (alkaline earth salts of phosphatidic acid), enzymes are incapable of catalyzing them under industrial conditions presently. Consequently, the industrial enzymatic degumming step has to be preceded by a chemical degumming step to arrive at degummed oil with a sufficiently low residual phosphorus content that can be physically refined. Future work needs to be aimed at the development of novel technology that facilitates hydrolysis of nonhydratable PLs and reduction of the process cost and environmental impact of degumming (Cowan 2013).

9.4 Enzyme Application in Lipid Modification

Oil and fat are mainly composed of triacylglycerols (TAGs), and the composition of TAGs is glycerol and three fatty acid molecules, which are usually different. It is generally known that the physical characteristics and nutritional properties of oil and fat depend on the type of TAGs, fatty acid composition, and positional distribution. In natural fats, acyl groups are distributed in a nonrandom pattern. They tend to concentrate on the specific "sn" position, which changes with different varieties, growth, environmental conditions, and parts of animals and plants. To improve the nutritional and functional properties of fats and oils, which is of great interest to food processors, lipid modification with various methods has been tried. With enzymes being involved in the modification, interesterification is the most typically used method, which includes three kinds of exchanges, that is, exchange of acyl groups between an ester and an acid (acidolysis), an ester and an alcohol (alcoholysis), and an ester and an ester (transesterification) (Figure 9.6).

Transesterification alters the positional distribution of fatty acids in the triacylglycerols. One example is the production of new TAGs similar to those in infant formula from blends of lard and soybean oil (Silva et al. 2009). The researchers found that after a lipase-catalyzed interesterification reaction with a lard–to–soybean oil ratio of 60:40 and 50:50, the ratios of linoleic acid to alpha linolenic acid were 13.1 and 13.9, respectively. And the ratios of polyunsaturated and saturated fatty acids increased to 0.9 and 1.2, respectively. These transesterified products showed potential application opportunities in infant formula because no significant alteration in the fatty acid profile of the starting blends was observed.

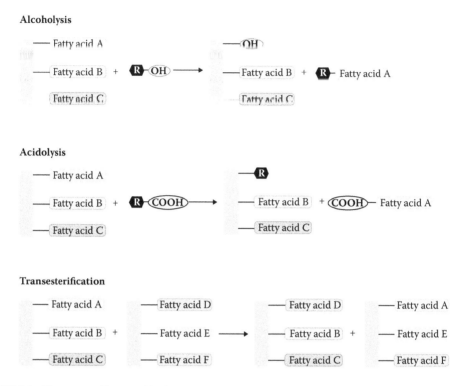

FIGURE 9.6   Three types of interesterification reactions of acylglycerols.

Alcoholysis is the reaction between an ester and an alcohol ester to produce new diacylglycerols and monoacylglycerols. Liu et al. (2012) attempted the production of diacylglycerols by glycerolysis of soybean oil. Immobilization of Lecitase®Ultra was employed in the study, and it catalyzed the glycerolysis reaction to produce diacylglycerols in a solvent-free system. Under the optimum conditions (glycerol/soybean oil mole ratio 10:1, initial water content 5 wt.%, and enzyme load 5 wt.%), a product with a 53.7 wt.% DAG content was obtained.

Acidolysis, the reaction of the transfer of an acyl group between an ester and an acid, has been used to produce human milk fat substitutes, cocoa butter substitutes, and modified fish oil products by the incorporation of new fatty acids or restructured to change the positions of fatty acids to improve their nutritional properties. Different starting materials were prepared to produce human milk fat (HMF) substitutes, such as lard, butter, palm stearin, and tripalmitin. HMF substitutes were produced by catalyzing acidolysis of high-melting palm stearin with fatty acids from rapeseed oil (Zou et al. 2012).

In past decades, there has been an increase in research efforts on substituting some chemical interesterification with enzymatic interesterification due to the inherent advantages associated with the enzymatic process. Compared with chemical interesterification, enzymatic reactions showed better specificity, less severe reaction conditions, less waste production, and higher economical attractiveness, considering immobilized enzymes can be reused.

9.4.1 Lipase and Its Specificity

Lipases (EC. 3.1.1.3) are key enzymes involved in fat digestion in vertebrates by converting insoluble triacylglycerols into more soluble products, fatty acids, and monoacylglycerols, which can easily be assimilated by the organism. Lipases are widely used in enzymatic interesterification processes and usually obtained from bacterial, yeast, fungal sources; the edible forestomach tissue of calves, kids, or lambs; and animal pancreatic tissues. Industrial lipases are also produced by the controlled fermentation of *Aspergillus niger* var., *Aspergillus oryzae* var., *Candida rugosa*, and *Rhizomucor miehei* as a powder or liquid. Modification of lipids is one of the typical applications of lipases in the food industry. Lipase-catalyzed reactions take place in the neat oil or in a nonpolar solvent.

All lipases contain an active serine in a consensus sequence G-X-S-X-G, which is also shown in serine proteases (Gly-Asp-Ser-Gly-Gly in the trypsin family and Gly-X-Ser-XAla in subtilisin) (Pabai et al. 1995). Recent structural analyses have shown conclusively that lipases contain the same constellation of the catalytic triad, Ser...His...Asp, present in all serine proteases although the topological position of the individual residue's side chain varies (Tanaka et al. 1993).

The lipase-catalyzed interesterification reaction showed superiority to chemical methods due to its specificity. Three major types of specificity can be distinguished. The first type involves positional or regiospecificity with most lipases being either 1,3-specific or fully random (nonspecific). The 1,3-specificity is the most common and is displayed by, for example, human gastric lipases as well as microbial lipases, such as from *Rhizomucor miehei*, *Thermomyces* sp., and various *Rhizopus* sp. The second type of specificity relates to the type of fatty acid involved. For example, certain *Cuphea* lipases show a distinct selectivity for short chain fatty acids. The third major lipase specificity is for certain types of acylglycerols. *Penicillium camembertii* lipase is known to act upon mono- and diglycerides only whereas other *Penicillium* lipases are reported to hydrolyze only triglycerides.

Nonspecific lipases show no positional or fatty acid specificity during interesterification. Interesterification with them after extended reaction times gives complete randomization of all fatty acids in all positions and gives the same products as chemical interesterification. Typical nonspecific lipases include lipases derived from *Candida cylindraceae*, *Corynebacterium acnes*, and *Staphylococcus aureus* (Soumanou et al. 2013).

Positional specificity is characterized as specificity toward ester bonds in positions sn-1,3 of the triacylglycerol. The inability of these lipases to act on position sn-2 on the triacylglycerol is generally caused by steric hindrance (Macrae and How 1988). An interesterification reaction using a 1,3-specific lipase will initially produce a mixture of triacylglycerols, 1,2- and 2,3-diacylglycerols, and free fatty acids (Macrae and How 1988). Lipases that are 1,3-specific include those from *Aspergillus niger*, *M. miehei*, *Rhizopus arrhizus*, and *Rhizopus delemar*. One interesting thing is that the specificity of individual lipases can change due to microenvironmental effects on the reactivity of functional groups or substrate molecules (Pabai et al. 1995). For example, lipase from *Candida parapsilosis* hydrolyzes the sn-2 position more rapidly than either of the sn-1 and sn-3 positions under aqueous conditions (Riaublanc et al. 1993). Positional specificity could bring out nutritionally improved interesterified fats and oils compared with chemical interesterified samples. As we all know, fish oils and some vegetables oils contain high degrees of essential PUFAs mostly located in sn-2 positions, and 2-monoacylglycerols are more easily absorbed than sn-1 or sn-3 monoacylglycerols in intestines. With a 1,3-specific lipase, the fatty acid composition of positions 1 and 3 can be changed to meet the targeted structural requirements while retaining the nutritionally beneficial essential fatty acids in position 2 (Ray and Bhattacharyya 1995).

Stereospecificity refers to the specificity that lipases differentiate the two primary esters sterically distinct at the sn-1 and sn-3 positions. When catalyzing with a stereospecific lipase, positions 1 and 3 are hydrolyzed at different rates. Stereospecificity is determined by the source of the lipase and the acyl groups, and can be influenced by lipid density at the interface and fatty acid chain length as well (Yu et al. 2009).

Fatty acid specificity occurs when many lipases are specific toward particular fatty acid substrates. Fatty acid chain length specificity has been found in many lipases with some being specific toward long-chain fatty acids and others being specific toward medium and short chain fatty acids. For example, porcine pancreatic lipase is specific toward short-chain fatty acids whereas lipase from *Penicillium cyclopium* is specific toward long-chain fatty acids. With interesterification reactions in organic media, lipases can also be specific toward certain alcohol species. Fatty acid specificity by certain lipases can be used in the production of short-chain fatty acids for use as dairy flavors and in the concentration of EPA and DHA in fish oils by lipases with lower activity toward these fatty acids.

The efficiency of lipase-catalyzed reactions depends not only on the specificity and other inherent properties, but also on environmental factors such as pH, the amount of water, solvent, temperature, and ratio of reactants. Among all these factors, the immobilization and reactor design have been gaining attention as to lipase-catalyzed reactions.

9.4.2 Immobilization and Reactors

Immobilized enzymes possess obvious advantages over free enzymes in reusability, rapid termination of reactions, lowered cost, controlled product formation, and ease of separation of the enzyme from the reactants and products. As to lipases, their selectivity and chemical and physical properties could also be improved during immobilization.

Extensive research on immobilization of lipases has been carried out, and most used methods are adsorption, entrapment, and covalent bonding. The easiest immobilization method of lipase in interesterification reactions is adsorption, which involves contacting an aqueous solution of the lipase with an organic or inorganic surface-active adsorbent. The process of adsorption can be accomplished through ion exchange or through hydrophobic or hydrophilic interactions and van der Waals interactions. Such factors as pH, temperature, solvent type, ionic strength, and protein and adsorbent concentrations would affect the efficiency and extent of immobilization. Common hydrophobic supports include polyethylene, polypropylene, styrene, and acrylic polymers whereas hydrophilic supports include Duolite, Celite, silica gel, activated carbon, clay, and Sepharose. The disadvantages of using hydrophilic supports include high losses of activity due to changes in conformation of the lipase, steric hindrance, and prevention of access of hydrophobic substrates. In addition, changes in pH, ionic strength, or temperature can cause desorption of lipase that has been adsorbed by ion exchange.

The effectiveness of the immobilization process is influenced by the internal structure of the support. Research showed that most of the enzyme is immobilized on the surface of the support if it contains only small pore structure, which prevents the occurrence of internal mass transfer limitations. If supports contain larger pore sizes, some lipases will be immobilized inside the pores, which can prevent access of the substrate to these lipase molecules (Ison et al. 1994). To evaluate the effectiveness of the immobilization process, the activity of an immobilized enzyme can be compared with the activity of free enzymes of equal amount determined under the same operating conditions. The effectiveness value can be used as a guide to the degree of inactivation of the enzyme caused by immobilization. For values close to 1.0, very little enzyme activity has been lost on immobilization whereas values much lower than 1 indicate high degrees of enzyme inactivation.

Performance of immobilized lipases, depend not only on the enzymes and the support material, but also on the handling and reaction conditions. It was found that freeze-drying of the immobilized enzyme could substantially reduce the moisture content of the immobilized lipase and dramatically improve its activity. Similarly, molecular sieves are able to reduce the amount of water in reaction systems and reduce the degree of hydrolysis by taking away the accumulated water in the reaction (Paludo et al. 2015).

For immobilized lipase, reactors can be batch or flow-through systems and can differ in the degree of mixing involved during the reaction. And the most common reactor systems include fixed bed, batch, continuous stirred tank, and membrane reactors (Figure 9.7). Two indexes, that is, volumetric activity and operational stability of the immobilized enzyme were often used to describe the productivity of the system. Volumetric activity is determined as the mass of product obtained per liter of reactor per hour, and operational stability is defined as the half-life of the immobilized enzyme.

A fixed bed reactor is a form of continuous flow reactor, in which the immobilized enzyme is packed in a column or as a flat bed, and the substrate and product streams are pumped in and out of the reactor at the same rate. The main advantages of fixed bed reactors include ease of scale-up, ease of operation, high efficiency, and low cost. In addition, it provides more surface area per unit volume than a membrane reactor system (Sheldon and van Pelt 2013). Before the reaction is catalyzed by immobilized lipase, the oil would go through a precolumn containing water-saturated silica or molecular sieves, allowing the oil to become saturated with sufficient but minimal water and to keep the progression of the interesterification reaction without undesired hydrolysis. In a fixed bed reactor, increasing residence time in the reactor can lead to increased product formation. However, complete conversion to products will never be achieved, and with an increase in product levels, a loss in productivity will occur. Although fixed bed reactors are more efficient than batch reactors, they are prone to fouling and compression. Therefore, dissolution of the oil in an organic solvent to reduce viscosity and removal of particulates from the substrate is necessary to run the system in a good state.

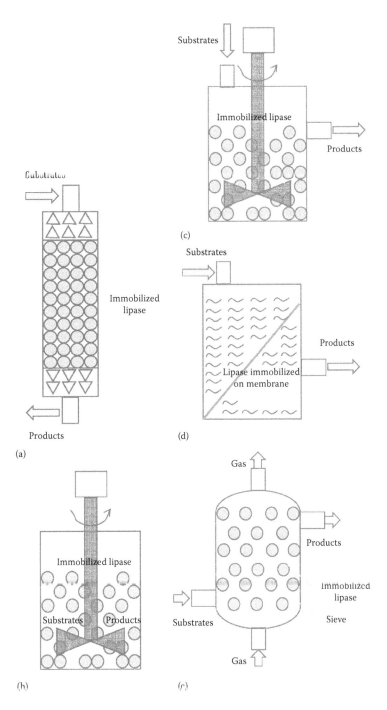

FIGURE 9.7 Typical reactors with immobilized lipase. (a) Fixed bed reactor, (b) stirred batch reactor, (c) continuous stirred tank reactor, (d) membrane reactor, and (e) fluidized bed reactor.

Another system, namely, the stirred batch reactor, is commonly used in laboratory experiments with lipase-catalyzed interesterification due to its simplicity and low cost. Compared with fixed bed reactors, stirred batch reactors do not require addition or removal of reactants and products except at the initial and final stages of the reaction. It is easy to build, and free enzymes can be used. However, a larger system or longer reaction times are required to achieve equivalent degrees of conversion in comparison with other systems, and side reactions can be significant (Kogej and Pavko 2001).

Continuous stirred tank reactors combine components of both fixed bed and batch reactors. There is an agitated tank into which reactants and products are added and removed at the same rate while continuous stirring is kept to eliminate mass transfer limitations encountered in fixed bed reactors. A continuous stirred tank reactor can be in the form of a tank with stirring from the top or bottom or a column with stirring accomplished by propellers attached to the sides of the column (Liu et al. 2008). Although they combine the advantages of both fixed bed reactors and stirred batch reactors, continuous stirred tank reactors are disadvantageous on higher power costs and increasing support particle breaking possibility.

Membrane reactors involve two-phase systems, in which the interface of two phases is at a membrane. Immobilization of enzymes onto semipermeable membranes is an attractive alternative for lipase-catalyzed interesterification reactions. With membrane reactor systems, pressure drops and fluid channeling reduced, effective diffusivity, enzyme stability, and membrane surface area-to-volume ratio are improved. In addition, the reaction and separation of substrates and products can be accomplished in one system (Giorno et al. 2007). This is especially important when water is produced during the esterification reaction. In a membrane, such as microporous polypropylene, the pores have dimensions of 0.075 by 0.15 mm, and the fibers have an internal diameter of 400 mm, providing 18 m^2 of surface area per gram of membrane. Immobilization of lipase on it can be accomplished by submerging the fibers in ethanol, rinsing them in buffer, then submerging them in lipase solution (Malcata et al. 1992).

Fluidized bed reactors are characterized in that the immobilized enzyme and support are kept suspended by the upward flow of substrate or gas at high flow rates. Although a quite low concentration of enzyme is used in such a large void volume, the fluidized bed reactors are very competent because channeling problems are eliminated and there is less change in pressure and less coalescence of emulsion droplets and no concentration gradients (Damnjanović et al. 2012). In a study, a gas lift reactor was adopted to produce a cocoa butter equivalent by interesterifying palm oil midfraction (Mojovic et al. 1994). The reaction in the gas lift reactor was more efficient than in a stirred batch reactor. Equilibrium was reached 25% earlier, and productivity was 2.8 times higher in the gas lift reactor.

9.4.3 Factors Affecting Lipase-Catalyzed Reaction

For enzymatic interesterification, the most influential factors include temperature, pH, water content, substrate composition, product composition, and surface active agents.

Similar to other enzymatic reactions, lipase-catalyzd interesterification would have a higher reaction rate with temperature, but decreases at very high temperatures due to irreversible denaturation of the enzyme. In food industry applications, due to the avoidance of organic solvents, the reaction temperatures are usually higher to keep the substrate in the liquid state. Thus, the chances of lipase denaturation increased. One solution is to substitute animal or plant lipases with extracellular microbial lipases, which are comparatively more thermostable (Mahadevan and Neelagund 2014). However, immobilization was a much-preferred choice, considering the advantages of it. Typically, immobilization fixes the enzyme in one conformation, restricts enzyme movement, and reduces the susceptibility of the enzyme to denaturation by heat. Thus, the operational stability is improved for lipases.

pH has a very significant effect on both enzyme stability and the catalyzed reaction. Different lipases could be catalytically active at a different pH due to the origin and the ionization state of residues in their active sites. The optimum pH for most lipases is 7~9 although lipases can be active over a wide range of acid and alkaline pHs (pH 4~10) (Sheldon and van Pelt 2013). Immobilization usually changes the pH optimum of lipases according to the partitioning of protons between the bulk phase and the microenvironment around the support and the restriction of proton diffusion by the support. When the lipase is immobilized on a polyanion matrix, the concentration of protons will be higher than that in the bulk phase, and the optimum pH of the enzyme will thus be shifted toward a more basic pH. This shifting of optimum pH occurs only in solutions with ionized support and low ionic strength systems. If protons are produced in the course of interesterification, the hydrogen ion concentration in the Nernst layer can be higher than in the bulk phase, thereby decreasing the pH in the vicinity of the lipase.

Among all enzymes applied in the food industry, lipases show much higher sensitivity to water content, or the dependence varies according to the source of the enzyme. Lipases from molds seem to be more tolerant to low water activity than bacterial lipases. The optimal water content for interesterification by different lipases ranges from 0.04% to 11% (w = v) although most reactions require water contents of <1% for effective interesterification (Gandhi et al. 2000). Generally, high water content leads the reaction equilibrium toward hydrolysis and low water content to ester synthesis. However, too low water activity would prevent all reactions from occurring because a certain amount of water is indispensable for lipases to keep the correct conformation. In a reaction system with an immobilized lipase, there is no aqueous shell surrounding the enzyme. Therefore, the oil phase must be well saturated with water to avoid diffusional limitations (Tweddell et al. 1998). Due to the accumulation of water during interesterification, the reaction equilibrium has to be manipulated to prevent ester hydrolysis other than ester synthesis reaction. Sabbani and Hedenstrom (2009) reported that by using a saturated salt solution in contact with the reaction mixture via the gas phase, the water produced during interesterification could be removed continuously. In another study, silicone tubing containing the salt solution was immersed in the reaction vessel, and water vapor was transferred out of the reaction system across the tubing wall and into the salt solution (Wehtje et al. 1997). In addition, adding molecular sieves to reactors near the end of the reaction or running the reaction under a vacuum can all be used to remove water while allowing the lipase to keep hydrated.

Substrate composition can have an effect on the lipase-catalyzed interesterification. Generally, the hydroxyl group in the sn-2 position has a negative inductive effect, so the hydrolysis rate of acylglycerols following the order triacylglycerols > diacylglycerols > monoacylglycerols (Foster and Berman 1981). Another factor related with lipase-catalyzed interesterification is steric hindrance, which usually possess a negative effect. If the composition of the substrate is such that it impedes access of the substrate to the active site, it is hard to improve the lipase activity. Conformation of the substrate can also affect the enzymatic reaction. Aliphatic chains and aromatic rings would be more easily accepted in the hydrophobic tunnel of the lipase than branched structures (Li et al. 2013). Oxidation of substrates, especially PUFAs, would bring an inhibitory effect on lipases, especially in reactions containing organic solvents. Considering this effect, refining of oils containing high levels of PUFAs is quite necessary before running interesterification reactions.

Product composition also could have a remarkable effect on lipase-catalyzed interesterification. A study showed that during the acidolysis of sunflower oil, MG and DG formation increased principally by the presence of excessive water formed during the reaction (Pacheco et al. 2010). During acidolysis of butter oil with undecanoic acid, concentrations of undecanoic acid was found to be greater than 250 mM and decreased the activity of porcine pancreatic lipase significantly (Elliott and Parkin 1991). In addition, free fatty acids with free or ionized carboxylic acid groups would acidify the microenvironment of aqueous phase surrounding the lipase and desorb water from the interface. Also, an increase of fatty acids in a reaction system could lead to more fatty acids partitioning into the surrounding water shell and thus limit access of substrates to the interface (Kuo and Parkin 1993).

Surface-active agents could be used during the immobilization of lipases. The presence of surface-active agents in lipase-catalyzed interesterification could have a complicated effect on reaction. In a microaqueous environment, the addition of lecithin or sugar esters as surface-active agents during the immobilization process can increase its activity by tenfold (Kurashige et al. 1993). In contrast, using surface-active agents to form an emulsion can dramatically decrease the rate of interesterification because they prevent contact between the lipase and substrate (Briand et al. 1995). Phospholipids that exist in oil as a minor component could have a negative effect on lipase activity. Due to the initial competition between phospholipids and triacylglycerols for the active site of the lipase, interesterification of triacylglycerols was inhibited in the reaction. Therefore, to prolong the half-life of immobilized lipases during interesterification, phospholipid content in oils must be controlled (Wang and Gordon 1991).

9.4.4 Typical Examples

Due to the economic disadvantage of enzymatic methods in production of commercial modified lipids, such as cocoa butter equivalent, margarines, and shortenings, there is still quite a long way to go before enzymatic interesterification replaces chemical methods. However, lipid modification with lipase has been gaining a lot of attention, especially in the development of reduced-calorie products, enriched lipids, and structured lipids. In addition, some research focuses on characterization of lipases with fatty acid specificities and potential applications of immobilized lipase in food industry. These are also important for oil related industries.

Commercial production of margarines with partial hydrogenation or chemical interesterification could generate a high content of trans fatty acids during these processes. Therefore, production of trans-free margarine by enzymatic interesterification is desirable and environmentally friendly. Zero-trans interesterified fats from camellia seed oil (CSO), palm stearin (PS), and coconut oil (CO) with three weight ratios (CSO/PS/CO, 50:50:10, 40:60:10, and 30:70:10) was prepared using Lipozyme TL IM (Ruan et al. 2014). Results showed that the interesterified products contained palmitic acid (34.28%–42.96%), stearic acid (3.96%–4.72%), oleic acid (38.73%–47.95%), linoleic acid (5.92%–6.36%) and total medium-chain fatty acids (MCFAs) (5.03%–5.50%). Compared with physical blends, triacylglycerols of OOO and PPP were decreased and formed new peaks of equivalent carbon number (ECN) 44 in the interesterified products. The product CPC3 showed a slip melting point of 36.8°C and a wide plastic range of solid fat content (SFC) (45.8%–0.4%) at 20°C–40°C. This research indicated that the zero-trans interesterified fats would have a potential functionality for margarine fats. Interesterification between tributyrin and methyl stearate in a solvent-free system for the production of low-calorie structured lipid (LCSL) with an efficient solid catalyst yielded more than 90% LCSL at 80°C within 1 h when the methyl stearate/tributyrin molar ratio of 2:1 was employed (Xie and Qi 2014). The obtained solid catalyst could be recovered easily and reused for several recycles with a negligible loss of activity. By using the solid base catalyst, an eco-friendly, more benign process for the interesterification reaction in a heterogeneous manner was developed.

Enzymatic acidolysis was applied to synthesize structured lipids (SLs) containing palmitic, oleic, and docosahexaenoic acids for possible use in infant formulas (Pande et al. 2013) using the substrates tripalmitin, extra virgin olive oil free fatty acids (EVOOFFA), and docosahexaenoic acid single-cell oil-free fatty acids (DHASCOFFA) in 1:1:1, 1:2:1, 1:3:2, 1:4:2, and 1:5:1 molar ratios. Reactions were carried out at 65°C for 24 h using Lipozyme® TL IM lipase. Results showed that SLs, SL132, SL142, and SL151 had desirable fatty acid distribution for infant formula use with nearly 60 mol% palmitic acid at the sn-2 position and oleic acid predominantly at the sn-1,3 positions. The total DHA content of SL132, SL142, and SL151 were 7.54, 6.72, and 5.89 mol%, respectively. The major TAG molecular species in the SLs were PPP, OPO, and PPO. The melting completion temperature of SL132 was 37.1°C, 35.2°C in SL142, and 32.9°C in SL151. The SLs synthesized have potential use in infant formulas.

One of the most outstanding advantages of lipase-catalyzed interesterification is fatty acid specialties. Nalder et al. (2014) synthesized a range of *para*-nitrophenol (*p*NP) acyl esters as a means to rapidly screen lipases for fatty acid selectivity using spectrophotometric detection. The chosen esters were based primarily on the most abundant fatty acids present in anchovy and tuna oils. *p*NP derivatives of C16:1 n-7, C18:1 n-9 (OA), C18:2 n-6 (LA), C18:3 n-3 (ALA), C20:5 n-3 (EPA), and C22:6 n-3 (DHA) were synthesized too. Storage stability of these *p*NP derivatives was shown to be at least 6 months, and all *p*NP derivatives, including those of EPA and DHA, were shown to be stable throughout the conditions of the assay. The authors applied the new assay substrates for the determination of fatty acid selectivity of five widely utilized lipases. Results showed that the lipase from *Candida rugosa* was the most selective in terms of omega-3 specificity, preferentially hydrolyzing all other medium-long chain substrates. Lipases from *Rhizomucor miehei* and *Thermomyces lanuginosa* also showed selectivity with a significant preference for saturated fatty acids. *Candida antarctica* lipase B and *Aspergillus niger* lipase were the least selective.

A recent review by DiCosimo et al. (2013) pointed out that immobilized lipase for food oil transesterification reached 10^5 tons/year. When including the immobilized lipase for chiral resolution of alcohols and amines and biodiesel production from triglycerides, the number could be even bigger. Exciting news from Alimentos Polar Comercial C.A., in Venezuela, is its new De Smet Ballestra Interzym Interesterification plant with a capacity of 80 tpd. The De Smet Interzym process utilizes Lipozyme® TL IM as an alternative to both chemical interesterification and hydrogenation. A pretreated blend of fats is pumped through a number of packed bed reactors (usually three or four) placed in series and kept at a temperature of around 70°C; typical flow rate is 1–2 kg oil per kg enzyme per hour (De Greyt and Dijkstra 2007). Using freshly deodorized oil blends as feed, 2.5–4.0 tons of oil was interesterified with 1 kg of immobilized enzyme, and the enzyme remains active for 2500–4000 h. The enzymatically interesterified oil does not need bleaching, only mild deodorization to remove some free fatty acid and off-flavors. Based on the reported enzyme productivity and enzyme pricing and the amount of sodium methylate catalyst used in a chemical interesterification process, the total costs of both interesterification processes were about equal.

REFERENCES

ATSDR of USA. 2011. "n-Hexane." Available at http://www.atsdr.cdc.gov/substances/toxsubstance.asp?toxid=68.

Badr, F. H., and M. Z. Sitohy. 1992. "Optimizing conditions for enzymatic extraction of sunflower oil." *Grasas Y Aceites* 43 (5):281–3.

Beisson, F., N. Ferté, S. Bruley, R. Voultoury, R. Verger, and V. Arondel. 2001. "Oil-bodies as substrates for lipolytic enzymes." *Biochimica et Biophysica Acta (BBA)—Molecular and Cell Biology of Lipids* 1531 (1–2):47–58.

Bocevska, M., D. Karlović, J. Turkulov, and D. Pericin. 1993. "Quality of corn germ oil obtained by aqueous enzymatic extraction." *Journal of the American Oil Chemists' Society* 70 (12):1273–7.

Briand, D., E. Dubreucq, and P. Galzy. 1995. "Functioning and regioselectivity of the lipase of candida-parapsilosis (ashford) langeron and talice in aqueous-medium—new interpretation of regioselectivity taking acyl migration into account." *European Journal of Biochemistry* 228 (1):169–75.

Campbell, K. A., and C. E. Glatz. 2009. "Mechanisms of aqueous extraction of soybean oil." *Journal of Agricultural and Food Chemistry* 57 (22):10904–12.

Chabrand, R. M., and C. E. Glatz. 2009. "Destabilization of the emulsion formed during the enzyme-assisted aqueous extraction of oil from soybean flour." *Enzyme and Microbial Technology* 45 (1):28–35.

Chabrand, R., H.-J. Kim, C. Zhang, C. E. Glatz, and S. Jung. 2008. "Destabilization of the emulsion formed during aqueous extraction of soybean oil." *Journal of the American Oil Chemists' Society* 85 (4):383–90.

Chi, Y., W. Zhang, R. Yang, X. Hua, and W. Zhao. 2014. "Characterization and destabilization of the emulsion formed during aqueous extraction of peanut oil." In *IFT Annual Meeting Abstract Book*. New Orleans, LA.

Cowan, D. 2013. "Enzyme processing." In *Edible Oil Processing*, edited by W. Hamm, R. J. Hamilton and G. Calliauw, 197–222. Chichester, UK: John Wiley & Sons.

Damnjanović, J. J., M. G. Žuža, J. K. Savanović, D. I. Bezbradica, D. Ž. Mijin, N. Bošković-Vragolović, and Z. D. Knežević-Jugović. 2012. "Covalently immobilized lipase catalyzing high-yielding optimized geranyl butyrate synthesis in a batch and fluidized bed reactor." *Journal of Molecular Catalysis B: Enzymatic* 75:50–9.

Dayton, C. L. G., and F. Galhardo. 2008. "Enzymatic degumming utilizing a mixture of PLA and PLC phospholipases." US20080182322.

De Greyt, W., and A. J. Dijkstra. 2007. "Fractionation and interesterificationin." In *Trans Fatty Acids*, edited by A. J. Dijkstra, J. R. J. Hamilton and W. Hamm, 181–202. Oxford, UK: Wiley-Blackwell.

De Maria, L., J. Vind, K. M. Oxenboll, A. Svendsen, and S. Patkar. 2007. "Phospholipases and their industrial applications." *Applied Microbiology and Biotechnology* 74 (2):290–300.

de Moura, J. M. L. N., and L. Johnson. 2009. "Two-stage countercurrent enzyme-assisted aqueous extraction processing of oil and protein from soybeans." *Journal of the American Oil Chemists' Society* 86 (3):283–9.

de Moura, J. M. L. N., K. Campbell, A. Mahfuz, S. Jung, C. E. Glatz, and L. Johnson. 2008. "Enzyme-assisted aqueous extraction of oil and protein from soybeans and cream de-emulsification." *Journal of the American Oil Chemists' Society* 85 (10):985–95.

Dickey, L. C., M. J. Kurantz, and N. Parris. 2008. "Oil separation from wet-milled corn germ dispersions by aqueous oil extraction and aqueous enzymatic oil extraction." *Industrial Crops and Products* 27 (3):303–7.

Dickey, L. C., M. J. Kurantz, N. Parris, A. McAloon, and R. A. Moreau. 2009. "Foam separation of oil from enzymatically treated wet-milled corn germ dispersions." *Journal of the American Oil Chemists Society* 86 (9):927–32.

DiCosimo, R., J. McAuliffe, A. J. Poulose, and G. Bohlmann. 2013. "Industrial use of immobilized enzymes." *Chemical Society Reviews* 42 (15):6437–74.

Dijkstra, A. J. 2011. "Enzymatic degumming." *Lipid Technology* 23 (2):36–8.

Dijkstra, A. J. 2013. "Degumming." In *Edible Oil Processing from a Patent Perspective*, edited by A. J. Dijkstra, 121 155. New York: Springer.

Dominguez, H., M. J. Nunez, and J. M. Lema. 1995a. "Aqueous processing of sunflower kernels with enzymatic technology." *Food Chemistry* 53 (4):427–34.

Dominguez, H., M. J. Nunez, and J. M. Lema. 1995b. "Enzyme-assisted hexane extraction of soya bean oil." *Food Chemistry* 54 (2):223–31.

Elliott, J. M., and K. L. Parkin. 1991. "Lipase-mediated acyl-exchange reactions with butteroil in anhydrous media." *Journal of the American Oil Chemists' Society* 68 (3):171–5.

Fang, X., and R. A. Moreau. 2014. "Extraction and demulsification of oil from wheat germ, barley germ, and rice bran using an aqueous enzymatic method." *Journal of the American Oil Chemists' Society* 91 (7):1261–8.

Foster, D. M., and M. Berman. 1981. "Hydrolysis of rat chylomicron acylglycerols: A kinetic model." *Journal of Lipid Research* 22 (3):506–13.

Frandsen, G. I., J. Mundy, and J. T. C. Tzen. 2001. "Oil bodies and their associated proteins, oleosin and caleosin." *Physiologia Plantarum* 112 (3):301–7.

Freitas, S. P., L. Hartman, S. Couri, F. H. Jablonka, and C. W. P. de Carvalho. 1997. "The combined application of extrusion and enzymatic technology for extraction of soybean oil." *Lipid/Fett* 99 (9):333–7.

Fullbrook, P. D. 1983. "The use of enzymes in the processing of oilseeds." *Journal of the American Oil Chemists' Society* 60 (2):476–8.

Gai, Q.-Y., J. Jiao, P.-S. Mu, W. Wang, M. Luo, C.-Y. Li, Y.-G. Zu, F.-Y. Wei, and Y.-J. Fu. 2013. "Microwave-assisted aqueous enzymatic extraction of oil from Isatis indigotica seeds and its evaluation of physico-chemical properties, fatty acid compositions and antioxidant activities." *Industrial Crops and Products* 45:303–11.

Gandhi, N. N., N. S. Patil, S. B. Sawant, J. B. Joshi, P. P. Wangikar, and D. Mukesh. 2000. "Lipase-catalyzed esterification." *Catalysis Reviews-Science and Engineering* 42 (4):439–80.

Giorno, L., E. D'Amore, R. Mazzei, E. Piacentini, J. Zhang, E. Drioli, R. Cassano, and N. Picci. 2007. "An innovative approach to improve the performance of a two separate phase enzyme membrane reactor by immobilizing lipase in presence of emulsion." *Journal of Membrane Science* 295 (1–2):95–101.

Hanmoungjai, P., D. L. Pyle, and K. Niranjan. 2001. "Enzymatic process for extracting oil and protein from rice bran." *Journal of the American Oil Chemists' Society* 78 (8):817–21.

Hanmoungjai, P., D. L. Pyle, and K. Niranjan. 2002. "Enzyme-assisted water-extraction of oil and protein from rice bran." *Journal of Chemical Technology and Biotechnology* 77 (7):771–6.

Hitchman, T. 2009. "Purifine® PLC: Industrial application in degumming and refining." In *AOCS Centenary Meeting & Expo*, 118. Orlando, FL.

Huang, A. H. 1996. "Oleosins and oil bodies in seeds and other organs." *Plant Physiology* 110 (4):1055–61.

Huang, S., M. L. Liang, Y. H. Xu, A. Rasool, and C. Li. 2014. "Characteristics and vegetable oils degumming of recombinant phospholipase B." *Chemical Engineering Journal* 237:23–8.

Ison, A. P., A. R. Macrae, C. G. Smith, and J. Bosley. 1994. "Mass-transfer effects in solvent-free fat interesterification reactions—influences on catalyst design." *Biotechnology and Bioengineering* 43 (2):122–30.

Jahani, M., M. Alizadeh, M. Pirozifard, and A. Qudsevali. 2008. "Optimization of enzymatic degumming process for rice bran oil using response surface methodology." *LWT—Food Science and Technology* 41 (10):1892–8.

Jiang, L. H., D. Hua, Z. Wang, and Xu S. Y. 2009. "Pilot-scale study on aqueous enzymatice extraction of oil and protein hydrolysates from peanut." *Food and Fermentation Industries* 35 (9):146–50 (in Chinese).

Jiang, L., D. Hua, Z. Wang, and S. Xu. 2010. "Aqueous enzymatic extraction of peanut oil and protein hydrolysates." *Food and Bioproducts Processing* 88 (2–3):233–8.

Jiang, F., J. Wang, I. Kaleem, D. Dai, X. Zhou, and C. Li. 2011. "Degumming of vegetable oils by a novel phospholipase B from Pseudomonas fluorescens BIT-18." *Bioresource Technology* 102 (17):8052–6.

Jiang, X. F., M. Chang, X. S. Wang, Q. Z. Jin, and X. G. Wang. 2014. "The effect of ultrasound on enzymatic degumming process of rapeseed oil by the use of phospholipase A(1)." *Ultrasonics Sonochemistry* 21 (1):142–8.

Johnson, L. A. 2007. "Recovery, refining, converting, and stabilizing edible fats and oils." In *Food Lipids Chemistry, Nutrition, and Biotechnology*, edited by C. C. Akoh and D. B. Min. Boca Raton, FL: CRC Press.

Keefe, S. F. O. 2007. "Nomenclature and classification of lipids." In *Food Lipids Chemistry, Nutrition, and Biotechnology*, edited by C. C. Akoh and D. B. Min, 1–38. Boca Raton, FL: CRC Press.

Kellens, M., and W. De Greyt. 2006. "Oil recuperation process." US20060030012 A1.

Kellens, M., and W. De Greyt. 2010. "Enzymatic oil recuperation process." US20100240917.

Kogej, A., and A. Pavko. 2001. "Laboratory experiments of lead biosorption by self-immobilized Rhizopus nigricans pellets in the batch stirred tank reactor and the packed bed column." *Chemical and Biochemical Engineering Quarterly* 15 (2):75–9.

Kuo, S. J., and K. L. Parkin. 1993. "Substrate preferences for lipase-mediated acyl-exchange reactions with butteroil are concentration-dependent." *Journal of the American Oil Chemists' Society* 70 (4):393–9.

Kurashige, J., N. Matsuzaki, and H. Takahashi. 1993. "Enzymatic modification of canola palm oil mixtures— Effects on the fluidity of the mixture." *Journal of the American Oil Chemists' Society* 70 (9):049–52.

Lamas, D. L., G. H. Crapiste, and D. T. Constenla. 2014. "Changes in quality and composition of sunflower oil during enzymatic degumming process." *LWT-Food Science and Technology* 58 (1):71–6.

Lamsal, B. P., and L. A. Johnson. 2007. "Separating oil from aqueous extraction fractions of soybean." *Journal of the American Oil Chemists' Society* 84 (8):785–92.

Lamsal, B. P., P. A. Murphy, and L. A. Johnson. 2006. "Flaking and extrusion as mechanical treatments for enzyme-assisted aqueous extraction of oil from soybeans." *Journal of the American Oil Chemists' Society* 83 (11):973–9.

Latif, S., and F. Anwar. 2009. "Effect of aqueous enzymatic processes on sunflower oil quality." *Journal of the American Oil Chemists' Society* 86 (4):393–400.

Latif, S., L. L. Diosady, and F. Anwar. 2008. "Enzyme-assisted aqueous extraction of oil and protein from canola (*Brassica napus* L.) seeds." *European Journal of Lipid Science and Technology* 110 (10):887–92.

Leng, Y. X., S. Y. Xu, Z. Wang, and R. J. Yang. 2006. "Aqueous enzymatic extraction of sunflower seed oil." *Food and Fermentation Industries* (in Chinese) 32 (10):127–31.

Li, X., S. Huang, L. Xu, and Y. Yan. 2013. "Conformation and catalytic properties studies of Candida rugosa Lip7 via enantioselective esterification of ibuprofen in organic solvents and ionic liquids." *The Scientific World Journal* 2013:364730.

Li, Y., L. Z. Jiang, Z. G. Zhang, H. B. Wu, J. Xu, and X. Wu. 2010. "Effect of cellulose degradation on soybean oil extraction yield through extrusion and expansion processing." *Transactions of the Chinese Society of Agricultural Machinery* (in Chinese) 41 (2):157–63.

Li, Y., Y. Zhang, X. N. Sui, Y. N. Zhang, H. X. Feng, and L. Z. Jiang. 2014. "Ultrasound-assisted aqueous enzymatic extraction of oil from perilla (Perilla frutescens L.) seeds." *Cyta-Journal of Food* 12 (1):16–21.

Liu, Q., C. Jia, J. M. Kim, P. Jiang, X. Zhang, B. Feng, and S. Xu. 2008. "Lipase-catalyzed selective synthesis of monolauroyl maltose using continuous stirred tank reactor." *Biotechnology Letters* 30 (3):497–502.

Liu, C. L., D. Li, T. Y. Wang, Y. H. Chang, and D. Y. Yu. 2011. "Aqueous enzymatic extraction of response surface optimization process of rice bran oil." *Food Industry* (in Chinese) 12:46–9.

Liu, N., Y. Wang, Q. Zhao, C. Cui, M. Fu, and M. Zhao. 2012. "Immobilisation of lecitase ultra for production of diacylglycerols by glycerolysis of soybean oil." *Food Chemistry* 134 (1):301–7.

Macrae, A. R., and P. How. 1988. "Rearrangement process." US 4719178.

Mahadevan, G. D., and S. E. Neelagund. 2014. "Thermostable lipase from Geobacillus sp. Iso5: Bioseparation, characterization and native structural studies." *Journal of Basic Microbiology* 54 (5):386–96.

Malcata, F. X., H. S. Garcia, C. G. Hill, and C. H. Amundson. 1992. "Hydrolysis of butteroil by immobilized lipase using a hollow-fiber reactor: Part I. Lipase adsorption studies." *Biotechnology and Bioengineering* 39 (6):647–57.

Mei, L., L. Wang, Q. Q. Li, J. N. Yu, and X. M. Xu. 2013. "Comparison of acid degumming and enzymatic degumming process for Silybum marianum seed oil." *Journal of the Science of Food and Agriculture* 93 (11):2822–8.

Mojovic, L., S. Silermarinkovic, G. Kukic, B. Bugarski, and G. Vunjaknovakovic. 1994. "Rhizopus-arrhizus lipase-catalyzed interesterification of palm oil midfraction in a gas-lift reactor." *Enzyme and Microbial Technology* 16 (2):159–62.

Moreau, R. A., D. B. Johnston, M. J. Powell, and K. B. Hicks. 2004. "A comparison of commercial enzymes for the aqueous enzymatic extraction of corn oil from corn germ." *Journal of the American Oil Chemists' Society* 81 (11):1071–5.

Murphy, D. J., I. Hernández-Pinzón, and K. Patel. 2001. "Role of lipid bodies and lipid-body proteins in seeds and other tissues." *Journal of Plant Physiology* 158 (4):471–8.

Nalder, T. D., S. Marshall, F. M. Pfeffer, and C. J. Barrow. 2014. "Characterisation of lipase fatty acid selectivity using novel omega-3 pNP-acyl esters." *Journal of Functional Foods* 6:259–69.

Pabai, F., S. Kermasha, and A. Morin. 1995. "Lipase from pseudomonas-fragi CRDA-323—Partial-purification, characterization and interesterification of butter fat." *Applied Microbiology and Biotechnology* 43 (1):42–51.

Pacheco, C., G. H. Crapiste, and M. E. Carrin. 2010. "Lipase-catalyzed acidolysis of sunflower oil: Kinetic behavior." *Journal of Food Engineering* 98 (4):492–7.

Paludo, N., J. S. Alves, C. Altmann, M. A. Z. Ayub, R. Fernandez-Lafuente, and R. C. Rodrigues. 2015. "The combined use of ultrasound and molecular sieves improves the synthesis of ethyl butyrate catalyzed by immobilized Thermomyces lanuginosus lipase." *Ultrasonics Sonochemistry* 22:89–94.

Pande, G., J. S. M. Sabir, N. A. Baeshen, and C. C. Akoh. 2013. "Synthesis of infant formula fat analogs enriched with DHA from extra virgin olive oil and tripalmitin." *Journal of the American Oil Chemists' Society* 90 (9):1311–8.

Perez, E. E., M. B. Fernández, S. M. Nolasco, and G. H. Crapiste. 2013. "Effect of pectinase on the oil solvent extraction from different genotypes of sunflower (Helianthus annuus L.)." *Journal of Food Engineering* 117 (3):393–8.

Qian, Z. J., Z. Wang, S. Y. Xu, R. J. Yang, and X. M. Zhang. 2004. "Aqueous enzymatic extraction of corn germ oil and recovering protein." *Journal of Wuxi University of Light Industry* (in Chinese) 23 (5):58–62.

Ray, S., and D. K. Bhattacharyya. 1995. "Comparative nutritional study of enzymatically and chemically interesterified palm oil products." *Journal of the American Oil Chemists' Society* 72 (3):327–30.

Ren, J. 2008. "Study on aqueous enzymatic extraction of oil from sunflower seed and the utilization of its protein." PhD dissertation, Jiangnan University.

Rhee, K. C., C. M. Cater, and K. F. Mattil. 1973. "Aqueous process for pilot plant-scale production of peanut protein concentrate." *Journal of Food Science* 38 (1):126–8.

Riaublanc, A., R. Ratomahenina, P. Galzy, and M. Nicolas. 1993. "Peculiar properties of lipase from candida-parapsilosis (ashford) langeron and talice." *Journal of the American Oil Chemists' Society* 70 (5):497–500.

Rosenthal, A., D. L. Pyle, and K. Niranjan. 1996. "Aqueous and enzymatic processes for edible oil extraction." *Enzyme and Microbial Technology* 19 (6):402–20.

Rosenthal, A., D. L. Pyle, and K. Niranjan. 1998. "Simultaneous aqueous extraction of oil and protein from soybean: Mechanisms for process design." *Food and Bioproducts Processing* 76 (4):224–30.

Rosenthal, A., D. L. Pyle, K. Niranjan, S. Gilmour, and L. Trinca. 2001. "Combined effect of operational variables and enzyme activity on aqueous enzymatic extraction of oil and protein from soybean." *Enzyme and Microbial Technology* 28 (6):499–509.

Ruan, X., X.-M. Zhu, H. Xiong, S.-Q. Wang, C.-Q. Bai, and Q. Zhao. 2014. "Characterisation of zero-trans margarine fats produced from camellia seed oil, palm stearin and coconut oil using enzymatic interesterification strategy." *International Journal of Food Science and Technology* 49 (1):91–7.

Sabbani, S., and E. Hedenstrom. 2009. "Control of water activity in lipase catalysed esterification of chiral alkanoic acids." *Journal of Molecular Catalysis B-Enzymatic* 58 (1–4):6–9.

Sarker, B. C., B. P. N. Singh, Y. C. Agrawal, and D. K. Gupta. 1998. "Optimization of enzyme pre-treatment of rapeseed for enhanced oil recovery." *Journal of Food Science and Technology-Mysore* 35 (2):183–6.

Sengupta, R., and D. K. Bhattacharyya. 1996. "Enzymatic extraction of mustard seed and rice bran." *Journal of the American Oil Chemists Society* 73 (6):687–92.

Sharma, A., S. K. Khare, and M. N. Gupta. 2001. "Enzyme-assisted aqueous extraction of rice bran oil." *Journal of the American Oil Chemists' Society* 78 (9):949–51.

Sharma, A., S. K. Khare, and M. N. Gupta. 2002. "Enzyme-assisted aqueous extraction of peanut oil." *Journal of the American Oil Chemists' Society* 79 (3):215–8.

Sheldon, R. A., and S. van Pelt. 2013. "Enzyme immobilisation in biocatalysis: Why, what and how." *Chemical Society Reviews* 42 (15):6223–35.

Silva, R. C., L. N. Cotting, T. P. Poltronieri, V. M. Balcao, D. B. de Almeida, L. A. G. Goncalves, R. Grimaldi, and L. A. Gioielli. 2009. "The effects of enzymatic interesterification on the physical-chemical properties of blends of lard and soybean oil." *LWT-Food Science and Technology* 42 (7):1275–82.

Sineiro, J., H. Dominguez, M. J. Nunez, and J. M. Lema. 1998. "Optimization of the enzymatic treatment during aqueous oil extraction from sunflower seeds." *Food Chemistry* 61 (4):467–74.

Soumanou, M. M., M. Pérignon, and P. Villeneuve. 2013. "Lipase-catalyzed interesterification reactions for human milk fat substitutes production: A review." *European Journal of Lipid Science and Technology* 115 (3):270–85.

Subrahmanyan, J., D. S. Bhatia, S. S. Kalbag, and N. Subramanian. 1959. "Integrated processing of peanut for the separation of major constituents." *JAOCS* 36:66–70.

Tanaka, Y., T. Funada, J. Hirano, and R. Hashizume. 1993. "Triglyceride specificity of Candida cylindracea lipase: Effect of docosahexaenoic acid on resistance of triglyceride to lipase." *Journal of the American Oil Chemists' Society* 70 (10):1031–4.

Tweddell, R. J., S. Kermasha, D. Combes, and A. Marty. 1998. "Esterification and interesterification activities of lipases from Rhizopus niveus and Mucor miehei in three different types of organic media: A comparative study." *Enzyme and Microbial Technology* 22 (6):439–45.

Tzen, J. T. C., Y. Z. Cao, P. Laurent, C. Ratnayake, and A. H. C. Huang. 1993. "Lipids, proteins, and structure of seed oil bodies from diverse species." *Plant Physiology* 101 (1):267–76.

U.S. Environmental Protection Agency. 2000. "Hexane." Available at http://www.epa.gov/ttn/atw/hlthef /hexane.html, accessed June 1, 2014.

Wang, Y. Y. 2005. "Studies on aqueous enzymatic extraction of oil and protein hydrolysates from peanuts." Jiangnan University.

Wang, Y. Q., and M. H. Gordon. 1991. "Effect of phospholipids on enzyme-catalyzed transesterification of oils." *Journal of the American Oil Chemists' Society* 68 (8):588–90.

Wang, Y., Z. Wang, S. Cheng, and F. Han. 2008. "Aqueous enzymatic extraction of oil and protein hydrolysates from peanut." *Food Science and Technology Research* 14 (6):533–40.

Wehtje, E., J. Kaur, P. Adlercreutz, S. Chand, and B. Mattiasson. 1997. "Water activity control in enzymatic esterification processes." *Enzyme and Microbial Technology* 21 (7):502–10.

Wu, J., L. A. Johnson, and S. Jung. 2009. "Demulsification of oil-rich emulsion from enzyme-assisted aqueous extraction of extruded soybean flakes." *Bioresource Technology* 100 (2):527–33.

Xie, W. L., and C. Qi. 2014. "Preparation of low calorie structured lipids catalyzed by 1,5,7-triazabicyclo 4.4.0 dec-5-ene(TBD)-functionalized mesoporous SBA-15 silica in a heterogeneous manner." *Journal of Agricultural and Food Chemistry* 62 (15):3348–55.

Yamamoto, N., and S. Tamura. 1999. "Textural properties, microstructure and sensory evaluation of cooked peanut compared with cooked soybean and kidney bean." *Journal of Home Economics of Japan* 50 (4):313–21.

Yang, H. P., S. Y. Wang, W. Song, J. Jin, and G. F. Zhu. 2004. "Study on extracting rice bran oil from rice bran by aqueous enzymatic method." *Food Science* (in Chinese) 25 (8):106–9.

Yu, L., Y. Xu, and X. Yu. 2009. "Purification and properties of a highly enantioselective l-menthyl acetate hydrolase from Burkholderia cepacia." *Journal of Molecular Catalysis B: Enzymatic* 57 (1–4):27–33.

Zhang, S.-B., Z. Wang, and S.-Y. Xu. 2007a. "Downstream processes for aqueous enzymatic extraction of rapeseed oil and protein hydrolysates." *Journal of the American Oil Chemists' Society* 84 (7):693–700.

Zhang, S. B., Z. Wang, and S. Y. Xu. 2007b. "Optimization of the aqueous enzymatic extraction of rapeseed oil and protein hydrolysates." *Journal of the American Oil Chemists' Society* 84 (1):97–105.

Zhang, S. B., Q. Y. Lu, H. Yang, Y. Li, and S. Wang. 2011. "Aqueous enzymatic extraction of oil and protein hydrolysates from roasted peanut seeds." *Journal of the American Oil Chemists' Society* 88 (5):727–32.

Zhang, S.-B., X.-J. Liu, Q.-Y. Lu, Z.-W. Wang, and X. Zhao. 2013. "Enzymatic demulsification of the oil-rich emulsion obtained by aqueous extraction of peanut seeds." *Journal of the American Oil Chemists' Society* 90 (8):1261–70.

Zhao, W., D. W. Wang, and Q. Li. 2010. "Optimal extraction processing of corn germ oil through water enzymolysis method." *Food Science* (in Chinese) 31 (24):206–9.

Zhao, X. Q., P. L. Sun, Y. Z. Fu, J. Wang, L. Q. Tian, and Q. Hou. 2014. "Effects of extrusion parameters on yield of rice bran oil in extracted by aqueous enzymatic method." *Journal of Agricultural Mechanization Research* (in Chinese) 8:134–8.

Zou, X.-Q., J.-H. Huang, Q.-Z. Jin, Y.-F. Liu, G.-J. Tao, L.-Z. Cheong, and X.-G. Wang. 2012. "Preparation of human milk fat substitutes from palm stearin with arachidonic and docosahexaenoic acid: Combination of enzymatic and physical methods." *Journal of Agricultural and Food Chemistry* 60 (37):9415–23.

10

Enzymes in Fruit Juice and Vegetable Processing

Luciana Francisco Fleuri, Clarissa Hamaio Okino Delgado, Paula Kern Novelli, Mayara Rodrigues Pivetta, Débora Zanoni do Prado, and Juliana Wagner Simon

CONTENTS

10.1 Fruit and Vegetable Global Overview

Fruits are appreciated for their organoleptic and nutritional characteristics as minerals, vitamins, sugars, and other beneficial components (Klavons et al. 1991). Worldwide production of fruit recorded growth from 420 million tons during 1989–1991 to 500 million tons in 1996, 724.5 million tons in 2009 and 728.4 million tons in 2010. The three largest fruit producers in the world are China, India, and Brazil, which contribute with 43.6% of the world's total production; however, they basically cater to their domestic markets (FAO 2012; IBGE 2012). It is estimated that 25% to 80% of harvested fruit is lost during transportation and storage due to its high perishability; and hence, they require special care in both postharvest storage and processing (Cardoso 2011). In this context, processing of fruits and vegetables has improved with the production of juices, pulps, purees, jelly, dried fruit, sweets, and jams, which allows uninterrupted consumption, waste reduction, and increased consumption of processed products (Kosseva 2009).

Regarding vegetables, production has increased to 94% from 1980 to 2004, and the annual average increase was 4.2% (E.U. 2007). As in the fruit chain, production of vegetables and greenery have many peculiarities from country to country. Moreover, the largest producers are China and India. In 2011, world production of sweet potatoes, pumpkins, potatoes, onions, lettuce, endive, garlic, eggplant, cucumbers, cauliflower, broccoli, carrots, turnips, and cabbage was 1.037 billion tons with China covering 40% and India 11% of total production, whereas the world production for other vegetables and greenery was 1.4 billion tons in 2011, totaling 41% for China and 11% for India (FAO 2012).

Enzymes are important throughout the life cycle of fruits and vegetables; however, they remain active even after harvest and may have undesirable effects on color, texture, taste, smell, and nutritional value (Bayludlill 2010). The final quality of fruit and vegetable products can be determined by manipulating the enzymes, disabling the activity of damaging ones and harnessing other beneficial enzymes.

The use of enzymes aids in the processing of fruits and vegetables, acting mainly in the hydrolysis of polysaccharides and facilitating the extraction of intracellular compounds. As the content of these biomolecules differs in each plant in composition and quantity, different enzymes are required for each process (Sreenath et al. 1994). The enzymatic process has advantages compared to the use of chemicals, such as higher specificity, mild reaction conditions, and less waste production (Devasena 2010). One of the goals of industrial processing of fruits and vegetables is to deactivate undesirable enzymes and eliminate microorganisms. However, this process usually causes loss of sensory and nutritional values, depending on the technique used (Pasha et al. 2014). Actions such as stripping, polishing, slicing, and grinding remove the plants' natural protection and cause injury, making them susceptible to dehydration, disintegration, and wilt. Mechanical damage causes exposition of the organic tissue to microorganisms in addition to an increase in respiration rate, ethylene production, and enzymatic browning (Rosen and Kader 1989). In conventional processing, such as canned and frozen vegetables, many of these problems are controlled by heat treatment and use of chemical preservatives, which promotes inactivation of enzymes and microorganism control. Other techniques are being developed to overcome sensory processing and nutrient deficit problems, such as controlled atmosphere packaging, edible films (Garcia and Barrett 2002), ultrasound (Terefe et al. 2012), pulsed electric fields (Terefe et al. 2013), and high pressure (Terefe et al. 2014).

In this chapter, the main enzymes used in fruit and vegetable processing and those involved in enzymatic browning are discussed.

10.2 Enzymes in the Processing of Fruits and Vegetables

During the ripening of vegetables, one of the most notable changes is softening, which is associated with alterations in the cell wall, middle lamellae, and membrane. Depending on the final product, vegetable texture can be handled in two ways: stated, if the end product is the vegetable in pieces, or minimally processed, if the final product is juice or oil. In both cases, the enzymes can be used or handled in the industry.

With the advances in enzyme information and technology, extracts from plant tissues, such as malt and papaya, or produced by bacteria, yeasts, and fungi can be applied in different industrial processes (Leadlay 1993). Table 10.1 presents some enzymes of industrial interest for fruits and vegetables as well as the microorganisms that produce them, their mode of action, and application in industry.

The fruit juice industry uses enzymes extensively because they provide higher yield, clarification, and improvement in the filtering process, which ensures a high quality for clarified juices. Table 10.2 shows the main enzymes used in various fruit juices and their technological applications.

Table 10.3 lists the enzymes used in plants for different purposes, including increasing the extraction yield of oils, carotenoids, and flavonoids, among other substances. The enzymes that are mostly used in the industry and new research developed in this field are dealt with in this chapter.

10.2.1 Pectinases

The pectin substances are glycoside macromolecules of high molecular weight formed by complexes of colloidal acid polysaccharides, composed of galacturonic acid residues linked by α-1,4 bonds, partially esterified by methyl and ester groups and partially or totally neutralized with one or more bases (ions of sodium, potassium, or ammonium) (Siew-Yin and Wee-Sim 2013). The American Chemical Society classified the pectin substances in protopectin, pectin acid, pectic acid, and pectin; the last three are partially soluble in water (Uenojo and Pastore 2007).

The term "pectin" is used for pectin acids soluble in water with variable degrees of methyl ester groups and degrees of neutralization capable of forming gels with sugars and acids under appropriate conditions

TABLE 10.1

Main Enzymes Used in Fruit and Vegetable Processing

	Main Microorganism Producer	Action	Technological Application	References
Pectinesterase	*Aspergillus* spp.	Remove methyl groups from galactose units of pectin	Increase firmness of vegetables and also used with pectinase	Croak and Corredig 2006
Pectinase	*Aspergillus* spp., *Penicillium funiculosum*	Pectin hydrolysis	Fruit juice clarification	Bruhlmann et al. 2000; Maktouf et al. 2014; Wang et al. 2013
Protopectinase	*Kluyveromyces fragilis, Galactomyces reesei, Trichosporon fragilis, Bacillus subtilis*	Catalyze pectin solubilization	Clarification and reduction of viscosity in fruit juices	Beg et al. 2000
Hemicellulase	*Aspergillus* spp., *Bacillus subtilis, Trichoderma reesei*	Hemicellulose hydrolysis	Used together with pectinase and cellulase. Help extraction of fruit juices, vegetable oils, and aromatic compounds	Chapla et al. 2010; Song et al. 2014; Wang et al. 2013
α-Amylase	*Aspergillus* spp., *Bacillus* spp., *Microbacterium imperiale*	Random hydrolyses α-1,4 bounds to rupture starch and produce maltose	Help in sugar production, for softness and increases volume of fruit juices and vegetables	Taniguchi and Honnda 2009
Glucoamylase	*Aspergillus niger, Rhizopus* spp.	Hydrolyze dextrin from starch in glucose	Fruit juice extraction and also used for corn syrup and glucose production	Lien and Man 2010

(Guo et al. 2014). They consist of axial structures of D-galacturonic units and contain molecules of L-ramnose, arabinose, galactose, and xylose as side chains. Pectins include protopectins, pectin acids, and pectic, and represent around 0.5% to 4.0% of the total weight of fresh fruits as well as constituents from 2% to 35% of the middle lamellae, which connects adjacent cells (Pifferi et al. 1989).

The pectin can be degraded by pectin lytic enzymes, which may be from vegetable or microbial origin. These enzymes are obtained from fruits and even fermentation of by-products of juice extraction; for example, mango by products were used as substrate for fermentation processes (Amid et al. 2013). The pectin lytic enzymes are widely produced by fermentation processes using a large variety of microorganisms, such as bacteria *Bacillus* spp. and *Pseudomonas* spp.; fungi *Penicillium* spp., *Aspergillus* spp.; and actinomyces *Amycolata* spp. and *Streptomyces* spp. (Bruhlmann et al. 2000; Maktouf et al. 2014; Wang et al. 2013).

The pectin lytic enzymes can be classified according to the type of degraded substrate or catalyzed reactions. There are basically three types of pectinases: (i) pectinesterase that catalyzes desertification or demethylation reactions, (ii) depolymerizing pectinases that are divided into hydrolases (polymethylgalacturonase and polygalacturonase) and lyases (polymethylgalacturonate lyase and polygalacturonate lyase), and (iii) protopectinases (Hoondal et al. 2002; Rehman et al. 2014).

The pectinesterase (polymethyl galacturonate esterase, PMGE, EC. 3.1.1.11) catalyzes the hydrolysis of methyl ester groups from pectin, releasing methanol and converting pectin in pectane (nonesterified polymer). They preferably act on the methyl ester group of the unit galacturonate close to a nonesterified unit, and they present optimum pH values ranging from 4.0 to 8.0 and optimum temperatures of 40°C to 50°C. They are widely marketed and provide protection, texture, and consistence for processed fruits and vegetables; in addition, they assist in the extraction and clarification of fruit juices (Croak and Corredig 2006).

TABLE 10.2

Main Enzymes Used in Fruit Juice Processing

Fruit Juice	Enzyme	Technological Application	References
Apple juice (*Pyrus malus* L.)	Amylase	Decrease starch concentration to improve durability	Carrin et al. 2004
Banana juice (*Musa sapientum* L.)	Pectinase	Decrease turbidity and viscosity	Lee et al. 2006
Blackcurrant juice (*Ribes nigrum* L.)	Pectinase, cellulase, and xylanase	Improve juice yield	Laaksonen et al. 2012
Blueberry (*Vaccinium myrtillus* L.)	Pectinase	Improve juice yield and anthocyanin level	Dinkova et al. 2014
Carambola or star fruit (*Carambola Averrhoa* L.)	Pectinase	Improve viscosity, turbidity, and color	Abdullah et al. 2007
Citrus juice (*Citrus sinensis* L. Osbeck)	Pectinase and cellulase	Decrease the turbidity of juice	Rai et al. 2004
Cherry juice (*Prunus avium* L.)	Pectinase and protease	Decrease turbidity and increase stability	Pinelo et al. 2010
Coffee (*Coffea Arabica* L.)	Pectinase and cellulase	Remove mucilage layer of coffee beans	Rani and Appaiah 2012
Cloudy ginkgo juice (*Ginkgo biloba* L.)	Amylase and protease	Reduce hydrolysis time to improve stability	Zhang et al. 2007
Elderberry juice (*Sambucus nigra* L.)	Pectinase	Improve juice yield and anthocyanin level	Landbo et al. 2007
Grape juice (*Vitis vinifera* L.)	Pectinase	Decrease turbidity and soluble solids	Sreenath and Santhanam 1991
Jicama juice (*Pachyrhyzus erosus* L.)	Pectinase, cellulase, and amylase	Improve juice yield	Lien and Man 2010
Lemon soft drink (*Citrus sinensis* L. Osbeck)	Pectinase	Clarification of peel extract to produce soft drinks	Novozymes 2013
Passion fruit juice (*Passiflora edulis* F.)	Amylase	Improve homogeneity and decrease juice turbidity	Okoth et al. 2000
Pineapple juice (*Ananas comosus* L.)	Pectinase and cellulase	Improve soluble solids and aromas	Sreenath et al. 1994
Pomegranate juice (*Punica granatum* L.)	Pectinase	Improve concentration of antioxidants and decrease turbidity	Rinaldi et al. 2013
Tea (*Camellia sinensis* (L) Kuntz)	Pectinase	Improve fermentation process	Chandini et al. 2011

The polymethylgalacturonase (PMG, EC. 3.2.1.64), which can be endo- or exo-PMG, hydrolyzes polymethylgalacturonates to oligomethylgalacturonates by the cleavage of bonds α-1,4. The endo-PMG catalyzes the random cleavage of α-1,4-glycosidic linkages of pectin, preferably highly esterified linkages. The exo-PMG promotes sequential hydrolysis of α-1,4-glycosidic pectin (Kashyap et al. 2001).

The polygalacturonases (PG, Endo EC. 3.2.1.15 and Exo EC. 3.2.1.67) hydrolyze α-1,4 glycoside bonds between two galacturonic acid residues. They can be endo-PG catalyzing the random hydrolysis of α-1,4-glucosidic bonds of pectin acid or exo-PG catalyzing the sequential hydrolysis of α-1,4-glycosidic linkages in pectin acids. Fungal polygalacturonases have high enzymatic activity and optimal activity at slightly acid pH and optimum temperatures between 30°C and 50°C (Kashyap et al. 2001).

The polymethylgalacturonate lyase (PMGL, EC. 4.2.2.10) catalyzes the hydrolysis of pectin by cleavage of hydrogen of the carbons in positions 4 and 5 of the aglycone portion of the pectin. Endo-PMGL catalyzes the random cleavage of α-1,4-glycosidic bonds within the pectin chain, and exo-PMGL catalyzes the sequential cleavage of bonds α-1,4-glycosidic of pectin from the chain ends. They exhibit optimal activity at acid pH and at temperatures between 40°C and 50°C (Hoondal et al. 2002).

The polygalacturonatelyase or pectinlyase (PGL, Endo EC. 4.2.2.2 and Exo EC. 4.2.2.9) catalyzes the cleavage of bonds α-1,4 of pectin acid by trans elimination. As they are, endo-PGL or exo-PGL, they

TABLE 10.3

Main Enzymes Used in Vegetable Products

Vegetable Product	Enzyme	Technological Application	References
Carrot (*Daucus carota* L.)	Pectinase	Improved nutritional properties as the content of polyphenols and flavonoids	Yoo et al. 2013
Chocolate (*Theobroma cacao* L.)	Amylase	Decrease viscosity of cacao syrup	Taniguchi and Honnda 2009
Date syrup (*Phoenix dactylifera* L,)	Cellulase and pectinase	Reduce turbidity and increase the extraction of soluble solids	Abbès et al. 2011
High-fructose corn syrup (HFCS—*Zea mays* L.)	Amylase	Conversion of glucose to fructose	Johnson et al. 2009
Olive oil (*Olea europaea* L.)	Pectinase, cellulase, and hemicellulase	Oil extraction from olive residue and improve soluble solids on oil extraction	Ranalli and Mattia 1997
Palm oil (*Elaeis guineenses* J.)	Pectinase, laccase, and cellulase	Carotenoid extraction from palm oil	Teixeira et al. 2013
Sunflower oil (*Helianthus annuus* L.)	Pectinase	Improve oil yield	Perez et al. 2013

catalyze the sequential cleavage, respectively, inside and outside the chain. They require Ca^{2+} and exhibit optimal activity at basic pH and at temperatures between 40°C and 50°C (Hoondal et al. 2002).

The protopectinases (EC. 3.2.1.99) solubilize protopectin. The protopectins are insoluble polysaccharides, which, after hydrolysis, form highly polymerized pectin-soluble substances. Based on their applications, they are mainly of two types: protopectinase type A (PPase-A), which reacts with the internal site, that is, the region of polygalacturonic acid of the protopectin, and protopectinase type B (PPase-B), which reacts with the external site or to the polysaccharide chains, which may be connected to the chains of polygalacturonic acid of the cell walls (Beg et al. 2000).

Pectins provide increased viscosity to the fruit pulps and also remain attached to the cellulose fibers, facilitating water retention. This viscosity makes pectin capable of forming gels, which is widely used in the jelly industry. However, it is considered a problem for extraction of pulp, causing fluid retention in the raw material and beverage turbidity (Guo et al. 2014; Renard et al. 1993).

The addition of pectinase facilitates hydrolysis of carbohydrates that allows better squeezing of fruit, providing increased extraction yield and enhancing the concentration of acids, flavorings, and colorants. Consequently, pectinases play a vital role in the food processing industries, especially in the production of beverages, which are widely used for extraction of oils, syrups, and starches (Rai et al. 2004; Yusof and Ibrahim 1994). These enzymes are used for reducing the texture of vegetables in order to facilitate processing. Frequently, combinations of pectinases, cellulases, and hemicellulases, known as macerating enzymes, are employed in juice production, in order to facilitate maceration, liquefaction, and clarification.

10.2.1.1 Pectinase Case Studies

Studies demonstrate that treatment of fruits with pectinolytic enzymes (PGL, PG, and PMGL) promote more than 96% reduction in viscosity and turbidity of juices (Maktouf et al. 2014). These enzymes are widely used in the fruit beverage industry.

Lien and Man (2010) tested commercial sources of enzymes in the production of jicama juice (*Pachyrhizus erosus* L.), and they obtained extraction percentages of 45.4%, 61.5%, 56.3%, 56.1%, 73.5%, and 55.9% for control, pectinase, cellulase, β-glucanase, α-amylase, and glucoamylase, respectively. However, the combination of pectinase, cellulase, and α-amylase promoted extraction of 92.7%. Ramos-de-la-Peña et al. (2012) tested the effect of concentration and incubation time with pectinase (Pectinex Ultra SP-L) in the jicama juice extraction, and the maximum extraction was 980 mL.kg^{-1} of roots after 10 h of treatment with 0.024 mL.g^{-1} of enzyme.

Addition of pectinases in the extraction of date syrup (*Phoenix dactylifera* L.) has increased the extraction yield and the concentration of reducing sugars, acids, and soluble dry matter. The higher concentration of reducing sugars prevented the crystallization of the syrup. Furthermore, addition of cellulase and pectinase brightens the syrup, and such characteristics are mostly appreciated by the consumer (Abbès et al. 2011).

The turbidity of fruit juices is considered undesirable, except for citrus juices, and it is the result of light scattering caused by substances in suspension by insoluble solids (usually pectin from cell walls). This turbidity is measured by viscosity, clarity, and concentration of insoluble solids and can be amended by clarification with centrifugation or micro- and ultrafiltration processes (Rai et al. 2004; Sarioglu et al. 2001).

The clarification process is related to degradation of pectin and subsequent precipitation and sedimentation of degraded compounds, usually involving the addition of pectinases, gelatin, silica, and betonite (Wang et al. 2013). The reduction in turbidity is achieved by electrostatic destabilization of pectin particles that are negatively charged. Pectin particles that cause turbidity are maintained in suspension as a result of the repulsive force between them. Pectinases in acid pH (common in fruit juices) are positively charged and, therefore, attract pectin, forming clusters of pectin in enzymes. Thus, these agglomerates are removed by filtration or centrifugation. The viscosity reduction is important to prevent pectin gelling during the concentration (Uenojo and Pastore 2007).

Pectinolytic enzyme preparations are also used in the clarification of citrus juices (especially lemon juice) and extraction of essential oils as well as in production of lemon peel extracts, which are used in the production of soft drinks (Novozymes 2013).

Among the main objectives of the oil industry is the optimization of the process of oil extraction with current focus on finding technologies that reduce energy consumption and are safe for the environment and workplace (Ramadan et al. 2008). In this context, the use of plant enzymes for pretreatments have expanded for further oil extraction as they assist in breakage and degradation of the cell wall structure, facilitating the release of oil and access of solvents (Latif and Anwar 2011).

Perez et al. (2013) examined the pretreatment with pectinase on the extraction of sunflower oil (*Helianthus annuus* L.) reaching up to 33% yield increase. Teixeira et al. (2013) tested the pretreatment with pectinase and cellulase on the extraction of carotene from palm oil (*Elaeis guineenses* J.) getting yield increases of up to 158%.

The maceration enzymes are used for the extraction of olive oil; however, mechanical maceration holds a considerable fraction of oil in the residue. Research has shown that enzyme combinations have better results than the use of isolated enzymes. Generally, commercial products contain pectinase, cellulase, and hemicellulase, for example, Olivex® (Fantozzi et al. 1977; Garcia et al. 2001), Cytolase 0® (Ranalli et al. 1998), Maxoliva® and Bioliva® (Ranalli et al. 2003).

The increase in productivity of the extraction of oil can depend on the commercial product and the olive cultivar. Ranalli et al. (2004) tested the Bioliva product in four olive cultivars and observed increases from 0.8% to 1.4%. Among the prepared enzyme, the increase in the extraction of olive oil was 1.2% with Bioliva, 1.37% with Maxoliva, and 1.44% with Cytolase 0 (Ranalli et al. 2003). According to the authors, although extraction with Bioliva is statistically lower than Maxoliva and Cytolase 0, this difference is not significant from the industrial point of view.

In addition to increased productivity, maceration enzymes have shown improvements in the nutritional and sensory properties of olive oil. The use of Bioliva increased incorporation of antioxidant components, such as phenol, tocopherols, carotenes, and xanthophylls (Ranalli et al. 2004).

Pectinases are important for the fermentation of tea, accelerating the fermentation process, improving the quality of the final product, and increasing the solids content. The addition of pectinases and tannases in the extraction of black tea increases the recovery of soluble solids (Chandini et al. 2011). Pectin enzymes are added in coffee bean processing to remove the mucilage layer of the grain, consisting of three quarters pectin substances. Cellulases and hemicellulases that are present in commercial preparations are sprayed on the grains, speeding up the brewing process. As the coffee bean treatment on a large scale with commercial enzymes is not economically viable, microbial pectin enzymes produced from the fermentation of waste mucilage are used (Rani and Appaiah 2012).

The protopectinases are less used in the industry than the maceration enzymes, but they have interesting effects. The protopectinases are a heterogeneous group of enzymes that catalyze the solubilization of pectin (Jayani et al. 2005). Yoo et al. (2013) demonstrated that protopectinase increased yield in the processing of carrot puree by 81% compared with only 56% of mechanical processing. Furthermore, protopectinase improved nutritional properties, as the content of polyphenols and flavonoids, by the maintenance of membrane integrity.

In contrast to what was previously explained, the consistency of vegetables may be an attractive effect in food processing. Enzymes, such as pectinesterase, can be used to maintain vegetables texture, and in addition, it is already found in commercial products. Jensen et al. (2005) reported an increase in pepper pieces' rigor with the use of pectin esterase. This enzyme has long been used for this purpose as shown by Hsu et al. (1965) in the production of canned tomatoes. In the industry, it is common to use calcium to maintain texture because calcium forms bonds between the free carboxyl groups of pectin chains, strengthening the cell wall. The combined treatment of cold bleaching to activate pectinesterase before immersion in calcium also helps to preserve the texture (Garcia and Barrett 2002).

10.2.2 Cellulases

Cellulose is a polysaccharide composed of several glucose units, and cellulases promote its hydrolysis. These enzymes are highly specific biocatalysts that act synergistically to release sugars, of which glucose has major industrial importance due to the possibility of its conversion into ethanol, sweeteners, phytohormones, and organic acids, among others. Steps involved in cellulose degradation by cellulases are still not fully understood, but cellulase is produced as a multienzyme system, including three enzymes that act synergistically in the hydrolysis of cellulose: β-1,4-endoglucanase (1,4-β-D-glucan 4-glucan hydrolase, EC. 3.2.1.4), cellobiohydrolase or exoglucanase (1,4-β-D-glucan cellobiohydrolase, EC. 3.2.1.91), and β-glucosidase (β-D-glucoside gluco-hydrolase, EC. 3.2.1.21) (Jecu 2000). The β-glucosidase enzyme participates in the global regulation of the cellulose hydrolysis because the enzymes endo- and exoglucanases are often inhibited by cellobiose. Therefore, β-glucosidase hydrolysis of cellobiose eventually reduces the inhibition of these two other enzymes (Oh et al. 1999).

The main commercial preparations of cellulases are obtained from filamentous fungi, such as *Aspergillus niger* (Cellulocast from Novozyme) and *Trichoderma reesei* (Megazyme). Cellulase-producing fungi include *Aspergillus, Trichoderma, Penicillium, Sporotrichum, Fusarium, Talaromyces, Thermoascus, Chaetomium, Humicola, Neocallimastix, Piromonas*, and *Sphaeromonas* (Bansal et al. 2012; Maalej-Achouri et al. 2012; Singhania et al. 2014; Sadaf and Khare 2014).

Cellulases are used in the beverage industry for the production of fruit juices and wine processing. They facilitate the extraction of juices and maceration for production of fruit nectars by breaking the cellulose chains that helps hold the liquid in the plant cells. Regarding the wine industry, β-glucosidase improves the extraction of pigments and flavoring substances present in grape skins. In addition, it breaks unpleasant-tasting compounds, releasing flavoring substances and improving beverage aroma and flavor (Juturu and Wu 2014).

10.2.2.1 Cellulase Case Studies

Pineapple juice is extracted by squeezing the fruits. The residual pulp obtained after pressing retains valuable substances, such as soluble solids and aromas that usually remain on these residues, and treatment with cellulase and pectinase makes possible the extraction of these compounds (Sreenath et al. 1994).

In the extraction of flavonoids from *Ginkgo biloba*, cellulase addition helps in the cell wall degradation and increases the solubility of the target compound in ethanol, increasing the extraction rate by more than 30% (Chen et al. 2011).

The extraction of syrup from *Phoenix dactylifera* L. usually originates low-quality products due to unpleasant texture. This problem can be lessened by the addition of enzymes, such as pectinases and cellulases that reduce turbidity and increase the extraction of soluble solids of interest (Abbès et al. 2011).

β-Glucosidase participates in reactions that release aroma compounds from glycoside precursors present in fruits and fermented products, improving sensory characteristics of the products in addition to producing bioactive molecules, such as isoflavones (Barrera-Islas et al. 2007; Riou et al. 1998). It also has the ability to hydrolyze anthocyanins, producing anthocyanidins and mono and disaccharides, which have low color and solubility and facilitates their removal. This ability is important for the juice industry because it prevents discoloration caused by anthocyanins during juice pasteurization (Palma-Fernandez et al. 2002; Villena et al. 2007). Moreover, they are used to hydrolyze compounds such as 4,5,7-trihydroxyflavanone-7-ramnoglicoside that are responsible for the bitterness in citrus juices (Riou et al. 1998).

10.2.3 Xylanases (Hemicellulases)

The hemicellulose xylan is the most abundant biopolymer in the world, representing about 30% of the renewable and organic carbon. Xylanases are present in superior plants, such as grasses, cereals, and trees (Cunha and Gandini 2010), covering a range of noncellulosic polysaccharides, such as D-xylose, D-mannose, D-glucose, L-arabinose, D-galactose, D-glucuronic, and D-galacturonic acid. Hemicelluloses are classified according to the principal sugar unit. Thus, when a polymer is hydrolyzed and xylose produced, it is called xylan (Collins et al. 2005). Most agricultural wastes contain 30% to 50% cellulose, 20% to 40% of hemicellulose, and 10% to 20% lignin based on their dry weight (Cunha and Gandini 2010).

Hemicellulases are a group of enzymes that degrade hemicellulose polymers, including endo-xylanase (endo-1,4-β-xylanase EC. 3.2.1.8), β-xylosidase (xylan-β-1,4-xylosidase, EC. 3.2.1.37), α-glucuronidase (α-glucosiduronase, EC. 3.2.1.139), α-arabinofuranosidase (α-L-arabinofuranosidase, EC. 3.2.1.55), arabinase (endo α-L-arabinase, EC. 3.2.1.99), acetyl xylan esterase (EC. 3.1.1.72), and feruloyl esterase (EC. 3.1.1.73) (Juturu and Wu 2013). Xylan and arabinoxylan polymers are hydrolyzed completely by the synergistic action of several xylanolytic enzymes: endo-1,4-β-D-xylanases, which randomly degrade β-D xylan linkages; β-D-xylosidases, that release xylose monomer from cleavage of the nonreducing end of xylo-oligosaccharides and xylobiose; and the other previously mentioned enzymes that degrade the hydrolysis products and/or more specific bonds (Terrasan et al. 2010).

These enzymes are produced by bacteria, fungi, algae, protozoa, nematodes, crustaceans, insects, and plants. Commercial xylanases are produced on a large scale in many countries, such as Japan, Finland, Germany, Ireland, Denmark, Canada, and the United States, especially by fermentation processes using fungi, yeast, actinomyces, and bacteria (Polizeli et al. 2005). Examples of microorganisms that produce these biocatalysts are *Aspergillus foetidus* TCC 4898, *Paenibacillus terrae* HPL-003, *Geobacillus stearothermophilus*, *Aspergillus niger*, *Humicola insolens*, and *Trichoderma reesei* (Song et al. 2014; Wang et al. 2013; Chapla et al. 2010; Bae et al. 2008).

Current studies on the production of xylanases have searched for an increase in productivity by developing improved microbial strains and more efficient fermentation techniques and recovery systems (Xu et al. 2005). Solid-state fermentation (SSF) is one of these techniques, which allows microorganism growth on substrates with low water activity. It has advantages, such as simple technologies and low cost, easy recovery of enzymes, utilization of simple and cheap substrates for fermentation, low power consumption, high volumetric productivity, and, often, high product yield (Fleuri et al. 2013; Chapla et al. 2010; Polizeli et al. 2005).

10.2.3.1 Xylanase Case Studies

Xylanases are used in the bread industry, clarification of fruit juice, wine and beer, and production of xylitol glucose in the candy industry (Zhao et al. 2013). Xylanases are used together with cellulases and pectinases for clarification and liquefaction in the fruit and juice industry (Biely 1985). The production of fruits and vegetables requires extraction, cleaning, and stabilization methods. In the 1930s, when the production of citrus juices began, yields were low due to poor filtration performance caused by turbidity. Increased knowledge of the chemical constituents of the fruit and the utilization of microbial enzymes helped solve these problems, improving yields of juice by liquefaction; stabilization of pulp; and higher recovery of aromas, essential oils, vitamins, minerals, and pigments as well as reducing the viscosity by hydrolyses of substances that are difficult to extract (Beg et al. 2001).

Biobleaching of palm fruit can be achieved by enzymes with the use of xylanases and laccases, resulting in improvements in optical properties, such as brightness and color (Martín-Sampedro et al. 2012).

Xylanases, in combination with endoglucanase, catalyze the hydrolysis of arabinoxylan and starch, which allows separation and isolation of gluten in wheat flour starch. Xylanase is also used for mucilage of coffee beans to improve the production process of roasted coffee (Wong et al. 1988; Wong and Saddler 1993).

10.2.4 Amylases

Starch is one of the largest natural reserves of polysaccharides, constituting a source of available energy (Sharma and Satyanarayana 2013). It is synthesized in plastids, present in leaves, and accumulates insoluble granules in superior and inferior plants. Starch is an important component of most food and also an essential factor for structure, consistency, and texture (Kuriki et al. 2006).

Amylolytic enzymes are differentiated by reaction mechanisms, catalytic activity, and structural characteristics. They include three families of glycoside hydrolases (GHs): GH 13 α-amylases (EC. 3.2.1.2), GH 14 β-amylases (EC. 3.2.1.3), and GH 15 glucoamylases (EC. 3.2.1.1). Amylases can also be classified according to their type of action as exo- and endo-amylases. Exo-amylases hydrolyze α-glucan successively into maltose and glucose whereas endo-amylase hydrolyzes α-glucan-forming oligosaccharides. The β-amylase hydrolyzes nonreducing terminals of starch and malto-oligosaccharides producing maltose units (Fazekas et al. 2013). Glucoamylase catalyzes hydrolytic release of glucose from the nonreducing end of the starch and oligosaccharides related. Amylolytic enzymes are present in a wide range of organisms, including animals, plants, and microorganisms. Microorganisms have been used as enzyme sources. Among them, *Aspergillus oryzae*, *A. niger*, *Bacillus amyloliquefaciens*, *B. circulans*, *B. licheniformis*, *B. stearothermophilus*, and *B. subtilis* are the most studied species for the production of α-amylase; *Bacillus* spp., *Pseudomonas* spp., and *Streptomyces* spp. for β-amylase; and *Pseudomonas amyloderamosa*, *Pseudomonas* sp., *Klebsiella* sp., *Bacillus brevis*, *B. licheniformis*, and *Bacillus acidopulluliticus* for glucoamylases (Taniguchi and Honnda 2009).

Usually, α-amylase occurs in plants, mammalian tissues, and microorganisms. Fungi α-amylase is thermosensitive, and it is inactivated at temperatures higher than 70°C (Macedo et al. 2005). α-Amylases account for about 30% of the enzymes in the world market and are applied in many industrial processes, such as starch saccharification, the textile industry, paper, brewing, baking, detergent, cakes, fruit juices, starch syrups, distilleries, preparation of drugs for stomach problems, and other pharmaceuticals (Roy et al. 2012; Van der Maarel et al. 2002). Amylases, which have optimum activity at acid pH, are primarily used to obtain glucose syrup and in the baking industry (Roy et al. 2013).

10.2.4.1 Amylase Case Studies

The α-amylase acts as an endo-enzyme, randomly fast decreasing the viscosity of the starch suspension, and, in the chocolate industry, it allows reducing the viscosity of cocoa syrup. This enzyme is also used in starch liquefaction processes in which, after gelatinization, the starch is hydrolyzed into dextrins. Compared to the chemical process, the enzymatic process has disadvantages, such as increased expenditure time, higher cost, and possible contamination; however, this product is vastly superior in quality, which enables the process (Taniguchi and Honnda 2009).

The saccharification process is the conversion of starch into other sugars, which is entirely performed by amylolytic enzymes to obtain a wide variety of sweetening syrups. High-fructose corn syrup (HFCS) is widely produced and marketed in the United States for its sweetness and solubility in beverages. It is produced by conversion of glucose to fructose with glucose isomerase enzymes obtained from bacterial sources (Johnson et al. 2009).

Apple juice is among the juices with higher starch concentrations, especially at the harvest beginning, with around 15% of this polysaccharide, which is responsible for turbidity, slow filtration, damage to the filtration membrane, and gelling after juice concentration (Helbert et al. 1996). Studies have shown that treatment with amylase neutralize the abovementioned problems (Carrin et al. 2004).

The addition of amylase in the production process of passion fruit juice increased homogeneity and decreased juice turbidity when added before pasteurization (Okoth et al. 2000).

One example for the use of enzymes in fruit juice is shown in Figure 10.1.

In the processing of fruit juices, many industries use enzymes to improve the final aspect of the product. The exact amount and which enzymes are used in the processing are normally trade secrets. However, as discussed in this chapter, enzymes are necessary to increase the amount of soluble solids. Thus, some fruit juice industries use approximately 3 to 20 g of commercial pectinase per 100 kg of fruit and from 0.2 to 2.0 g of commercial cellulase per 100 kg of fruit with temperatures ranges from 30°C to 50°C in the auxiliary tank. In the clarification stage, from 1.5 to 3.0 g of commercial pectinase per 100 kg of fruit and from 0.5 to 2.0 g of commercial amylase per 100 kg of fruit are used with temperatures from 25°C to 45°C. Enzyme addition in the auxiliary tank aims to remove the insoluble pectin and in the clarification stage aims to reduce the viscosity, degrade residual starch, and facilitate the flocculation of insoluble molecules. This process is used in many types of fruit juices, such as apple, grape, and citrus. The details presented in Table 10.4 show the decrease in residues obtained after centrifugation of commercial juices, due to the effect of enzyme action on clarification.

Figure 10.2 summarizes the use of different enzymes in the processing of apple juice.

FIGURE 10.1   Enzymes added in juice processing (general flow).

TABLE 10.4

Apple Juice Visual Aspect before and after Centrifugation, Showing the Residual Accumulation on the Bottom of the Tube

Juice	Description	Before Centrifugation	After Centrifugation
	–Granny smith and Red delicious variety		
	–Pasteurization		
	–Pasteurization		
	–Fuji and Gala variety		
	–Produced at Santa Catarina, Brazil		
	–Crushed fruit		
	–Centrifugation (low temperature)		
	–Enzymes addition		
	–Clarification		
	–Pasteurization		
	–Vacuum filtering		

10.3 Enzymes Involved in Flavor and Aromas of Fruits and Vegetables

The perception of flavor is determined by both taste and aroma of the food. Taste is determined by the contents of sugars, organic acids, phenols, tannins, and other compounds, such as terpenes and carotenoids. However, most important flavor characteristics are related to aromatic volatile compounds. These compounds are complex and include alcohols, aldehydes, esters, ketones, and lactones, among others (Beaulieu and Baldwin 2002). For example, C6 and C9 compounds that are alcohols and aldehydes are

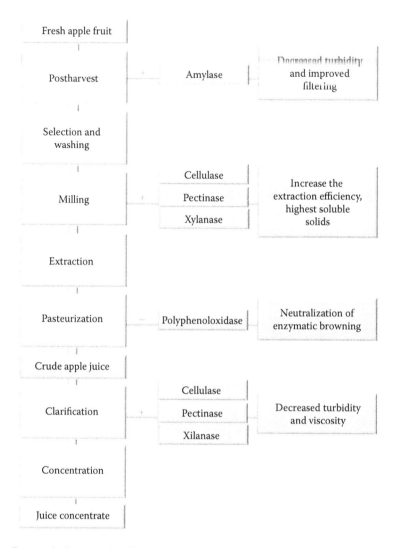

FIGURE 10.2 Enzymes in the processing of apple juice ("+" = addition and "−" = subtraction).

widely used as additives in food due to the aroma with "green touch," a characteristic of freshly cut grass and fresh fruit (Schwab et al. 2008).

The volatile compounds can be synthesized from various substrates, such as fatty acids, amino acids, and carotenoids (Goff and Klee 2006). At least four enzymes are involved in the synthesis of volatile compounds from fatty acids: lipoxygenase, hydro peroxide lyase, 3Z, 2E-enal isomerase, and alcohol dehydrogenase; moreover the main enzyme among these is the lipoxygenase (Schwab et al. 2008). The lipoxygenase (linoleate: oxygen oxidoreductase, EC. 1.13.11.12) catalyzes the oxidation of polyunsaturated fatty acids with cis, cis-1,4-pentadiene (Siedow 1991). The main application of these enzymes has been in the extraction from plant sources, such as soy (Whitehead et al. 1995), banana (Kuo et al. 2006), tomato (Cass et al. 2000), and olive (Akacha and Gargouri 2009), and subsequently, they reacted with fatty acids to produce volatile compounds.

In addition to enzymes directly related to the volatile compounds, it was demonstrated that Bioliva increased the sensory score of olive oil due to an increased release of pleasant volatile compounds as trans-2-hexenal and a lower release of unpleasant volatile compounds (Ranalli et al. 2004).

The flavor in citrus fruit juices can be enzymatically modified through limoninase and naringinase enzymes as they degrade agents responsible for the bitter taste, the limonin and naringin, respectively (Ribeiro et al. 2010). Limonoate dehydrogenase prevents formation of limonin because it catalyzes the oxidation of its precursor lactone A-ring to 17-dehydrolimonoate, a nonbitter derivative that cannot be converted into limonin (Humanes et al. 1997). In addition to the use of enzymes, the bitterness of citrus fruit juices can be reduced by adsorbing polymers. Amberlite XAD-16HP and Dowex Optipore L285 resins were observed to be effective in reducing the bitterness of orange juice although Dowex Optipore L285 induced other modifications on juice, such as reduction of titratable acidity (TA) and increased content of total soluble solids (TSS), the ratio of TSS to TA, and pH (Kola et al. 2010).

10.4 Enzymes Acting in Fruit and Vegetable Browning

Among the changes in the color of fruits and vegetables, enzymatic browning is what causes major economic loss, and its prevention during processing and storage is a major problem in the food industry (Terefe et al. 2014). The main enzymes involved in vegetable browning are polyphenol oxidase (PPO), peroxidase (POD), and β-glucosidase (Zabetakis et al. 2000) with greater relevance for PPO and POD. The phenomenon is caused by oxidation of phenolic compounds to quinones with polymerization reactions leading to production of melanin, which present a dark color (Marshall et al. 2000).

It is known that after harvest some fruits suffer fast enzymatic browning. This process occurs due to enzymatic factors, discussed earlier in this chapter. In order to delay the enzymatic browning, some antibrowning agents may be used; they are normally solutions into which the pieces of fruit or vegetable are immersed. The most commonly used antibrowning agents are ascorbic acid, citric acid, and $CaCl_2$ among others; they may be used alone or in combination with other chemicals.

Figure 10.3 shows the effect of some chemical solutions as treatments to reduce or prevent browning in potatoes (*Solanum tuberosum* L.) and Yacon (*Smallanthus sonchifolius*). Ascorbic acid is widely used as an antibrowning agent, as a reducing agent acting on the enzymatic reaction of the PPO. In the same manner, citric acid can be used as an antibrowning reducing agent as well. Some halides are known as PPO inhibitors, so the use of calcium chloride ($CaCl_2$) can help reduce vegetable browning, acting on pH and as a chelating agent. As already reported by many authors (Ghidelli et al. 2013; Severini et al. 2003), combined treatments seemed to be more effective to prevent fresh-cut vegetables from browning, followed by $CaCl_2$ and acid reducing agents.

The use of these agents is a strategy often used by food manufacturers to increase the shelf life of fruits and vegetables that are peeled or sliced.

10.4.1 Polyphenol Oxidases

Polyphenol oxidase (PPO, EC. 1.10.3.1) was first identified in mushrooms in 1856 by Schoenbein, and in 1896, Bourquelot and Bertrand observed the appearance of a dark substance in mushrooms, which was designated as tyrosinase (Fatibello-Filho and Vieira 2002; Ramírez et al. 2003; Scott 1975). Polyphenol oxidase belongs to the group of oxido-reductases, and it has copper as a prosthetic group (Mendonça and Guerra 2003). This enzyme causes oxidation of phenol substances that are derived from secondary products containing a phenol group or a functional hydroxyl group attached to one aromatic ring. They are produced by plants with a wide variety of functions (Taiz and Zeiger 2009). PPO is present in high concentrations in mushrooms, potato tubers, peaches, apples, bananas, yerba mate, coffee beans, tobacco leaves, microorganisms, and invertebrates (Eisenmenger and Reyes-De-Corcuera 2009; Reed 1975; Tauber 1950).

Polyphenol oxidases are enzymes with a binuclear active site formed by two copper atoms, capable of introducing oxygen in the ortho position of the hydroxyl group in one aromatic ring, followed by oxidation of diphenol to quinone. PPO nomenclature differentiates between monophenol oxidase (tyrosinase, EC. 1.14.18.1) and catechol oxidase or o-diphenol:oxygen oxidoreductase (EC. 1.10.3.2) (Mayer 2006). PPO receives an assortment of names, depending on its substrate, as laccase, cresolase, catecholase, and

Potato (*Solanum tuberosum* L.)

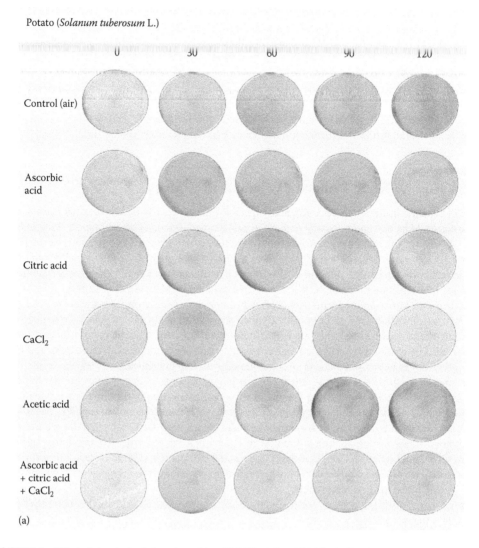

(a)

FIGURE 10.3   Effect of chemical solutions (ascorbic acid 0.4%, citric acid 0.4%, CaCl₂ 0.2%, acetic acid 0.4%) as treatments to reduce or prevent browning in potatoes (a) (*Solanum tuberosum* L.).

(Continued)

tyrosinase. PPO catalyzes the oxidation of monophenol (p-cresol, tyrosine, phenol) to diphenol and the oxidation of diphenols (catechol, L-dopa, dopamine, adrenalin) to benzoquinones (Fatibello-Filho and Vieira 2002).

According to Goupy et al. (1995) and Ramírez et al. (2003), benzoquinones are highly reactive with O_2, sulfhydryl, amines, amino acids, and proteins. Furthermore, several products are generated in reactions, such as melanin, which presents colors, such as yellow, red, brown, and black, characterizing enzymatic browning coloration. Moreover, the oxidation of phenolic compounds by polyphenol oxidase is the most responsible reaction for the enzymatic browning in foods during harvesting, handling, storage, and processing (Eisenmenger and Reyes-De-Corcuera 2009; Núñez-Delicado et al. 2007).

PPO can be present as a soluble fraction or ion bound to the membrane in plants. Histochemical techniques show that these enzymes are located in the chloroplasts where the PPO gene is encoded

Yacon (*Smallanthus sonchifolius* [Poepp & Endl.] H.)

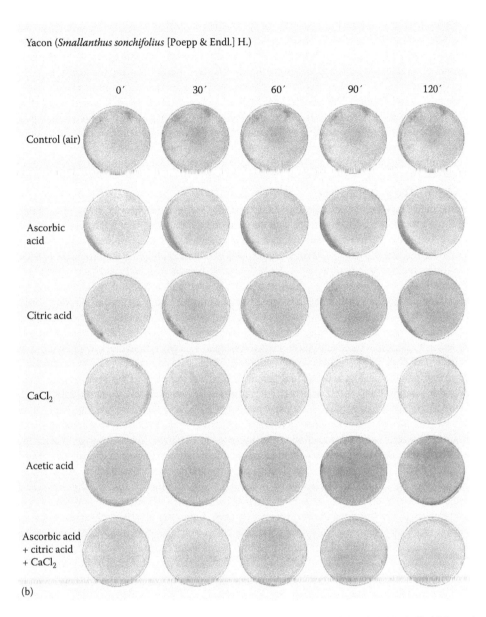

(b)

FIGURE 10.3 (CONTINUED) Effect of chemical solutions (ascorbic acid 0.4%, citric acid 0.4%, $CaCl_2$ 0.2%, acetic acid 0.4%) as treatments to reduce or prevent browning in yacon (b) (*Smallanthus sonchifolius*).

in the nucleus and translated in the cytosol, in which pro-PPO is formed and then transported to the chloroplast to be cleaved by one protease, bringing it into its active form (Martinez and Whitaker 1995). Furthermore, the PPO distribution in several parts of fruits and vegetables may be different, and the relationship between the soluble PPO and ion bound varies depending on the maturation stage (Vámos-Vigyázó 1981).

Researchers have shown that apple PPO is encoded by a multiple gene, the expression of which is regulated by tissue damage (Kim et al. 2001). Constabel and Ryan (1998) evaluated the response of PPO to damage in several plants, especially tomatoes, tobacco, and ornamental trees. In another study with aspen, Constabel et al. (2000) also observed a significant expression of PPO response to mechanical damage in both damaged and nondamaged leaves.

Observations related to the enzymatic activity in peaches subjected to mechanical damage demonstrated the behavior of PPO over 48 h of storage. It was found that in intact tissues, the activity of PPO showed a slight and gradual decrease during the 48-h period. However, in damaged tissues, the behavior was different, with increased activity, peaking around 24 h after injury (Tourino et al. 1993).

Studies conducted to characterize the PPO between fresh, frozen, canned, and processed industrial peaches demonstrated that for frozen peaches the optimum temperature was 20°C and after 30 min with 80% activity retained at temperatures ranging from 15°C to 40°C. As for canned and processed peaches, latent PPO was negative. Belluzzo et al. (2009) tested several inhibitors of PPO in peaches and concluded that ascorbic acid, β-mercaptoethanol, sodium metabisulfite, and cysteine were effective.

Enzymatic browning is controlled by both processing techniques, such as freezing and heating, as by inhibitors that can be antioxidants, acidulants, and chelating agents, among others, both from natural and synthetic sources (Marshall et al. 2000).

Among the many PPO inhibitors investigated, only a few are used in industry, due to problems with unwanted flavors (off-flavor), food security, and sustainability. The most commonly used additives are ascorbic acid and sulfites. However, sulfites can cause allergies, and their use on fresh products was banned in the United States by the Food and Drug Administration, leading the food industry to seek new alternatives for its replacement (Loizzo et al. 2012). Much has been researched regarding new PPO inhibitors, and good reviews were posted by Loizzo et al. (2012), Chang (2009), and Kim and Uyama (2005).

The natural compound 4-hexylresorcinol has been researched in preventing enzymatic browning. Guerrero-Beltrán et al. (2005) observed that the utilization of 4-hexylresorcinol with ascorbic acid or cysteine caused reduction in the activity of PPO and changes in color of mango puree. However, the isolated use of 4-hexylresorcinol was not effective. Arias et al. (2007) concluded that 4-hexylresorcinol acts as a mixed inhibitor because it influences both the V_{max} (reaction rate) and the K_m (kinetics constant) of PPO. The combined use of 4-hexylresorcinol, L-cysteine, and kojic acid inhibited 89.2% of the activity of PPO in apple juice after 24 h (Iyidogan and Bayindirli 2004). Kojic acid is another natural compound widely studied recently; it is even used as a standard inhibitor in comparison with new inhibitors. Kojic acid acts to chelate transition metal ions, such as Fe^{3+} and Cu^{2+} in addition to impound free radicals (Kim and Uyama 2005). Whereas for the monophenol activity, their inhibitory effect is competitive, the inhibition is mixed for diphenol activity (Chang 2009).

Proteases have also been used for prevention of enzymatic browning. Proteases are enzymes that catalyze the hydrolysis of proteins that are degraded to peptides and amino acids (Molina et al. 2013). It is suggested that protease inhibits PPO activity by proteolysis or by binding to specific sites for activation. Another suggested mechanism of action is related to the presence of sulfhydryl groups (such as cysteine) (Garcia and Barrett 2002). In one of the first protease studies, Labuza et al. (1992) observed a decrease in apple browning with the use of papain and in potato pieces as well. Perera et al. (2010) observed that the use of pineapple juice along with a high-pressure technique reduced browning and PPO activity in apple pieces. The inhibitory effect of pineapple juice was attributed to the presence of bromelin and sulfhydryl groups or malic and citric acids. The prevention of browning by pineapple juice was also observed in pieces of banana (Chaisakdanugull et al. 2007). However, Tochi et al. (2009) evaluated the browning of apple juice in the presence of bromelin, L-cysteine, and ascorbic acid and concluded that bromelin was not effective.

Recently, a new inhibitor of PPO was extracted from German cockroaches (*Blattella germanica*) and tested on purified PPOs from apple and potato (Grotheer et al. 2012). The inhibitor reduced the activity of potato PPO from 15% to 25% and apple PPO from 60% to 70%. The authors suggested that the inhibitor was a protein with no competitive action, but the results are not conclusive. Tsukamoto et al. (1992) extracted PPO inhibitors from insects and isolated three compounds of low molecular weight (2850 Da; 3100 Da, and 3350 Da) from domestic fly pupae (*Musca domestica* L.). The inhibitors were stable to heat and to pH and exhibited competitive inhibition.

Ramos et al. (2013) observed that inhibitors, such as sodium dodecyl sulfate, Na-EDTA, sodium bisulfite, and ascorbic acid reduced peroxidase activity and browning in roots of cassava processed until the sixth day of storage.

However, the polyphenol oxidase has positive action in the processing of tea, coffee, cacao, apples, cider, plums, black raisins, black figs, and sapote (Ramírez et al. 2003). Their action could result in interference in the flavor and color of food as occurs in black tea; in reducing bitterness and astringency of products, such as cocoa; and in the formation of aldehydes from amino acids.

10.4.2 Peroxidases

Peroxidases (POD, EC. 1.11.1.7), isolated in 1898 by Linossier, are found in plants, microorganisms, and animals. They are oxidase enzymes that use hydrogen peroxide (H_2O_2) as a catalyst for oxidation of monophenols, diphenols, polyphenols, aminophenols, and others (Fatibello-Filho and Vieira 2002).

POD is thermostable and may have its activity recovery in plant systems after heat treatment also used as a parameter for bleaching efficiency (Aguero et al. 2008; Gonçalves et al. 2007). POD has industrial application in wastewater treatment, fine chemical biosynthesis, and paper treatment, among others (Eisenmenger and Reyes-De-Corcuera 2009). It is related to undesirable changes in flavor, color (browning), texture, and nutritional value in foods (Gonçalves et al. 2007). It also shows beneficial changes related to texture due to the formation of intercrossed protein links (Duarte-Vázquez et al. 2000).

POD uses hydrogen peroxide as an electron acceptor and can utilize a wide range of substrates, leading to formation of radicals that can react subsequently, non-enzymatically, with other substrates (Oort 2010). Both enzymes PPO and POD are of great importance in plant sources, such as for the formation of quinones and tannins, which show, respectively, antimicrobial and biopesticide functions. Furthermore, they form compounds with proteins, which will act as a physical barrier against the pathogens ingress (Campos and Silveira 2003).

The browning of sugar cane juice after extraction is due to PPO and POD enzymes that oxidize phenolic compounds present in the juice (Bucheli and Bobinson 1994; Qudsieh et al. 2002). While analyzing grape cultivars, it was observed that peroxidase extracts demonstrated similar activity both for the soluble fraction and for the bound fraction, and for polyphenol oxidase, the highest activity was observed for cultivar Ruby (about 70%) (Freitas et al. 2008). According to Cenci (1994), during storage of grapes, changes in the content of PPO and POD enzymes, associated with senescence of tissues, may occur, increasing peroxidase activity and impairing grape quality although thermostable behavior of PPO and POD for these varieties was demonstrated to be nonlinear (Freitas et al. 2008).

Activities of POD and PPO in enzyme extracts of grape varieties Benitaka and Ruby decreased with time of exposure and increased with temperature. Thus, higher enzyme inactivation was observed at 85°C with 10 min exposure time; however, the thermal treatments were not enough to complete inactivation of the enzymes. For jellies and juices, temperature also caused decrease in PPO and POD activities, but it was not sufficient for complete inactivation of these enzymes (Freitas et al. 2008).

Studies with guavas showed that the activity of PPO and POD decreased with temperature and time of heat exposure but were not fully inactivated (Zanatta et al. 2006). This may occur due to the presence of thermostable isoenzymes that interfere in the complete inactivation of these enzymes (Alvim and Clemente 1998). Pineapple cultivar IAC "gomo-de-mel" when submitted to different temperatures showed reduction in peroxidase enzyme activity. This can be observed after 120 s at a temperature of 90°C (Brito et al. 2007).

Other studies have shown that apple peel from the Fuji cultivar had higher enzymatic activity both for PPO and POD when compared to its pulp. The same could be observed for the Gala cultivar. Related to the heat treatment in concentrated extracts of pulp and peel of both cultivars, decreased activity of PPO to its total inactivation after 10 min of exposure to temperature of 75°C, which demonstrated what was not observed for POD, which had its activity decreased but not inactivated (Valderram et al. 2001).

Research has shown that the activity of PPO in the soluble and ionic forms in broccoli were higher in extracts of fresh broccoli than in bleached broccoli. Regarding soluble POD, the activity was lower in bleached broccoli, about one ninth less than the activity found in fresh broccoli, which indicates that bleaching was partially effective in denaturation of these enzymes. As for the ion-bound POD, considerable differences were not observed for the two types of broccoli (Lopes and Clement 2002).

10.5 Future Perspectives

The use of enzymes in the food industry has increased significantly, generating more added value to the final product and an increase in production while reducing costs (Ribeiro et al. 2010). The enzyme-assisted extraction is also an ecofriendly extraction method for the production of a variety of bioactive molecules, saving processing time and energy, and potentially providing a more reproducible extraction process on the commercial scale (Puri et al. 2012), which demonstrates its importance for the future markets.

The integration of enzymes in food processes is a well-established approach, but evidence clearly shows that dedicated research efforts are consistently being made as to make this application of biological agents more effective and/or diversified (Fernandes 2010). To use enzymes most effectively, it is important to understand their catalytic property and mode of action, optimal operational conditions, and which enzyme or enzyme combination is appropriate for the plant material selected (Puri et al. 2012).

Some recent studies have been made for some species and their products in order to improve their production as seen in jicama juice (*Pachyrhyzus erosus* L.) (Lien and Man 2010; Ramos-de-la-Peña et al. 2012), date syrup (*Phoenix dactylifera* L.) (Abbès et al. 2011), and palm oil (*Elaeis guineenses* J.) (Teixeira et al. 2013), but future investigations are needed to expand the currently available enzymatic processes, in particular to further enhance the yields of bioactive compounds (Puri et al. 2012). The future challenge is to define the most appropriate and economical methods of enzymatic use and establish protocols for each plant and product.

The cost of the enzyme is of prime importance in order to realize the full potential of the extraction strategy. The costs are expected to decrease as technology and techniques advance and as cheaper growth substrates are explored for producing enzymes (Puri et al. 2011). Solid-state fermentation (SSF) is one promising technique, presenting advantages when carried out with different agricultural and kitchen waste residues that adds value by decreasing the cost of enzyme production, reducing the quantity of solid waste, and boosting the environmentally friendly management of agricultural and domestic waste, including corn cobs, carrot peelings, composite, grass, leaves, orange peelings, pineapple peelings, potato peelings, rice husk, sugarcane baggage, saw dust, wheat bran, and wheat straw, which are increasing as a result of the rising population (Bansal et al. 2011; da Silva et al. 2005).

Also, the research needs to achieve a higher scale as much of the work has been carried out on laboratory scale, and there has been little work at pilot scale and full-scale investigations on the use of enzymes. Pilot scale research would facilitate better evaluation of the technology and its constraints and opportunities (Puri et al. 2011).

REFERENCES

Abbès, F.; Bouaziz, M. A.; Blecker, C.; Masmoudi, M.; Attia, H.; Besbes, S. Date syrup: Effect of hydrolytic enzymes (pectinase/cellulase) on physicochemical characteristics, sensory and functional properties. *Food Science and Technology* 44 (2011):1827–1834.

Abdullah, A. G. L.; Sulaiman, N. M.; Aroua, M. J.; Noor, M. M. Response surface optimization of conditions for clarification of carambola fruit juice using a commercial enzyme. *Journal of Food Engineering* 81 (2007):65–71.

Aguero, M. V.; Ansorena, M. R.; Roura, S. I.; Del Valle, C. E. Thermal inactivation of peroxidase during blanching of butternut squash. *Food Science and Technology* 41 (2008):401–407.

Akacha, N. B.; Gargouri, M. Enzymatic synthesis of green notes with hydroperoxide-lyase from olive leaves and alcohol-dehydrogenase from yeast in liquid/gas reactor. *Process Biochemistry* 44 (2009):1122–1127.

Alvim, K. C.; Clemente, E. Estudo da termoestabilidade de peroxidases extraídas da polpa e casca de mexerica (*Citrus deliciosa*). *Acta Scientiarum* 20 no. 2 (1998):201–205.

Amid, M.; Manap, M. Y. A.; Mustaf, S. Purification of pectinase from mango (*Mangifera indica* L. cv. Chokanan) waste using an aqueous organic phase system: A potential low cost source of the enzyme. *Journal of Chromatography B* 931 (2013):17–22.

Arias, E.; González, J.; Peiró, J. M.; Oria, R.; Lopez-Buesa, P. Browning prevention by ascorbic acid and 4-hexylresorcinol: Different mechanisms of action on polyphenol oxidase in the presence and in the absence of substrates. *Journal of Food Science: Food Chemistry and Toxicology* 72 no. 9 (2007):464–470.

Bae, H. J.; Kim, H. J.; Kim, Y. S. Production of a recombinant xylanase in plants and its potential for pulp biobleaching applications. *Bioresource Technology* 99 (2008):3513–3519.

Bansal, N.; Tewari, R.; Gupta, J. K.; Soni, S. K.; Soni, R. A novel strain of *Aspergillus niger* producing a cocktail of industrial depolymerising enzymes for the production of second generation biofuels. *BioResources* 6 (2011):552–569.

Bansal, N.; Tewari, R.; Soni, R.; Soni, S. K. Production of cellulases from *Aspergillus niger* NS-2 in solid state fermentation on agricultural and kitchen waste residues. *Waste Management* 32 (2012):1341–1346.

Barrera-Islas, G. A.; Ramos-Valdivia, A. C.; Salgado, L. M.; Ponde-Noyola, T. Characterization of beta-glucosidase produced by a high specific growth-rate mutante of *Cellulomonas flavigena*. *Current Microbiology* 54 (2007):266–270.

Bayindirli, A. "Introduction to enzymes." In: *Enzymes in Fruit and Vegetable Processing. Chemistry and Engineering Applications*, 1–18. Boca Raton, FL: CRC Press, 2010.

Beaulieu, J. C.; Baldwin, E. A. "Flavor and aroma of fresh-cut fruits and vegetables," In: Lamikanra, O., ed., *Fresh-Cut Fruits and Vegetables: Science, Technology, and Market*, 391–425. Boca Raton, FL: CRC Press, 2002.

Beg, Q. K.; Bhushan, B.; Kapoor, M.; Hoondal, G. S. Effect of amino acids on production of xylanase and pectinase from a *Streptomyces* sp. QG-11–3. *World Journal of Microbiology and Biotechnology* 16 (2000):211–213.

Beg, Q.; Kapoor, M.; Mahajan, L.; Hoondal, G. Microbial xylanases and their industrial applications: A review. *Applied Microbiology and Biotechnology*, 56 (2001):326–338.

Belluzzo, A. S. F.; Fleuri, L. F.; Macedo, J. A.; Macedo, G. A. Characterization of Biuti peach polyphenoloxidase. *Food Science and Biotechnology* 18 (2009):878–883.

Biely, P. Microbial xylanolytic systems. *Trends Biotechnology* 3 (1985):286–290.

Brito, C. A. K.; Sato, H. H.; Spironello, A.; Siqueira, W. J. Abacaxi IAC gomo-de-mel (*Ananas comosus* (L.) Merril): Característica da polpa e da peroxidase do suco. *Boletim CEPPA* 25 no. 2 (2007):257–266.

Bruhlmann, F.; Leupin, M.; Erismann, K. H.; Fiechter, A. Enzymatic degumming of ramie bast fibers. *Journal of Biotechnology* 76 (2000):43–50.

Bucheli, C. S.; Robinson, S. P. Contribution of enzymatic browning to color in sugarcane juice. *Journal of Agricultural and Food Chemistry* 42 no. 2 (1994):257–261.

Campos, A. D.; Silveira, E. M. L. Metodologia para determinação da peroxidase e da polifenoloxidase em plantas. *Embrapa-Communicado Técnico* 87 (2003):1–3.

Cardoso, F. S. N. "Frutas e hortaliças" In: *Matérias primas alimentícias—Composição e controle de qualidade*, 6–24. Rio de Janeiro: Editora Guanabara Koogan, 2011.

Cass, B. J.; Schade, F.; Robinson, C. W.; Thompson, J. E.; Legge, R. L. Production of tomato flavor volatiles from a crude enzyme preparation using a hollow-fiber reactor. *Biotechnology and Bioengineering* 67 (2000):372–377.

Cenci, S. A. Ácido naftalenoacético (ANA) e cloreto de cálcio na précolheita de uva niágara rosada (*Vitis labrusca* L. X *Vitis vinifera* L.): Avaliação do potencial de conservação no armazenamento. Lavras: These in Food Science—Escola Superior de Agricultura de Lavras, 1994, 109 pp.

Chaisakdanugull, C.; Theerakulkait, C.; Wrolstad, R. E. Pineapple juice and its fractions in enzymatic browning inhibition of banana (Musa [AAA Group] Gros Michel). *Journal of Agriculture and Food Chemistry* 55 (2007):4252–4257.

Chandini, S. K.; Jaganmohan-Rao, L.; Gowthaman, M. K.; Haware, D. J.; Subramanian, R. Enzymatic treatment to improve the quality of black tea extracts. *Food Chemistry* 127 (2011):1039–1045.

Chang, T. S. An updated review of tirosinase inhibitors. *International Journal of Molecular Sciences* 10 (2009):2440–2475.

Chapla, D.; Divechab, J.; Madamwara, D.; Shaha, A. Utilization of agro-industrial waste for xylanase production by *Aspergillus foetidus* MTCC 4898 under solid state fermentation and its application in saccharification. *Biochemical Engineering Journal* 49 (2010):361–369.

Chen, S.; Xing, X. H.; Huang, J. J.; Xu, M. S. Enzyme-assisted extraction of flavonoids from Ginkgo biloba leaves: Improvement effect of flavonol transglycosylation catalyzed by *Penicillium decumbens* cellulose. *Enzyme and Microbial Technology* 48 (2011):100–105.

Collins, T.; Gerday, C.; Feller, G. Xylanases, xylanase families and extremophilic xylanases. *FEMS Microbiology* 29 (2005):3–23.

Constabel, C. P.; Ryan, C. A. A survey of wound- and methyl jasmonate-induced leaf polyphenol oxidase in crop plants. *Phytochemistry* 47 no. 4 (1998):507–511.

Constabel, C. P.; Yip, L.; Patton, J. J.; Christopher, M. E. Polyphenol oxidase from hybrid poplar. Cloning and expression in response to wounding and herbivory. *Plant Physiology* 124 (2000):285–295.

Croak, S.; Corredig, M. The role of pectin in orange juice stabilization: Effect of pectin methylesterase and pectinase activity on the size of cloud particles. *Food Hydrocolloids* 20 (2006):961–965.

Cunha, A. G.; Gandini, A. Turning polysaccharides into hydrophobic materials: A critical review. Part 2. Hemicelluloses, chitin/chitosan, starch, pectin and alginates. *Cellulose*, 17 (2010):1045–1065.

da Silva, R., Lago, E. S., Merheb, C. W., Macchione, M. M.; Park, Y. K.; Gomes, E. Production of xylanase and CMCase on solid state fermentation in different residues by *Thermoascus aurantiacus* Miehe. *Brazilian Journal of Microbiology* 36 (2005):235 241.

Devasena, T. *Enzymology*. New Delhi: Oxford, 2010.

Dinkova, R.; Heffels, P.; Shikov, V.; Weber, F.; Schieber, A.; Mihalev, K. Effect of enzyme-assisted extraction on the chilled storage stability of bilberry (*Vaccinium myrtillus* L.) anthocyanins in skin extracts and freshly pressed juices. *Food Research International* 65 (2014):35–41.

Duarte-Vázquez, M. A.; García-Almendárez, B. G.; Raragalado, C.; Whitaker, J. R. Purification and partial characterization of theree turnip (*Brassica napus* L. var. esculenta D.C.). *Journal of Agricultural and Food Chemistry* 48 (2000):1574–1579.

Eisenmenger, M. J.; Reyes-De-Corcuera, J. I. High pressure enhancement of enzymes: A review. *Enzyme and Microbial Technology* 45 (2009):331–347.

European Comission—Agricultural commodity markets past developments fruit and vegetables—An analysis of consumption, production and trade based on statistics from the Food and Agriculture Organisation (FAO). Available at http://ec.europa.eu/agriculture/analysis/tradepol/worldmarkets/fruitveg/072007_en.pdf, accessed December 2013.

Fantozzi, P.; Petruccioli, G.; Montedoro, G. Trattamenti con additivi enzimatici alle paste di oliva sottoposte ad estrazione per pressione unica: Influenze delle cultivars, dell'epoca di raccolta e della conservazione. *Rivista Italiana delle Sostanze Grasse* 54 (1977):381–388.

FAO. FAOSTAT, 2012. Available at http://faostat.fao.org/, accessed December 2013.

Fatibello-Filho, O.; Vieira, I. C. Uso analítico de tecidos e de extratos brutos vegetais como fonte enzimática. *Química Nova* 25 no. 3 (2002):455–464.

Fazekas, E.; Szabó, K.; Kandra, L.; Gyémánt, G. Unexpected mode of action of sweet potato β-amylase on maltooligomer substrates. *Biochimica et Biophysica Acta (BBA)—Proteins and Proteomics* 1834 no. 10 (2013):1976–1981.

Fernandes, P. Enzymes in food processing: A condensed overview on strategies for better biocatalysts. *Enzyme Research* 2010 (2010):1–19.

Fleuri, L. F.; Lima, G. P. P. Polissacarídeos: Obtenção e Aplicação na Indústria de Alimentos. In: Pastore, G. M., Bicas, J. L., Maróstica-Junior, M. R., eds., *Biotecnologia de Alimentos da série Ciência, Tecnologia, Engenharia de Alimentos e Nutrição*. São Paulo: Atheneu, 2013, pp. 297–317.

Freitas, A. A.; Francelin, M. F.; Hirata, G. F.; Clemente, E.; Schmidt, F. L. Atividades das enzimas peroxidase (POD) e polifenoloxidase (PPO) nas uvas das cultivares benitaka e rubi e em seus sucos e geleias. *Ciência e Tecnologia de Alimentos* 28 (2008):172–177.

Garcia, A.; Brenes, M.; Moyano, M. J.; Alba, J.; García, P.; Garrido, A. Improvement of phenolic compound content in virgin olive oils by using enzymes during malaxation. *Journal of Food Engineering* 48 (2001):189–194.

Garcia, E.; Barrett, D. M. "Preservative treatments for fresh-cut fruits and vegetables." In: Lamikandra, O., ed., *Fresh-Cut Fruits and Vegetables: Science, Technology, and Market*, 267–303. Boca Raton, FL: CRC Press, 2002.

Ghidelli, C.; Rojas-Argudo, C.; Mateos, M.; Pérez-Gago, M. B. Effect of antioxidants in controlling enzymatic browning of minimally processed persimmon 'Rojo Brillante.' *Postharvest Biology and Technology*, 86 (2013):487–493.

Goff, S. A.; Klee, H. J. Plant volatile compounds: Sensory cues for health and nutritional value? *Science* 311 (2006):815–819.

Gonçalves, E. M.; Pinheiro, J.; Abreu, M.; Brandão, T. R. S.; Silva, C. L. M. Modelling the kinetics of peroxidase inactivation, colour and texture changes of pumpkin (*Cucurbita maxima* L.) during blanching. *Journal of Food Engineering* 81 (2007):693–701.

Goupy, P.; Amiot, M. J.; Richard-Forget, R.; Duprat, F.; Aubert, S.; Nicolas, J. Enzymatic browning of model solutions and apple phenolic extracts by apple polyphenoloxidase. *Journal of Food Science* 60 no. 3 (1995):497–502.

Grotheer, P.; Valles, S.; Simonne, A.; Kim, J. M.; Marshall, M. R. Polyphenol oxidase inhibitor(s) from german cockroach (*Blattella germanica*) extract. *Journal of Food Biochemistry* 36 (2012):292–300.

Guerrero-Beltrán, J. A.; Swanson, B. G.; Barbosa-Cánovasa. Inhibition of polyphenoloxidase in mango puree with 4-hexylresorcinol, cysteine and ascorbic acid. *LWT—Food Science and Technology* 38 (2005):625–630.

Guo, X.; Zhao, W.; Pang, X.; Liao, X.; Hu, X.; Wu, J. Emulsion stabilizing properties of pectins extracted by high hydrostatic pressure, high-speed shearing homogenization and traditional thermal methods: A comparative study. *Food Hydrocolloids* 35 (2014):217–225.

Helbert, W.; Schulein, M.; Henrissat, B. Electron microscopic investigation of the diffusion of *Bacillus licheniformis* amylase into corn starch granules. *International Journal of Biological Macromolecules* 19 (1996):165–169.

Hoondal, G. S.; Tiwari, R. P.; Tewari, R.; Dahiya, N.; Beg, Q. K. Microbial alkaline pectinases and their industrial applications: A review. *Applied Microbiology and Biotechnology* 59 (2002):409–418.

Hsu, C. P., Deshpande, S. N., Desrosier, N. W. Role of pectin methylesterase in firmness of canned tomatoes. *Journal of Food Science* 30 (1965):583–588.

Humanes, L.; López-Ruiz, A.; Merino, M. T.; Roldán, J. M.; Diez, J. Purification and characterization of limonoate dehydrogenase from *Rhodococcus fascians*. *Applied Environment and Microbiololy* 63 (1997):3385–3389.

Instituto Brasileiro de Geografia e Estatística (IBGE), Brazil, 2012. Available at www.ibge.gov.br/home/estatistica /economia/agropecuaria/, accessed December 2013.

Iyidogan, N. F.; Bayindirli, A. Effect of L-cysteine, kojic acid and 4-hexylresorcinol combination on inhibition of enzymatic browning in Amasya apple juice. *Journal of Food Engineering* 62 no. 3 (2004):299–304.

Jayani, R. S.; Saxena, S.; Gupta, R. Microbial pectinolytic enzymes: A review. *Process Biochemistry* 40 (2005):2931–2944.

Jecu, L. Solid-state fermentation of agricultural wastes for endoglucanase production. *Industrial Crops Products* 11 (2000):1–5.

Jensen, M.; Petersen, B. R.; Adler-Nissen, J. Enzymatic firming of processed red pepper by means of exogenous pectinesterase. *Food Biotechnology* 18 no. 2 (2005):217–227.

Johnson, R.; Padmaja, G.; Moorthy, S. N. Comparative production of glucose and high fructose syrup from cassava and sweet potato roots by direct conversion techniques. *Innovative Food Science & Emerging Technologies* 10 no. 4 (2009):616–620.

Juturu, V.; Wu, J. C. Insight into microbial hemicellulases other than xylanases: A review. *Journal of Chemical Technology and Biotechnology* 88 (2013):353–363.

Juturu, V.; Wu, J. C. Microbial cellulases: Engineering, production and applications. *Renewable and Sustainable Energy Reviews* 33 (2014):188–203.

Kashyap, D. R.; Vohra, P. K.; Chopra, S.; Tewari, R. Applications of pectinases in the commercial sector: A review. *Bioresource Technology* 77 (2001):215–227.

Kim, J. H.; Seo, Y. S.; Kim, J. E.; Sung, S.-K.; Song, K. J.; Na, G.; Kim, W. T. Two polyphenol oxidases are differentially expressed during vegetative and reproductive development and in response to wounding in the Fuji apple. *Plant Science* 161 no. 6 (2001):1145–1152.

Kim, Y. J.; Uyama, H. Tyrosinase inhibitors from natural and synthetic sources: Structure, inhibition mechanism and perspective for the future. *Cellular and Molecular Life Sciences* 62 (2005):1707–1723.

Klavons, J. A.; Bennett, R. D.; Vannier, S. H. Nature of protein constituent of orange juice cloud. *Agriculture and Food Chemistry* 39 no. 9 (1991):1545–1548.

Kola, O.; Kaya, C.; Duran, H.; Altan, A. Removal of limonin bitterness by treatment of ion exchange and adsorbent resins. *Food Science and Biotechnology* 19 no. 2 (2010):411–416.

Kosseva, M. R. "Processing of food wastes." In: Taylor, S. L., ed., *Advances in Food and Nutrition Research*, 58–60. Dublin: Elsevier Inc., 2009.

Kuo, J. M.; Hwang, A.; Yeh, D. B.; Pan, M. H.; Tsai, M. L.; Pan, B. S. Lipoxygenase from banana leaf: Purification and characterization of an enzyme that catalyzes linoleic acid oxygenation at the 9-position. *Journal of Agriculture and Food Chemistry* 54 no. 8 (2006):3151–3156.

Kuriki, T.; Takata, H.; Yanase, M.; Ohdan, K.; Fujii, K.; Terada, Y.; Takaha T.; Hondoh, H.; Matsuura, Y.; Imanaka, T. The concept of the α-amylase family: A rational tool for interconverting glucanohydrolases/ glucanotransferases, and their specificities. *Journal Applied Glycoscience* 53 (2006):155–161.

Laaksonen, O.; Sandell, M.; Nordlund, E.; Heinio, R. L.; Malinen, H. L.; Jaakkola, M.; Kallio, H. The effect of enzymatic treatment on blackcurrant (Ribes nigrum) juice. *Food Chemistry* 130 (2012):31–41.

Labuza, T. P.; Lillemo, J. H.; Taoukis, P. S. Inhibition of polyphenol oxidase by proteolytic enzymes. *Fruit Processing* 2 no. 1 (1992):9–13.

Landbo, A. K.; Kaack, K.; Meyer, A. S. Statistically designed two step response surface optimization of enzymatic prepress treatment to increase juice yield and lower turbidity of elderberry juice. *Innovative Food Science and Emerging Technologies* 8 (2007):135–142.

Latif, S.; Anwar, F. Aqueous enzymatic sesame oil and protein extraction. *Food Chemistry* 125 no. 2 (2011):679–684.

Leadlay, P. F. *An Introduction to Enzyme Chemistry.* Cambridge: The Royal Society of Chemistry, 1993, 82 pp.

Lee, W. C.; Yusof, S.; Hamid, N. S. A.; Baharin, B. S. Optimizing conditions for enzymatic clarification of banana juice using response surface methodology (RSM). *Journal of Food Engineering* 73 (2006):55–63.

Lien, N. L. P.; Man, L. V. V. Application of commercial enzymes for jicama pulp treatment in juice production. *Science and Technology Development* 13 (2010):64–76.

Loizzo, M. R.; Tundis, R.; Menichini, F. Natural and synthetic tyrosinase inhibitors as antibrowning agents: An update. *Comprehensive Reviews in Food Science and Food Safety* 11 (2012):378–398.

Lopes, A. S.; Clemente, E. Minerais e enzimas oxidativas em brócolis (*Brassica oleracea* L. cv. Itálica) minimamente processado. *Acta Scientiarum* 24 no. 6 (2002):1615–1618.

Maalej-Achouri, I.; Guerfali, M.; Belhaj-Ben, I. R.; Gargouri, A.; Belghith, H. The effect of *Talaromyces thermophilus* cellulase-free xylanase and commercial laccase on lignocellulosic components during the bleaching of kraft pulp. *International Biodeterioration and Biodegradation* 75 (2012):43–48.

Macedo, G. A.; Pastore, G. M.; Sato, H. H.; Park, Y. K. *Bioquímica Experimental de Alimentos.* São Paulo: Varela, 2005.

Maktouf, S.; Neifarb, M.; Driraa, S. J.; Bakloutic, S.; Fendria, M.; Châabouni, S. E. Lemon juice clarification using fungal pectinolytic enzymes coupled to membrane ultrafiltration. *Food and Bioproducts Processing* 92 no. 1 (2014):14–19.

Marshall, M. R.; Kim, J.; Wei, C. *Enzymatic Browning in Fruits, Vegetables and Seafoods.* Washington, DC: FAO, 2000.

Martín-Sampedro, R.; Rodríguez, A.; Ferrer, A.; García-Fuentevilla, L. L.; Eugenio, M. E. Biobleaching of pulp from oil palm empty fruit bunches with laccase and xylanase. *Bioresource Technology* 110 (2012):371–378.

Matinez-Valverde, I.; Periago, M. J.; Provan, G.; Chesson, A. Phenolic compounds, lycopene and antioxidant activity in commercial varieties of tomato (*Lycopersicum esculentum*). *Journal of the Science Food and Agriculture* 82 (2002):323–330.

Mayer, A. Polyphenol oxidases in plants and fungi: Going places? A review. *Phytochemistry* 67 no. 21 (2006):2318–2331.

Mendonça, S. C.; Guerra, N. B. Métodos físicos e químicos empregados no controle do escurecimento enzimático de vegetais. *Boletim Sociedade Brasileira de Ciência e Tecnologia de Alimentos* 37 no. 2 (2003):113–116.

Molina, G.; Prazeres, J. N.; Ballus, C. A. "Aplicação de enzimas na indústria de alimentos." In: Pastore, G. M.; Bicas, J. L.; Maróstica Júnior, M. R., eds., *Biotecnologia de Alimentos*, 343–365. São Paulo: Ediora Atheneu, 2013.

Novozymes Enzymes at work. 2013. Available at www.novozymes.com/en/about-us/brochures/documents /enzymes_at_work.pdf, accessed September 2013.

Núñez-Delicado, E.; Serrano-Megías, M. S.; Pérez López, A. J.; López-Nicolás, J. M. Characterization of polyphenol oxidase from napoleon grape. *Food Chemistry* 100 (2007):108–114.

Oh, K.; Hamada, K.; Saito, M.; Lee, H. J.; Matsuoka, H. Isolation and properties of an extracellular β-glicosidase from a filamentous fungus *Cladosporium resinae*, isolated from querosene. *Bioscience, Biotechnology and Biochemistry* 63 no. 2 (1999):281–287.

Oh, K. B.; Hamada, K.; Saito, M.; Lee, H. J.; Matsuoka, H. Isolation and properties of an extracellular β-glucosidase from a filamentous fungus, *Cladosporium resinae*, isolated from kerosene. *Bioscience, Biotechnology and Biochemistry* 63 (1999):281–287.

Okoth, M. W.; Kaahwa, A. R.; Imungi, J. K. The effect of homogenisation, stabiliser and amylase on cloudiness of passion fruit juice. *Food Control* 11 (2000):305–311.

Oort, M. "Enzymes in bread making." In: Whitehurst, R. J.; Oort, M., eds., *Enzymes in Food Technology*, 103–143. Chichester: Wiley-Blackwell, 2010.

Palma-Fernandez, E. R. D.; Gomes, E.; Silva, R. Purification and characterization of two β-glicosidases from thermophilic fungus *Thermonascus aurantiacus* Miehe. *Folia Microbiology* 47 (2002):685–690.

Pasha, I.; Saeed, F.; Sultan, M. T.; Khan, M. R.; Rohi, M. Recent developments in minimal processing: A tool to retain nutritional quality of food. *Critical Reviews in Food Science and Nutrition* 54 no. 3 (2014):340–351.

Perera, N.; Gamage, T. V.; Wakeling, L.; Gamlath, G. G. S.; Versteeg, C. Colour and texture of apples high pressure processed in pineapple juice. *Innovative Food Science and Emerging Technologies* 11 (2010):39–46.

Perez, E. E.; Fernández, M. B.; Nolasco, S. M.; Crapiste, G. H. Effect of pectinase on the oil solvent extraction from different genotypes of sunflower (*Helianthus annuus* L.) *Journal of Food Engineering* 117 (2013):393–398.

Pifferi, P. G.; Buska, G.; Manenti, I.; Lo Presti, A.; Spagna, G. Immobilization of pectinesterase on c-alumina for the treatment of juices. *Belgian Journal of Food Chemistry and Biotechnology* 44 (1989):173–182.

Pinelo, M.; Zeuner, B.; Meyer, A. S. Juice clarification by protease and pectinase treatments indicates new roles of pectin and protein in cherry juice turbidity. *Food and Bioproducts Processing* 88 (2010):259–265.

Polizeli, M. L. T. M.; Rizzatti, A. C. S.; Monti, R.; Terenzi, H. F.; Jorge, J. A.; Amorim, S. S. Xylanases from fungi: Properties and industrial applications. *Applied Microbiology and Biotechnology* 67 (2005):577–591.

Puri, M.; Sharma, D.; Tiwari, A. K. Downstream processing of stevioside and its potential applications. *Biotechnology Advances* 29 (2011):781–791.

Puri, M.; Sharma, D.; Barrow, C. J. Enzyme-assisted extraction of bioactives from plants. *Trends in Biotechnology* 30 no. 1 (2012):37–44.

Qudsieh, H. Y. M.; Yusof, S.; Osman, A.; Rahman, R. A. Effect of maturity on chlorophyll, tannin, color and polyphenol oxidase (PPO) activity of sugarcane juice (*Saccharum officinarum* var. Yellow cane). *Journal of Agricultural and Food Chemistry* 50 (2002):1615–1618.

Rai, P.; Majumdar, G. C.; Das Gupta, S.; De, S. Optimizing pectinase usage in pretreatment of mosambi juice for clarification by response surface methodology. *Journal of Food Engineering* 64 (2004):397–403.

Ramadan, M. F.; Sitohy, M. Z.; Moersel, J. T. Solvent and enzyme-aided aqueous extraction of goldenberry (*Physalis peruviana* L.) pomace oil: Impact of processing on composition and quality of oil and meal. *European Food Research Technology* 226 (2008):1445–1458.

Ramírez, E. C.; Whitaker, J. R.; Virador, V. M. "Polyphenoloxidase." In: Whitaker, J. R.; Voragen, A. G. J.; Wong, D. W. S., eds., *Handbook of Food Enzymology*, 509–524. New York: Marcel Dekker Inc., 2003.

Ramos, P. A. S.; Sediyama, T.; Viana, A. E. S.; Pereira, D. M.; Finger, F. L. Efeito de inibidores da peroxidase sobre a conservação de raízes de mandioca *in natura*. *Brazilian Journal of Food Technology* 16 no. 2 (2013):116–124.

Ramos-de-la-Peña, A. M.; Renard, C. M. G. C.; Wicker, L.; Montañes, J.; Reyes-Veja, M. L.; Voget, C.; Contreras-Esquivel, J. C. Enzymatic liquefaction of jicama (*Pachyrhizus erosus*) tuberous roots and characterization of the cell walls after processing. *Food Science and Technology* 49 (2012):257–262.

Ranalli, A.; De Mattia, G.; Ferrante, M. L. The characteristics of percolation olive oils produced with a new processing enzyme aid. *International Journal of Food Science and Technology* 32 (1998):247–258.

Ranalli, A.; Malfattp, A.; Pollastri, L.; Contento, S.; Lucera, L. Analytical quality and genuineness of enzyme-extracted virgin olive oil. *Journal of Food Quality* 26 (2003):149–164.

Ranalli, A.; Mattia, G. D. Characterisation of olive oil produced with a new enzyme processing aid. *Journal of American Oil Chemist's Society* 74 (1997):1105–1113.

Ranalli, A.; Lucera, L.; Contento, S.; Simone, N.; Del Re, P. Bioactive constituents, flavors and aromas of virgin oils obtained by processing olives with a natural enzyme extract. *European Journal of Lipid Science Technology* 106 (2004):187–197.

Rani, M. U.; Appaiah, K. A. A. *Gluconacetobacter hansenii* UAC09-mediated transformation of polyphenols and pectin of coffee cherry husk extract. *Food Chemistry* 130 (2012):243 247.

Reed, G. "General characteristics of enzymes." In: Reed, G., ed., *Enzymes in Food Processing*, 15–19. New York: Academic Press, 1975.

Rehman, H. U.; Nawaz, M. A.; Aman, A.; Baloch, A. H.; Qader, S. A. U. Immobilization of pectinase from Bacillus licheniformis KIBGE-IB21 on chitosan beads for continuous degradation of pectin polymers. 3 (2014):282–287.

Renard, C. M. G. C.; Thibault, J. F.; Voragen, A. G. J.; Pilnik, V. D. B. W. Studies on apple protopectin VI: Extraction of pectins from apple cell walls with rhamnogalacturonase. *Carbohydrate Polymers* 22 (1993):203–210.

Ribeiro, D. S.; Henrique, S. M. B.; Oliveira, L. S.; Macedo, G. A.; Fleuri, L. F. Enzymes in juice processing: A review. *International Journal of Food Science and Technology* 45 (2010):635–641.

Rinaldi, M.; Caligiani, A.; Borgese, R.; Palla, G.; Barbanti, D.; Massani, R. The effect of fruit processing and enzymatic treatments on pomegranate juice composition, antioxidant activity and polyphenols content. *Food Science Technology* 53 (2013):355–359.

Riou, C.; Salmon, J. M.; Vallier, I. Z.; Gunata, Z.; Barre, P. Purification, characterization and substrate specificity of a novel highly glucose-tolerant β-glicosidase from *Aspergillus orzyzae*. *Applied Microbiology and Biotechnology* 64 (1998):3607–3614.

Rosen, J. C.; Kader, A. A. Postharvest physiology and quality maintenance of sliced pear and strawberry fruits. *Journal of Food Science* 54 (1989):656–659.

Roy, J. K.; Rai, S. K.; Mukherjee, A. K. Characterization and application of a detergent-stable alkaline alpha-amylase from *Bacillus subtilis* strain AS-S01a. *International Journal Biological Macromolecules* 50 (2012):219–229.

Roy, J. K.; Borah, A.; Mahantab, C. L.; Mukherjeea, A. K. Cloning and overexpression of raw starch digesting α-amylase gene from *Bacillus subtilis* strain AS01a in *Escherichia coli* and application of the purified recombinant-amylase (AmyBS-I) in raw starch digestion and baking industry. *Journal of Molecular Catalysis B: Enzymatic* 97 (2013):118–129.

Sadaf, A.; Khare, S. K. Production of *Sporotrichum thermophile* xylanase by solid state fermentation utilizing deoiled Jatropha curcas seed cake and its application in xylooligosachharide synthesis. *Bioresource Technology* 153 (2014):126–130.

Sarioglu, K.; Demir, N.; Acar, J.; Mutlu, M. The use of commercial pectinase in the fruit juice industry, Part 2: Determination of the kinetic behaviour of immobilized commercial pectinase. *Journal of Food Engineering* 47 (2001):271–274.

Schwab, W.; Davidovich-Rikanati, R.; Lewinsohn, E. Biosynthesis of plant-derived flavor compounds. *The Plant Journal* 54 (2008):712–732.

Scott, D. "Oxidoreductases." In: Reed, G., ed., *Enzymes in Food Processing*, 222–254. New York: Academic Press, 1975.

Severini, C.; Baiano, A.; Pilli, T. D.; Romaniello, R.; Derossi, A. Prevention of enzymatic browning in sliced potatoes by blanching in boiling saline solutions. *Lebensm.-Wiss. u.-Technol.* 36 (2003):657–665.

Sharma, A.; Satyanarayana, T. Microbial acid-stable α-amylases: Characteristics, genetic engineering and applications. *Process Biochemistry* 48 (2013):201–211.

Siedow, J. N. Plant lipoxygenase: Structure and function. *Annual Review of Plant Physiology and Plant Molecular Biology* 42 (1991):145–188.

Siew-Yin, C.; Wee-Sim, C. Effect of extraction conditions on the yield and chemical properties of pectin from cocoa husks. *Food Chemistry* 141 (2013):3752–3758.

Singhania, R. R.; Saini, J. K.; Saini, R.; Adsul, M.; Mathur, A.; Gupta, R.; Tuli, D. K. Bioethanol production from wheat straw via enzymatic route employing *Penicillium janthinellum* cellulases. *Bioresource Technology* 169 (2014):490–495.

Song, H. Y.; Lim, H. K.; Kim, D. I.; Lee, K. I.; Hwang, I. T. A new bi-modular endo-β-1,4-xylanase KRICT PX-3 from whole genome sequence of *Paenibacillus terrae* HPL-003. *Enzyme and Microbial Technology* 54 (2014):1–7.

Sreenath, H. K.; Santhanam, K. The use of commercial enzymes in white grape juice clarification. *Journal of Fermentation and Bioengineering* 73 no. 3 (1991):241–243.

Sreenath, H. K.; Sudarshanakrishna, K. R.; Santhanam, K. Improvement of juice recovery from pineapple pulp/residue using cellulases and pectinases. *Journal of Fermentation and Bioengineering* 78 no. 6 (1994):486–488.

Taiz, L.; Zeiger, E. *Fisiologia Vegetal*, 4a ed. Porto Alegre: Artmed, 2009.

Taniguchi, H.; Honnda, Y. Amylases chapter. In *Applied Microbiology: Industrial Amylases*. Japan: Elsevier Inc., 2009.

Tauber, H. *The Chemistry and Technology of Enzymes*. New York: John Wiley & Sons, 1950.

Teixeira, C. B.; Macedo, G. A.; Macedo, J. A.; Silva, L. H. M.; Rodrigues, A. M. C. Simultaneous extraction of oil and antioxidant compounds from oil palm fruit (*Elaeis guineensis*) by an aqueous enzymatic process. *Bioresource Technology* 129 (2013):575–581.

Terefe, N. S.; Buckow, R.; Versteeg, C. Quality related enzymes in plant based products: Effects of novel food processing technologies, Part 3: Ultrasonic processing. *Critical Reviews in Food Science and Nutrition* (2012). doi:10.1080/10408398.2011.586134.

Terefe, N. S.; Buckow, R.; Versteeg, C. Quality related enzymes in plant based products: Effects of novel food processing technologies, Part 2: Pulsed electric field processing. *Critical Reviews in Food Science and Nutrition* (2013). doi.10.1080/10408398.2012.701253.

Terefe, N. S.; Buckow, R.; Versteeg, C. Quality related enzymes in plant based products: Effects of novel food processing technologies, Part 1: High-pressure processing. *Critical Reviews in Food Science and Nutrition* 54 (2014):24–63.

Terrasan, C. R. F.; Temer, B.; Duarte, M. C. T.; Carmona, E. C. Production of xylanolytic enzymes by *Penicillium janczewskii*. *Bioresource Technology* 101 (2010):4139–4143.

Tochi, B. N.; Wang, Z.; Xu, S. Y.; Zhang, W. Effect of Stem Bromelain on the browning of apple juice. *American Journal of Food Technology* 4 no. 4 (2009):146–153.

Tourino, M. C. C.; Chitarra, A. B.; Gavilanes, M. C. Injúria mecânica em tecidos de frutos de pessegueiros (*Prununs persica* [L.] Batsch): Mecanismos de cura. *Boletim da Sociedade Brasileira de Ciência e Tecnologia de Alimentos* 27 no. 2 (1993):69–78.

Tsukamoto, T.; Ichimaru, Y.; Kanegae, N.; Watanabe, K.; Yamaura, I.; Katsura, Y.; Funatsu, M. Identification and isolation of endogenous insect phenoloxidase inhibitors. *Biochemical and Biophysical Research Communications* 184 no. 1 (1992):86–92.

Uenojo, M.; Pastore, G. M. Pectinases: Aplicações industriais e perspectivas. *Química Nova* 30 no. 2 (2007):1–14.

Vámos-Vigyázó, L. Polyphenol oxidase and peroxidase in fruits and vegetables. *Critical Reviews in Food Science and Nutrition* 15 no. 1 (1981):49–127.

Van der Maarel, M. J. E.; Van der Veen, B.; Uitdehaag, J. C. M.; Leemhuis, H.; Dijkhuizen, L. Properties and applications of starch-converting enzymes of the alpha-amylase family. *Journal of Biotechnology* 94 (2002):37–155.

Villena, M. A.; Iranzo, J. F. U.; Parez, A. I. B. β-glicosidase activity in wine yeasts: Application in enology. *Enzyme Microbiology and Technology* 40 (2007):420–425.

Wang, B.; Cheng, F.; Lub, Y.; Ge, W.; Zhang, M.; Yue, B. Immobilization of pectinase from *Penicillium oxalicum* F67 ontomagnetic cornstarch microspheres: Characterization and application in juice production. *Journal of Molecular Catalysis B: Enzymatic* 97 (2013):137–143.

Whitehead, M.; Muller, B. L.; Dean, C. Industrial use of soybean lipoxygenase for the production of natural green note flavor compounds. *Cereal Foods World* 40 (1995):193–197.

Wong, K. K. Y.; Saddler, J. N. "Applications of hemicellulases in the food, feed, and pulp and paper industries." In: Coughlan, M. P.; Hazlewood, G. P., eds., *Hemicelluloses and Hemicellulases*, 127–143. London: Portland Press, 1993.

Wong, K. K. Y.; Tan, L. U. L.; Saddler, J. N. Multiplicity of β-1,4-xylanase in microorganisms, functions and applications. *Microbiology Review* 52 no. 3: (1988):305–317.

Xu, Z. H.; Bail, Y. L.; Xu, X.; Shi, J. S.; Tao, W. I. Production of alkali-tolerant cellulose free xylanase by *Pseudomonas* sp. UN024 with wheat bran as the main substrate. *World Journal of Microbiology and Biotechnology* 21 (2005):575–581.

Yoo, J. K.; Lee, J. H.; Cho, H. Y.; Kim, J. G. Change of antioxidant activities in carrots (*Daucus carota* var. sativa) with enzyme treatment. *Journal of the Korean Society of Food Science and Nutrition* 42 no. 2 (2013):262–267.

Yusof, S.; Ibrahim, N. Quality of soursop juice after pectinase enzyme treatment. *Food Chemistry* 51 (1994):8–88.

Zabetakis, I.; Leclerc, D.; Kajda, P. The effects of high hydrostatic pressure on the strawberry anthocyanins. *Journal of Agriculture and Food Chemistry* 48 (2000):2749–2754.

Zanatta, C. L.; Zotarelli, M. F.; Clemente, E. Peroxidase (POD) e Polifenoloxidase (PPO) em polpa de goiaba (*Psidium guajava* R.). *Ciência e Tecnologia de Alimentos* 26 no. 3 (2006):705–708.

Zhang, H.; Wang, Z.; Xu, S. Y. Optimization of processing parameters for cloudy ginkgo (*Ginkgo biloba* Linn.) juice. *Journal of Food Engineering* 80 (2007):1226–1232.

Zhao, L.; Meng, K.; Shi, P.; Bai, Y.; Luo, H.; Huang, H.; Wang, Y.; Yang, P.; Yao, B. A novel thermophilic xylanase from *Achaetomium* sp. Xz-8 with high catalytic efficiency and application potentials in the brewing and other industries. *Process Biochemistry* 48 (2013):1879–1885.

11

Enzymes in Beverage Processing

Gustavo Molina, Fabiano Jares Contesini, Hélia Harumi Sato, and Gláucia Maria Pastore

CONTENTS

11.1 Introduction

The beverage industry is an important segment of the food industry and is greatly highlighted because the necessity for water by people is not sufficiently satisfied by drinking plain water, and in this case, the majority prefers flavored drinks. There is a great variety of drinks, such as natural fluids, including the juices of crushed fruits; infusions made by steeping leaves or seeds in hot or cold water; concoctions manufactured by the soft drink industry; and products of brewers and vintners. The preferences for which beverages are consumed depend on social and cultural habits added to the massive influence from advertising (Wiseman 2002).

Beverages can be classified as alcoholic and nonalcoholic drinks based on the alcohol content in the drink. An alcoholic beverage is a drink containing ethanol, popularly known as alcohol, and they are classified as beers, wines, and distilled spirits. Nonalcoholic beverages are technically beverages that contain less than 0.5% alcohol by volume, and there are nonalcoholic versions of some alcoholic beverages. They include juices of fruits and vegetables, milk, carbonated and uncarbonated soft drinks, coffee, and tea. Soft drinks contain carbonated water, sweetening, and flavoring agents. The sweetening agents are normally sugar, high-fructose corn syrup, or a sugar substitute. In addition, in some cases, soft drinks may also contain caffeine or fruit juice. Among soft drinks, they are called sodas, pop, coke, soda pop, fizzy drink, or carbonated beverages. Another type of nonalcoholic drinks is syrups that are thick, viscous liquids consisting primarily of a solution of sugar in water and containing a large amount of dissolved sugars (Shyam Kumar and Chandrasekaran 2012; Chandrasekaran 2012).

Both alcoholic and nonalcoholic beverages represent a remarkable share of the market in the food industry worldwide. In this case, beer, which is one of the most popular alcoholic beverages, holds a strategic economic position with production exceeding 1.34 billion hectoliters in 2002. In addition, beer is the fifth most consumed beverage in the world after tea, carbonates, coffee, and milk, and continues to be a popular drink with an average consumption of 23 L per person per year (Fillaudeau et al. 2006). The consumption of beer in Germany is about 151 L per person/per year, and American breweries producing approximately 156,900 million barrels of beer a year. More specifically, each barrel is the equivalent of 117 L or approximately 31 gallons (http://www.madehow.com/Volume-2/Beer.html).

On the other hand, wine production is one of the oldest endeavors of mankind, and the Egyptians and the Assyrians were known to make wine from grapes as early as 3500 BC. The basic process has remained unaltered for centuries in spite of the fact that some new wine products have been developed during recent periods. The process for producing wine includes destemming, crushing, pressing, fermentation, racking, and bottling (Shyam Kumar and Chandrasekaran 2012).

Among nonalcoholic beverages, there are soft drinks that consist primarily of carbonated water, sugar, and flavorings. Approximately 200 nations consume sweet, sparkling soda with an annual consumption of more than 34 billion gallons. In addition, soft drinks rank as America's favorite beverage segment, representing 25% of the total beverage market (http://manufacturingtodaynigeria.com/index.php/cover).

Beverage production includes some important steps, such as extraction of fruits and maceration, which involves the denaturation of components of the cell wall: cellulose, hemicellulose, and pectin. Hence, cellulases, hemicellulases, and pectinases are required for obtaining better quality in several types of beverages, including beers and wines. Enzymes are important biocatalysts that present a broad range of industrial applications. Their efficiency is mostly attributed to specificity for the substrates, which leads to a yield of high value-added products with minimal formation of by-products and their use in mild reaction conditions. Among the sources of enzymes, microorganisms correspond to a highlighted group due to the feasibility of production and the higher concentration of protein expression in the fermentative processes. Among the enzymes, the most used in the beverage industry are proteases, amylases, pectinases, and cellulases, mainly applied for clarification, aroma enhancement, and to support fermentation processes, among many other uses.

In this chapter, various aspects of enzymes used in different stages of beverage production, including alcoholic and nonalcoholic drinks, are presented. Additionally, different aspects of beverage production are also dealt with in order to show the various processes and products as well as the great potential of this industry.

11.2 Alcoholic Beverages

11.2.1 General Aspects

Alcoholic beverages are socially and culturally appreciated worldwide. They include fermented beverages, such as beers and wines, and distilled spirits, including liqueurs, vodka, and cachaça. Maceration and fermentation of grains and fruits are present in all these beverage preparations and there are steps that need special treatment to guarantee acceptable organoleptic properties to justify their commercialization. Furthermore, some studies report functional properties of beverages, such as wines (Staško et al. 2008) and beers (Gasowski et al. 2004).

Exogenous enzymes have relevant applications in the beverage industry, because during alcoholic beverage production, there is a necessity for use of enzymes in the extraction of important compounds that may present remarkable properties, including functional properties and sensory properties that are dependent on the phenolic compounds, oligossacharides, and aroma, including ketones and other compounds as summarized in Figures 11.1 and 11.2. Some enzymes and their applications in alcoholic beverage production are summarized in Table 11.1.

One group that is truly relevant for this application are the enzymes capable of lysing the cell wall of grains and grapes, resulting in the releasing of many compounds, including polyphenols and other

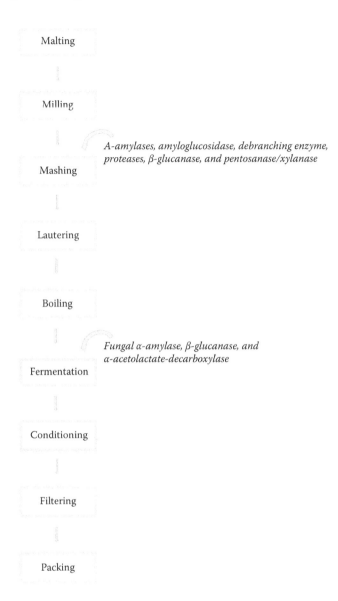

Malting

Milling

Mashing

A-amylases, amyloglucosidase, debranching enzyme, proteases, β-glucanase, and pentosanase/xylanase

Lautering

Boiling

Fermentation

Fungal α-amylase, β-glucanase, and α-acetolactate-decarboxylase

Conditioning

Filtering

Packing

FIGURE 11.1 Diagram showing the exact stage of application of exogenous enzymes during beer production.

functional compounds with antioxidant activity. They include cellulases, hemicellulases, and pectino-lytic enzymes that are used as highly concentrated commercial preparations. In addition, these enzymes are important for improving filterability and viscosity of beverages. Other enzymes are related to the aroma compounds of beers, such as acetolactate decarboxylases, or to improve fermentation processes, helping yeasts, providing nitrogen, which is the case of proteases. Table 11.2 shows details of commercial enzymes used in brewing and wine production.

11.2.2 Brewing

Beer production can be divided into two major stages: main fermentation and maturation. During fermentation, the fermentable sugars in the substrate for the fermentation, the wort, is metabolized. On the other side, during the maturation, flavor, carbon dioxide content, haze stability, attenuation, and other properties of the beer are carefully adjusted (Hough et al. 1982). Taking into account the advantages of shortening these steps in the brewing process, beer fermentation is a very important target for new

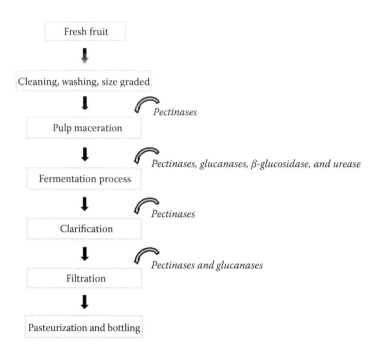

FIGURE 11.2 Diagram showing the exact stage of application of exogenous enzymes during wine production.

TABLE 11.1

Enzymes and Their Use in the Production of Alcoholic Beverages

Enzyme	Beverage	Application	References
Lichenase	Beer	Hydrolysis of lichenan	Chaari et al. 2014
Acetolactate decarboxylase	Beer	Decomposition of acetolactate	Godtfredsen et al. 1984
Pectinase	Wine	Improvement of aromatic profile	Servili et al. 1992
β-Glucanase	Wine	Influence on the autolysis of yeast	Torresi et al. 2014
Pectinases, cellulases, hemicellulases	Distilled spirit	Hydrolysis of bound monoterpenes	Fundira et al. 2002

biotechnological techniques, such as the use of immobilized yeast systems (Mensour et al. 1996) for shortening the time of fermentation and the utilization of several enzymes and mutant yeast (Liu et al. 2008; Scheffler and Bamforth 2005) for the reduction of time necessary for maturation.

The enzyme acetolactate decarboxylase has an important application in brewing (Bamforth 2009). The rate-limiting step in flavor maturation of beer in particular and in beer maturation in general is the conversion of acetolactate to diacetyl, and this has an important role in maturation (Godtfredsen and Ottesen 1982). The presence of this compound results in an undesirable flavor in beer when it is present in quantities greater than approximately 0.1 ppm (Hough et al. 1982). Yet diacetyl is enzymatically converted into acetoin and also into 2,3-butanediol by yeast cells during maturation. The addition of acetolactate decarboxylases to fermenting wort or maturing beer can entirely omit this by the decomposition of acetolactate (Eskin and Fereidoon 2013; Godtfredsen and Ottesen 1982; Godtfredsen et al. 1984).

Godtfredsen and Ottesen (1982) isolated an α-acetolactate decarboxylase from *Enterobacter aerogenes* strain 1033 and observed that, after the addition of the enzyme to freshly fermented beer, α-acetolactate and α-aceto-α-hydroxybutyrate were effectively removed in 24 h to an inferior level of taste threshold of the corresponding volatile diacetyl, diketones, and 2,3-pentanedione. This is very interesting because other important properties of the beer were not affected, and the beer matured in the

TABLE 11.2

Details of Commercial Enzymes Used in Brewing and Wine Production

Commercial Enzyme	Scientific Names	Vendor	References
Maturex®ª	α-Acetolactate decarboxylase	Novozymes	http://www.novozymes.com
Ceramix® PLUS*ª	Cellulases, α-amylases, glucanases, neutral protease, xilanase, and pentotanases	Novozymes	http://www.novozymes.com
Neutrase®ª	Protease	Novozymes	http://www.novozymes.com
Flavourzyme®ª	Peptidase	Novozymes	http://www.novozymes.com
Palkolase H1ᵇ/ª	Amylase	Maps Enzyme	http://www.mapsenzymes.com
Protamex®ª	Protease	Novozymes	http://www.novozymes.com
α-Amylase from barley maltª	α-Amylase	Sigma-Aldrich	http://www.sigmaldrich.com
Brewer Clarex®ª	Protease	DSM food specialties	http://www.dsm.com
Vinozym®Vintage FCE Gᵇ	Cinamil esterase	Novozymes	http://www.novozymes.com
Novoclair® Speedᵇ	Polygalacturonase	Novozymes	http://www.novozymes.com
VinoTaste® Proᵇ	β-Glucanases	Novozymes	http://www.novozymes.com
Vinoflow® Maxᵇ	Pectinase	Novozymes	http://www.novozymes.com
Rohavin® MX	Glucosidase	AB Enzymes GmbH	http://www.abenzymes.com

ª Enzymes used for beer production.
ᵇ Enzymes used for wine production.

presence of α-acetolactate decarboxylase was considered to have an equally satisfactory quality when compared with conventionally prepared beer. In addition, Dulieu et al. (2000) set up a simple model that focused on α-acetolactate and diacetyl profiles during beer fermentation. They observed that encapsulated α-acetolactate decarboxylase allows the acceleration of beer fermentation as efficiently as free α-acetolactate decarboxylase. This is relevant due to the possibility of reuse of the immobilized enzyme, which is interesting from the economic point of view.

One important use of enzymes applied in any step of brewing is in barley cell wall lysis. Barley is malted before its use in brewing for synthesizing enzymes capable of degrading starch as well as allowing the synthesis of the enzymes capable of removing the cell wall components enclosing the starch in the starchy endosperm (Briggs 1998). Although in well-modified malt, most of the cell walls from the starchy endosperm of barley are removed, in the case of extensively modified malt, there are some residual polysaccharides, for instance, β-glucan and arabinoxylan that can either be in the form of undegraded structure or as solubilized but incompletely hydrolyzed material (Bamforth 1994). Therefore, the brewer has few tools to deal with residual unmodified or poorly modified regions of endosperm, or even completely unmodified endosperm, including the use of low temperature mash protocols. However, the use of enzymes is a very interesting alternative technique, considering their efficiency and specificity regarding lysis of cell walls.

A commercial enzyme preparation containing β-glucanases and xylanases was investigated (Scheffler and Bamforth 2005) for its ability to reduce the viscosity of mash and increase extract recovery when applied for the degradation of cells from starch endosperm. A preparation containing mainly xylanase increased wort viscosity when added in lower levels to mash, probably by releasing β-glucan from its entrapment by arabinoxylan within the cell walls of barley. On the other hand, the glucanases showed a greater impact than xylanases on viscosity. Therefore, the combinations of β-glucanase and xylanase had the greatest impact on extract yield and viscosity.

Zhao et al. (2012) studied an extracellular β-1,4-glucanase isolated from the culture filtrate of *Phialophora* sp. G5 and cloned its encoding gene. Native enzyme presented optimal activity in the range of 55°C–60°C and the pH range of 4.5–5.0 as well as superior properties than most fungal β-1,4-glucanases, including stability in acidic and alkaline conditions (pH 2.0–9.0) and strong resistance against pepsin and trypsin, retaining 89% and 94% activity, respectively. The enzyme was capable of decreasing the viscosity of barley bean feed and mash and increasing the filtration rate of mash, which makes this enzyme a good candidate for utilization in brewing industries.

Amylases are another important enzyme applied in brewing. According to Goode et al. (2005), the addition of a commercial α-amylase preparation and pH adjustment during the brewery mashing process resulted in rheological changes. The authors observed clear correlations between the level of amylase present in mashes and the rheological data points representing primary grain/starch swelling and subsequent starch digestibility. Furthermore, the presence of smaller starch granules resulted in an increase of viscosity in secondary starch gelatinization, which was clearly correlated with the level of mash amylase. It could be noticed that some parameters, such as grain components, amylolytic enzymes, and mash pH, collectively influence viscosity profiles during the mashing process.

The nitrogenous compounds used by brewer's yeast are free amino nitrogen (FAN), such as the sum of the individual wort amino acids, ammonium ions, and small peptides (di- and tripeptides). Adequate levels of these compounds result in efficient yeast cell growth and, as a consequence, desirable fermentation performance. Proteases are important enzymes that have been studied to overcome this problem because, although wort contains abundant proteins and longer chain peptides, brewer's yeast is not able to assimilate these compounds, considering that cells hardly secrete proteases during fermentation.

Lei et al. (2013) reported that unavailable nitrogen from two types of high-gravity worts (20°P and 24°P) was utilized by adding three food-grade commercial protease preparations, Neutrase, Flavorzyme, and Protamex, at the beginning of the fermentation process. It was observed that protease supplementation significantly increased the free amino nitrogen level and hence the amount of cell suspension in the later stages of fermentation. More specifically, the best fermentation performance with regard to significantly improved wort fermentability, higher ethanol yield, and flavor volatile formation were obtained using Flavorzyme. The authors also observed that the foam of final beers produced by protease addition was as stable as that of the control at each of the corresponding gravities.

Piddocke et al. (2011) studied the effect of the addition of the multicomponent protease enzyme Flavourzyme and its influence on the metabolism of the brewer's yeast strain Weihenstephan 34/70 using transcriptome and metabolome analysis. The authors observed that on one side there was a significantly higher impact of protease addition for maltose syrup-supplemented fermentations. On the other side, the addition of glucose syrup to increase the gravity in the wort resulted in increased glucose repression, which resulted in the inhibition of amino acid uptake and therefore inhibited the effect of the protease addition.

11.2.3 Wine

Red and white wines are fermented beverages highly appreciated worldwide because they present a very complex composition, leading to highlighted properties. They are the consequence of an important specific group of compounds, such as polyphenols, anthocyanins, oligo- and polysaccharides, and volatile compounds. To obtain the best wine properties, efficient treatments must be carried out in all the wine production steps, including maceration and clarification. Within this context, some enzymes can be used in different steps of winemaking. They can be used to clarify and filter the must and wine, which results in an increase of their physical–chemical stability and a strengthening of the aromatic profile or color of wines produced from certain varieties of grape (Boulton et al. 2002). Relevant characteristics of different wines are presented in Table 11.3.

The maceration step needs enzymatic preparations that contain pectinases, cellulases, and hemicellulases, which result in an improvement in the yield of the must after pressing and a more efficient extraction in addition to an improvement in aging due to the increase of tannins and proantocyanidines (Rogerson et al. 2000). The pectinase preparations used in red winemaking are mainly obtained from *Aspergillus* sp. They are applied to maximize the extraction of free-run juice during maceration and to influence clarification and filtration (Canal-Llaubères 1993). These enzymes act on the pectic substances commonly present in the structural polysaccharides in the middle lamella and primary grape cell wall (Pinelo et al. 2006). On the other hand, glycosidase activity increases and improves the varietal aromas of the wine, breaking the bonds that connect many of the nonodiferous sugar-bonded aromatic compounds of the grapes. This increases the free volatile fraction and therefore influences the sensory improvement (Cabaroglu et al. 2003; Palomo et al. 2005).

TABLE 11.3

Relevant Characteristics of Different Wines

Grape	Source of Material	Flavor Characteristics
Cabernet Sauvignon	California (USA), Bordeaux and Pomerol (France), and Italy's Tuscan coast region	Full, tannic wines with notes of black currant and cassis
Merlot	France and Italy	Medium to full-bodied wines with flavors of black cherry, plum, and tobacco
Monastrell or Mourvèdre	Rhone Valley and Provence (France), Jumila and Yecla (Spain), California (USA), and the South of Australia	Earthy and savory reds
Albariño	Galicia (Spain) and Vinho Verde (Portugal)	Apple, pear, and/or citrus nuances
Malbec	Southwest of France and Argentina	Structured, robust wines, with notes of blackberry, plum, and leather

Source: Robinson, J. *Oxford Companion to Wine* (3rd Edition). London: Oxford University Press, 2006; Lewin, B. *Wine Myths and Reality.* Dover: Vendange Press, 2010.

Phenolics and anthocyanins are extremely important compounds, responsible for red wine color and some sensory characteristics. Many factors, including grape variety and maturity, soil, and climate influence the presence of these compounds in wine (González-Sanjosé et al. 1990). They are mainly presented in the berry skin, which justifies efforts in the development of different strategies to help the rupture of the berry skin cell walls to facilitate the liberation of phenolic compounds. The use of enzymes has a relevant impact on the polyphenol content and composition. According to some studies, an increase in the anthocyanin levels was observed (Kelebek et al. 2007), and some studies reported a decrease in the anthocyanin and phenolic compounds (Borazan et al. 2013).

The effect of using commercial pectinolytic macerating enzyme preparation has been evaluated as a possible alternative to cold prefermentative maceration. It was observed that in the 2008 vintage, when the wines were macerated with enzymes, the richest content of phenolic and anthocyanic compounds was obtained (Ortega-Heras et al. 2012).

Ducasse et al. (2011) studied the effect of macerating enzymes (pectinases) on polysaccharide and polyphenol of Merlot wines. The authors observed that enzyme-treated wines presented more *Rhamnogalacturonan* II and less polysaccharide rich in arabinose and galactose over the vintages. In addition, the wine polyphenol composition was also modified after treatment with the enzymes. Furthermore, Favre et al. (2014) applied different maceration techniques to evaluate the polyphenol composition of Tannat red wines and found that maceration using enzymes increased proanthocyanidins (up to 2250 mg/L).

The polysaccharidic fraction of wines possesses polysaccharides rich in galactose and arabinose, such as type II arabinogalactan-proteins and arabinans, rhamnogalacturonans coming from the pecto-cellulosic cell walls of grape berries (Vidal et al. 2003), and mannoproteins (MPs) released by the yeasts during fermentation (Llaubères et al. 1987). The enzymatic treatment can modify the structure and amounts of polysaccharides released in wines (Ayestaran et al. 2004), which is highly important. They also play a role in the colloidal stability of wines because they aggregate with tannins.

Oligosaccharides have only recently been shown to occur in wine, and more efforts are necessary to study their importance to wines. These molecules find medicinal, food, and agricultural applications, and they play a role in plant defense responses (Darvill and Albersheim 1984; Elleuch et al. 2011; Qiang et al. 2009). Regarding the importance of oligosaccharides in wine quality, these compounds have physicochemical properties, such as chelation of cations (Cescutti and Rizzo 2001). In a study, Carignan and Merlot wines presented approximately 300 mg of oligosaccharides/L. These oligosaccharides are structurally related to plant cell wall polysaccharides. After the analysis of the structures of oligosaccharides, short chains of galacturonic acid with a degree of polymerization between 2 and 6 and a short chain of rhamnogalacturonan-oligomers constituted by the repeats of rhamnose and galacturonic acid, arising from smooth regions or hairy regions of pectin, respectively, but also 4-OMe-oligo-glucuronoxylan oligosaccharides produced from hemicellulose were observed (Ducasse et al. 2010).

In another study, similar structures have been isolated from other wines. Forty-five complex free oligosaccharides in red and white wines, Grignolino and Chardonnay, respectively, were isolated and characterized, reaching concentrations of approximately 100 mg/L. Oligosaccharides were hexose-oligosaccharides, xyloglucans, and arabinogalactans, probably a natural by-product of the degradation of cell wall polysaccharides (Dordiga et al. 2012).

When Merlot wines were treated with macerating enzymes, differences in the total oligosaccharide concentration were observed. In the majority of cases, enzyme-treated wines contained lower amounts of oligosaccharides compared to the control. Therefore, some recent studies reported the application of different enzymes on the release of oligosaccharides (Ducasse et al. 2011).

Apolinar-Valiente et al. (2014) studied the release of oligosaccharides in Monastrell wine according to the geographical origin of the grapes. The use of a commercial enzyme preparation with polygalacturonase, pectinesterase pectin lyase, and β-glucanase activity modified the arabinose/galactose and the rhamnose/galacturonic acid ratios in the wines Cañada Judío and Albatana terroirs. The arabinose/galactose/rhamnose ratio in the wines Cañada Judío, Albatana, and Chaparral-Bullas terroirs were also modified. It was observed that the "terroir" impacts the effect of commercial enzyme treatment on wine oligosaccharide composition.

Aroma compounds are another important group responsible for sensory properties of wines, and they can be very much influenced with the use of enzymes. The flavor profile of a wine is composed of the primary aromas, characteristic of the variety of the grape, secondary aromas that are originated from yeasts during the fermentation process, and tertiary aromas developed during the maturing process of the wine in the bottle or oak barrels (Piñeiro et al. 2006). Among the volatile components, there are C13-norisoprenoids, monoterpenes, aliphatic alcohols, and benzene derivatives. The volatile compounds that are in free forms contribute directly to the aroma, but in much higher concentrations, there are nonvolatile forms (bound to sugars). They accumulate in the grape during the maturation process (Günata et al. 1985). The efficient hydrolysis of these compounds can be achieved with β-glucosidases.

Albariño wines have been studied, and both aromatic extracting and clarifying enzymes on the aroma were evaluated and compared. Different commercial pectinolytic preparations were used—Endozym cultivar, Depectil clarification, and Rapidase Xpress—and it was observed that all the wines, independent of the type of enzyme, presented different aromatic characteristics in comparison to the control wines. However, the wines obtained from the enzyme applied during maceration (Endozym cultivar) showed the highest contents in varietal compounds and benzene derivatives, ethyl esters, or phenylethyl acetate. When maceration and clarification enzymes (Endozym cultivar, Depectil clarification, and Rapidase Xpress) were applied in combination, the release of aromas by the glycosidase enzyme during the pellicular contact was observed. These results indicated that the glycosidase enzymatic treatment was apparently effective for the improvement of the aroma of Albariño wines (Armada et al. 2010).

Sieiro et al. (2014) applied a recombinant endopolygalacturonase of *Kluyveromyces marxianus* for the aroma enhancement of Albariño and observed that there was a significant increase of the total compounds responsible for the aroma, different from the effect when using a commercial pectic enzyme. After characterization, wines made by using the recombinant enzyme presented greater richness and diversity regarding the number of aromatic compounds present, which were citric, balsamic, spicy, and floral aromas.

11.2.4 Distilled Spirits

Fruit spirits are highly consumed in different countries such as the United States and European countries. They are high-value alcoholic beverages that can be commercially obtained by the distillation of fermented fruit mash, juice, or pomace. Despite the fact that the use of enzymes in alcoholic beverages is focused on fermented ones, here we discuss some studies on distilled spirits.

Fundira et al. (2002) studied the clarification of marula fruit (*Sclerocarya berria* sub. caffra) juice using commercial enzymes to improve the yield and observed an increase in yield of up to 12% in juice treated with the enzyme Rapidase as well as a 15-fold improvement in juice clarity and an increase in total terpenes after treatment with prefermentation processing enzymes. Afterward, postfermented marula wine was treated with enzymes as an attempt to hydrolyze bound monoterpenes, resulting in an increase in the free monoterpenes of at least 92%. As a result, both positive and negative effects on the flavor of the juice, wine, and distillate were observed using different enzymes.

As discussed earlier, pectinolytic enzymes are widely applied in the beverage industry, including in the preparation of distilled spirits. They are liquefaction enzymes applied during fermentation of fruit mashes to improve the yield of ethanol and the ability to pump the mash. These enzymes not only hydrolyze pectin, but also hydrolyze the methyl ester in pectin (pectinesterase activity), resulting in methanol formation, and in some cases, it can lead to distilled products containing concentrations of methanol above the legal limit, demanding methanol monitoring if these enzymes are used for commercial products, in the case of some fruits (Andraous et al. 2004). Moreover, methanol, for instance, has been reported as being associated with harmful health effects, including fatigue, headache, nausea, visual impairment, or complete blindness (Merck Index 2001).

Zhang et al. (2011) investigated the influence of pectinase treatment on fruit spirits produced from apple mash, juice, and pomace and observed that after fermentation and distillation of apple mash, juice, and pomace methanol, ethanol, *n*-propanol, isobutanol, and isoamyl alcohol were identified as the major alcohols in all the apple spirits. In addition, there were significant differences between the methanol concentrations of pectinase-treated and nonpectinase-treated apple spirits. In this case, apple pomace yielded significantly higher methanol concentrations than apple mash and juice.

Cellulases are also applied in the production of distilled spirits. In a study, the influence of cellulases on the flavor substances, such as ethanol, aldehydes, organic acids, and esters in the Chinese product Daqu liquor, was evaluated. It was observed that the addition of cellulases to fermenting grains resulted in a very low content of aldehydes and fusel oil in liquor. Yet the content of four main ethyl esters and ethyl propionate as well as relative acids were higher. On evaluation of the sensory properties it was observed that the addition of cellulases highlighted the Daqu liquor style features (Li et al. 2011).

11.3 Nonalcoholic Beverages

11.3.1 General Aspects

Fruit and vegetable juices are rich in minerals, vitamins, and other components that are beneficial to human health, so they are considered to be of high nutritional value (Maktoufa et al. 2014). Traditional juices, such as orange, grape, tomato, and fruit blends, are well established in developed countries. Nowadays, minor juices, especially tropical juices and juice products, have attracted consumers' attention, which has called for new strategies that facilitate their production and commercialization (Abdullah et al. 2007).

The fruit pulp consists of cells, and polysaccharides constitute the main structural polymers of the cell wall (90%–100%). The cell wall composition depends on the fruit species and evolves as a function of the agronomic and climatic conditions, the fruit maturity, and the fruit storage mode and duration (Aehle 2004; Mac Cann and Roberts 1994).

Due to the presence of some colloids, consisting mainly of polysaccharides (pectin, cellulose, hemicellulose, lignin, and starch), protein, tannin, metals, and others, some juices acquire a cloudy appearance after extraction, which consumers often resent (Maktoufa et al. 2014; Vaillant et al. 2001). In fact, cloudiness stemming primarily from the presence of pectin is one of the major problems encountered during the preparation of fruit juices. Pectin comprises a long chain of α-D-galacturonic acid units linked by alpha 1-4 bonds. In solution, this polymer has a viscous appearance as well as high gelling power (Masmoudi et al. 2008). Pectins can be associated with plant polymers and the cell debris, generating a fiber-like molecular structure. To minimize juice turbidity while maintaining its organoleptic and sensory characteristics, it is essential to conduct a clarification step (Maktoufa et al. 2014). However, cloudiness might be difficult to remove, and enzymatic depectinization may be necessary (Abdullah et al. 2007).

Pectin is an essential structural fruit component. In combination with hemicellulose, it binds single cells to form the fruit tissues. The pectin chains are almost exclusively composed of D-galacturonic acid units partially esterified with methanol. In immature fruit, pectin is mostly insoluble. As the fruit ripens, some of the pectic substances gradually break down in the cell walls of the skin and flesh to yield materials constituted by polysaccharides. The general term "pectic substances" covers not only pectins but just about everything resulting from the degradation processes that involve pectin and which occur upon fruit ripening, including soluble forms (Taylor 2005).

As fruit matures, it becomes softer, less acidic, and sweeter, therefore, the juice processor must take all these changes into account. Apples, in particular, are best processed prior to their fully ripe state because solubilized pectin and softened fruit tissues will seriously affect the juice separation efficiency and afford low yields. Other fruits, such as berry fruits, need to be fully ripened if optimal flavor is to be achieved. Hence, if the resulting juice is to be clarified, enzymatic treatment will be required at some stage, to break pectin down and enable precipitation or sedimentation of the resulting pectic substances. If the final product is to be a juice concentrate, it is particularly important that no pectin be left available to "jam" the operation (Taylor 2005).

The pectin content specific to the fruit species impacts juice production. The right pectinase balance, which is necessary during the treatment, will depend on the fruit. For example, in apples, orange peel, grapes, pineapples, and strawberries the pectin content (wt %) varies from 0.5 to 1.6, from 3.5 to 5.5, from 0.1 to 0.4, from 0.04 to 0.1, and from 0.5 to 0.7, respectively. Depending on their chemical form, pectins are categorized as either soluble or insoluble fibers, which cannot be absorbed by the human digestive tract (Mieszczakowska-Frąc et al. 2012). Nevertheless, enzymes can modify these fibers into short polysaccharide fragments that humans can absorb. Enzymatic pectin degradation diminishes raw juice viscosity, thereby increasing the juice yield (Voragen et al. 1992) and improving production efficiency. Most commercial enzyme preparations contain pectinases, cellulases, and hemicellulases at various ratios, and they act in different ways. During production of a juice that tends to be cloudy, it is essential that the enzyme break pectin into short fragments to obtain a cloud-stable juice (Beldman et al. 1997).

The first application of enzymes in the fruit juice industry consisted of using pectinases for apple juice clarification, which took place in the 1930s. This procedure reduced juice viscosity after pectin breakdown, improving the overall quality of the industrial apple juice. Later, treatment of red berry juice with pectinases furnished a better quality product after depectinization (Aehle 2004). Due to their wide spectrum of industrial applications, pectinases are recognized as one of the most important classes of enzymes in juice and nonalcoholic beverage technology (Uenojo and Pastore 2007). Nowadays, most of the enzymes applied in fruit maceration are obtained from *Aspergillus japonicus*, *Aspergillus niger*, *Mucor flavus*, *Aspergillus awamori*, *Penicillium italicum*, *Bacillus macerans*, *Erwinia carotovora*, *Saccharomyces pastorianus*, *Saccharomyces cerevisiae*, and *Kluyveromyces marxianus* (Rodrigues 2012).

The main goal of using enzymes, such as pectinases and amylases, to degrade fruit pectin and starch during the clarification stage is to prevent postbottling haze formation while still ensuring juice concentration, smaller storage volume, cheaper transportation, concentrate stability, and higher juice yield (Grassin and Fauquembergue 1996a).

The various developments in the area of nonalcoholic beverages have prompted a very qualified industry. Indeed, enzyme suppliers provide fruit juice producers with tailor-made enzyme preparations blended on the basis of fruit composition. This guarantees optimal quality and stability of the products, shorter process time, and larger plant capacity. In this sense, enzyme technology devoted to the nonalcoholic beverage industry has added value to raw materials and reduced the amount of waste (Aehle 2004).

11.3.2 Fruit Juice Industry

11.3.2.1 Application of Important Enzymes

11.3.2.1.1 Pectinases

Many microorganisms produce enzymes that degrade fruit cell walls. In general, commercial pectinases applied in the fruit juice industry come from selected strains of *Aspergillus* sp., *Penicillium* sp., and others. On the basis of their action toward pectins, pectinases are defined and classified as (i) pectin lyase (EC. 4.2.2.10), a pectin depolymerase of the endotype that has great affinity for long, highly methylated

TABLE 11.4

Important Enzymes for the Fruit Juice Industry and Their Functions

Enzyme	EC.	Function	Application
Maceration enzymes: Pectinases, cellulases, and hemicellulases	–	They hydrolyze pectin and soluble components of the cell wall, reduce viscosity, and maintain fruit juice texture.	Improved extraction of fruit juices and release of aromas, polysaccharides, and pigments
Pectin lyase	4.2.2.10	It cleaves pectin into oligosaccharides without the action of esterase.	Fruit juice clarification
Pectin methyl esterase	3.1.1.11	It removes methyl esters and releases methanol, enabling polygalacturonase to digest the pectin.	Fruit pulp production
Polygalacturonase	3.2.1.15	It randomly digests pectins and hydrolyzes polygalacturonan in the cell walls.	Fruit juice clarification and high-viscosity purées
Hemicellulases	3.2.1.8	They hydrolyze cell wall polysaccharides.	Production of nonclarified vegetable juices
Xylanases	3.2.1.8	They hydrolyze xylan and arabinoxylan.	Fruit juice clarification
Cellulases	3.2.1.4	They hydrolyze cell wall polysaccharides.	Juice extraction

Source: Uenojo, M., and G. M. Pastore, *Química Nova*, 30, 2, 388–394, 2007; Rodrigues, S. Enzyme maceration. In *Advances in Fruit Processing Technologies*, 235–246. London: CRC Press, 2012.

chains and which acts by β-elimination of methylated α-1,4 homogalacturonan, to form C4–C5 unsaturated oligo-uronides; (ii) pectin methylesterase (EC. 3.1.1.11), which removes methoxyl groups from pectin to give methanol and less highly methylated pectin; and (iii) polygalacturonase (EC. 3.2.1.15), which exists in endo- and exo-forms, both of which act only on pectin with a degree of esterification of less than 50%–60%. Endo-polygalacturonase generally reduces viscosity more markedly as compared with exo-polygalacturonase (Aehle 2004). Table 11.4 lists the main enzymes applied in the juice industry and their functions.

Numerous studies have reported that depectinization using enzymes such as pectinases could effectively clarify fruit juices (Alvarez et al. 1998; Ceci and Lozano 1998; Chamchong and Noomhorm 1991; Grassin and Fauquembergue 1996a,b; Vaillant et al. 2001; Yusof and Ibrahim 1994). Pectinases hydrolyze pectin and cause pectin–protein complexes to flocculate. The juice resulting from this pectinase treatment will have less pectin and lower viscosity, which should facilitate the subsequent filtration processes. With the aid of enzymatic pretreatment using pectinases, it is possible to degrade pectin and increase the permeate flux during fruit juice microfiltration, ultrafiltration, and reverse osmosis (Abdullah et al. 2007).

Concerning soft fruits, the usual way to apply enzymes is to dose them at the recommended level (e.g., 0.1%) into the prewarmed mash, followed by vigorous mixing and standing for a recommended period at a constant temperature. For example, a typical pectinase will require a standing period of 1.5 h in the case of black currants and other soft fruits. To determine the exact time, it is necessary to take a series of mash samples during the enzymation stage, press or filter off some of the juice, and treat it with excess alcohol (e.g., 40 ml of single strength juices and 60 ml of alcohol in a 100-ml measuring cylinder). If present, dissolved pectin will be thrown out of the solution as an insoluble gel; if the pectin is fully degraded, it will form a flocculent precipitate of pectic substances that will settle at the bottom of the measuring cylinder (Taylor 2005).

11.3.2.1.2 Amyloglucosidases

Additionally, the fruit juice industry employs amylases and amyloglucosidases to process fruits that contain starch, such as unripe apples at harvest time. Several important amylases exist in fungal strains, such as *Aspergillus niger* (Fogarty and Kelly 1980). In general, it is possible to apply these enzymes after starch gelatinization above 75°C, to prevent postbottling haze formation due to starch retrogradation. In some cases, hemicellulases also find application in the juice industry because they can hydrolyze galactans, arabinogalactans, xylans, and xyloglucans (Aehle 2004).

Pome fruits sometimes require the activity of other enzymes. Fruit picked before maturity and ripened under controlled atmospheric conditions in a cool storage might retain starch from the unripe fruit. This starch can form a gel during juice processing, culminating in precipitation and haze effects in the final product. Amylases help overcome this issue because they can cleave any residual starch (Taylor 2005).

11.3.2.1.3 Cellulases

Cellulases can accelerate color removal during fruit processing. In recent years, these enzymes have found successful application in the "total" liquefaction of plant tissues during processing, obviating the need for a press while increasing yields.

11.3.2.1.4 α-L-rhamnosidase

The use of conventional chemical methods, such as acid hydrolysis or alkaline cleavage to prepare minor flavonoids inevitably leads to side reactions. In this context, the bioconversion of many compounds by a specific type of glycosyl hydrolase is advantageous due to the selectivity and mildness of the reaction conditions (Da Silva et al. 2013). The enzyme α-L-rhamnosidase (EC. 3.2.1.40) specifically cleaves terminal α-L-rhamnose from a large number of natural products including hesperidin, naringin, rutin, quercitrin, terpenyl glycosides, and many other natural glycosides containing terminal α-L-rhamnose (Monti et al. 2004; Yadav et al. 2010).

Only two commercial preparations of α-L-rhamnosidases, naringinase, and hesperidinase are available, and both originate from fungal sources. Hesperidinase exists in *Aspergillus niger* and *Penicillium* species; naringinase is obtained from *Penicillium decumbens*. All these preparations display significant β-D-glucosidase activity, catalyzing the hydrolysis of terminal nonreducing residues into β-D-glucosides and releasing glucose (Yadav et al. 2010).

α-L-rhamnosidase has become a biotechnologically important enzyme due to its use in a variety of processes, such as debittering of citrus fruit juices (Busto et al. 2007), manufacture of prunin from naringin (Roitner et al. 1984), enhancement of wine aromas by the enzymatic hydrolysis of terpenyl glycosides containing L-rhamnose (Spagna et al. 2000), and de-rhamnosylation of many L-rhamnose-containing steroids, such as diosgenin, desglucoruscin, and ginsenosides-Rg2, which yield de-rhamnosylated products, are of clinical importance (Monti et al. 2004).

11.3.2.1.5 Tannases

Tannin acyl hydrolases, commonly referred to as tannases (EC. 3.1.1.20), are inducible enzymes produced by fungi, yeast, and bacteria (Ferreira et al. 2013). The main feature of tannases has mostly been their activity on complex polyphenolics and their ability to hydrolyze the ester bond (galloyl ester of an alcohol moiety) and the depside bond (galloyl ester of gallic acid) of substrates, such as tannic acid, epicatechin gallate, epigallocatechin gallate, and chlorogenic acid (García-Conesa et al. 2001). Battestin and coworkers (2008) studied the tannase from *Paecilomyces variotti*, which mediated the hydrolysis of the green tea polyphenol epigalocatechin gallate to generate epicatechin and gallic acid. As a consequence, the antioxidant activity of the tea extract increased after the enzymatic reaction. Similarly, another study employed tannase obtained from *Paecilomyces variotti* to biotransform orange juice polyphenols and modify their biological activity. Tannase extract was able to modify the polyphenolic composition of the orange juice and act on naringin and hesperidin to remove glycosides (Ferreira et al. 2013).

11.3.2.2 Juice Production: From Research to Industry

11.3.2.2.1 Orange Juice

Orange juice is the most consumed fruit juice worldwide (Taylor 2005). The development of frozen concentrated orange juice started in 1940. In certain countries, the legislation does not permit the use of enzymes during premium juice production. Nevertheless, enzymes find many other applications therein: They raise the yield of solids recovery during pulp washing, thereby facilitating concentrated juice fabrication and improving essential oil recovery from the peel, debittering, and juice clarification (Grassin and Fauquembergue 1996b). Additionally, it is possible to treat the whole orange with a 2% pectinase solution, to digest the albedo, followed by fruit peeling, rinsing, cooling, and packaging (Aehle 2004).

11.3.2.2.2 Apple Juice

The second most commercialized juice worldwide is apple juice, both in clear and concentrated forms. About 54×10^6 tons of this raw material is gathered per year, and 10% is processed to obtain mostly clear concentrate. The United States, Poland, Argentina, Italy, Chile, Germany, and China the major producers. Meanwhile, 2% of the raw material is processed as cloudy juice, cider, and applesauce (Grassin and Fauquembergue 1996b). In this sense, many advances have been made in the field of apple juice manufacture, and the use of enzymes in this industry is crucial.

Figure 11.3 shows the utilization of enzymes during the production stages of generic fruit juice (adapted from Toepfl et al. 2014). During juice fabrication, enzymes can participate in two different stages: (i) they can be mixed with the fruits during initial press for juice release, and (ii) they can be used to catalyze juice depectinization. Aehle (2004) presented a complete flow chart for apple juice production. During juice storage, endogenous apple pectinases slowly transform insoluble protopectin into soluble pectin. At the same time, endogenous apple amylases slowly degrade starch into glucose and consume it during postharvest metabolism. This raises the soluble pectin content, which could reach values from 0.5 to 5 g/kg in an overripe apple. Considering that apple pectin is highly methylated, commercial enzymes must contain a high concentration of pectinlyase or pectin methylestearase in association with polygalacturonase and other enzymes. The use of commercial pectinases from *Aspergillus* sp. in apple mash is also necessary because the activities of the endogenous enzymes are too low to promote the desired effect (Aehle 2004).

In fact, the main application of enzymes is to reduce the turbidity due to the presence of pectin. One liter of apple juice with 13% dry matter can contain approximately 2–5 g of pectin, depending on the

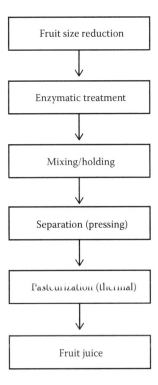

FIGURE 11.3 Synthetic scheme of the production stages of a generic fruit juice using enzymes during the process. (Adapted from Toepfl, S., Siemer, C., Saldãna-Navarro, G., and V. Heinz. In *Emerging Technologies for Food Processing*, 93–108. Oxford: Academic Press, 2014.)

fruit maturity. Arabanase activity prevents haze formation after juice concentration, and juice clarification should occur after enzymatic depectinization. Pectinases partially hydrolyze the pectin gel, which
results in electrostatic aggregation of oppositely charged particles, flocculation of the cloud, and complete juice clarification (Taylor 2005).

11.3.2.2.3 Other Fruit Juices

In the same way, the production of several clear and concentrated juices, such as those from black currant, raspberry, and strawberry, demands enzymatic maceration and depectinization. In general, these
juices have high pectin content, approximately 7 g L^{-1} as compared with 0.5 g L^{-1} in apple juice. Pectin
hairy regions remain as a soluble colloid in the juice; hemicelluloses tend to bind to phenolics and proteins during processing and storage, forming irreversibly linked brown complexes that enzymes can no
longer break down. This makes juice clarification, filtration, and concentration difficult (Aehle 2004).

Meanwhile, red berry juice can be produced in two ways: First, a single-step process involves simultaneous maceration and depectinization. Here, pectinases help to improve juice and color extraction while
retaining the organoleptic properties of the fruit. Prior to maceration with enzymes, the pulp sometimes
requires heating to 90°C to inhibit fruit oxidases. The other is a two-stage process that consists of fruit
pulp enzymatic maceration followed by a second addition of enzymes for juice depectinization at low
temperatures to prevent aroma loss and promote fabrication of high-quality juices (Aehle 2004).

Recent developments have allowed for commercialization of juices based on tropical fruits, such as
star fruit (or carambola), which includes enzymatic treatment in its processing (Abdullah et al. 2007).
Additionally, an array of tropical fruits find application in the manufacture of juices of commercial interest, namely apricot, peach, kiwi, mango, guava, papaya, and banana, which dismiss the need for enzymes
during processing (Aehle 2004). These fruits are mainly processed to purée and stored before further
processing to cloudy or clear juice, and the problem related to viscosity can be lessened with the use of
pectinases and amylases during clear juice production (Grassin and Fauquembergue 1996b).

Pectin (0.1%) is among the constituents involved in lemon juice opalescence. The cloudy lemon juice that
emerges after juice extraction contains particles consisting of pectin and proteins that remain in suspension
bound to citral and lemon flavor. In the past, clarification of this juice took place upon addition of large
amounts of bentonite or sulfur dioxide; nowadays, this process involves acidic pectinases (Aehle 2004).

Clarified juices (from which enzymatic treatment has actively removed pectin as an integral part of
the process) can be filtered bright and then either pasteurized or rendered sterile by means of membrane
filters. These filters can directly eliminate yeasts and mold. Citrus juice pasteurization is conducted at
temperatures above 95°C to eliminate the undesirable effects of pectolytic enzymes, particularly pectin esterase. These enzymes act by demethoxylating pectin, which generates cross-linking between the
resulting polygalacturonic acid molecules, gelation effects, and loss of cloud stability (Taylor 2005).

In addition to the most well-known processes and practical applications that have been successfully
developed in the industry, in recent years several strategies have arisen, aiming to improve the available
products and to allow the marketing of new fruit juices by using enzymes.

In this perspective, a number of studies on the optimization of enzymatic pretreatment to clarify fruit
juices have been reported, especially for tropical fruit juices (Lee et al. 2006; Rai et al. 2004; Sin et al.
2006). A number of factors, such as incubation time, incubation temperature, and enzyme concentration,
affect the enzymatic treatment that is necessary to hydrolyze pectic substances (Baumann 1981; Lee et
al. 2006; Rai et al. 2004; Sin et al. 2006).

Lately, the use of membrane processes such as microfiltration (MF) and ultrafiltration (UF) during
juice clarification has gained prominence over some conventional treatments, such as diatomaceous
earth, paper filters, and bentonite (Jiao et al. 2004; Todisco et al. 1998; Wu et al. 1990). However, the
major disadvantage of this process is membrane fouling during permeation. This results from retention of some components over the membrane surface, which rapidly reduces flux (Cassano et al. 2007;
Espamer et al. 2006). Enzymatic treatment of the juice with hemicellulases, phenol oxidase, and pectinases, in particular, can help to overcome this drawback—the enzymes degrade the colloidal particles
before the ultrafiltration step, thereby reducing juice viscosity via insoluble pectin depolymerization
(Kilara 1982; Kilara and Van Buren 1989). Commercial preparations of pectinases, which include such
enzymes as pectin lyases, polygalacturonase, and pectin esterase, are often used (Abdullah et al. 2007).

Many studies have described fruit juice depectinization by pectinases as an efficient strategy to abate turbidity (Kashyap et al. 2001; Landbo et al. 2007). Success will depend on various physicochemical factors, such as incubation time, enzyme concentration, and incubation temperature (Abdullah et al. 2007).

Recent advances in enzyme biochemistry, materials science, and computational technology have contributed to investigations into the chemical and biological modification of enzymes, to obtain enhanced stability and catalytic performance against adverse conditions that deactivate the enzyme (Gassara-Chatti et al. 2013).

Another promising approach is to incorporate enzymes into polymeric nanostructures (Kim et al. 2008). This is a flexible method to design polymeric structures; it enables the tailoring of enzymes for countless applications, such as nonaqueous catalysis, bionano devices, intelligent molecular machines, and artificial cells (Börner 2009).

It is possible to prepare a hydrogel using different concentrations of gelatin, pectin, and CMC to entrap ligninolytic enzymes (MnP, laccase, and LiP) into matrices. This process takes place under mild conditions and does not damage the native structure of the enzyme. These hydrogels find application in juice clarification, but it is crucial to select the best formulation. Results have shown that enzyme entrapment significantly ($p < 0.05$) improved the thermal stability of enzymes at 4°C and 75°C. An increase in particle size and viscosity of the formulation rendered the enzyme more stable. Polyphenol reduction and enhanced clarity in a mixture of berry and pomegranate juices was more significant ($p > 0.05$) upon treatment with encapsulated enzymes as compared with free enzymes. Hence, enzymatic treatment constituted an efficient alternative for juice clarification (Gassara-Chatti et al. 2013).

Some authors have tested immobilization of the enzyme papain and found that it displayed enhanced activity, better tolerance to variations in the medium pH and temperature, improved storage stability, and good reusability as compared with the free enzyme. Both the free and immobilized enzymes effectively clarified pomegranate juice (Mosafa et al. 2013).

Enzymatic reactions have been carried out using commercial rhamnosidases (hesperidinase and naringinase) and β-D-glucosidase using an experimental design as well as statistical tools to analyze the data for each enzyme and to assess the independent variables (pH and temperature) that significantly affected enzymatic activity (dependent variable). DPPH and FRAP assays helped to evaluate the antioxidant activity before and after bioconversion. Aliquots of control and enzyme-treated samples were taken at different times and analyzed by UPLC–MS. The antioxidant activity of both treated juices was higher than that of the untreated juices, confirming that the bioconversion reaction conditions efficiently incremented the antioxidant activity. Conduction of the hesperidinase reaction for 4 h converted 60% of hesperidin into hesperetin in orange juice. The enzymatic treatment also enhanced the antioxidant capacities of the glycosylated juice. These results represent a step forward when it comes to using α-L-rhamnosidases to produce functional beverages by way of de-glycosylation of their flavonoid glycosides (Da Silva et al. 2013).

The evolution of enzyme technology has allowed processors to obtain higher juice yield (productivity) together with better quality of the finished products. The use of specific enzymes adapted to fruit processing improves color stability and turbidity, increasing the shelf life of juices and concentrates. Additionally, the diversity and the high specificity of commercial enzymes pave the way for the development of new processes and new types of fruit-derived products.

Response surface methodology (RSM) has aided simultaneous analysis of the effects of enzymatic treatment conditions, such as the incubation time, incubation temperature, and enzyme concentration, on the physical characteristics of the juice, including turbidity, clarity, viscosity, and color (Abdullah et al. 2007). In this case, the study employed a two factor central composite design to establish the optimum conditions for the enzymatic treatment to clarify carambola (star fruit) juice. The enzyme concentration was the most important factor affecting the characteristics of the star fruit juice. The recommended enzymatic treatment condition was enzyme at 0.10% concentration at 30°C for 20 min (Abdullah et al. 2007).

11.3.3 Other Products of Nonalcoholic Beverage Industries

The book *Enzymes in Food and Beverage Processing* published earlier (Ory and Allen 1977) dealt with the application and use of enzymes during the manufacture of nonalcoholic products, such as black tea and instant tea, coffee, and milk. Indeed the main nonalcoholic beverages include juices of fruits and vegetables, milk, carbonated and uncarbonated soft drinks, coffee, and tea (Shyam Kumar and Chandrasekaran 2012 ; Chandrasekaran 2012). In fact, the use of enzymes in these nonalcoholic products has remained an active research subject in recent years. Although a majority of studies on applications of enzymes in the nonalcoholic beverage industry were focused on fruit juices, a few other beverages deserve attention in regard to application of enzymes during some steps of their production process.

11.3.3.1 Milk

In this context, the study of enzyme application in milk is a key specialization within both the fields of biochemistry and dairy science (Kelly and Fox 2006). The utilization of chymosin in the dairy industry during the coagulation of milk is well understood, and the involvement of enzymes is an important factor for the quality and safety of milk and milk derivatives. Despite this potential, the volume of enzyme used in the dairy sector is low (Law and Goodenough 1995). Several enzymes are used in the milk processing industry that could allow the development of nonalcoholic beverages with different properties from this raw material (Law 2010). They include acid proteinases for milk coagulation; proteinases and peptidases for production of hypoallergenic milk-based foods; lipases for structurally modified milk fat products and flavor enhancement; beta galactosidase for lactose-reduced whey products, and lactoperoxidase for cold sterilization of milk.

The latest research aimed at using enzymes to obtain products that are more targeted to health benefits and improvement in their functional properties. Dietary proteins, in particular milk proteins, contain bioactive peptides that may modulate different bodily functions, such as digestive, cardiovascular, immune, and nervous systems, and therefore contribute to the maintenance of consumer health (Haque et al. 2009). These compounds have already been reported in fermented milk products, such as yogurt, sour milk, or kefir, and the main strategy to obtain such compounds are related to the proteolytic system of lactic acid bacteria (LAB), food-grade enzymes, or a combination of both to release the functional peptides from the milk proteins directly in the fermented milk products (Hafeez et al. 2014).

Many gastrointestinal enzymes, such as pepsin, trypsin, and chymotrypsin, and microbial enzymes, such as thermolysin and alcalase, have been successfully employed to release biologically active peptides from the milk proteins (Choi et al. 2012; Hafeez et al. 2014), greatly increasing the nutritional quality of the original products. In the same perspective, selected microorganisms were recently evaluated for their proteolytic activity and capability to produce fermented milk enriched with ACE-inhibitory (ACEI) peptides (Chaves-López et al. 2014).

11.3.3.2 Tea

After water, tea (*Camellia sinensis*) is the second most consumed beverage in the world (Borrelli et al. 2004), and black tea is the major part of the tea consumed (Lea 1995), followed by green tea. An earlier study showed the production of a traditional nonalcoholic beverage from the leaves of *Camellia sinensis*. In this process, the leaves are picked and cut or bruised to promote enzymatic degradation. The mixture was then heated to interrupt this reaction (to approximately 70°C–95°C) once the desired point was reached and undergoes a fermentation step (Sanderson 1972).

Another important approach is related to the use of enzymes for accurate assessment and validation of levels of bioactive tea catechins in new products. Extraction and analysis of physiologically significant tea catechins from complex food matrices are complicated by strong association of tea catechins with macronutrients, such as proteins. Thus, Ferruzzi and Green (2006) investigated the recovery of tea catechins from dairy matrices and evaluate pepsin treatment as an enzymatic step to enhance catechin recovery from milk and other protein-rich formulations. Authors observed the recovery of total catechins was highest for pepsin-treated samples (89%–102%), followed by methanol deproteination (78%–87%), and acid precipitation (20%–74%) with values decreasing with increased milk content.

The use of tannase and various cell wall-digesting enzymes has been attempted at various stages of green tea processing, improving the extraction efficiency of polyphenols and enhancing the antioxidant capacity of the extract. The various enzymatic methods proposed for processing green tea were recently presented by Murugesh and Subramanian (2014).

11.3.3.3 Ready-to-Drink (RTD) Tea

Another important application of enzymes is during the decreaming step for the production of ready-to-drink (RTD) tea. During production, the development of haze and formation of tea cream influences the production and quality of the product. Tannase is the most commonly employed enzyme (tannin acyl hydrolase) aimed at improving cold-water extractability/solubility and decreasing tea cream formation as well as improving clarity. Additionally, enzyme treatments have been attempted at three stages of black tea processing, namely, enzymatic treatment to green tea and conversion to black tea, enzymatic treatment to black tea followed by extraction, and enzymatic clarification of the extract (Chandini et al. 2013).

11.3.3.4 Coffee

Aiming to improve flavor and reduce bitterness in the resulting roast and ground coffee product, the authors presented a process for treating green or partially roasted coffee beans with enzymes. The patent deposited by Small and Asquith (1990) showed that the green or partially roasted beans were treated with a solution containing cell wall-digesting, cell storage component-digesting, or phenol oxidase enzymes under a pressure of at least approximately 250 psi, and then they were dried and roasted. Murthy and Naidu (2011) reported the treatment of pectinase produced by coffee pulp and application of the same on demucilage of coffee pulp, indicating waste recycling with value addition, which is also economical for the coffee industry. The authors proved that the isolated pectinase, when applied in Robusta coffee, enabled a reduction of the time in demucilization, reducing the pH and sugars released.

11.3.3.5 Functional Foods

Enzymes may also be used for the development of new products to meet consumer demand for functional foods. In this perspective, the company SternEnzym® created a functional nonalcoholic drink based on cereals, resulting in a product with antioxidants, vitamins, and other health-promoting substances. A mixture of commercial enzymes (Optizym BA, Optizym HC, Optizym BP, Optizym GA, Optizym A, Betamalt) are applied in this product aiming to confer sweetening from starch breakdown, viscosity from pentosan breakdown, mouthfeel, and digestibility, from glucan and protein breakdown, respectively (http://www.stern-enzym.de, accessed on October 15, 2014). The process proposed by this enzyme company is outlined in Figure 11.4.

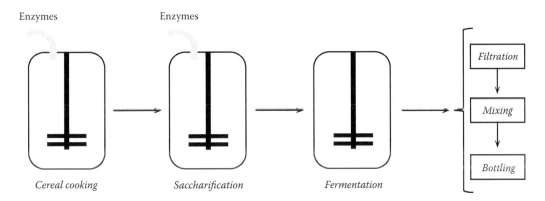

FIGURE 11.4 Simplified process of SternEnzym for the production of a functional nonalcoholic drink based on cereals. (All rights for the product and process are reserved to SternEnzym.)

11.4 Concluding Remarks

The beverage industry is an important segment of the food industry, and in recent years, this industry has sought the evolution of processes as well as developing new products to meet consumer demand. The application of exogenous enzymes finds a very relevant field in some steps of alcoholic beverage production, in which there is the necessity for extraction of important compounds that may present remarkable properties, including functional properties and sensory properties related to the concentration of these compounds. Additionally, consumer demand has encouraged new technologies for the production and commercialization of nonalcoholic beverages, as in the case of fruit juices, which in consequence have developed based on the potential application of enzymes in various processes. In fact, the use of specific enzymes adapted to fruit processing improves color stability and turbidity, increasing the shelf life of juices and concentrates. As discussed herein, the evolution of enzyme technology has allowed processes with higher beverage yield (productivity) together with better quality of the finished products. Additionally, the diversity and the high specificity of commercial enzymes pave the way for the development of new processes and new types of both alcoholic and nonalcoholic products, meeting the requirements of these industrial segments that are experiencing strong growth in recent years.

REFERENCES

Aehle, W. *Enzymes in Industry: Production and Applications*. Weinheim, Germany: Wiley-VCH Verlag GmbH & Co. KGaA, 2004.

Álvarez, S., Alvarez, R., Riera, F. A., and Coca, P. J. "Influence of depectinization on apple juice ultrafiltration." *Colloids and Surfaces A: Physicochemical and Engineering Aspects* 138 no. 2–3 (1998): 377–382.

Andraous, J. I., Claus, M. J., Lindemann, D. J., and Berglund, K. A. "Effect of liquefaction enzymes on methanol concentration of distilled fruit spirits." *American Journal of Enology and Viticulture* 55 no. 2 (2004): 199–201.

Apolinar-Valiente, R., Williams, P., Mazerolles, G. et al. "Effect of enzyme additions on the oligosaccharide composition of Monastrell red wines from four different wine-growing origins in Spain." *Food Chemistry* 156 (2014): 151–159.

Armada, L., Fernández, E., and Falqué, E. "Influence of several enzymatic treatments on aromatic composition of white wines." *LWT—Food Science and Technology* 43 no. 10 (2010): 1517–1525.

Ayestaran, B., Guadalupe, Z., and León, D. "Quantification of major grape polysaccharides (*Tempranillo* v.) released by maceration enzymes during the fermentation process." *Analytica Chimica Acta* 513 no. 1 (2004): 29–39.

Bamforth, C. W. "β-Glucan and β-glucanases in malting and brewing: Practical aspects." *Brewers Digest* 69 (1994): 12–21.

Bamforth, C. W. "Current perspectives on the role of enzymes in brewing." *Journal of Cereal Science* 50 no. 3 (2009): 353–357.

Battestin, V., Macedo, G. A., and de Freitas, V. "Hidrolysis of epigallocatechin gallate using a tannase from *Paecilomyces variotii*." *Food Chemistry* 108 no. 1 (2008): 228–233.

Baumann, J. W. "Application of enzymes in fruit juice technology." In *Enzymes and Food Processing*, 129–147. London: Applied Science Publication, 1981.

Beldman, G., Muter, M., van den Broek, L. A. M. et al. "Classical and novel pectin degrading enzymes: Mode of action and application aspects." *European Seminar Enzymes for Food* (1997): 2–16.

Borazan, A. A., and Bozan, B. "The influence of pectolytic enzyme addition and prefermentative mash heating during the winemaking process on the phenolic composition of Okuzgozu red wine." *Food Chemistry* 138 no. 1 (2013): 389–395.

Bordiga, M., Travaglia, F., Meyrand, M. et al. "Identification and characterization of complex bioactive oligosaccharides in white and red wine by a combination of mass spectrometry and gas chromatography." *Journal of Agricultural and Food Chemistry* 60 no. 14 (2012): 3700–3707.

Börner, H. G. "Strategies exploiting functions and self-assembly properties of bioconjugates for polymer and materials sciences." *Progress in Polymer Science* 34 no. 9 (2009): 811–851.

Borrelli, F., Capasso, R., Russo, A., and Ernst, E. "Systematic review: Green tea and gastrointestinal cancer risk." *Alimentary Pharmacology and Therapeutics* 19 no. 5 (2004): 497–510.

Boulton, R. B., Singleton, V. L., Bisson, L. F., and Kunkee, R. E. *Teoría y práctica de la elaboración del vino.* Zaragoza: Editorial Acribia, 2002.

Briggs, D. E. *Malts and Malting.* London: Blackie Academic & Professional, 1998.

Busto, M. D., Meza, V., Ortega, N., and Perez-Matcos, M. "Immobilization of naringinase from *Aspergillus niger* CECT 2088 in poly(vinyl alcohol) cryogels for the debittering of juices." *Food Chemistry* 104 no. 3 (2007): 1177–1182.

Cabaroglu, T., Selli, S., Canbas, A., Lepoutre, J. P., and Günata, Z. "Wine flavor enhancement through the use of exogenous fungal glycosidases." *Enzyme and Microbial Technology* 33 no. 5 (2003): 581–587.

Canal-Llaubères, R. M. "Enzymes in winemaking." In *Wine Microbiology and Biotechnology*, 477–506. Washington, DC: Hardwood Academic Publishers, 1993.

Cassano, A., Donato, L., and Drioli, E. "Ultrafiltration of kiwifruit juice: Operating parameters, juice quality and membrane fouling." *Journal of Food Engineering* 79 no. 2 (2007): 613–621.

Ceci, L., and Lozano, J. "Determination of enzymatic activities of commercial pectinases for the clarification of apple juice." *Food Chemistry* 61 no. 1–2 (1998): 237–241.

Cescutti, P., and Rizzo, R. "Divalent cation interactions with oligo-galacturonides." *Journal of Agricultural and Food Chemistry* 49 no. 7 (2001): 3262–3267.

Chaari, F., Belghith-Fendri, L., Blibech, M. et al. "Biochemical characterization of a lichenase from *Penicillium occitanis* Pol6 and its potential application in the brewing industry." *Process Biochemistry* 49 no. 6 (2014): 1040–1046.

Chamchong, M., and Noomhorm, A. "Effect of pH and enzymatic treatment on microfiltration and ultrafiltration of tangerine juice." *Journal of Food Process Engineering* 14 no. 1 (1991): 21–34.

Chandini, S. K., Subramanian, R., and Rao, L. J. "Application of enzymes in the production of RTD black tea beverages: A review." *Critical Reviews in Food Science and Nutrition* 53 (2013): 180–197.

Chandrasekaran, M. *Valorization of Food Processing By-Products.* London: CRC Press, 2012.

Chaves-López, C., Serio, A., Paparella, A. et al. "Impact of microbial cultures on proteolysis and release of bioactive peptides in fermented milk." *Food Microbiology* 42 (2014): 117–121.

Choi, J., Sabikhi, L., Hassan, A., and Anand, S. "Bioactive peptides in dairy products." *International Journal of Dairy Technology* 65 no. 1 (2012): 1–12.

Da Silva, C. M. G., Contesini, F. J., Sawaya, A. C. H. F. et al. "Enhancement of the antioxidant activity of orange and lime juices by flavonoid enzymatic de-glycosylation." *Food Research International* 52 no. 1 (2013): 308–314.

Darvill, A. G., and Albersheim, P. "Phytoalexins and their elicitors—A defense against microbial infection in plants." *Annual Review of Plant Biology* 155 (1984): 507–516.

Ducasse, M.-A., Williams, P., Meudec, E., Cheynier, V., and Doco, T. "Isolation of Carignan and Merlot red wine oligosaccharides and their characterization by ESI-MS." *Carbohydrate Polymers* 79 (2010): 747–754.

Ducasse, M.-A., Williams, P., Canal-Llauberes, R.-M., Mazerolles, G., Cheynier, V., and Doco, T. "Effect of macerating enzymes on the oligosaccharide profiles of Merlot red wines." *Journal of Agricultural and Food Chemistry* 59 no. 12 (2011): 6558–6567.

Dulieu, C., Moll, M., Boudrant, J., and Poncelet, D. "Improved performances and control of beer fermentation using encapsulated α-acetolactate decarboxylase and modeling." *Biotechnology Progress* 16 no. 6 (2000): 958–965.

Elleuch, M., Bedigian, D., Roiseux, O., Besbes, S., Blecker, C., and Attia, H. "Dietary fibre and fibre-rich by-products of food processing: Characterisation, technological functionality and commercial applications: A review." *Food Chemistry* 124 no. 2 (2011): 411–442.

Eskin, M. N. A., and Fereidoon, S. *Biochemistry of Foods.* London: Academic Press, 2013.

Espamer, L., Pagliero, C., Ochoa, A., and Marchese, J. "Clarification of lemon juice using membrane process." *Desalination* 200 no. 1–3 (2006): 565–567.

Favre, G., Peña-Neira, Á., Baldi, C. et al. "Low molecular-weight phenols in Tannat wines made by alternative winemaking procedures." *Food Chemistry* 158 no. 1 (2014): 504–512.

Ferreira, L. R., Macedo, J. A., Ribeiro, M. L., and Macedo, G. A. "Improving the chemopreventive potential of orange juice by enzymatic biotransformation." *Food Research International* 51 no. 2 (2013): 526–535.

Ferruzzi, M. G., and Green, R. J. "Analysis of catechins from milk–tea beverages by enzyme assisted extraction followed by high performance liquid chromatography." *Food Chemistry* 99 no. 3 (2006): 484–491.

Fillaudeau, L., Blanpain-Avet, P., and Daufin, G. "Water, wastewater and waste management in brewing industries." *Journal of Cleaner Production* 14 no. 5 (2006): 463–471.

Fogarty, W. M., and Kelly, C. T. "Amylases, amyloglucosidases and related glucanases." In *Microbial Enzymes and Bioconversion: Economic Microbiology*, 115–170. London: Academic Press, 1980.

Fundira, M., Blom, M., Pretorius, I. S., and Van Rensburg, P. "Comparison of commercial enzymes for the processing of marula pulp, wine, and spirits." *Journal of Food Science* 67 no. 6 (2002): 2346–2351.

García-Conesa, M.-T., Østergaard, P. B., Kauppinen, S., and Williamson, G. "Hydrolysis of diethyl diferulates by tannase from *Aspergillus oryzae*." *Carbohydrate Polymers* 44 no. 4 (2001): 319–324.

Gasowski, B., Leontowicz, M., Leontowicz, H. et al. "The influence of beer with different antioxidant potential on plasma lipids, plasma antioxidant capacity, and bile excretion of rats fed cholesterol-containing and cholesterol-free diets." *The Journal of Nutritional Biochemistry* 15 no. 9 (2004): 527–533.

Gassara-Chatti, F., Brar, S. K., Ajila, C. M., Verma, M., Tyagi, R. D., and Valero, J. R. "Encapsulation of ligninolytic enzymes and its application in clarification of juice." *Food Chemistry* 137 no. 1–4 (2013): 18–24.

Godtfredsen, S. E., and Ottesen, M. "Maturation of beer with α-acetolactate decarboxylase." *Carlsberg Research Communications* 47 no. 2 (1982): 93–102.

Godtfredsen, S. E., Rasmussen, A. M., Ottesen, M., Rafn, P., and Peitersen, N. "Occurrence of α-acetolactate decarboxylases among lactic acid bacteria and their utilization for maturation of beer." *Applied Microbiology and Biotechnology* 20 no. 1 (1984): 23–28.

González-Sanjosé, M. L., Santamaría, G., and Díez, C. "Anthocyanins as parameters for differentiating wines by grape variety, wine-growing region and winemaking methods." *Journal of Food Composition and Analysis* 3 no. 1 (1990): 54–56.

Goode, D. L., Ulmer, H., Arendt, M., and Elke, K. "Model studies to understand the effect of amylase additions and pH adjustment on the rheological behaviour of simulated brewery mashes." *Journal of the Institute of Brewing* 111 no. 2 (2005): 153–164.

Grassin, C., and Fauquembergue, P. "Application of pectinases in beverages." In *Pectins and Pectinases*, 453–462. Amsterdam: Elsevier Science, 1996a.

Grassin, C., and Fauquembergue, P. "Fruit juices." In *Industrial Enzymology*. London: Macmillan, 1996b.

Günata, Y. Z., Bayonove, C., Baumes, R., and Cordonnier, R. "The aroma of grapes: I. Extraction and determination of free and glycosidically bound fractions of some grape aroma components." *Journal of Chromatography* 331 (1985): 83–90.

Hafeez, Z., Cakir-Kiefer, C., Roux, E., Perrin, C., Miclo, L., and Dary-Mourot, A. "Strategies of producing bioactive peptides from milk proteins to functionalize fermented milk products." *Food Research International* 63 (2014): 71–80.

Haque, E., Chand, R., and Kapila, S. "Biofunctional properties of bioactive peptides of milk origin." *Food Reviews International* 25 no. 1 (2009): 28–43.

Hough, U. S., Brigs, D. E., Stevens, R., and Young, T. W. *Malting and Brewing Science*. London: Chapman and Hall, 1982.

Jiao, B. L., Cassano, A., and Drioli, E. "Recent advances on membrane processes for the concentration of fruit juices: A review." *Journal of Food Engineering* 63 no. 3 (2004): 303–324.

Kashyap, D. R., Vohra, P. K., Chopra, S., and Tewari, R. "Applications of pectinases in the commercial sector: A review." *Bioresource Technology* 77 no. 3 (2001): 215–227.

Kelebek, H., Canbas, A., Cabaroglu, T., and Selli, S. "Improvement of anthocyanin content in the cv. Okuzgozu wines by using pectolytic enzymes." *Food Chemistry* 105 no. 1 (2007): 334–339.

Kelly, A. L., and Fox, P. F. "Indigenous enzymes in milk: A synopsis of future research requirements." *International Dairy Journal* 16 no. 6 (2006): 707–715.

Kilara, A. "Enzymes and their uses in the processed apple industry: A review." *Process Biochemistry* 23 (1982): 35–41.

Kilara, A., and Van Buren, J. P. "Clarification of apple juice." In Processed Apple Products (Ed. Downing, D.L.), 83–95. New York: Van NostrandReinhold, 1989.

Kim, J., Grate, J. W., and Wang, P. "Nanobiocatalysis and its potential applications." *Trends in Biotechnology* 26 no. 11 (2008): 639–646.

Landbo, A.-K., Kaack, K., and Meyer, A. S. "Statistically designed two step response surface optimization of enzymatic prepress treatment to increase juice yield and lower turbidity of elderberry juice." *Innovative Food Science & Emerging Technologies* 8 no. 1 (2007): 135–142.

Law, B. A. "Enzymes in dairy product manufacture." In *Enzymes in Food Technology* (Eds. R.J. Whitehurst and M. van Oortm), pp. 88–102. Oxford: Blackwell Publishing Ltd., 2010.

Law, B. A., and Goodenough, P. W. "Enzymes in milk and cheese production." In *Enzymes in Food Processing*, 114–143. Glasgow: Blackie Academic & Professional, 1995.

Lea, A. G. H. "Enzymes in the production of beverages and fruit juices." In *Enzymes in Food Processing*, 223–249. Glasgow: Blackie Academic & Professional, 1995.

Lee, W. C., Yusof, S., Hamid, N. S. A., and Baharin, B. S. "Optimizing conditions for enzymatic clarification of banana juice using response surface methodology (RSM)." *Journal of Food Engineering* 73 no. 1 (2006): 55–63.

Lei, H., Zhao, H., and Zhao, M. "Proteases supplementation to high gravity worts enhances fermentation performance of brewer's yeast." *Biochemical Engineering Journal* 77 no. 15 (2013): 1–6.

Lewin, B. *Wine Myths and Reality*. Dover: Vendange Press, 2010.

Li, X., Wu, S., and Zhang, Z. "Investigation on the influence of cellulase on the flavor substances of the Daqu liquor." *China Brewing* 30 no. 6 (2011): 80–83.

Liew Abdullah, A. G., Sulaiman, N. M., Aroua, M. K., and Megat Mohd Noor, M. J. "Response surface optimization of conditions for clarification of carambola fruit juice using a commercial enzyme." *Journal of Food Engineering* 81 no. 1 (2007): 65–71.

Liu, Z., Zhang, G., and Sun, Y. "Mutagenizing brewing yeast strain for improving fermentation property of beer." *Journal of Bioscience and Bioengineering* 106 no. 1 (2008): 33–38.

Llaubères, R.-M., Dubourdieu, D., and Villettaz, J.-C. "Exocellular polysaccharides from *Saccharomyces* in wine." *Journal of the Science of Food and Agriculture* 41 no. 3 (1987): 277–286.

Mac Cann, M. C., and Roberts, K. "Plant cell walls: Murals and mosaics." *Agro-Food Industry Hi-Technology* 5 (1994): 43–46.

Maktoufa, S., Neifarb, M., Driraa, S. J., Bakloutic, S., Fendri, M., and Châabouni, S. E. "Lemon juice clarification using fungal pectinolytic enzymes coupled to membrane ultrafiltration." *Food and Bioproducts Processing* 92 no. 1 (2014): 14–19.

Masmoudi, M., Besbes, S., Chaabouni, M. et al. "Optimization of pectin extraction from lemon by-product with acidified date juice using response surface methodology." *Carbohydrate Polymers* 74 (2008): 185–192.

Mensour, N. A., Margaritis, A., Briens, C. L., Pilkington, H., and Russell, I. "Application of immobilized yeast cells in the brewing industry." *Progress in Biotechnology* 11 (1996): 661–671.

Mieszczakowska-Frąc, M., Markowski, J., Zbrzeźniak, M., and Płocharski, W. "Impact of enzyme on quality of black currant and plum juices." *LWT—Food Science and Technology* 49 no. 2 (2012): 251–256.

Monti, D., Pisvejcová, A., Kren, V., Lama, M., and Riva, S. "Generation of an alpha-L-rhamnosidase library and its application for the selective derhamnosylation of natural products." *Biotechnology and Bioengineering* 87 no. 6 (2004): 763–771.

Mosafa, L., Moghadam, M., and Shahedi, M. "Papain enzyme supported on magnetic nanoparticles: Preparation, characterization and application in the fruit juice clarification." *Chinese Journal of Catalysis* 34 no. 10 (2013): 1897–1904.

Murthy, P. S., and Naidu, M. "Improvement of Robusta coffee fermentation with microbial enzymes." *European Journal of Applied Sciences* 3 no. 4 (2011): 130–139.

Murugesh, C. S., and Subramanian, R. "Applications of enzymes in processing green tea beverages: Impact on antioxidants." In *Processing and Impact on Antioxidants in Beverages*, 99–108. Oxford: Academic Press, 2014.

Ortega-Heras, M., Pérez-Magariño, S., and González-Sanjosé, M. L. "Comparative study of the use of maceration enzymes and cold pre-fermentative maceration on phenolic and anthocyanic composition and colour of a Mencía red wine." *LWT—Food Science and Technology* 48 no. 1 (2012): 1–8.

Ory, R. L., and Allen, J. S. A. *Enzymes in Food and Beverage Processing*. Washington: ACS Symposium Series, 1977.

Palomo, E. S., Hidalgo, M. C. D.-M., González-Viñas, M. A., and Pérez-Coello, M. S. "Aroma enhancement in wines from different grape varieties using exogenous glycosidases." *Food Chemistry* 92 no. 4 (2005): 627–635.

Piddocke, M. P., Fazio, A., Vongsangnak, A. et al. "Revealing the beneficial effect of protease supplementation to high gravity beer fermentations using '-omics' techniques." *Microbial Cell Factories* 10 no. 27 (2011): 1–27.

Piñeiro, Z., Natera, R., Castro, R., Palma, M., Puertas, B., and Barroso, C. G. "Characterization of volatile fraction of monovarietal wines: Influence of winemaking practices." *Analytica Chimica Acta* 563 no. 1–2 (2006): 165–172.

Pinelo, M., Arnous, A., and Meyer, A. S. "Upgrading of grape skins: Significance of plant cell-wall structural components and extraction techniques for phenol release." *Trends in Food Science and Technology* 17 no. 11 (2006): 579–590.

Qiang, X., Lie, C. Y., and Bing, W. Q. "Health benefit application of functional oligosaccharides." *Carbohydrate Polymers* 77 no. 3 (2009): 435–441.

Rai, P., Majumdar, G. C., DasGupta, S., and De, S. "Optimizing pectinase usage in pretreatment of mosambi juice for clarification by response surface methodology." *Journal of Food Engineering* 64 no. 3 (2004): 397–403.

Robinson, J. *Oxford Companion to Wine* (3rd Edition). London: Oxford University Press, 2006.

Rodrigues, S. "Enzyme maceration." In *Advances in Fruit Processing Technologies*, 235–246. London: CRC Press, 2012.

Rogerson, F. S. S., Vale, E., Grande, H. J., and Silva, M. C. M. "Alternative processing of port-wine using pectolytic enzymes." *Ciencia Y Tecnología Alimentaria* 2 no. 5 (2000): 222–227.

Roitner, M., Schalkhammer, T., and Pittner, F. "Preparation of prunin with the help of immobilized naringinase pretreated with alkaline buffer." *Applied Biochemistry and Biotechnology* 9 (1984): 483–488.

Sanderson, G. W. "The chemistry of tea and tea manufacturing." In *Recent Advances in Phytochemistry* (Eds. Runeckles and Tso, T.S.), pp. 247–316. London: Academic Press, 1972.

Scheffler, A., and Bamforth, C. W. "Exogenous b-glucanases and pentosanases and their impact on mashing." *Enzyme and Microbial Technology* 36 (2005): 813–817.

Servili, M., Begliomini, A. L., Montedoro, G., Petruccioli, M., and Federici, F. "Utilisation of a yeast pectinase in olive oil extraction and red wine making processes." *Journal Science of Food and Agriculture* 58 no. 2 (1992): 253–260.

Shyam Kumar, R., and Chandrasekaran, M. 2012. "Beverages." In *Valorization of Food Processing By-Products*, Ed. M. Chandrasekaran, 584–609. Boca Raton, FL: CRC Press, Taylor & Francis Group.

Sieiro, C., Villa, T. G., Silva, A. F., García-Fragaa, B., and Vilanova, M. "Albariño wine aroma enhancement through the use of a recombinant polygalacturonase from *Kluyveromyces marxianus*." *Food Chemistry* 145 (2014): 179–185.

Sin, H. N., Yusof, S., Hamid, N. S. A., and Rahman, R. "Optimization of enzymatic clarification of sapodilla juice using response surface methodology." *Journal of Food Engineering* 73 no. 4 (2006): 313–319.

Small, L. E., and Asquith, T. N. Process for treating coffee beans with enzyme-containing solution under pressure to reduce bitterness. US4904484 A, 11 abr., 1990.

Spagna, G., Barbagallo, R. N., Martino, A., and Pifferi, P. G. "A simple method for purifying glycosidases: Alpha-l-rhamnopyranosidase from *Aspergillus niger* to increase the aroma of Moscato wine." *Enzyme and Microbial Technology* 27 no. 7 (2000): 522–530.

Staško, A., Brezová, V., Mazúr, M., Čertík, M., Kaliňák, M., and Gescheidt, G. "A comparative study on the antioxidant properties of Slovakian and Austrian wines." *LWT—Food Science and Technology* 41 no. 10 (2008): 2126–2135.

SternEnzym. Available at http://www.stern-enzym.de, accessed on October 15, 2014.

Taylor, B. "Fruit and juice processing." In *Chemistry and Technology of Soft Drinks and Fruit Juices*, 35–66. Ashurst: Blackwell Publishing Ltd., 2005.

The Merck Index. "An Encyclopedia of Chemicals, Drugs, and Biologicals." New Jersey, USA: Whitehouse Station, 2001.

Todisco, S., Tallarico, P., and Drioli, E. "Modelling and analyses of the effects of ultrafiltration on the quality of freshly squeezed orange juice." *Italian Food and Beverage Technology* 12 (1998): 3–8.

Toepfl, S., Siemer, C., Saldāna-Navarro, G., and Heinz, V. "Overview of pulsed electric fields processing for food." In *Emerging Technologies for Food Processing*, 93–108. Oxford: Academic Press, 2014.

Torresi, S., Frangipane, M. T., Garzillo, A. M. V., Massantini, R., and Contini, M. "Effects of a β-glucanase enzymatic preparation on yeast lysis during aging of traditional sparkling wines." *Food Research International* 55 (2014): 83–92.

Uenojo, M., and Pastore, G. M. "Pectinases: Aplicações industriais e perspectivas." *Química Nova* 30 no. 2 (2007): 388–394.

Vaillant, F., Millan, A., Dornier, M., Decloux, M., and Reynes, M. "Strategy for economical optimisation of the clarification of pulpy fruit juices using crossflow microfiltration." *Journal of Food Engineering* 48 no. 1 (2001): 83–90.

Vidal, S., Williams, P., Doco, T., Moutounet, M., and Pellerin, P. "The polysaccharides of red wine: Total fractionation and characterization." *Carbohydrate Polymers* 54 no. 4 (2003): 439–447.

Voragen, A. G. J., Schols, H. A., and Beldman, G. "Tailor-made enzymes in fruit juice processing." *Flüssiges Obst* 59 (1992): 404–410.

Wiseman, G. *Nutrition and Health*. Boca Raton, FL: Taylor & Francis, 2002.

Wu, M. L., Zall, R. R., and Tzeng, W. C. "Microfiltration and ultrafiltration comparison of apple juices clarification." *Journal of Food Science* 55 no. 4 (1990): 1162–1163.

Yadav, V., Yadav, P. K., Yadav, S., and Yadav, K. D. S. "α-L-rhamnosidase: A review." *Process Biochemistry* 45 no. 8 (2010): 1226–1235.

Yusof, S., and Ibrahim, N. "Quality of soursop juice after pectinase enzyme treatment." *Food Chemistry* 51 no. 1 (1994): 83–88.

Zhang, H., Woodams, E. E., and Hang, Y. D. "Influence of pectinase treatment on fruit spirits from apple mash, juice and pomace." *Process Biochemistry* 46 no. 10 (2011): 1909–1913.

Zhao, J., Shi, P., Yuan, T. et al. "Purification, gene cloning and characterization of an acidic β-1,4-glucanase from *Phialophora* sp. G5 with potential applications in the brewing and feed industries." *Journal of Bioscience and Bioengineering* 114 no. 4 (2012): 379–384.

12

Enzymes in Flavors and Food Additives

Renu Nandakumar and Mamoru Wakayama

CONTENTS

12.1 Introduction

In the modern food industry, various types of microbial enzymes have been used for the synthesis of important food additives, improvement of food production processes, and food properties. Historically, in Denmark in 1874, Christian Hansen produced the rennet-containing enzyme chymosin for cheese production on a commercial basis. In 1894, Jokichi Takamine commercialized the diastatic enzyme produced by *Aspergillus oryzae* as Takadiastase. Pectinase and papain were industrially used for the production of fruit juice and beer for the removal of insoluble pectin and protein at the turn of the twentieth century. Since then, many distinctive enzymes have been explored for their application potential for enhancing the efficiency of various industrial-scale food production processes and to improve the food quality. For example, various amylases, glucose isomerases, and cyclomaltodextrin glucanotransferases in the production of starch sugars; proteases, peptidases, lipases, glutaminases, and chimosins in the fermentation and brewery industries; and naringinase, tannase, and pectinase in the beverage industry.

Among flavors and food additives, sweeteners are one of the most important food additives, which are indispensable to home cooking and the food-service and processed-food industries throughout the world.

From ancient times onward, sucrose derived from sugarcane and sugar beets is an important sweetener. But cultivation of these sugar-producing plants is possible only in limited regions; sugarcane can grow under a highly warm climate, and sugar beets can grow in a relatively cold climate. In addition to the limitations imposed by climatic requirements for their cultivation, purification of sucrose from sugarcane and sugar beets is expensive and time consuming. Since the 1980s, sweeteners called "starch sugars" have been proposed as an alternative to sucrose, and they have now captured a major share of the sugar market around the world. Many enzymes with different reaction mechanisms for starch degradation have been developed and were used in the production of starch-derived sweeteners with various functions.

On the other hand, umami compounds are also important flavors and food additives and, in particular, in high demand for seasonings. L-Glutamate is a typical flavor-enhancing amino acid. Historically, Dr. Kikunae Ikeda found L-glutamate was responsible for the savory-enhancing properties of kelp-like seaweed, *konbu* in Japanese, *Laminaria japonica* and developed the manufacturing process of seasoning (Ajinomoto) containing L-glutamate from *konbu*. At present, L-glutamate in amounts of more than two million tons per year is produced by a fermentation method using *Corynebacterium glutamicum*. The other well-known umami compounds are mononucleotides, such as inosinate and guanylate. The former is a main umami component of dried bonito flakes, and the latter is a main umami of dried mushroom. Recently, in addition to these conventional flavors, peptides with new functions, serving "*kokumi*," with a very rich taste and enhancing salty taste, have been reported.

In this chapter, we particularly focus on the functional sweeteners synthesized using processing enzymes and umami compounds, including new functional peptides as well as conventional flavors and the enzymes used for synthesis of them. We also deal with the enzymes that control food texture and quality.

12.2 Sweeteners

12.2.1 High-Fructose Corn Syrup

Among the starch-derived sugars, high-fructose corn syrup (HFCS) is the most prominent as an alternative to sucrose. HFCS is produced via the following process that involves three steps: The first step is preparation of dextrin from starch by the action of thermostable α-amylase (α-1,4-endoglucanase) from *Bacillus licheniformis* (dextrinization process). The second step involves the preparation of α-D-glucose from dextrin formed in the dextrinization process by the action of glucoamylase, which produces α-D-glucose by catalyzing the hydrolysis of α-1,4-glycoside linkage at the nonreducible end and isoamylase, which catalyzes the hydrolysis of α-1,6-linkage at the branched point of dexrin (saccharification process). The third step in the process is the preparation of isomerized sugar from α-D-glucose formed in the saccharification process by the action of glucose isomerase, which catalyzes the interconversion between glucose and fructose (isomerization process). The resultant isomerized sugar is generally called high-fructose corn syrup (HFCS) and mainly consists of glucose and fructose at an approximate ratio of 50:50. HFCS is widely used as a sweetener to refreshing drinks, bread, confectionary, pickles, flavoring, and alcoholic drinks. The schematic process of HFCS production is shown in Figure 12.1.

α-Amylase (α-1,4-endoglucanase), which is used in the dextrinization process is highly thermostable. The enzyme from *B. licheniformis* is hyperthermostable, and some mutant enzymes can retain full enzyme activity at 100°C (Declerck et al. 2003; Nielsen and Borchert 2000). Isoamylase or pullulanase, which catalyzes the hydrolysis of α-1,6-linkage at the branched point of dextrin, have been commercially used (Kuriki and Imanaka 1989; Kusano et al. 1990; Li et al. 2013; Urlaub and Woeber 1975). Glucose isomerases from *Streptomyces* sp. have been used extensively for isomerization of glucose (Chen et al. 1979; Rasmussen et al. 1994). For example, Sweetzyme, the commercial product of Novo Nordisk Co. is the preparation from *Streptomyces murinus* (Bandlish et al. 2002). Isomerases immobilized with ion-exchange resins, such as DEAE-cellulose, have been used in industrial processes. On the other hand, maltose is produced from dextrin by the action of β-amylase, which catalyzes the hydrolysis of dextrin at the nonreducing end and releases maltobiose. β-Amylases from plants such as wheat, barley, and soybeans have been widely used in the food industry (Daba et al. 2013; Kang et al. 2004; Rejzek et al. 2011;

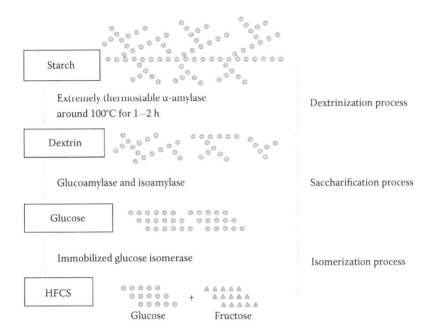

FIGURE 12.1 Schematic process of HFCS production.

Vinje et al. 2011). The sweetness of maltose is about 35% of that of sucrose, but its taste is smoother than sucrose. Therefore, there is an increasing demand for maltose for applications in foods such as starch syrup, chocolate, ice cream, and confectionary.

12.2.2 Coupling Sugar

Cyclomaltodextrin glucanotransferase or cyclodextrin glucosyltransferase (CGTase) catalyzes the formation of α-1,4-glycosyl linkage by acting on starch or dextrin and is often used for the synthesis of cyclodextrin, a cyclic compound consisting of 6-8 α-D-glucoses (Son et al. 2008). Because cyclodextrin has unique characteristics that can accommodate chemicals such as volatile compounds, drugs, nutrients, and pigments in the ring space, it has widely been used in the fields of medicine, cosmetics, food, and agro-chemistry (Martin Del Valle 2004). CGTase catalyzes the transfer reaction of glucose or maltooligosyl residue from starch to glucose moiety of sucrose to generate maltosylsucrose. A series of maltosylsucrose consisting of one to several glucoses bound to the nonreducible end of sucrose with α-1,4-linkage is formed with monosaccharides and oligosaccharides. This sugar mixture is called "coupling sugar," which has a sweet quality similar to that of sucrose and beneficial as an easily digestible food. Its sweetness is about 60% of that of sucrose. However, compared to sucrose, maltosylsucrose, one of the main ingredients of coupling sugar, was reported to have high anti-cariogenicity. In addition, coupling sugar is also known to inhibit glucosyltransferase responsible for the formation of an insoluble α 1,3 glucan layer in which *Streptococcus mutans*, a major cariogenic pathogen, preferentially inhabits (Pleszczynska et al. 2010). So far, no deleterious effects of its use in foods have been reported. In Japan, coupling sugar is widely employed in the manufacture of candies, cookies, chocolates, etc. The chemical structure of erlose, one of the main ingredients of coupling sugar, is shown in Figure 12.2.

Palatinose, also known as isomaltulose, is a disaccharide that is synthesized from sucrose via an intramolecular transferring reaction catalyzed by α-glucosyltransferase from *Protaminobacter rubrum* (Neto and Menao 2009). This compound consists of glucose and fructose with α-1,6-linkage (Figure 12.3). Its sweetness is about 40% of that of sucrose, but because it exhibits high anti-cariogenicity like coupling sugar, its demand in the food industry is steadily growing (Crittenden and Playne 1996).

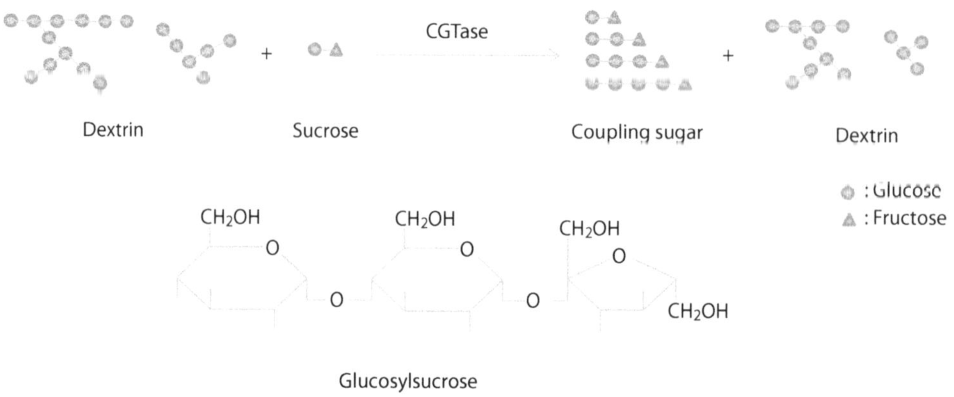

FIGURE 12.2   Coupling sugar. (From Crittenden, R. G., and M. J. Playne, *Trends in Food Sci.*, 7, 353–360, 1996.)

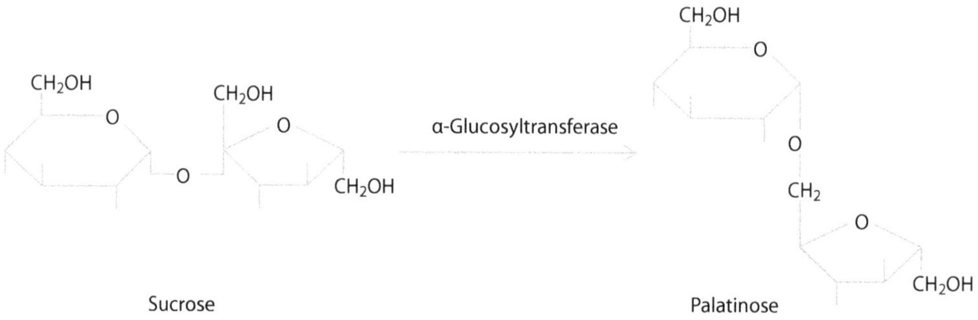

FIGURE 12.3   Palatinose. (From Crittenden, R. G., and M. J. Playne, *Trends in Food Sci.*, 7, 353–360, 1996.)

12.2.3 Fructo-Oligosaccharide

Fructo-oligosaccharide (FOS), also known as oligofructose or oligofructan, is a mixture of oligosaccharides with the general structure Glu-(Fru)$_n$ or (GF$_n$) with n ranging from 1 to 5. It is synthesized by the transfructosylation reaction of a β-fructosidase from *Aspergillus niger* on sucrose. The unique structural feature of FOS is that D-fructose residues linked by β(2→1) bonds with a terminal α(1→2) linked D-glucose. The chemical structures of the main components of commercial FOS, kestose (GF2), and nystose (GF3) are shown in Figure 12.4. Due to its unique structure, FOS resists hydrolysis by salivary and intestinal digestive enzymes and has a lower caloric value, contributing to the dietary fiber content. FOS is used as an additive to yogurt and other dairy products.

FOS is sometimes used in combination with other sweeteners for improvement of their sweetness profile and aftertaste. FOS also stimulates the growth of the beneficial *Bifidobacterium species*, contributing to the maintenance and improvement of intestinal environment (Saulnier et al. 2007). Physiological functions of saccharides are summarized in Table 12.1.

12.2.4 Trehalose

Trehalose is a nonreducing disaccharide, consisting of two glucoses with α,α-1,1-linkage, and is widely found in nature (Figure 12.5). In fungi and yeasts, trehalose is thought to be one of the reserve carbon sources (Winkler et al. 1991). In general, it is well known that trehalose has a protective function against tissue damage from desiccation and freezing in plants and nematodes (Behm 1997; Mueller et al. 1995). Trehalose does not react with amino acids in the Maillard reaction, and its sweetness is about 45% of that of sucrose. Trehalose has several applications as a sweetener and a stabilizing agent in foods, cosmetics,

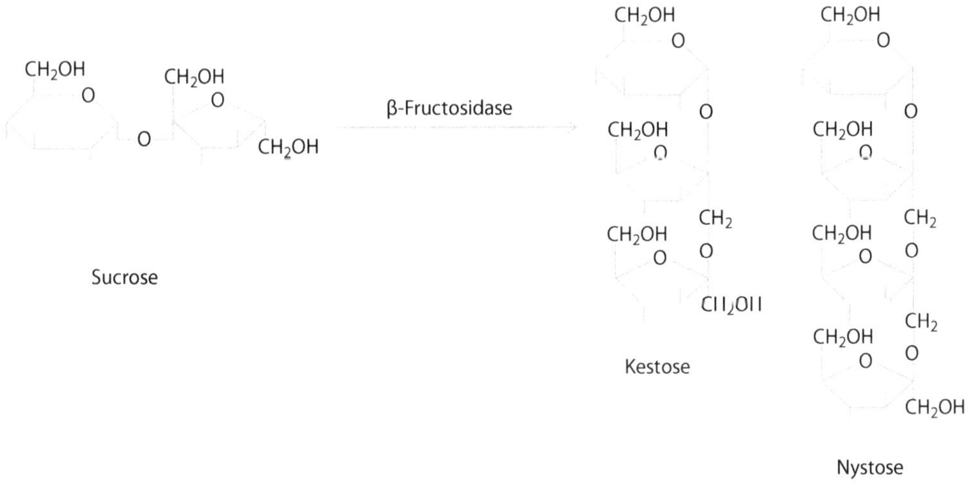

FIGURE 12.4 Fructo-oligosaccharide. (From Crittenden, R. G., and M. J. Playne, *Trends in Food Sci.*, 7, 353–360, 1996.)

TABLE 12.1

Physiological Function of Saccharides

Saccharide	Structure	Oral Microbe			Small Intestine	Large Intestine
		Acid Production	Polysaccharide Production	Inhibitory Effect for Polysaccharide Production	Assimilation	Stimulation of Profitable Bacteria Growth
Sucrose	G^1-^2F	++	+	−	+	
Glucose	G	++	−	−	+	
Fructose	F	++	−	−	+	
Maltose	G^1-^4G	+	−	+	+	
Isomaltose	G^1-^6G	−	−	+	+	
Isomaltulose	G^1-^6F	−	−	+	−	+
Glucosylsucrose	G^1-$^4G^1$-^2F	−	−	+	+	
Maltosylsucrose	G^1-$^4G^1$-$^4G^1$-^2F	−	−	+	+	
Kestose	G^1-$^2F^1$-^2F	−	−	+	−	
Nystose	G^1-$(^2F^1)_2$-^2F	−	−	+	−	+
Lactose	Gal^1-^4G	−	−	−	+	+
Lactulose	Gal^1-^4F	−	−	−	−	+

Source: Modified from Komaki, T. "Knowledge of Enzyme Application (in Japanese)." ver 4, Saiwai Shobo (2000).

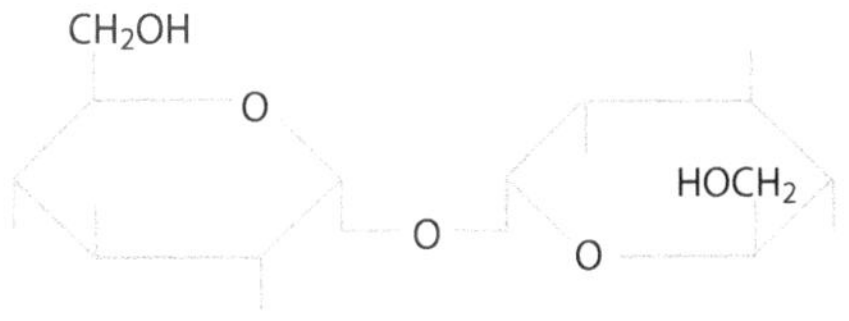

FIGURE 12.5 Trehalose.

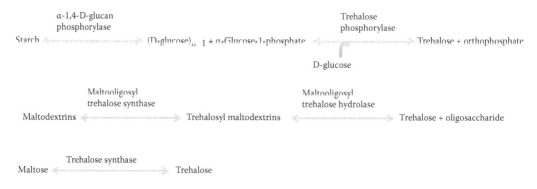

FIGURE 12.6 Biosynthetic pathways of trehalose. (Modified from Schiraldi, C. et al., *Trends Biotechnol.*, 20, 10, 420–425, 2000.)

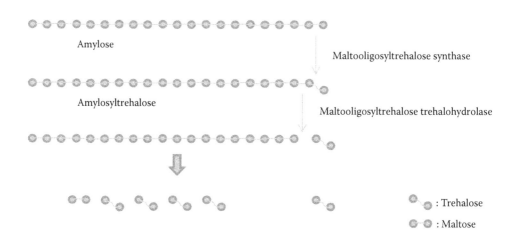

Trehalose production

FIGURE 12.7 Trehalose. (From Nakada, T. et al., *Biosci. Biotech. Biochem.*, 59, 2215–2218, 1995.)

and drugs (Ohtake and Wang 2011). The biosynthetic pathways of trehalose are summarized in Figure 12.6. Enzymatic methods are generally not cost-effective for the industrial scale production of trehalose due to their inadequate yield except for the enzymatic system of maltooligosyl trehalose synthase (MTAase) and maltooligosyl trehalose trehalohydrolase (MTHase) developed by Hayashihara Institute Corp (Maruta et al. 1995; Nakada et al. 1995a,b). This production scheme (Figure 12.7) has resulted in significant yields of trehalose especially by not utilizing any high-energy compounds such as UDP-glucose and glucose 6-phosphate.

12.2.5 Galacto-Oligosaccharide

β-Galactosidase catalyzes the hydrolysis of the galactosyl moiety from the nonreducing end of various galacosyloligosaccharides. It is used in the dairy industry to hydrolyze lactose, Gal-(β1,4)-Glc, the main saccharide of milk. Microbial β-galactosidases, such as those from *Kluyveromyces lactis*, were commonly used in the preparation of lactose-hydrolyzed milk for treating lactose intolerance (Mateo et al. 2004). In addition, hydrolysis of lactose to galactose and glucose by β-galactosidase prevents the lactose crystallization in frozen and condensed milk products and also intensifies the sweetness.

β-galactosidase can also catalyze transgalactosylation, in which a galactose moiety is transferred to the saccharides, forming different galacto-oligosaccharides (GOS) (Gosling et al. 2010). GOS can reduce

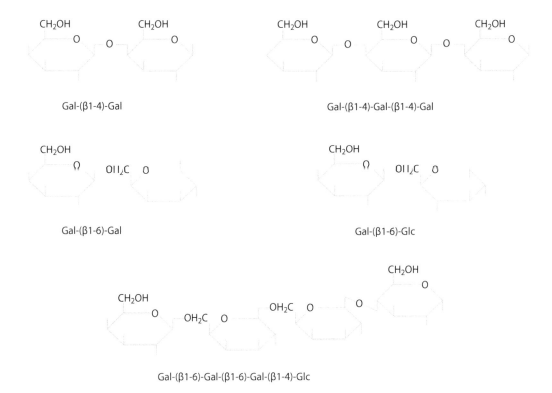

FIGURE 12.8 Galacto-oligosaccharide. (From Gosling, A. et al., *Food Chem.*, 121, 307–318, 2010.)

the level of blood serum cholesterol, improve mineral absorption, and suppress colon cancer development (Vulevic et al. 2013). Production of GOS by microbial β-galactosidases, such as Gal-(β1,4)-Gal, Gal-(β1,4)-Gal-(β1,4)-Gal, Gal-(β1,6)-Gal, Gal-(β1,6)-Glc, and Gal-(β1,6)-Gal-(β1,6)-Gal-(β1,4)-Glc has been reported (Figure 12.8).

12.2.6 Psicose

D-Psicose, a C-3 epimer of D-fructose, is one of the rare sugars because of its low natural abundance. Recently, various physiological roles of D-psicose have been reported: inhibition of α-glycosidase, suppression of hepatic lipogenic enzyme activity, and blood glucose elevation (Hayashi et al. 2010; Hossain et al. 2011; Iida et al. 2008) and received increased attention to its application in food and medicine. A new method for the large-scale production of D-psicose has been developed. The Izumori group has reported the preparation of D-psicose from D-fructose using immobilized D-tagatose 3-epimerase (D-TE) from *Pseudomonas* sp. ST-24 (Itoh et al. 1995; Takeshima et al. 2000). It was also shown that by coupling immobilized D-TE and D-xylose isomerase, D-psicose can be produced directly from D-glucose formed by the same process for isomerized sugar production (Figure 12.9).

12.2.7 Aspartame

In recent years, artificial sweeteners, including noncaloric sweeteners and anticariogenic sweeteners, have been proposed as an alternative to natural, calorie-containing sweeteners to lower the risk for diabetes, obesity, and cardiovascular diseases. For example, sugar substitutes, such as sucralose, saccharin, acesulfame potassium, and aspartame, are frequently used as food additives in various foods and soft drinks alone or in combination with natural sweeteners. Peptide-type nonsugar sweeteners, such as aspartame, have been widely used as ingredients in bottled/canned beverages, such as coffee, tea, sports drinks, and processed and pet foods. Aspartame, α-aspartyl-phenylalanine methylester, has been

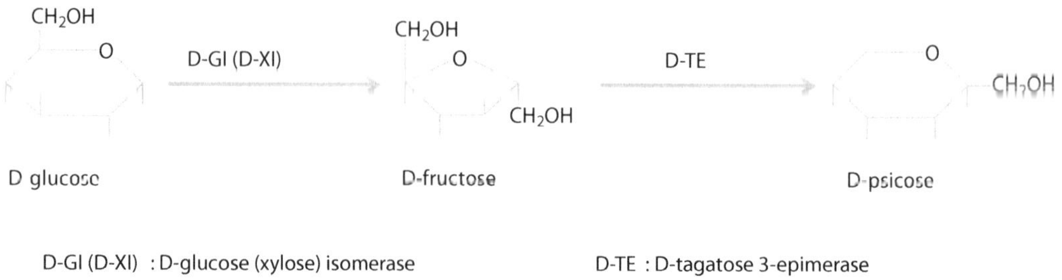

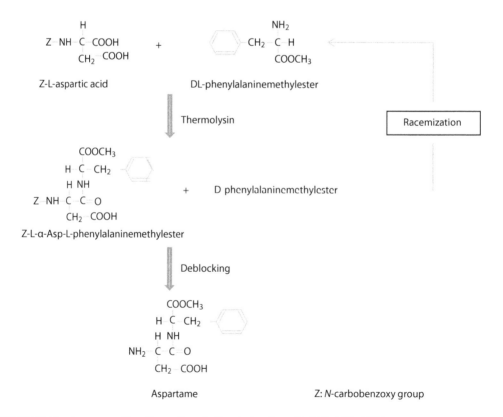

D-GI (D-XI) : D-glucose (xylose) isomerase   D-TE : D-tagatose 3-epimerase

FIGURE 12.9   Psicose. (Modified from Itoh, H. et al., *J. Biosci. Bioeng.*, 80, 101–103, 1995.)

synthesized by both organic chemical reactions and enzymatic reactions. For the enzymatic synthesis of aspartame thermolysin, a neutral protease from *Bacillus thermoproteolyticus*, which is reported to exhibit an exponential increase in enzymatic activity in the presence of increasing NaCl concentration, has been used. This enzyme can catalyze the reverse reaction of hydrolysis of peptide bonds in the mixture of *N*-carbobenzoxy-L-aspartic acid and DL-phenylalanine methylester (Erbeldinger et al. 2000; Inouye et al. 2007; Isowa et al. 1979; Miyanaga et al. 1995; Nagayasu et al. 1994; Ogino et al. 2010). Finally, the *N*-carbobenzoxy-protecting group is removed chemically, and unreacted D-phenylalanine methyl ester is converted to its racemic form, DL-phenylalanine methylester (Figure 12.10). Aspartame is a high-intensity artificial sweetener whose sweetness degree is about 200 times higher than that of sucrose.

FIGURE 12.10   Aspartame. (From Isowa, Y. et al., *Tetrahedron Lett.*, 28, 2611–2612, 1979.)

12.3 Glycosides

12.3.1 Ascorbic Acid Glycoside

Ascorbic acid, also called vitamin C, is a water-soluble compound, which is essential for the growth and maintenance of the body. Ascorbic acid serves as a strong antioxidant that can scavenge reactive oxygen species, such as superoxide anions and other oxidizing substances that can damage DNA. Ascorbic acid is known to have antiaging and anticancer properties and lowers the overall risk of developing heart diseases and osteoarthritis. A severe form of vitamin C deficiency is known as scurvy. Ascorbic acid being a natural antioxidant, it easily undergoes oxidation to dehydroascorbic acid, consequently losing its antioxidant activity. Glycosylation of ascorbic acid into two different types of glycosyl ascorbic acid by two different enzymes was proposed to improve its oxidative stability. Glycosylated products include 6-O-α-D-glucopyranosyl-L-ascorbic acid (6-GA) synthesized from ascorbic acid and maltose by the action of α-glucosidase from *Aspergillus niger* (McCleary and Gibson 1989) and 2-O-α-D-glucopyranosyl-L-ascorbic acid (2-GA) by the action of CGTase from *Bacillus* sp. (Bae et al. 2002; Han et al. 2012, 2013). The chemical structures of 6-GA and 2-GA are shown in Figure 12.11.

12.3.2 Narindine

Citrus juice is one of the most popular juices in the world. However, in most cases, its taste can be too bitter, and the bitterness is primarily due to naringin present in the citrus fruits. To minimize the bitterness, the debittering enzyme, naringinase (α-L-rhamnosidase), which breaks naringin into prunin and rhamnose is frequently used in the commercial production process of citrus juice (Bram and Solomons 1965; Puri and Banerjee 2000; Puri et al. 2005). For effective debittering, β-glucosidase, which catalyzes hydrolysis of prunin to glucose and naringenin, flavorless flavanone, is used together with naringinase (Figure 12.12).

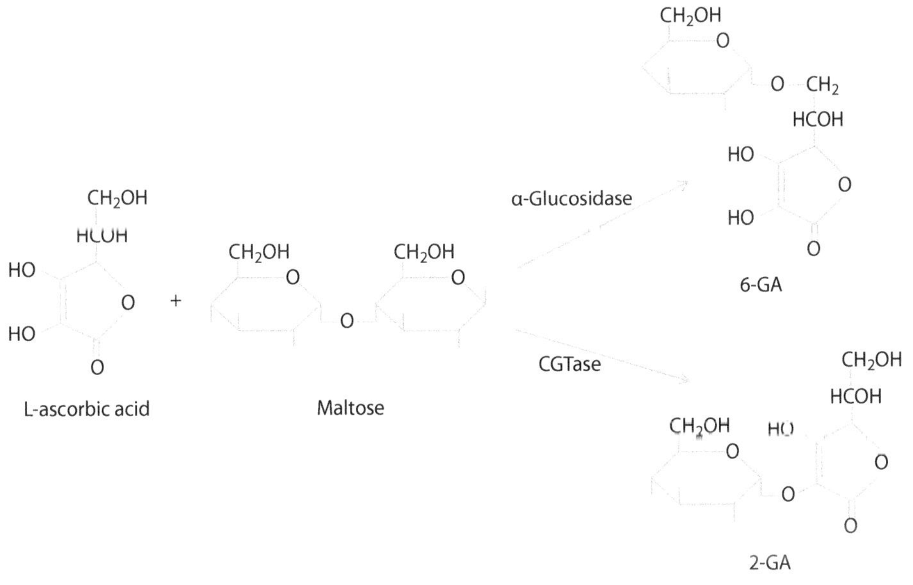

FIGURE 12.11   Ascorbic acid glycoside. (From Han, R. et al., *Appl. Microbiol. Biotechnol.*, 95, 313–320, 2012.)

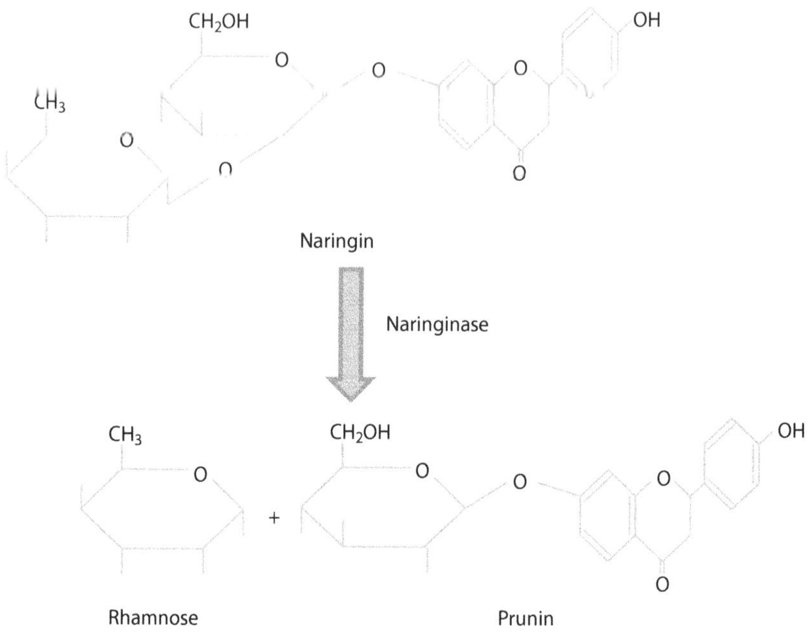

FIGURE 12.12   Naringin. (From Eiji Ichijima, *Chemistry of Enzyme* [in Japanese], Asakura Shoten, 1995.)

12.4 Umami

12.4.1 Glutamic Acid

L-Glutamic acid is an acclaimed flavor-enhancing amino acid and widely used as an important sea-soning in home cooking, food service, and processed-food industries around the world. In the early part of this century, L-glutamic acid was prepared by its extraction from seaweed. In the late 1960s, industrial production of L-glutamic acid by fermentation technology was developed mainly by using *Corynebacterium glutamicum*. Ajinomoto, a popular home-cooking seasoning, is monosodium gluta-mate. On the other hand, L-glutamic acid in fermented foods is derived from a hydrolysate of proteins present in raw materials. L-Glutamic acid can also be generated by the hydrolysis of L-glutamine in the protein hydrolysate; however, in the absence of glutaminase, most of the L-glutamine liberated is converted to tasteless pyroglutamic acid (Figure 12.13). The content of L-glutamic acid in Japanese soy sauce is one important criterion for the determination of its quality. Glutaminase from *Aspergillus oryzae*, which is commonly used for soy sauce fermentation, is shown to be inhibited by the high salt concentration in the soy sauce fermentation process (Yano et al. 1988). Therefore, salt-tolerant glutamin-ases have been investigated; especially those from *Micrococcus luteus*, *Bacillus subtilis*, and *Aspergillus oryzae* have been extensively studied for their application in soy sauce fermentation (Masuo et al. 2005;

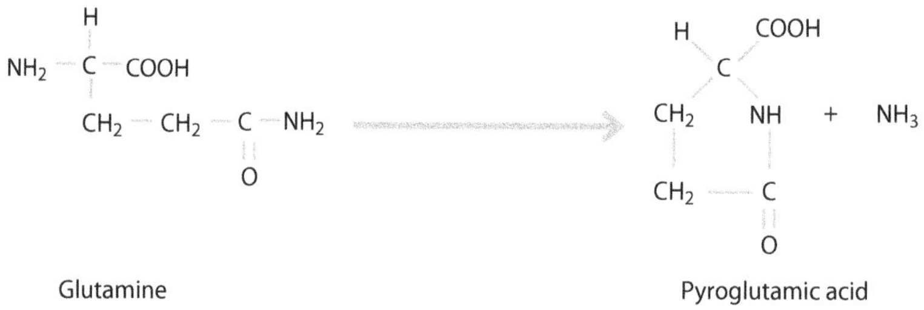

FIGURE 12.13   Pyroglutamic acid. (From Nandakumar, R. et al., *J. Mol. Cat. B: Enzymatic*, 23, 87–100, 2003.)

Moriguchi et al. 1994; Shimizu et al. 1991). The glutaminase from *B. subtilis* is currently available as a commercial product.

12.4.2 Theanine

Theanine, γ-glutamylethylamide, is a nonprotein-derived amino acid that was first isolated from green tea leaves in the 1940s (Sakato 1949). It is considered to be a unique amino acid in nature because its occurrence appears to be limited to the *Camellia* genus, particularly the tea-producing plant, *C. sinensis* (Juneja et al. 1999). Theanine is known for its distinctive umami taste; therefore, teas with a high theanine content are generally considered to be of higher quality. Beneficial effects of theanine include improved relaxation, improvement of memory and learning ability, strengthening of the immune system, prevention of cancer and vascular diseases, and suppression of weight gain (Kimura et al. 2007; Kurihara et al. 2007; Owen et al. 2008; Zhang et al. 2001; Zheng et al. 2004). Theanine can be produced from glutamate and ethylamine using a combination reaction of bacterial glutamine synthetase with a sugar fermentation reaction of baker's yeast as an ATP-regenerating system has been reported (Nandakumar et al. 2003; Wakisaka et al. 1998) (Figure 12.14). On the other hand, glutaminase from *Pseudomonas nitroreducens* has been found to catalyze a γ-glutamyl transfer reaction in a mixture containing a γ-glutamyl donor and γ-glutamyl acceptor (Abelian et al. 1993). This enzymatic method for theanine production has been industrially developed using glutaminase from *P. nitroreducens* (Figure 12.15).

γ-Glutamyltranspeptidase (GGT), which catalyzes the transfer of γ-glutamyl moiety from γ-glutamyl compounds, such as glutathione, to amino acids and peptides as well as the hydrolysis of γ-glutamyl compounds, has been reported for the enzymatic synthesis of theanine (Suzuki et al. 2002a). Other

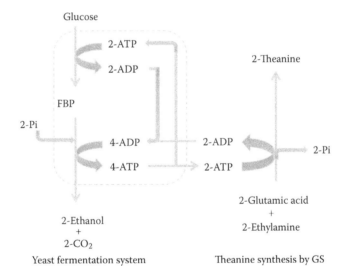

FIGURE 12.14 Theanine synthesis by Gln synthetase. (From Nandakumar, R. et al., *J. Mol. Cat. B: Enzymatic*, 23, 87–100, 2003.)

$$\text{Glutamine} + \text{Ethylamine} \longrightarrow \text{Theanine} + NH_3$$

Glutamine — Ethylamine — Theanine

FIGURE 12.15 Theanine synthesis by glutaminase. (From Nandakumar, R. et al., *J. Mol. Cat. B: Enzymatic*, 23, 87–100, 2003.)

physiologically important γ-glutamyl compounds, such as γ-glutamyl taurine and γ-glutamyl DOPA, have also been synthesized by an enzymatic method using GGT (Kumagai et al. 1988). It has been shown that γ-glutamylization using GGT effectively reduced the bitter taste of several amino acids and increased their preference (Suzuki et al. 2002b).

12.4.3 *Kokumi* Peptides

Kokumi compounds might be defined as chemicals that are nearly tasteless themselves but can enhance mouthfeel and complexity and induce more long-lasting savory taste sensation when combined with savory foods (Dunkel et al. 2007). As *kokumi* compounds, γ-glutamyl compounds, such as γ-Glu-Leu, γ-Glu-Val, γ-Glu-Cys-Gly (glutathione), and γ-Glu-Cys-β-Ala, have been reported in beans, Gouda cheese, and fish sauces (Dunkel et al. 2007; Kuroda et al. 2012; Toelstede et al. 2009). There is currently great interest in *kokumi* compounds because they can modify the basic tastes, such as sweet, salty, and umami, when they are added to beverages or food. Recently, it has been reported that the calcium-sensing receptor (CaSR) is a receptor for *kokumi* compounds, which serve as CaSR agonists (Maruyama et al. 2012; Ohsu et al. 2010). Among those γ-glutamyl compounds, γ-Glu-Val-Gly has been determined to be a potent *kokumi* peptide by CaSR assay and sensory evaluation (Table 12.2) (Kuroda et al. 2012). Because the contents of γ-Glu-Val-Gly in foodstuff are very low, an extraction method from foods is not available. Therefore, the large-scale production of γ-Glu-Val-Gly is dependent on synthetic methods: overall chemical synthesis or semichemical synthesis combined with enzymes, such as γ-glutamyltranspeptidase, γ-glutamylcysteine synthetase, and glutathione synthase. As described in theanine synthesis, microbial γ-glutamyltranspeptidases are highly expected to be potent enzymes for various *kokumi* compounds, including γ-Glu-Val-Gly (Suzuki et al. 2002a).

12.4.4 Salty Peptides

Aspartame is one of the most famous artificial sweeteners. This sweet compound is a dipeptide composed of Asp and Phe-OMe as described in previous Section 12.2.7. Dipeptides such as Leu-Phe and Arg-Leu have been known as bitter peptides. Tada et al. (1984) and Seki et al. (1990) reported that ornityl dipeptides, such as ornityl-β-alanine, ornithyltaurine, and ornithyl-Lys, exhibited a salty taste and the significance of pH of the peptide solution for salty taste. Recently, Leu-Ser, Leu-Ile, Arg-Phe, and Arg-Arg have been introduced as new salty taste-enhancing peptides (Arai et al. 2013; Schindler et al. 2011). These peptides will be useful as an alternative to sodium chloride, which is thought to be undesirable for hypertensives. Among these salty taste-modulating peptides, the enzymatic synthesis of dipeptides, such as Leu-Ser has been reported using L-amino acid ligase (Arai et al. 2013). L-Amino

TABLE 12.2

Contents of γ-Glu-Val-Gly in Commercial Fish Sauces

Samples	Country of Origin	Contents of γ-Glu-Val-Gly (mg/dL)
Nampura A	Thailand	0.27
Nampura B	Thailand	0.12
Nampura C	Thailand	0.31
Nampura D	Thailand	0.20
Nampura E	Thailand	0.23
Nuoc Mum A	Vietnam	1.20
Nuoc Mum B	Vietnam	1.26
Nuoc Mum C	Vietnam	1.23
Nuoc Mum D	Vietnam	1.06
Nuoc Mum E	Vietnam	1.04
Yu-lu	China	0.11
Shottsuru	Japan	0.28

Source: Modified from Kuroda, M. et al., *J. Agric. Food Chem.* 60: 7291–7296 (2012).

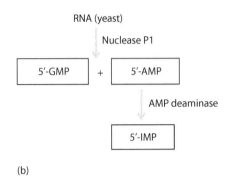

FIGURE 12.16 Nucleic acid-based flavor. (From Eiji Ichijima, *Chemistry of Enzyme* [in Japanese], Asakura Shoten, 1995.)

acid ligase might be a potent enzyme for synthesis of these dipeptides, such as L-amino acid ester acyl transferase (Yokozaki and Hara 2005).

12.4.5 IMP and GMP

In addition to amino acid-based flavor compounds, such as L-glutamic acid, L-aspartic acid, and L-theanine, nucleic acid-based flavor compounds, 5′-guanylic acid (5′-GMP) and 5′-inosinic acid (5′-IMP), have been used from ancient times (Figure 12.16a). 5′-GMP, the flavor component of dried mushrooms, and 5′-IMP, the flavor component of dried bonito, are indispensable for traditional Japanese foods. Figure 12.16b shows the schematic process for the industrial production of 5′-GMP and 5′-IMP from yeast RNA. In this process, two enzymes are employed as catalysts: a nuclease P1 from *Penicillium citrium* (Ying et al. 2006) and an AMP deaminase from *A. oryzae* (Minato et al. 1965).

12.5 Enzymes Controlling Food Texture and Quality

12.5.1 Pectinase

Pectin is one of the important polysaccharides used in the food industry. For example, it is used as a gelling agent and a natural texturizer to enhance the taste and appearance of processed foods, strawberry jam, marmalade, etc. On the other hand, soluble pectins in fruit juices causes hazing or cloudiness to juices, such as apple, grapefruit, lime, lemon, and orange. Although cloudy juices, especially from tropical fruits, have a growing market, clear fruit juices, especially from apples and grapes, are preferred by some consumers, and it is desirable to make it an international standard. The clarification process usually involves the enzymatic degradation of fruit pectin using pectinases, a heterogeneous group of enzymes catalyzing the depolymerization of pectin (Alkorta et al. 1998). These enzymes are classified into pectin esterase, pectin lyase, and polygalacturonase (Alana et al. 1990; Glinka and Liao 2011; Kant et al. 2013). Pectinolytic enzymes from *Aspergillus niger* are commonly used in the food industry (Dinu et al. 2007). Figure 12.17 illustrates the mode of pectin degradation by enzymes.

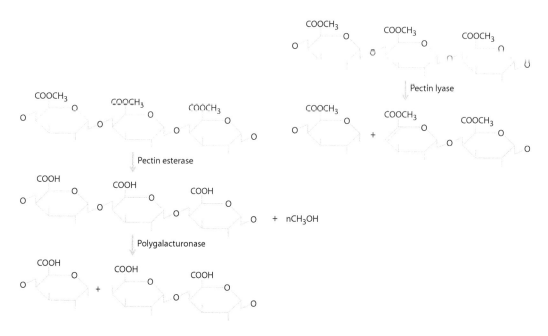

FIGURE 12.17   Pectinase. (From Eiji Ichijima, *Chemistry of Enzyme* [in Japanese], Asakura Shoten, 1995.)

12.5.2 Transglutaminase

Formation of cross-links between proteins is significant from the point of view of improvement of food properties, such as firmness, viscosity, elasticity, and water-binding capacity. Transglutaminases, which catalyze the cross-linking reaction between proteins, have been reported in microbial sources, such as *Streptomyces* sp., *Bacillus* sp., and *Corynebacterium* sp. (Placido et al. 2008; Umakoshi et al. 2011; Zhang et al. 2012). Transglutaminase has been widely used in various food industries, including cheese manufacture, other dairy products, meat processing, and the bakery industry. The enzyme catalyzes the cross-linkage between proteins by transferring the γ-carboxyamide group of glutamine residue of one protein to the ε-amino group of lysine residue of the same protein or another protein (Figure 12.18). Transglutaminase catalyzes deamination reaction of γ-carboxyamide group of glutamine residue of protein in the absence of free amine groups resulting in a hydrolysis reaction. The enzyme that specifically catalyzes the hydrolysis of the γ-carboxyamide group of glutamine residue of protein is called protein glutaminase. This enzyme is used to improve properties of proteins, such as solubility, emulsification, gelation, sensitivity to proteases, reduction of allergenicity, and reactivity toward transglutaminase.

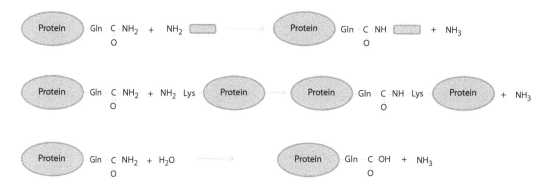

FIGURE 12.18   Transglutaminase. (From Eiji Ichijima, *Chemistry of Enzyme* [in Japanese], Asakura Shoten, 1995.)

12.5.3 Asparaginase

Asparaginase catalyzes the hydrolysis of asparagine to aspartic acid and ammonia and is widely distributed in nature from microorganisms to mammals (Figure 12.19). The enzyme is considered to play a significant role in asparagine metabolism in cells. Asparaginases from various microbial sources have been studied (Derst et al. 2000; Harms et al. 1991; Kotzia and Labrou 2005; Yao et al. 2005); in particular, the enzymes from *Escherichia coli* and *Erwinia chrysanthemi* have been extensively investigated with respect to their antitumor activity. They have been employed for many years as effective drugs for the treatment of acute lymphocytic leukemia (Clavell et al. 1986). Aspartic acid, a product of the hydrolysis reaction of asparaginase, is a major dietary supplement known to aid in liver detoxification and increase resistence to fatigue as well as to have a sour and savory taste. So far, most of the focus on asparaginase was as an anticancer drug for treatment of leukemia, and not much attention has been given to its potential applications in the food industry.

Recently, non-negligible amounts of acrylamide were detected in processed foods, such as fried potatoes, cereals, bread, and roasted coffee (Tareke et al. 2002). Because acrylamide is an extremely toxic and carcinogenic compound and a potential human health hazard, various ways to improve food production processes and to minimize acrylamide content have been proposed. Asparagine and reducing sugars are considered to be the main precursors for acrylamide formation in foods (Zyzak et al. 2003). It has been shown that the reducing sugars (glucose, fructose) are the limiting factors in acrylamide formation in potatoes, and asparagine appears to be the limiting factor in cereal products (Amrein et al. 2004). The proposed mechanism of acrylamide formation is shown in Figure 12.20. It has been shown

FIGURE 12.19 Asparaginase.

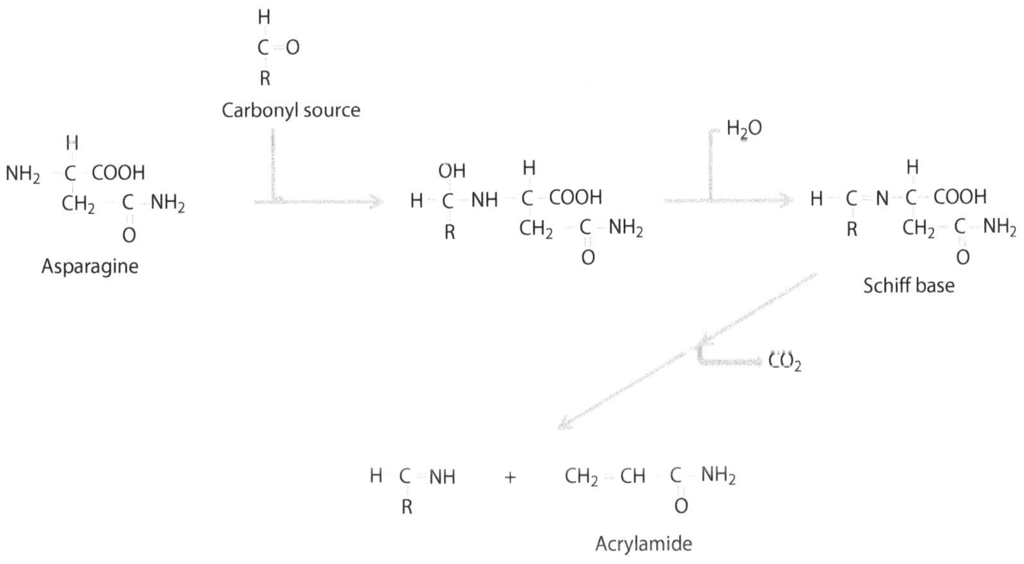

FIGURE 12.20 Proposed mechanism of acrylamide formation. (Modified from Zyzak, D. V. et al., *J. Agric. Food Chem.*, 51, 4782–4787 [2003].)

that the treatment of raw food material with microbial asparaginase effectively reduces the amount of L-asparagine, simultaneously suppressing the generation of acrylamide in processed foods, such as fried potato (Amrein et al. 2004; Zyzak et al. 2003). Asparaginases from *A. oryzae* and *Bacillus subtilis* have been used for acrylamide mitigation (Hendriksen et al. 2009; Onishi et al. 2011; Pedreschi et al. 2008; Yano et al. 2008). In the near future, many asparaginases from food-grade organisms are expected to be employed in industrial-scale food production processes.

REFERENCES

Abelian, V. H., T. Okubo, K. Mutoh, D. C. Chu, M. Kim, and T. Yamamoto. 1993. A continuous production method for theanine by immobilized *Pseudomonas nitroreducens* cells. *J. Ferment. Bioeng.* 76:195–198.

Alana, A., I. Alkorta, J. B. Dominguez, M. J. Llama, and J. L. Serra. 1990. Pectin lyase activity in a *Penicillium italicum* strain. *Appl. Environ. Microbiol.* 56:3755–3759.

Alkorta, I., C. Garbisu, M. J. Llama, and J. L. Serra. 1998. Industrial applications of pectic enzymes: A review. *Process Biochem.* 33:21–28.

Amrein, T. M., B. Schoenbaechler, F. Escher, and R. Amado. 2004. Acrylamide in gingerbread: Critical factors for formation and possible ways for reduction. *J. Agric. Food Chem.* 52:4282–4288.

Arai, T., Y. Arima, S. Ishikura, and K. Kino. 2013. L-Amino acid ligase from *Pseudomonas syringae* producing Tabtoxin can be used for enzymatic synthesis of various functional peptides. *Appl. Environ. Microbiol.* 79:5023–5029.

Bae, H. K., S. B. Lee, C. S. Park et al. 2002. Modification of ascorbic acid using transglycosylation activity of *Bacillus stearothermophilus* maltogenic amylase to enhance its oxidative stability. *J. Agric. Food Chem.* 50:3309–3316.

Bandlish, R. K., H. J. Michael, K. L. Epting, C. Vieille, and R. M. Kelly. 2002. Glucose-to-fructose conversion at high temperature with xylose (glucose) isomerase from *Streptomyces murinus* and two hyperthermophilic *Thermotoga* species. *Biotechnol. Bioeng.* 80:185–194.

Behm, C. A. 1997. The role of trehalose in the physiology of nematodes. *Int. J. Parasitol.* 27:215–229.

Bram, B., and G. L. Solomons. 1965. Production of the enzyme naringinase by *Aspergillus niger*. *Appl. Microbiol.* 13:842–845.

Chen, W. P., A. W. Anderson, and Y. W. Han. 1979. Extraction of glucose isomerase from *Streptomyces flavogriseus*. *Appl. Environ. Microbiol.* 37:324–331.

Clavell, L. A., R. D. Gelber, H. J. Cohen et al. 1986. Four agent induction and intensive asparaginase therapy for treatment of childhood acute lymphoblastic leukemia. *New Engl. J. Med.* 315:657–663.

Crittenden, R. G., and M. J. Playne. 1996. Production, properties and application of food-grade oligosaccharides. *Trends Food Sci.* 7:353–360.

Daba, T., K. Kojima, and K. Inouye. 2013. Interaction of wheat β-amylase with maltose and glucose as examined by fluorescence. *J. Biochem.* 154:85–92.

Declerck, N., M. Machius, P. Joyet, G. Wiegand, R. Huber, and C. Gaillardin. 2003. Hyperthermostabilization of *Bacillus licheniformis* α-amylase and modulation of its stability over a 50°C temperature range. *Protein Eng.* 16:287–293.

Derst, C., J. Henseling, and K. H. Roehm. 2000. Engineering the substrate specificity of *Escherichia coli* asparaginase II. Selective reduction of glutaminase activity by amino acid replacements at position 248. *Protein Sci.* 9:2009–2017.

Dinu, D., M. T. Nechifor, G. Stoian, M. Costache, and A. Dinischiotu. 2007. Enzymes with new biochemical properties in the pectinolytic complex produced by *Aspergillus niger* MIUG 16. *J. Biotechnol.* 131:128–137.

Dunkel, A., J. Koester, and T. Hofmann. 2007. Molecular and sensory characterization of γ-glutamyl peptides as key contributors to the kokumi taste of edible beans (*Phaseolus vulgaris* L.). *J. Agric. Food Chem.* 55:6712–6719.

Erbeldinger, M., A. J. Mesiano, and A. J. Russell. 2000. Enzymatic catalysis of formation of Z-aspartame in ionic liquid—An alternative to enzymatic catalysis in organic solvents. *Biotechnol. Prog.* 16:1129–1131.

Glinka, E. M., and Y. C. Liao. 2011. Purification and partial characterization of pectin methylesterase produced by *Fusarium asiaticum*. *Fungal Biol.* 115:1112–1121.

Gosling, A., G. W. Stevens, A. R. Barber, S. E. Kentish, and S. L. Gras. 2010. Recent advances refining galac-tooligosaccharide production from lactose. *Food Chem.* 121:307–318.

Han, R., L. Liu, J. Li, G. Du, and J. Chen. 2012. Functions, applications and production of 2-O-D-glucopyranosyl-L-ascorbic acid. *Appl. Microbiol. Biotechnol.* 95:313–320.

Han, R., L. Liu, H. D. Shin, R. R. Chen, G. Du, and J. Chen. 2013. Site-saturation engineering of lysine 47 in cyclodextrin glycosyltransferase from *Paenibacillus macerans* to enhance substrate specificity towards maltodextrin for enzymatic synthesis of 2-O-D-glucopyranosyl-L-ascorbic acid (AA-2G). *Appl. Microbiol. Biotechnol.* 97:5851–5860.

Harms, E., A. Wehner, M. P. Jennings, K. J. Pugh, I. R. Beacham, and K. H. Roehm. 1991. Construction of expression systems for *Escherichia coli* asparaginase II and two-step purification of the recombinant enzyme from periplasmic extracts. *Protein Expr. Purif.* 2.144–150.

Hayashi, N., T. Iida, T. Yamada et al. 2010. Study on the postprandial blood glucose suppression effect of D-psicose in borderline diabetes and the safety of long-term ingestion by normal human subjects. *Biosci. Biotechnol. Biochem.* 74:510–519.

Hendriksen, H. V., B. A. Kombrust, P. R. Ostergaard, and M. A. Stringer. 2009. Evaluating the potential for enzymatic acrylamide mitigation in a range of food products using an asparaginase from *Aspergillus oryzae*. *J. Agric. Food Chem.* 57:4168–4176.

Hossain, M. A., S. Kitagaki, D. Nakano et al. 2011. Rare sugar D-psicose improves insulin sensitivity and glucose tolerance in type 2 diabetes Otsuka Long-Evans Tokushima Fatty (OLETF) rats. *Biochem. Biophys. Res. Commun.* 405:7–12.

Iida, T., Y. Kishimoto, Y. Koshikawa et al. 2008. Acute D-psicose administration decrease the glycemic responses to an oral maltodextrin tolerance test in normal adults. *J. Nutr. Sci. Vitaminol.* 54:511–514.

Inouye, K., M. Kusano, Y. Hashida, M. Minoda, and K. Yasukawa. 2007. Engineering, expression, purification, and production of recombinant thermolysin. *Biotechnol. Annu. Rev.* 13:43–64.

Isowa, Y., M. Ohmori, T. Ichikawa, and K. Mori. 1979. The thermolysin-catalyzed condensation reactions of N-substituted aspartic and glutamic acids with phenylalanine alkly esters. *Tetrahedron Lett.* 28:2611–2612.

Itoh, H., T. Sato, and K. Izumori. 1995. Preparation of D-psicose from D-fructose by immobilized D-tagatose 3-epimerase. *J. Biosci. Bioeng.* 80:101–103.

Juneja, L. R., D. C. Chu, T. Okubo, Y. Nagano, and H. Yokogoshi. 1999. L-Theanine: A unique amino acid of green tea and its relaxation effect in humans. *Trends Food Sci. Technol.* 10:199–204.

Kang, Y. N., M. Adachi, S. Utsumi, and B. Mikami. 2004. The roles of glu186 and glu380 in the catalytic reaction of soybean beta-amylase. *J. Mol. Biol.* 18:1129–1140.

Kant, S., A. Vohra, and R. Gupta. 2013. Purification and physicochemical properties of polygalacturonase from *Aspergillus niger* MTCC 3323. *Protein Expr. Purif.* 87:11–16.

Kimura, K., M. Ozeki, L. R. Juneja, and H. Ohira. 2007. L-Theanine reduces psychological and physiological stress responses. *Biol. Psychol.* 74:39–45.

Komaki, T. 2000. "Knowledge of Enzyme Application (in Japanese)" ver 4, Saiwai Shobo.

Kotzia, G. A., and N. E. Labrou. 2005. Cloning, expression and characterization of *Erwinia carotovora* L-asparaginase. *J. Biotechnol.* 119:309–323.

Kumagai, H., T. Echigo, H. Suzuki, and T. Tochikura. 1988. Distribution, formation and stabilization of yeast glutathione S-transferase. *Agric. Biol. Chem.* 52:1377–1382.

Kurihara, S., S. Shibahara, H. Arisaka, and Y. Akiyama. 2007. Enhancement of antigen-specific immunoglobulin G production in mice by co-administration of L-cysteine and L-theanine. *J. Vet. Med. Sci.* 69:1263–1270.

Kuriki, T., and T. Imanaka. 1989. Nucleotide sequence of the neopullulanase gene from *Bacillus stearothermophilus*. *J. Gen. Microbiol.* 135:1521–1528.

Kuroda, M., Y. Kato, J. Yamazaki et al. 2012. Determination and quantification of γ-glutamyl-valyl-glycine in commercial fish sauces. *J. Agric. Food Chem.* 55:7291–7296.

Kusano, S., S. Takahashi, D. Fujimoto, and Y. Sakano. 1990. Effects of reduced malto-oligosaccharides on the thermal stability of pullulanase from *Bacillus acidpullulyticus*. *Carbohydr. Res.* 199:83–89.

Li, Y., D. Niu, L. Zhang, Z. Wang, and G. Shi. 2013. Purification, characterization and cloning of a thermotolerant isoamylase produced from *Bacillus* sp. CICIM 304. *J. Ind. Microbiol. Biotechnol.* 40:437–446.

Martin Del Valle, E. M. 2004. Cyclodextrins and their uses: A review. *Process Biochem.* 39:1033–1046.

Maruta, K., T. Nakada, M. Kubota et al. 1995. Formation of trehalose from maltooligosaccharides by a novel enzymatic system. *Biosci. Biotechnol. Biochem.* 59:1829–1834.

Maruyama, Y., R. Yasuda, M. Kuroda, and Y. Eto. 2012. *Kokumi* substances, enhancers of basic tastes, induce responses in calcium-sensing receptor expressing taste cells. *PLoS One* 7:e34489.

Masuo, N., K. Yoshimune, K. Ito, K. Matsushima, Y. Koyama, and M. Moriguchi. 2005. *Micrococcus luteus* K-3-type glutaminase from *Aspergillus oryzae* RIB40 is salt-tolerant. *J. Biosci. Bioeng.* 100:576–578.

Mateo, C., R. Monti, B. C. Pessela et al. 2004. Immobilization of lactase from *Kluyveromyces lactis* greatly reduces the inhibition promoted by glucose. Full hydrolysis of lactose in milk. *Biotechnol. Prog.* 20:1259–1262.

McCleary, B. V., and T. S. Gibson. 1989. Purification, properties, and industrial significance of transglucosidase from *Aspergillus niger. Carbohydr. Res.* 185:147–162.

Minato, S., T. Tagawa, and K. Nakanishi. 1965. Studies on nonspecific adenosine deaminase from Takadiastase. I. Purification and properties. *J. Biochem.* 58:519–525.

Miyanaga, M., T. Tanaka, T. Sakiyama, and K. Nakanishi. 1995. Synthesis of aspartame precursor with an immobilized thermolysin in mixed organic solvents. *Biotechnol. Bioeng.* 46:631–635.

Moriguchi, M., K. Sakai, R. Tateyama, Y. Furuta, and M. Wakayama. 1994. Isolation and characterization of salt-tolerant glutaminase from marine *Micrococcus luteus. J. Ferment. Bioeng.* 77:621–625.

Mueller, J., T. Boller, and A. Wiemken. 1995. Trehalose and trehalase in plants: Recent developments. *Plant Sci.* 112:1–9.

Nagayasu, T., M. Miyanaga, T. Tanaka, T. Sakiyama, and K. Nakanishi. 1994. Synthesis of dipeptide precursors with an immobilized thermolysin in ethyl acetate. *Biotechnol. Bioeng.* 43:1108–1117.

Nakada, T., K. Maruta, K. Tsusaki et al. 1995a. Purification and properties of a novel enzyme, maltooligosyl trehalose synthase, from *Arthrobacter* sp. Q36. *Biosci. Biotechnol. Biochem.* 59:2210–2214.

Nakada, T., K. Maruta, H. Mitsuzumi et al. 1995b. Purification and properties of a novel enzyme, maltooligosyl trehalose trehalohydrolase, from *Arthrobacter* sp. Q36. *Biosci. Biotechnol. Biochem.* 59:2215–2218.

Nandakumar, R., K. Yoshimune, M. Wakayama, and M. Moriguchi. 2003. Microbial glutaminase: Biochemistry, molecular approaches and applications in the food industry. *J. Mol. Cat. B; Enzymatic* 23:87–100.

Neto, P. O., and P. T. P. Menao. 2009. Isomaltulose production from sucrose by *Protaminobacter rubrum* immobilized in calcium alginate. *Biores. Technol.* 100:4252–4256.

Nielsen, J. E., and T. V. Borchert. 2000. Protein engineering of bacterial α-amylase. *Biochim. Biophys. Acta* 1543:253–274.

Ogino, H., S. Tsuchiyama, M. Yasuda, and N. Doukyu. 2010. Enhancement of the aspartame precursor synthetic activity of an organic solvent-stable protease. *Protein Eng. Des. Sel.* 23:147–152.

Onishi, Y., S. Yano, J. Thongsanit, K. Takagi, K. Yoshimune, and M. Wakayama. 2011. Expression in *Escherichia coli* of a gene encoding type II L-asparaginase from *Bacillus subtilis*, and characterization of its properties. *Ann Microbiol.* 61:517–524.

Ohsu, T., Y. Amino, H. Nagasaki et al. 2010. Involvement of the calcium-sensing receptor in human taste perception. *J. Biol. Chem.* 285:1016–1022.

Ohtake, S., and Y. J. Wang. 2011. Trehalose: Current use and future applications. *J. Pharm. Sci.* 100:2020–2053.

Owen, G. N., H. Parnell, E. A. D. Bruin, and J. A. Rycroft. 2008. The combined effects of L-theanine and caffeine on cognitive performance and mood. *Nutr. Neurosci.* 11:193–198.

Pedreschi, F., K. Kaack, and K. Granby. 2008. The effect of asparaginase on acrylamide formation in French fries. *Food Chem.* 109:386–392.

Placido, D., C. G. Fernandes, A. Isidro, M. A. Carrondo, A. O. Henriques, and M. Archer. 2008. Auto-induction and purification of *Bacillus subtilis* transglutaminase (Tgl) and its preliminary crystallographic characterization. *Protein Expr. Purif.* 59:1–8.

Pleszczynska, M., A. Wiater, and J. Szczodrak. 2010. Mutanase from *Paenibacillus* sp. MP-1 produced inductively by fungal α-1,3-glucan and its potential for the degradation of mutan and *Streptococcus mutans* biofilm. *Biotechnol. Lett.* 32:1699–1704.

Puri, M., and U. C. Banerjee. 2000. Production, purification, and characterization of the debittering enzyme naringinase. *Biotechnol. Adv.* 18:207–217.

Puri, M., A. Banerjee, and U. C. Banerjee. 2005. Optimization of process parameters for the production of naringinase by *Aspergillus niger* MTCC 1344. *Process Biochem.* 40:195–201.

Rasmussen, H., T. la Cour, J. Nyborg, and M. Schuelein. 1994. Structure determination of glucose isomerase from *Streptomyces murinus* at 2.6 Å resolution. *Acta Crystallogr. D Biol. Crystallogr.* 50:124–131.

Rejzek, M., C. E. Stevenson, A. M. Southard et al. 2011. Chemical genetics and cereal starch metabolism: Structural basis of the non-covalent and covalent inhibition of barley β-amylase. *Mol. Biosyst.* 7:718–730.

Sakato, Y. 1949. The chemical constituents of tea: III. A new amide theanine. *Nippon Nogei Kagakukaishi* 23:262–267 (in Japanese).

Saulnier, D. M., D. Molenaar, W. M. deVos, G. R. Gibson, and S. Kolida. 2007. Identification of prebiotic fructo-oligosaccharide metabolism in *Lactobacillus plantarum* WCFS1 through microarrays. *Appl. Environ. Microbial.* 73:1753–1765.

Schindler, A., A. Dunkel, F. Staehler et al. 2011. Discovery of salt taste enhancing arginyl dipeptides in protein digests and fish sauces by means of sensomics approach. *J. Agric. Food Chem.* 59:12578–12588.

Seki, T., Y. Kawasaki, M. Tamura et al. 1990. Further study on the salty peptide ornithyl-β-alanine. Some effects pH and additive ions on the saltiness. *J. Agric. Food Chem.* 38.23–29.

Shimizu, Y., A. Ueyama, and K. Goto. 1991. Purification and characterization of glutaminase from *Bacillus subtilis* GT strain. *J. Brew. Soc. Japan* 66:441–446.

Son, Y. J., R. Chan-Su, P. Yong-Cheol, S. So-Yeon, L. Yoon-Seung, and S. Jin-Ho. 2008. Production of cyclo-dextrins in ultrafiltration membrane reactor containing cyclodextrin glycosyltransferase from *Bacillus macerans. J. Microbiol. Biotechnol.* 18:725–729.

Suzuki, H., S. Izuka, N. Miyakawa, and H. Kumagai. 2002a. Enzymatic production of theanine, an "umami" component of tea, from glutamine and ethylamine with bacterial γ-glutamyltranspeptidase. *Enzyme Microbial. Technol.* 31:884–889.

Suzuki, H., Y. Kajimoto, and H. Kumagai. 2002b. Improvement of the bitter taste of amino acids through the transpeptidation reaction of bacterial γ-glutamyltranspeptidase. *J. Agric. Food Chem.* 50:313–318.

Tada, M., I. Shinoda, and H. Okai. 1984. L-Ornithyltaurine, a new salty peptide. *J. Agric. Food Chem.* 32:992–996.

Takeshima, K., A. Suga, and K. Izumori. 2000. Mass production of D-psicose from D-fructose by a continuous bioreactor system using immobilized D-tagatose 3-epimerase. *J. Biosci. Bioeng.* 90:453–455.

Tareke, E., P. Rydberg, P. Karlsson, S. Eriksson, and M. Tornqvist. 2002. Analysis of acrylamide, a carcinogen formed in heated foodstuffs. *J. Agric. Food Chem.* 50:4998–5006.

Toelstede, S., A. Dunkel, and T. Hofmann. 2009. A series of kokumi peptides impart the long-lasting mouthful-ness of matured Gouda cheese. *J. Agric. Food Chem.* 57:1440–1448.

Umakoshi, M., T. Hirasawa, C. Furusawa, Y. Takenaka, Y. Kikuchi, and H. Shimizu. 2011. Improving protein secretion of a transglutaminase-secreting *Corynebacterium glutamicum* recombinant strain on the basis of ^{13}C metabolic flux analysis. *J. Biosci. Bioeng.* 112:595–601.

Urlaub, H., and G. Woeber. 1975. Identification of isoamylase, glycogen-debranching enzyme, from *Bacillus amyloliquefaciens. FEBS Lett.* 57:1–4.

Vinje, M. A., D. K. Willis, S. H. Duke, and C. A. Henson. 2011. Differential expression of two β-amylase genes (*Bmy*1 and *Bmy*2) in developing and mature barley grain. *Planta* 233:1001–1010.

Vulevic, J., A. Juric, G. Tzortzis, and G. R. Gibson. 2013. A mixture of trans-galactooligosaccharides reduces markers of metabolic syndrome and modulates the fecal microbiota and immune function of overweight adults. *J. Nutr.* 143:324–331.

Wakisaka, S., Y. Ohshima, M. Ogawa, T. Tochikura, and T. Tachiki. 1998. Characterization and efficiency of glutamine production by coupling of a bacterial glutamine synthetase reaction with the alcoholic fermen-tation system of Baker's yeast. *Appl. Environ. Microbiol.* 64:2952–2957.

Winkler, K., I. Kienle, M. Burgert, J. C. Wagner, and H. Holzer. 1991. Metabolic regulation of the trehalose content of vegetative yeast. *FEBS Lett.* 291:269–272.

Yano, T., M. Ito, K. Tomita, H. Kumagai, and T. Tochikura. 1988. Purification and properties of glutaminase from *Aspergillus oryzae. J. Ferment. Technol.* 66:137–143.

Yano, S., R. Minato, J. Thongsanit, T. Tachiki, and M. Wakayama. 2008. Overexpression of type I asparaginase of *Bacillus subtilis* in *Escherichia coli*, rapid purification and characterization of recombinant type I L-asparaginase. *Ann. Microbiol.* 58:711–716.

Yao, M., Y. Yasutake, H. Morita, and I. Tanaka. 2005. Structure of the type I L-asparaginase from the hyperther-mophilic archaeon *Pyrococcus horikoshii* at 2.16 Å resolution. *Acta Cryst. D* 61:294–301.

Ying, G. Q., S. Lu-E, Y. Yi, Z. X. Tang, and J. S. Chen. 2006. Production, purification and characterization of nuclease p1 from *Penicillium citrinum. Process Biochem.* 41:1276–1281.

Yokozaki, K., and S. Hara. 2005. A novel and efficient enzymatic method for the production of peptides from unprotected starting materials. *J. Biotechnol.* 115:211–220.

Zhang, G., Y. Miura, and K. Yagasaki. 2001. Inhibitory effects of theanine and sera from theanine-fed rats on receptor-mediated cancer cell invasion beneath mesothelial-cell monolayers. *Cytotechnology* 36:195–200.

Zhang, L., L. Zhang, H. Yi et al. 2012. Enzymatic characterization of transglutaminase from *Streptomyces mobaraensis* DSM 40587 in high salt and effect of enzymatic cross-linking yak milk proteins on functional properties of stirred yogurt. *J. Dairy Sci.* 95:3559 3568.

Zheng, G., K. Sayama, T. Okubo, L. R. Juneja, and L. Oguni. 2004. Anti-obesity effects of three major components of green tea, catechins, caffeine and theanine, in mice. *In Vivo* 18:55–62.

Zyzak, D. V., R. A. Sanders, M. Stojanovic et al. 2003. Acrylamide formation mechanism in heated foods. *J. Agric. Food Chem.* 51:4782–4787.

13

Enzymes in Milk, Cheese, and Associated Dairy Products

Utpal Roy, Ravinder Kumar, Sanjeev Kumar, Monica Puniya, and Anil K. Puniya

CONTENTS

13.1 Introduction

Natural biocatalysts or enzymes were originally obtained from living cells in the twentieth century, and later on, molecular characterization was carried out. Urease was the first enzyme invented by James B. Sumner in 1926 that started the actual journey of enzyme application. Applications of enzymes are diverse, and they are contributing in pharmaceutical, food, beverage, leather, meat, and environmental applications. Apart from these, enzymes have numerous commercial applications in the food industry. The commercialization required enormous efforts to optimize large-scale production and deep characterization. All of these processes relied on enzymes produced by spontaneously growing microorganisms, or preexisting enzymes are added in preparations obtained from different sources, such as calves, rumen, papaya, and grape. The huge ratio of enzymes currently used in industries are hydrolytic in nature, and the major application is to degrade natural components or in biotransformation (Klik et al. 2002).

Among the various food sectors, dairy is an important industry as dairy foods are popular and common (for instance, cheese, yogurt, etc.) worldwide whereas other foods (e.g., tofu, miso, tempeh, idli, pizza, etc.) remain geographically region-specific. The per capita consumption of dairy foods is increasing year by year. The production and quality improvement of milk and milk products involve various enzymes (chymosin, lactase, lipase, etc.). The use of enzymes in dairy is very old as rennet has been used since 6000 BC for cheese production. After the rennet, proteinases, lipases, beta-galactosidases, superoxide dismutase, glucose oxidase, and sulphydryl oxidase are other important dairy enzymes (Kumar et al. 2010; Law 2002; Neelakantan et al. 1999). The enzymes used in the dairy sector may be from microbial or nonmicrobial origins. Rennet is presently produced by both animal and microbial origin.

Several microbes, for example, *Rhizomucor pusillus, R. miehei, Endothia parasitica, Aspergillus oryzae*, and *Irpex lactis*, are used extensively for rennet production in cheese manufacture. Calf and buffalo chymosin has been successfully produced by introducing the animal chymosin gene into the microbial systems (Kumar et al. 2010; Mohanty et al. 1999; Vallejo et al. 2012). In this chapter, applications of enzymes in the dairy industry, particularly in milk coagulation, for cheese production, and other therapeutic dairy product formulation, for example, low-lactose milk, are discussed.

13.2 Cold Active Enzymes: Utility in Dairy Industry

Different kinds of enzymes have been recovered from organisms living in different habitats, for example, hot water springs, Antarctica, high-altitude hills, rivers, and marine conditions. Among these, cold-region microbes have proven potential in various industrial processes. Numerous microorganism bacteria, yeasts, unicellular algae, and fungi (*Polaromonas, Pseudoalteromonas, Psychroflexus, Moraxella, Psychrobacter, Arthrobacter, Micrococcus, Bacillus, Halorubrum, Methanogenium, Penicillium, Cryptococcus, Candida*, and *Cladosporium*) have showed potential for colonizing in cold environments and therefore have developed adaptations enabling them to compensate for the adverse effects of low temperatures (Gerday et al. 2000; Joshi and Satyanarayana 2013; Marx et al. 2007). Microorganisms demonstrated their strange ability to adapt to extreme cold or hot environments by modifying their own biocatalyst, and therefore, enzymes from psychrophiles have become conspicuous candidates for various food and industrial applications. Additionally, productions of metabolites of psychrophilic microorganisms carried out at low temperatures always remain beneficial as they lower the chances of contamination of mesophilic microbes.

The psychrophilic microbes have produced several enzymes that act at low temperatures. The enzyme having high catalytic efficiency with reduced activation energy at low temperatures is known as a cold-adapted enzyme. Reduced activation energy results in high catalytic activity, and this characteristic of cold-active enzymes is imparted by the native or flexibility of the protein structure (Deming 2002). Enzymes (cold-active enzymes) from species living at low temperatures are adapted to perform their catalytic function at low temperatures (viz. 0°C–10°C), and therefore, they are more suitable for low-temperature industrial applications (Spiwok et al. 2007). The enzymes produced by cold-adapted microbes have an economical edge by being more productive than mesophilic or thermophilic homologues at low temperatures, thereby helping energy savings. On the other hand, it has been observed that enzymes that originated from mesophilic and thermophilic microbes can be stable and may exhibit a native conformation or structure at low temperatures; however, they lose their catalytic efficiency or show poor enzymatic activity. The cold-adapted enzymes offer another advantage of avoidance of unwanted chemical reactions that may occur at elevated temperatures particularly for heat-sensitive substrates (Jeon et al. 2009). Cold-active enzymes have different conformational and structural properties, such as reduced core hydrophobicity, but higher surface hydrophobicity; poor secondary structure; weak interdomain and intersubunit interaction, and longer loops. The amino acid composition of these enzymes also varies as compared to mesophilic and thermophilic originated enzymes, that is, a high concentration of glycine residue, low ratio of arginine to lysine, and low amount of proline in the loop while being higher in α-helices. These enzymes possess weak metal-binding sites, a weak disulfide bridge, small electrostatic bonding, and high structural entropy in the nonfolded state (Cavicchioli and Siddiqui 2006; Cavicchioli et al. 2011; Joshi and Satyanarayana 2013; Saunders et al. 2003; Siddiqui and Cavicchioli 2006). For instance, cold-adapted α-amylase from *Pseudoalteromonas haloplanktis* and citrate synthase revealed increased local structural flexibility at the active site and gives these enzymes more efficiency at low temperatures in unique ways (Bjelic et al. 2008; Siddiqui et al. 2005; Spiwok et al. 2007). Similarly, xylanase obtained from the Antarctic bacterium *Pseudoalteromonas haloplanktis* showed greater activity in comparison with enzymes of the mesophilic family (Collins et al. 2003).

These natural characteristics of cold-adapted enzymes bear significance for the food and dairy industry in which it is essential to decimate spoilage and change in nutritional value and flavor of the original heat-labile substrates and products (Cavicchioli et al. 2002; Gerday et al. 2000; Margesin et al. 2003; Russell 1998; Tutino et al. 2009). A cold-adapted lipase from *Pseudomonas* strain P38 found use in nonaqueous biotransformation for the synthesis in n-heptane of the flavoring compound butyl caprylate and cheese ripening (Tan et al. 1996).

Cold-active enzymes from various sources demonstrate their potential in cheese making in different ways. Rennet, commonly used in cheese preparation, helps the coagulation process. Rennet includes several enzymes, such as rennin and protease. Rennin and protease work in combination to facilitate curding and precipitation, which separates the milk from the curds. They break down the caseins, which help keep milk in its liquid form (Hildebrandt et al. 2009; Hoyoux et al. 2001; Wang et al. 2008). β-Galactosidase is also used to reduce lactose responsible for severe induced intolerances in approximately two thirds of the world's population. Cold-active β-galactosidases have been used to treat dairy products under mild conditions to avoid spoilage and changes in the taste and nutritional value, and cold-active β-galactosidases could be inactivated at a low temperature without heat treatment (Margesin and Feller 2010; Margesin and Schinner 1994). All of these enzymes improve the yield of dairy production processes in multiple ways.

The major drawback in the use of cold-active enzymes for some applications is their weak thermostability, and protein engineering is now potentially emerging to combat this challenge to overcome this thermolability problem (Margesin and Feller 2010; Margesin and Schinner 1999).

13.3 Enzymes in Dairy Industry

There are various enzymes (more than 60) that are used in different dairy food production processes. A few important enzymes (Table 13.1) along with their applications are discussed in this section of the chapter.

13.3.1 Lactase

Lactose is an important disaccharide present in milk at about 4.7% (w/v) concentration. Although milk is considered to be a complete food, a large fraction of the world's adult population is fraught with lactose intolerance and has difficulty in consuming milk and dairy products. Milk consumption in lactose-intolerant subjects, including children, has been reported to cause severe tissue dehydrating diarrhea, bloating, stomach cramps, and even death. Therefore, the hydrolysis of lactose in milk helps prevent digestive problems caused by lactose intolerance. Lactose hydrolysis offers several advantages; it enhances the sweetness of the hydrolyzed milk and improves scoop and richness of cream in ice cream, yogurt, and frozen desserts (Kosseva et al. 2009; Oliveira et al. 2011; Panesar et al. 2006, 2010).

To hydrolyze the lactose, the lactase or β-galactosidase (EC. 3.2.1.23) enzyme is used, which helps to combat the problem of lactose intolerance. Lactase catalyzes the reaction and breaks the β-glycosidic bond in lactose to yield galactose and glucose. A number of microorganisms known to produce lactase enzymes are *Aspergillus niger*, *A. oryzae*, *Kluveromyces fragilis*, and *K. lactis*. Each microbe has its own metabolic system of production of lactase, for instance, *Aspergillus* spp. secretes β-galactosidase into the extracellular medium whereas *Kluyveromyces* spp. produces intracellular β-galactosidase. Hence, the breakdown of lactose in *Kluyveromyces* spp. is carried out in the intracellular pathway. Permease helps in the transportation of lactose, and afterward, it is hydrolyzed by lactose to yield glucose and galactose (Domingues et al. 2010). The metabolite produced by microbes should be safe, and the lactase produced by the abovementioned microorganisms has GRAS status, which makes it suitable in human food-related applications. Lactase produced by these fungal strains exhibits maximum catalytic efficiency in the pH range 2.5–5.4 and up to 50°C in temperature (González-Siso 1996; Kosseva et al. 2009; Panesar et al. 2006), and yeast-originated enzymes revealed maximum activity in the neutral pH range (6.0–7.0). Thus, lactase produced by yeast is more favorable for hydrolysis of milk and sweet whey (Kosseva et al. 2009; Oliveira et al. 2011; Panesar et al. 2006; Yang and Silva 1995; Zadow 1984). Commonly, lactase is used to hydrolyze the acid whey left over after the production of fresh or soft cheese.

Moreover, hydrolyzing lactose into glucose and galactose has a number of applications, particularly for lactose utilization present in a sufficient concentration in whey. Hydrolysis of lactose offers a possible solution to the disposal problem posed by cheese whey when the product turns several times sweeter,

TABLE 13.1

Enzymes Used in Dairy Industries

Enzymes	Enzyme Class	Source(s)	Applications	Reference(s)
Proteinases and peptidases	EC. 3.4.23.23	Lactic acid bacteria (*Lactobacillus acidophillus*, *Lb. plantarum*, *Lb. delbrueckii* sp. *bulgaricus*, *Lb. lactis*, and *Lb. helveticus*)	Dairy-based flavoring preparations, for example, cheese and milk	Ardo et al. 1989; Chapline and Buckc 1990; Coulson et al. 1991; Lynch et al. 1999; Neelakantan et al. 1999
Lysozyme	EC. 3.2.1.17	Hen egg white or *Micrococcus lysodeikticus*, egg albumin, cow milk, human milk, papaya juice	Nitrate replacer for washed-curd cheeses and cheeses with eyes (e.g., Emmental)	Law 2002; Law and Goodenough 1995; Liburdi et al. 2014
Lipase	EC. 3.1.1.3	*Rhizopus arrhizus*, *R. miehei*, *P. roqueforti*, *P. candidum*, and *A. niger*	Cheese flavor, production of cocoa butter, and emulsifier	Benjamin and Pandey 1998; Ghanem 2007; Haki and Rakshit 2003; Joseph et al. 2008; Law 2002; Pandey et al. 1999; Verma et al. 2012
Lactoperoxidase	EC. 1.11.1.7	Milk, saliva, and colostrum	Cold sterilization of milk: milk replacers for calves	Law 2002; Law and John 1981; Neelakantan et al. 1999; Reiter and Marshall 1979
Lactase or β-galactosidase	EC. 3.2.1.23	*Aspergillus niger*, *A. oryzae*, *Kluveromyces fragilis*, and *K. lactis*	Lactose-reduced whey products, galacto-oligosaccharides production, low-lactose milk, whey hydrolysis for ethanol production	González-Siso 1996; Gosling et al. 2010; Mahmoud and Helmy 2009; Oliveira et al. 2011; Panesar et al. 2006; Park and Oh 2010; Yang and Silva 1995
Glucose oxidase	EC. 1.1.3.4	*Aspergillus niger*, *P. glaucum*, and *P. amagasakiense*	Preservation of dairy foods	Neelakantan et al. 1999; Ramachandran et al. 2006
Bovine rennet (chymosin, pepsin)	EC. 3.4.23.4, EC. 3.4.23.1	Animal (calf, goat, sheep, rabbit, cow, and buffalo), Plant microbes (*R. miehei*, *A. oryza*, *Mucor pussillus*, *M. meihei*, and *Irpex lactis*)	Clotting in cheese preparation, cheddar cheese, cottage cheese, cream cheese, cream cheese spread, cream cheese spread with (naming the added ingredients)	Fernandez 2010; Fox et al. 1996; Kumar et al. 2010; Mohanty et al. 1999; Neelakantan et al. 1999; Noseda et al. 2014;

and whey may be concentrated to microbiologically secured syrup (70%, w/v). In cheese making, the supernatant remaining after coagulation of milk proteins contains a good amount of lactose, and whey cannot be easily disposed of in the environment because of high oxygen demand. Around 160 million tons of whey is left over after cheese production annually, worldwide. High levels of BOD (30–50 g/L) and COD (60–80 g/L) are needed to degrade the whey (Guimarães et al. 2010). Lactose present in whey is the sole reason for this high COD and BOD value. β-Galactosidases have been proven to be the best alternate to overcome this bio burden and to explore the new application of whey-based dairy products. In recent years, hydrolyzed whey has been used to produce several value-added food products as well other important molecules (Oliveira et al. 2011). For instance, *Saccharomyces cerevisiae* is used to produce ethanol from hydrolyzed whey although catabolite repression-resistant mutants must be applied in order to avoid glucose–galactose diauxy (Bailey et al. 1982; Terrell et al. 1984).

Moreover, hydrolyzed whey syrup can be used in various dairy foods. Hydrolyzed whey syrup is produced from whey permeate at temperatures of 35°C to 50°C, the by-product of cheese making and casein production. The hydrolysis step can be on the whey itself or on the whey permeate derived from an ultrafiltration plant used to make whey protein concentrate. The ultrafiltered from permeate still contains some whey protein but is enriched with lactose (Law 2002). Afterward, it is allowed to be concentrated by up to 15%–20% total solids, demineralized by ion exchange through nanofiltration, then heated according to the type of lactase treatment to be used. The hydrolysis of whey can be achieved using the *K. lactis* immobilized enzyme reactor or *Aspergillus*-originated β-galactosidases. However, both processes produce sufficient free glucose and galactose to make the product sweet, and this (sweetening) property is enhanced by evaporation of the hydrolysate to 60% TS to make the final syrup (Law 2002). The whey syrup is made sticky by the high concentration of glucose and galactose and thus not completely dried to powdered form, but it is sold and used in that form. This hydrolyzed whey syrup is used in ice cream, milk desserts, and sauces. The syrup is also an excellent caramel ingredient and as a sweetener in cereal bars (Law 2002).

Generally, for such processes, immobilized β-galactosidase is used, and various immobilization strategies have been adopted, such as fiber entrapment, gel entrapment, etc., using sepharose, sodium alginate, polyacrylamide gel, cellulose acetate glass, and others. Immobilization of lactase with cellulose sheets has been shown to be one of most successful and suitable strategies.

The role of β-galactosidase in cheese ripening has also been shown. Trials were carried out to accelerate the Ras cheese ripening. Pretreatment of milk was carried out with β-galactosidase enzyme. It was seen that flavor strength and liberation of nitrogen compounds, free amino acids, and free fatty acids was increased in cheese prepared from milk treated with β-galactosidase. Interestingly, the ripening period (2 months) of cheese was reduced in comparison with control (4 months) cheese (Farahat et al. 1985).

β-Galactosidases with transgalactosylation activities are highly attractive candidates for the production of added-value lactose derivatives. In fact, galacto-oligosaccharides (GOS) are increasingly finding application in functional foods, namely as low-calorie sweeteners in fermented milk products and beverages. The use of β-galactosidases for the production of GOS has been recently reviewed (Gosling et al. 2010; Park and Oh 2010). Moreover, the recombinant thermostable β-galactosidase possessed a high level of transgalactosylation activity in parallel with the hydrolysis of lactose in milk. Oligosaccharides are no longer considered to be the undesired products they were in the past (Gosling et al. 2010; Shukla and Wierzbicki 1975). Nowadays, low-lactose milk with GOS is considered to be a value-added food product. Indeed, GOS produced via the transgalactosylation activity of β-galactosidase are utilized as growth-promoting substrates of bifidobacteria in the human intestine (Park and Oh 2010). Also, this lactose needs to be removed from the industrial pollutant (Shukla and Wierzbicki 1975).

Organisms such as *A. niger*, *A. oryzae*. *Candida pseudotropicalis*, and *K. lactis* are the major sources of industrial lactase. Lactase applications are in batch and immobilized enzyme technology, favoring the *Aspergillus* and *Kluyveromyces* sources, respectively. Lactase has been made available in tablet form (e.g., Lactaid®) to treat lactose intolerance. These tablets deliver active lactase to the gut to degrade lactose and reduce intolerance symptoms.

13.3.2 Lipases

Lipases (triacylglycerol acylhydrolases, EC. 3.1.1.3) are serine hydrolases. The enzymatic mechanism of lipase involves the hydrolysis of triacylglycerols to generate free fatty acids, which may be diacylglycerols and monoacylglycerols along with glycerol. Lipases have emerged as one of the leading biocatalysts or bioaccelerators with proven potential for contributing to the multibillion-dollar underexploited lip-tech bioindustry and are used in both in situ lipid metabolism and ex situ multifaceted industrial application (Aravindan et al. 2007; Benjamin and Pandey 1998; Pandey et al. 1999; Verma et al. 2012). Apart from the various applications in pharmaceutical, food, and other related industries, lipase has several applications in dairy food processing. Most lipases are of fungal and bacterial origin. *R. miehei* lipase has been used for Italian cheeses, and lipase from *P. roqueforti* and *P. candidum* has also shown satisfactory enzymatic activity. Lipases are exclusively used for the hydrolysis of milk fat in the dairy industry. Lipolysis of cheese is one of the important applications of this enzyme; for example, lipolysis enhances the flavor of Swiss cheese. Another classic example of lipase application is the peppery flavor of blue cheese. The lipolytic activity of this enzyme yields fatty acids of short chain and methyl ketones, which impart this flavor in blue cheese. Moreover, "palatase" (a form of lipase) prepared from the pancreas of bovine and porcine, *M. miehei, A. niger, P. roqueforti*, and *Rhizopus arrhizus* are used in NOVO processing to produce medium-age cheese, which is used in soups, dips, and dressings of foods. This enzyme is also used to enhance the various cheese (Swiss, Cheddar, blue) attributes, such as flavors, texture, and ripening (Tomasini et al. 1993). The addition of a lipase such as palatase from *M. miehei* and *A. niger*, can potentially influence generation of flavor in the Manchego cheese variety manufactured in central Spain from Manchega ewe's and cow's milk (Fernández-García et al. 1999; Law 2002; Ordóñez et al. 1997). Moreover, the combination of lipases from different sources should be judiciously chosen to control bitterness or development of typical flavor that is critically influenced by lipolytic activity and the ratios of free fatty acid release (Aravindan et al. 2007; Law 2001; Neelakantan et al. 1999; Verma et al. 2012).

Another type of lipase, named phospholipase, also has various applications in the dairy industry, especially in cheese production. Phospholipids are a small quantity (0.5%) of total milk lipids, and their role is in the stabilization of milk fat to avoid coalescence. Apart from this, phospholipids also have various functions in dairy foods, for instance, as coemulsifiers. So to increase the dairy food quality and modification of attributes of phospholipids, the enzyme phospholipases are used. It has been reported that fungal-originated phospholipase decreases the fat loss in whey and also enhances the yield of Mozzarella cheese (Ipsen 2013). It has been shown that hydrolysis of phospholipids enhances the yield of cheese. Phospholipase is also used to improve the foaming quality of whey protein along with enhanced heat stability (Casado et al. 2012; Ipsen 2013).

13.3.3 Rennet

13.3.3.1 Source of Rennet

The majority of microbial rennet is derived from bacterial and fungal strains. Microbial sources were explored in the late 1980s due to the scarcity of animal-originated rennet and to promote vegetarian-friendly cheese. Some microbial rennets are derived from microorganisms that naturally produce chymosin, which is responsible for the coagulation of milk. Numerous microbes have been reported as potential sources of rennet alternates. *A. oryza, Mucor pussilus, M. meihei*, and *Irpex lactis* have been reported as sources of microbial renin. Among the various microbial origins, *Rhizomucor miehei* has been shown to produce a good-quality coagulant (Law 2002; Mohanty et al. 1999; Rao and Dutta 1981). Approximately 50% of world cheese producers are using microbial-originated rennet. Yet animal rennet is dominant due to its special attributes, such as better flavor and cheese texture. On the other hand, microbial rennet sometime causes low yield, flavor, and poor texture. Other types of microbial rennet have been altered in a laboratory to produce chymosin (acid proteinase) by means of the same genes found in a calf's stomach (Kumar et al. 2010; Law 2002; Mohanty et al. 1999; Noseda et al. 2014; Vallejo et al. 2012).

13.3.3.2 Cheese

Cheese is the most famous dairy product throughout the world. Rennet is used to produce cheese. It is composed of many enzymes (rennin and chymosin, EC. 3.4.23.4), which assist baby ruminants to digest milk after suckling. Rennet is a crude extract obtained from the fourth stomach of young unweaned ruminants, such as goats, cows, and sheep, produced as an inactive precursor, prorennin. It is assumed that many animal species, for example, rabbit, goat, sheep, and buffalo, are also potential sources of rennet and need more exploration of these targets for commercial purposes (Mohanty et al. 1999; Rao and Dutta 1981). Approximately 80%–90% of rennet is chymosin (Kumar et al. 2010). The coagulation of milk is carried out by calf rennet by targeting k-casein. The enzyme that targets α- and β-casein for curd formulation results in poor yield of cheese. Apart from chymosin, the other constituent of calf rennet is pepsin (EC. 3.4.23.1). It is believed that pepsin also enhances the ripening attributes of cheese.

In additional to animal sources, many plants (e.g., *Cynara cardunculus*) have also been reported to produce proteinase, which has the ability to coagulate milk. It is also seen that enzymes obtained from plant origin are efficient enough for commercial purposes. However, the flavor of the end product is not comparable with a product prepared from animal-originated enzyme cheese (Almeida et al. 2015).

Rennet causes coagulation of the milk that is imperative for cheese production. Prorennin is converted to rennin by its autocatalysis or by the action of pepsin. It is used on a large scale in the dairy sector to manufacture a quality curd along with pleasant flavor. This enzyme acts at a specific position of the milk protein. It cleaves a single peptide bond in k-casein and yields nonsoluble para-k-casein and C-terminal glycopeptides (Law 2002; Rao et al. 1998). In milk, caseins are secreted, and their self-associate aggregation is called micelles. The alpha and beta caseins present in the micelles are kept away from precipitating by their interactions with k-casein. In essence, k-casein normally keeps the majority of milk protein soluble and prevents it from spontaneous coagulation. In the presence of the enzymes, it gets away from the interaction and forms a clot. The coagulation is done by the enzyme in a two-step process. During the first step, the enzyme produces insoluble para-casein and peptide-soluble glycopeptide, and in the second step, precipitation of para-casein takes place by Ca^{2+}. Chymosin acts on k-casein whereas the other portion of casein remains less attacked. The catalytic efficiency of rennin depends on calcium (Kumar et al. 2010; Law 2002). In the absence of calcium, the para-k-casein gets linked with casein, which is labile to calcium, and avoids precipitation. In the presence of calcium, para-k-casein and calcium-sensitive casein formulate clots. The cleavage target site of chymosin in k-casein lies in between phenylalanine and methionine at the 105–106 position, and the hydrolysis of this bond yields pentapeptide. Chymosin proteolytically digests and inactivates k-casein, converting it into para-kappa-casein and a smaller protein called macropeptide. Para-k-casein does not have the ability to stabilize the micellar structure, and the calcium-insoluble caseins precipitate, forming a curd.

Various types of rennet have been used for coagulation of milk. Phillipinos prepare Kesong puti cheese using buffalo rennet. Another type of rennet (lamb rennet) as compared to calf rennet, produces a much sharper flavor, for example, in Kefalyotri cheese, a hard and salty cheese popular in Greece and Cyprus. Pig or porcine rennet contains pepsin, which is used in combination with calf rennet in equal concentrations. Rennet from calves shows high catalytic activity in the pH range of 6.2–6.4, and pepsin remains highly active at pH 1.7–2.3. Both the enzymes help each other during the coagulation of milk. To find out the best coagulation of milk, different enzyme ratios were used during cheese production, for example, calf rennet/bovine pepsin (50:50 or 25:75 or vice versa). A similar strategy was carried out with a rennet, calf rennet, and porcine pepsin combination. The rennet from microbial sources, that is, bacterial, fungal, fungal and pepsin, and bacterial and pepsin, have also been used in different combinations.

During cheese preparation, rennet mixed with milk to yield coagulum and some of the rennet remains trapped in the cheese and helps break down the cheese structure as the cheese ages. The major enzyme in rennet is renin, and the presence of other enzymes depends on the source of origin. Hence, the presence or absence of enzymes in rennet can affect the ripening of cheese. For instance, generation of flavor during the ripening of different cheeses, for example, Gouda, Cheddar, and Edam, depends on the presence of proteolysis enzymes (Fox et al. 1996). Moreover, chymosin obtained from *M. miehei* is thermostable but causes bitterness in cheese in comparison with heat-labile enzymes. The treatment of enzymes with peracids and hydrogen peroxides or with other oxidizing agents can overcome this problem of off-flavors in cheese as it decreases the thermal ability of the enzymes.

Rennet, along with other enzymes, enhances the flavor quality. Enzyme combinations Flavourzyme (a fungal protease/peptidase complex) entrapped in calcium alginate, (Anjani et al. 2007) k-carragenan, and milk fat (Kailasapathy and Lam 2005) showed attractive results of cheese ripening. Further, immobilization of Flavourzyme (bacterial proteases and Palatase) in liposomes yields quick ripening of Cheddar cheese (Fernandez 2010; Kheadr et al. 2003). Roncal cheese prepared by renneting at high temperature (32°C–37°C) using the same strategy as adopted for Manchego cheese revealed the sharp and piquant flavor after 4 months of ripening (Nuñez et al. 1991; Ordonez et al. 1980).

13.3.4 Proteinases and Peptidases

The processing of cheese involves several other enzymes. Proteinases (EC. 3.4.23.23) are significant enzymes that are used to increase the ripening of cheese in combination with different enzymes. Proteolysis is a desirable attribute to impart flavor and texture in different types of cheese. Proteinase breaks milk protein into small peptides, and their further degradation liberates free amino acids, which are imperative for cheese flavor and ripening. Plasmin, rennet, and proteinases from starter bacteria, or even nonstarter, are the key proteinases used during the processing of cheese (Khadidja et al. 2012; Neelakantan et al. 1999; Soda 1993; Wilkinson 1993; Wilkinson et al. 2002). Proteinases and peptidases from the starter lactic acid bacteria and adventitious nonstarter lactic acid bacterial culture break down casein (αs1, αs2, and κ-casein) into smaller peptides responsible for flavor and soft texture development in Cheddar cheese (Lynch et al. 1999). These proteinases, when used with neutral proteinases, generally enhance cheese flavor. It has been noted that acid proteases, when used solely, increase the bitterness of cheese. Further, it is also documented that lipase in combination with proteinase impart low bitterness, improved cheese flavor, and acidity in cheese.

Dairy fermented products are generally produced with the help of lactic acid bacteria (LAB). The growth of LAB in milk depends on proteolytic or proteinase enzymes. These enzymes are also important for flavor enhancement in fermented dairy products. Amino-peptidases are crucial for flavor development of dairy products during fermentation as these enzymes are able to liberate amino acid from peptides. On the other hand, proteinases break proteins into peptides, ultimately into amino acids, that are useful for the growth of LAB. The proteinases of LAB are composed of proteinases and peptidase (endo, amino, tri). The microorganisms, namely, *Lactobacillus acidophillus*, *Lb. plantarum*, *Lb. delbrueckii* sp. *bulgaricus*, *Lb. lactis*, and *Lb. helveticus,* use the proteolytic system for growth during fermentation of milk. Numerous investigations have shown Cheddar flavor generation, increase in proteolysis, and flavor after the addition of lactobacilli (Khadidja et al. 2012; Lane and Fox 1996; Lynch et al. 1999). It has been reported that lacto-bacilli treated with heat can liberate aminopeptidase to produce debittering in cheese ripening at the initial stage to achieve the desired flavor (Ardo and Mansson 1990; Ardo et al. 1989; Soda 1993). In another study, aminopeptidase accelase composed of six aminopeptidases from *Lactococcus lactis* showed a decrease in the ripening period of flat cheese and enhanced the flavor of cheese (Coulson et al. 1991).

13.3.5 Lactoperoxidase

Lactoperoxidase (EC. 1.11.1.7) has antimicrobial activity against Gram-negative bacteria. This enzyme is naturally present in milk, saliva, and colostrum. Lactoperoxidase converts thiocyanate into hypo-thiocyanate by an oxidation reaction in the presence of hydrogen peroxide. Hypothiocyanate acts as a bactericidal component (Law 2002; Law and John 1981; Seifu et al. 2005). Generally, the level of natu-ral hydrogen peroxide remains low in raw milk that does not induce lactoperoxidase. Hence, commer-cially, hydrogen peroxide is supplemented to activate the lactoperoxidase enzyme system. This method of reduction of microbial load using lactoperoxidase is named cold sterilization, which is very successful in regions in which heat treatment methods are not prevalent. This process is even successful to eradicate microbes at 4°C (Law 2002; Reiter and Marshall 1979; Seifu et al. 2005).

13.3.6 Lysozyme

Lysozyme (EC. 3.2.1.17) is bactericidal in action and widespread. It is used to protect Gouda, Grana Padano, and different varieties of hard and semihard cheeses from the defect "late blowing" caused by *Clostridium tyrobutyricum*. This microbe forms spores that cannot be removed by pasteurization of milk. So lysozyme is used as an alternative to remove the vegetative cells and spores of *Clostridium tyro-butyricum* (Law and Goodenough 1995; Liburdi et al. 2014). Lysozyme remains active for a long dura-tion in cheese. Sometimes, lysozyme suppresses the growth of LAB also, which is necessary for cheese processing. However, LAB is less sensitive to lysozyme. Lysozyme is also used in yogurt to arrest the growth of *Listeria monocytogenes*. Moreover, this enzyme has some limitations, which make it uncom-mon in fermented dairy products. Second, fermented products have low pH that can affect the growth of undesired microflora (Law 2002; Law and Goodenough 1995; Liburdi et al. 2014).

13.3.7 Glucose Oxidase

Glucose oxidase (EC. 1.1.3.4) belongs to the oxo-reductase class of the enzymes. It catalyzes oxidation of glucose and converts it into D-glucono-δ-lactone and hydrogen peroxide. Glucose oxidase is generally used for preservation of dairy foods (Ramachandran et al. 2006). The major applications of this enzyme are glucose and oxygen removal and generation of acid and hydrogen peroxide. The enzyme catalase further uses this generated hydrogen peroxide and helps in preservation. Apart from this, glucose oxidase has been used in production of cottage and mozzarella cheese along with acids. Generally, HCl and/or lactic acid are used as acidogens for these cheese varieties. Glucose oxidase removes glucose and oxygen and lowers the pH of milk, which leads to an increase in coagulum. The yield of coagulum was higher when glucose oxidase was used along with acidogens (Neelakantan et al. 1999; Ramachandran et al. 2006).

13.3.8 Other Enzymes Used in Dairy Products

Enzymes, such as transglutaminase, sulphydryl oxidase, catalase, and superoxide dismutase, are used in different dairy food applications. But the role of these enzymes is limited (Neelakantan et al. 1999). Catalase, along with glucose oxidase, is helpful in preservation of dairy foods, and the enzyme superoxide dismutase is used as an antioxidant. Sulphydryl oxidase is used to remove sulphydryl compounds from heat-treated milk, which cause off-flavor. Moreover, transglutaminase is used for the gelation of caseins and whey proteins. Additionally, this enzyme also reduces the syneresis in acid milk gels and enhances texture with the shelf life of yogurt (Neelakantan et al. 1999).

13.4 Conclusions

The use of enzymes in the dairy sector is witnessing rapid increase day by day for a variety of applications. The worldwide production of enzymes for dairy food manufacturing is huge. But enzyme producers are focusing on the production of few enzymes. Nevertheless, advancement in technology and processes is poised to attract manufacturers for production of different kinds of enzymes for making dairy products. Many novel enzyme genes are getting identified, and enzymes with new and exciting or improved properties are likely to be discovered or evolved. It can be assumed that production of dairy enzymes will be at its height as demand for value-added food is increasing and act as an inducer for the production of enzymes.

REFERENCES

Almeida, C.M., Gomes, D., Faro, C. and Simões, I. 2015. Engineering a cardosin B-derived rennet for sheep and goat cheese manufacture. *Applied Microbiology and Biotechnology* 99(1):269–281.

Anjani, K., Kailasapathy, K. and Phillips, M. 2007. Microencapsulation of enzymes for potential application in acceleration of cheese ripening. *International Dairy Journal* 17(1):79–86.

Aravindan, R., Anbumathi, P. and Viruthagiri, T. 2007. Lipase applications in food industry. *Indian Journal of Biotechnology* 6:141–158.

Ardo, Y. and Mansson, H.L. 1990. Heat treated lactobacilli develop desirable aroma in low-fat cheese. *Scandinavian Dairy Information* 4(1):38–40.

Ardo, T., Larsson, P., Mansson, L. and Hedenberg, A. 1989. Studies on peptidolysis during early maturation and its influence on low-fat cheese quality. *Milchwissenschaf't* 44:485–490.

Bailey, R.B., Benitez, T. and Woodward, A. 1982. *Saccharomyces cerevisiae* mutants resistant to catabolite repression—Use in cheese whey hydrolysate fermentation. *Applied and Environmental Microbiology* 44:631–639.

Benjamin, S. and Pandey, A. 1998. *Candida rugosa* lipases: Molecular biology and versatility in biotechnology. *Yeast* 14(12):1069–1087.

Bjelic, S., Brandsdal, B.O. and Aqvist, J. 2008. Cold adaptation of enzyme reaction rates. *Biochemistry* 47:10049–10057.

Casado, V., Martín, D., Torres, C. and Reglero, G. 2012. Phospholipases in food industry: A review. *Methods in Molecular Biology* 861:495–523.

Cavicchioli, R. *and* Siddiqui, K.S. 2006. Cold-adapted enzymes. In: *Enzyme Technology*, eds. A. Pandey, C. Webb, C.R., Soccol and C. Larroche, 615–638. New York: Springer Science.

Cavicchioli, R., Siddiqui, K.S., Andrews, D. and Sowers, K.R. 2002. Low-temperature extremophiles and their applications. *Current Opinion in Biotechnology* 13:253–261.

Cavicchioli, R., Charlton, T., Ertan, H., Omar, S.M., Siddiqui, K.S. and Williams, T.J. 2011. Biotechnological uses of enzymes from psychrophiles. *Microbial Biotechnology* 4:449–460.

Chaplin, M.F. and Bucke, C. 1990. *Enzyme Technology*. Cambridge: Cambridge University Press.

Collins, T., Meuwis, M.A., Gerday, C. and Feller, G. 2003. Activity, stability and flexibility in glycosidases adapted to extreme thermal environments. *Journal of Molecular Biology* 328:419–428.

Coulson, J., Pawlett, D. and Wivell, R. 1991. Making high quality low-fat cheese. *Dairy, Industries International* 56(7):31–32.

Deming, J.D. 2002. Psycrophiles and polar regions. *Current Opinion in Microbiology* 5:301–309.

Domingues, L., Guimarães, P.M. and Oliveira, C. 2010. Metabolic engineering of *Saccharomyces cerevisiae* for lactose/whey fermentation. *Bioengineered Bugs* 1:164–171.

Farahat, S.M., Rabie, A.M., Baky, A.A.A., El-Neshawy, A.A. and Mobasher, S. 1985. β-Galactosidase in the acceleration of Ras cheese ripening. *Nahrung* 29:247–254.

Fernandez, P. 2010. Enzymes in food processing: A condensed overview on strategies for better biocatalysts. *Enzyme Research*. Article ID 862537, 19 pp.

Fernández-García, E., Tomillo, J. and Nuñez, M. 1999. Effect of added proteinases and level of starter culture on the formation of biogenic amines in raw milk Manchego cheese. *International Journal of Food Microbiology* 52:189–196.

Fox, P.F., Wallace, J.M., Morgan, S., Lynch, C.M., Niland, E.J. and Tobin, J. 1996. Acceleration of cheese ripening. *Antonie van Leeuwenhoek* 70:271–297.

Gerday, C., Aittaleb, M., Bentahir, M., Chessa, J.P., Claverie, P., Collins, T., D'Amico, S., Dumont, J., Garsoux, G. and Georges, F. 2000. Cold-adapted enzymes: From fundamentals to biotechnology. *Tibetech* 18:103–107.

Ghanem, E.H., AlSayeed, H.A. and Saleh, K.M. 2000. An alkalophilic thermostable lipase produced by a new isolate of *Bacillus alcalophilus*. *World Journal of Microbiology and Biotechnology* 16:459–464.

González-Siso, M.I. 1996. The biotechnological utilization of cheese whey: A review. *Bioresource Technology* 57(1):1–11.

Gosling, A., Stevens, G.W., Barber, A.R., Kentish, S.E. and Gras, S.L. 2010. Recent advances refining galactooligosaccharide production from lactose. *Food Chemistry* 121:307–318.

Guimarães, P.M., Teixeira, J.A. and Domingues, L. 2010. Fermentation of lactose to bio-ethanol by yeasts as part of integrated solutions for the valorisation of cheese whey. *Biotechnology Advance* 28(3):375–384.

Haki, G.D. and Rakshith, S.K. 2003. Developments in industrially important thermostable enzymes: A review. *Bioresource Technology* 89:17–34.

Hildebrandt, P., Wanarska, M. and Kur, J. 2009. A new cold-adapted β-D-galactosidase from the Antarctic Arthrobacter sp. 32c—Gene cloning, overexpression, purification and properties. *BMC Microbiology* 9:151.

Hoyoux, A., Jennes, I., Dubois, P., Genicot, S., Dubail, F., François, J.M., Baise, E., Feller, G. and Gerday, C. 2001. Cold-adapted beta-galactosidase from the Antarctic psychrophile *Pseudoalteromonas haloplanktis*. *Applied and Environmental Microbiology* 67:1529–1535.

Ipsen, R. 2013. Use of phospholipases to modify phospholipid functionality in dairy processing. *Journal of Dairy Science* 96(1):202 (E-supplement).

Jeon, J., Kim, J.T. and Kang, S. 2009. Characterization and its potential application of two esterases derived from the Arctic sediment metagenome. *Marine Biotechnology* 11:307–316.

Joseph, B., Ramteke, P.W. and Thomas, G. 2008. Cold active microbial lipases: Some hot issues and recent developments. *Biotechnology Advances* 26:457–470.

Joshi, S. and Satyanarayana, T. 2013. Biotechnology of cold-active proteases. *Biology* 2:755–783.

Kailasapathy, K. and Lam, S.H. 2005. Application of encapsulated enzymes to accelerate cheese ripening. *International Dairy Journal* 15:929–939.

Khadidja, B., Salima, R., Halima, Z.K. and Nour Eddine, K. 2012. Specific aminopeptidases of indigenous *Lactobacillus brevis* and *Lactobacillus plantarum*. *African Journal of Biotechnology* 11(88):15438–15445.

Kheadr, E.E., Vuillemard, J.C. and Ei-Deeb, S.A. 2003. Impact of liposome-encapsulated enzyme cocktails on Cheddar cheese ripening. *Food Research International* 36(3):241–252.

Kirk, O., Borchert, T.V. and Fuglsang, C.C. 2002. Industrial enzyme applications. *Current Opinion in Biotechnology* 13(4):345–351.

Kosseva, M.R., Panesar, P.S., Kaur, G. and Kennedy, J.F. 2009. Use of immobilised biocatalysts in the processing of cheese whey. *International Journal of Biological Macromolecules* 45:437–447.

Kumar, A., Grover, S., Sharma, J. and Batish, V.K. 2010. Chymosin and other milk coagulants: Sources and biotechnological interventions. *Critical Review in Microbiology* 30(4):243–258.

Lane, C.N. and Fox, P.F. 1996. Contribution of starter and adjunct lactobacilli to proteolysis in Cheddar cheese during ripening. *International Dairy Journal* 6:715–728.

Law, B.A. 2001. Controlled and accelerated cheese ripening: The research base for new technologies. *International Dairy Journal* 11:383–398.

Law, B.A. 2002. Enzymes in the manufacture of dairy products. In: *Enzymes in Food Technology*, eds. R.J. Whitehurst and B.A. Law, 91–108. Boca Raton, FL: CRC Press.

Law, B.A. and John, P. 1981. Effect of LP bactericidal system on the formation of the electrochemical proton gradient in *E. coli. Microbiological Letter* 10:67–70.

Law, B.A. and Goodenough, P.W. 1995. Enzymes in milk and cheese production. In: *Enzymes in Food Technology*, 2nd edition, eds. G.A. Tucker and L.F.G. Woods, 114–143. Glasgow: Blackie Academic & Professional.

Liburdi, K., Benucci, I. and Esti, M. 2014. Lysozyme in wine: An overview of current and future applications. *Comprehensive Reviews in Food Science and Food Safety* 13:1062–1073.

Lynch, C.M., Muir, D.D., Banks, J.M., McSweeny, P.L.H. and Fox, P.F. 1999. Influence of adjunct cultures of *Lactobacillus paracasei* ssp. *paracasei* or *Lactobacillus plantarum* of Cheddar cheese ripening. *Journal of Dairy Science* 82:1618–1628.

Mahmoud, A.R.D. and Helmy, A.W. 2009. A novel cold-active and alkali-stable β-glucosidase gene isolated from the marine bacterium *Martelella mediterrânea. Australian Journal of Basic and Applied Sciences* 3:3808–3817.

Margesin, R. and Schinner, F. 1994. Properties of cold-adapted microorganisms and their potential role in biotechnology. *Journal of Biotechnology* 33(1):1–14.

Margesin, R. and Schinner, F. 1999. *Biotechnological Applications of Cold-adopted Organisms*. New York: Springer Science.

Margesin, R. and Feller, G. 2010. Biotechnological applications of psychrophiles. *Environmental Technology* 31:835–844.

Margesin, R., Gander, S., Zacke, G., Gounot, A.M. and Schinner, F. 2003. Hydrocarbon degradation and enzyme activities of cold-adapted bacteria and yeasts. *Extremophiles* 7:451–458.

Marx, J.C., Collins, T., D'Amico, S., Feller, G. and Gerday, C. 2007. Cold-adapted enzymes from marine Antarctic microorganisms. *Marine Biotechnology (NY)* 9:293–304.

Mohanty, A.K., Mukhopadhyay, U.K., Grover, S. and Batish, V.K. 1999. Bovine chymosin: Production by rDNA technology and application in cheese manufacture. *Biotechnology Advances* 17(2–3):205–217.

Neelakantan, S., Mohanty, A.K. and Kaushik, J.K. 1999. Production and use of microbial enzymes for dairy processing. *Current Science* 77(1):143–148.

Noseda, D.G., Blasco, M., Recúpero, M. and Galvagno, M.A. 2014. Bioprocess and downstream optimization of recombinant bovine chymosin B in *Pichia* (*Komagataella*) *pastoris* under methanol-inducible AOXI promoter. *Protein Expression and Purification* 104:85–91.

Nuñez, M., Guillen, A.M., Rodriguez-Marin, M.A., Marilla, M.A., Gaya, P. and Medina, M. 1991. Accelerated ripening of ewes' milk Manchego cheese: The effect of neutral proteinases. *Journal of Dairy Science* 74(12):4108–4118.

Oliveira, C., Guimarães, P.M.R. and Domingues, L. 2011. Recombinant microbial systems for improved β-galactosidase production and biotechnological application. *Biotechnology Advances* 29:600–609.

Ordonez, J.A., Masso, J.A., Marmol, M.P. and Ramos, M. 1980. Contribution to the study of Roncal cheese. *Lait* 60:283–294.

Ordóñez, A.I., Ibañez, F.C., Torre, P. and Barcina, Y. 1997. Formation of biogenic amines in Idiazábal ewe's-milk cheese: Effect of ripening, pasteurization and starter. *Journal of Food Protection* 60:1371–1375.

Pandey, A., Benjamin, S., Soccol, C.R., Nigam, P., Krieger, N. and Soccol, V.T. 1999. The realm of microbial lipases in biotechnology. *Biotechnology and Applied Biochemistry* 29:119–131.

Panesar, P.S., Panesar, R., Singh, R.S., Kennedy, J.F. and Kumar, H. 2006. Microbial production, immobilization and applications of β-D-galactosidase. *Journal of Chemical Technology and Biotechnology* 81:530–543.

Panesar, P.S., Kumari, S. and Panesar, R. 2010. Potential applications of immobilized β-galactosidase in food processing industries. *Enzyme Research* 10:1–16. Article ID 473137, 16 pp.

Park, A.R. and Oh, D.K. 2010. Galacto-oligosaccharide production using microbial beta-galactosidase: Current state and perspectives. *Applied Microbiology and Biotechnology* 85:1279–1286.

Ramachandran, S., Fontanille, P., Pandey, A. and Larroche, C. 2006. Gluconic acid: Properties, applications and microbial production *Food Technology and Biotechnology* 44(2):185–195.

Rao, S. and Dutta, S.M. 1981. Extraction of rennet from abomasum of suckling buffalo calves. *Indian Journal of Dairy Science* 34(2):235–237.

Rao, M.B., A.M. Tanksale, M.S. and Ghatge, V.V. 1998. Deshpande molecular and biotechnological aspects of microbial proteases. *Microbiology and Molecular Biology Reviews* 62:597–635.

Reiter, B. and Marshall, V.M. 1979. Bactericidal activity of the lactoperoxidase system against psychrotrophic *Pseudomonas* spp. in raw milk. In: *Cold Tolerant Microbes in Spoilage and the Environment*, eds. A.D. Russel and R. Fuller, 153–164. London: Academic Press.

Russell, N.J. 1998. Molecular adaptations in psychrophilic bacteria: Potential for biotechnological applications. *Advances in Biochemical Engineering/Biotechnology* 61:1–21.

Saunders, N.F.W., Thomas, T., Curmi, P.M., Mattick, J.S., Kuczek, E., Slade, R., Davis, J., Franzmann, P.D., Boone, D., Rusterholtz, K., Feldman, R., Gates, C., Bench, S., Sowers, K., Kadner, K., Aerts, A., Dehal, P., Detter, C., Glavina, T., Lucas, S., Richardson, P., Larimer, F., Hauser, L., Land, M. and Cavicchioli, R. 2003. Mechanisms of thermal adaptation revealed from the genomes of the Antarctic Archaea *Methanogenium frigidum* and *Methanococcoides burtonii*. *Genome Research* 13:1580–1588.

Seifu, E., Buys, E.M. and Donkin, E.F. 2005. Significance of the lactoperoxidase system in the dairy industry and its potential applications: A review. *Trends in Food Science & Technology* 16(4):137–154.

Shukla, T.P. and Wierzbicki, L.E. 1975. Beta-galactosidase technology: A solution to the lactose problem. *Critical Review in Food Science and Nutrition* 5:325–356.

Siddiqui, K.S. and Cavicchioli, R. 2006. Cold-adapted enzymes. *Annual Review of Biochemistry* 75:403–433.

Siddiqui, K.S., Feller, G., D'Amico, S., Gerday, C., Giaquinto, L. and Cavicchioli, R. 2005. The active site is the least stable structure in the unfolding pathway of a multidomain cold-adapted alpha-amylase. *Journal of Bacteriology* 187:6197–6205.

Soda, M.E. 1993. Accelerated maturation of cheese. *International Dairy Journal* 3:531–544.

Spiwok, V., Lipovová, P., Skálová, T., Dusková, J., Dohnálek, J., Hasek, J., Russell, N.J. and Králová, B. 2007. Cold-active enzymes studied by comparative molecular dynamics simulation. *Journal of Molecular Modeling* 13:485–497.

Tan, S., Owusu, A.R.K. and Knapp, J. 1996. Low temperature organin phase biocatalysis using cold-adapted lipase from psychrotrophic *Pseudomonas* P38. *Food Chemistry* 57:415–418.

Terrell, S.L., Bernard, A. and Bailey, R.B. 1984. Ethanol from whey-continuous fermentation with a catabolite repression-resistant *Saccharomyces cerevisiae* mutant. *Applied and Environmental Microbiology* 48:577.

Tomasini, A., Bustillo, G. and Lebeault, J. 1993. Fat lipolysed with a commercial lipase for the production of Blue cheese flavour. *International Dairy Journal* 3:117–127.

Tutino, M.L., di Prisco, G., Marino, G. and de Pascale, D. 2009. Cold-adapted esterases and lipases: From fundamentals to application. *Protein and Peptide Letter* 16:1172–1180.

Vallejo, J.A., Ageitos, J.M., Poza, M. and Villa, T.G. 2012. Short communication: A comparative analysis of recombinant chymosins. *Journal of Dairy Science* 95(2):609–613.

Verma, N., Thakur, S. and Bhatt, A.K. 2012. Microbial lipases: Industrial applications and properties. *International Research Journal of Biological Sciences* 1(8):88–92.

Wang, Q.F., Hou, Y.H., Xu, Z., Mioa, J.-N. and Li, G.-Y. 2008. Purification and properties of an extracellular cold-active protease from the psychrophilic bacterium *Pseudoalteromonas* sp. NJ276. *Biochemical Engineering Journal* 38(3):362–368.

Wilkinson, M.G. 1993. Acceleration of ripening in cheese. In: *Cheese Chemistry, Physics and Microbiology*, Volume 1, General aspects, 2nd edition, ed. P.F. Fox, 523–555. London: Chapman and Hall.

Wilkinson, M.G., van den Berg, G. and Law, B.A. 2002. Technological properties of commercially available enzyme preparations other than coagulants. *Bulletin IDF* 371:16–19.

Yang, S.T. and Silva, E.M. 1995. Novel products and new technologies for use of a familiar carbohydrate, milk lactose. *Journal of Dairy Science* 78:2541–2562.

Zadow, J.G. 1984. Lactose properties and uses. *Journal of Dairy Science* 67:2654–2679.

14

Enzymes in Meat Processing

**Cinthia Baú Betim Cazarin, Gláucia Cariclo Lima,
Juliana Kelly da Silva, and Mário Roberto Maróstica Jr.**

CONTENTS

14.1 Introduction

The consumption of meat and meat products since the last decade is quickly rising, and consumers' choices are generally based on availability, price, and tradition (Delgado 2003). The world per capita consumption of livestock products in 1964–1966 was 24.2 kg per year and may reach 45.3 kg per year in 2030 (WHO 2002).

The search for healthier products along with the increase in awareness of environmental impact have led to the reduction of waste and the use of growth promoters and additives (Pearson and Dutson 1997). Consequently, to attend to the consumers' demands, the meat industry has been developing enzymatic methods to add value and improve the tenderization and texture as well as the efficiency of carcass utilization (Lantto et al. 2009).

Animal feed has great impact on the meat composition, but there are other aspects that influence meat quality, such as animal species, breed, and method of production (Lantto et al. 2009). The amounts of muscle, bone, fat, and edible offal or inedible by-products obtained from the meat industry can vary due to slaughtering methods and cultural reasons. However, waste implys high costs, driving research to the development of methods to restructure low-valued cuts and by-products, improving appearance and texture and increasing market value (Lantto et al. 2009; Marques et al. 2010).

In this way, the uses of enzyme in the meat industry have grown, and enzymes used in meat treatment are presented in Figure 14.1. In this chapter, the advantages of enzyme use in meat processing are addressed, mainly focused on nutritional improvements and technological and environmental aspects.

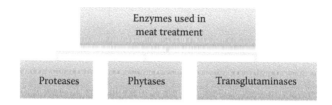

FIGURE 14.1   Enzymes used in meat treatment.

14.2 Application of Proteases in Meat Products

Tenderness is the major organoleptic characteristic and a determinant of the quality and acceptability of meats and their products. This attribute is important because the meat is cooked (consumed only this way for most meat types) and is strongly related to product acceptance (Bekhit et al. 2014b; Miller et al. 2001). However, only about 10% of the meat in a carcass is constituted of prime cuts that have high tenderness (Polkinghorne et al. 2008).

Meat processing generates a significant amount of by-products and wastes every day. There are some systems to treat these residues and by-products, but at most times, they are underutilized. The use of enzymes to improve tenderness is an alternative to increasing the use of waste from the meat industry, making it a good alternative for commercial products (Toldrá et al. 2012).

Tenderness is an important sensory property, especially for red meat, which has a characteristic toughness, unlike other types of meat, such as chicken or fish (Bekhit et al. 2014a). Shackelford et al. (2001) observed that tenderness is the primary determinant of satisfaction among consumers in the United States. In addition, consumers are inclined to pay more for tender meat (Feldkamp et al. 2005; Lusk et al. 2001; Shackelford et al. 2001). Proteases are used to improve tenderness, mainly in the non prime meat cuts, adding value and making them interesting to the industry (Bekhit et al. 2014b).

The toughness of the meat is determined by two important factors: connective tissue (structural proteins that provide support for the muscles) and rigor mortis (postmortem shortening of the muscle). The structure and/or the amount of different collagens and elastin in connective tissue will influence the meat toughness (Lepetit 2008). Endogen enzyme (calpain) can degrade postmortem myofibrillar proteins, improving the meat tenderness; however, there is another endogen enzyme (calpastin) that inactivates the previous. The concentration of these enzymes in the meat varies among the species and determines the range of the meat tenderness (Cheret et al. 2007).

The age of the animal can influence the degree of tenderness so that older animal muscles shows stronger connective tissue and less proteolytic enzymes to tenderize the meat postmortem (Veiseth et al. 2004). The tenderness of this type of meat can be greatly improved by the use of exogenous enzymes (Bekhit et al. 2014a).

Postmortem interventions used in fresh meat tenderization can be classified into three categories—physical, chemical, and enzymatic—based on their mode of action (Bekhit et al. 2014a). Physical or mechanical actions cause alteration in the connective tissue and the myofibrillar protein networks and stimulate proteolysis through direct interactions between endogenous enzymes, cofactors, and substrates, which can reduce the palpability. Chemical interventions, although improving the texture of meat and the palatability, sometimes increase the amount of added salt and can become unhealthy (Bekhit et al. 2014b; Sun 2009). In this context, the use of exogenous enzymes by infusion, marination, or injection can be one enzymatic technique to achieve the tenderization (Bekhit et al. 2014b). A scheme of the treatment used for meat tenderization is shown in Figure 14.2.

The exogenous enzymes used herein are proteases derived from plant, microbial, and animal sources. They act on myofibrillar proteins and connective tissue to produce a more fragmented and disintegrated structure (Bekhit et al. 2014b). The main enzymes used in meat processing are shown in Table 14.1, and some commercial producers of enzymes are shown in Table 14.2.

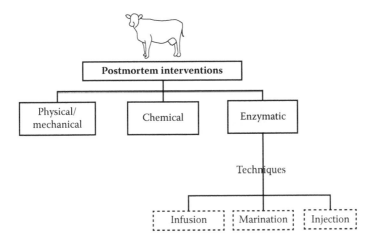

FIGURE 14.2 Scheme of treatment used for meat tenderization.

TABLE 14.1

Main Proteolytic Enzymes Used in Meat Processing

Source	Enzyme	Source	Action	Site of Action Proteolytic
Plants	Papain	Papaya (*Carica papaya*)	Hydrolysis of a wide range of bonds (peptide, amide, ester, thiol ester, and thiono ester bonds)	Actomyosin or collagen
	Bromelain	Pineapple (*Ananas comosus*)	Hydrolysis of a wide range of bonds (peptide, amide, ester, thiol ester, and thiono ester bonds)	Actomyosin or collagen
	Ficin	Fig (*Ficus glabrata, Ficus anthelmintica,* and *Ficus laurifolia*)	Hydrolysis of a wide range of bonds (peptide, amide, ester, thiol ester, and thiono ester bonds)	Actomyosin or collagen
	Actinidin	Kiwi fruit (*Actinidia deliciosa*)	Hydrolysis of a wide range of bonds (peptide, amide, ester, thiol ester, and thiono ester bonds)	Actomyosin or collagen
	Cucumin	Kachri fruit (*Cucumis pubescens*)	Hydrolysis of a wide range of bonds (peptide, amide, ester, thiol ester, and thiono ester bonds)	Actomyosin or collagen
	Zingibain	Ginger (*Zingiber officinale*)	Hydrolysis of a wide range of bonds (peptide, amide, ester, thiol ester, and thiono ester bonds)	Actomyosin or collagen
Micro-organisms	Collagenase aspartic protease	*Clostridium histolyticum*	Peptidases	Elastin or actomyosin
	Fungal protease EPg222	*Aspergillus oryzae*	Peptidases	Elastin or actomyosin
	Thermophile enzyme E	*Bacillus* strain E A.1	Peptidases	Elastin or actomyosin
	A.1 protease 4-1.A protease	*Thermus* strain Rt4-1.A	Peptidases	Elastin or actomyosin
	Caldolysin	*Thermus* strain T-35	Peptidases	Elastin or actomyosin
	Elastase	*Bacillus* sp. EL31410	Peptidases	Elastin or actomyosin
	Collagenase	*Vibrio* B-30	Peptidases	Elastin or actomyosin
Animal	Porcine pancreatin	Pork	Elastase	Component myofibrillar

Source: Bekhit, A. A., D. L. Hopkins, G. Geesink, A. A. Bekhit, and P. Franks, *Critical Reviews in Food Science and Nutrition*, 54, 8, 1012–1031, 2014.

TABLE 14.2

Examples of Some Commercial Enzymes Used in Meat Processing

Protease	Commercial Name	Stated Activity Units/mg[a]
Papain	Panol® 300 Purified Papain	>300 MCU
Bromelain	Enzeco® Bromelain 240	228–276 MCU
Ficin	Enzeco® Ficin 260	250–300 MCU
Aspergillus	Enzeco® Fungal Protease Concentrate	400 HUT
Aspergillus	Enzeco® Fungal Protease 400	400 HUT
Bacillus	Enzeco® Neutral Bacterial Protease 160 K	152–184 PC

Source: Calkins, C. R., and G. Sullivan, *Adding Enzymes to Improve Beef Tenderness* 2007 [cited 29/09/2014]. Available at http://www.beef research.org/CMDocs/BeefResearch/PE_Fact_Sheets/Adding_Enzymes _to_Improve_Beef_Tenderness.pdf.

Note: HUT = hemoglobin unit; MCU = milk clot unit; PC = proteolytical unit.

The effects of using enzymes in meat to obtain tenderness in the product can be achieved by two mechanisms. First, the enzymes can catalyze breakdown of protein covalent bonds, generating smaller peptide fragments or amino acids, altering the meat structure. The second mechanism involves the use of some enzymes that may promote the formation of new or different covalent bonds in the native protein structure (transglutaminases), which can change the tenderness of the meat (Weiss et al. 2010). In this section, the enzymes that can break down the protein covalent bonds are addressed.

Papain, bromelain, and ficin are the most studied plant enzymes used to obtain postmortem meat tenderization (Bekhit et al. 2014a; Payne 2009). They are endopeptidases with low substrate specificity and are able to catalyze the hydrolysis of a wide range of bonds (peptide, amide, ester, and thiol ester and thiono ester bonds) (Schwimmer 1981). These enzymes exerted a proteolytic effect on myosin and other myofibrillar proteins reflecting in meat tenderness (Melendo et al. 1996).

14.2.1 Papain

Papain was the first sulfhydryl enzyme discovered. This enzyme is extracted from the latex of the papaya plant (*Carica papaya*), whose natural role is to protect the fruit against insects (Konno 2011). Papain has been shown to have a broad-spectrum enzymatic activity over range of pH (5–8) and optimal temperature (65°C) (Bekhit et al. 2014a). The enzyme contains six sulfhydryls and one free cysteine, which is part of the active site. The enzyme has seven sites for recognizing substrate amino acid residues, and it hydrolyses amides of arginine and lysine readily and glutamine, histidine, glycine, and tyrosine at reduced rates (Dubey et al. 2007). The enzyme is stable for months at storage temperatures below 4°C and can be inactivated above 90°C (Bekhit et al. 2014a). Papain is used to tenderize meat; however, it can overtenderize the meat surface, which limits its use (Marques et al. 2010). Scanning electron micrographs of intramuscular connective tissue prepared from proteolytic enzyme-treated meats is shown in Figure 14.3.

14.2.2 Bromelain

Bromelain is a mixture of cysteine proteases, extracted from the stem or fruit of the *Ananas comosus* plant (Bromeliacea family). The proteolytic activity of the bromelain extract from the stem is lower than the extract from the fruits (Barrett et al. 2004). Different bromelain purification techniques have been investigated (Nadzirah et al. 2013). The enzymatic activity spectrum is slightly less than that of papain with detected proteolytic activity on synthetic peptides at pH 5.0–7.0 and an optimal temperature of 50°C (Rowan et al. 1990). Pure bromelain is stable when stored at −20°C (Rowan et al. 1990). In a study

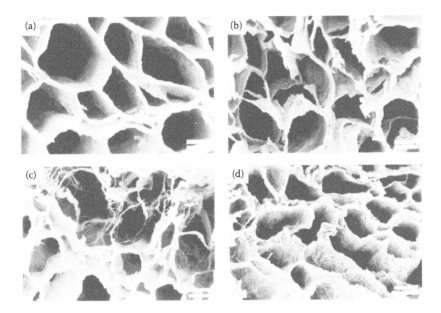

FIGURE 14.3 Scanning electron micrographs of intramuscular connective tissue prepared from enzyme-treated meats. The intramuscular connective tissues were prepared from meats treated with proteolytic enzymes after osmotic dehydration and stored for 24 h at 34°C. Each scale mark indicates 20 mm. (a) Control, (b) treated with papain, (c) treated with proteinase from *A. sojae*, and (d) treated with proteinase from *A. oryzae*. (Reprinted from Gerelt, B., Ikeuchi, Y., Suzuki, A., *Meat Science*, 56, 3, 311–318, 2000, with permission from Elsevier.)

on the effect of bromelain extract (BE) in tenderizing muscle foods, breaking tissue fibers, on a microstructural level, degradation of the cell membranes and in the connections between the sarcolemma and the myofibrils was observed (Ketnawa and Rawdkuen 2011). Bromelain was effective in tenderizing the meat, but there is some limitation in its use, as with papain, because the enzyme can promote overtenderization (Marques et al. 2010). On the other hand, bromelain generally generates a better flavor than papain (Kim and Taub 1991). The appearance of meat treated with different protease enzymes is shown in Figure 14.4.

14.2.3 Ficin

Ficin is obtained from the latex of *Ficus glabrata*, *Ficus anthelmintica*, and *Ficus laurifolia*, popularly known as "fig" (Gaughran 1976). The maximum activity of ficin is obtained within a pH range of 5–8 and a temperature range of 45°C–55°C. The effect of ficin on meat protein was evaluated on the quality attributes of sausage, and the results indicated that some quality attributes of meat products can be improved by the use of this enzyme (Ramezani et al. 2003).

14.2.4 Actinidin

Actinidin is another potential meat tenderizer, a less-researched cysteine protease present in kiwifruit (*Actinidia deliciosa*) (Liu et al. 2011). Actinidin is more stable in high pHs (7–10) than other enzymes, and its optimal activity is at 58°C–62°C (Yamaguchi et al. 1982). Infusing lamb carcasses prerigor in kiwifruit juice extract improved meat tenderness in the loin and hind leg region and maintained the color shelf life, which is usually degraded when a plant-derived enzyme is used (Liu et al. 2011). There is increased interest in this protease mainly because of its effects in the tenderization process, which is less extensive than the other proteases (Bekhit et al. 2014a).

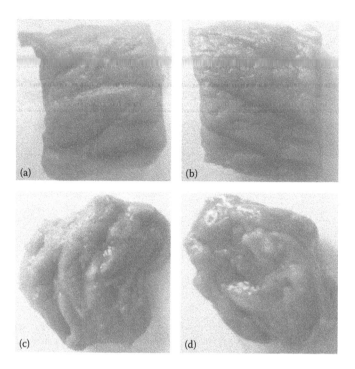

FIGURE 14.4 Appearance of meat treated with different enzymes. MCP-01 (b), bromelain (c), or papain (d) with 10 U of caseinolytic activity were incubated with beef at 4°C for 72 h. The beef dipped in distilled water was used as a control (a). The pictures were taken by a Nikon camera. (Reprinted from Zhao, G. Y., Zhou, M. Y., Zhao, H. L., Chen, X. L., Xie, B. B., Zhang, X. Y., He, H. L., Zhou, B. C., Zhang, Y. Z., *Food Chemistry*, 134, 4, 1738–1744, 2012, with permission from Elsevier.)

14.2.5 Zingibain

Zingibain is obtained from ginger (*Zingiber officinale*) and has a maximum activity at pH 6–7 and at 60°C. The enzyme has more specificity to collagen compared with actomyosin (Bekhit et al. 2014a). Use of ginger extract in meat tenderization was observed to promote better sensory qualities in relation to appearance, flavor, tenderness, and acceptability than the enzymes from *Cucumis trigonus* Roxb (Kachri) (Naveena et al. 2004). The addition of ascorbate increased the enzyme half-life, increasing its attractiveness as a commercial product (Adulyatham and Owusu-Apenten 2005). Ginger extract has the advantage of avoiding lipid oxidation and microbial growth, which keep meat qualities during storage (Mendiratta et al. 2000; Naveena et al. 2004).

14.2.6 Cucumin

Cucumin is obtained from kachri fruit, a melon variety fruit, which is available in India, West Pakistan, Afghanistan, and Persia. In these places, the kachri is traditionally used as a food-tenderizing agent (Maiti et al. 2008; Naveena et al. 2004). Buffalo meat has a lower cholesterol content than beef but has a higher toughness that decreases sensory characteristics (Naveena et al. 2004). In order to improve the consumption of this meat, previous tenderization is necessary. The use of cucumin (kachri fruit extract) promises to be effective in the tenderization of this meat (Naveena et al. 2004).

14.2.7 Fungal Protease

Microbial enzymes are another class of proteases with greater specificity to substrates. The production of this kind of enzyme has been stimulated due to biotechnology aspects that permit the production of large amounts with high purity and because their production does not depend on the seasonality of the raw material (Bekhit et al. 2014a). The use of EPg222, a fungal extracellular protease isolated from *Penicillium chrysogenum*, showed changes in the texture of whole pieces of pork loin (Benito et al. 2003).

14.2.8 Elastase

Elastase is an enzyme that breaks down elastin in connective tissue. Elastase from *Bacillus* sp. EL31410 was observed to contribute to tenderization of beef. After 4 h of treatment with 1% elastase solution, meat hardness was observed to be 30% decreased. Elastase showed better juiciness and flavor parameters compared to papain (Qihe et al. 2006). Gastric elastases have been recovered from marine and freshwater species, such as carp, catfish, and Atlantic cod, with the ability to rapidly degrade elastin taste (Weiss et al. 2010).

14.2.9 Animal Proteases

Animal proteases are less used to tenderize meat. In a recent study, injection of pancreatin on beef semitendinosus tended to improve overall tenderness, and the results showed that taste was not affected (Pietrasik et al. 2010). Pancreatin is more effective in muscle in which the myofibrillar component is the main contributor to toughness, and in muscles with high connective tissue content, its action is limited (Pietrasik and Shand 2010).

14.3 Application of Protease Inhibitors in Meat Products

Protease inhibitors have been studied to avoid the overtenderization problem. Plants, especially legumes, are strong sources of protease inhibitors, but the successful use of these substances depends on some factors: effective levels of use, consumer acceptability, effects on the meat sensory quality, and cost (Bekhit et al. 2014a). Ascorbic acid has the ability to up- and downregulate the activity of the proteases (papain, bromelain, actinidin, and zingibain) (Ha et al. 2014).

14.4 Production of Bioactive Peptides

The use of peptidases in meat is not limited to the food industry. Proteolytic activity during meat processing generates a large amount of different peptides and free amino acids. Current research has shown the potential biologic action in these peptide fractions (Toldrá et al. 2012), especially if the waste could be used to generate high-value products, such as those bioactive peptides. Bioactive peptides can exert health benefits, such as antihypertensive action and antioxidant or opioid activity, among others (Arihara 2006; Arihara and Ohata 2010; Toldrá et al. 2012).

The inhibition of the angiotensin I-converting enzyme (ACE), the enzyme involved in increasing the mechanism of blood pressure, probably constitutes the most relevant bioactivity (Toldrá et al. 2012). The ACE-I inhibitory activity of peptides derived from the hydrolysis of porcine muscle proteins by the action of *Lactobacillus sakei* CRL1862 and *Lactobacillus curvatus* CRL705 was investigated at 30°C for 36 h. Peptides with ACE inhibitory activity were generated, and these peptides could have great promise in the development of novel therapeutics and as a functional food for preventing hypertension (Castellano et al. 2013; Toldrá et al. 2012). In fact, some peptides derived from enzymatic hydrolysis exerted a significant blood pressure decrease in rats after just 8 h of administration (Escudero et al. 2012). A few peptides with relevant ACE-inhibitory activity tested in vitro and in vivo against spontaneous hypertensive rats were reported from chicken collagen hydrolyzate obtained with a protease from *Aspergillus oryzae* (Saiga et al. 2008).

A study showed that a peptide fraction from pork meat obtained by papain action had antithrombotic activity in vivo in doses of 210 and 70 mg/kg^{-1} (Shimizu et al. 2009). Other peptides have showed bioactive effects, such as opioid (Froidevaux et al. 2008; Pihlanto-Leppälä 2000; Piot et al. 1992),

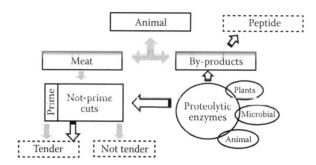

FIGURE 14.5   Use of the proteolytic enzymes in meat and meat processing by-product.

antimicrobial (Kitts and Weiler 2003; Pellegrini 2003), antioxidant (Chang et al. 2007), and immuno-regulatory (Gill et al. 2000) action.

The application of enzymes in meat products and by-products is summarized in Figure 14.5.

14.5 Phytase as Feed Additive in Meat Animal Production

Enzymes are added into animal feed to improve its utilization, achieve better production performances, and reduce the environmental impact of excessive phosphorus excretion (Perić et al. 2011). In plant seeds, phytate forms a complex with phosphorus and calcium, making them unavailable for absorption (Bedford and Partridge 2010); in addition, it can bind to amino acids and proteins promoting the inhibition of digestive enzymes and decreasing the nutritive value of food (Dvořáková 1998). Phytase (myo-inositol hexaphosphate phosphohydrolase) is the enzyme used to hydrolyze the phytate molecule and release phosphorus (Lan et al. 2010; Lim and Lee 2009; Selle et al. 2009).

Pigs, poultry, and fish do not produce the phytase enzyme. This enzyme deficiency is corrected by phytase addition in the animal feed mainly to release these minerals, improving the absorption and decreasing inorganic phosphorus supplementation (Lei et al. 2013). The addition of phytase in feed improves the phosphorus digestibility into 60% to 80%, dose-dependent (Lei et al. 2013), and prevents soil phosphorus accumulation through less excretion. The presence of phytate in the diet depresses endogenous carbohydrase activity (Liu et al. 2008). However, recent studies have shown the benefits of phytase supplementation in the improvement of energy and protein and amino acid digestibility (Bedford and Partridge 2010).

The use of enzymes in animal feed generally aims to replace some nutrients for those with lower quality and reduce feed costs or to improve animal growth, egg production, and feed conversion (Bedford and Partridge 2010).

Phytase is produced by different microorganisms, such as *Aspergillus ficuum* (Jafari-Tapeh et al. 2012), *Aspergillus niger* (Bhavsar et al. 2011), *Bacillus subtilis* (Singh et al. 2013), *Alcaligenes* sp. (Vijayaraghavan et al. 2013), and *Penicillium purpurogenum* (Awad et al. 2014), etc.

The costs of fish meal are increasingly expensive, leading to the current research focus looking for alternatives to produce fish feed formulation that replaces fish meal with cheaper and more readily available vegetable protein sources (Marques et al. 2010).

Diet supplementation of *O. niloticus* fingerlings with microbial phytase at 1000 U kg^{-1} improved the growth performance by reducing the effect of phytic acid and releasing the chelated protein and minerals of plant-based diets (Hassaan et al. 2013). Tahoun et al. (2009) observed that a diet containing distiller's dried grain with soluble supplemented phytase improved growth and feed utilization parameters in Nile tilapia (*Oreochromis niloticus*) fingerlings until a dose of 150 mg/kg^{-1}. Increases in mineral content in the *Cyprinus carpio* fingerling carcasses was observed with a diet supplemented with phytase in 1000 UF/kg^{-1} (Rocha et al. 2010).

The uses of microbial phytase in broiler chicken production resulted in higher body weight gain, improved feed conversion ratio, and increased protein content in the breast and thigh muscles, contributing to the high quality of chicken meat (Kliment et al. 2012). In addition, the chickens supplemented with microbial phytase improved the meat quality and the profitability of egg production (Amin and Hamidi 2013; Gao et al. 2013; Perić et al. 2011).

Cost reduction is possible using phytase in animal feed. The addition of phytase (500 U/kg^{-1}) in corn–soybean pig feed with reduced calcium and phosphorus content promoted increased amino acid and energy digestibility, showing that it is possible to reduce the concentrations of amino acids and energy in diets supplemented with phytase (Johnston et al. 2004). In this way, feeds formulated with corn, soybean bran, defatted rice bran, and soybean oil for growing swine supplemented with phytase (759 U/kg^{-1}) showed that it is possible to eliminate the traditional phosphorus sources in the feed (Moreira et al. 2003).

Briefly, the effects of phytase use in animal feed are related to improvement of meat quality, cost reduction, and decreased danger of accumulation of environmental phosphorus from the manure.

14.6 Transglutaminase: Use in the Manufacture of Meat Products

Transglutaminase (TGase; protein-glutamine γ-glutamyltransferase, EC. 2.3.2.13) is an enzyme that catalyzes an acyl-transfer reaction between the carboxyamide group of peptide-bound glutamine residues (acyl donors) and a variety of primary amines (acyl acceptors), including the amino group of lysine residues in certain proteins. TGase also catalyzes the deamidation of glutamine residues in the absence of amine substrates, using water molecules as acyl acceptors (Motoki and Seguro 1998).

Until the end of the 1980s, commercial transglutaminase could only be obtained from animal tissues, most commonly from guinea pig liver. Therefore, the rare source, complicated procedure, and high cost of the enzyme production reduced its wide application in food processing. However, recently, the production of transglutaminase by microorganisms is well established. Thus, food industries widely use this enzyme nowadays. The microbial transglutaminase is obtained mainly from a variant *Streptoverticillium mobaraense* (Zhu and Tramper 2008).

The optimum pH of microbial transglutaminase activity is around 5 to 8 although it expresses some enzymatic activity even at pH 4 or 9. The optimum temperature for enzymatic activity is 50°C, but it still has some activity at 10°C or at temperatures just above the freezing point (Motoki and Seguro 1998).

TGase-catalyzed protein cross-linking is able to improve the functional characteristics of protein food, such as texture, flavor, and shelf life. In addition, the enzyme can also reduce the allergenicity of certain foods (Zhu and Tramper 2008), such as milk proteins, and produce food proteins of higher nutritional value through cross-linking of different proteins containing complementary limiting essential amino acids (Zhu et al. 1995). TGase affects the gelling of concentrated solutions of proteins, such as soybean proteins, milk proteins, beef, pork, chicken, and fish gelatin and myosin (Motoki and Seguro 1998).

In fact, numerous studies about quality characteristics were done, such as functional and textural properties of meat products manufactured with transglutaminase. The use of TGase in meat processing includes development of methods for restructuring of low-value cuts and trimmings to improve their appearance, flavor, and texture, in addition to enhancing market value. Restructuring treatment usually involves size reduction, reforming, and binding (Zhu et al. 1995). Indeed, TGase improves the firmness, gel elasticity of meat products, water-holding capacity, and heat stability, avoiding some undesirable attributes, such as high viscosity and excessive meat adhesiveness (Ahhmed et al. 2009). Thus, diverse meat products have been manufactured by using TGase, such as sausages of chicken and beef (Ahhmed et al. 2007), ham (de Avila et al. 2010; Fulladosa et al. 2009), chicken meatballs (Tseng et al. 2000), beef (Carballo et al. 2006), fish (Zhang et al. 2013), and steak (Canto et al. 2014).

The effect of the addition of different levels of TGase (0.05% and 0.10% of TGase and an emulsion without TGase) to pork meat was determined (Herrero et al. 2008). In addition to TGase, sodium caseinate, and maltodextrin were also used in all formulations. Results of the study indicated differences in texture profile parameters among systems with or without enzymes. Hardness, springiness, and cohesiveness were higher in meat systems supplemented with TGase (0.05% and 0.10%) than in the system with no enzyme addition. This study also showed the occurrence of secondary structural changes in meat proteins by using TGase, and significant correlations were found between these secondary structural changes in meat proteins and their textural properties (Herrero et al. 2008).

TGase was shown to be effective in improving gel strength in chicken and beef sausages (Ahhmed et al. 2007). Addition of the enzyme significantly increased the breaking strength values in treated samples of each meat type compared to nontreated ones even though beef reacted most strongly at 80°C. The study suggested that the functional properties of TGase make it a very good additive to improve texture and gelation of sausages. However, it increases the values of breaking strength in meat species. The breaking strength is highly related to the protein complexity force, due to the interaction of proteins after adding the enzyme. The beneficial functional effects of TGase are primarily involved with the protein compounds and reformations, especially with the myosin heavy chain (MHC) (Ahhmed et al. 2007). Thus, TGase reacts differently depending on the kind of meat. Later, the same group investigated the factors that cause differences in the improvements of gel strength and ε-(γ-glutamyl) lysine content in chicken and beef myofibrillar proteins after addition of microbial transglutaminase (Ahhmed et al. 2009). They raised some factors that might be involved in TGase reaction in different meat species, such as variation in muscle physiology and morphogenesis, the identity of free amino acids, the amount and distance between transferable amino acids, and the amount of TGase inhibitors (Ahhmed et al. 2009).

TGase alone seems to have little effect on binding strength in the meat system (Kuraishi et al. 1997). The presence of salt is necessary once it improves solubilization of myofribillar proteins as the exudate is a binding agent. This exudate is a good substrate for cross-linking reactions by TGase. However, due to a negative effect on product taste and the health demands of less salt usage, researchers have replaced sodium chlorate to sodium caseinate (Kuraishi et al. 1997) and other nonmeat ingredients (Colmenero et al. 2005). Thus, some work was carried out using a TGase and sodium caseinate combination in emulsion (Carballo et al. 2006; Colmenero et al. 2005; Kuraishi et al. 1997). Sodium caseinate was shown to be a good substrate for TGase, facilitating cross-linking and promoting a more stable gel matrix.

TGase has also been used in studies aimed at the development of healthier meat products, such as those with low-salt content as well as phosphate-free. The effect of transglutaminase level, salt, and different processing conditions of restructured cooked pork shoulder were investigated in order to produce a phosphate-free product with reduced salt content (Dimitrakopoulou et al. 2005). Six treatments of restructured cooked pork shoulder with two levels of salt (2% and 1%) and three levels of transglutaminase (0%, 0.075%, and 0.15%) were prepared. TGase level had no effect on chemical composition or physicochemical parameters. However, concerning sensory attributes, TGase level positively affected the consistency and the overall acceptability of the restructured cooked pork shoulder. The study concluded that transglutaminase can be used at a concentration of 0.15% to produce phosphate-free restructured cooked pork shoulder with a reduced salt level (1%) and acceptable sensory attributes (Dimitrakopoulou et al. 2005).

The application of a microbial transglutaminase/sodium caseinate (MTG/C) system on three meat species (pork, chicken, lamb) was investigated (Carballo et al. 2006). The effect of MTG/C systems on meat batter characteristics (water binding and textural properties of raw and cooked products) was determined in the presence of NaCl (1.5 g/100 g^{-1}) and sodium tripolyphosphate (0.5 g/100 g^{-1}), and storage time (96 h at 3°C). Samples prepared from pork and lamb with only MTG/C (salt-free) had the highest cooking loss values. Hardness and chewiness tended to be higher in cooked samples containing MTG/C than in samples containing only salts. However, the efficiency of the MTG/C system as a texture conditioner of cooked products varied with the meat source. The MTG/C system was less effective in cooked lamb and more effective in cooked pork.

Another study compared the effects of combinations of MTG and various non meat ingredients (caseinate, potassium chloride, and wheat fiber) used as salt replacers with the effects of sodium chloride on the physicochemical properties, such as cooking loss, emulsion stability, texture, and color of frankfurters with walnuts added. Caseinate, potassium chloride, or fiber combined with MTG resulted in harder, springier, and chewier frankfurters with better water- and fat-binding properties (emulsion stability and cooking loss) when compared with those with TGase addition only. The combination of TGase and caseinate showed the highest lightness and the lowest redness values. However, frankfurters with sodium chloride had a harder, springier, and chewier gel/emulsion network with lower cooking loss than those that were sodium chloride-free (Colmenero et al. 2005).

The functional properties of heat-induced gels prepared from MTG-treated porcine myofibrillar protein (MP) containing sodium caseinate with or without konjac (a plant grown in China, Korea, Taiwan, Japan, and southeast Asia) flour under different salt concentrations (0.1, 0.3, and 0.6 M sodium chloride) were evaluated (Dondero et al. 2006). It was observed that the combination of konjac flour (KF) and transglutaminase improved the gel strength at 0.1 and 0.3 M sodium chloride when compared to non-transglutaminase controls. Moreover the combination of TGase and KF improved the gelling and water-binding properties of MP gels at an intermediate level of salt (0.3 MNaCl).

Low salt chicken meatballs were formulated using TGase (Tseng et al. 2000). Gel strength and yield of meatballs increased as the concentration of TGase in the batter increased from 0.05% to 1%. The addition of TGase did not adversely affect the product color. Scanning electron microscopy images showed higher order in meat gels compared to those that were TGase free. In another investigation, in the physicochemical and sensory attributes of low-sodium restructured caiman steaks containing microbial transglutaminase (MTG) and salt replacers (KCl and $MgCl_2$), it was observed that the formulations named T-3 (0.75% NaCl + 1% MTG + 0.75% $MgCl_2$) and T-4 (0.75% NaCl + 1% MTG + 0.375% KCl + 0.375% $MgCl_2$) can be employed as suitable strategies of sodium chloride reduction on restructured caiman steaks (Canto et al. 2014). The study suggested that MTG improved the texture (instrumental and sensory) and could be used as a replacer of sodium chloride (KCl and $MgCl_2$), improving cooking yield, succulence, and consumer acceptance.

Other treatments have also been associated with TGase for gel quality improvement on meat products. The addition of TGase and application of high pressure (HP) at 40°C on chicken meat batters with egg proteins with reduced content of sodium chloride and without phosphates resulted in gels with more cutting force, hardness, and chewiness than samples subjected only to pressure or those obtained by traditional heat treatment (Trespalacios and Pla 2007). TGase-treatment under high hydrostatic pressure (HHP) also led to greater enhancement of heat-induced porcine plasma gel properties compared to control samples (Fort et al. 2008). TGase combined with other ingredients seems to be a promising additive to improve the quality of meat products, mainly their texture characteristics. Therefore, this is a way to improve the use of animal carcasses to maximize the yield of marketable meat products.

14.7 Conclusion

Meat and its products are much consumed worldwide, and tenderness is the main characteristic considered by consumers. The supply chain starts in the choice of the animal feed; followed by the use of machinery and equipment; and the incorporation of intermediate products to the final product, which should provide good quality to meet the demands of the consumer market. In this way, strategies to improve meat tenderness could increase the value of not-prime cuts, making possible the use of by-products from the meat industry and reducing the waste and environmental impact generated in this supply chain. The use of enzymes in meat technology is growing and promise desirable advantages in terms of costs and quality of the final product.

However, although these natural products have been shown to be effective in improving meat quality, it is important to determine some optimal conditions to prevent the overtenderization and meat sensory quality loss. There are many natural sources of these enzymes, especially by-products from the food industry; for example, bromelain can be extracted from pineapple peel. The use of these enzymes can be profitable for the meat industry in the future in addition to contributing to sustainable management of environmental problems due to meat industry waste.

REFERENCES

Adulyatham, P., and R. Owusu-Apenten. 2005. Stabilization and partial purification of a protease from ginger rhizome (*Zingiber officinale* Roscoe). *Journal of Food Science* 70 (3):C231–C234.

Ahhmed, A. M., S. Kawahara, K. Ohta, K. Nakade, T. Soeda, and M. Muguruma. 2007. Differentiation in improvements of gel strength in chicken and beef sausages induced by transglutaminase. *Meat Science* 76 (3):455–462.

Ahhmed, A. M., T. Nasu, D. Q. Huy, Y. Tomisaka, S. Kawahara, and M. Muguruma. 2009. Effect of microbial transglutaminase on the natural actomyosin cross-linking in chicken and beef. *Meat Science* 82 (2):170–178.

Amin, M. R., and E. N. B. Hamidi. 2013. Effect of phytase supplementation on the performance of bab-cock-380 layer hens. *Journal of Tropical Resources and Sustainable Science* 1 (1):36–41.

Arihara, K. 2006. Strategies for designing novel functional meat products. *Meat Science* 74 (1):219–229.

Arihara, K., and M. Ohata. 2010. Functional meat products. In *Handbook of Meat Processing*, edited by F. Toldrá. Ames, IA: Wiley-Blackwell.

Awad, G. E. A., M. M. I. Helal, E. N. Danial, and M. A. Esawy. 2014. Optimization of phytase production by Penicillium purpurogenum GE1 under solid state fermentation by using Box–Behnken design. *Saudi Journal of Biological Sciences* 21 (1):81–88.

Barrett, A. J., N. D. Rawlings, and J. F. Woessner. 2004. *Handbook of Proteolytic Enzymes*. Amsterdam: Elsevier Academic Press.

Bedford, M. R., and G. G. Partridge. 2010. *Enzymes in Farm Animal Nutrition*, 2nd ed. Cambridge: CAB International.

Bekhit, A. A., D. L. Hopkins, G. Geesink, A. A. Bekhit, and P. Franks. 2014a. Exogenous proteases for meat tenderization. *Critical Reviews in Food Science and Nutrition* 54 (8):1012–1031.

Bekhit, A. E. A., A. Carne, M. Ha, and P. Franks. 2014b. Physical interventions to manipulate texture and ten-derness of fresh meat: A review. *International Journal of Food Properties* 17 (2):433–453.

Benito, M. J., M. Rodriguez, R. Acosta, and J. J. Cordoba. 2003. Effect of the fungal extracellular protease EPg222 on texture of whole pieces of pork loin. *Meat Science* 65 (2):877–884.

Bhavsar, K., V. R. Kumar, and J. M. Khire. 2011. High level phytase production by *Aspergillus niger* NCIM 563 in solid state culture: Response surface optimization, up-scaling, and its partial characterization. *Journal of Industrial Microbiology & Biotechnology* 38 (9):1407–1417.

Calkins, C. R., and G. Sullivan. 2014. *Adding Enzymes to Improve Beef Tenderness 2007* (cited September 29, 2014). Available at http://www.beefresearch.org/CMDocs/BeefResearch/PE_Fact_Sheets/Adding _Enzymes_to_Improve_Beef_Tenderness.pdf.

Canto, A. C., B. R. Lima, S. P. Suman et al. 2014. Physico-chemical and sensory attributes of low-sodium restructured caiman steaks containing microbial transglutaminase and salt replacers. *Meat Science* 96 (1):623–632.

Carballo, J., J. Ayo, and F. J. Colmenero. 2006. Microbial transglutaminase and caseinate as cold set binders: Influence of meat species and chilling storage. *LWT-Food Science and Technology* 39 (6):692–699.

Castellano, P., M.-C. Aristoy, M. Á. Sentandreu, G. Vignolo, and F. Toldrá. 2013. Peptides with angiotensin I converting enzyme (ACE) inhibitory activity generated from porcine skeletal muscle proteins by the action of meat-borne *Lactobacillus*. *Journal of Proteomics* 89:183–190.

Chang, C.-Y., K.-C. Wu, and S.-H. Chiang. 2007. Antioxidant properties and protein compositions of porcine haemoglobin hydrolysates. *Food Chemistry* 100 (4):1537–1543.

Cheret, R., C. Delbarre-Ladrat, M. de Lamballerie-Anton, and V. Verrez-Bagnis. 2007. Calpain and cathepsin activities in post mortem fish and meat muscles. *Food Chemistry* 101 (4):1474–1479.

Colmenero, F. J., M. J. Ayo, and J. Carballo. 2005. Physicochemical properties of low sodium frankfurter with added walnut: Effect of transglutaminase combined with caseinate, KCl and dietary fibre as salt replac-ers. *Meat Science* 69 (4):781–788.

de Avila, M. D. R., J. A. Ordonez, L. de la Hoz, A. M. Herrero, and M. L. Cambero. 2010. Microbial transglu-taminase for cold-set binding of unsalted/salted pork models and restructured dry ham. *Meat Science* 84 (4):747–754.

Delgado, C. L. 2003. Rising consumption of meat and milk in developing countries has created a new food revolution. *The Journal of Nutrition* 133 (11):3907S–3910S.

Dimitrakopoulou, M. A., J. A. Ambrosiadis, F. K. Zetou, and J. G. Bloukas. 2005. Effect of salt and trans-glutaminase (TG) level and processing conditions on quality characteristics of phosphate-free, cooked, restructured pork shoulder. *Meat Science* 70 (4):743–749.

Dondero, M., V. Figueroa, X. Morales, and E. Curotto. 2006. Transglutaminase effects on gelation capacity of thermally induced beef protein gels. *Food Chemistry* 99 (3):546–554.

Dubey, V. K., M. Pande, B. K. Singh, and M. V. Jangannadham. 2007. Papain-like proteases: Applications of their inhibitors. *African Journal of Biotechnology* 6 (9):1077–1086.

Dvořáková, J. 1998. Phytase: Sources, preparation and exploitation. *Folia Microbiologica* 43 (4):323–338.

Escudero, E., F. Toldra, M. A. Sentandreu, H. Nishimura, and K. Arihara. 2012. Antihypertensive activity of peptides identified in the in vitro gastrointestinal digest of pork meat. *Meat Science* 91 (3):382–384.

Feldkamp, T. J., T. C. Schroeder, and J. L. Lusk. 2005. Determining consumer valuation of differentiated beef steak quality attributes. *Journal of Muscle Foods* 16 (1):1–15.

Fort, N., T. C. Lanier, P. M. Amato, C. Carretero, and E. Saguer. 2008. Simultaneous application of microbial transglutaminase and high hydrostatic pressure to improve heat induced gelation of pork plasma. *Meat Science* 80 (3):939–943.

Froidevaux, R., M. Vanhoute, D. Lecouturier, P. Dhulster, and D. Guillochon. 2008. Continuous preparation of two opioïd peptides and recycling of organic solvent using liquid/liquid extraction coupled with aluminium oxide column during haemoglobin hydrolysis by immobilized pepsin. *Process Biochemistry* 43 (4):431–437.

Fulladosa, E., X. Serra, P. Gou, and J. Arnau. 2009. Effects of potassium lactate and high pressure on transglutaminase restructured dry-cured hams with reduced salt content. *Meat Science* 82 (2):213–218.

Gao, C. Q., C. Ji, J. Y. Zhang, L. H. Zhao, and Q. G. Ma. 2013. Effect of a novel plant phytase on performance, egg quality, apparent ileal nutrient digestibility and bone mineralization of laying hens fed corn–soybean diets. *Animal Feed Science and Technology* 186 (1–2):101–105.

Gaughran, E. R. 1976. Ficin: History and present status. *Pharmaceutical Biology* 14 (1):1–21.

Gerelt, B., Y. Ikeuchi, and A. Suzuki. 2000. Meat tenderization by proteolytic enzymes after osmotic dehydration. *Meat Science* 56 (3):311–318.

Gill, H. S., F. Doull, K. J. Rutherfurd, and M. L. Cross. 2000. Immunoregulatory peptides in bovine milk. *British Journal of Nutrition* 84:S111–S117.

Ha, M., A. El-Din Bekhit, and A. Carne. 2014. Effects of l- and iso-ascorbic acid on meat protein hydrolyzing activity of four commercial plant and three microbial protease preparations. *Food Chemistry* 149:1–9.

Hassaan, M. S., M. A. Soltan, H. M. Agouz, and A. M. Badr. 2013. Influences of calcium/phosphorus ratio on supplemental microbial phytase efficiency for Nile tilapia (*Oreochromis niloticus*). *The Egyptian Journal of Aquatic Research* 39 (3):205–213.

Herrero, A. M., M. I. Cambero, J. A. Ordonez, L. De la Hoz, and P. Carmona. 2008. Raman spectroscopy study of the structural effect of microbial transglutaminase on meat systems and its relationship with textural characteristics. *Food Chemistry* 109 (1):25–32.

Jafari-Tapeh, H., Z. Hamidi-Esfahani, and M. H. Azizi. 2012. Culture condition improvement for phytase production in solid state fermentation by *Aspergillus ficuum* using statistical method. *ISRN Chemical Engineering* 2012:5.

Johnston, S. L., S. B. Williams, L. L. Southern et al. 2004. Effect of phytase addition and dietary calcium and phosphorus levels on plasma metabolites and ileal and total-tract nutrient digestibility in pigs. *Journal of Animal Science* 82 (3):705–714.

Ketnawa, S., and S. Rawdkuen. 2011. Application of bromelain extract for muscle foods tenderization. *Food and Nutrition Sciences* 2 (5):393–401.

Kim, H.-J., and I. A. Taub. 1991. Specific degradation of myosin in meat by bromelain. *Food Chemistry* 40 (3):337–343.

Kitts, D. D., and K. Weiler. 2003. Bioactive proteins and peptides from food sources. Applications of bioprocesses used in isolation and recovery. *Current Pharmaceutical Design* 9 (16):1309–1323.

Kliment, M., M. Angelovičová, and S. Nagy. 2012. The effect of microbial phytase on broiler chicken production and nutritional quality of meat. *Scientific Papers Animal Science and Biotechnologies* 45 (1):46–50.

Konno, K. 2011. Plant latex and other exudates as plant defense systems: Roles of various defense chemicals and proteins contained therein. *Phytochemistry* 72 (13):1510–1530.

Kuraishi, C., J. Sakamoto, K. Yamazaki, Y. Susa, C. Kuhara, and T. Soeda. 1997. Production of restructured meat using microbial transglutaminase without salt or cooking. *Journal of Food Science* 62 (3):488.

Lan, G. Q., N. Abdullah, S. Jalaludin, and Y. W. Ho. 2010. In vitro and in vivo enzymatic dephosphorylation of phytate in maize–soya bean meal diets for broiler chickens by phytase of *Mitsuokella jalaludinii*. *Animal Feed Science and Technology* 158 (3–4):155–164.

Lantto, R., K. Kruus, E. Puolanne, K. Honkapää, K. Roininen, and J. Buchert. 2009. Enzymes in meat processing, in Enzymes in Food Technology, 2nd ed. (eds. R. J. Whitehurst and M. van Oort), Wiley-Blackwell, Oxford, UK. doi: 10.1002/9781444309935.ch12

Lei, X. G., J. D. Weaver, E. Mullaney, A. H. Ullah, and M. J. Azain. 2013. Phytase, a new life for an "old" enzyme. *Annual Review of Animal Biosciences* 1:283–309.

Lepetit, J. 2008. Collagen contribution to meat toughness: Theoretical aspects. *Meat Science* 80 (4):960–967.

Lim, S.-J., and K.-J. Lee. 2009. Partial replacement of fish meal by cottonseed meal and soybean meal with iron and phytase supplementation for parrot fish *Oplegnathus fasciatus*. *Aquaculture* 290 (3–4):283–289.

Liu, N., Y. J. Ru, F. D. Li, and A. J. Cowieson. 2008. Effect of diet containing phytate and phytase on the activity and messenger ribonucleic acid expression of carbohydrase and transporter in chickens. *Journal of Animal Science* 86 (12):3432–3439.

Liu, C., Y. L. Xiong, and G. K. Rentfrow. 2011. Kiwifruit protease extract injection reduces toughness of pork loin muscle induced by freeze–thaw abuse. *LWT-Food Science and Technology* 44 (10):2026–2031.

Lusk, J. L., J. A. Fox, T. C. Schroeder, J. Mintert, and M. Koohmaraie. 2001. In-store valuation of steak tenderness. *American Journal of Agricultural Economics* 83 (3):539–550.

Maiti, A. K., S. S. Ahlawat, D. P. Sharma, and N. Khanna. 2008. Application of natural tenderizers in meat—A review. *Agricultural Reviews* 29 (3):226–230.

Marques, A. Y., M. R. Marostica, and G. M. Pastore. 2010. Some nutritional, technological and environmental advances in the use of enzymes in meat products. *Enzyme Research* 2010:480923.

Melendo, J. A., J. A. Beltrán, I. Jaime, R. Sancho, and P. Roncalés. 1996. Limited proteolysis of myofibrillar proteins by bromelain decreases toughness of coarse dry sausage. *Food Chemistry* 57 (3):429–433.

Mendiratta, S. K., A. S. R. Anjaneyulu, V. Lakshmanan, B. M. Navenna, and G. S. Bisht. 2000. Tenderizing and antioxidant effect of ginger extract on sheep meat. *Journal of Food Science and Technology* 37 (6):651–655.

Miller, M. F., M. A. Carr, C. B. Ramsey, K. L. Crockett, and L. C. Hoover. 2001. Consumer thresholds for establishing the value of beef tenderness. *Journal of Animal Science* 79 (12):3062–3068.

Moreira, J. A., D. M. S. S. Vitti, M. A. da Trindade Neto, and J. B. Lopes. 2003. Phytase enzyme in diets containing defatted rice bran for growing swine. *Scientia Agricola* 60:631–636.

Motoki, M., and K. Seguro. 1998. Transglutaminase and its use for food processing. *Trends in Food Science & Technology* 9 (5):204–210.

Nadzirah, K. Z., S. Zainal, A. Noriham, and I. Normah. 2013. Efficacy of selected purification techniques for bromelain. *International Food Research Journal* 20 (1):43–46.

Naveena, B. M., S. K. Mendiratta, and A. S. R. Anjaneyulu. 2004. Tenderization of buffalo meat using plant proteases from *Cucumis trigonus* Roxb (Kachri) and *Zingiber officinale* roscoe (Ginger rhizome). *Meat Science* 68 (3):363–369.

Payne, C. T. 2009. Enzymes. In *Ingredients in Meat Products*, edited by R. Tarté. New York: Springer.

Pearson, A. M., and T. R. Dutson. 1997. *Production and Processing of Healthy Meat, Poultry and Fish Products, Vol. 11, Advances in Meat Research*. London: Blackie Academic & Professional.

Pellegrini, A. 2003. Antimicrobial peptides from food proteins. *Current Pharmaceutical Design* 9 (16):1225–1238.

Perić, L., K. Sartowska, N. Milošević, M. Đukić-Stojčić, S. Bjedov, and N. Nikolova. 2011. The effect of enzymes on the economics of poultry meat and egg production. *Macedonian Journal of Animal Science* 1 (1):113–117.

Pietrasik, Z., and P. J. Shand. 2010. Bovine biceps femoris is resistant to tenderization by lower-salt moisture enhancement or enzyme addition. *Canadian Journal of Animal Science* 90 (4):495–503.

Pietrasik, Z., J. L. Aalhus, L. L. Gibson, and P. J. Shand. 2010. Influence of blade tenderization, moisture enhancement and pancreatin enzyme treatment on the processing characteristics and tenderness of beef semitendinosus muscle. *Meat Science* 84 (3):512–517.

Pihlanto-Leppälä, A. 2000. Bioactive peptides derived from bovine whey proteins: Opioid and ace-inhibitory peptides. *Trends in Food Science & Technology* 11 (9–10):347–356.

Piot, J.-M., Q. Zhao, D. Guillochon, G. Ricart, and D. Thomas. 1992. Isolation and characterization of two opioid peptides from a bovine hemoglobin peptic hydrolysate. *Biochemical and Biophysical Research Communications* 189 (1):101–110.

Polkinghorne, R., J. Philpott, A. Gee, A. Doljanin, and J. Innes. 2008. Development of a commercial system to apply the Meat Standards Australia grading model to optimise the return on eating quality in a beef supply chain. *Australian Journal of Experimental Agriculture* 48 (11):1451–1458.

Qihe, C., H. Guoqing, J. Yingchun, and N. Hui. 2006. Effects of elastase from a *Bacillus* strain on the tenderization of beef meat. *Food Chemistry* 98 (4):624–629.

Ramezani, R., M. Aminlari, and H. Fallahi. 2003. Effect of chemically modified soy proteins and ficin-tenderized meat on the quality attributes of sausage. *Journal of Food Science* 68 (1):85–88.

Rocha, C. B., J. L. F. Pouey, S. R. N. Piedras, D. B. S. Enke, and J. M. Fernandes. 2010. Fitase na dieta de alevinos de carpa húngara: Desempenho e características de carcaça. *Arquivo Brasileiro de Medicina Veterinária e Zootecnia* 62:1462–1468.

Rowan, A. D., D. J. Buttle, and A. J. Barrett. 1990. The cysteine proteinases of the pineapple plant. *Biochemical Journal* 266 (3):869–875.

Saiga, A., K. Iwai, T. Hayakawa et al. 2008. Angiotensin I-converting enzyme-inhibitory peptides obtained from chicken collagen hydrolysate. *Journal of Agricultural and Food Chemistry* 56 (20):9586–9591.

Schwimmer, S. 1981. *Source Book of Food Enzymology*. Westport, CT: AVI Publishing Co.

Selle, P. H., A. J. Cowieson, and V. Ravindran. 2009. Consequences of calcium interactions with phytate and phytase for poultry and pigs. *Livestock Science* 124 (1–3):126–141.

Shackelford, S. D., T. L. Wheeler, M. K. Meade, J. O. Reagan, B. L. Byrnes, and M. Koohmaraie. 2001. Consumer impressions of Tender Select beef. *Journal of Animal Science* 79 (10):2605–2614.

Shimizu, M., N. Sawashita, F. Morimatsu et al. 2009. Antithrombotic papain-hydrolyzed peptides isolated from pork meat. *Thrombosis Research* 123 (5):753–757.

Singh, N. K., D. K. Joshi, and R. K. Gupta. 2013. Isolation of phytase producing bacteria and optimization of phytase production parameters. *Jundishapur Journal of Microbiology* 6 (5):e6419.

Sun, X. D. 2009. Utilization of restructuring technology in the production of meat products: A review. *Cyta-Journal of Food* 7 (2):153–162.

Tahoun, A. M., H. A. Abo-State, and Y. A. Hammouda. 2009. Effect of adding commercial phytase to DDGS based diets on the performance and feed utilization of Nile Tilapia (*Oreochromis niloticus*) Fingerlings. *American-Eurasian Journal of Agricultural & Environmental Sciences* 5 (4):550–555.

Toldrá, F., M. C. Aristoy, L. Mora, and M. Reig. 2012. Innovations in value-addition of edible meat by-products. *Meat Science* 92 (3):290–296.

Trespalacios, P., and R. Pla. 2007. Synergistic action of transglutaminase and high pressure on chicken meat and egg gels in absence of phosphates. *Food Chemistry* 104 (4):1718–1727.

Tseng, T. F., D. C. Liu, and M. T. Chen. 2000. Evaluation of transglutaminase on the quality of low-salt chicken meatballs. *Meat Science* 55 (4):427–431.

Veiseth, E., S. D. Shackelford, T. L. Wheeler, and M. Koohmaraie. 2004. Factors regulating lamb longissimus tenderness are affected by age at slaughter. *Meat Science* 68 (4):635–640.

Vijayaraghavan, P., R. R. Primiya, and S. G. P. Vincent. 2013. Thermostable alkaline phytase from *Alcaligenes* sp. in improving bioavailability of phosphorus in animal feed: In vitro analysis. *ISRN Biotechnology* 2013:6.

Weiss, J., M. Gibis, V. Schuh, and H. Salminen. 2010. Advances in ingredient and processing systems for meat and meat products. *Meat Science* 86 (1):196–213.

WHO (World Health Organization). 2002. Diet, nutrition and the prevention of chronic diseases: Global and regional food consumption patterns and trends. In *Nutrition Health Topics*, edited by J. W. F. E. Consultation. Geneva: WHO—World Health Organization.

Yamaguchi, T., Y. Yamashita, I. Takeda, and H. Kiso. 1982. Proteolytic enzymes in green asparagus, kiwi fruit and miut: Occurrence and partial characterization. *Agricultural and Biological Chemistry* 46 (8):1983–1986.

Zhang, F., L. Fang, C. Wang et al. 2013. Effects of starches on the textural, rheological, and color properties of surimi-beef gels with microbial transglutaminase. *Meat Science* 93 (3):533–537.

Zhao, G. Y., M. Y. Zhou, H. L. Zhao et al. 2012. Tenderization effect of cold-adapted collagenolytic protease MCP-01 on beef meat at low temperature and its mechanism. *Food Chemistry* 134 (4):1738–1744.

Zhu, Y., and J. Tramper. 2008. Novel applications for microbial transglutaminase beyond food processing. *Trends in Biotechnology* 26 (10):559–565.

Zhu, Y., J. Bol, A. Rinzema, and J. Tramper. 1995. Microbial transglutaminase—A review of its production and application in food processing. *Applied Microbiology and Biotechnology* 44 (3–4):277–282.

15

Enzymes in Seafood Processing

P. V. Suresh, T. Nidheesh, and Gaurav Kumar Pal

CONTENTS

15.1 Introduction

The term "seafood" is commonly used to describe a group of biologically diverse edible animals (without mammals), consisting of fish (finfish) and shellfish, whether of fresh water, estuarine, or marine habitats (FAO 2013; Venugopal 2006). Further, the word "seafood" is applied to edible animals both harvested from aquatic sources by fishing and produced by farming. Aquatic animals, such as frogs and turtles, that are served as food and eaten by humans are also considered to be seafood (FAO 2013). Edible seaweeds are also seafood and are widely eaten as sea vegetables all over the world, particularly in Asian countries (FAO 2013). In most countries, the bulk quantity of seafood harvested is sold fresh for local consumption. However, a major portion of seafood harvested is processed in some way or another throughout the world, owing to its extremely perishable nature or export potential or to produce a range of products (processed and ready to eat) with different textures and flavors. In general, commercial seafood processing recovers only 20% to 50% as edible portions, and the remaining parts (50%–80%) are discarded as "nonedible" by-products (Guerard 2007; Suresh and Prabhu 2012). Normally, these by-products are rich sources of different valuable components, such as protein, oils and lipids, bioactive peptides, pigments, flavors, chitin, collagen, vitamins, minerals, enzymes, etc., useful for various applications in many fields (Suresh and Prabhu 2012).

TABLE 15.1

Various Applications of Enzymes in Seafood Processing Industries

Field of Enzyme Application	Unit Operations in Which Enzymes Are Used
Seafood process development	Removal of scales from fish
	Removal of skin from fish
	Removal of skin from squid and tenderization of squid
	Peeling of shrimp and shucking of clam
	Ripening of salted fish
	Production of fish sauce (acceleration of fish sauce production)
	Caviar production
	Production of surimi and restructured fishery products
Seafood by-product processing	Recovery of protein and protein hydrolysate from seafood by-products
	Recovery of oil, lipids, and fatty acids from seafood by-products
	Recovery of chitin and its derivatives from seafood processing by-products
	Recovery of flavor from seafood by-products
	Recovery of pigment from seafood by-products
	Recovery of calcium and minerals from seafood by-products
	Recovery of collagen from seafood by-products
	Recovery of bioactive peptides from seafood by-products
	Recovery of bioactive compounds from other seafood resources
Other applications in seafood processing	Extension of shelf life of fishery products
	Removal of off-odor and fishy taste
	Retention of color of cooked shrimp and crab
	Reduction in the viscosity of stick water
Analytical methods in seafood processing	Assessment of seafood quality and freshness
	Direct enzymatic methods for quality evaluation
	Detection of frozen fishery products
	Identification of fish species and its adulteration

Enzymatic methods have become an essential and integral part of the processes used by the seafood industry to produce a large and diversified range of products as well as to improve the quality of finished products. This is mainly due to the highly specific nature of enzymes. Enzymes are highly active at very low concentrations under mild conditions of pH and temperature, which results in fewer unwanted side effects and by-products in the production process. Traditional applications of enzymes in seafood processing have been limited to very few products, such as fish protein hydrolysate, fish sauce, or cured herring, etc., based on the action of endogenous proteases present in the fish (Diaz-López and García-Carreño 2000). Due to the advancement of enzyme technology, additional applications of enzymes from different sources in seafood processing have emerged recently.

Applications of enzymes in the seafood industry include descaling of fish, deskinning of fish, ripening of salted fish, production of fish sauce, production of surimi, removal of the exoskeleton from shellfish, extraction of fish oil, and preparation of ω-3 polyunsaturated fatty acid (ω-3 PUFAs or n-3 PUFAs)-enriched marine oil, flavor production, extraction of chitin and carotenoproteins from shellfish processing discards, production of fish protein hydrolysates, etc. Enzymes have also been used as components of biosensors for assessing the quality as well as the freshness of seafood. In addition, a number of enzymes have also been used as constituents of different methods developed for identifying the seafood species and its adulterations. Various enzymatic methods used in seafood processing are summarized in Table 15.1.

15.2 Enzymes in Seafood Processing: Sources and Applications

The traditional applications of enzymes in seafood processing have been limited to very few products/ processes such as fish protein hydrolysate, fish sauce, or cured herring, etc., based on the action of endogenous proteases (Diaz-López and García-Carreño 2000). Recently, additional applications of enzymes from different sources in seafood processing have emerged due to the advancements in biotechnology and enzyme technology. These include the improvement of traditional applications by using exogenous enzymes to accelerate the process, production of other products, and the use of enzymes to improve the production processes or as alternative processes in seafood production. Applications of enzymes in seafood processing can be classified broadly into two main groups: (1) enzymes in seafood process development and (2) enzymes in seafood by-product processing. In addition, a number of enzymatic methods have been developed for checking the quality and freshness of seafood (raw and processed), species identification of seafood and to check seafood (raw and processed) adulterations (Figure 15.1).

Enzymes commonly used in seafood processing are presented in Table 15.2. The enzymes used in seafood processing are commercially obtained from microorganism (natural isolates or recombinants), plant, and animal (terrestrial or marine) sources. Microbial enzymes are often more useful than enzymes derived from plants or animals because of the great variety of catalytic activities available, the high yield, ease of genetic manipulation, regular supply due to absence of seasonal fluctuations, and rapid growth of microorganisms on inexpensive culture media. Among the various enzymes, protease and transglutaminase are the most important enzymes used in seafood processing. Enzymes from seafood by-products (marine source enzymes), particularly from fish viscera, and their role in seafood processing have been discussed by a number of researchers (Benjakul et al. 2010a; Diaz-López and García-Carreño 2000; Shahidi and Kamil 2001; Venugopal 2015).

15.2.1 Proteases

Proteases are widely utilized as biotechnological tools in seafood product development, such as deskinning, descaling, caviar production, tenderization of squid, ripening of salted fish, fish sauce, etc. In addition, proteases are also extensively examined for the processing of seafood by-products to make hydrolysates from seafood processing by-products and underutilized seafood and to extract valuable components, such as seafood flavor, chitin, pigment, oil and fats, etc. The role of proteases in seafood processing has been discussed by several investigators (Benjakul et al. 2010a; Diaz-López and García-Carreño 2000; Klomklao 2008; Shahidi and Kamil 2001; Venugopal 2015; Venugopal et al. 2000).

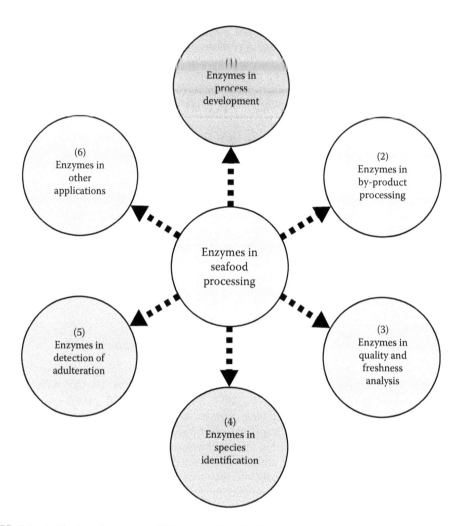

FIGURE 15.1   Application of enzymes in different areas in seafood processing: schematic presentation.

15.2.2 Transglutaminase

Transglutaminase (TGase, glutaminyl-peptide:amine γ-glutamyltransferase, EC. 2.3.2.13) is an enzyme capable of catalyzing acyl transfer reactions by cross-linking, introducing proteins, such as myosin, peptides, and primary amines (Helena et al. 2008; Kieliszek and Misiewicz 2014; Morrissey and Okada 2007) and also by forming hydrophobic interactions (Park 2005). It is well documented that TGase catalyzes covalent bonding between the ε-amino group of lysyl residues and the γ-carboxamide group of glutaminyl residues of adjacent proteins (Uresti et al. 2006). It is a calcium-dependent endogenous enzyme that is generally found in the muscle tissue of fish and other animals. Microbial transglutaminase (MTGase) is an extracellular calcium-independent TGase produced by a number of microorganisms, such as *Streptoverticillium* sp., *Bacillus subtilis*, and *Streptomyces* sp. MTGase is active over a wide range of pH and temperatures (0°C–70°C). The finding of calcium-independent TGase from *Streptoverticillium mobaraense* boosted the TGase investigation and utilization in seafood product development because of the better availability of this enzyme (Aidaroos et al. 2011; Helena et al. 2008; Yu et al. 2008). MTGase prepared from *Streptoverticillium* sp. is the only transglutaminase product available commercially as a food enzyme preparation (Kuraishi et al. 2001). TGase (Kuraishi et al. 2001) and MTGase and their applications in food processing have been reviewed (Kieliszek and Misiewicz 2014).

TGase and MTGase are widely examined as biotechnological tools in seafood processing, especially for the setting of minced fish muscle in surimi production (Kuraishi et al. 2001; Park 2005). Potential

TABLE 15.2

Various Enzymes and Their Potential Applications in Seafood Processing

Enzymes	Probable Application in Seafood Processing
Proteases	Removal of scales from fish
	Removal of skin from fish
	Removal of skin from squid and tenderization of squid
	Peeling of shrimp and shucking of clam
	Ripening of salted fish
	Production of fish sauce (acceleration of fish sauce production)
	Caviar production
	Reduction in the viscosity of stick water
	Recovery of protein and protein hydrolysate from seafood by-products
	Recovery of oil, lipids, and fatty acids from seafood by-products
	Recovery of chitin from seafood processing by-products
	Recovery of flavor from seafood by-products
	Recovery of pigment from seafood by-products
	Recovery of calcium and minerals from seafood by-products
	Recovery of collagen from seafood by-products
	Recovery of bioactive peptides from seafood by-products
Transglutaminase	Surimi and restructured fishery products
	Fish meat film formations
	Fish meat mince formulations
	Binding of ingredients with fish meat
	Raw and processed fish egg products
	Finfish texture modification
	Shark fin processing
	Gelatin and collagen bond formation
	Reduction of drip in thawed frozen fish
	Freeze-texturization of fishery products
Lipases	Enrichment of ω-3 PUFA
	Isolation of oil and fats from seafood products
	Evaluation of seafood quality and freshness
Other Enzymes	
Chitinases	*N*-acetyl-D-glucosamine preparation
Glucose oxidase	Preservation of seafood
	Retention of color of cooked shrimp and crab
Lysozyme	Preservation of seafood
Catalase	Preservation of seafood
Urease	Removal of off-odor and fishy taste
	Evaluation of seafood quality and freshness
Xanthine oxidase	Evaluation of seafood quality and freshness
Nucleotide phosphorylase	Evaluation of seafood quality and freshness
Glutamine dehydrogenase	Evaluation of seafood quality and freshness
Putrescine oxidase	Evaluation of seafood quality and freshness
Diamine oxidase	Evaluation of seafood quality and freshness
Glycerol dehydrogenase	Evaluation of seafood quality and freshness
Restriction endonucleases	Identification of fish species and detection of adulteration
Taq polymerase	Identification of fish species and detection of adulteration

uses of TGase and MTGase in seafood processing include in surimi and surimi-based fish analogues; fish meat paste formulations; binding of ingredients such as soybean, milk proteins, egg white, etc., with fish meat; fish meat sheet formation; raw and processed fish egg products; finfish texture modification, involving gelatin and collagen bond formation; shark fin processing; reduction of drip in thawed frozen fish; freeze-texturization of fishery products, etc. (Venugopal et al. 2000). This enzyme is also examined for the processing of seafood by-products to make biodegradable gelatin film (Benjakul et al. 2010a).

15.2.3 Lipases

Lipolytic enzymes consist of two major groups, the lipases, which are triacylglycerol acylhydrolases (EC. 3.1.1.3), and the phospholipases A_1 (3.1.1.32) and A_2 (3.1.1.4), which are phosphoglyceride acyl hydrolases. Although phospholipases C (3.1.4.3) and D (3.1.4.4) are not acylhydrolases, they are nonetheless commonly included as lipolytic enzymes. Lipolytic enzymes are involved in the breakdown and thus in the mobilization of lipids within the cells of individual organisms as well as in the transfer of lipids from one organism to another (Beisson et al. 2000; Fariha et al. 2006). Triacylglycerols are found wildly in plants, animals, and microorganisms. The commercial sources of these enzymes include animal tissue, microorganisms, and seafood by-products. Lipases differ greatly with respect to their origins (which can be bacterial, fungal, mammalian, etc.) and their properties, and they can catalyze the hydrolysis or synthesis of a wide range of different carboxylic esters and liberate organic acids and glycerols. All these enzymes show high specific activity toward glyceridic substrates (Fariha et al. 2006). Lipases are widely exploited as biotechnological tools for the isolation of oil and fats from seafood by-products (and underutilized seafood) as well as in the preparation and concentration of ω-3 PUFA-enriched marine oils (Guerard 2007; Wanasundara 2011).

15.3 Enzymes in Seafood Process Development

15.3.1 Enzymes in Removal of Scales from Fish

Descaling of fish is one of the major fish processing operations due to the demand for fish fillets with skin on but without scales as well as in the leather industry in which fish skins without scales are frequently used (Benjakul et al. 2010a; Shahidi and Kamil 2001; Venugopal et al. 2000). Usually, the descaling of fish is performed manually or by mechanical scrubbing and scraping methods. However, these processes are very tedious as fish have large amounts of scales. Further, there are a number of problems with these processes, including incomplete removal of scales, partial damage or tearing of the skin, loss of the color and shiny appearance associated with skin, damage to muscle texture, etc. These resulted in the development of enzymatic methods to remove scales from fish of several species (Benjakul et al. 2010a; Venugopal 2015). The enzymatic method includes mainly three steps: denaturation and loosening of the mucous layer and outer protein structures of the skin, enzymatic degradation of the outer skin structures, and washing off the scales with water jets (Benjakul et al. 2010a; Shahidi and Kamil 2001; Venugopal et al. 2000).

15.3.2 Enzymes in Removal of Skin from Fish

The common method of removing skin from fish fillets is purely mechanical, but an automated machine in operation tears the skin off the flesh. The ease of the deskinning process varies greatly among fish species (Benjakul et al. 2010a; Venugopal et al. 2000; Vilhelmsson 1997). A number of enzymatic processes have been developed as an alternative to physical and mechanical methods in the processing of skin removal. The enzymatic method for removal of fish skin is generally based on the solubilization of collagenous skin tissue without degrading the muscle tissue. Endogenous proteolytic enzymes from seafood as well as exogenous enzymes from other sources can be applied for skin removal from fish (Benjakul et al. 2010a; Shahidi and Kamil 2001). Proteases are the principal enzymes used for skin removal from fish, but they are usually mixed with carbohydrases in order to facilitate skin removal. Enzymatic methods have been developed for deskinning of various fish species, including ocean perch, haddock, tuna, herring, and skate wings (Benjakul et al. 2010a; Shahidi and Kamil 2001).

15.3.3 Enzymes in Removal of Skin from Squid and Tenderization of Squid

The squid outer surface is protected by double-layered skin membranes. The eating quality of squid is highly determined by the degree of deskinning of squid due to its characteristic and rather tough texture (Shahidi and Kamil 2001; Venugopal et al. 2000). In general, the removal of squid skin requires a pretreatment in 5% salt solution at 45°C for 10 min to soften the tough outer protective layer. The treatment also activates endogenous enzymes, causing a softening of the skin. However, this conventional mechanical deskinning method removes only the pigmented outer squid skin, leaving the rubbery inner membrane intact. Enzymatic deskinning processes for removing the double-layered skin of squid using proteolytic enzymes from squid intestines have been reported. In addition, tenderization of squid meat by enzymatic treatment has been developed (Shahidi and Kamil 2001; Venugopal et al. 2000).

15.3.4 Enzymes in Peeling of Shrimp and Shucking of Clams

Shrimp processing (peeling) is one of the major agro-industries in tropical and subtropical countries based on aquaculture and marine fishing and has attained great importance during the last couple of decades due to the ever-increasing demand for processed shrimp in the international market (Ghorbel-Bellaaj et al. 2011; Kumar and Suresh 2014; Nidheesh and Suresh 2015a; Suresh 2012). The process of loosening the shells from the muscle tissue of shrimp prior to peeling by the use of a mixture of carbohydrases and cellulases derived from *Aspergillus niger* was reported (Venugopal et al. 2000). Shucking of clams by an enzymatic process has also been reported (Venugopal et al. 2000).

15.3.5 Enzymes in Ripening of Salted Fish

The use of pelagic fish species for salting is a very ordinary practice in several European countries. The salted anchovy (*Engraulis encrasicholus*) is a traditional heavily salted food product in Mediterranean countries with a high nutritional value that contains approximately between 14% and 15% of NaCl (Llorente et al. 2007). Salting is a conventional preservation technique for herring (*Clupea harengus*) in many European countries, for instance, in Norway, Finland, Denmark, Iceland, Holland, Russia, and Germany (Rahaman 2014). The ripening of fish is traditionally carried out by storing fish (gutted or ungutted) together with salt and probably sugar in barrels for several months in order to develop the characteristic changes in texture and flavor (Rahaman 2014; Shahidi and Kamil 2001; Venugopal et al. 2000). It is a complex enzymatic conversion of main components of the raw material and subsequent changes of the products due to their autolysis, which is responsible for the organoleptic properties of the final product. The ripening process appears to be principally endogenous proteolytic, leading to a degradation of muscle proteins, resulting in a soft textured product, taste-active peptides, and free amino acids (Rahaman 2014; Shahidi and Kamil 2001). Fish tissue cathepsins as well as enzymes from microorganisms have also been reported to be responsible for autolysis. Most researchers believe that the process of ripening salted fish is attributable mainly to a proteolytic reaction caused by digestive enzymes released from pyloric caeca and other intestines. Therefore, in the process of ripening headed and gutted fish, special enzymatic substances or internal organs of other animals rich in enzymes are often used (Kołakowski and Bednarczyk 2003). Accelerated fermentation of salted fish by the incorporation of exogenous protease preparation has been documented (Shahidi and Kamil 2001; Venugopal et al. 2000).

15.3.6 Enzymes in Production of Fish Sauce

Fish sauce is a traditional fermented fish product, which is used as an important source of protein in several Southeast Asian countries (Benjakul et al. 2010a). It is also consumed in Europe and North America. Some examples of such products are *nuoc mam* in Vietnam and Cambodia, *nam-pla* in Thailand, *patis* in the Philippines, *uwoshoyu* in Japan, and *ngapi* in Burma (Diaz-López and García-Carreño 2000). It is a clear liquid product produced by the hydrolysis of fish protein in the presence of a high salt concentration. It is prepared by adding 20%–30% marine salt to fish (small and commercially unimportant fish) such as anchovies, sardines, or round scad, and permitting them to hydrolyze in closed containers (Shahidi and Kamil 2001). For complete hydrolysis of fish tissue and flavor development, the fermentations were carried out for 3 months to 2 years at approximately 30°C–35°C (Klomklao et al. 2006). During incubation, endogenous halotolerant enzymes present in the muscle gradually degrade the fish tissue in order to form a clear, amber-colored liquid with a high content of free amino acids and excellent flavor (Shahidi and Kamil 2001). Due to the high salt content, in general the growth of bacteria is negligible.

Because fish sauce fermentation requires a long period of time, several methods have been proposed by modifying the fermentation conditions to accelerate the process and reduce the production time. These accelerated methods have accomplished an improved liquefaction rate (Benjakul et al. 2010a, Shahidi and Kamil 2001). However, the products from the accelerated processes at alkaline condition have shown inferior flavor and bitter taste. Acceleration of the protein hydrolysis in traditional fish sauce by the addition of exogenous enzymes under controlled conditions was reported. Plant proteinases, such as papain, bromelain, and ficin, were used to shorten the fish sauce fermentation time. These plant-derived enzymes are cysteine proteases, and most are active under low acid conditions. Addition of these exogenous proteinases decreases the liquefaction time, but the flavor characteristic of the final product was normally inferior to the traditional product (Benjakul et al. 2010a; Shahidi and Kamil 2001; Venugopal 2015). The enzymes recovered from marine animals have also been successfully used in the acceleration of fermentation for the production of high-quality fish sauce. The fermentation time of capelin sauce supplemented with squid hepatopancreatic proteases was shortened to 6 months in addition to recording higher acceptability than traditional fish sauce or other commercial protease supplements (Shahidi and Kamil 2001). Fermentation of squid meat with added proteolytic enzymes was reported (Shahidi and Kamil 2001). Supplementing the spleen of skipjack tuna (*Katsuwonus pelamis*) accelerated liquefaction of sardines (*Sardinella gibbosa*) for fish sauce production (Klomklao et al. 2006).

15.3.7 Enzymes in Caviar Production

Traditionally *caviar* refers only to the riddles and cured roe of the sturgeon (*Acipenser* sp. and *Huso huso*) (Shahidi and Kamil 2001; Vilhelmsson 1997). However, in recent years, the term "caviar" is also used to refer to the treated roe of other less-expensive fish species, such as catfish (*Parasilurus astus*), herring (*Clupea harengus*), capelin, lumpfish, and skipjack (Shahidi and Kamil 2001; Vilhelmsson 1997). The preparation of caviar starts with the riddling process, which consists of separating the roe (egg) from the roe sack (ovaries) of the fish at the moment of slaughter. The riddling process in the preparation of caviar is a difficult task, which is carried out either mechanically or manually. The conventional process of separation of roe has several disadvantages, such as being labor-intensive, giving a poor yield, having a short shelf life due to severe mechanical damage to the eggs, etc. (Shahidi and Kamil 2001; Vilhelmsson 1997). Various enzymatic methods have been examined for the production of superior-quality caviar from different fish species (Diaz-López and García-Carreño 2000).

15.3.8 Enzymes in Surimi and Restructured Fishery Products

The word "surimi" is derived from the Japanese verb *suru*, which means "to mince," and it refers to minced fish muscle tissue that has been refined by repeated washing with water. It is widely used in many countries, particularly Japan, for the development of textured seafood analogues, such as crab legs, shrimp, etc. In Japan, surimi is principally used in the manufacture of various types of heat-gelled products, such as *itatsuki*, *kamaboko*, *chikuwa*, *hanpen*, and *satsuma-age*. Surimi is considered to be a valuable, functional proteinaceous ingredient, similar to the use of soybean protein concentrates, in a variety of food products (Benjakul et al. 2010a; Diaz-López and García-Carreño 2000; Shahidi and Kamil 2001).

Surimi forms thermo-irreversible gels upon heating, constituted by a three-dimensional protein network formed mainly of actomyosin. The characteristic rheological properties of surimi are largely based on the gel-forming ability of the fish myofibrillar proteins, myosin being the most important. The textural characteristics developed during gelation are expressed as gel strength, which is the primary determinant for surimi quality and price. The setting of surimi is induced by the addition of exogenous TGases. Generally, a little amount of salt is added to enhance the protein solubility and cross-linking of myofibrillar proteins in surimi (Benjakul et al. 2010a; Diaz-López and García-Carreño 2000; Shahidi and Kamil 2001). It is reported that the salt requirement can be overcome by incorporation of MTGase (Jongjareonrak et al. 2006; Ramirez et al. 2007; Téllez-Luis et al. 2002). MTGase also affects the low temperature gelation of fish protein sols leading to textural modifications (Benjakul et al. 2008). MTGase has been proven as a potential gel-enhancing agent, which improves the gel strength of surimi produced from low-quality lizardfish. MTGase was reported to be a powerful gel enhancer of mince from lizardfish stored in ice at different temperatures (Benjakul et al. 2008), and it has been used widely in restructured fish products to reduce the salt required due to its ability in protein cross-linking (Benjakul et al. 2010a). Protein additives, sodium ascorbate, and MTGase influenced the texture and color of red tilapia surimi gel (Duangmal and Taluengphol 2010). To enhance the gel strength of surimi with MTGase added, a number of ingredients involving protein additive, protease inhibitors as well as hydrocolloids have been used in conjuction (Benjakul et al. 2010a). The addition of MTGase in sardine surimi eliminated the adverse effect of Bambara groundnut protein isolate (BGPI) on muscle protein gelation and produced a rigid mixed protein gel (Kudre et al. 2013). Furthermore, the addition of BGPI prepared with heat treatment in the presence of EDTA in combination with MTGase in sardine surimi exhibited the highest gel strengths with negligible lipid oxidation and beany flavor (Kudre et al. 2014).

Restructured fishery products are processed from minced and/or chopped muscle, usually with added ingredients, to make products with a new appearance and texture. Development of a new generation of fishery products called analogues or substitutes has been witnessed over the last 30 years and most of them are formulated essentially from surimi (Helena et al. 2005). MTGase and alginate as additives were used in cold gelification of minced hake (*Merluccius capensis*) muscle (Helena et al. 2005). Restructured fish product from white croaker (*Micropogonias furnieri*) mince was prepared using MTGase (Gonçalve and Passos 2010). The restructured fillet developed has advantages when compared to traditional fillets, such as absence of spine and less flavor intensity (wash cycles). Chitosan and MTGase influenced the gel-forming ability of horse mackerel (*Trachurus* spp.) muscle under high pressure (Carmen et al. 2005).

Unlike fish species, squid mantle muscle has been characterized by a low gel-forming capacity because of its high protease activity, which causes preferential degradation of myosin during heating (Park et al. 2005). MTGase would be an effective additive for inducing the setting effect on squid muscle thermal gelation because it is fully active in the absence of calcium and at higher temperatures (40°C–50°C). MTGase and starch influenced the thermal gelation of salted squid muscle paste (Park et al. 2005).

15.4 Enzymes in Seafood By-Product Processing

Finfish as well as shellfish processing by-products are prospective sources of a number of valuable components with a wide range of potential applications in various fields including food, feed, bio-medicine, fine chemicals, agriculture, the environment, pharmaceuticals, etc. These valuable components from seafood processing by-products include chitin; pigment; enzymes; bioactive peptides; minerals, such as calcium, collagen, lipid, and long-chain ω-3 PUFAs, etc. Some of these components have been extracted and marketed commercially (Suresh and Prabhu 2012). In recent years, enzymatic processes have emerged as an integral part of seafood by-product processing not only as an attractive substitute to chemical, physical, and mechanical methods, but also as these enzymatic tools are eco-friendly (Guerard 2007; Rustad 2003; Shahidi and Kamil 2001; Suresh and Prabhu 2012).

15.4.1 Enzymes in the Recovery of Protein and Protein Hydrolysate from Seafood By-Products

Seafood processing by-products are rich sources of functionally active and nutritive marine proteins with various potential applications (Okada et al. 2008; Suresh and Prabhu 2012). Proteins from seafood by-products are recovered chiefly in the form of protein hydrolysates, which are derived from proteins as peptides of various sizes after chemical or enzymatic hydrolysis (Guerard 2007; Kristinsson and Rasco 2000a). The protein hydrolysates have a wide range of potential applications, including use as flavor enhancers, functional ingredients, or as nutritional additives to foods of low protein quality (Sachindra et al. 2011). The hydrolysates, extracted from seafood by-products, have become popular in the food industry due to the high protein content (Córdova-Murueta et al. 2007). Marine protein hydrolysates are rich sources of biologically active peptides with considerable potential in pharmacology and/or as growth-stimulating agents in animal feeds (Gildberg and Stenberg 2001).

Hydrolysis of protein of seafood by-products can be achieved by both the digestive enzymes of the fish itself as well as by the addition of enzymes of external sources (Rustad 2003). Enzymatic hydrolysis possesses a greater advantage when compared with chemical hydrolysis or autolysis caused by the endogenous enzymes in fish and shellfish. A number of methods for recovering marine protein as hydrolysates from seafood by-products of different origin with added enzymes have been attempted (Benjakul et al. 2010a). The main principle of the methods is similar in most cases with minor alteration, depending on the target compounds and functionalities. In general, the method of production of protein hydrolysates from seafood by-products (or underutilized seafood species) involves mixing of crushed raw materials with enzymes and incubation in a reactor for a specified time under optimum conditions for the enzyme activity followed by inactivation of enzymes by heating or pH adjustment. The hydrolyzed liquor is then separated by filtration and centrifugation and concentrated by dehydration or drying (Guerard 2007; Okada et al. 2008; Suresh and Prabhu 2012). A generalized method for preparation of protein hydrolysates by enzymatic process from seafood by-products is presented in Figure 15.2.

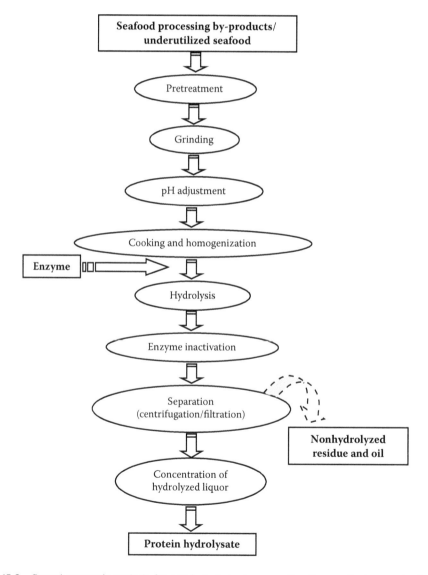

FIGURE 15.2 General enzymatic methods for protein hydrolysate preparation from seafood processing by-products or underutilized seafood. (Adapted from Suresh, P. V. and G. N. Prabhu, Scafood. In *Valorization of Food Processing Byproducts*, Ed. Chandrasekaran, M. CRC Press, Boca Raton, FL, pp. 685–736, 2012.)

By-products from several fish species have been exploited for commercial production of protein hydrolysate with the most common source being fillet by-products (Okada et al. 2008; Suresh and Prabhu 2012). By-products from shellfish processing, such as squid (Ono et al. 2004) and shrimp (Bhaskar et al. 2007) as well as jellyfish (Zhuang et al. 2009) are possible sources of high-quality protein hydrolysates. Arctic and Antarctic krill have also been reported as sources of protein and protein hydrolysate (Sountarna et al. 2007).

Protein hydrolysate with specific functionalities and/or biological properties can be prepared by using specific enzymes and hydrolysis conditions (Kristinsson 2007; Sachindra et al. 2011). The most common commercial proteases used for seafood protein hydrolysate preparation include those from microbial sources, such as Alcalase, Neutrase, Flavourzyme, Umamizyme, Protamex, etc.; from plant sources, such as papain, bromelain, ficin, etc.; and from animal sources, such as pepsin, trypsin, chymotrypsin, etc. (Benjakul et al. 2010a; Guerard 2007). The use of enzymes from microbial sources for protein hydrolysis offers a number of advantages, such as a wide variety of available catalytic activities and greater pH and temperature stability (Guerard et al. 2002). Among various commercial proteases, Alcalase, an alkaline enzyme produced from *Bacillus licheniformis* and developed by Novozymes, has been reported to be one of the most effective enzymes for hydrolysis of fish proteins (Aspmo et al. 2005; Bhaskar et al. 2007).

15.4.2 Enzymes in the Recovery of Oil, Lipids, and Fatty Acids from Seafood By-Products

Seafood processing by-products (head, gut, liver, etc.) are potential sources of marine lipid-based compounds (Amit et al. 2010; Mbatia 2011; Shahidi 2007). The lipid-based compounds that can be recovered from seafood processing by-products include oils, ω-3 fatty acids, phospholipids, squalene, vitamins, cholesterol, etc. (Suresh and Prabhu 2012). The beneficial effects of fish oils on human health are well recognized and documented (Berge and Barnathan 2005; Venugopal 2009). ω-3 PUFAs are essential fatty acids (FAs) that cannot be synthesized in the human body. The main source of ω-3 PUFAs are marine products, and they are synthesized by phytoplankton and algae, transferred through the food web and incorporated into the lipids of fish and marine mammals (Shahidi 2008). Fish oils are still considered to be the least expensive natural source of ω-3 PUFAs (Guerard 2007). The occurrence of ω-3 PUFAs, especially eicosapentaenoic acid (EPA, 20:5 n-3) and docosahexaenoic acid (DHA, 22:6 n-3) in fish oils are responsible for their health benefits (Osborn and Akok 2002; Shahidi and Alasalvar 2011).

The enzymatic methods used for the recovery of oil from seafood by-products can be classified into two different categories: (i) protease-catalyzed hydrolysis and (ii) lipase-catalyzed hydrolysis and esterification (Guerard 2007; Mbatia 2011; Wanasundara 2011). Several commercial enzymes from animal, plant, or microbial sources can be used for this purpose, such as pepsin, papain, Alcalase, Neutrase, Corolase PN-L, Corolase 7089, Flavourzyme, and Protamex (Kahveci and Xu 2011; Kristinsson and Rasco 2000b; Liaset et al. 2000; Linder et al. 2005; Mbatia 2011; Okada and Morrissey 2007).

The protease-catalyzed hydrolysis method is used to extract the ω-3 PUFAs from the starting raw material. It is recognized that oil extracted from seafood processing by-products with enzymatic hydrolysis is similar to that produced by conventional methods (Dumay et al. 2004; Linder et al. 2002; Sathivel et al. 2003). Commercial proteases from plants (papain), from animals (chymotrypsin), and from microbes (Protamex, Flavourzyme, Neutrase, and Alcalase) were used for the extraction of oil and lipids from fish sources by protease-catalyzed hydrolysis (Amit et al. 2010; Dumay et al. 2004; Laset et al. 2003; Swapna et al. 2011a). Approximately 77% of total lipids with high EPA and DHA content were achieved from salmon frames by extraction using Protamex at 55°C, pH 6.5 for 60 min (Laset et al. 2003). Commercial proteases have been used to release oil from salmon (*Salmo salar*) heads resulting in improved yields as compared to yields obtained after heat treatment (Gbogouri et al. 2006). The use of commercial proteases, bromelain, and Protex 30L for oil recovery from Nile perch (*Lates niloticus*) and salmon (*Salmo salar*) heads was evaluated (Mbatia et al. 2010b). The maximum oil yield was obtained when hydrolysis was performed with Protex 30L at 55°C without pH adjustment or water addition. They reported that an oil yield of 11.2% and 15.7% of wet weight was obtained from Nile perch and salmon heads, respectively, compared to 13.8% and 17.6%, respectively, obtained using solvent extraction (Mbatia et al. 2010a). The recovery of lipids and protein simultaneously by hydrolysis of visceral waste of Indian carp was attempted using four different commercial proteases (Protease-P-Amano6, Alcalase, Protex, and Neutrase) (Swapna et al. 2011a). The hydrolysis using fungal protease (Protease-P-Amano6) resulted in maximum lipid recovery (74.9%) followed by Alcalase (61.7%). Hydrolysis of seafood by-products using proteolytic enzyme enhanced the fat extraction (Dumay et al. 2004). Different commercial proteases (Alcalase, Neutrase, and Flavourzyme™) were tested for their ability to release the oil content of salmon heads, and the amount of oil (17%) obtained after 2 h was close to that obtained by the chemical extraction method (20%) (Linder et al. 2005).

In general, the quantity of ω-3 fatty acids in the natural sources is very low and may not provide the required amount of these fatty acids as such for therapeutic purposes (Osborn and Akok 2002). The main method for concentration of ω-3 PUFAs are chromatographic separation, fractional or molecular distillation, low-temperature crystallization, supercritical fluid extraction, and urea complexation, and they have been recently reviewed elsewhere (Rubio-Rodríguez et al. 2010). Lipase-catalyzed hydrolysis, on the other hand, has been proven to be a straightforward and selective process for production of ω-3 PUFA concentrates (Kahveci and Xu 2011). In order to concentrate n-3 fatty acids, the lipase-catalyzed hydrolysis and esterification of existing oils can be applied (Guerard 2007). Many different lipases specifically from both fungi and yeast have been employed for the purpose. Preparation of ω-3 fatty acid concentrates using lipase from salmon by-products have been reported (Linder et al. 2002). Enrichment of ω-3 PUFAs in the glyceride fraction of salmon oil by *Candida rugosa* lipase catalyzed hydrolysis has been reported (Kahveci and Xu 2011). The researchers found that the total ω-3 PUFA content in the product was 38.71% (mol. %), more than double the initial level. Lipase-catalyzed hydrolysis was employed for the production of ω-3 PUFA concentrate from oil extracted from Pacific sardines (*Sardinops sagax*) (Okada and Morrissey 2007). Nonspecific lipases used in the study were significantly better than 1,3-specific lipases at concentrating ω-3 PUFAs from sardine oil. Lipolysis of the salmon head oil with Novozym SP398 yielded a mixture of free fatty acids and acylglycerols (24 h, 45% hydrolysis) (Linder et al. 2005). Enrichment of EPA and DHA in the glyceride fraction was carried out by hydrolysis of oils extracted from Nile perch viscera with lipases from *Candida rugosa*, *Thermomyces lanuginosus*, and *Pseudomonas cepacia* (Mbatia et al. 2010b). *T. lanuginosus* lipase gave better DHA enrichment than *C. rugosa* lipase but showed low DHA recoveries and was ineffective in EPA enrichment. Lipase from *C. rugosa* could thus be better in catalyzing hydrolysis of Nile perch viscera oil for enrichment of both DHA and EPA in acylglycerols. The authors confirmed that *T. lanuginosus* lipase would be the enzyme of choice if a high ratio of DHA to EPA is desired (Mbatia et al. 2010b). All the enzymatic processes used for production of ω-3 PUFA concentrates are reported to be safer and more efficient than any other methods because the enzymatic reactions, regardless of whether they involve hydrolysis or esterification, can be conducted under ambient temperatures, normal pressure, and in a nitrogen-protected environment (Wanasundara 2011).

15.4.3 Enzymes in the Recovery of Chitin and Its Derivatives from Seafood Processing By-Products

Chitin is a linear amino polysaccharide of β-1, 4 linked *N*-acetyl-D-glucosamine. It is the second most abundant and renewable biopolymer on earth, next only to cellulose, and is available to the extent of more than 10 gigatons annually (Tharanathan and Kittur 2003). Chitin and its derivatives are interesting biomolecules with unique properties that offer a wide range of potential industrial and biological applications in the field (Nidheesh and Suresh 2015a,b; Tharanathan and Kittur 2003).

The solid by-products of crustacean (shrimp, crab, lobster, krill, etc.) processing provide an important source of industrial production of chitin. Chitin present in the crustacean by-products is tightly associated with proteins; minerals, such as calcium carbonate; and lipids, including pigments. Therefore, the process of chitin preparation from shellfish by-products involves removal of minerals (demineralization), proteins (deproteinization), and pigments (decolorization) (Suresh and Prabhu 2012). The conventional chemical chitin purification process is extremely hazardous, energy-consuming, and damaging to the environment, using high concentrations of strong acid and alkali as well as resulting in a product with inferior quality (Nidheesh and Suresh 2015a,b). Alternative methods of chitin extraction from crustacean by-products using enzymes have been attempted, and these alternative processes, using milder treatment, resulted in chitin with consistent quality (Bhaskar et al. 2011; Suresh and Prabhu 2012). A variety of enzymatic procedures for deproteinization of decalcified crustacean by-products have been developed (Bhaskar et al. 2010; Gildberg and Stenberg 2001).

Chitin oligomers are produced by partial hydrolysis of pure chitin or crude crustacean processing by-products using acid or enzymes. The enzymatic methods are reported to have a number of advantages over acid hydrolysis (Suresh et al. 2011). Recently, the enzymatic production of chitin oligomers from crustaceans, such as shrimp and crab by-products using thermoactive chitinases of *Penicillium monoverticillium* CFR2 was reported (Suresh et al. 2011).

N-acetyl-D-glucosamine is the basic unit of many biologically important poly and oligosaccharides, glycoproteins, and glycolipids. These molecules have attracted much attention owing to their potential therapeutic and health benefits (Suresh and Kumar 2012; Venugopal 2009). Chitin extracted from the crustacean by-products are the commercial source of *N*-acetyl-D-glucosamine. Enzymatic processes have been attempted for the production of *N*-acetyl-D-glucosamine from crustacean chitin. Chitinolytic enzymes from microbes (Suresh and Kumar 2012), animals such as squid (Matsumiya 2004), and non-chitinase commercial enzymes (Sukwattanasinitt et al. 2002) have been used for production of *N*-acetyl-D-glucosamine from different chitin substrates.

15.4.4 Enzymes in Recovery of Flavor from Seafood By-Products

Flavor extracts derived from seafood by-products are very popular items, mainly in Asian markets, and good-quality seafood flavors are in high demand (Lee 2007). They are used in products such as seafood sauces, chowders, soups, bisques, instant noodles, snacks, cereal-based extrusion products like shrimp chips, and surimi-based products such as crab and shrimp analogues. The complex flavor of seafood is composed of equally important taste and aroma active components (Kim and Cadwallader 2011; Suresh and Prabhu 2012). The production of natural seafood flavor extracts from processing by-products has been a commercial practice in France and Japan (Lee 2007).

Enzymatic hydrolysis of seafood processing by-products using commercial proteases is currently employed for the production of most commercial natural seafood flavor extracts over other methods, such as fermentation, due to the advantages, such as high yield, good quality with less off-flavor generated, and control of flavor characteristics through variation of enzyme reactions (Lee 2007; Suresh and Prabhu 2012). Enzymatic hydrolysis has a significant effect on both taste and aroma of seafood flavor extracted. However, with the enzymatic techniques, the key consideration is the selection of the right type of enzyme for a given raw material. In a comparative study using different commercial proteases, such as Protamex, Flavourzyme, Alcalase, Neutrase, Optimase, and HT-Proteolytic 200, for seafood flavor production, it was found that Protamex produced from *Bacillus* sp. and Flavourzyme from *Aspergillus oryzae* are the best enzymes for the purpose (Yang and Lee 2000). Lobster flavors from lobster canning by-products consist of residual head meat; clam flavor from sea clam (*Spisula solidissima*) processing by-products consist of residual meat, belly, and viscera; and fish flavor from red hake (*Urophycis chuss*) processing by-products consist of head, gut, and frame, which have been extracted successfully using commercial proteases (Benjakul et al. 2010a; Lee 2007).

15.4.5 Enzymes in Recovery of Pigment from Seafood By-Products

Carotenoids are responsible for the bright red, orange, and yellow coloration of the flesh and skins of fish, such as salmonids (e.g., salmon and trout), and the exoskeletons of crustaceans (Simpson 2007). The carotenoids extracted from crustacean by-products have potential commercial application as a source of pigmentation in aquaculture and poultry as colorants for food, drugs, and cosmetics (Sachindra et al. 2011; Simpson 2007; Suresh and Prabhu 2012).

Crustacean, such as shrimp, lobster, crabs, crayfish, and krill, by-products are the main source of natural carotenoids, principally astaxanthin (Sachindra et al. 2007, 2011). In crustaceans, the carotenoids occur in free and/or in complex form, such as carotenoprotein, and its total quantity was found to vary depending on species. In order to extract carotenoids as the stable carotenoprotein complex, enzymatic techniques using proteases have been attempted by various researchers (Cremades et al.

2003; Holanda and Netto 2006; Klomklao et al. 2009; Sila et al. 2012, 2014). Extraction of carotenoprotein from shrimp by-products using EDTA and the proteolytic enzyme trypsin have been
tested. In a comparative evaluation of different enzymes for the recovery of carotenoprotein from
brown shrimp shells, trypsin showed maximum recovery of carotenoid (55%), followed by pepsin
and papain with 50% recovery of the pigment during the same period (Chakrabarti 2002). Klomklao
et al. (2009) used trypsin isolated from bluefish (*Pomatomus saltatrix*) pyloric caeca on the recovery of carotenoprotein from black tiger shrimp (*Penaeus monodon*) shells and compared bluefish
and bovine trypsins toward carotenoprotein recovery; the hydrolytic activity of bluefish trypsin was
found to be similar to that obtained by trypsin from the bovine pancreas. Isolation of carotenoproteins
from shrimp by-products using the trypsin purified from the viscera of Tunisian barbel (*Barbus
callensis*) have been reported (Sila et al. 2012). Furthermore, functional properties and in vitro antioxidative activity of the peptidic fraction of carotenoproteins from shrimp (*Parapenaeus longiros-
tris*) by-products generated by enzymatic treatment with Alcalase@ were evaluated (Sila et al. 2014).
Recovery of the carotenoprotein component from shrimp (*Xiphopenaeus kroyeri*) processing by-products
by enzymatic hydrolysis have been attempted (Holanda and Netto 2006). Carotenoprotein from
shells of Pacific white shrimp (*Litopenaeus vannamei*) was extracted with the aid of proteases from
hepatopancreas of the same species (Senphan et al. 2014). They observed that the recovery of carotenoprotein increased with increasing protease levels and hydrolysis times but principally reached a
plateau at 120 min of reaction.

15.4.6 Enzymes in Recovery of Calcium and Minerals from Seafood By-Products

Fish bone comprises a significant part of seafood processing by-products and is considered to be a potential source of minerals, particularly calcium. Due to its high mineral content, fish bone material derived
from processing of large fish can be considered a suitable natural source for extracting calcium for use
in food and supplements (Kim and Jung 2007; Suresh and Prabhu 2012; Venugopal 2006). Enzymatic
modification and fortification of bones from processing by-products of different seafood species for use
as a functionally active calcium supplement or fortifier and other minerals have been attempted (Kim
and Jung 2007). Skeletal frames from industrial processing of Hoki (*Johnius belengerii*) were digested
by a heterogeneous proteolytic enzyme extracted from the intestine of Bluefin tuna (*Thunnus thynnus*) in
order to utilize the bone to produce nutraceuticals with a high calcium bioavailability (Jung et al. 2005).
The bone from the Hoki (*J. belengerii*) frame was recovered using a proteolytic enzyme isolated from
mackerel intestine (Kim et al. 2003).

15.4.7 Enzymes in Recovery of Collagen from Seafood By-Products

Collagen is one of the major structural components of both vertebrates and invertebrates and is found
widely in skin, bone, tendon, and cartilage (Regenstein and Zhou 2007). Gelatine is a class of protein
fractions that have no existence in nature but are derived from the parent protein collagen by denaturation (Sachindra et al. 2011). Collagen and gelatine find applications in food, biomedical, etc., industries. In recent years, marine-derived collagen and gelatine have received much interest, partially due
to the requirements for *kosher* and *halal* food product development and consumers' concerns about
bovine spongiform encephalopathy (mad cow disease) in products from land-based animal collagen
(Regenstein and Zhou 2007). Marine-derived collagen can be isolated from finfish by-products, such as
skin, scale, bone, cartilage, and swim bladder, as well as different by-products of shellfish (cuttlefish,
octopus, squid, jellyfish, and sea urchin) (Benjakul et al. 2010a; Jongjareonrak et al. 2005; Krishnan and
Perumal 2013; Liu et al. 2010; Regenstein and Zhou 2007; Suresh and Prabhu 2012).

Collagens can be classified into salt-soluble collagen, acid-soluble collagen, and pepsin-soluble collagen based on the preparation methods. Collagen is generally extracted using an acid solubilization process, in which noncollagenous matter, pigment, and lipid are first removed by alkali solution (Benjakul et
al. 2010a). However, the acid extraction process gives a low yield of collagen. Enzymatic extraction with
limited pepsin proteolysis is widely used for an increased yield of collagen from seafood by-products,
and it can be applied alone or right after the acid extraction process (Nagai and Suzuki 2002; Regenstein

and Zhou 2007). In a study (Nagai et al. 2002), the yield of pepsin-solubilized collagen was found to be higher (44.7%) than acid-solubilized collagen (10.7%) from Ocellate Puffer fish (*Takifugu rubripes*) skin. A 50% yield of pepsin solubilized from the outer skin of the paper nautilus using 10% pepsin (w/v) was observed (Nagai and Suzuki 2002). Pepsin-solubilized collagen was isolated from the skin of grass carp (*Ctennopharyngodon idella*) with a yield of 35% (dry weight basis) (Zhang et al. 2007). Acid and pepsin-solubilized collagens from the skin of Brown stripe red snapper (*Lutjanus vitta*) was isolated and characterized (Jongjareonrak ct al. 2005).

Collagen from the outer skin of cuttlefish (*Sepia lycidas*) was extracted by the acidic process, followed by pepsin digestion (Nagai et al. 2001). In this process, the initial acid extraction of the cuttlefish outer skin with acetic acid yielded only 2% (dwb) of collagen. With further digestion of the residue with 10% (w/v) pepsin, a solubilized collagen was obtained with a yield of 35% (dwb). Pepsin-solubilized collagen from giant red sea cucumbers (*Parastichopus* sp.) was isolated and characterized (Liu et al. 2010). Acid-solubilized and pepsin-solubilized collagens were prepared from jellyfish (*Chrysaora quinquecirrha*) from the southeast coast of India. In this preparation, the yield of collagen was 0.48% acid solubilized and 1.28% pepsin solubilized on the basis of lyophilized dry weight (Krishnan and Perumal 2013).

Pepsin from marine sources is another promising enzyme for collagen isolation, and it can be extracted or recovered from seafood processing by-products, especially fish viscera (Benjakul et al. 2010a). Application of fish pepsin for the isolation of collagen from the skin of different fish species resulted in increased collagen yield (Benjakul et al. 2010b). Enhanced recovery of collagen from bigeye snapper skin was achieved using the pepsin from bigeye snapper (Nalinanon et al. 2007). The yields of collagen from bigeye snapper extraction for 48 h with acid and with bigeye snapper pepsin (20 kU/g skin) were 5.3% and 18.8% (dry wt. basis), respectively. After presoaking in acid for 24 h, followed by collagen extraction with bigeye snapper pepsin at the level of 20 kU/g skin for 48 h, the yield increased to 19.8%, which was greater than that of collagen extracted using porcine pepsin at the same level (13.0%) (Nalinanon et al. 2007). Pepsin from the stomach of tuna (albacore, skipjack, and tonggol) was used for the extraction of collagen from the skin of threadfin bream (Nalinanon et al. 2008). The yield and integrity of pepsin-solubilized collagen from threadfin bream skin using pepsin from different tuna species differed greatly. The yield of collagen increased by 1.84- to 2.31-fold, and albacore pepsin showed comparable extraction efficiency to porcine pepsin. Tuna pepsin-solubilized collagen from thread bream had a similar protein configuration to acid-solubilized collagen, which can be categorized as type I collagen while degradation of α and β components took place when pepsin from skipjack tuna was used. Albacore pepsin demonstrated similar extraction efficiency to porcine pepsin and had no adverse effect on the integrity of the resulting collagen. Thus, albacore pepsin could be used for collagen extraction from thread beam skin (Nalinanon et al. 2008). Central composite design (CCD) of response surface methodology (RSM) was used for the optimization of collagen extraction from yellowfin tuna dorsal skins (Woo et al. 2008). The optimal conditions were 0.92N (NaOH concentration), 24 h (NaOH treatment time), 0.98% (w/v) (pepsin concentration), and 23.5 h (digestion time). The predicted collagen content under optimal conditions was 26.7%, and the actual experimental collagen content was 27.1%. In a study, we found yields of collagen from scales of Indian carp (*Catla* sp.) with acid and with commercial pepsin (1:1000 U) at 0.5% (w/v) level were 1.5% ± 0.34% and 1.8% ± 0.12% (dry wt. basis), respectively (unpublished data).

Most of the collagen produced commercially is converted to gelatine, owing to the wider industrial uses of gelatin (Suresh and Prabhu 2012). TGase can be used for cross-linking proteins for the development of biodegradable and edible gelatin films (Benjakul et al. 2010a).

15.4.8 Enzymes in Recovery of Bioactive Peptides from Seafood By-Products

Food-derived bioactive peptides have potential regulatory roles in human bodies beyond their basic role as nutrient sources (Hartmann and Meisel 2007; Venugopal 2006). Seafood proteins have gained much attention as an attractive source of bioactive peptides (peptides derived from seafood protein hydrolysates) due to the abundance of raw materials in the form of processing discards (Guerard 2007; Samaranayaka and Li-Chan 2011). Enzymatic hydrolysis has been the most common method of producing bioactive peptides from seafood protein (Suresh and Prabhu 2012). Bioactive peptides generally depend on the source of the parent protein and the sequence of hydrolysis when different enzymes are used (Benjakul et al. 2010a).

Use of exogenous enzymes is preferred in most cases due to the reduction in time required to obtain a similar degree of hydrolysis as well as better control of the hydrolysis to obtain more consistent molecular weight profile and peptide composition (Samaranayaka and Li-Chan 2011). Low molecular weight peptides with high calcium binding from pepsinolytic hydrolysates of Alaska pollock backbone (Jung et al. 2006) and hoki frame (Jung and Kim 2007) have been isolated and characterized. The effect of different types of enzymes (Alcalase, α-chymotrypsin, Neutrase, pepsin, papain, and trypsin) on the antioxidant activity of tuna backbone hydrolysate was reported (Je et al. 2007). Among the different hydrolysates obtained, peptic hydrolysate showed the highest antioxidant activity. Protein hydrolysates from yellow stripe trevally, prepared using Flavourzyme, had a higher antioxidant activity compared with hydrolysates prepared using Alcalase (Klompong et al. 2007). Protein hydrolysates with different antioxidant activity from round scad muscle Flavourzyme and Alcalase were prepared (Thiansilakul et al. 2007). They also observed that protein hydrolysate prepared using Flavourzyme exhibited greater antioxidant activity than protein hydrolysate prepared using Alcalase. Antihypertensive peptides, such as angiotensin converting enzyme (ACE) inhibitory peptides, have been identified from hydrolysates prepared from various finfish and shellfish processing by-products (He et al. 2007; Liu et al. 2011; Murry and Fitz-Gerald 2007). It was reported that hydrolysates prepared using Protames and SM98011 protease from *Bacilllus* sp. exhibited a lower IC_{50} when compared with hydrolysates prepared using other enzymes (He et al. 2007). Marine clam, *Meretrix casta* (Chemnitz), protein hydrolysates prepared from different organs (body, foot, and viscera), using commercial enzymes (pepsin, trypsin, and papain), were reported to have antioxidant activity (Nazeer et al. 2013). Sowmya et al. (2014) prepared antioxidant-rich protein isolate by enzymatic hydrolysis of shrimp by-product.

15.4.9 Enzyme in Recovery of Bioactive Compounds from Other Seafood Resources

Marine seaweeds, such as Irish moss, laver, and kelp, are important food sources in many Asian countries, such as Japan, China, and Korea, etc. (Senanayake et al. 2011). Seaweeds are rich sources of flavor and various bioactive compounds, such as iodine, phlorotannins, glutathione, and carotenoid pigment fucoxanthin and also carbohydrates, such as alginates and algenic acid (Jeon et al. 2011; Suresh and Prabhu 2012; Venugopal 2006). Extensive research has been carried out on the enzymatic isolation of some of these compounds from seaweeds (Chang-Bum et al. 2004; Heo et al. 2005; Jae-Young et al. 2009; Jeon et al. 2011).

Seven brown seaweeds (*Ecklonia cava, Ishige okamurae, Sargassum fullvelum, S. horneri, S. coreanum, S. thunbergii, Scytosipon lomentaria*) were enzymatically hydrolyzed with five carbohrases (Viscozyme, Celluclast, Amyloglucosidase, Termamyl, and Ultraflo) and five proteases (Protamex, Kojizyme, Neutrase, Flavourzyme, and Alcalase) from Novozyme Co. (Novozyme Nordisk, Bagsvaerd, Denmark) (Heo et al. 2005). The enzymatic seaweed extracts exhibited more prominent effects in hydrogen peroxide scavenging activity (approximately 90%) compared to the other scavenging (1,1-diphenyl-2-pricrylhydrazyl, free radical, superoxide anion, hydroxyl radical) activities, and the activity of enzymatic extracts was even higher than that of the commercial antioxidants. The Ultraflo and Alcalase extracts of *S. horneri* were dose-dependent and thermally stable. Moreover, these two enzymatic extracts strongly inhibited DNA damage (approximately 50%) (Heo et al. 2005). Enzyme-assisted extraction of *Undaria pinnatifida* was performed using five proteases (Alcalase, Flavourzyme, Neutrase, Trypsin, and Protamex) and six carbohrases (Amyloglucosidase, Dextrozyme, Maltogenase, Promozyme, Viscozyme, and Celluclast) in order to acquire extracts rich in antioxidants (Jae-Young et al. 2009). They evaluated the antioxidant potential of the extracts using electron spin resonance spectroscopy on 1,1-diphenyl-2-picrylhydrazyl, hydroxyl, and superoxide radicals. Enzymatic extracts of *U. pinnatifida* exhibited strong radical scavenging activity on 1,1-diphenyl-2-picrylhydrazyl and hydroxyl radical, and activity increased with increasing extract concentration (Jae-Young et al. 2009). The brown seaweed *S. lomentaria* was enzymatically hydrolyzed to prepare water-soluble extracts by five carbohrases (Amyloglucosidase, Celluclast, Termamyl, Ultraflo, and Viscozyme) and five proteases (Alcalase, Flavourzyme, Kojizyme, Neutrase, and Protamex) (Chang-Bum et al. 2004). The enzymatic extracts of *S. lomentaria* exhibited strong scavenging activity on hydroxyl, alkyl, and 1,1-diphenyl-2-picrylhydrazyl radical, and the activity increased with the increment of concentration of the enzymatic extracts. The authors reported that scavenging activity enzymatic extracts were similar to the scavenging values on hydroxyl and alkyl radicals and lower than the scavenging value on 1,1-diphenyl-2-picrylhydrazyl radical compared with vitamin C used as a reference, respectively (Chang-Bum et al. 2004).

15.5 Enzymes in Other Applications in Seafood Processing

15.5.1 Enzymes in Extension of Shelf Life of Fishery Products

Fresh as well as processed seafoods are highly perishable due to various reasons, such as by the action of endogenous enzymes, microbial contamination, and oxidation of lipid compounds. Shelf life of seafood products stored under different conditions can be extended by using exogenous enzymes, such as glucose oxidase and lysozyme, which control the growth of spoilage-causing microorganisms. The potential of glutathione peroxidase to prevent oxidative deterioration in fish muscle during handling, storage, and processing has been suggested (Venugopal et al. 2000). Further, an extensive review is also available on the potentials of exogenous enzymes on the preservation of seafood (Venugopal et al. 2000).

15.5.2 Enzymes in Removal of Off-Odor and Fishy Taste

Meat of Elasmobranch (sharks and rays) species contains significant amounts of urea, which adversely influences the consumer acceptance of these products. Treatment of shark meat with soybean flour, which is rich in urease, has been attempted to degrade urea in the fish meat (Venugopal et al. 2000).

15.5.3 Retention of Color of Cooked Shrimp and Crab

Precooked frozen shrimp have a pleasant pink color due to their content of carotenoids. This red color is easily oxidized to yellow, which can be prevented by giving the cooked shrimp a dip treatment in glucose oxidase. The enzyme may also be included in the glazing solution prior to freezing of the shellfish. The treatment was also good for color retention of cooked crab meat (Venugopal et al. 2000).

15.5.4 Enzymes in Reduction in the Viscosity of Stick Water

Stick water is an aqueous by-product obtained during the manufacture of fish meal. The fish by-products or waste is cooked and then pressed to squeeze out the moisture. The pressed fish solids are then dried into fish meal. The pressed liquid is processed in order to recover more fish solids and fish oil. The remaining liquid is called *stick water*. In addition to protein, it contains amino acids, vitamins, and minerals and has excellent nutritional value. It can be sold as a feedstuff labeled as condensed or dried fish solubles. The protein content of stick water presents a problem during evaporation/concentration. This problem can be solved by reducing the viscosity of the stick water by enzymatic processes using protease treatment (Venugopal et al. 2000, 2015). A patent that describes the process for preparing a stick water hydrolysate using commercial proteolytic enzymes, preferably Alcalase®, is available (Simonsen 2003).

15.6 Seafood By-Product Sources of Enzymes

As marine organisms have adapted to diverse and extreme environmental conditions (high salt concentration, low or high temperatures, high pressure, etc.), the enzymes from these organisms may differ from those of terrestrial organisms in their characteristic properties (Suresh and Prabhu 2012; Venugopal 2009). Thus, marine-derived enzymes have received much attention for their potential applications in many fields.

Seafood processing by-products offer a renewable source of various industrially as well as economically important enzymes. Extraction of enzymes from seafood processing by-products may improve the economics of the seafood processing industry while minimizing the environmental impact of by-product disposal (Shahidi and Kamil 2001; Venugopal et al. 2000; Zhou et al. 2011). Various methods, such as ensilage, membrane technology, ohmic heating, precipitation, aqueous two-phase systems, and chromatography, have been used over the years to recover enzymes economically from seafood processing by-products (Morrissey and Okada 2007; Sachindra et al. 2011). Seafood processing by-products, principally viscera, can serve as one economically viable source for marine enzymes. Morrissey and Okada (2007) stated that recovery of enzymes from seafood by-products could help in the environmental and ethical concerns surrounding the discards and improve the bottom line for seafood industries wishing to exploit new technologies and markets.

Extractable enzymes from seafood by-products include digestive proteinases, extracellular gastric proteinases, chitinases, lipases, phospholipases, TGases, polyphenoloxidase, and others. Among the various enzymes from seafood by-products, digestive proteolytic enzymes have received considerable attention over the last two to three decades due to the availability of raw materials, such as viscera, and their high rate of enzymatic activity (Morrissey and Okada 2007; Shahidi and Kamil 2001; Venugopal 2015). Finfish processing by-products, especially viscera, constitute a large proportion of digestive enzymes, including collaginases, trypsin, pepsin, chymotrypsin, elastase, and carboxylpeptidase (Sachindra et al. 2011). In addition to the digestive proteases from visceral by-products, the proteolytic enzymes found in muscle cells as well as other organs of seafood, such as the hepatopancreas in shellfish, can be extracted from wash water of fish and shellfish processing. Collagenolytic enzymes have been extracted and characterized from various seafood processing by-products (Daboor et al. 2012; Kim et al. 2002). The unusual catalytic activities of marine proteases have been useful for their applications in much food processing (Simpson 2000; Shahidi and Kamil 2001; Swapna et al. 2011b).

In addition to digestive enzymes, a number of other industrially important enzymes have been extracted from finfish and shellfish by-products, such as lipases, chitinases, alkaline phosphatase, hyaluronidase, and β-acetyl-D-glucosaminidase (Sachindra et al. 2011). Chitinases are enzymes that hydrolyze β-(1,4)-N-acetylglucosaminide linkages in the chitin polymer, which are widely distributed in various groups of animals. It is a complex enzyme, classified into two groups, namely endochitinases (EC. 3.2.1.14) and β-N-acetylhexosaminidases (EC. 3.2.1.52), the actions of which are known to be synergistic and consecutive (Suresh and Chandrasekaran 1999). Chitinolytic enzymes have been extracted and characterized from processing by-products of shellfish, such as squid, cuttlefish, clam, shrimp, and so on, and finfish, such as red sea bream, cod, Japanese eel, silver croaker, and so on (Mana et al. 2009; Matsumiya 2004).

Seafood by-products are considered to be an important source of lipid-degrading enzymes. Lipases have been extracted and characterized from by-products of cod, salmon, rohu, oil sardine, mullet, Indian mackerel, and red sea bream (Swapna et al. 2011b). Alginate lyases are an important group of carbohydrate-degrading enzymes with wide applications, especially in the recovery of biomolecules from seaweeds and in algal biotechnology for the isolation of algal protoplast. Alginate lyase has been isolated from gut, gland, and heptopancrease of marine mollusks (Wong et al. 2000). Polyphenoloxidases are widely distributed in nature and is responsible for dark pigments in fruits, vegetables, and crustacean species. They are generally found in shellfish and their by-products, including shrimp, lobster, and cuttlefish (Morrissey and Okada 2007). This enzyme has a major role in seafood processing and in postharvest quality of shellfish. TGases have been purified and characterized from walleye pollock by-products. In addition to the abovementioned enzymes, other enzymes, such as myosin ATPase, acid and alkaline phosphatase, lipoxygenase, amylase, α-glucosidase and β-glucosidase, cellulase, xylanase, and lysozyme, have been isolated and purified from seafood processing by-products (Morrissey and Okada 2007).

15.7 Conclusion

The world seafood industry plays a significant role in providing protein-rich and healthy food to a significant part of the world's populace. The seafood industries continue to grow throughout the world along with the increasing demand for processed seafood to satisfy the requirement of an emerging society that attempts to cope with the current lifestyle. Compared with other sectors of food processing, enzyme technology in seafood processing is still in its early stages. Enzymatic seafood processing is generally overlooked due to the complex diversity, highly perishable nature, and complex processing steps of seafood and related economic constraints. Even though the use of endogenous enzymes in the seafood industry is well established, the adoption of new technologies in recent years has resulted in little progress in this sector. The advent of enzyme technology through modern biotechnology certainly has changed the processing of seafood product development. Further, enzymatic processes have become an integral part of seafood by-product processing, and the application of such methods also enhanced the recovery of additional value-added compounds from seafood processing discards. Owing to its unique qualities, the enzymatic approach is also considered to be an alternative solution to circumvent the environmental problems caused by the mechanical and chemical processing of seafood. In this context, some of the potentials of enzymes in seafood product development, by-product processing, and related processes that can be applied for easier processing and development of high-quality seafood products were discussed in this chapter, highlighting the impeccable role of enzymes in fish processing. Overall, the recent technological achievements explained herein represent important contributions of enzyme technology to the sector of seafood and the progress is promising. However, as compared with other fields of food processing, enzymatic seafood processing is inadequate with respect to the aquatic realm, which is very rich with largely available resources of food products that are of prospective interest to mankind.

REFERENCES

Aidaroos, H. I., Du, G., and Chen, J. 2011. Microbial fed-batch production of transglutaminase using ammonium sulphate and calcium chloride by *Streptomyces hygroscopicus*. *Biotechnology, Bioinformatics and Bioengineering* 1(2): 173–178.

Amit, K. R., Swapna, H. C., Bhaskar, N., Halami, P. M., and Sachindra, N. M. 2010. Effect of fermentation ensilaging on recovery of oil from fresh water fish viscera. *Enzyme and Microbial Technology* 46: 9–13.

Aspmo, S. I., Horn, S. J., and Eijsink, V. G. 2005. Hydrolysates from Atlantic cod (*Gadus morhua* L.) viscera as components of microbial growth media. *Process Biochemistry* 40(12): 3714–3722.

Beisson, F., Arondel, V., and Verger, R. 2000. Assaying *Arabidopsis* lipase activity. *Biochemical Society Transactions* 28: 773–775.

Benjakul, S., Phatcharat, S., Tammatinna, A., Visessanguan, W., and Kishimura, H. 2008. Improvement of gelling properties of lizardfish mince as influenced by microbial transglutaminase and fish freshness. *Journal of Food Science* 73: 239–246.

Benjakul, S., Klomklao, S., and Simpson, B. K. 2010a. Enzymes in fish processes. In *Enzymes in Food Technology*, R. J. Whitehurst and M. V. Oort (eds.), 2nd Edition, pp. 211–235. Chichester, West Sussex, UK: Wiley-Blackwell.

Benjakul, S., Thiansilakul, Y., Visessanguan, W., Roytrakul, S., Kishimura, H., Prodpran, T., and Meesane, J. 2010b. Extraction and characterization of pepsin-solubilised collagens from the skin of bigeye snapper (*Priacanthus tayenus* and *Priacanthus macracanthus*). *Journal of the Science of Food and Agriculture* 90(1): 132–138.

Berge, A., and Barnathan, G. 2005. Fatty acids from lipids of marine organisms: Molecular biodiversity, roles as biomarkers, biologically active compounds, and economical aspects. In *Marine Biotechnology I, Series: Advances in Biochemical Engineering/Biotechnology*, Y. Le Gal and R. Ulber (eds.), Vol. 96, pp. 49–126. Berlin Heidelberg: Springer-Verlag.

Bhaskar, N., Suresh, P. V., Sakhare, P. Z., and Sachindra, N. M. 2007. Shrimp biowaste fermentation with *Pediococcus acidolactici* CFR2182: Optimization of fermentation conditions by response surface methodology and effect of optimized conditions on deproteination/demineralization and carotenoid recovery. *Enzyme and Microbial Technology* 40(5): 1427–1434.

Bhaskar, N., Suresh, P. V., Sakhare, P. Z., and Sachindra, N. M. 2010. Yield and chemical composition of fractions from fermented shrimp biowaste. *Waste Management and Research* 28: 64–70.

Bhaskar, N., Sachindra, N. M., Suresh, P. V., and Mahendrakar, N. S. 2011. Microbial reclamation of fish industry by-products. In *Aquaculture Microbiology and Biotechnology*, D. Montet and R. C. Ray (eds.), pp. 249–276. Eneld, NH: Science Publishers.

Carmen, G.-G. M., Pilar, M., Teresa, S. M., and Miriam, P.-M. 2005. Effect of chitosan and microbial transglutaminase on the gel forming ability of horse mackerel (*Trachurus* spp.) muscle under high pressure. *Food Research International* 38: 103–110.

Chakrabarti, R. 2002. Carotenoprotein from tropical brown shrimp shell waste by enzymatic process. *Food Biotechnology* 16(1): 81–90.

Chang-Bum, A., Jeon, Y. J., Kang, D. S., Shin, T. S., and Jung, B. M. 2004. Free radical scavenging activity of enzymatic extracts from a brown seaweed *Scytosiphon lomentaria* by electron spin resonance spectrometry. *Food Research International* 37: 253–258.

Córdova-Murueta, J. H., Navarrete-del-Toro, M. D. L. Á., and García Carreño, F. L. 2007. Concentrates of fish protein from by catch species produced by various drying processes. *Food Chemistry* 100(2): 705–711.

Cremades, O., Parrado, J., Alvarez-Osorio, M., Jover, M., Collantes de Teran, L., Gutierrez, J., and Bautista, J. 2003. Isolation and characterization of carotenoproteins from crayfish (*Procambarus clarkii*). *Food Chemistry* 82: 559–566.

Daboor, S. M., Budge, S. M., Ghaly, A. E., Brooks, M. S., and Dave, D. 2012. Isolation and activation of collagenase from fish processing waste. *Advances in Bioscience and Biotechnology* 3: 191–203.

Diaz-López, M., and García-Carreño, F. L. 2000. Applications of fish and shellfish enzymes in food and feed products. In *Seafood Enzymes*, N. F. Haard and B. K. Simpson (eds.), pp. 571–618. New York: Marcel Dekker.

Duangmal, K., and Taluengphol, A. 2010. Effect of protein additives, sodium ascorbate, and microbial transglutaminase on the texture and colour of red tilapia surimi gel. *International Journal of Food Science and Technology* 45: 48–55.

Dumay, O., Tari, P. S., Tomasini, J. A., and Mouillot, D. 2004. Functional groups of lagoon fish species in Languedoc-Roussillon, southern France. *Journal of Fish Biology* 64(4): 970–983.

FAO. 2013. *Yearbook of Fishery Statistics*. Statistics and Information Service, Caracalla 00153, Rome: Food and Agriculture Organization of the United Nations. Available at http://www.fao.org/fishery/statistics/en.

Fariha, H., Aamer, A. S., and Abdul, H. 2006. Industrial applications of microbial lipases. *Enzyme and Microbial Technology* 39: 235–251.

Gbogouri, G. A., Linder, M., Fanni, J., and Parmentier, M. 2006. Analysis of lipids extracted from salmon (*Salmo salar*) heads by commercial proteolytic enzymes. *European Journal of Lipid Science and Technology* 108: 766–775.

Ghorbel-Bellaaj, O., Hmidet, N., Jellouli, K., Younes, I., Maâlej, H., Hachicha, R., and Nasri, M. 2011. Shrimp waste fermentation with *Pseudomonas aeruginosa* A2: Optimization of chitin extraction conditions through Plackett-Burman and response surface methodology approaches. *International Journal of Biological Macromolecules* 48(4): 596–602.

Gildberg, A., and Stenberg, E. 2001. A new process for advanced utilization of shrimp waste. *Process Biochemistry* 36: 809–812.

Gonçalve, A. A., and Passos, M. G. 2010. Restructured fish product from white croaker (*Micropogonias furnieri*) mince using microbial transglutaminase. *Brazilian Archives of Biology and Technology* 53(4): 987–995.

Guerard, F. 2007. Enzymatic methods for marine by-products recovery. In *Maximizing the Value of Marine By-Products*, F. Shahidi (ed.), pp. 107–136. Cambridge: Woodhead Publishing Ltd., CRC Press.

Guerard, F., Guimas, L., and Binet, A. 2002. Production of tuna waste hydrolysates by a commercial neutral protease preparation. *Journal of Molecular Catalysis B: Enzymatic* 19: 489–498.

Hartmann, R., and Meisel, H. 2007. Food-derived peptides with biological activity: From research to food applications. *Current Opinion in Biotechnology* 18(2): 163–169.

He, H. L., Chen, X. L., Wu, H., Sun, C. Y., Zhang, Y. Z., and Zhou, B. C. 2007. High throughput and rapid screening of marine protein hydrolysate enriched in peptides with angiotensin-I-converting enzyme inhibitory activity by capillary electrophoresis. *Bioresource Technology* 98: 3499–3505.

Helena, S. N., Luciane, V. S., Cristiano, P. P., and Sandra, T. 2005. Biosensor based on xanthine oxidase for monitoring hypoxanthine in fish meat. *American Journal of Biochemistry and Biotechnology* 1(2): 85–89.

Helena, M. M., Carballo, J., and Borderías, A. J. 2008. Influence of alginate and microbial transglutaminase as binding ingredients on restructured fish muscle processed at low temperature *Journal of the Science of Food and Agriculture* 88: 1529–1536.

Heo, S. J., Park, E. J., Lee, K. W., and Jeon, Y. J. 2005. Antioxidant activities of enzymatic extracts from brown seaweeds. *Bioresource Technology* 96(14): 1613–1623.

Holanda, H. D., and Netto, F. M. 2006. Recovery of components from shrimp (*Xiphopenaeus kroyeri*) processing waste by enzymatic hydrolysis. *Journal of Food Science* 71: 298–303.

Jae-Young, J., Pyo-Jam, P., Eun-Kyung, K., Jung-Suk, P., Ho-Dong, Y., Kwang-Rae, K., and Chang-Bum, A. 2009. Antioxidant activity of enzymatic extracts from the brown seaweed *Undaria pinnatifida* by electron spin resonance spectroscopy. *LWT—Food Science and Technology* 42: 874–878.

Je, J. Y., Zhong-Ji, Q., Byun, H.-G., and Se-Kwon, K. 2007. Purification and characterization of an antioxidant peptide obtained from tuna backbone protein by enzymatic hydrolysis. *Process Biochemistry* 42: 840–846.

Jeon, Y.-J., Wijesinghe, W. A. J. P., and Kim, S.-K. 2011. Enzyme-assisted extraction and recovery of bioactive components from seaweeds. In *Handbook of Marine Macroalgae: Biotechnology and Applied Phycology*, S.-K. Kim (ed.). Chichester, UK: John Wiley & Sons, Ltd.

Jongjareonrak, A., Benjakul, S., Visessanguan, W., Nagai, T., and Tanaka, M. 2005. Isolation and characterisation of acid and pepsin-solubilised collagens from the skin of brown stripe red snapper (*Lutjanus vitta*). *Food Chemistry* 93. 475–484.

Jongjareonrak, A., Benjakul, S., Visessanguan, W., and Tanaka, M. 2006. Skin gelatin from bigeye snapper and brown stripe red snapper: Chemical compositions and effect of microbial transglutaminase on gel properties. *Food Hydrocolloids* 20: 1216–1222.

Jung, W. K., and Kim, S. K. 2007. Calcium-binding peptide derived from pepsinolytic hydrolysates of hoki (*Johnius belengerii*) frame. *European Food Research and Technology* 224(6): 763–767.

Jung, W. K., Park, P. J., Byun, H. G., Moon, S. H., and Kim, S. K. 2005. Preparation of hoki (*Johnius belengerii*) bone oligophosphopeptide with a high affinity to calcium by carnivorous intestine crude proteinase. *Food Chemistry* 91: 333–340.

Jung, W. K., Karawita, R., Heo, S. J., Lee, B. J., Kim, S. K., and Jeon, Y. J. 2006. Recovery of a novel Ca-binding peptide from Alaska pollack (*Theragra chalcogramma*) backbone by pepsinolytic hydrolysis. *Process Biochemistry* 41: 2097–2100.

Kahveci, D., and Xu, X. 2011. Repeated hydrolysis process is effective for enrichment of omega 3 polyunsaturated fatty acids in salmon oil by *Candida rugosa* lipase. *Food Chemistry* 129(4): 1552–1558.

Kieliszek, M., and Misiewicz, A. 2014. Microbial transglutaminase and its application in the food industry. A review. *Folia Microbiology* 59: 241–250.

Kim, S. K., and Jung, V. K. 2007. Fish and bone as a calcium source. In *Maximizing the Value of Marine By-Products*, F. Shahidi (ed.), pp. 328–339. Cambridge: Woodhead Publishing Ltd., CRC Press.

Kim, H., and Cadwallader, K. R. 2011. Instrument analysis of seafood flavour. In *Handbook of Seafood Quality, Safety and Health Application*, C. Alasalvar, F. Shahidi, K. Miyashita and U. Wanasundara (eds.), pp. 50–62. Chichester, West Sussex, UK: Wiley-Blackwell.

Kim, S. K., Park, P. J., Kim, J. B., and Shahidi, F. 2002. Purification and characterization of a collagenolytic protease from the filefish, *Novoden modestrus*. *Journal of Biochemistry and Molecular Biology* 35(2): 165–171.

Kim, S. K., Park, P. J., Byun, H. G., Je, J. Y., Moon, S. H., and Kim, S. H. 2003. Recovery of fish bone from hoki (*Johnius belengerii*) frame using a proteolytic enzyme isolated from mackerel intestine. *Journal Food Biochemistry* 27(3): 255–266.

Klomklao, S. 2008. Digestive proteinases from marine organisms and their applications. *The Songklanakarin Journal of Science and Technology* 30(1): 37–46.

Klomklao, S., Benjakul, S., Visessanguan, W., Kishimura, H., and Simpson, B. K. 2006. Effects of the addition of spleen of skipjack tuna (*Katsuwonus pelamis*) on the liquefaction and characteristics of fish sauce made from sardine (*Sardinella gibbosa*). *Food Chemistry* 98: 440–452.

Klomklao, S., Benjakul, S., Visessanguan, W., Kishimura, H., and Simpson, B. K. 2009. Extraction of carotenoprotein from black tiger shrimp shells with the aid of bluefish trypsin. *Journal of Food Biochemistry* 33(2): 201–217.

Klompong, V., Benjakul, S., Kantachote, D., and Shahidi, F. 2007. Antioxidative activity and functional properties of protein hydrolysate of yellow stripe trevally (*Selaroides leptolepis*) as influenced by the degree of hydrolysis and enzyme type. *Food Chemistry* 102(4): 1317–1327.

Kołakowski, E., and Bednarczyk, B. 2003. Changes in headed and gutted baltic herring during immersed salting in brine with the addition of acetic acid part 2. Intensity of proteolysis. *Electronic Journal of Polish Agricultural Universities* 6(1): 10. Available at http://www.ejpau.media.pl/volume6/issue1/food/art-10.html.

Krishnan, S., and Perumal, P. 2013. Preparation and biomedical characterization of jellyfish (*Chrysaora quinquecirrha*) collagen from southeast coast of India. *International Journal of Pharmacy and Pharmaceutical Sciences* 5(3): 698–701.

Kristinsson, H. G. 2007. Aquatic food protein hydrolysates. In *Maximizing the Value of Marine By-Products*, F. Shahidi (ed.), pp. 229–248. Cambridge: Woodhead Publishing Ltd., CRC Press.

Kristinsson, H. G., and Rasco, B. A. 2000a. Fish protein hydrolysates: Production, biochemical, and functional properties. *Critical Reviews in Food Science and Nutrition* 40(1): 43–81.

Kristinsson, H. G., and Rasco, B. A. 2000b. Biochemical and functional properties of Atlantic salmon (*Salmo salar*) muscle proteins hydrolyzed with various alkaline proteases. *Journal of Agricultural and Food Chemistry* 48(3): 657–666.

Kudre, T. G., and Benjakul, S. 2013. Combining effect of microbial transglutaminase and Bambara groundnut protein isolate on gel properties of surimi from sardine (*Sardinella albella*). *Food Biophysics* 8(4): 240–249.

Kudre, T. G., and Benjakul, S. 2014. Effects of Bambara groundnut protein isolates and microbial transglutaminase on textural and sensorial properties of surimi gel from sardine (*Sardinella albella*). *Food and Bioprocess Technology* 7(6): 1570–1580.

Kumar, P. A., and Suresh, P. V. 2014. Biodegradation of shrimp biowaste by marine *Exiguobacterium* sp. CFR26M and concomitant production of extracellular protease and antioxidant materials: Production and process optimization by response surface methodology. *Marine Biotechnology* 16: 202–218.

Kuraishi, C., Yamazaki, K., and Susa, Y. 2001. Transglutaminase: Its utilization in the food industry. *Food Reviews International* 17(2): 221–246.

Laset, B., Julshamn, K., and Espe, M. 2003. Chemical composition and theoretical nutritional evaluation of the produced fractions from enzymatic hydrolysis of salmon frames with Protamex. *Process Biochemistry* 38: 1747–1759.

Lee, C. M. 2007. Seafood flavor from processing by-products. In *Maximizing the Value of Marine By-Products*, F. Shahidi (ed.), pp. 304–327. Cambridge: Woodhead Publishing Ltd., CRC Press.

Liaset, B., Lied, E., and Espe, M. 2000. Enzymatic hydrolysis of by-products from the fish-filleting industry; chemical characterisation and nutritional evaluation. *Journal of the Science of Food and Agriculture* 80(5): 581–589.

Linder, M., Matouba, E., Fanni, J., and Parmentier, M. 2002. Enrichment of salmon oil with n-3 PUFA by lipolysis, filtration and enzyme re-esterification. *European Journal of Lipid Science Technology* 104: 455–462.

Linder, M., Fanni, J., and Parmentier, M. 2005. Proteolytic extraction of salmon oil and PUFA concentration by lipases. *Marine Biotechnology* 7(1): 70–76.

Liu, Z., Oliveira, A. C. M., and Su, Y. C. 2010. Purification and characterization of pepsin-solubilized collagen from skin and connective tissue of giant red sea cucumber (*Parastichopus californicus*). *Journal of Agriculture Food Chemistry* 58: 1270–1274.

Liu, Z., Chen, D., Su, Y. C., and Zeng, M. Y. 2011. Optimization of hydrolysis conditions for the production of the angiotensin-I converting enzyme inhibitory peptides from sea cucumber collagen hydrolysates. *Journal of Aquatic Food Product Technology* 20(2): 222–232.

Llorente, H. R., Altonaga, Z. M., Peral, D. I. I., Ibargüen, S. M., and Gartzia, P. I. 2007. Salting dynamics for anchovy (*Engraulis encrasicholus*) with salt replacers. Proceedings of European Congress of Chemical Engineering (ECCE-6), Copenhagen, September 16–20.

Mana, I., Kouji, M., Atsushi, M., and Masahiro, M. 2009. Purification and characterization of chitinase from the stomach of silver croaker *Pennahia argentatus*. *Protein Expression and Purification* 65(2): 214–222.

Matsumiya, M. 2004. Enzymatic production of *N*-acetyl-D-glucosamine using crude enzyme from the liver of squids. *Food Science and Technology Research* 10(3): 296–299.

Mbatia, B. 2011. Valorisation of fish waste biomass through recovery of nutritional lipids and biogas. Doctoral thesis, Lund University, Sweden. Available at http://lup.lub.lu.se.

Mbatia, B., Adlercreutz, D., Adlercreutz, P., Mahadhy, A., Mulaa, F., and Mattiasson, B. 2010a. Enzymatic oil extraction and positional analysis of omega-3 fatty acids in Nile perch and salmon heads. *Process Biochemistry* 45: 815–819.

Mbatia, B., Adlercreutz, P., Mulaa, F., and Mattiasson, B. 2010b. Enzymatic enrichment of omega-3 polyunsaturated fatty acids in Nile perch (*Lates niloticus*) viscera oil. *European Journal of Lipid Science and Technology* 112: 977–984.

Morrissey, M. T., and Okada, T. 2007. Marine enzymes from seafood by-products. In *Maximizing the Value of Marine By-Products*, F. Shahidi (ed.), pp. 374–389. Cambridge: Woodhead Publishing Ltd., CRC Press.

Murry, B. A., and Fitz-Gerald, R. J. 2007. Angiotensin converting enzyme inhibitory peptides derived from food protein: Biochemistry, bioactivity and production. *Current Pharmaceutical Design* 53: 773–791.

Nagai, T., and Suzuki, N. 2002. Preparation and partial characterization of collagen from paper nautilus (*Argonauta argo*, Lin) outer skin. *Food Chemistry* 76: 149–153.

Nagai, T., Yamashita, E., Taniguch, K., Kanamori, N., and Suzuki, N. 2001. Isolation and characterization of collagen from the outer skin waste material of cuttlefish (*Sepia lycidas*). *Food Chemistry* 72: 425–429.

Nagai, T., Araki, Y., and Suzuki, N. 2002. Collagen of the skin of ocellate puffer fish (*Takifugu rubripes*). *Food Chemistry* 78: 173 177.

Nalinanon, S., Benjakul, S., Visessanguan, W., and Kishimura, H. 2007. Use of pepsin for collagen extraction from the skin of bigeye snapper (*Priccanthus tayenus*). *Food Chemistry* 104: 593–601.

Nalinanon, S., Benjakul, S., Visessanguan, W., and Kishimura, H. 2008. Tuna pepsin: Characteristics and its use for collagen extraction from the skin of threadfin bream (*Nemipterus* spp.). *Journal of Food Science* 73: 413 419.

Nazeer, R. A., Divya Prabha, K. R., Sampath Kumar, N. S., and Ganesh, R. J. 2013. Isolation of antioxidant peptides from clam, *Meretrix casta* (Chemnitz). *Journal of Food Science and Technology* 50(4): 777–783.

Nidheesh, T., and Suresh, P. V. 2015a. Optimization of conditions for isolation of high quality chitin from shrimp processing raw byproducts using response surface methodology and its characterization. *Journal of Food Science and Technology* 52(6): 3812–3823.

Nidheesh, T., and Suresh, P. V. 2015b. Functional polysaccharide from shellfish by-products. In *Fish Processing Byproducts-Quality Assessment and Applications*, N. S. Mahendrakar and N. M. Sachindra (eds.), pp. 299–345. Houston, TX: Studium Press LLC.

Okada, T., and Morrissey, M. T. 2007. Production of ω-3 polyunsaturated fatty acid concentrate from sardine oil by lipase-catalyzed hydrolysis. *Food Chemistry* 103(4): 1411–1419.

Okada, T., Hosokawa, M., Ono, S., and Miyashita, K. 2008. Enzymatic production of marine-derived protein hydrolysates and their bioactive peptides for use in foods and nutraceuticals. In *Biocatalysis and Bioenergy*, C. T. Hou and J. F. Shaw (eds.), pp. 491–519. Hoboken, NJ: John Wiley & Sons, Inc.

Ono, S., Kasai, D., Sugano, T., Ohba, K., and Takahasi, K. 2004. Production of water soluble antioxidative plastein from squid heptopancrease. *Journal of Oleo Science* 53: 267–273.

Osborn, H. T., and Akok, C. C. 2002. Structured lipids-novel fats with medical, nutraceutical, and food applications. *Comprehensive Reviews in Food Science and Food Safety* 1: 93–103.

Park, J. W. 2005. Surimi seafood: Products, market, and manufacturing. In *Surimi and Surimi Seafood*, Park, J. W. (ed.), pp. 375–433. Boca Raton, FL: Taylor & Francis.

Park, S.-H., Cho, S.-Y., Kimura, M., Nozawa, H., and Seki, N. 2005. Effects of microbial transglutaminase and starch on the thermal gelation of salted squid muscle paste. *Fisheries Science* 71: 896–903.

Rahaman, Md. Atikur. 2014. Effect of processing parameters on salting of herring. Thesis, Norwegian University of Science and Technology, Uppsala, Sweden. Available at http://www.diva-portal.org/smash/get/diva2:743707/FULLTEXT01.pd.

Ramirez, J. A., del Angel, A., Uresti, R. M., Velazquez, G., and Vazquez, M. 2007. Low-salt restructured fish products using low-value fish species from the Gulf of Mexico. *International Journal of Food Science and Technology* 42: 1039–1045.

Regenstein, J. M., and Zhou, P. 2007. Collagen and gelatin from marine byproducts. In *Maximizing the Value of Marine By-Products*, F. Shahidi (ed.), pp. 279–299. Cambridge: Woodhead Publishing Ltd., CRC Press.

Rubio-Rodríguez, N., Beltrán, S., Jaime, I., Sara, M., Sanz, M. T., and Carballido, J. R. 2010. Production of omega-3 polyunsaturated fatty acid concentrates: A review. *Innovative Food Science and Emerging Technologies* 11: 1–12.

Rustad, T. 2003. Utilization of marine by-products. *Electronic Journal of Environmental, Agricultural and Food Chemistry* 2: 458–463.

Sachindra, N. M., Bhaskar, N., Siddegowda, G. S., Sathisha, A. D., and Suresh, P. V. 2007. Recovery of carotenoids from ensilaged shrimp waste. *Bioresource Technology* 98: 1642–1646.

Sachindra, N. M., Bhaskar, N., Hosokawa, M., and Miyashita, K. 2011. Value addition of seafood processing discards. In *Handbook of Seafood Quality, Safety and Health Application*, C. Alasalvar, F. Shahidi, K. Miyashita and U. Wanasundara (eds.), pp. 390–398. Chichester, West Sussex, UK: Wiley-Blackwell.

Samaranayaka, A. G. P., and Li-Chan, E. C. Y. 2011. Bioactive peptides from seafood and their health effects. In *Handbook of Seafood Quality, Safety and Health Application*, C. Alasalvar, F. Shahidi, K. Miyashita and U. Wanasundara (eds.), pp. 485–493. Chichester, West Sussex, UK: Wiley-Blackwell.

Sathivel, S., Prinyawiwatkul, W., King, J. M., Grimm, C. C., and Lloyd, S. 2003. Oil production from catfish viscera. *Journal of the American Oil Chemists' Society* 8: 377–382.

Senanayake, S. P. J. N., Ahmed, N., and Fichtali, J. 2011. Neutraceutical and bioactive from marine algae. In *Handbook of Seafood Quality, Safety and Health Application*, C. Alasalvar, F. Shahidi, K. Miyashita and U. Wanasundara (eds.), pp. 456–463. Chichester, West Sussex, UK: Wiley-Blackwell.

Senphan, T., Benjakul, S., and Kishimura, H. 2014. Characteristics and antioxidative activity of carotenoprotein from shells of Pacific white shrimp extracted using hepatopancreas proteases. *Food Bioscience* 5: 54–63.

Shahidi, F. 2007. Marine oils from seafood waste. In *Maximizing the Value of Marine By-Products*, F. Shahidi (ed.), pp. 258–278. Cambridge: Woodhead Publishing Ltd., CRC Press.

Shahidi, F. 2008. *Bioactives from Marine Resources*, pp. 24–34, Washington, DC: ACS Publications.

Shahidi, F., and Kamil, J. Y. V. A. 2001. Enzymes from fish and aquatic invertebrates and their application in the food industry. *Trends in Food Science and Technology* 12: 435–464.

Shahidi, F., and Alasalvar, C. 2011. Marine oils and other marine nutraceuticals. In *Handbook of Seafood Quality, Safety and Health Application*, C. Alasalvar, F. Shahidi, K. Miyashita and U. Wanasundara (eds.), pp. 444–454. Chichester, West Sussex, UK: Wiley-Blackwell.

Sila, A., Nasri, R., Jridi, M., Balti, R., Nasri, M., and Bougatef, A. 2012. Characterisation of and its application for recovery of carotenoproteins from shrimp wastes. *Food Chemistry* 132: 1287–1295.

Sila, A., Sayari, N., Balti, R., Martinez-Alvarez, O., Nedjar-Arroume, N., Nasri, M., and Bougatef, A. 2014. Biochemical and antioxidant properties of peptidic fraction of carotenoproteins generated from shrimp by-products by enzymatic hydrolysis. *Food Chemistry* 148: 445–452.

Simonsen, P. S. 2003. Stick water hydrolyzate and a process for the preparation thereof. European Patent EP 0990393 A1.

Simpson, B. K. 2000. Digestive proteinases from marine animals. In *Seafood Enzymes: Utilization and Influence on Postharvest Seafood Quality*, N. F. Haard and B. K. Simpson (eds.), pp. 531–540. New York: Marcel Dekker.

Simpson, B. K. 2007. Pigments from by-products of seafood processing. In *Maximizing the Value of Marine By-Products*. F. Shahidi (ed.), pp. 413–432. Cambridge: Woodhead Publishing Ltd., CRC Press.

Sountama, J., Kiessling, A., Melle, W., Waagbo, R., and Olsen, R. E. 2007. Protein from Northern krill (*Thysanoessa inermis*), Antarctic krill (*Euphausia superba*) and the Arctic amphipod (*Themisto libellula*) can partially replace fish meal in diets to Atlantic salmon (*Salmo salar*) without affecting product quality. *Aquaculture Nutrition* 13: 50–58.

Sowmya, R., Ravikumar, T. M., Vivek, R., Rathinaraj, K., and Sachindra, N. M. 2014. Optimization of enzymatic hydrolysis of shrimp waste for recovery of antioxidant activity rich protein isolate. *Journal of Food Science and Technology* 51(11): 3199–3207.

Sukwattanasinitt, M., Zhu, H., Sashiwa, H., and Aiba, S. 2002. Utilization of commercial non-chitinase enzymes from fungi for preparation of 2-acetamido-2-deoxy-D-glucose from β-chitin. *Carbohydrate Research* 337: 133–137.

Suresh, P. V. 2012. Biodegradation of shrimp processing bio-waste and concomitant production of chitinase enzyme and N-acetyl-D-glucosamine by marine bacteria: Production and process optimization. *World Journal Microbiology and Biotechnology* 28: 2945–2962.

Suresh, P. V., and Chandrasekaran, M. 1999. Impact of process parameters on chitinase production by an alkalophilic marine *Beauveria bassiana* in solid state fermentation. *Process Biochemistry* 43(3): 257–267.

Suresh, P. V., and Kumar, P. K. A. 2012. Enhanced degradation of α-chitin materials prepared from shrimp processing byproduct and production of N-acetyl-D-glucosamine by thermoactive chitinases from soil mesophilic fungi. *Biodegradation* 23: 597–607.

Suresh, P. V., and Prabhu, G. N. 2012. Seafood. In *Valorization of Food Processing Byproducts*, M. Chandrasekaran (ed.), pp. 685–736. Boca Raton, FL: CRC Press.

Suresh, P. V., Kumar, P. K. A., and Sachindra, N. M. 2011. Thermoactive β-N-acetylhexosaminidase production by a soil isolate of *Penicillium monoverticillium* CFR 2 under solid state fermentation: Parameter optimization and application for N-acetyl chitooligosaccharides preparation from chitin. *World Journal of Microbiology and Biotechnology* 27: 1435–1447.

Swapna, C. H., Bijinu, B., Amit, K. R., and Bhaskar, N. 2011a. Simultaneous recovery of lipids and proteins by enzymatic hydrolysis of fish industry waste using different commercial proteases. *Applied Biochemistry and Biotechnology* 164: 115–124.

Swapna, C. H., Amit, K. R., Sachindra, M. N., and Bhaskar, N. 2011b. Seafood enzymes and their potential industrial application. In *Handbook of Seafood Quality, Safety and Health Application*, C. Alasalvar, F. Shahidi, K. Miyashita and U. Wanasundara (eds.), pp. 522–532. Chichester, West Sussex, UK: Wiley-Blackwell.

Téllez-Luis, S. J., Uresti, R. M., Ramírez, J. A., and Vázquez, M. 2002. Low-salt restructured fish products using microbial transglutaminase. *Journal of Food Science and Agriculture* 82: 953–959.

Tharanathan, R. N., and Kittur, F. S. 2003. Chitin—The undisputed biomolecule of great potential. *Critical Reviews in Food Science and Nutrition* 43(1): 61–87.

Thiansilakul, Y., Benjakul, S., and Shahidi, F. 2007. Compositions, functional properties and antioxidative activity of protein hydrolysates prepared from round scad (*Decapterus maruadsi*). *Food Chemistry* 103: 1385–1394.

Uresti, R. M., Velázquez, G., Vázquez, M., Ramírez, J. A., and Torres, J. A. 2006. Effects of combining microbial transglutaminase and high pressure processing treatments on the mechanical properties of heat-induced gels prepared from arrowtooth flounder (*Atheresthes stomias*). *Food Chemistry* 94: 202–209.

Venugopal, V. 2006. *Seafood Processing: Adding Value through Quick Freezing Retortable Packing and Cook-Chilling*. Boca Raton, FL: Taylor & Francis, CRC Press.

Venugopal, V. 2009. *Marine Products for Healthcare: Functional and Bioactive Nutraceutical Compounds from the Ocean*. Boca Raton, FL: Taylor & Francis, CRC Press.

Venugopal, V. 2015. Fish industry byproducts as source of enzymes and application of enzymes in seafood processing. In *Fish Processing Byproducts—Quality Assessment and Applications*, N. S. Mahendrakar and N. M. Sachindra (eds.), pp. 136–171. Houston, TX: Studium Press LLC.

Venugopal, V., Lakshmanan, R., Doke, S. N., and Bongirwar, D. R. 2000. Enzymes in fish processing, biosensors and quality control. *Food Biotechnology* 14: 21–27.

Vilhelmsson, O. 1997. The state of enzyme biotechnology in the fish processing industry. *Trends in Food Science and Technology* 8: 266–270.

Wanasundara, U. 2011. Preparative and industrial-scale isolation and purification of omega-3 polyunsaturated fatty acids from marine sources. In *Handbook of Seafood Quality, Safety and Health Application*, C. Alasalvar, F. Shahidi, K. Miyashita and U. Wanasundara (eds.), pp. 464–475. Chichester, West Sussex, UK: Wiley-Blackwell.

Wong, T. Y., Preston, L. A., and Schiller, N. L. 2000. Alginate lyase: A review of major sources and enzyme characteristic structure-function analysis, biological roles and applications. *Annual Review Microbiology* 54: 289–340.

Woo, J.-W., Yu, S.-J., Chob, S.-M., Leea, Y.-B., and Kim, S.-B. 2008. Extraction optimization and properties of collagen from yellowfin tuna (*Thunnus albacares*) dorsal skin. *Food Hydrocolloids* 22: 879–887.

Yang, Y., and Lee, C. M. 2000. Enzyme-assisted bio-production of lobster flavor from the process by-product and its chemical and sensory properties. In *Seafood in Health and Nutrition-Transformation in Fisheries and Aquaculture: Global Perspectives*, F. Shahidi (ed.), pp. 169–193. St. John's, NL, Canada: Science Tech Publishing Co.

Yu, Y. J., Wu, S. C., Chan, H. H., Chen, Y. C., Chen, Z. Y., and Yang, M. T. 2008. Overproduction of soluble recombinant transglutaminase from *Streptomyces netropsis* in *Escherichia coli*. *Applied Microbiology and Biotechnology* 81: 523–532.

Zhang, Y., Liu, W., Li, G., Shi, B., Miao, Y., and Wu, X. 2007. Isolation and partial characterization of pepsin-soluble collagen from the skin of grass carp (*Ctenopharyngodon idella*). *Food Chemistry* 103: 906–912.

Zhou, L., Budge, S. M., Ghaly, A. E., Brooks, M. S., and Dave, D. 2011. Extraction, purification and characterization of fish chymotrypsin: A review. *American Journal of Biochemistry and Biotechnology* 7(3): 104–123.

Zhuang, Y., Xue, Z., and Ba-Fang, L. 2009. Optimization of antioxidant activity by response surface methodology in hydrolysates of jellyfish (*Rhopilema esculentum*) umbrella collagen. *Journal of Zhejiang University Science B* 10(8): 572–579.

Section III

Advances in Food Grade
Enzyme Biotechnology

16

Enzymes in Synthesis of Novel Functional Food Ingredients

Amit Kumar Jaiswal and Samriti Sharma

CONTENTS

16.1 Introduction

The philosophy "Let food be thy medicine and medicine be thy food" was embraced by Hippocrates, the father of medicine, about 2500 years ago. This meant that the food should contain nutrition in amounts capable of meeting the energy and nutritional requirements of an individual that would ultimately help develop a strong resistance to most common infections that were then prevalent (Hasler 1998). However, with the advent of modern drug therapy, there has been a drastic change in this philosophy. This is mainly due to the decreased realization of food in health care (Hasler 2000; Milner 1999). It was also an era in which much emphasis was given to preferences and tastes over nutrition as such foods were supposed to be a major contributor to the economic growth of a country (Hasler 1998). It was in the first half of the twentieth century that increased incidences of diseases related to over-nutrition made the health care sector realize the negative impact of prolonged consumption of such foods. This led to an increased wave of awareness in the form of public health guidelines, emphasizing the importance of a balanced nutritious diet. This included a diet low in salt and saturated fats and high in vegetables, fruits, whole grains, and legumes, which help reduce the risks of chronic diseases such as cancer, diabetes, etc. This need led to the birth of "functional foods."

Functional foods stand as a new category of remarkably promising foods that possess the properties of low cholesterol, antioxidants, anticancer, antiaging, etc. These properties render functional foods quite appealing, and subsequently, there has been increased demand for these products (Diplock et al. 1998; Mohamed 2014). Functional foods are currently a favorite topic among researchers, dieticians, and producers and consumers of the foods. To date, a wide range of definitions of functional foods exists across the world. For instance, according to Diplock et al (1999), functional foods are those foods that can be accepted by the human body in normal amounts as part of our daily dietary needs and not as any medication. In general terms, any food that provides energy and meets the nutritional requirement of an individual on a daily basis is a functional food. Table 16.1 summarizes some of the most commonly used definitions of functional foods. Based on these definitions, it can be concluded that any food can be considered a functional food if it (a) provides health benefits, (b) constitutes a part of the normal diet, and (c) possesses nutritional functions.

In today's era, food is not only intended to satisfy hunger and provide necessary nutrients essential for the well-being of a human body, but it also plays a crucial role in the prevention of nutrition-related diseases and improves the physical and mental state of its consumers (Siro et al. 2008). Generally, functional foods are often confused with nutraceuticals; both are intended to promote health through the use of food and its components, but they are totally different terms.

"Nutraceuticals," as the name suggests, are a combination of two well-known terms: "nutrients" and "pharmaceuticals." Functional foods, on the other hand, are normal food products with certain specific functions aimed at the well-being of an individual. The two terms also differ in their modes of consumption as the former is dose dependent whereas the latter is a part of the regular diet. However, the two terms have always been a topic of debate as they can be used interchangeably at times. According to El Sohaimy (2012), functional foods for one could serve as nutraceuticals for another. This means that, when food is consumed to provide the required amounts of proteins, vitamins, and minerals to the body

TABLE 16.1

Most Commonly Used Definitions of Functional Foods

Author/Organizations	Definitions
Food Safety Authority Ireland (FSAI), Ireland	"Functional food is a food with characteristics that can help achieve or maintain good health in addition to providing basic nutrition."
British Nutrition Foundation (BNF), UK	"Functional food is a food that delivers additional or enhanced benefits over and above their basic nutritional value and includes broad range of products."
European Food Information Council (EUFIC)	"Functional food is a food intended to be consumed as a part of normal diet and contain biologically active components which offer potential of enhanced health or reduced risk of disease."
The Commonwealth Scientific and Industrial Research Organisation (CSIRO), Australia	"Any food or food component that may provide demonstrated physiological benefits or reduce the risk of chronic diseases above and beyond basic nutritional functional, is a functional food."
Agriculture and Agri-Food Canada	"Functional foods are foods enhanced with bioactive ingredients and which have demonstrated health benefits."
Food and Nutrition Board (FNB) of National Academy of Sciences, USA	"Functional foods are those that encompasses potentially healthful products including any modified food or food ingredient that may provide a health benefit beyond that of traditional nutrients that it contains."
Food for Specified Health Uses (FOSHU), Japan	"Foods which are based on knowledge between food and food components and health expected to have certain health benefits, and have been licensed to bear a label claiming that a person using them for specified health use may expect to obtain the health use through the consumption thereof."
Iowa State University, USA	"Foods that have been linked to health benefits."
International Food Information Council (IFIC), Washington, USA	"Foods that may provide health benefit beyond basic nutrition."
Health Canada, Ontario, Canada	"Functional foods as products that resemble traditional foods but possess demonstrated physiological benefits."

for its healthy survival, it is said to be a functional food. However, when this functional food is used to prevent or treat a disease or disorder, it is said to be a nutraceutical (El Sohaimy 2012).

Because the two terms are still evolving, several gaps seem to exist. Their producers wish to provide enough information on the health claims of these products. In addition, the governmental regulating bodies also face a tough time classifying these products as they occupy a position in between regular foods and drugs. Functional foods and nutraceuticals are still not legally recognized in the United States (Hasler and Brown 2009). In the European Union, on the other hand, functional food is regulated under general food law regulation (Regulation [EC] No. 178/2002) and under Regulation (EC) No. 1924/2006, which covers claims referring to health in order to avoid any confusion among its consumers (Koch et al. 2014). However, lack of a single universally accepted guideline not only fails to draw a clear line between functional foods and nutraceuticals, but it also leaves the companies entering the health food market in confusion and complicates the import and export of similar products worldwide.

16.2 Classification of Functional Foods and Ingredients

In the absence of uniform legal status, "functional foods" is often considered to be a marketing term across the globe. This is mainly because of the lack of a single widely accepted definition available, which thereby fails to draw a clear boundary between what can and what cannot be considered functional foods. However, based on the definitions mentioned in Table 16.1 as accepted in different parts of the world, functional foods can be broadly categorized as the following:

- Naturally occurring foods, which are the conventional foods containing naturally occurring bioactive components
- Processed foods, which have been modified by any of the following ways: addition of a component, removal of a component, modification of components, modification of bioavailability, or foods produced by the combination of any of the above methods (Henry 2010)

Table 16.2 gives examples of the most commonly known functional foods along with the main functional ingredient and the health benefits of these foods. Most of these naturally occurring functional foods have been known to mankind for their properties and have therefore been an important part of the diet. For example, tomatoes, both in raw form and processed form are one of the most widely consumed vegetables across the world. They are known to be rich in potassium, folate, and vitamins, such as A, C, and E. In addition, they are also one of the richest sources of lycopene. Research in the past has identified the role of lycopene in inhibition of proliferation, antiandrogen, and antigrowth factor effects, such as conditions of cancer (De Marzo et al. 2007; Wang et al. 2003). Epidemiological studies have shown that tomato polyphenols are involved in physiological properties such as anti-inflammatory, antimicrobials, antiviral effects, etc. (Navarro-González et al. 2011). Flavonols found in tomatoes are an essential component for healthy skin (Canene-Adams et al. 2005; Stewart et al. 2000).

Epidemiological studies suggest that diets rich in Brassica vegetables are associated with a lower risk of several diseases, such as atherosclerosis, stroke, cancer, diabetes, arthritis, and aging, which could be due to their richness in phenolic acids, flavonoids, isothiocyanates, and glucosinolates (Cartea and Velasco 2008; Jaiswal et al. 2011). Inflammatory and antioxidation properties found in watermelon makes it another suitable choice as a functional food for obtaining health benefits of increased arginine availability, reducing serum concentration of cardiovascular risk factors, and also helping in the dietary management of metabolic syndrome in diabetes mellitus type 2 (formerly noninsulin-dependent diabetes mellitus) and obesity (Kim et al. 2014; Wu et al. 2007).

Berries such as strawberries, cranberries, and blueberries; coconut; mushrooms; soybeans; and pomegranates have also been recognized for their functional components of proven health benefits (Basu et al. 2014; Lasekan 2014). Some of these are even under clinical trials for their role in the treatment of cancer and other diseases (Gu et al. 2013). These are some of the examples of naturally occurring foods with potential health benefits and considered as functional foods. In addtion, these foods are rich in several

TABLE 16.2

Some of the Most Commonly Available Naturally Occurring Foods with Functional Components and Potential Health Benefits

Functional Foods	Functional Components	Potential Benefits	References
Tomatoes	Lycopene	Prostate health	Cortés-Olmos et al. (2014)
Berries	Phenolic acids, flavonoids, anthocyanin	Effective in the treatment of urinary tract infection and seasonal influenza	Lasekan (2014)
Pomegranate	Polyphenols such as punicalagins, punicalins, gallagic acid, ellagic acid	Effective against cardiovascular disease, diabetes, prostate cancer	Johanningsmeier and Harris (2011)
Broccoli	Glucosinolate	Prevention of cancer	Jeffery et al. (2003)
Citrus	Flavanones	Neutralizes free radicals, reduced risk of some cancers	Gattuso et al. (2007)
Soybeans	Isoflavones	Lowers LDL and total cholesterol	Hasler (1998)
Fish oils	Omega 3-fatty acids	Reduced risk of cardiovascular disease	Holub and Holub (2004)
Wheat bran, corn bran, fruit skin	Insoluble fiber	Reduced risk of breast and colon cancer, promotes healthy digestive tract	Reddy et al. (2000); Vitaglione et al. (2008)
Carrots, sweet potatoes, apricots, cantaloupes, peaches, dark green leafy vegetables	β-carotene	Most potent pro-vitamin A form, maintains healthy eyes; vision cycle, reduces the risk of cancers and heart diseases	Hasler (1998, 2002); Ishida et al. (2000); Schieber et al. (2001); Stacewicz-Sapuntzakis et al. (2001)
Onions, apples, tea, red lettuce	Flavonols	Reduced risks of cardiovascular diseases; prevent oxidative damage to cells, lipids and DNA; promote bone health, prevent osteoporosis; possess anti-inflammatory properties	Liu (2003); Rice-Evans (2001); Schieber et al. (2001)
Celery, oregano, parsley	Flavones	Effective against different types of cancers and osteoporosis	Shahidi (2009)
Flaxseed, sesame, whole grain wheat bread, rice, cashew nuts	Lignans	Reduces cancer of reproductive organs and colon cancer, improves cardiovascular health	Kamal-Eldin et al. (2011)
Kale, spinach, corn, egg, asparagus	Leutin, Zeaxanthin	Maintenance of eye health	Halsted (2003); Hasler (2000)
Oat bran, oat meal, oat flour, barley, rye	β-glucan	Reduce risk of coronary heart diseases	Brennan and Cleary (2005)
Psyllium seed husk, peas, beans, apple, citrus fruits	Soluble fibers	Effective in treatment for coronary heart diseases and different cancers	Brennan and Cleary (2005)
Cauliflower, broccoli, broccoli sprouts, cabbage	Sulforaphane	Enhance detoxification of certain undesirable compounds	Robbins et al. (2005)
Fish, red meat, whole grains, garlic	Selenium	Neutralizes free radicals, supports maintenance of immune and prostate health	Ferrari (2007)
Apple, pears, citrus fruits, coffee	Caffeic acid, feluric acid	Promotes eye health and heart health	Liu (2003); Rice-Evans (2001); Schieber et al. (2001)

types of oligosaccharides and offer ingredients that could be extracted for commercial production of functional foods.

Processed foods, on the other hand, are the foods that have been modified either by the addition or removal of certain components in order to increase the potential of the food product in providing health benefits to its consumers. Table 16.3 illustrates some of the most commonly available processed functional foods in the market along with their functional ingredients, added functional components, and potential health benefits.

Based on their source of origin, functional foods can also be categorized as the following:

- Plant-derived functional foods, which include foods containing active components derived from plants
- Animal-derived functional foods, which include foods containing functional components derived from animals, such as dairy products and fish oil

Foods derived from plants provide a great source of certain ingredients that have long been known for their positive impact on the health of an individual. These fruits, vegetables, nuts, legumes, whole grains, tea, coffee, etc. contain ingredients that serve as bioactive compounds, such as polyphenolic compounds, flavonoids, plant sterols and stanols, and vitamins C, E, and carotenoids (Alissa and Ferns 2012). Certain plant secondary metabolites such as glucosinolates and capsaicinoids are other examples of bioactive compounds of which the former has been found to be associated with a reduction of carcinomas in lungs, stomach, colon, and rectum, whereas the latter is supposed to provide relief from pain by acting on the peripheral part of the sensory nervous system. Although certain studies do indicate some allergic reactions associated with some of these compounds, their potential as functional ingredients cannot be avoided (Lavecchia et al. 2013).

Animal-derived functional foods can broadly be divided into two main categories as meat and eggs and dairy products. Meat and eggs have long been a part of the major diet in several civilizations. Meat

TABLE 16.3

Some of the Most Commonly Available Processed Functional Foods Along with Their Functional Ingredients, Added Functional Components, and Potential Health Benefits

Functional Foods	Functional Components	Potential Benefits	References
Orange juice with added vitamin D, juices with calcium	Vitamin D	Improved bone health	Dawson-Hughes et al. (1997)
Yogurt with probiotics	Probiotics	Improved gastrointestinal digestion	Roberfroid (2000)
Bread and cereals with added fiber	Fibers	Decreases the problem of constipation and reduces the risk of cancers	Charalampopoulos et al. (2002); Pharmaceutiques (1995)
Margarine fortified with plant sterols	Plant sterols and phytosterols	Reduces cholesterol	Berger et al. (2004)
Grains with folic acid	Folic acid	Reduce risk of heart diseases and neural tube birth defects	Alissa and Ferns (2014); De Wals et al. (2007)
Juices with added fibers	Fibers	Reduces risks of certain cancers and heart diseases, cholesterol, hypertension, and constipation	Brennan and Cleary (2005)
Beverages and salad dressings with antioxidants	Antioxidants	Promote overall growth	Hasler (1998)
Sport bars and drinks	Varies with types	Extra source of energy	
Eggs with Omega 3 fatty acids	Omega 3 fatty acids	Reduces risk of heart diseases	Holub and Holub (2004)

and meat-derived products are major sources of bioactive compounds, such as iron, zinc, vitamins, and minerals. Fish oil, for instance, is one of the richest sources of long-chain fatty acids, such as DHA and EPA. In certain cases, meats and eggs have been enriched with selenium to increase the functionality of the diet due to its proven role in metabolism. The dairy industry is one of the key sectors of the food industry throughout the world. Milk and milk-derived products are the richest sources of conjugated linoleic acids that provide protection from various types of cancer (Kralik et al. 2012). Probiotic products rely greatly on the dairy industry.

In general, government agencies, such as the U.S. Food and Drug Administration (FDA), control the key decisions regarding the regulation, safety, and marketing of food products. In the case of functional foods, lack of a formal definition issued by these regulating bodies gives a freedom to the manufacturers to choose the market for such products. In such cases, functional foods are currently marketed as the following:

- Enriched foods: This category includes foods that are produced by the addition of one or more nutrients that have previously been lost during their processing.
- Fortified foods: These are foods that have been prepared by the addition of one or more nutrients that impart a specific functionality.
- Enhanced foods: These are prepared by the addition of one or more active ingredients by modification or indirect methods.
- Altered foods: These are produced by the removal, reduction, or replacement of a deleterious component by some beneficial ingredient.

Usually, functional foods are those foods that contain at least one ingredient that provides an important function in improving health and/or reducing the chances of disease, such as diabetes, cancer, hypertension, and several cardiovascular diseases (Granato et al. 2010). These ingredients may include prebiotics, probiotics, antioxidants, carotenoids, lycopene, etc. As discussed earlier, the natural functional foods include the class of conventional foods that are unmodified whole foods, such as fruits and vegetables. Processed foods, on the contrary, include foods that have been modified by any of the processes, such as fortification, enrichment, or enhancement of conventional foods. These include calcium-fortified juices, folate-enriched breads, or foods enhanced with bioactive components, such as plant sterols (Granato et al. 2010).

Efforts are also being made to modify foods using biotechnological approaches to improve the nutritional value of foods, thereby creating a new field of foods in the market. These include foods with increased omega 3-fatty acids or decreased trans-fatty acid content (Granato et al. 2010). Some of the most common products that are found in markets throughout the world include probiotics, prebiotics, functional cereals, functional meat, functional eggs, etc. Despite their development in all food markets, functional food products are not homogeneously scattered over all segments of the food industry. Some of the major products include confectionery, dairy, soft drinks, bakery, and the baby food market (Siro et al. 2008).

16.3 Functional Food Market, Trends, and Major Products

16.3.1 Market Trends of Functional Foods

Over the past few years, the concept of functional foods has gained momentum, which is on account of an increased understanding of the relationship between nutrition and health. Despite their existence in a wide range of foods, they fail to attain uniform distribution in all segments of the growing market (Siro et al. 2008). Technical obstacles, lack of proper legislation, and consumer willingness are some of the main reasons that make the development and commerce of functional foods a complex, expensive, and risky task (Siro et al. 2008). This could be because of the lack of a globally accepted definition of functional foods. This creates confusion in the minds of its consumers regarding functional food products. As a result of this, consumers prefer products such as yogurt, cereals, and juices as healthy functional foods that are well known (Annunziata and Vecchio 2011).

Since the birth of the concept of functional foods, Japan has pioneered the development of this sector of the food industry, generating interest and awareness in the regions of the United States and Europe. These countries took no time in realizing the potential of functional foods in reducing the cost of health care and as a source of commercial revolution in the food industry. It is estimated that global functional food markets contribute US$33 billion to the overall food industry with the United States offering the largest market segment, followed by Europe and Japan (Siro et al. 2008). Due to ambiguity about the functional food products across the globe, certain other studies estimate a much higher value of US$61 billion to the functional food market. Despite these variations in the share of functional foods, they are supposed to contribute to 90% of the sales from the three main markets of the United States, Europe, and Japan (Siro et al. 2008).

Countries such as Germany, France, the Netherlands, and the United Kingdom constitute the major functional food markets of Europe with Hungary, Poland, and Russia being some new emerging ones since 2005 (Annunziata and Vecchio 2011). However, a high level of divergence exists between the Eastern and Western countries of the world with respect to the nature of functional foods. In the Eastern world, for instance, Foods for Specific Health Use (FOSHU) labeled functional foods are accepted as first-generation functional foods and preferred for their function over taste. On the contrary, emphasis is laid on adding functionality to traditional foods rather than considering them as a separate class (Siro et al. 2008). Not only this, but variation also exists within different states of Europe based on food traditions and cultural heritage with higher numbers of consumers in the northern and central parts as compared to the Mediterranean regions (Annunziata and Vecchio 2011).

Despite the efforts by the food industries throughout the globe, functional foods still suffer great losses. This was mainly because most companies developed a single product rather than throwing a range of products into the market, thereby providing the customers with fewer options to choose from (Menrad 2003).

16.3.2 Industries Involved in Functional Food Production

The market for functional foods is very unevenly distributed across the world. Certain types of functional foods are much more favored among consumers. This is mainly influenced by their use in certain cultures that dates back several centuries. However, increased cost of health care and awareness among consumers regarding the state of well-being and a healthy lifestyle are playing major roles in the growth of the present functional foods market. Functional products are mainly launched in dairy, beverages, bakery, confectionery, and the baby food market. The European market currently is flooded by products promising gut health, specifically probiotic-based products. Menrad (2003) divided key suppliers of functional food products into the following categories:

- Multinational food companies
- Pharmaceutical companies and companies involved in the production of dietary supplements
- National category leaders
- Small and medium-sized companies
- Retail companies
- Suppliers of functional ingredients

Major multinational companies involved in the production of functional foods are Yakult Honsa Co. Ltd., Danone, Unilever, Nestle, Kellogg, and Quaker Oats. Table 16.4 demonstrates major functional foods and the industries involved worldwide. Yakult Honsa Co. Ltd., established in 1930, is a Japan-based company that introduced the concept of probiotic-based dairy products for the first time. Its first product was Yakult, which was brought to the market in 1935 and is a fermented mixture of skim milk with a specific strain of *Lactobacillus casei* Shirota. Following its success in the Japanese market, the product is now marketed in 31 different countries, such as Australia, New Zealand, India, Indonesia, Vietnam, America, the Philippines, Thailand, South Korea, Singapore, Hong Kong, and China. Mexico serves as one the largest Yakult selling markets.

TABLE 16.4

Major Functional Food Brands Worldwide

Functional Foods	Products	Companies
Dairy products	Probiotic yogurt/drinks	Dannon (France)
		CoolBrands International (Canada)
		Yakult (Japan)
		National Foods (Pakistan)
	Cholesterol lowering yogurt	Dannon (France)
	Omega 3 yogurt and milk	Jalna Dairy Foods (Australia)
		PB Foods (UK)
	Fortified milk	Borden (USA)
		PB Foods (UK)
		Murray Goulbum (Australia)
Bakery products	Whole grain and fortified breads	Sara Lee (USA)
		George Weston (Australia)
	Breads with omega 3 fatty acids	Wegmans Food Markets (USA)
		Arnold Foods (USA)
		George Weston (Australia)
	High-fiber cookies	RD Foods (USA)
		Quaker (UK)
	High-fiber white bread	Quality Bakers (Australia)
		George Weston
Cereal products	Heart-healthy cereals	Quaker (UK)
		Kraft Foods (USA)
		Kellogg (USA)
		Uncle Toby's (Australia)
	Calcium-fortified cereal bars	Quaker (UK)
		Kellogg (USA)
	High-fiber cereal bars	Kellogg (USA)
	Heart-healthy cereal bars	Nature Valley
	Whole grain pasta	Barilla (Italy)
		Kraft (USA)
	Ready-to-eat energy treats	Kellogg (USA)
	Ready-to-eat health bars	Kellogg (USA)
	Ready-to-eat biscuits	Sanitarium
	Ready-to-eat staples	Kellogg (USA)
	Ready-to-eat bran	Kellogg (USA)
Soy	Soy milk	White Wave (USA)
		Sanitarium (Australia)
		Parmalat (Italy)
Beverages	Fruit juices and juice blends	Tropicana (USA)
		Minute Maid (USA)
		Ocean Spray (USA)
		National Foods (Pakistan)
		Campbell Soup (USA)
	Enhanced waters	Gatorade (USA)
		Energy Brands (USA)
		PepsiCo (USA)
		Coca-Cola (USA)
	Teas with antioxidants	Tetley (UK)
		Lipton (UK)

(Continued)

TABLE 16.4 (CONTINUED)

Major Functional Food Brands Worldwide

Functional Foods	Products	Companies
Meat, fish, eggs	Canned fish with omega 3	Star-Kist (USA)
		Chicken of the Sea (USA)
		Heinz (USA)
	Fish oil supplements	Bumble Bee and Leiner Health (USA)
	Frozen fish with omega 3	Simplot (Australia)
	Omega 3-DHA/lutein enriched eggs	Gold Circle Farms (USA)
		Eggland's Best (USA)
		Pace Farms (Australia)
		Farm Pride (Australia)

Danone is another leading name in the field of probiotic products. This French company deals with a wide range of dairy products, cereals, baby foods, and yogurts sold under brand names such as Actimel, Activia, YoCrunch, and Dannon (the United States). This French-based multinational company has several joint ventures with companies such as Britannica (India), Yakult (Japan), Bright Dairy (China), etc. Another well-known brand available in the market is Becel ProActiv, which is a margarine-based UniLever product claiming to reduce cholesterol levels. The brand is sold in countries such as Belgium, Brazil, Bulgaria, Canada, Estonia, Denmark, Finland, Germany, Greece, the Netherlands, Portugal, Romania, the United Kingdom, Ireland, Spain, Poland, Australia, New Zealand, and South Africa, where it is sold under the name Flora.

Novartis, GlaxoSmithKline, Johnson & Johnson, and Abbott Laboratories are some of the renowned pharmaceutical companies also involved in the production of a range of functional food. Novartis launched a project named AVIVA to market a line of functional food products in 1999 in Austria and Switzerland, which was later withdrawn in the year 2001 owing to lesser consumer willingness to pay more for such foods. The Altus Food Company was another joint venture of Novartis and Quaker Oats to target the functional food market in the United States, Canada, and Mexico. GlaxoSmithKline contributes to the functional food market through its products such as Ribena, which is a fruit-based soft drink available in both carbonated and noncarbonated forms claiming to be fortified with vitamins such as A and E; and Lucozade, which is a series of energy and soft drinks. Retail stores such as ALDI and LIDL have come up with their own label brands, especially in the dairy sector, in recent years across Europe.

However, it has been seen that multinational companies lead the sales of functional food products, which could be attributed to their well-established R&D departments and specific in-house resources and expertise in the domain of functional food technology and research (Menrad 2003). Also the history of success of these companies helps develop a positive image among consumers, which aids in the product development and marketing required for these functional food products. Small- and medium-scale companies, on the other hand, lack the adequate know-how, intensive R&D, and investment required for the launch of such products (Menrad 2003).

16.3.3 Major Functional Foods in the Market

16.3.3.1 Probiotics

According to the FDA, probiotics are defined as "live microorganisms that, when administered, confer a health benefit on the host." The most commonly used microorganisms are *Lactobacillus* and *Bifidobacterium*, which have been proven to exist in the human intestinal tract (Siro et al. 2008). Probiotics with these and several other bacteria have long been found to provide a healthy gut. Probiotics help enhance nutrient availability, reduce lactose intolerance, prevent and treat allergies and eczema, and treat irritable bowel syndrome, infections caused by *Helicobacter pyroli* (Drago et al. 2013; Godin 1998). With a view to fulfill the proposed health claims, a wide range of products serves as carriers of probiotic strains. These include dairy-based products, such as sweet drinks, flavored milk, cheeses, and yogurt;

soy-based products, such as soy milk and soy cream cheese; and juice-based products, such as tomato juice, carrot juice, orange juice, etc. (Nagpal et al. 2012). These products are marketed under different trade names throughout the world.

Due to their years-long use in several cultures and a positive image among consumers, probiotics hold a major share of the functional food market. Among the probiotic markets, Europe represents the largest and fastest-growing market, followed by Japan (Yamaguishi et al. 2011). It accounts for a share of >1.4 billion euros with the biggest sales of 1 billion euros achieved by yogurts and desserts, followed by probiotic milk (Saxelin 2008). Current market leaders for probiotic-based functional foods include Danone, Yakult, Nestle, etc. The probiotic market generated US$15.9 billion in 2008 and was projected to reach US$28.8 billion by 2015 (Yamaguishi et al. 2011).

16.3.3.2 Prebiotics

Prebiotics were first introduced by Gibson and Roberfroid (1995). Prebiotic means "a non-digestible food ingredient that beneficially affects the host by selectively stimulating the growth of one or a limited number of bacteria in the colon." These food ingredients include inulin, fructo-oligosaccharides (FOS), trans-galactosylated oligosaccharides, and soybean oligosaccharides that, for instance, promote the growth of bacteria such as *bifidobacterium* and *lactobacilli* (Manning and Gibson 2004; Schrezenmeir and de Vrese 2001). Other similar prebiotic food ingredients include glycooligosaccharides, glucooligosaccharides, lactulose, lactitol, maltooligosaccharides, xylooligosaccharides, stachyose, raffinose, and sucrose thermal oligosaccharides (Patterson and Burkholder 2003). Improved flora inside the gut thus results in increased resistance to pathogenic bacteria, lower levels of blood ammonia, increased immune response activation, and a reduced risk of cancer (Manning and Gibson 2004). One of the most widely used prebiotics is inulin-type fructans. Inulins are "polydisperse carbohydrate material consisting mainly of β (2-1) fructosyl-fructose links."

Prebiotic products have been classified on two main criteria: (i) their resistance to gastric juices, hydrolysis by the enzymes, and absorption by the gastrointestinal tract and (ii) their fermentation by microorganisms present in the intestine and selective stimulation of the growth of intestinal microflora associated with health and well-being (Roberfroid 2007). Current prebiotic sales are US$110 million, which is expected to double in the next 5 years. Thirty-five percent of this market is contributed by inulin-based products, 25% by mannan oligosaccharides, and 10% by fructan oligosaccharides.

16.3.3.3 Functional Drinks

Functional drinks are another subsector of functional foods currently known in the market. It includes nonalcoholic beverages that are fortified/enhanced to provide health benefits of improved heart health, immunity, digestion, bone health, etc. Some of the common products in this category include functional drinks with herbal extracts, high-efficiency sport drinks, cholesterol-lowering drinks, eye-health drinks, bone-health drinks, and ready-to-drink teas (Whitehead 2008). Studies have shown that fruit drinks containing β-glucan can reduce serum concentrations of LDL cholesterol, and reduction in non-HDL and LDL cholesterol was observed in the case of soy drinks enriched with plant sterols, thereby providing a solution to manage hypercholesterolemia (Weidner et al. 2008). Other types of functional drinks are ACE drinks, in which antioxidant vitamins A, C, and E are added to fruit- and vegetable-based soft drinks. In order to achieve a desirable value of vitamin A in drinks, compounds such as retinol, retinyl acetate, retinyl palmitate, and β-carotene are added to them (Viñas et al. 2013). Several fruit juices, such as orange juice, have also been fortified with sterol esters to attain potential health benefits (Alemany-Costa et al. 2012).

16.3.3.4 Functional Meat

Functional meat is another category of functional food that has been brought to the market in order to achieve increased health benefits. This has been done while keeping in mind the increasing demand of meat and expansion of the meat industry throughout the world. In developing countries, meat has always been seen as the food of the rich. But with increased industrialization, more and more countries are entering the list of developed countries. It is also reported that the consumption of meat in developing countries has increased over the past decades. Thus, it provides an opportunity to follow practices that can improve the functionality of the existing products available in the market, thereby bringing a new term of *functional meat* to its consumers. Recent research has shed light on the presence of several meat-based bioactive ingredients, such as carnosine, anserine, L-carnitine, and conjugated linoleic acid (Arihara 2006).

The functional quality of meat and meat-based products could be improved by several means, for example, modification of carcass composition at the animal production stage and altering the meat raw materials in addition to reformulating meat products by reducing fat content; modifying fatty acid profile; reducing cholesterol, calories, sodium content, and nitrites; and incorporating functional ingredients of both plant and animal origin (Jiménez-Colmenero et al. 2001). Conventional meat products have been combined with fish or sea-related products to balance the omega-6/omega-3 polyunsaturated fatty acids (PUFAs) in the diet and obtain a synergic combination of antioxidants in order to increase the intake of antioxidants considering their role in reduction of free radicals from the body (Reglero et al. 2008). Market studies have identified Europe to be a great potential platform for meat products containing fiber (López and Pérez-Alvarez 2008).

16.3.3.5 Designer Eggs

Attempts have been made to produce designer eggs that will revolutionize the functional food market in the future. As the name suggests, these are eggs enriched with linolenic acid (a precursor of DHA) to provide protection against fatal ischemic heart disease (Brouwer et al. 2004). This is achieved by enriching the hen's diet with flaxseeds or linseeds, resulting in the enrichment of the egg yolk with alpha-linolenic acid. Another alternative is to enhance the levels of n-3 in the egg by including preformed DHA in the diet.

16.3.3.6 Bakery Products

Bakery products constitute one of the largest sectors of the food industry due to increased dependence of consumers around the world. They thus serve as a good option to introduce ingredients that can improve the functionality of the currently available bakery products. Bakery products possess wheat flour, semolina, and water as the main constituents. High starch content in these ingredients is an excellent source of energy, proteins, and fatty acids. They also contain certain essential vitamins, minerals, antioxidants, and phytochemicals. Efforts in the past have been made to enrich bread with omega 3 PUFAs to increase essential fatty acid intake. In order to produce a bakery product with additional functions, technologies such as Super Micro Atomization Retention Technology (SMART) are required to produce SuperCoat omega 3 products (Kadam and Prabhasankar 2010). Another study shows that bread can be enriched with microencapsulated tuna oil that increases DHA (Yep et al. 2002). Rice bran, which is usually either discarded or used no more than as an animal feed, is now being utilized to add functional properties to bread. This is targeted to introduce the nutraceutical properties of rice bran into bread (Lima et al. 2002).

β-Glucan is a type of soluble fiber that has gained importance in cereals as an active functional ingredient. Research has already provided facts supporting its role in insulin resistance; reducing hypertension and obesity; lowering serum cholesterol levels; and stimulating the immune system, anticoagulants, antimutagenic, and antitumorigenic properties (Mantovani et al. 2008; Wood 2007). The level of β-glucans varies from 1% in wheat grains to 5% 11% in barley. Considering the health benefits of β-glucans, breads and pasta have been enriched with these ingredients in order to produce another category of functional foods, known as functional cereals (Brennan and Cleary 2005). These cereals will thus serve as fermentable substrates, promoting growth of microflora inside the gut.

16.4 Methods of Processing of Functional Foods

In order to prepare a desired functional food product, several production steps are necessary in order to produce a food product from raw ingredients; change the physical and chemical appearance of the product; ensure food safety, constant quality, and shelf life; and add or remove one or another component to achieve the desired functional benefits. In general, heat treatment is the most widely used method for several solid and liquid products to make them palatable with an added bioavailability of nutrients, denaturized proteins, and modified carbohydrates and starches and to develop desired flavors, aroma, color, and texture or to inactivate microorganisms to reach required sterility. However, as a consequence of heat treatment, there are certain degrees of loss in nutritional contents of food products. Fruits and vegetables are a major source of functional ingredients, such as polyphenols, flavonoids, and isothiocyanates, etc., and are generally extracted using organic solvents, which may be toxic and are not environmentally friendly.

In recent years, several alternative strategies have been employed for the processing of functional foods, which preserve inherent properties of raw material intact as well as extract higher content of functional ingredients from plant materials as compared to conventional methods. In order to achieve this, several novel food processing methods have been developed and employed in the food industries; for example, ultrasonication uses high-frequency short pulses to disrupt the cell wall of microorganisms, thereby leading to its inactivation. Ultrasonication treatment has been reported to increase the shelf life of food products and also increase bioactive extraction (Briones-Labarca et al. 2015). High hydrostatic pressure (HHP) uses high pressure in the range of 100–600 MPa for food processing. As this is a nonthermal technology, sensory and nutritional attributes of the product remain virtually unaffected, thus yielding products with better quality than those processed using traditional methods. It has the ability to inactivate microorganisms as well as enzymes responsible for shortening the life of a product. In addition, it can modify the functional properties of components, such as proteins, which, in turn, can lead to the development of new products (Briones-Labarca et al. 2015).

Pulse electric field (PEF) processing is a nonthermal food processing technology, which uses a series of short and high-voltage pulses to pasteurize liquid food. This process renders microorganisms inactive and also acts upon enzymes, such as peroxidases and polyphenol oxidases, which cause enzymatic browning and color changes in food. It also aids in the extraction of nutraceuticals from plant parts (San Martin et al. 2002; Segovia et al. 2015). However, these processing methods have several drawbacks, such as being expensive, tedious to operate, being hazardous, having high processing costs, and requiring stringent process control operations. These problems demand a novel and optimized method for the production of functional foods. Enzyme-assisted production of functional food is an age-old concept whose extensive applications have recently come into the limelight with the advancement of biotechnology (Puri et al. 2012).

Enzymes are biological catalysts with high substrate specificity. They are highly precise in the reaction they catalyze and can perform in mild conditions in aqueous solutions. Because of their biological nature, enzymes are ideal for the extraction of new food ingredients for functional food applications and can also be applied in a range of functional food product development. The use of enzymes in the food industry also provides a cleaner and greener alternative as opposed to conventional chemical reaction processes currently employed in nutraceutical production (Meyer 2010). The ability of enzymes to cleave biomolecules at specific sites results in better yields of the desirable compounds as compared to physical and chemical processing methods.

16.5 Enzymes in Synthesis of Functional Foods

These days, enzymes are isolated from living cells, which led to their pilot-scale production and wider application in the food industry. Microorganisms are the most important source of commercial enzymes. Microorganisms do not contain the same enzymes as plants or animals, but usually, it is found that they produce a similar enzyme that will catalyze the desired reaction. There is much emphasis focused on the isolation and identification of novel microbial strains for the production of desired enzymes or strains derived from mutation and genetic engineering (Afifi et al. 2014; Liu et al. 2014); however, in the industry, a large number of microbial enzymes come from a very limited number of genera, and *Aspergillus* and *Bacillus* are the most predominate species.

Enzymes that are used in the food industry are supposed to conform to certain safety guidelines before they can be commercially exploited. An overall safety assessment of each enzyme preparation intended for use in food or food processing must be performed. This assessment should include an evaluation of the safety of the production organism, the enzyme component, side activities, the manufacturing process, and the consideration of dietary exposure (FAO 2001). All the enzymes from animal origin and preparation of enzymes must comply with meat inspection requirements and be handled in accordance with good hygienic practice, and plant-origin enzyme preparations must contain components that leave no residues harmful to health in the processed finished food under normal conditions of use (FAO 2001). However, according to safety regulations for enzymes obtained from microbial strains, the following issues were considered: the pathogenicity and toxicity of the production strain, allergies and irritations, interaction of enzymes from other food components, products of enzymatic reactions, carcinogenic and mutagenic effects, and the effect of food enzymes on consumption. The primary focus when assessing a production strain is its toxigenic potential. Toxigenic potential is defined as the ability of a microorganism to produce chemicals (toxins) that can cause food poisoning.

The fungal-origin enzymes should be evaluated for toxicity; they should not contain significant amounts of mycotoxins that are known to be synthesized by production strains. Source microorganisms must be discrete and stable strains or variants that have been taxonomically characterized to enable them to be assigned unique identities as the sources of the enzyme preparations that are the subject of individual specifications (Landry et al. 2003). The reference or production strain number may be included in individual specifications. In addition, culture media used for the growth of microbial sources must consist of components that leave no residues harmful to health in the processed finished food under normal conditions of use (FAO 2001).

A single class of enzyme can be used for different purposes. The primary question that needs to be answered is the objective that has to be achieved: Is the enzyme going to be used for reduction of viscosity; change the texture or flavor of food; or remove certain characteristics, such as bitterness, from beverage products. Enzymes may be selected based on the nature of the substrate they act upon, such as starch, protein, and fats. Another aspect that should be evaluated during enzyme selection is the overall conditions of the food or the beverage, such as pH, temperature, moisture content, reaction time necessary, and other processing conditions. The source of the enzyme also determines its operational efficiency; for example, fungal enzymes tend to be more active in acidic pH, and bacterial, yeast, and animal sources show higher activity in alkaline conditions. It is also necessary to study the side reactions of the enzymes as well. For example, fungal proteases used in flour production also exhibit amylase activity (Guo and Xu 2005).

16.6 Application of Enzymes in Food Industry: Case Studies

16.6.1 Enzyme-Mediated Synthesis of Nondigestible Oligosaccharides

According to IUPAC nomenclature, oligosaccharides are saccharides that contain 3–10 sugar moieties. These oligosaccharides are low molecular weight carbohydrates and can be classified as digestible and nondigestible based on their physiological properties. Roberfroid (1997) defined nondigestible oligosaccharides (NDOs) as "an oligomeric carbohydrate, the osidic bond of which is in a spatial configuration that makes it resistant to the hydrolytic activity of the intestinal digestive enzymes. However, it is still sensitive to hydrolysis by enzymes of the colonic bacteria which then ferment its monomers to produce short chain carboxylic acids, gases and cellular energy for metabolic activities, growth and proliferation."

Nondigestible oligosaccharides have long been known for their functional properties (Roberfroid 1997). They possess certain physiological and physicochemical properties that make them behave as dietary fibers and prebiotics (Mussatto and Mancilha 2007). A diet enriched with NDOs helps improve the microbiology of the gut by encouraging the growth of healthy microflora inside the gut. They do so by promoting the growth of certain specific microorganisms, such as *Bifidobacterium* and *Lactobacillus* over other pathogenic bacteria followed by their predominance in the gastrointestinal tract as observed through their increased occurrence in the feces (Pharmaceutiques 1995). It is believed that the configuration of the osidic bond prevents the hydrolysis of the NDOs in the upper part of the gastrointestinal tract. These later undergo the metabolic process known as fermentation in the caeco-colon, serving as an indirect energy substrate. It is due to this fermentation that the pH of the gut decreases, thereby preventing the development of bacterial strains such as *E. coli*, *Clostridia*, etc. (Delzenne and Roberfroid 1994). It also helps improve the absorption of minerals from the intestine and lipid metabolism, prevent cancer of certain types, and develop a strong immune system (Swennen et al. 2006).

Based on the type of monosaccharide unit attached, nondigestible saccharides can be glucooligosaccharides, galactooligosaccharides, fructooligosaccharides, and xylooligosaccharides (Mussatto and Mancilha 2007). In addition, other compounds known as NDOs are lactulose, isomaltulose, raffinose, cyclodextrins, and glycosylsucrose (Mussatto and Mancilha 2007). These NDOs can be obtained by several methods, such as (i) direct extraction from natural sources, (ii) chemical synthesis from disaccharides or polysaccharides, and (iii) enzymatic methods (Mussatto and Mancilha 2007). Oligosaccharides can be prepared from lactose enzymatically by (i) transgalactosylation with lactose as the galactosyl donor and β-galactosidase as the catalyst synthesizing galactooligosaccharide, (ii) transgalactosylation of fructose and sucrose with β-galactosidase as the catalyst yielding lactulose and lactosucrose, and (iii) transfructosylation or transglucosylation of lactose with sucrose as the glycosyl donor and levansucrase or dextransucrase as the calayst forming lactosucrose and glucosylactose, respectively (Gänzle 2012).

Of all the abovementioned NDOs, galactooligosaccharides (GOS) are the most commonly used and known prebiotics. GOS are composed of 2–10 galactose units linked to a terminal glucose unit by β (1-4) and β (1-6) bonds. Recent studies have shown that GOS poses a positive impact on immunity and calcium absorption in the body (Rastall 2013). GOS is an intermediate product of the lactose hydrolysis by a hydrolase enzyme, β-galactosidase, which attacks the o-glucosyl group of lactose (Torres et al. 2010). Thus, in processes involving the hydrolysis of lactose, GOS is often considered to be an undesirable by-product. However, considering the prebiotic properties of GOS, there has been an increased interest in the commercial production of GOS by various food and pharmaceutical sectors (Torres et al. 2010).

GOS are most commonly prepared by enzymatic methods (Gänzle 2012; Torres et al. 2010). Glycosyltransferase (EC. 2.4) and glycoside hydrolase (EC. 3.2.1) are the two enzymes used to catalyze the production of GOS (Torres et al. 2010). Glycosyltransferase does so by catalyzing the formation of glycosidic linkage through transfer of a saccharide from donor to the acceptor molecule; glycoside hydrolase, on the contrary, catalyzes the hydrolysis of a glycoside bond by two mechanisms, depending upon the stereochemistry as the (i) inverting mechanism or (ii) retaining mechanism (Davies and Henrissat 1995). However, factors such as poor availability, high prices, and choice of sugars as the substrates restrain glycosyltransferases to be used for commercial production of GOS.

Glycoside hydrolase, on the other hand, is preferred for producing GOS commercially due to its increased availability (Torres et al. 2010). As mentioned above, glycoside hydrolases are generally known for their hydrolyzing property; however, in certain cases, retaining glycoside hydrolases can be used to synthesize glycosidic linkages by the process known as transglycosylation (Van Den Broek and Voragen 2008). The glycoside hydrolase consists of GH1, GH2, GH35, and GH42 families based on the types of substrates whereas the catalytic mechanism remains the same (Gänzle 2012; Van Den Broek and Voragen 2008).

β-Galactosidase is by far the most commonly known glycoside hydrolase that catalysizes the formation of GOS from lactose. β-Galactosidase of the GH2 family has higher lactase hydrolysis activity and transglycosylation activity than GH42 (Van den Broek et al. 2005). However, a competition always remains between both mechanisms. In order to favor transglycosylation over hydrolysis to promote the synthesis of galactooligosaccharides, it is necessary to optimize reaction parameters, such as pH, temperature, substrate concentration, etc. (Van Den Broek and Voragen 2008). In some cases, site-directed mutagenesis has been applied to enhance the feasibility of the reaction. Also, GH42 β-galactosidase from *Lactobacilli* and *Bifidobacteria* are inhibited by high concentrations of lactose and thus less preferred for transgalactosylation.

16.6.2 Role of Enzymes in the Synthesis of Functional Meat

The meat industry is one of the world's largest growing food sectors. With increasing demands for meat and meat-based products and competition, tremendous efforts and contributions have been made in the field of research to increase the quality of these products. Meat is a great source of fatty acids, minerals, dietary fiber, antioxidants, and bioactive peptides (Decker and Park 2010). Considering the increased demand for healthier food options, functional meat offers a great potential both for producers and consumers. For instance, muscle foods are a major source of bioactive compounds, such as iron, zinc, conjugated linoleic acids, and vitamin B (Jiménez-Colmenero et al. 2001).

Meat and meat-based products can be modified in the following ways to improve their functional properties: (i) addition of ingredients and (ii) removal of harmful compounds. One such example is the addition of olive oils to traditional Spanish sausage to replace up to 30% of the pork back fat (Fernández-Ginés et al. 2005; Muguerza et al. 2001). The functional value of meat can be improved by adding functional compounds, such as conjugated linolenic acid, vitamin E, n3 fatty acids, and selenium in the animal diet or ingredients such as vegetable proteins, dietary fibers, herbs, spices, and lactic acid bacteria during the processing of meat (Zhang et al. 2010). In addition, recent work also demonstrates the generation of peptides from meat and meat products during the processes of fermentation, curing and aging, and enzymatic hydrolysis (Zhang et al. 2010).

Research in the past has identified peptides released from animal and plant proteins that offer attributes of antimicrobial properties, blood pressure-lowering effects, cholesterol-lowering effects, antioxidant properties, immunomodulatory effects, etc., to a human body that is beyond normal nutrition (Hartmann and Meisel 2007). These peptides, often known as bioactive peptides, remain inactive within their parent protein and are released inside the gastrointestinal tract or during the processing of foods containing these peptides (Korhonen and Pihlanto 2003). Lafarga and Hayes (2014) described bioactive peptides as 2–30 amino acid sequences that impart a positive health effect to consumers. This section, however, focuses on meat-based bioactive compounds. Some of the examples of such compounds include conjugated linoleic acid, carnosine, anserine, L-carnitine, glutathione, taurine, creatine, etc. (Arihara 2006). Bioactive peptides can be produced (i) during processing of meat and meat-based products by methods of fermentation and enzymatic hydrolysis and (ii) directly from meat proteins, followed by their use to enrich the meat (Arihara 2006; Jang and Lee 2005; Saiga et al. 2003).

Meat has largely been used to synthesize bioactive peptides by enzymatic methods. ACE inhibitory peptides are one such group of bioactive peptides synthesized from meat proteins using enzymes. ACE or angiotensin-converting enzyme is found in many tissues and body fluids in the human body, and it regulates the blood pressure and maintains the electrolyte balance by converting inactive angiotensin I into vasoconstricting angiotensin II, thereby increasing the blood pressure of the body (Meisel et al. 2006). Thus, ACE-inhibitory peptides have been found to block this conversion of angiotensin I to angiotensin II and help maintain blood pressure (Ogbru and Marks 2008). Although several drugs that inhibit this conversion are present in the market, the need to find natural solutions to this problem of hypertension gave birth to the research of ACE-inhibitory peptides. Due to their smaller size, these peptides can easily cross the digestive epithelial barrier and reach the blood vessels, thereby providing a natural cure (Yust et al. 2003). Currently, ACE peptides have largely been isolated from whey proteins, fish, pork, chicken meat, chicken eggs, and beef (Ghassem et al. 2011; Jang and Lee 2005; Saiga et al. 2003; Vercruysse et al. 2005). This has mainly been achieved by enzymatic hydrolysis of the meat. Based on the type of meat used, a wide range of enzymes has been used for hydrolysis, such as trypsin, chymotripsin, pepsin, pancreatin, thermolysin, and proteinase A. However, a correct combination of enzymes is advised in order to attain high yields.

16.7 Concluding Remarks

Functional foods stand as a new category of foods with numerous health benefits, such as low cholesterol, high antioxidants, and anticancer and antiaging properties, etc. In recent years, demand for such food is increasing, which could be due to consumer awareness of foods and beverages with health-promoting nutrients. Several methods are in practice in the food industry for the processing of these functional foods, such as conventional heat treatment or chemically mediated extraction of nutraceuticals from plant parts. In recent years, numerous novel technologies were employed, such as high hydrostatic pressure, pulsed electric field, ultrasonication, etc.; however, these processing methods have several drawbacks, such as being expensive, tedious to operate, having high processing costs, and requiring stringent process control operations. Enzymes are environmentally friendly and nontoxic and do not require expensive technology and operating expertise. Their use offers the potential for many exciting applications in the improvement of foods, and they are commonly used in the production of nutraceuticals and in the synthesis of a range of functional foods. However, there is still a long way to go in recognizing the full potential of enzymes in food and other industrial applications; for example, a range of increasingly sophisticated enzymes for specific functional food applications should be developed. In addition, various properties of enzymes, such as thermostability, specificity, and catalytic efficiency, should be improved to withstand extreme conditions, such as such as high temperature, pH, etc., for several industrial processes.

REFERENCES

Afifi, A. F., Abo-Elmagd, H. I., & Housseiny, M. M. 2014. Improvement of alkaline protease production by Penicillium chrysogenum NRRL 792 through physical and chemical mutation, optimization, characterization and genetic variation between mutant and wild-type strains. *Annals of Microbiology, 64*(2), 521–530.

Alemany-Costa, L., González-Larena, M., García-Llatas, G., Alegría, A., Barberá, R., Sánchez-Siles, L. M., & Lagarda, M. J. 2012. Sterol stability in functional fruit beverages enriched with different plant sterol sources. *Food Research International, 48*(1), 265–270.

Alissa, E. M., & Ferns, G. A. 2012. Functional foods and nutraceuticals in the primary prevention of cardiovascular diseases. *Journal of Nutrition and Metabolism, 2012*, 569486.

Alissa, E. M., & Ferns, G. A. 2014. Potential cardio-protective effects of functional foods. In: Noomhorm, A., Ahmad I. and Anal A. K. (Eds). *Functional Foods and Dietary Supplements: Processing Effects and Health Benefits*, pp. 463–487. John Wiley & Sons, Chichester, UK.

Annunziata, A., & Vecchio, R. 2011. Functional foods development in the European market: A consumer perspective. *Journal of Functional Foods, 3*(3), 223–228.

Arihara, K. 2006. Strategies for designing novel functional meat products. *Meat Science, 74*(1), 219–229.

Basu, A., Nguyen, A., Betts, N. M., & Lyons, T. J. 2014. Strawberry as a functional food: An evidence-based review. *Critical Reviews in Food Science and Nutrition, 54*(6), 790–806.

Berger, A., Jones, P., & Abumweis, S. S. 2004. Plant sterols: Factors affecting their efficacy and safety as functional food ingredients. *Lipids in Health and Disease, 3*(5), 907–919.

Brennan, C. S., & Cleary, L. J. 2005. The potential use of cereal $(1 \rightarrow 3, 1 \rightarrow 4)$-β-D-glucans as functional food ingredients. *Journal of Cereal Science, 42*(1), 1–13.

Briones-Labarca, V., Plaza-Morales, M., Giovagnoli-Vicuña, C., & Jamett, F. 2015. High hydrostatic pressure and ultrasound extractions of antioxidant compounds, sulforaphane and fatty acids from Chilean papaya (*Vasconcellea pubescens*) seeds: Effects of extraction conditions and methods. *LWT-Food Science and Technology, 60*(1), 525–534.

Brouwer, I. A., Katan, M. B., & Zock, P. L. 2004. Dietary α-linolenic acid is associated with reduced risk of fatal coronary heart disease, but increased prostate cancer risk: A meta-analysis. *The Journal of Nutrition, 134*(4), 919–922.

Canene-Adams, K., Campbell, J. K., Zaripheh, S., Jeffery, E. H., & Erdman, J. W. 2005. The tomato as a functional food. *The Journal of Nutrition, 135*(5), 1226–1230.

Cartea, M. E., & Velasco, P. 2008. Glucosinolates in Brassica foods: Bioavailability in food and significance for human health. *Phytochemistry Reviews, 7*(2), 213–229.

Charalampopoulos, D., Wang, R., Pandiella, S., & Webb, C. 2002. Application of cereals and cereal components in functional foods: A review. *International Journal of Food Microbiology, 79*(1), 131–141.

Cortés-Olmos, C., Leiva-Brondo, M., Roselló, J., Raigón, M. D., & Cebolla-Cornejo, J. 2014. The role of traditional varieties of tomato as sources of functional compounds. *Journal of the Science of Food and Agriculture, 94*(14), 2888–2904.

Davies, G., & Henrissat, B. 1995. Structures and mechanisms of glycosyl hydrolases. *Structure, 3*(9), 853–859.

Dawson-Hughes, B., Harris, S. S., Krall, E. A., & Dallal, G. E. 1997. Effect of calcium and vitamin D supplementation on bone density in men and women 65 years of age or older. *New England Journal of Medicine, 337*(10), 670–676.

Decker, E. A., & Park, Y. 2010. Healthier meat products as functional foods. *Meat Science, 86*(1), 49–55.

Delzenne, N. M., & Roberfroid, M. 1994. Physiological effects of non-digestible oligosaccharides. *LWT-Food Science and Technology, 27*(1), 1–6.

De Marzo, A. M., Platz, E. A., Sutcliffe, S., Xu, J., Grönberg, H., Drake, C. G., & Nelson, W. G. 2007. Inflammation in prostate carcinogenesis. *Nature Reviews Cancer, 7*(4), 256–269.

De Wals, P., Tairou, F., Van Allen, M. I., Uh, S.-H., Lowry, R. B., Sibbald, B., & Crowley, M. 2007. Reduction in neural-tube defects after folic acid fortification in Canada. *New England Journal of Medicine, 357*(2), 135–142.

Diplock, A., Charuleux, J.-L., Crozier-Willi, G., Kok, F., Rice-Evans, C., Roberfroid, M., & Vina-Ribes, J. 1998. Functional food science and defence against reactive oxidative species. *British Journal of Nutrition, 80*(S1), S77–S112.

Diplock, A. T., Aggett, P. J., Ashwell, M., Bornet, F., Fern, E. B., & Roberfroid, M. D. 1999. Scientific concepts in functional foods in Europe: Consensus document. *British Journal of Nutrition (United Kingdom), 81,* S1–S27.

Drago, L., Toscano, M., & Pigatto, P. 2013. Probiotics: Immunomodulatory properties in allergy and eczema. *Giornale italiano di dermatologia e venereologia: Organo ufficiale, Societa italiana di dermatologia e sifilografia, 148*(5), 505–514.

El Sohaimy, S. 2012. Functional foods and nutraceuticals-modern approach to food science. *World Applied Sciences Journal, 20*(5), 691–708.

FAO. 2001. General specifications and considerations for enzyme preparations used in food processing. 57th session. *Compendium of Food Additive Specifications.* FAO Food and Nutrition Paper, 52.

Fernández-Ginés, J. M., Fernández-López, J., Sayas-Barberá, E., & Pérez-Alvarez, J. 2005. Meat products as functional foods: A review. *Journal of Food Science, 70*(2), R37–R43.

Ferrari, C. K. 2007. Functional foods and physical activities in health promotion of aging people. *Maturitas, 58*(4), 327–339.

Gänzle, M. G. 2012. Enzymatic synthesis of galacto-oligosaccharides and other lactose derivatives (hetero-oligosaccharides) from lactose. *International Dairy Journal, 22*(2), 116–122.

Gattuso, G., Barreca, D., Gargiulli, C., Leuzzi, U., & Caristi, C. 2007. Flavonoid composition of citrus juices. *Molecules, 12*(8), 1641–1673.

Ghassem, M., Arihara, K., Babji, A. S., Said, M., & Ibrahim, S. 2011. Purification and identification of ACE inhibitory peptides from Haruan (*Channa striatus*) myofibrillar protein hydrolysate using HPLC–ESI–TOF MS/MS. *Food Chemistry, 129*(4), 1770–1777.

Gibson, G. R. & Roberfroid, M. B. 1995. Dietary modulation of the human colonic microbiota introducing the concept of prebiotics. *Journal of Nutrition* 125(6), 1401–1412.

Goldin, B. R. 1998. Health benefits of probiotics. *The British Journal of Nutrition, 80*(4), S203–S207.

Granato, D., Branco, G. F., Cruz, A. G., Faria, J. D. A. F., & Shah, N. P. 2010. Probiotic dairy products as functional foods. *Comprehensive Reviews in Food Science and Food Safety, 9*(5), 455–470.

Gu, J., Ahn-Jarvis, J. H., Riedl, K. M., Schwartz, S. J., Clinton, S. K., & Vodovotz, Y. 2013. Characterization of black raspberry functional food products for cancer prevention human clinical trials. *Journal of Agricultural and Food Chemistry* 62(18), 3997–4006.

Guo, Z., & Xu, X. 2005. New opportunity for enzymatic modification of fats and oils with industrial potentials. *Organic & Biomolecular Chemistry, 3*(14), 2615–2619.

Halsted, C. H. 2003. Dietary supplements and functional foods: 2 sides of a coin? *The American Journal of Clinical Nutrition, 77*(4), 1001S–1007S.

Hartmann, R., & Meisel, H. 2007. Food-derived peptides with biological activity: From research to food applications. *Current Opinion in Biotechnology, 18*(2), 163–169.

Hasler, C. M. 1998. Functional foods: Their role in disease prevention and health promotion. *Food Technology—Champaign then Chicago, 52,* 63–147.

Hasler, C. M. 2000. The changing face of functional foods. *Journal of the American College of Nutrition, 19*(Suppl. 5), 499S–506S.

Hasler, C. M. 2002. Functional foods: Benefits, concerns and challenges—A position paper from the American Council on Science and Health. *The Journal of Nutrition, 132*(12), 3772–3781.

Hasler, C. M., & Brown, A. C. 2009. Position of the American Dietetic Association: Functional foods. *Journal of the American Dietetic Association, 109*(4), 735–746.

Henry, C. 2010. Functional foods. *European Journal of Clinical Nutrition, 64*(7), 657–659.

Holub, D. J., & Holub, B. J. 2004. Omega-3 fatty acids from fish oils and cardiovascular disease. *Molecular and Cellular Biochemistry, 263*(1), 217–225.

Ishida, H., Suzuno, H., Sugiyama, N., Innami, S., Tadokoro, T., & Maekawa, A. 2000. Nutritive evaluation on chemical components of leaves, stalks and stems of sweet potatoes (*Ipomoea batatas* poir). *Food Chemistry, 68*(3), 359–367.

Jaiswal, A. K., Rajauria, G., Abu-Ghannam, N., & Gupta, S. 2011. Phenolic composition, antioxidant capacity and antibacterial activity of selected Irish Brassica vegetables. *Natural Products Communications, 6*(9), 1299–1304.

Jang, A., & Lee, M. 2005. Purification and identification of angiotensin converting enzyme inhibitory peptides from beef hydrolysates. *Meat Science, 69*(4), 653–661.

Jeffery, E., Brown, A., Kurilich, A., Keck, A., Matusheski, N., Klein, B., & Juvik, J. 2003. Variation in content of bioactive components in broccoli. *Journal of Food Composition and Analysis, 16*(3), 323–330.

Jiménez-Colmenero, F., Carballo, J., & Cofrades, S. 2001. Healthier meat and meat products: Their role as functional foods. *Meat Science, 59*(1), 5–13.

Johanningsmeier, S. D., & Harris, G. K. 2011. Pomegranate as a functional food and nutraceutical source. *Annual Review of Food Science and Technology, 2*, 181–201.

Kadam, S., & Prabhasankar, P. 2010. Marine foods as functional ingredients in bakery and pasta products. *Food Research International, 43*(8), 1975–1980.

Kamal-Eldin, A., Moazzami, A., & Washi, S. 2011. Sesame seed lignans: Potent physiological modulators and possible ingredients in functional foods and nutraceuticals. *Recent Patents on Food, Nutrition & Agriculture, 3*(1), 17–29.

Kim, C. H., Park, M. K., Kim, S. K., & Cho, Y. H. 2014. Antioxidant capacity and anti-inflammatory activity of lycopene in watermelon. *International Journal of Food Science & Technology, 49*(9), 2083–2091.

Koch, A., Brandenburger, S., Türpe, S., & Birringer, M. 2014. The need for a legal distinction of nutraceuticals. *Food and Nutrition Sciences, 5*, 905–913.

Korhonen, H., & Pihlanto, A. 2003. Food-derived bioactive peptides—Opportunities for designing future foods. *Current Pharmaceutical Design, 9*(16), 1297–1308.

Kralik, G., Kušec, G., Grčević, M., Ðurkin, I., & Kralik, I. 2012. Animal products as coventional and functional food—An overwiev. *Acta Agriculturae Slovenica, 3*, 18.

Lafarga, T., & Hayes, M. 2014. Bioactive peptides from meat muscle and by-products: Generation, functionality and application as functional ingredients. *Meat Science, 98*(2), 227–239.

Landry, T. D., Chew, L., Davis, J. W., Frawley, N., Foley, H. H., Stelman, S. J., & Hanselman, D. S. 2003. Safety evaluation of an α-amylase enzyme preparation derived from the archaeal order *Thermococcales* as expressed in *Pseudomonas fluorescens* biovar I. *Regulatory Toxicology and Pharmacology, 37*(1), 149–168.

Lasekan, O. 2014. Exotic berries as a functional food. *Current Opinion in Clinical Nutrition & Metabolic Care, 17*(6), 589–595.

Lavecchia, T., Rea, G., Antonacci, A., & Giardi, M. T. 2013. Healthy and adverse effects of plant-derived functional metabolites: The need of revealing their content and bioactivity in a complex food matrix. *Critical Reviews in Food Science and Nutrition, 53*(2), 198–213.

Lima, I., Guraya, H., & Champagne, E. 2002. The functional effectiveness of reprocessed rice bran as an ingredient in bakery products. *Food/Nahrung, 46*(2), 112–117.

Liu, R. H. 2003. Health benefits of fruit and vegetables are from additive and synergistic combinations of phytochemicals. *The American Journal of Clinical Nutrition, 78*(3), 517S–520S.

Liu, Z., Liu, L., Österlund, T., Hou, J., Huang, M., Fagerberg, L., & Nielsen, J. 2014. Improved production of a heterologous amylase in saccharomyces cerevisiae by inverse metabolic engineering. *Applied and Environmental Microbiology, 80*(17), 5542–5550.

Manning, T. S., & Gibson, G. R. 2004. Prebiotics. *Best Practice & Research Clinical Gastroenterology, 18*(2), 287–298.

Mantovani, M. S., Bellini, M. F., Angeli, J. P. F., Oliveira, R. J., Silva, A. F., & Ribeiro, L. R. 2008. β-Glucans in promoting health: Prevention against mutation and cancer. *Mutation Research/Reviews in Mutation Research, 658*(3), 154–161.

Meisel, H., Walsh, D., Murray, B., & FitzGerald, R. 2006. ACE inhibitory peptides. In: Mine, Y. and Shahidi, F. (Eds.). *Nutraceutical Proteins and Peptides in Health and Disease*, pp. 269–315. CRC Press, Boca Raton, FL.

Menrad, K. 2003. Market and marketing of functional food in Europe. *Journal of Food Engineering, 56*(2), 181–188.

Meyer, A. S. 2010. Enzyme technology for precision functional food ingredient processes. *Annals of the New York Academy of Sciences, 1190*(1), 126–132.

Milner, J. 1999. Functional foods and health promotion. *The Journal of Nutrition, 129*(7), 1395S–1397S.

Mohamed, S. 2014. Functional foods against metabolic syndrome (obesity, diabetes, hypertension and dyslipidemia) and cardiovasular disease. *Trends in Food Science & Technology, 35*(2), 114–128.

Muguerza, E., Gimeno, O., Ansorena, D., Bloukas, J., & Astiasarán, I. 2001. Effect of replacing pork back fat with pre-emulsified olive oil on lipid fraction and sensory quality of Chorizo de Pamplona—A traditional Spanish fermented sausage. *Meat Science, 59*(3), 251–258.

Mussatto, S. I., & Mancilha, I. M. 2007. Non-digestible oligosaccharides. A review. *Carbohydrate Polymers, 68*(3), 587–597.

Nagpal, R., Kumar, A., Kumar, M., Behare, P. V., Jain, S., & Yadav, H. 2012. Probiotics, their health benefits and applications for developing healthier foods: A review. *FEMS Microbiology Letters, 334*(1), 1–15.

Navarro-González, I., García-Valverde, V., García-Alonso, J., & Periago, M. 2011. Chemical profile, functional and antioxidant properties of tomato peel fiber. *Food Research International, 44*(5), 1528–1535.

Ogbru, O. 2008. ACE Inhibitors (Angiotensin Converting Enzyme Inhibitors). In: Marks, J. (Ed.). Medical and Pharmacy. MedicineNet, Inc. Available at http://www.medicinenet.com/ace_inhibitors/article.htm.

Patterson, J., & Burkholder, K. 2003. Application of prebiotics and probiotics in poultry production. *Poultry Science, 82*(4), 627–631.

Pérez-Alvarez, J. A. 2008. Overview of meat products as functional foods. In: J. F. López (Ed.). *Technological Strategies for Functional Meat Products Development*, pp. 1–17. Transworld Research Network, Kerela, India.

Pharmaceutiques, U. D. L. 1995. Dietary modulation of the human colonie microbiota: Introducing the concept of prebiotics. *The Journal of Nutrition, 125*, 1401–1412.

Puri, M., Sharma, D., & Barrow, C. J. 2012. Enzyme-assisted extraction of bioactives from plants. *Trends in Biotechnology, 30*(1), 37–44.

Rastall, R. A. 2013. Gluco and galacto-oligosaccharides in food: Update on health effects and relevance in healthy nutrition. *Current Opinion in Clinical Nutrition & Metabolic Care, 16*(6), 675–678.

Reddy, B. S., Hirose, Y., Cohen, L. A., Simi, B., Cooma, I., & Rao, C. V. 2000. Preventive potential of wheat bran fractions against experimental colon carcinogenesis: Implications for human colon cancer prevention. *Cancer Research, 60*(17), 4792–4797.

Reglero, G., Frial, P., Cifuentes, A., García-Risco, M. R., Jaime, L., Marin, F. R., Palanca, V., Ruiz-Rodríguez, A., Santoyo, S., Señoráns, F. J., Soler-Rivas, C., Torres, C., and Ibañez, E. 2008. Meat-based functional foods for dietary equilibrium omega-6/omega-3. *Molecular Nutrition & Food Research, 52*(10), 1153–1161.

Rice-Evans, C. 2001. Flavonoid antioxidants. *Current Medicinal Chemistry, 8*(7), 797–807.

Robbins, R. J., Keck, A.-S., Banuelos, G., & Finley, J. W. 2005. Cultivation conditions and selenium fertilization alter the phenolic profile, glucosinolate, and sulforaphane content of broccoli. *Journal of Medicinal Food, 8*(2), 204–214.

Roberfroid, M. 1997. Health benefits of non-digestible oligosaccharides. In: Kritchevsky, D. & Bonfield, C. (Eds.). *Dietary Fiber in Health and Disease*, pp. 211–219. Plenum Press, New York, USA.

Roberfroid, M. B. 2000. Prebiotics and probiotics: Are they functional foods? *The American Journal of Clinical Nutrition, 71*(6), 1682s 1687s.

Roberfroid, M. 2007. Prebiotics: The concept revisited. *The Journal of Nutrition, 137*(3), 830S–837S.

Saiga, A., Okumura, T., Makihara, T., Katsuta, S., Shimizu, T., Yamada, R., & Nishimura, T. 2003. Angiotensin I-converting enzyme inhibitory peptides in a hydrolyzed chicken breast muscle extract. *Journal of Agricultural and Food Chemistry, 51*(6), 1741–1745.

San Martin, M., Barbosa-Cánovas, G., & Swanson, B. 2002. Food processing by high hydrostatic pressure. *Critical Reviews in Food Science and Nutrition, 42*(6), 627–645.

Saxelin, M. 2008. Probiotic formulations and applications, the current probiotics market, and changes in the marketplace: A European perspective. *Clinical Infectious Diseases, 46*(Suppl. 2), S76–S79.

Schieber, A., Stintzing, F., & Carle, R. 2001. By-products of plant food processing as a source of functional compounds—Recent developments. *Trends in Food Science & Technology, 12*(11), 401–413.

Schrezenmeir, J., & de Vrese, M. 2001. Probiotics, prebiotics, and synbiotics—Approaching a definition. *The American Journal of Clinical Nutrition, 73*(2), 361s–364s.

Segovia, F. J., Luengo, E., Corral-Pérez, J. J., Raso, J., & Almajano, M. P. 2015. Improvements in the aqueous extraction of polyphenols from borage (*Borago officinalis* L.) leaves by pulsed electric fields: Pulsed electric fields (PEF) applications. *Industrial Crops and Products*. doi:10.1016/j.indcrop.2014.11.010.

Shahidi, F. 2009. Nutraceuticals and functional foods: Whole versus processed foods. *Trends in Food Science & Technology, 20*(9), 376–387.

Siro, I., Kapolna, E., Kapolna, B., & Lugasi, A. 2008. Functional food. Product development, marketing and consumer acceptance—A review. *Appetite, 51*(3), 456–467.

Stacewicz-Sapuntzakis, M., Bowen, P. E., Hussain, E. A., Damayanti-Wood, B. I., & Farnsworth, N. R. 2001. Chemical composition and potential health effects of prunes: A functional food? *Critical Reviews in Food Science and Nutrition, 41*(4), 251–286.

Stewart, A. J., Bozonnet, S., Mullen, W., Jenkins, G. I., Lean, M. E., & Crozier, A. 2000. Occurrence of flavonols in tomatoes and tomato-based products. *Journal of Agricultural and Food Chemistry, 48*(7), 2663–2669.

Swennen, K., Courtin, C. M., & Delcour, J. A. 2006. Non-digestible oligosaccharides with prebiotic properties. *Critical Reviews in Food Science and Nutrition, 46*(6), 459–471.

Torres, D. P., Gonçalves, M. D. P. F., Teixeira, J. A., & Rodrigues, L. R. 2010. Galacto-oligosaccharides: Production, properties, applications, and significance as prebiotics. *Comprehensive Reviews in Food Science and Food Safety, 9*(5), 438–454.

Van Den Broek, L. A., & Voragen, A. G. 2008. Bifidobacterium-glycoside hydrolases and (potential) prebiotics. *Innovative Food Science & Emerging Technologies, 9*(4), 401–407.

Van den Broek, L. A., Hinz, S. W., Beldman, G., Doeswijk-Voragen, C. H., Vincken, J.-P., & Voragen, A. G. 2005. Glycosyl hydrolases from Bifidobacterium adolescentis DSM20083. An overview. *Le Lait, 85*(1–2), 125–133.

Vercruysse, L., Van Camp, J., & Smagghe, G. 2005. ACE inhibitory peptides derived from enzymatic hydrolysates of animal muscle protein: A review. *Journal of Agricultural and Food Chemistry, 53*(21), 8106–8115.

Viñas, P., Bravo-Bravo, M., López-García, I., & Hernández-Córdoba, M. 2013. Quantification of β-carotene, retinol, retinyl acetate and retinyl palmitate in enriched fruit juices using dispersive liquid–liquid microextraction coupled to liquid chromatography with fluorescence detection and atmospheric pressure chemical ionization-mass spectrometry. *Journal of Chromatography A, 1275*, 1–8.

Vitaglione, P., Napolitano, A., & Fogliano, V. 2008. Cereal dietary fibre: A natural functional ingredient to deliver phenolic compounds into the gut. *Trends in Food Science & Technology, 19*(9), 451–463.

Wang, S., DeGroff, V. L., & Clinton, S. K. 2003. Tomato and soy polyphenols reduce insulin-like growth factor-I-stimulated rat prostate cancer cell proliferation and apoptotic resistance in vitro via inhibition of intracellular signaling pathways involving tyrosine kinase. *The Journal of Nutrition, 133*(7), 2367–2376.

Weidner, C., Krempf, M., Bard, J.-M., Cazaubiel, M., & Bell, D. 2008. Cholesterol lowering effect of a soy drink enriched with plant sterols in a French population with moderate hypercholesterolemia. *Lipids in Health and Disease, 7*(1), 35–42.

Whitehead, J. 2008. 12 Functional drinks containing herbal extracts. In: Ashurst, P. R. (Ed.). *Chemistry and Technology of Soft Drinks and Fruit Juices*, pp. 300–335. John Wiley & Sons, Oxford, UK.

Wood, P. J. 2007. Cereal β-glucans in diet and health. *Journal of Cereal Science, 46*(3), 230–238.

Wu, G., Collins, J. K., Perkins-Veazie, P., Siddiq, M., Dolan, K. D., Kelly, K. A., Meininger, C. J. 2007. Dietary supplementation with watermelon pomace juice enhances arginine availability and ameliorates the metabolic syndrome in Zucker diabetic fatty rats. *The Journal of Nutrition, 137*(12), 2680–2685.

Yamaguishi, C. T., Spier, M. R., Lindner, J. D. D., Soccol, V. T., & Soccol, C. R. 2011. Current market trends and future directions. In: Liong, M. T. (Ed.), *v*, pp. 299–319. Springer-Verlag, Berlin.

Yep, Y. L., Li, D., Mann, N. J., Bode, O., & Sinclair, A. J. 2002. Bread enriched with microencapsulated tuna oil increases plasma docosahexaenoic acid and total omega-3 fatty acids in humans. *Asia Pacific Journal of Clinical Nutrition, 11*(4), 285–291.

Yust, M. A. M., Pedroche, J., Giron-Calle, J., Alaiz, M., Millán, F., & Vioque, J. 2003. Production of ace inhibitory peptides by digestion of chickpea legumin with alcalase. *Food Chemistry, 81*(3), 363–369.

Zhang, W., Xiao, S., Samaraweera, H., Lee, E. J., & Ahn, D. U. 2010. Improving functional value of meat products. *Meat Science, 86*(1), 15–31.

17

Enzymes in Processing of Nutraceuticals

Fahad M. A. Al-Hemaid and Muthusamy Chandrasekaran

CONTENTS

17.1 Introduction

Natural remedies have been the choice of people since ancient times for the prevention and treatment of a variety of common ailments and diseases. With the evolution of modern medicine and the discovery of antibiotics, natural remedies have become secondary although traditional treatments, such as ayurveda, siddha, naturopathy, etc., still persist with natural herbs and natural compounds as the soul of their medicines. Nevertheless, recently, there has been resurgence of interest and emergence of medicinal foods that contribute to the maintenance of well-being, enhancement of health, modulating immunity, and consequent prevention as well as treatment of specific diseases. The best-known physiological benefits or protection against chronic disease include antioxidant activity (resveratrol, flavonoids), cancer prevention (broccoli, fiddleheads), reduction of hypercholesterolemia (soluble dietary fibers), and protection against cardiovascular disease (α-linolenic acid). These so-called *medicinal foods* are termed *nutraceuticals* nowadays.

According to the European Nutraceutical Association, nutraceuticals are "naturally derived bioactive compounds that are found in foods, dietary supplements and herbal products, and have health promoting, disease preventing, or medicinal properties." The term "nutraceutical" was originally defined, by Stephen L. DeFelice, founder and chairman of the Foundation for Innovation in Medicine (DeFelice 1992), as a combination of the words "nutrition" and "pharmaceutical," referring to nutritional products that have effects that are relevant to health (Figure 17.1). In contrast to pharmaceuticals, nutraceuticals are not synthetic substances or chemical compounds formulated for specific indications. According to Ames et al. (1993), nutraceuticals, a term that combines the words "nutrition" (a nourishing food or food

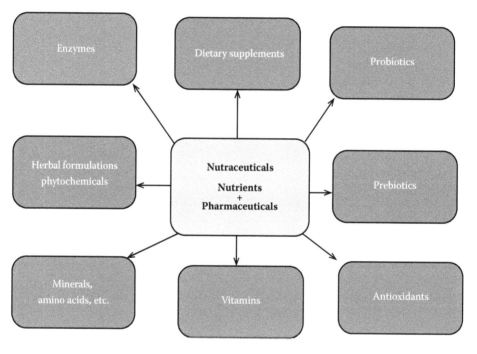

FIGURE 17.1 Concept and classes of nutraceuticals.

TABLE 17.1

Categories of Natural Food Sources Used as Nutraceuticals

Categories of Naturally Available Food Sources	Examples
Dietary fiber	Cereals and grains
Probiotics	*Lactobacillus acidophilus* *Bifidobacterium lactis*
Prebiotics	Digestive enzymes, fructo oligosaccharides (FOS), soybean oligosaccharides, inulin
Polyunsaturated fatty acids	Omega 3 fatty acids
Antioxidant vitamins	Vitamins A, C, E
Phytochemicals	Polyphenols, flavonoids, isoflavonoids anthocyanidins, phytoestrogens, terpenoids, carotenoids, lycopene, limonoids, phytosterols, glucosinolates, polysaccharides, ellagic acid (EA) and ellagitannins (ET)

component) and "pharmaceutical" (a medical drug), are foods or food products that provide health and medical benefits, including the prevention and treatment of disease whereas, according to Zeisel (1999), nutraceuticals are diet supplements that deliver a concentrated form of a presumed bioactive agent from a food, presented in a nonfood matrix, and used with the purpose of enhancing health in dosages that exceed those that could be obtained from normal foods. In the past few years, many food bioactive constituents have been commercialized in the form of pharmaceutical products (pills, capsules, solutions, gels, liquors, powders, granulates, etc.) that incorporate food extracts or phytochemical-enriched extracts to which a beneficial physiological function has been directly or indirectly attributed. These ranges of products cannot be truly classified as "food," and this new hybrid term between nutrients and pharmaceuticals—"nutraceuticals"—has been coined to designate them (Espin et al. 2007). Table 17.1 lists the various categories of natural food sources used as nutraceuticals.

Nutraceutical products may range from dietary supplements; isolated nutrients (e.g., vitamins, minerals, coenzyme Q, carnitine, and botanicals, such as ginseng and gingko biloba); herbal products (Table 17.2);

TABLE 17.2
Common Herbs Used as Nutraceuticals and Their Health Benefits

Common Name of Herb	Biological Name and Part Used	Constituents	Health Benefits
Garlic	Dried bulbs of *Allium sativum* (Liliaceae)	Alliin and allicin	Anti-inflammatory, antibacterial, antigout, nervine tonic
Maiden hair tree	Leaves of *Ginkgo biloba* (Ginkgoaceae)	Ginkgolide and bilobalide	PAF antagonist, memory enhancer, antioxidant
Ginger	Rhizomes of *Zingiber officinale* (Zingiberaceae)	Zingiberene and gingerols	Stimulant, chronic bronchitis, hyperglycemia and throat ache
Echinacea	Dried herb of *Echinacea purpurea* (Asteraceae)	Alkylamide and echinacoside	Anti-inflammatory, immunomodulator, antiviral
Ginseng	Dried root of *Panax ginseng* (Araliaceae)	Ginsenosides and panaxosides	Stimulating immune and nervous system and adaptogenic properties
Licorice	Dried root of *Glycyrrhiza glabra* (Leguminosae)	Glycyrrhizin and liquirtin	Anti-inflammatory and antiallergic, expectorant
St. John's wort	Dried aerial part of *Hypericum perforatum* (Hypericaceae)	Hypericin and hyperforin	Antidepressant, against HIV and hepatitis c virus
Turmeric	Rhizome of *Curcuma longa* (Zingiberacae)	Curcumin	Anti-inflammatory, antiarthritic, anticancer, and antiseptic
Onion	Dried bulb of *Allium cepa* Linn. (Liliaceae)	Allicin and alliin	Hypoglycemic activity, antibiotic and antiatherosclerosis
Valeriana	Dried root of *Valeriana officinalis* Linn. (Valerianaceae)	Valerenic acid and valerate	Tranquillizer, migraine and menstrual pain, intestinal cramps, bronchial spasm
Aloes	Dried juice of leaves *Aloe barbadensis* Mill. (Liliaceae)	Aloins and aloesin	Dilates capillaries, anti-inflammatory, emollient, wound-healing properties
Goldenseal	Dried root of *Hydrastis canadensis* (Ranunculaceae)	Hydrastine and berberine	Antimicrobial, astringent, antihemorrhagic, treatment of mucosal inflammation
Senna	Dried leaves of *Cassia angustifolia* (Leguminosae)	Sennosides	Purgative
Asafoetida	Oleo gum resin of *Ferula assafoetida* L. (Umbelliferae)	Ferulic acid and umbellic acid	Stimulant, carminative, expectorant
Bael	Unripe fruits of *Aegle marmelos* Corr. (Rutaceae)	Marmelosin	Digestive, appetizer, treatment of diarrhea and dysentery
Brahmi	Herbs of *Centella asiatica* (Umbelliferae)	Asiaticoside and madecassoside	Nervine tonic, spasmolytic, antianxiety

Source: Chauhan, B., Kumar, G., Kalam, N., Ansari, S. H., *J Adv Pharm Tech Res*, 4, 4–8, 2013.

processed foods, such as cereals, soups, and beverages; functional foods (e.g., bran, oats, omega 3s, prebiotics, plant sterols, and stanols); medicinal foods (e.g., health bars with added medications); and specific diets to genetically engineered foods.

A large list of nutraceuticals containing phytochemicals from foods is now available on the market. For example, the carotenoid lycopene, alliaceae (garlic, onion) extracts containing sulfur derivatives (i.e., alliin and allicin), glucosinolate extracts, and phytosterol extracts are widely commercialized products. Some of the most common phytochemicals found in the nutraceutical market are polyphenols, such as anthocyanins, proanthocyanidins, flavonols, stilbenes, hydroxycinnamates, coumarins, ellagic acid (EA) and ellagitannins (ETs), isoflavones, lignans, etc. (Espin et al. 2007). Some examples of marketed nutraceutical products are presented in Table 17.3.

Although the modern nutraceutical market began to develop in Japan during the 1980s, now it has global presence, and recently the nutraceutical industry has grown into a global market concomitant with the advent of modern biotechnologies. In fact, at present, nutraceuticals represent the fastest growing segment of the food industry.

The search for less expensive treatments and the undesirable aggressive side effects often associated with synthetic molecules have contributed to the popularity of nutraceuticals. The global nutraceuticals product market reached $142.1 billion in 2011 and is expected to reach $204.8 billion by 2017, growing at a CAGR of 6.3%, according to a new market report from Transparency Market Research, Albany, NY. Asia-Pacific (including Japan) is expected to have the second largest market share after North America by 2017. North America enjoyed the highest market share for nutraceutical products at $56.4 billion in 2011. Health-conscious consumers in the United States have been driving success in the region. Global growth has been spurred primarily by dietary supplements. In this market, protein and peptides are expected to grow 6.6% during the forecasted period (2012–2017) (http://www.nutraceuticalsworld .com/issues/2013-09/view_industry-news/nutraceuticals-market-to-reach-2048-billion-by-2017/#sthash .U2iq2NK2.dpuf).

Discovery and production of nutraceuticals over pharmaceuticals have become the favored options of pharmaceutical and biotech companies. Thus, pharmaceutical and biotech companies, such as Monsanto, American Home Products, Dupont, BioCorrex, Abbott Laboratories, Warner-Lambert, Johnson & Johnson, Novartis, Metabolex, Scio-tech, Genzyme Transgenic, PPL Therapeutics, Unigen, and Interneuron, have committed to the discovery of nutraceuticals (Palthur et al. 2010).

Nutraceuticals are all set to emerge as a major industry in the new millennium, and in this context, the role of enzymes in the processing of phytochemicals as nutraceuticals from natural plant

TABLE 17.3

List of Some Marketed Nutraceutical Products

Product	Category	Content	Manufacturers
Calcirol	Calcium supplement	Calcium and vitamins	Cadila Healthcare Limited, Ahmedabad, India
GRD (Growth, revitality, and development)	Nutritional supplement	Proteins, vitamins, minerals, and carbohydrates	Zydus Cadila Ltd., Ahmedabad, India
Proteinex®	Protein supplement	Predigested proteins, vitamins, minerals, and carbohydrates	Pfizer Ltd., Mumbai, India
Coral calcium	Calcium supplement	Calcium and trace minerals	Nature's Answer, Hauppauge, NY, USA
Chyawanprash	Immune booster	Amla, ashwagandha, pippali	Dabur India Ltd.
Omega woman	Immune supplement	Antioxidants, vitamins, and phytochemicals (e.g., lycopene and resveratrol)	Wassen, Surrey, UK
Celestial Healthtone	Immune booster	Dry fruit extract	Celestial Biolabs Limited
Amiriprash (Gold)	Good immunomodulator	Chyawanprash Avaleha, Swarnabhasma, and RasSindur	UAP Pharma Pvt. Ltd.

Source: Chauhan, B., Kumar, G., Kalam, N., Ansari, S. H., *J Adv Pharm Tech Res*, 4, 4–8, 2013.

sources is discussed in this chapter because enzyme processing is an ecofriendly and cost-effective technology.

17.2 Nutraceuticals: Phytochemicals, Sources, Characteristics

Phytochemicals are nonessential nutrients and are mainly produced by plants to provide them protection. Phytochemicals of nutraceutical importance are bioactive constituents that may provide medical health benefits, including the prevention and/or treatment of diseases and physiological disorders. They, either alone and/or in combination, are reported to have specific pharmacological effects on human health as anti-inflammatory, antiallergic, antioxidant, antibacterial, antifungal, antispasmodic, chemopreventive, hepatoprotective, hypolipidemic, neuroprotective, antiaging, analgesic, immuno-modulator, carminative, and hypotensive agents; treatment for diabetes, osteoporosis, DNA damage, cancer, and heart diseases; to act to induce apoptosis or a diuretic effect; to stimulate the CNS; and to protect from UVB-induced carcinogenesis, psychotic diseases, ulcers, and microbial, viral, and parasitic infections (Dillard and German 2000; Nichenametla et al. 2006; Packer and Weber 2001; Prakash and Kumar 2011; Prakash et al. 2012).

Phytochemicals include polyphenols, flavonoids, isoflavonoids, anthocyanidins, phytoestrogens, terpenoids, carotenoids, limonoids, phytosterols, glucosinolates, and polysaccharides. A majority of foods, such as whole grains, beans, fruits, vegetables, and herbs, contain phytochemicals of nutraceutical importance. A list of phytochemicals of nutraceutical importance, their sources, and probable health effects is presented in Table 17.4.

17.2.1 Polyphenols

Polyphenols are among the most abundant antioxidants in the human diet and represent an important group of bioactive compounds in foods (Dueñas et al. 2005). The most popular polyphenols are the flavonoids, a family composed of flavones, flavonols, and anthocyanins, among others. They are present in fruits, vegetables, cereals, olive oil, dried vegetables, chocolate, and beverages, such as coffee, tea, and wine. The major sources of dietary polyphenols are cereals, legumes (barley, corn, nuts, oats, rice, sorghum, wheat, beans, and pulses), oilseeds (rapeseed, canola, flaxseed, and olive seeds), fruits (apple, grape, pear, cherry, and various berries), vegetables, and beverages (fruit juices, tea, coffee, cocoa, beer, and wine) (Cieślik et al. 2006; Katalinic et al. 2006; Kaul and Kapoor, 2001; Prakash and Kumar 2011; Scalbert et al. 2005).

The major constituent of tea polyphenols are Flavan-3-ols also known as flavonols (catechin, epicatechin, catechingallate, and epigallo-catechingallate), flavanols (quercetin, kaempferol, and their glycosides), flavones (vitexin, isovintexin), and phenolic acids (gallic acid, chlorogenic acid). Caffeic acid in the form of caffeoyl esters and coumaric acids are common in apples, pears, and grapes. Chlorogenic acid is rich in apples and pears. Gallic acid is rich in grapes. Apples contain high levels of quercetin among fruits. Citrus fruits are major sources of flavonones, and hesperidin is found in abundance (120–250 mg/L) in orange juice. Quercetin occurs in its glycosylated form as rutin in fruits and vegetables, and particularly, onions are its rich source (Anagnostopoulou et al. 2006; Prakash et al. 2007; Singh et al. 2009).

Phenols protect plants from oxidative damage and have been studied extensively as antioxidant protectants for human beings and play a beneficial role in reducing the risk of coronary heart disease, diabetes, hypertension, and some types of cancer (Andjelković et al. 2006; Arts and Hollman 2005; Cieślik et al. 2006; Miller and Larrea 2002; Prakash and Kumar 2011; Willcox et al. 2004).

17.2.2 Anthocyanins

Anthocyanins are pigments of fruits, such as cherries, plums, strawberries, raspberries, blackberries, and red currants, and their content varies from 0.15 to 4.5 mg/g in fresh berries. Both anthocyanins and anthocyanin-rich berries or derived extracts have been reported to exhibit a wide range of protective effects with potential benefits for humans that include visual capacity, brain cognitive function, obesity,

TABLE 17.4

List of Phytochemicals of Nutraceutical Importance, Their Sources, and Probable Health Effects

Phytochemical	Source	Health Benefit	References
Dietary polyphenols, flavones, flavonols	Vegetables: dried vegetables; cereals: barley, corn, nuts, oats, rice, sorghum, wheat, beans, and pulses; olive oil: oilseeds (rapeseed, canola, flaxseed, and olive seeds); fruits (apple, grape, pear, cherry, and various berries); fruit juices; tea, coffee, cocoa, chocolate, beer, and wine	As antioxidant protectants for human beings, play beneficial role in reducing the risk of coronary heart disease, diabetes, hypertension, and some types of cancer	Cieślik et al. 2006; Karalini et al. 2006; Kaul and Kapoor 2001; Prakash and Kumar 2011; Scalbert et al. 2005
Tea polyphenols: catechin, epicatechin, catechingallate epigallo-catechingallate, quercetin, kaempferol and their glycosides, vitexin, isovintexin, gallic acid	Tea	Antioxidants	
Caffeic acid in the form of caffeoyl esters and coumaric acids	Apples, pears, and grapes		
Chlorogenic acid	Apples and pears		
Gallic acid	Grapes		
Quercetin	Apples		
Flavonones and hesperidin	Citrus fruits, orange juice		
Anthocyanins	Cherries, plums, strawberries, raspberries, blackberries, and red currant	Visual capacity, brain cognitive function, obesity, ulcer protection, cardiovascular risk, cancer prevention	Espin et al. 2007
Anthocyanidins	Fruits and flowers	Antioxidative and antimutagenic	Prakash and Kumar 2011
Proanthocyanidins	Grapes, apples, strawberries, beans, nuts, cocoa, wine	Antioxidant activity of plasma, decrease of LDL cholesterol fraction, vasodilatation, decrease of blood pressure, maintenance of endothelium function	Williamson and Manach 2005
	Grapeseed extracts and cocoa-derived products, such as milk drinks and other beverages, snack bars, chocolate	Strengthens and repairs connective tissue, multiple sclerosis, teeth and gums, reduces allergic responses, enhances capillary strength and vascular function, reduces blood pressure and cholesterol levels, prevents cancer, strengthens the immune system, improves vision, reduces skin aging and loss of elasticity	Espin et al. 2007; Heiss et al. 2005, 2007; Schroeter et al. 2006; Wang-Polagruto et al. 2006

(Continued)

TABLE 17.4 (CONTINUED)

List of Phytochemicals of Nutraceutical Importance, Their Sources, and Probable Health Effects

Phytochemical	Source	Health Benefit	References
Flavonoids	Soybean and its products, legumes, berries, whole grains, and cereals	Antioxidant, anti-inflammatory, antiallergic, hepatoprotective, antithrombotic, antiviral, and anticarcinogenic	Middleton et al. 2000; Stalikas 2007
Flavanones: naringin and hesperidin	Grapefruit and oranges	Antioxidant, hypocholesterolemic, hypoglycemic, prevention of bone loss, and antitumor osteoporosis, or cancer	Espin et al. 2007; Garg et al. 2001; Nanach and Donovan 2004
Eriodictyol and its glycoside eriocitrin	Lemon peel		Garg et al. 2001
Resveratrol (3,5,4′-trans-trihydroxystilbene)	Grape pomace extracts, root of the Japanese knotweed *Polygonum cuspidatum*	Prevent or reduce cancer and cardiovascular diseases	Aensi et al. 2002; Bradamante et al. 2004; Jang et al. 1997
Phytoestrogens: genistein and daidzein, coumestrol	Soybeans	Preventing and treating cancer and osteoporosis; hormone replacement therapy (HRT); amelioration of postmenopausal symptoms, such as hot flashes; cardiovascular diseases; cognitive function; breast and prostate cancer	Espin et al. 2007; Howes et al. 2006; Ikeda et al. 2006; Lee et al. 2005; Verheus et al. 2007; Williamson-Hughes et al. 2006
Terpenoids: isoprenoids, tocotrienols and tocopherols	Fruits, vegetables, and grains	Antioxidant activity	Nichenametla et al. 2006; Prakash et al. 2004; Prakash and Kumar 2011; Sahl 2005
Carotenoids: beta-carotene, lycopene, lutein, and zeaxanthin	Fruits and vegetables	Protect against uterine, prostate, breast, colorectal, and lung cancers	Cartea and Velasco 2008; Prakash et al. 2004; Prakash and Kumar 2011; Sharma et al. 2005
Limonoids: limonene	Citrus fruit, orange peel oil	Provide chemotherapeutic activity by inhibiting Phase I enzymes and inducing Phase II detoxification enzymes in the liver; inhibits pancreatic carcinogenesis	Cartea and Velasco 2008; Prakash et al. 2004; Prakash and Kumar 2011; Sharma et al. 2005
Glucosinolates	Cruciferous vegetables, broccoli, and cauliflower	Activators of liver detoxification enzymes	Al-Azzawi and Wahab 2010
Oligo- or monosaccharides	Mushrooms, fruits, vegetables, legumes, and certain cereals and grains, such as oats and barley	Prevention of cancers, obesity, cardiovascular diseases, inhibition of tumor growth, neutralize the side effects of chemotherapy and radiation, normalize blood pressure, aid in balancing blood sugar, combat autoimmune disease, act as an anti-inflammatory, lower cholesterol and blood lipids, improve liver function, and increase calcium absorption	Choi and Choi 2012; Lee et al. 2003; Ooi and Liu 2000; Sulkowska-Ziaja et al. 2011; Warrand 2006

ulcer protection, cardiovascular risk, and cancer prevention (Espin et al. 2007). Hence, the berries and anthocyanins are being increasingly exploited as nutraceuticals and dietary supplements. Single-berry extracts (e.g., bilberry or wild blueberry) or natural combinations of various berries (blends of blueberry, strawberry, cranberry, wild bilberry, elderberry, and raspberry extracts) and also in combination with other food components are commercialized as powders, capsules, or tablets with a declared content of anthocyanins ranging from <1% to >25% or tablets containing from 40 mg up to 250 mg of anthocyanins. These anthocyanin-based nutraceuticals or supplements are marketed for their high level of antioxidant capacity (Espin et al. 2007).

17.2.3 Anthocyanidins

Anthocyanidins are water-soluble flavonoids that are aglycones of anthocyanins, and they are among the principal pigments in fruits and flowers. Aglycons and glycosides have antioxidative and antimutagenic properties in vivo (Prakash and Kumar 2011).

17.2.4 Proanthocyanidins

Proanthocyanidins are widespread throughout the plant kingdom and the second most abundant natural phenolics after lignin. They have become part of the human diet upon consumption of fruits (grapes, apples, strawberries, etc.), beans, nuts, cocoa, and wine (http://www.nal.usda.gov/fnic/foodcomp/Data /PA/PA.pdf). They are reported to have effects on the vascular system, including an increase in the antioxidant activity of plasma, a decrease of LDL cholesterol fraction and oxidative stress-derived substances, an improvement of endothelium vasodilatation, a decrease in blood pressure, maintenance of endothelium function, etc. (Williamson and Manach 2005); strengthening and repair of connective tissue; aid patients with multiple sclerosis; strengthen teeth and gums; reducing allergic responses by minimizing histamine production; enhancing capillary strength and vascular function; reducing blood pressure and cholesterol levels; enable prevention of cancer; strengthening the immune system; increasing peripheral circulation; improving vision; and reducing skin aging and loss of elasticity (Espin et al. 2007). These activities have been mainly reported in grapeseed extracts and cocoa-derived products, such as milk drinks and other beverages, snack bars, chocolate, etc. (Espin et al. 2007; Heiss et al. 2005, 2007; Schroeter et al. 2006; Wang-Polagruto et al. 2006). Procyanidin-rich products are one of the most common nutraceuticals on the market, and the most popular are those based on grapeseed extracts, which are usually sold as "95% procyanidins standardized extract" pills or capsules.

17.2.5 Flavonoids

The major active nutraceutical ingredients in plants are flavonoids. They are known for their antioxidant anti-inflammatory, antiallergic, hepatoprotective, antithrombotic, antiviral, and anticarcinogenic activities. Flavonoids are natural compounds showing high physiological activities in therapies for inflammation, heart disease, and cancer (Middleton et al. 2000; Stalikas 2007). They occur as aglycones, glycosides, and methylated derivatives. Flavonoids have a structure similar to that of estrogen and exert both estrogenic and antiestrogenic effects and provide possible protection against bone loss and heart diseases. The precursors of these substances are mainly found in Leguminosae and are abundant in soybean and its products, legumes, berries, whole grains, and cereals.

17.2.6 Flavanones

Aglycones naringenin and hesperetin are the most representative flavanones, and their corresponding glycosides, naringin and hesperidin, are abundant in grapefruit and oranges, respectively (Garg et al. 2001; Manach and Donovan 2004), whereas eriodictyol and its glycoside eriocitrin are abundant in lemon peel (Garg et al. 2001). These compounds exhibit a wide range of biological and pharmacological activities, including antioxidant, hypocholesterolemic, hypoglycemic, prevention of bone losses, and antitumor, indicating potential beneficial effects in humans against diseases, such as cardiovascular diseases, diabetes, osteoporosis, or cancer (Espin et al. 2007). Supplements containing flavanones, such as hesperidin or naringenin, as the main components are less represented in the current market of nutraceuticals compared to isoflavones or anthocyanin-containing products. At present, most flavanone-containing supplements are prepared from citrus fruit extracts, marketed mostly as citrus bioflavonoid complexes, and often mixed with large quantities of vitamin C and a blend of other flavonoids, such as flavonols. Some tablets that contain the flavanone hesperidin are also available, but it is also found mixed with other compounds, such as the flavone diosmin, or even mixed with enzymes, such as the proteolytic enzyme bromelain, apparently to aid in the absorption of hesperidin (Espin et al. 2007).

17.2.7　Resveratrol

Resveratrol (3,5,4′-trans-trihydroxystilbene) has been described as a compound that can prevent or reduce a wide range of diseases, such as cancer (Asensi et al. 2002; Jang et al. 1997), cardiovascular diseases, and ischemic damage (Bradamante et al. 2004) as well as increase the resistance to stress and prolong the lifespan of various organisms from yeast (Howitz et al. 2003) to vertebrates (Baur et al. 2006; Valenzano et al. 2006). The biological activities mentioned above have been detailed in several publications, including some reviews in which many of the main mechanisms of action of this stilbene, including inhibition of ornithine decarboxylase and cyclo-oxygenases; inhibition of angiogenesis; cell cycle alteration; cell death promotion; free radical scavenging capacity that prevents lipid peroxidation; inhibition of platelet aggregation; vasodilatation; estrogenicity/antiestrogenicity; antibacterial, antiviral, and antihelminthic; increase of the cognitive capacity; sirtuin activation; neuroprotection, etc. (Bau and Sinclair 2006; Delmas et al. 2005; Espin et al. 2007; Signorelli and Ghidoni 2005) have been described.

　　Resveratrol-containing nutraceuticals are often prepared from *Vitis vinifera* extracts or grape pomace extracts. Resveratrol is a phytoalexin, and therefore, the normal levels of this compound in grapes or derived products (i.e., wine) are very low and largely variable. Many of the current resveratrol-containing nutraceuticals are enriched in this compound by adding purified resveratrol extracted from the root of the Japanese knotweed *Polygonum cuspidatum*. Resveratrol supplements are mainly sold as capsules or pills with different contents of the compound from a few milligrams to 500 mg per capsule (Espin et al. 2007).

17.2.8　Isoflavones

Isoflavones are flavonoids belonging to the so-called phytoestrogens and are one of the most investigated polyphenols so far. They have the potential for preventing and treating cancer and osteoporosis and in the amelioration of postmenopausal symptoms, such as hot flashes and osteoporosis (Espin et al. 2007; Howes et al. 2006; Ikeda et al. 2006; Williamson-Hughes et al. 2006). Other important biological activities are related to effects on cardiovascular diseases, cognitive function, and breast and prostate cancer (Lee et al. 2005; Verheus et al. 2007). Soybeans are an unusually concentrated source of isoflavones, including genistein and daidzein, and soy is the major source of dietary isoflavones. The isoflavones of soy have received considerable attention owing to their binding to the estrogen receptor class of compounds, thus representing an activity of a number of phytochemicals termed phytoestrogens. Genistein inhibits the growth of most hormone-dependent and independent cancer cells in vitro, including colonic cancer cells. Isoflavone-based nutraceuticals are one of the most widely tested polyphenol supplements so far, and there are many isoflavone-based supplements commercially available. They are mostly prepared from fermented or unfermented concentrated soybean extracts or red clover extracts and are sold in different forms: pills, tablets, extracts, etc. (Espin et al. 2007).

17.2.9　Ellagic Acid (EA) and Ellagitannins (ETs)

Pomegranate juice is currently recognized as one of the most powerful in vitro antioxidant foods. This remarkable activity has been associated to ETs, such as punicalagin, that are characteristic of this fruit (Gil et al. 2000). The antioxidant activity and the punicalagin content have been suggested as the possible mediators of the different health effects reported for pomegranate juice. These effects include protection against cardiovascular diseases (decrease in atherosclerosis risk factors, such as hypertension, platelet aggregation, oxidative stress, and blood lipid profiles) (Aviram et al. 2000, 2002, 2004; Rosenblat et al. 2006) and cancer prevention (Pantuck et al. 2006). The main natural sources of ETs and EA supplements are berry extracts (mainly red raspberry). Berry extracts can be commercialized either as a source of anthocyanins or as a source of EA, depending on the extraction procedure. Another important source of EA supplements is pomegranate extract. Most commercially available nutraceuticals contain a mixture of compounds because they are usually prepared from raw extracts from different food products (Espin et al. 2007).

17.2.10 Phytoestrogens

Phytoestrogens are nonsteroidal phytochemicals quite similar in structure and function to gonadal estrogen hormone. On the basis of chemical structure, phytoestrogens can be classified as flavonoids, isoflavonoids, coumestans, stilbenes, and lignans. They occur in either plants or their seeds. Soybean is rich in isoflavones whereas the soy sprout is a potent source of coumestrol, the major coumestan. They offer an alternative therapy for hormone replacement therapy (HRT) with beneficial effects on the cardiovascular system and may even alleviate menopausal symptoms. They are potential alternatives to the synthetic selective estrogen receptor modulators (SERMs), which are currently applied in HRT. They have antioxidant effects due to their polyphenolic nature, are anticarcinogenic, can modulate steroid metabolism or of detoxification enzymes, can interfere with calcium-transport, and have favorable effects on lipid and lipoprotein profiles (Prakash and Gupta 2011; Sakamoto et al. 2010).

17.2.11 Terpenoids

The terpenes, also known as isoprenoids, are the largest class of phytonutrients in green foods and grains. They have a unique antioxidant activity in their interaction with free radicals. They react with free radicals by partitioning themselves into fatty membranes by virtue of their long carbon side chain. The most studied terpene antioxidants are the tocotrienols and tocopherols found naturally in whole grains, and they have effects on cancer cells. The tocotrienols are effective apoptotic inducers for human breast cancer cells. The impact of a diet of fruits, vegetables, and grains on reduction of cancer risk may be explained by the actions of terpenes in vivo (Nichenametla et al. 2006; Prakash et al. 2004; Prakash and Kumar 2011; Stahl 2005).

17.2.12 Carotenoids

Carotenoids are a group of lipophilic molecules including the major components β-carotene, lycopene, lutein, and zeaxanthin. Although β-carotene, a provitamin A, accumulates in the skin providing a golden yellow color, lutein and zeaxanthin accumulate preferentially in the macula lutea, where they protect the retina against oxidative damage from UV light. Carotenoids, present in fruits and vegetables, are highly pigmented (yellow, orange, and red), and they are incorporated into the yolk of eggs when consumed by birds. Carotenoids comprise two types of molecules, carotenes and xanthophylls. Carotenes are tissue-specific in their biological activity, and β-carotene has vitamin A activity. β-carotene, lycopene, and lutein, protect against uterine, prostate, breast, colorectal, and lung cancers. They may also protect against the risk of digestive tract cancer. The xanthophyll types of carotenoids offer protection to other antioxidants, and they may exhibit tissue-specific protection. Zeaxanthin, cryoptoxanthin, and astaxanthin are members of the xanthophyll group (Cartea and Velasco 2008; Prakash et al. 2004; Prakash and Kumar 2011; Sharma et al. 2003). Carotenoids are globally promoted as an added value to food products for maintaining good health, and it is predicted that the carotenoid market worldwide will reach US$1.2 billion in 2015 (http://www.nutraceuticalsworld.com /contents/view_breakingnews/2010-10-28/carotenoids-market-to-exceed-1-billion-by-2015). The food industry has been investing in the extraction and concentration of carotenoids from natural sources, and several patents are available worldwide on the subject (Riggi 2010). There are more than 700 identified carotenoids, largely distributed in the animal and vegetable kingdoms; the ones that are important to human health are α-carotene, β-carotene, β-cryptoxanthin, lycopene, lutein, and zeaxanthin (Rodriguez-Amaya et al. 2008). Among the identified carotenes, the most important to the human body is β-carotene, which is the main precursor of vitamin A (Albuquerque et al. 2003). Other carotenes (α-carotene, γ-carotene, and β-cryptoxanthin) also have provitamin A activity although it is reduced to approximately 50% as compared to the β-carotene activity (Rodriguez-Amaya et al. 2008). Xanthophylls comprise cryptoxanthin, lutein, and zeaxanthin. Lutein and zeaxanthin have been found to prevent macular degeneration of the human retina and to act in the reduction of cataract risks (Rodriguez-Amaya et al. 2008).

Carotenoids from fruits, vegetables, and animal products are usually fat-soluble and are associated with lipid fractions (Su et al. 2002). Carotenoids have been extracted and enriched mainly from carrots, tomato waste products, and also from other vegetable sources.

17.2.13 Limonoids

Limonoids are terpenes present in citrus fruit. They provide chemotherapeutic activity by inhibiting Phase I enzymes (cytochrome P450 enzymes) and inducing Phase II detoxification enzymes (UDP-glucuronlytransferases [UGTs], glutathione-S-transferases [GSTs], sulfotransferases [SULTs]) involved in the metabolic detoxification process in the liver. D-Limonene, the most common monocyclic monoterpene, found within orange peel oil, inhibits pancreatic carcinogenesis induced in experimental models and also provides protection to lung tissue (Cartea and Velasco 2008; Prakash et al. 2004; Prakash and Kumar 2011; Sharma et al. 2003).

17.2.14 Phytosterols

Phytosterols are another important terpene subclass. Two sterol molecules that are synthesized by plants are β-sitosterol and its glycoside. In animals, these two molecules exhibit anti-inflammatory, antineoplastic, antipyretic, and immunomodulating activity. Phytosterols were reported to block inflammatory enzymes, for example, by modifying the prostaglandin pathways in a way that protected platelets. Phytosterols compete with cholesterol in the intestine for uptake and aid in the elimination of cholesterol from the body. Saturated phytosterols appear to be more effective than unsaturated compounds in decreasing cholesterol concentrations in the body. Their actions reduce serum or plasma total cholesterol and low-density lipoprotein (LDL) cholesterol. Competition with cholesterol for absorption from the intestine is not unexpected as the structure of plant sterols is similar to that of cholesterol (Dillard and German 2000; Prakash and Kumar 2011).

17.2.15 Glucosinolates

Glucosinolates are present in cruciferous vegetables and are activators of liver detoxification enzymes. Consumption of cruciferous vegetables offers a phytochemical strategy for providing protection against carcinogenesis, mutagenesis, and other forms of toxicity of electrophiles and reactive forms of oxygen. The sprouts of certain crucifers, including broccoli and cauliflower, contain higher amounts of glucoraphanin (glucosinolate of sulforaphane) than do the corresponding mature plants. Crucifer sprouts may protect against the risk of cancer more effectively than the same quantity of mature vegetables of the same variety (Al-Azzawi and Wahab 2010; Cartea and Velasco 2008; Prakash et al. 2007; Traka and Mithen 2009).

17.2.16 Polysaccharides

Oligo- or monosaccharides have nutraceutical properties such that they are used in food as fat replacers, dietary fibers, or prebiotics. Their physiological activities include prevention of cancers, obesity, and cardiovascular diseases; inhibition of tumor growth; neutralization of the side effects of chemotherapy and radiation; normalizing blood pressure; aiding in balancing blood sugar; combat of autoimmune disease, acting as an anti-inflammatory, lowering cholesterol and blood lipids; improving liver function; and increasing calcium absorption. Many mushrooms; fruits; vegetables; legumes; and certain cereals and grains, such as oats and barley, have been reported to contain beneficial polysaccharides. These polysaccharides are used for deriving oligo- or monosaccharides through controlled enzymatic hydrolysis or physical methods (Choi and Choi 2012; Lee et al. 2003; Ooi and Liu 2000; Sulkowska-Ziaja et al. 2011, Warrand 2006).

17.3 Extraction of Nutraceuticals from Plants

The commonly used methods for the extraction of nutraceuticals from sources are the conventional liquid–liquid or solid–liquid extraction. Traditional solid–liquid extraction methods, such as Soxhlet extraction, which have been used for many decades, are very time-consuming and require relatively large quantities of solvents (Luque de Castro and Garcia-Ayuso 1998). The large amount of solvent used not only increases operating costs but also causes additional environmental problems.

Various novel and advanced techniques, including ultrasound-assisted extraction (Vinatoru 2001), microwave-assisted extraction (Kaufmann and Christen 2002), supercritical fluid extraction (Lang and Wai 2001; Marr and Gamse 2000; Meireles and Angela 2003), and accelerated solvent extraction (Kaufmann and Christen 2002; Smith 2002), which are fast and efficient for extracting chemicals from solid plant matrices have been developed for the extraction of nutraceuticals from plants in order to shorten the extraction time, decrease the solvent consumption, increase the extraction yield, and enhance the quality of extracts. These techniques have the possibility of working at elevated temperatures and/or pressures in addition to significantly reducing the time of extraction (Wang and Weller 2006). However, novel extraction techniques have only been found in a limited field of applications.

Gas chromatography (GC) and high-performance liquid chromatography (HPLC) are the most widespread separation tools because of their versatility and easy operation, but capillary electrophoresis (CE) has also proven to be effective. These techniques are typically used in a combined way with mass spectrometry (MS), nuclear magnetic resonance spectroscopy (NMR), Fourier-transform infrared (FTIR) spectroscopy, and other detection methods. Size polarity, thermal stability, and other properties of the samples will affect the choice of the separation technique (GC, HPLC, CE) and the type of detector employed (MS, NMR, FTIR, flame ionization, fluorescence). Further, an important aspect with respect to nutraceuticals is the need for appropriate, fast, and cost-effective extraction processes able to isolate the compounds of interest. In this regard, advanced extraction techniques, such as pressurized liquid extraction and QuEChERS (quick, easy, cheap, effective, rugged, and safe), have demonstrated superior performance over conventional liquid–liquid methods (Mondello 2013).

There is a need to evolve economically and ecologically feasible extraction technologies for deriving nutraceuticals from plant materials. As a consequence, enzyme- and instant controlled pressure drop-assisted extractions were tried as previous steps during extractions, which help to release the compounds from the matrix. Enzymes have been used particularly for the treatment of plant material prior to conventional methods of extraction. Studies indicated that the exploitation of enzymes in the industry for extracting plant bioactives for their application in food is a promising field. Consequently, enzyme-based extractions have become the subject of research and development of novel ecofriendly technologies. Enzyme-based extraction techniques have the potential to be commercially attractive because there exists a market for ecofriendly extraction methods for production of a variety of bioactives. The enzyme-assisted extraction of natural compounds can save processing time and energy in addition to providing a more potential and reproducible extraction process at the commercial scale (Puri et al. 2012). These technologies could lead to augmentation of an innovative approach in the coming years to increase the production of specific compounds for use as nutraceuticals or as ingredients in the design of functional foods.

17.4 Enzyme-Assisted Extraction Processes

Enzymic processing for complete extraction of nutraceutical chemicals without the use of solvents is an alternative and attractive option in developing ecofriendly and cost-effective technology. Pretreatment of raw material employing enzymes results in the reduction of extraction time, minimizes the usage of solvents, and leads to enhancement in yield and quality of product (Meyer 2010; Sowbhagya and Chitra 2010). The schematic diagram presented in Figure 17.2 shows the role of enzymes in extraction of biomolecules of nutraceutical importance from plant resources. However, prior knowledge of the cell wall composition of the raw materials is necessary for appropriate selection of suitable enzymes. Further, minimal solvent use during extraction is important, particularly for both environmental reasons and complying with regulatory stipulations, indenting a "greener" technology option compared to traditional nonenzymatic extraction.

According to Puri et al. (2012), enzyme-assisted extraction of bioactive compounds from plants has potential commercial and technical limitations: (i) The cost of enzymes is relatively expensive for processing large volumes of raw material; (ii) currently available enzyme preparations cannot completely hydrolyze plant cell walls, limiting extraction yields of compounds, including the extraction of stevioside; and (iii) enzyme-assisted extraction can be difficult to scale up to industrial scale because enzymes behave differently as environmental conditions, such as the percentage of dissolved oxygen, temperature, and nutrient availability vary. Nevertheless, enzyme-based extraction has the potential to facilitate an increase in extraction yields and enhance product quality by enabling the use of milder processing conditions, such as lower extraction temperatures, provided the limitations are overcome through research and development efforts.

Unlike other nonthermal processes, such as high hydrostatic pressure (HP), compressed carbon dioxide (cCO_2), SC-CO_2, and high electric field pulses (HELP), enzyme-assisted extraction can readily be tested on the laboratory scale (Puri et al. 2012). The major advantage of using enzymes for extraction of

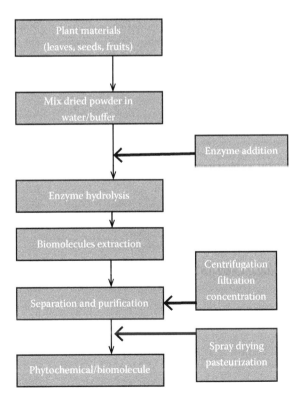

FIGURE 17.2 Enzyme-assisted isolation of biomolecules of nutraceutical importance from plant resources.

nutraceuticals and bioactive substances from natural resources is that it is possible to select and apply enzymes with specific functions and employ optimal process conditions, such as temperature and concentration, toward obtaining maximal yield. Despite the fact that enzymes normally function at an optimal temperature, they can be used over a broad range of temperatures, enabling economic process in addition to contributing to product quality. Effective enzymic digestion of plant cell walls and release of chemical substances from cells resulting in significant increase in extraction yields is achieved if the substrate particle size is reduced prior to enzymatic treatment. Further, it was noted that in enzyme-assisted aqueous extraction, the enzymes can rupture the polysaccharide–protein colloid in the cell wall, creating an emulsion that interferes with extraction, and hence, nonaqueous systems are preferable for some materials because they minimize the formation of polysaccharide–protein colloid emulsions (Concha et al. 2004).

The successful application of enzymes for the extraction of a variety of products, including the extraction of carotenoids from marigold flower or tomato skin (Barzana et al. 2002; Dehghan-Shoar et al. 2011), vanillin from vanilla green pods (Ruiz-Teran et al. 2001), polysaccharide from sterculia (Wu et al. 2005), and oil from grapeseed (Passos et al. 2009) and polyphenols (Yang et al. 2010), indicates that enzymes can also be useful for the extraction of bioactive compounds from other plant sources. Use of an enzyme system at 0.1% w/w in the processing of pectic polysaccharide for

TABLE 17.5

Extraction of Bioactive Components from Various Plant Sources Employing Enzymes

Bioactive Compounds	Source	Enzymes Used	References
Oil	Grapeseed	Cellulase, protease, xylase, and pectinase	deMaria et al. 2007
Carotenoids	Marigold flower	Viscozyme, pectinex, neutrase, corolase, and HT-proteolytic	Wu et al. 2005
Volatile oil	Mandarin peel	Xylan-degrading enzymes	Mishra et al. 2005
Carotene	Carrot pomace	Pectinex Ultra SP-L	Stoll et al. 2003
Lycopene	Tomato	Pancreatin	Dehghan-Shoar et al. 2011
	Tomato	Cellulase and pectinase	Choudhari and Ananthanarayan 2007
Capsaicin	Chile	Cellulase, hemicellulase, and pectinase	Sampathu et al. 2006
Colorant	Pitaya	Pectinolytic, hemicellulolytic, and cellulolytic enzymes	Schweiggert et al. 2009
Anthocyanin	Grape skin	Pectinex BE3-L	Muñoz et al. 2004
Sugar	Grapefruit peel waste	Cellulase and pectinase	Wilkins et al. 2007
Oligosaccharide	Rice bran	Cellulase	Patindol et al. 2007
Inulin	Jerusalem artichoke	Inulinase	Singh et al. 2006
Starch	Cassava	Pectinase enzyme	Dzogbefia et al. 2008
Pectin	Pumpkin	Xylanase, cellulase, β-glucosidase, endopolygalacturonase, and pectinesterase	Ptichkina et al. 2008
Vanillin	Vanilla green pods	β-glucosidase and pectinase	Ruiz-Teran et al. 2001
Flavonoid (naringin)	Kinnow peel	Recombinant rhamnosidase	Puri et al. 2011
Phenolics	Citrus peel	Celluzyme MX	Li et al. 2006
Proteins	Lentils and white beans	Glucoamylases	Bildstein et al. 2008
Polyphenols	Grape pomace	Pectinolytic	Kammerer et al. 2005
Catechins	Tea beverage	Pepsin	Ferruzzi and Green 2006
Lignans	Flax	Cellulase and glycosidase	Renouard et al. 2010
Soluble fiber	Carrot pomace	Cellulase-rich crude preparation	Yoon et al. 2005

enhancing extraction of an antioxidant led to increased extraction from 1.7 to 7.4 g/kg of raw material dry weight (Gan and Latiff 2010). An increase of phenolic compounds (25.90%–39.72%) and sugars (12–14 g/L) have been observed after enzyme-assisted extraction from citrus peel and grape pomace (Kammerer et al. 2005; Li et al. 2006). Defatted grapeseed meal is high in phenolic antioxidants, and enzyme-assisted oil extraction gave a 59.4% yield improvement when compared with a nonenzymatic oil extraction process (Tobar et al. 2005). It was also observed that enzymes can be used to disrupt the pectin–cellulose complex in citrus peel and enhance flavonoid (naringin) production (Puri et al. 2011).

However, there is a need to optimize the process parameters pH, time, temperature, and enzyme concentration for each specific process that influences enzyme-assisted release of bioactive substances (Puri et al. 2012). Enzyme-assisted extraction methodology for the extraction of some bioactive components from various plant sources is summarized in Table 17.5.

17.5 Enzymes Used in Extraction of Nutraceuticals

Various enzymes are employed for the extraction of nutraceuticals and functional food molecules from natural sources. Among the various enzymes, polysaccharases, particularly cellulases, pectinases, and hemicellulase, are employed to disrupt the structural integrity of the plant cell wall and enhance the extraction of desired bioactive molecules and nutraceutical substances from plants. These enzymes hydrolyze cell wall components and increase cell wall permeability, which consequently results in higher extraction yields of molecules. In fact, enzyme-assisted extraction of natural functional compounds from plants has been widely investigated in recent years for its advantages in easy operation, high efficiency, and ecofriendly outcome. Most of the studies employed cellulase, pectinase, and β-glucosidase to hydrolyze and degrade plant cell wall constituents to improve the release of intracellular contents (Barzana et al. 2002; Lavecchia and Zuorro 2008; Salgado-Roman et al. 2008; Sowbhagya and Chitra 2010).

17.6 Case Studies

17.6.1 Phenolic Compounds

Commercial pectinolytic enzymes decreased particle size from 500–1000 mm to <125 mm and increased the phenolic yield from 1.6- to fivefold in pomace (Landbo and Meyer 2001). The effect of *Thermobifida fusca* cellulase on apple peel produced an improvement in the yield of phenols and reduced sugar production and antioxidant capacity. Approximately 60 mg of reducing sugar equivalent was produced per gram of apple peel when treated with cellulase enzyme compared with only 20 mg using a nonenzymatic extraction method (Kim et al. 2005).

Candida antarctica lipase was used for lipophilizing phenolic extracts from *Sorbus aucuparia* (L.) using octadecanol as the alkyl donor. The lipophilized extract was then assessed for its antioxidant activity in refined, bleached, and deodorized rapeseed oil during accelerated storage at 65°C and frying at 180°C. It was observed that rapeseed oil fortified with lipophilized extract recorded significantly higher storage stability with a 43% decrease in peroxide value after 7 days of storage compared to rapeseed control without extract. Further, the lipophilized extract exhibited markedly better protection against thermo-oxidative degradation during frying indicating that lipophilization of phenolic compounds enhanced their protective activity in frying oil besides improving the functionality of the prepared food (Aladedunye et al. 2015).

17.6.2 Flavanoids

Enzymes have been used to increase flavonoid release from plant material while minimizing the use of solvents and heat (Kaur et al. 2010). The solubility of the target compound is a matter of concern during extraction with solvents because low solubility in the extractant leads to low extraction yield and requires a large amount of solvents, which largely impedes the economic efficiency in the industry. In this context, the enzyme-assisted extraction approach contributes to improvement in the extraction not only from enhanced cell wall degradation, but also from the increased solubility of target compounds in the extractant. Plant flavonoids generally have poor solubility in mild solvents, such as ethanol water solution (Chen et al. 2011). *Ginkgo biloba*, a Chinese medicinal plant well known for its high content of flavonoids (24%) and small amount of terpene lactones (6%), contains three predominant flavonol aglycones, quercetin, kaempferol, and isorhamnetin (Ahlemeyer and Krieglstein 2003; Gray et al. 2007). Extract from *G. biloba* is among the most popular phytomedicines and herbal dietary supplements (Tang et al. 2007). A commercial cellulase improved extraction of flavonoids from *G. biloba* leaves. Enzymes from *Penicillium decumbens* resulted in far better degradation of powdered dried leaves than *Trichoderma reesei* cellulase and *Aspergillus niger* pectinase. The extraction yield under optimized conditions was 28.3 mg/g dry weight, which was 102% higher than extraction without enzymes. The *P. decumbens* cellulase was found to improve extraction of the three predominant flavonoids from ginkgo leaves through flavonol transglycosylation in *G. biloba*. The enzymatic bifunctionality was utilized for cell degradation and target product transformation in the extraction of natural compounds from plants. With *P. decumbens* cellulase, even the cell wall degradation activity was seen at the same level as with pectinase, and there was another 0.7-fold increase in the extraction yield due to flavonol transglycosylation. Further, it was also observed that the flavonol transglycosylation rate was about 50% under the optimal conditions, indicating that when half of the flavonol aglycones were enzymatically glycosylated into more soluble glycosides, the extraction yield was increased by at least 70% (Chen et al. 2011). The *P. decumbens* cellulase was demonstrated to be a highly efficient enzyme to assist the extraction of flavonoid compounds in mild solvents, such as ethanol–water, because its high activity not only facilitated cell wall degradation, but also enabled transglycosylation of flavonol aglycones into more polar glycosides, which have higher solubility in the extractant. This study indicated that enzyme-assisted extraction of plant secondary metabolites could be an alternative approach for enzyme-facilitated cell wall degradation along with targeting enzymatic modification of the properties of target compounds. However, it is imperative to find an enzyme that could modify the properties, such as solubility, of target compounds for more efficient extraction without damaging their bioactivity. In this context, use of bifunctional enzymes or mixed enzyme preparations could help in achieving even higher extraction performance (Chen et al. 2011). Although the enzyme-assisted approach has largely improved the extraction rate, due to the intrinsic low polarity of flavonol aglycones, the extraction after enzymatic treatment still has to be performed with solvents, such as hexane, acetone, and butanone, in order to achieve high productivity (Strati and Oreopoulou 2011).

17.6.3 Lycopene

Consumption of fresh and processed tomato products is reported to be associated with a reduced risk of some types of cancer (Ellinger et al. 2006) and a lower incidence of ischemic heart disease (Willcox et al. 2003), primarily attributed to the antioxidant activity of tomato-based products because increased oxidative stress is a common pathogenetic factor for the above mentioned diseases (Elahi et al. 2009; Visconti and Grieco 2009). Tomatoes contain a wide variety of antioxidants, including vitamin E, ascorbic acid, carotenoids, flavonoids, and phenolics (Borguini and Torres 2009). In addition to being a powerful antioxidant, lycopene, the carotenoid responsible for the deep red color of ripe tomatoes, is believed to have important biological properties, including induction of apoptosis, inhibition of cell proliferation, and increase in gap-junctional communication, and consequently play a role in disease prevention (Rao et al. 2006; Seren et al. 2008; van Breeman and Pajkovic 2008).

Tomato pomace (tomato peels and seeds), the solid residue remaining after the industrial processing of tomatoes (Del Valle et al. 2006), constitutes about 3%–5% of the total weight of processed tomatoes, which, according to the estimates provided by the World Processing Tomato Council (WPTC 2010), leads to annual production volumes of more than 1,200,000 tons worldwide. At present, considerable amounts of tomato pomace are produced in the world and disposed of as solid waste or offsite. Tomato pomace is a very rich source of lycopene, which progressively accumulates in the outer skin layer of tomatoes during the ripening process (Zuorroa et al. 2011). At the end of ripening, lycopene levels in the skin are about five times higher than in the pulp (Sharma and Maguer 1996). Hence, the exploitation of this material as a source of natural lycopene could not only provide economic benefits, but also could contribute to environmental protection. Despite these advantages, lycopene recovery from this material is not straightforward as revealed by the low yields achievable with conventional solvent extraction procedures (Strati and Oreopoulou 2011). Hence, cell wall-degrading enzymes were used as a means for improving lycopene extraction from tomato skins. Studies revealed that the release of lycopene from cellulase- or pectinase-treated tomato skins was two- or threefold higher, respectively, compared to that from the untreated material. Further, enzyme-aided extraction of lycopene from tomato tissues using cellulases and pectinases under optimized conditions resulted in a significant increase (206%) in lycopene yield versus control experiments (Choudhari and Ananthanarayan 2007). Similarly, lycopene-assisted pancreatin digestion of tomato skin provided a 2.5-fold increase in yield. A digestion step prior to extraction by solvents was necessary to efficiently extract lycopene from the raw material (Dehghan-Shoar et al. 2011).

Further, it was also inferred that mixed hydrolytic enzyme preparations are more appropriate than individual enzymes because commercial food-grade enzyme preparations with pectinolytic, cellulolytic, and hemicellulolytic activities were capable of significantly enhancing lycopene recovery from tomato skins (Lavecchia and Zuorro 2008) and tomato paste (Zuorro and Lavecchia 2010). The observed enhancement in lycopene extractability was attributed to the fact that the tomato peel tissue is rich in cellulose, hemicellulose, and pectin (Gross 1984) and that the preparations used have cellulolytic, hemicellulolytic, and pectinolytic activities, which could have facilitated cell-wall polysaccharide degradation and increased solvent penetration and lycopene dissolution.

17.6.4 Anthocyanins

Enzymes facilitate increase in the yield of extraction of polyphenols and anthocyanins from black currant juice. Pectinase and protease were employed for the extraction of anthocyanins from black currant juice press residues (Landbo and Meyer 2001). More anthocyanin was extracted during the vinification process when enzymes were applied to grape skins (Muñoz et al. 2004). In another study, extraction of anthocyanins from berries was studied, and the enzyme was found to effectively hydrolyze glycosides to their corresponding aglycones (Puupponen-Pimia et al. 2008).

17.6.5 Response Surface Methodology (RSM) for Optimization of Process Parameters for Extraction

The traditional one-factor-at-a-time approach to the process of enzyme-assisted extraction optimization is time-consuming and can ignore the interactions among various factors. Response surface methodology (RSM) (Myers and Montgomery 2002), a powerful tool for optimization and control purposes, enables evaluation of several process parameters, such as time, temperature, enzyme type, and concentration. It has been successfully applied for developing, improving, and optimizing biochemical and biotechnological processes related to food systems, including production of pectic polysaccharide (Wu et al. 2005), enzymes (Puri et al. 2010), and phenolic antioxidants from fruits (Pompeu et al. 2009).

RSM was employed for determining the optimal extraction conditions and the characteristics of the product obtained by the use of an enzyme preparation with multiple hydrolytic activities. Tomato skins were pretreated by a food-grade enzyme preparation with pectinolytic and cellulolytic activities and then subjected to hexane extraction. It was found that the extraction conditions T = 30°C, extraction time = 3.18 h, and enzyme load = 0.16 kg/kg were observed as optimal. Further, pretreatment of tomato skins by the 50:50 Pec–Cel preparation resulted in an eight- to 18-fold increase in extraction yields. These facts, together with the comparatively low cost of commercial food-grade enzyme preparations, lend strong support to the possible implementation of the process on an industrial scale. These results strongly advocated the use of cell wall-degrading enzymes as an effective means for recovering lycopene from tomato waste (Zuorro et al. 2011).

17.7 Conclusion and Future Perspectives

Among the emerging food processing industries, the nutraceutical industry is witnessing rapid growth at a rate far exceeding the expansion in the food and pharmaceutical industries owing to the consumer demand for health food. The expanding nutraceutical market clearly indicates that end users are seeking minimally processed food with extra nutritional benefits and organoleptic value. This development, in turn, is reflected in terms of expansion in the global nutraceutical markets. Within the stringent environmental regulations and deep concern for environmental protection from excess use of chemicals in food processing, there is a growing concern and need for ecofriendly and cost-effective processing technologies. In this context, exploitation of enzymes in industry for extracting plant bioactives for their application in food is a promising field. As has been shown in this chapter, enzyme-assisted extraction of biological molecules of nutraceutical importance from various plant sources has already marked its beginning, and it has been noted that it can save processing time and energy and potentially provide a more reproducible extraction process at the commercial scale. However, intensive research and development activities are warranted in the future to expand the currently available enzymatic processes, in particular to further enhance the yields of bioactive compounds. In fact, food and pharmaceutical scientists speculate that enzyme applications represent another exciting frontier in nutraceutical development and processing. Enzymes have been underemployed, and hence, they have to be effectively harnessed in the coming future. Further, the effect of enzyme processing on the biological availability and effectiveness of nutraceuticals also has to be evaluated to ensure effectiveness of the enzyme technology in nutraceutical production.

REFERENCES

Ahlemeyer, B., Krieglstein, J. 2003. Neuroprotective effects of *Ginkgo biloba* extract. *Cell. Mol. Life Sci.* 60, 1779–1792.

Aladedunye, F., Niehaus, K., Bednarz, H., Thiyam-Hollander, U., Fehling, E., Matthaus, B. 2015. Enzymatic lipophilization of phenolic extract from rowanberry (*Sorbus aucuparia*) and evaluation of antioxidative activity in edible oil. *LWT—Food Sci. Technol.* 60, 56–62.

Al-Azzawi, F., Wahab, M. 2010. Effectiveness of phytoestrogens in climacteric medicine. *Ann. N.Y. Acad. Sci.* 1205, 262–267.

Albuquerque, M. L. S., Guedes, I., Alcantara, Jr., P., Moreira, S. G. C. 2003. Infrared absorption spectra of Buriti (*Mauritia flexuosa* L.) oil. *Vib. Spectrosc.* 33, 127–131.

Ames, B. N., Mark, K. S., Tory, M. H. 1993. Oxidants, antioxidants and the degenerative disease of aging. *Proc. Natl. Acad. U.S.A.* 90, 7915–7922.

Anagnostopoulou, M. A., Kefalas, P., Papageorgiou, V. P., Assimopoulou, A. N., Boskou, D. 2006. Radical scavenging activity of various extracts and fractions of sweet orange peel (*Citrus sinensis*). *Food Chem.* 94, 19–25.

Andjelković, M., Camp, J. V., Meulenaer, B. D., Depaemelaere, G., Socaciu, C., Verloo, M., Verhe, R. 2006. Iron-chelation properties of phenolic acids bearing catechol and galloyl groups. *Food Chem.* 98, 23–31.

Arts, I., Hollman, P. 2005. Polyphenols and disease risk in epidemiologic studies. *Am. J. Clin. Nutr.* 81, 317S–325S.

Asensi, M., Medina, I., Ortega, A., Carretero, J., Bano, M. C., Obrador, E., Estrela, J. M. 2002. Inhibition of cancer growth by resveratrol is related to its low bioavailability. *Free Radic. Biol. Med.* 33, 387–398.

Aviram, M., Dornfeld, L., Rosenblat, M., Volkova, N., Kaplan, M., Colemann, R., Hayek, T., Presser, D., Fuhrman, B. 2000. Pomegranate juice consumption reduces oxidative stress, atherogenic modifications to LDL, and platelet aggregation: Studies in humans and in atherosclerotic apolipoprotein E-deficient mice. *Am. J. Clin. Nutr.* 71, 1062–1076.

Aviram, M., Dornfeld, L., Kaplan, M. et al. 2002. Pomegranate juice flavonoids inhibit low-density lipoprotein oxidation and cardiovascular diseases: Studies in atherosclerotic mice and in humans. *Drugs Exp. Clin. Res.* 28, 49–62.

Aviram, M., Rosenblat, M., Gaitini, D. et al. 2004. Pomegranate juice consumption for 3 years by patients with carotid artery stenosis reduces common carotid intima-media thickness, blood pressure and LDL oxidation. *Clin. Nutr.* 23, 423–433.

Barzana, E., Rubio, D., Santamaria, R. I., Garcia-Correa, O., Garcia, F., Ridaura Sanz, V. E., López-Munguía, A. 2002. Enzyme-mediated solvent extraction of carotenoids from marigold flower (*Tagetes erecta*). *J. Agric. Food Chem.* 50, 4491–4496.

Bau, J. A., Sinclair, D. A. 2006. Therapeutic potential of resveratrol: The in vivo evidence. *Nat. Rev. Drug Discov.* 5, 493–506.

Baur, J. A., Pearson, K. J., Price, N. L. et al. 2006. Resveratrol improves health and survival of mice on a high-calorie diet. *Nature* 444, 337–342.

Bildstein, M., Lohmann, M., Hennigs, C., Krause, A., Hilz, H. 2008. An enzyme based extraction process for the purification and enrichment of vegetable proteins to be applied in bakery products. *Eur. Food Res. Technol.* 228, 177–186.

Borguini, R. G., Torres, E. A. F. D. 2009. Tomatoes and tomato products as dietary sources of antioxidants. *Food Rev. Int.* 25, 313–325.

Bradamante, S., Barenghi, L., Villa, A. 2004. Cardiovascular protective effects of resveratrol. *Cardiovasc. Drug Rev.* 22, 169–188.

Cartea, M. E., Velasco, P. 2008. Glucosinolates in Brassica foods: Bioavailability in food and significance for human health. *Phytochem. Rev.* 7, 213–229.

Chauhan, B., Kumar, G., Kalam, N., Ansari, S. H. 2013. Current concepts and prospects of herbal nutraceutical: A review. *J. Adv. Pharm. Technol. Res.* 4, 4–8.

Chen, S., Xing, X.-H., Huang, J.-J., Xu, M.-S. 2011. Enzyme-assisted extraction of flavonoids from Ginkgo biloba leaves: Improvement effect of flavonol transglycosylation catalysed by *Penicillium decumbens* cellulase. *Enzyme Microbiol. Technol.* 48, 100–105.

Choi, B. D., Choi, Y. J. 2012. Nutraceutical functionalities of polysaccharides from marine invertebrates. *Adv. Food Nutr. Res.* 65, 11–30.

Choudhari, S. M., Ananthanarayan, L. 2007. Enzyme aided extraction of lycopene from tomato tissues. *Food Chem.* 102, 77–81.

Cieślik, E., Greda, A., Adamus, W. 2006. Contents of polyphenols in fruits and vegetables. *Food Chem.* 94, 135–142.

Concha, J., Soto, C., Chamy, R., Zúñiga, M. E. 2004. Enzymatic pretreatment on rose-hip oil extraction: Hydrolysis and pressing conditions. *J. Am. Oil Chem. Soc.* 81, 549–552.

DeFelice, S. L. 1992. The nutraceutical initiative: A recommendation for U.S. Economic and regulatory reforms. *Genet. Eng. News* 12, 13–15.

Dehghan-Shoar, Z., Hardacre, A. K., Meerdink, G., Brennan, C. S. 2011. Lycopene extraction from extruded products containing tomato skin. *Int. J. Food Sci. Technol.* 46, 365–371.

Delmas, D., Rebe, C., Micheau, O., Athias, A., Gambert, P., Grazide, S., Laurent, G., Latruffe, N., Solary, E. 2005. Resveratrol: Preventing properties against vascular alterations and ageing. *Mol. Nutr. Food Res.* 49, 377–395.

Del Valle, M., Cámara, M., Torija, M.-E. 2006. Chemical characterization of tomato pomace. *J. Sci. Food Agric.* 86, 1232–1236.

deMaria, L., Vind, J., Oxenbøll, K. M., Svendsen, A., Patkar, S. 2007. Phospholipases and their industrial applications: Mini review. *Appl. Microbiol. Biotechnol.* 74, 290–300.

Dillard, C. J., German, J. B. 2000. Phytochemicals: Nutraceuticals and human health. *J. Sci. Food Agric.* 80, 1744–1756.

Dueñas, M., Fernández, D., Hernández, T., Estrella, I., Muñoz, R. 2005. Bioactive phenolic compounds of cowpeas (*Vigna sinensis* L.). Modifications by fermentation with natural microflora and with *Lactobacillus plantarum* ATCC 14917. *J. Sci. Food Agric.* 85, 297–304.

Dzogbefia, V. P., Ofosu, G. A., Oldham, J. H. 2008. Evaluation of locally produced Saccharomyces cerevisiae pectinase enzyme for industrial extraction of starch from cassava in Ghana. *Sci. Res. Essay* 3, 365–369.

Elahi, M. M., Kong, Y. X., Matata, B. M. 2009. Oxidative stress as a mediator of cardiovascular disease. *Oxid. Med. Cell. Longev.* 2, 259–269.

Ellinger, S., Ellinger, J., Stehle, P. 2006. Tomatoes, tomato products and lycopene in the prevention and treatment of prostate cancer: Do we have the evidence from intervention studies? *Curr. Opin. Clin. Nutr. Metab. Care* 9, 722–727.

Espin, J. C., Garcia-Conesa, M. T., Tomas-Barbera, F. A. 2007. Nutraceuticals: Facts and fiction. *Phytochemistry* 68, 2986–3008.

Ferruzzi, M. G., Green, R. J. 2006. Analysis of catechins from milk–tea beverages by enzyme assisted extraction followed by high performance liquid chromatography. *Food Chem.* 99, 484–491.

Gan, C. Y., Latiff, A. A. 2010. Extraction of antioxidant pecticpolysaccharide from mangosteen (*Garcinia mangostana*) rind: Optimization using response surface methodology. *Carbohydr. Polym.* 83, 600–607.

Garg, A., Garg, S., Zaneveld, L. J., Singla, A. K. 2001. Chemistry and pharmacology of the citrus bioflavonoid hesperidin. *Phytother. Res.* 15(8), 655–669.

Gil, M. I., Tomas-Barberán, F. A., Hess-Pierce, B., Holcroft, D. M., Kader, A. A. 2000. Antioxidant activity of pomegranate juice and its relationship with phenolic composition and processing. *J. Agric. Food Chem.* 48, 4581–4589.

Gray, D., LeVanseler, K., Pan, M., Waysek, E. H. 2007. Evaluation of a method to determine flavonol aglycones in Ginkgo biloba dietary supplement crude materials and finished products by high-performance liquid chromatography: Collaborative study. *J. AOAC Int.* 90, 43–53.

Gross, K. C. 1984. Fractionation and partial characterization of cell walls from normal and non-ripening mutant tomato fruit. *Physiol. Plant* 62, 25–32.

Heiss, C., Kleinbongard, P., Dejam, A., Perre, S., Schoeter, H., Sies, H., Kelm, M. 2005. Acute consumption of flavanol-rich cocoa and the reversal of endothelial dysfunction in smokers. *J. Am. Coll. Cardiol.* 46, 1276–1283.

Heiss, C., Finis, D., Kleinbongard, P., Hoffmann, A., Rassaf, T., Kelm, M., Sies, H. 2007. Sustained increase in flow-mediated dilation after daily intake of high-flavanol cocoa drink over 1 week. *J. Cardiovasc. Pharmacol.* 49, 74–80.

Howes, L. G., Howes, J. B., Knight, D. C. 2006. Isoflavone therapy for menopausal flushes: A systematic review and meta-analysis. *Maturitas* 55, 203–211.

Howitz, K. T., Bitterman, K. J., Cohen, H. Y. et al. 2003. Small molecule activators of sirtuins extend *Saccharomyces cerevisiae* lifespan. *Nature* 425, 191–196.

Ikeda, Y., Iki, M., Morita, A., Kajita, E., Kagamimori, S., Kagawa, Y., Yoneshima, H. 2006. Intake of fermented soybeans, natto, is associated with reduced bone loss in postmenopausal women: Japanese population-based osteoporosis (JPOS) study. *J. Nutr.* 136, 1323–1328.

Jang, M. S., Cai, E. N., Udeani, G. O. et al. 1997. Cancer chemopreventive activity of resveratrol, a natural product derived from grapes. *Science* 275, 218–220.

Kammerer, D., Claus, A., Schieber, A., Carle, R. 2005. A novel process for the recovery of polyphenols from grape (*Vitis vinifera*) pomace. *J. Food Sci.* 70, 157–163.

Katalinic, V., Milos, M., Kulisic, T., Jukic, M. 2006. Screening of 70 medicinal plant extracts for antioxidant capacity and total phenols. *Food Chem.* 94, 550–557.

Kaufmann, B., Christen, P. 2002. Recent extraction techniques for natural products: Microwave-assisted extraction and pressurized solvent extraction. *Phytochem. Anal.* 13, 105–113.

Kaul, C., Kapoor, H. C. 2001. Antioxidants in fruits and vegetables—The millenniums health. *Int. J. Food Sci. Technol.* 36, 703–725.

Kaur, A., Singh, S., Singh, R. S., Schwarz, W. H., Puri, M. 2010. Hydrolysis of citrus peel naringin by recombinant a-L-rhamnosidase from *Clostridium stercorarium*. *J. Chem. Technol. Biotechnol.* 85, 1419–1422.

Kim, Y. J., Kim, D. O., Chun, O. K., Shin, D. H., Jung, H., Lee, C. Y., Wilson, D. B. 2005. Phenolic extraction from apple peel by cellulases from *Thermobifida fusca*. *J. Agric. Food Chem.* 53, 9560–9565.

Landbo, A. K., Meyer, A. S. 2001. Enzyme-assisted extraction of antioxidative phenols from black currant juice press residues (*Ribes nigrum*). *J. Agric. Food Chem.* 49, 3169–3177.

Lang, Q., Wai, C. M. 2001. Supercritical fluid extraction in herbal and natural product studies—A practical review. *Talanta* 53, 771–782.

Lavecchia, R., Zuorro, A. 2008. Improved lycopene extraction from tomato peels using cell-wall degrading enzymes. *Eur. Food Res. Technol.* 228, 153–158.

Lee, B. C., Bae, J. T., Pyo, H. B., Choe, T. B., Kim, S. W., Hwang, H. J., Yun, J. W. 2003. Biological activities of the polysaccharides produced from submerged culture of the edible Basidiomycete *Grifola frondosa*. *Enzyme Microb. Technol.* 32, 574–581.

Lee, J. H., Park, C. H., Jung, K. C., Rhee, H. S., Yang, C. H. 2005. Negative regulation of beta-catenin/Tcf signaling by naringenin in AGS gastric cancer cell. *Biochem. Biophys. Res. Commun.* 335(3), 771–776.

Li, B. B., Smitha, B., Hossain, M. M. 2006. Extraction of phenolics from citrus peels: II. Enzyme-assisted extraction method. *Sep. Purif. Technol.* 48, 189–196.

Luque de Castro, M. D., Garcia-Ayuso, L. E. 1998. Soxhlet extraction of solid materials: An outdated technique with a promising innovative future. *Anal. Chim. Acta* 369, 1–10.

Manach, C., Donovan, J. L. 2004. Pharmacokinetics and metabolism of dietary flavonoids in humans. *Free Radic. Res.* 38(8), 771–785.

Marr, R., Gamse, T. 2000. Use of supercritical fluids for different processes including new developments—A review. *Chem. Eng. Process.* 39, 19–28.

Meireles, A., Angela, M. 2003. Supercritical extraction from solid: Process design data (2001–2003). *Curr. Opin. Solid State Mater. Sci.* 7, 321–330.

Meyer, A. S. 2010. Enzyme technology for precision functional food ingredients processes. *Ann. N.Y. Acad. Sci.* 1190, 126–132.

Middleton, E., Kandaswami, C., Theoharides, T. C. 2000. The effects of plant flavonoids on mammalian cells: Implications for inflammation, heart disease, and cancer. *Pharmacol. Rev.* 52, 673–751.

Miller, N. J., Larrea, M. B. R. 2002. Flavonoids and other plant phenols in the diet: Their significance as antioxidants. *J. Nutr. Environ. Med.* 12, 39–51.

Mishra, D., Shukla, A. K., Dixit, A. K., Singh, K. 2005. Aqueous enzymatic extraction of oil from mandarin peels. *J. Oleo Sci.* 54, 355–359.

Mondello, L. 2013. Nutraceuticals and separations. *Anal. Bioanal. Chem.* 405(13), 4589–4590.

Muñoz, O., Sepulveda, M., Schwartz, M. 2004. Effects of enzymatic treatment on anthocyanic pigments from grapes skin from Chilean wine. *Food Chem.* 87, 487–490.

Myers, R. H., Montgomery, D. C., eds. 2002. *Response Surface Methodology: Process and Product Optimization Using Designed Experiments*, 3rd ed. Wiley, New York, p. 824.

Nichenametla, S. N., Taruscio, T. G., Barney, D. L., Exon, J. H. 2006. A review of the effects and mechanism of polyphenolics in cancer. *Crit. Rev. Food Sci. Nutr.* 46, 161–183.

Ooi, V. C. E., Liu, F. 2000. Immunomodulation and anti-cancer activity of polysaccharide-protein complexes. *Curr. Med. Chem.* 7, 715–729.

Packer, L., Weber, S. U. 2001. *The Role of Vitamin E in the Emerging Field of Nutraceuticals*. Marcel Dekker, New York, pp. 27–43.

Palthur, M. P., Palthur, S. S., Chitta, S. K. 2010. Nutraceuticals: A conceptual definition. *Int. J. Pharm. Pharm. Sci.* 2(3), 19–27.

Pantuck, A. J., Leppert, J. T., Zomorodian, N. et al. 2006. Phase II study of pomegranate juice for men with rising prostate-specific antigen following surgery or radiation for prostate cancer. *Clin. Cancer Res.* 12, 4018–4026.

Passos, C. P., Yilmaz, S., Silva, C. M., Coimbra, M. A. 2009. Enhancement of grape seed oil extraction using a cell wall degrading enzyme cocktail. *Food Chem.* 115, 48–53.

Patindol, J., Wang, I., Wang, Y.-J. 2007. Cellulase-assisted extraction of oligosaccharides from defatted rice bran. *J. Food Sci.* 72, C517–C521.

Pompeu, D. R., Silva, E. M., Rogez, H. 2009. Optimisation of the solvent extraction of phenolic antioxidants from fruits of Euterpeu oleracea using response surface methodology. *Bioresour. Technol.* 100, 6076–6082.

Prakash, D., Gupta, C. 2011. Role of phytoestrogens as nutraceuticals in human health. A review. *Biotechnol. Indian J.* 5(3), 1–8.

Prakash, D., Kumar, N. 2011. Cost effective natural antioxidants. In: R. R. Watson, J. K. Gerald and V. R. Preedy, eds. *Nutrients, Dietary Supplements and Nutraceuticals*. Humana Press, Springer, New York, pp. 163–188.

Prakash, D., Dhakarey, R., Mishra, A. 2004. Carotenoids: The phytochemicals of nutraceutical importance. *Ind. J. Agric. Biochem.* 17, 1–8.

Prakash, D., Singh, B. N., Upadhyay, G. 2007. Antioxidant and free radical scavenging activities of phenols from onion (*Allium cepa*). *Food Chem.* 102, 1389–1393.

Prakash, D., Gupta, C., Sharma, G. 2012. Importance of phytochemicals in nutraceuticals. *J. Chin. Med. Res. Dev. (JCMRD)* 1(3), 70–78.

Ptichkina, N. M., Markina, O. A., Rumyantseva, G. N. 2008. Pectin extraction from pumpkin with the aid of microbial enzymes. *Food Hydrocolloids* 22, 192–195.

Puri, M., Kaur, A., Singh, R. S., Singh, A. 2010. Response surface optimization of medium components for naringinase production from Staphylococcus xylosus MAK2. *Appl. Biochem. Biotechnol.* 19, 162–181.

Puri, M., Kaur, A., Schwarz, W. H., Singh, S., Kennedy, J. F. 2011. Molecular characterization and enzymatic hydrolysis of naringin extracted from kinnow peel waste. *Int. J. Biol. Macromol.* 48, 58–62.

Puri, M., Sharma, D., Barrow, C. J. 2012. Enzyme-assisted extraction of bioactives from plants. *Trends Biotechnol.* 30(1), 37–44. doi:10.1016/j.tibtech.2011.06.014.

Puupponen-Pimia, R., Nohynek, L., Ammann, S., Oksman-Caldentey, K. M., Buchert, J. 2008. Enzyme-assisted processing increases antimicrobial and antioxidant activity of bilberry. *J. Agric. Food Chem.* 56, 681–688.

Rao, A. V., Raym, M. R., Rao, L. G. 2006. Lycopene. *Adv. Food Nutr. Res.* 51, 99–164.

Renouard, S., Hanoa, C., Corbina, C. et al. 2010. Cellulase-assisted release of secoisolariciresinol from extracts of flax (*Linum usitatissimum*) hulls and whole seeds. *Food Chem.* 122, 679–687.

Riggi, E. 2010. Recent patents on the extraction of carotenoids. *Recent Pat. Food Nutr. Agric.* 2(1), 75–82.

Rodriguez-Amaya, D. B., Kimura, M., Godoy, H. T., Amaya-Farfan, J. 2008. Updated Brazilian database on food carotenoids: Factors affecting carotenoid composition. *J. Food Compos. Anal.* 21, 445–463.

Rosenblat, M., Hayek, T., Aviram, M. 2006. Anti-oxidative effects of pomegranate juice (PJ) consumption by diabetic patients on serum and on macrophages. *Atherosclerosis* 187, 363–371.

Ruiz-Teran, F., Perez-Amador, I., López-Munguia, A. 2001. Enzymatic extraction and transformation of gluco-vanillin to vanillin from vanilla green pods. *J. Agric. Food Chem.* 49, 5207–5209.

Sakamoto, T., Horiguchi, H., Oguma, E. 2010. Effects of diverse dietary phytoestrogens on cell growth, cell cycle and apoptosis in estrogen-receptor-positive breast cancer cells. *J. Nutr. Biochem.* 21, 856–864.

Salgado-Roman, M., Botello-Alvarez, E., Rico-Martínez, R., Jiménez-Islas, H., Cárdenas-Manríquez, M., Navarrete-Bolaños, J. L. 2008. Enzymatic treatment to improve extraction of capsaicinoids and carotenoids from chili (*Capsicum annuum*) fruits. *J. Agric. Food Chem.* 56, 10012–10018.

Sampathu, S. R., Naidu, M. M., Sowbhagya, H. B., Naik, J. P., Krishnamurthy, N. 2006. A process for making chilli oleoresin of improved quality. US Patent No. 7097867.

Scalbert, A., Manach, C., Morand, C., Remesy, C. 2005. Dietary polyphenols and the prevention of diseases. *Crit. Rev. Food Sci. Nutr.* 45, 287–306.

Schroeter, H., Heiss, C., Balzer, J. et al. 2006. (-)-Epicatechin mediates beneficial effects of flavanol-rich cocoa on vascular function in humans. *Proc. Natl. Acad. Sci. U.S.A.* 103, 1024–1029.

Schweiggert, R. M., Villalobos-Gutierrez, M. G., Esquivel, P., Carle, R. 2009. Development and optimization of low temperature enzyme-assisted liquefaction for the production of colouring foodstuff from purple pitaya (*Hylocereus* sp.). *Eur. Food Res. Technol.* 230, 269–280.

Seren, S., Lieberman, R., Bayraktar, U. D., Heath, E., Sahin, K., Andic, F., Kucuk, O. 2008. Lycopene in cancer prevention and treatment. *Am. J. Ther.* 15, 66–81.

Sharma, S. K., Maguer, L. M. 1996. Lycopene in tomatoes and tomato pulp fractions. *Ital. J. Food Sci.* 2, 107–113.

Sharma, G., Singh, R. P., Chan, D. C. F., Agarwal, R. 2003. Silibinin induces growth inhibition and apoptotic cell death in human lung carcinoma cells. *Anticancer Res.* 23, 2649–2655.

Signorelli, P., Ghidoni, R. 2005. Resveratrol as an anticancer nutrient: Molecular basis, open questions and promises. *J. Nutr. Biochem.* 16, 449–466.

Singh, R. S., Dhaliwal, R., Puri, M. 2006. Production of inulinase from Kluyveromyces marxianus YS-1 using root extract of *Asparagus racemosus*. *Proc. Biochem.*, 41, 1703–1707.

Singh, B. N., Singh, B. R., Singh, R. L., Prakash, D., Singh, D. P., Sharma, B. K., Upadhyay, G., Singh, H. B. 2009. Polyphenolics from various extracts/fractions of red onion (*Allium cepa*) peel with potential anti-oxidant and antimutagenic activities. *Food Chem. Toxicol.* 47, 1161–1167.

Smith, R. M. 2002. Extractions with superheated water. *J. Chromatogr. A* 975, 31–46.

Sowbhagya, H. B., Chitra, V. N. 2010. Enzyme-assisted extraction of flavorings and colorants from plant materials. *Crit. Rev. Food Sci. Nutr.* 50, 146–161.

Stahl, W. 2005. Bioactivity and protective effects of natural carotenoids. *Biochim. Biophys. Acta* 1740, 101–107.

Stalikas, C. D. 2007. Extraction, separation, and detection methods for phenolic acids and flavonoids. *J. Sep. Sci.* 30, 3268–3295.

Stoll, T., Schweiggert, U., Schieber, A., Carle, R. 2003. Process for the recovery of a carotene rich functional food ingredient from carrot pomace by enzymatic liquification. *Innov. Food Sci. Emerg. Technol.* 4, 415–423.

Strati, I. F., Oreopoulou, V. 2011. Effect of extraction parameters on the carotenoid recovery from tomato waste. *Int. J. Food Sci. Technol.* 46, 23–29.

Su, Q., Rowley, K. G., Balazs, N. D. H. 2002. Carotenoids: Separation methods applicable to biological samples. *J. Chromatogr. B.* 781, 393–418.

Sulkowska-Ziaja, K., Karczewska, E., Wojtas, I., Budak, A., Ena muszy-ska, B., Ekiert, H. 2011. Isolation and biological activities of polysaccharide fractions from mycelium of *Sarcodon imbricatus* Karst. (Basidiomycota) cultured *in vitro*. *Acta Pol. Pharm. Drug Res.* 68, 143–145.

Tang, J., Sun, J., Zhang, Y., Li, L., Cui, F., He, Z. 2007. Herb–drug interactions: Effect of Ginkgo biloba extract on the pharmacokinetics of theophylline in rats. *Food Chem. Toxicol.* 45, 2441–2445.

Tobar, P., Moure, A., Soto, C., Chamy, R., Zúñiga, M. E. 2005. Winery solid residue revalorization into oil and antioxidant with nutraceutical properties by an enzyme assisted process. *Water Sci. Technol.* 51, 47–52.

Traka, M., Mithen, R. 2009. Glucosinolates, isothiocyanates and human health. *Phytochem. Rev.* 8, 269–282.

Valenzano, D. R., Terzibasi, E., Genade, T., Cattaneo, A., Domenici, L., Cellerino, A. 2006. Resveratrol prolongs lifespan and retards the onset of age-related markers in a shortlived vertebrate. *Curr. Biol.* 16, 296–300.

van Breemen, R. B., Pajkovic, N. 2008. Multitargeted therapy of cancer by lycopene. *Cancer Lett.* 269, 339–351.

Verheus, M., van Gils, C. H., Keinan-Boker, L., Grace, P. B., Bingham, S. A., Peeters, P. H. M. 2007. Plasma phytoestrogens and subsequent breast cancer risk. *J. Clin. Oncol.* 25, 648–655.

Vinatoru, M. 2001. An overview of the ultrasonically assisted extraction of bioactive principles from herbs. *Ultrason. Sonochem.* 8(3), 303–313.

Visconti, R., Grieco, D. 2009. New insights on oxidative stress in cancer. *Curr. Opin. Drug Discov. Dev.* 12, 240–245.

Wang, L., Weller, C. L. 2006. Recent advances in extraction of nutraceuticals from plants. *Trends Food Sci. Technol.* 17, 300–312.

Wang-Polagruto, J. F., Villablanca, A. C., Polagruto, J. A. et al. 2006. Chronic consumption of flavanol-rich cocoa improves endothelial function and decreases vascular cell adhesion molecule in hype cholesterol-emic postmenopausal women. *J. Cardiovasc. Pharmacol.* 47, S177–S186.

Warrand, J. 2006. Healthy Polysaccharides. *Food Technol. Biotechnol.* 44, 355–370.

Wilkins, M. R., Widmer, W. W., Grohmann, K., Cameron, R. G. 2007. Hydrolysis of grapefruit peel waste with cellulase and pectinase enzymes. *Bioresour. Technol.* 98, 1596–1601.

Willcox, J. K., Catignani, G. L., Lazarus, S. 2003. Tomatoes and cardiovascular health. *Crit. Rev. Food Sci. Nutr.* 43, 1–18.

Willcox, J. K., Ash, S. L., Catignani, G. L. 2004. Antioxidants and prevention of chronic diseases. *Crit. Rev. Food Sci. Nutr.* 44, 275–295.

Williamson, G., Manach, C. 2005. Bioavailability and bioefficacy of polyphenols in humans. II. Review of 93 intervention studies. *Am. J. Clin. Nutr.* 81, 243S–255S.

Williamson-Hughes, P. S., Flickinger, B. D., Messina, M. J., Empie, M. W. 2006. Isoflavone supplements containing predominantly genistein reduce hot flash symptoms: A critical review of published articles. *Menopause-J. North Am. Menopause Soc.* 13, 831–839.

WPTC. 2010. World production estimate as of October 18, 2010, Release # 33. Available at http://www.wptc.to/releases/releases18.pdf (accessed January 14, 2010).

Wu, Y., Cui, S. W., Tang, J., Gu, X. 2005. Optimization of extraction process of crude polysaccharides from boat-fruited sterculia seeds by response surface methodology. *Food Chem.* 105, 1599–1605.

Yang, Y. C., Li, J., Zu, Y.-G., Fu, Y.-J., Luo, M., Wu, N., Liu, X.-L. 2010. Optimisation of microwave-assisted enzymatic extraction of corilagin and geraniin from Geranium sibiricum Linne and evaluation of antioxidant activity. *Food Chem.* 122, 373–380.

Yoon, K. Y., Cha, M., Shin, S. R., Kim, K. S. 2005. Enzymatic production of a soluble-fibre hydrolyzate from carrot pomace and its sugar composition. *Food Chem.* 92, 151–157.

Zeisel, S. H. 1999. Regulation of "nutraceuticals." *Science* 285, 185–186.

Zuorro, A., Lavecchia, R. 2010. Mild enzymatic method for the extraction of lycopene from tomato paste. *Biotechnol Biotechnol Equipment*, 24, 1854–1857.

Zuorro, A., Fidaleo, M., Lavecchi, R. 2011. Enzyme-assisted extraction of lycopene from tomato processing waste. *Enzyme Microb. Technol.* 49, 567–573.

18

Enzyme-Mediated Novel Biotransformations in Food and Beverage Production

Pedro Fernandes

CONTENTS

18.1 Introduction

The use of the catalytic action of enzymes in the production of goods for food and feed is deeply rooted in mankind. Thus, the most ancient reports on biotransformation/fermentation date back to 6000 BC, involving beer production (Buchholz and Collins 2013; Gurung et al. 2013). This close trend between enzymes and food and feed endured, ultimately including relevant milestones in enzyme technology, such as the isomerization of glucose to fructose, which, combined with immobilization and merged in a cascade of amylase-catalyzed reactions, led to the production of high fructose corn syrup, one of the largest volume operations solely involving enzymes, or the production of enantiomerically pure amino acids for use as food additives or as building blocks for the production of sweeteners (Bornscheuer et al. 2012; Liese et al. 2006; Wenda et al. 2011). These are some of the well-known applications of enzymes in the food and feed industry, part of a universe that corresponds to the largest fraction of industrial enzymes. Thus, a recent review states that, on its own, the revenues of the enzyme market for food and beverage in 2011 were $1.2 billion with a prospective value of $1.8 billion by 2016 (Adrio and Demain 2014). On top of this, the market for feed enzymes presented revenue of $780 million by 2012 and is expected to reach $1.2 billion by 2018 (Anonymous 2014). It is therefore understandable that the use of enzymes for practical uses increases, and different methodologies based on their use are constantly being designed and tested. The present work aims to present some of these, based on application in given goods and applications within food and feed.

18.2 Sugars and Sweeteners

18.2.1 Stevioside Derivatives

A particular area of the food industry in which enzyme activity is of particular interest is in the production of sugars and sweeteners. There is a particular focus on sugars and sweeteners with high sweetening power and low caloric value (Purkayastha 2014). Among these is stevioside, the most abundant steviol glycoside from stevia, with a sweetening power 300-fold that of sucrose but also possessing some bitterness and undesirable aftertaste. These latter features can, however, be overcome through glycosylation of stevioside (Figure 18.1).

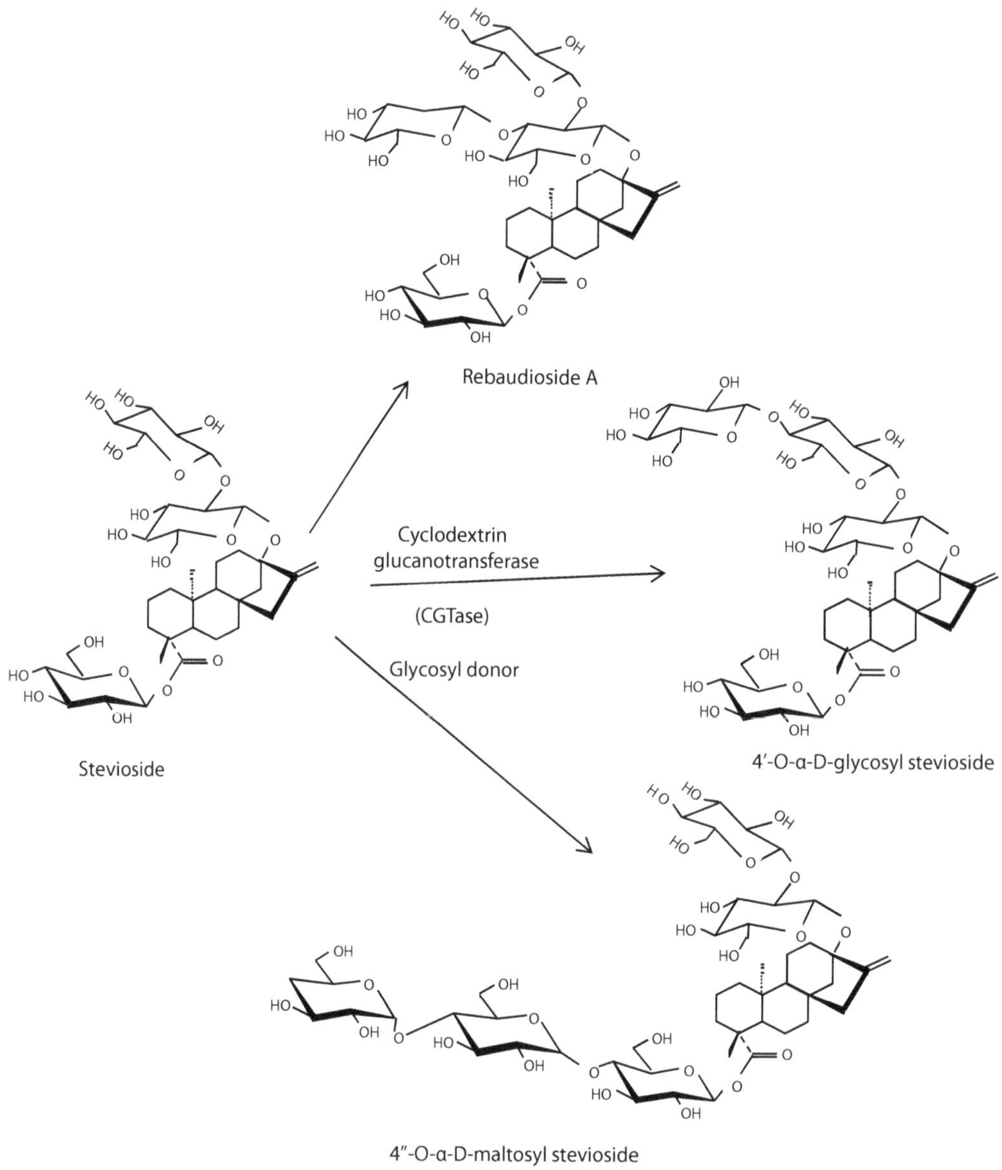

FIGURE 18.1   Some representative examples of stevioside derivatives obtained through cyclodextrin glucanotransferase action.

Actually, rebaudioside A, the second most prominent steviol glycoside, is actually a β-1,3-monoglucosyl stevioside, with which the unwanted organoleptic features are mitigated and the sweetening power is enhanced up to 1.5-fold (Chatsudthipong and Muanprasat 2009). Moreover, it has been typically accepted that an increase in sweetening power is favored by mono- and diglycosylation at the 13-O-β-sophorosyl moiety of stevioside. On the other hand, glycosylation at the 19-O-glucosyl moiety is unwanted because it leads to a decrease in intensity of sweetness (Tanaka 1997). Thus, the use of enzymatic formulation with fitting glycosylation activity, together with an adequate glucose donor, should provide an adequate setup for the production of a high-intensity sweetener with minimal bitterness and astringency of after-taste. Based on this concept, cyclodextrin glucanotransferases (CGTase) isolated from *Bacillus stearo-thermophilus* B-5076 and *B. macerans* BIO-4m were effectively used for the transglycosylation of a 1:1 (w/w) mixture of stevioside and rebaudioside, using dextrose equivalent of 0.15–0.3 thinned starch as the donor. Incubation of a 40% solution of glycosides and starch (1:1, w/w) at 55°C to 60°C for 48 h led to a mixture of seven high molecular weight derivatives of the steviol glycosides, displaying a sweetness intensity 170-fold higher than that of sucrose. This was further enhanced by incubating the reaction products in the presence of an α-amylase so that mono- and diglycosylated forms were obtained (Kochikyan et al. 2006). The concerted use of two enzymes was also reported by Lee and coauthors for the production of rebaudioside A from stevioside contained in liquors from processed stevia leaves. Thus, a β-1,3-glucosyl transferase was combined with a β-1,3-glucosyl oligosaccharide from curdlan or laminarin (no further disclosed) as a donor, stevioside as an acceptor, and a β-1,3-glucanase to break down the β-1,3 glucose contained in the oligosaccharide. Incubation at 50°C for 5 h allowed the production of rebaudio-side A (Lee et al. 2011).

The nature of the glucose donor may clearly condition the outcome of the biotransformation. Thus, the use of β-cyclodextrin rather than starch yielded nine derivatives although proper adjustment of the enzyme content favors the production of two 1,4 transglycosylated products, 4'-O-α-D-glycosyl stevio-side and 4″-O-α-D-maltosyl stevioside, that display the preferred organoleptic properties (Abelyan et al. 2004). Actually, this specific biotransformation was later used as a test bed to assess the possibility of conjugation of enzymatic and physical methods, namely microwave irradiation, for an improved process. The authors validated their approach because under similar operational conditions, but with incubation under microwave conditions (2450 MHz, 300 W), the biotransformation was completed in 1 min, yielding 4'-O-α-D-glycosyl stevioside and 4″-O-α-D-maltosyl stevioside in relative proportions of 66% and 24%, respectively, and 90% of total product yield. The enzymatic biotransformation on its own lasted for 12 h and allowed for the production of 4'-O-α-D-glycosyl stevioside and 4″-O-α-D-maltosyl stevioside in relative proportions of 46% and 24%, and total product yield of 70% (Jaitak et al. 2009). Data suggest a positive effect of microwave irradiation in enzyme performance—a feature that is not always observed. Despite the encouraging results, no further significant developments on the use of this combined methodology in stevioside transglycosylation have been reported although similarly positive observations were recently published on the transglycosylation of rutin (Sun et al. 2011). Also, within this trend to couple physical and enzymatic methods, ultrasonic assistance to starch hydrolysis catalyzed by α-amylase was shown to speed up starch retrogradation, and the reducing sugars titer was higher than in the purely enzymatic hydrolysis of starch. In addition, amylopectin is debranched; therefore, the rela-tive amount of amylose is increased, and the polymerization degree decreases as does the viscosity of the solution. This behavior was related to the ultrasound rupturing the macromolecular chains of starch, decreasing starch particle size, and damaging its structure by opening channels and providing pathways for water diffusion and favoring enzyme action (Hu et al. 2013). Mechanical effects and mass transfer enhancement resulting from cavitation created by ultrasound irradiation have also been advantageously combined with pectinesterase activity as the core of an ecofriendly methodology to extract genipin from genipap (Ramos-de-la-Peña et al. 2014). Genipin is a natural colorant used in beverages and desserts. In this particular case, and unlike previous methodologies with which physical and enzymatic events occur simultaneously, the pectinesterase was added after the ultrasonic-assisted extraction to separate genipin from pectin and proteins. While the ultrasound promoted the disruption of the plant cell wall, allowing for the cell content to be released, including genipin and pectin, efficient demethylation of the pectin molecule was carried out through the action of the pectinesterase. The latter action promoted the polyelectrolyte complex formation between pectic polysaccharides and proteins, preventing the typical cross-linking of genipin.

Highly substituted steviosides are unwanted because they do not possess adequate sweetening quality. Thus, Li and coworkers developed a strategy for the production of a mixture composed mostly of mono- and di-glycosylated steviosides using a commercial cyclodextrin glucanotransferase (Toruzyme 3.0 L) as a catalyst and cornstarch hydrolysate as a glucose donor (Li et al. 2013). The authors were able to establish that their goal was favored by operating at 60°C because it enhanced yield in the two products as well as hydrolysis of unwanted compounds, in a solution of 20 gL^{-1} of stevioside and cornstarch (1:1, w/w) at pH 6.0 and an enzyme load not exceeding 40 U g^{-1} stevioside. Further increase of the enzyme load to 50 U g^{-1} stevioside slightly favored the production of triglycosylated derivates but allowed for stevioside conversion of 77% after 5 h of operation (Li et al. 2013). Starch and β-cyclodextrins have been the favored glycosyl donors for tranglycosylation processes involving the use of CGTases as biocatalysis, but details on the mechanism of the process were unknown. Recently, however, Lu and Xia offered a significant contribution toward the understanding of the transglucosylation pathways of stevioside with diverse donors, including starch, α- and β-cyclodextrins, with the commercial Toruzyme 3.0 L as biocatalyst. The authors were able to establish that the monosaccharide- and disaccharide-tested fructose, glucose, sucrose, and maltose failed as effective glycosyl donors. Thus, negligible conversion of stevioside was observed, but for maltose. Even in this case poor transglycosylation activity was observed, suggesting side reactions. On the other hand and under similar operating conditions, cyclodextrins and starches allowed a transglucosylation yield of up to 80% in 5 h. The suggested pathways differed in each case because cyclodextrins were considered to perform transglycosylation through coupling so that intermediates of reducing sugar are produced, followed by disproportionation with stevioside. Transglycosylation involving starches are believed to combine the cyclodextrin pathway and the hydrolytic pathway of starches (Lu and Xia 2014).

Current efforts toward the enzymatic glycosylation of stevioside mostly rely on the use of CGTase and oligo/polysaccharides as glucose donors after past attempts with pullanase, isomaltase, and dextrin saccharase (Li et al. 2013). However, in a recent publication, biotransformation of stevioside catalyzed by an alternansucrase from *Leuconostoc citreum* SK24.002 was successfully achieved using sucrose, a cheap substrate, as a donor. The highest transglycosylation yield from a 10 gL^{-1} stevioside concentration was about 44%, achieved at 20°C and pH 5.4 after 24 h of incubation using a sucrose concentration of 10 gL^{-1}. The product composition contained a dispersion of mono-di- and tri-glycosylated steviosides and their isomers (Musa et al. 2014).

A slightly different and somehow opposite approach toward the production of food components from stevioside has been suggested and implemented, and it relies on the selective cleavage of the β-1,2-glycosidic linkage of the sophorosyl moiety at C13 of stevioside to produce rubusoside (Figure 18.2)

β-Glucosidase

Stevioside   Rubusoside

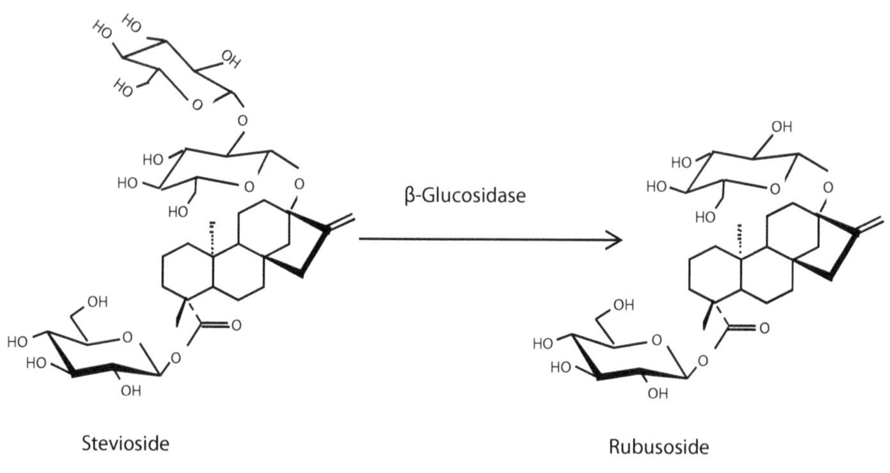

FIGURE 18.2 Hydrolysis of stevioside to rubusoside through β-glucosidase action.

(Wan et al. 2012). Although rubusoside is only 80% as sweet as stevioside (Ko et al. 2012), it is also useful as a sweetening enhancer and displays antiangiogenic and antiallergic activities and solubilizing properties, thus favoring its use in different formulations in food and beverages (Ko et al. 2012; Prakash et al. 2011). Moreover, rubusoside is about tenfold more expensive than stevioside, further making the biotransformation clearly appealing (Ko et al. 2012). A β-galactosidase from *Aspergillus* sp. was shown to allow for such cleavage. Actually, a stevioside conversion of 98% corresponding to a rubusoside yield of 91% was obtained after 72 h of incubation at 60°C of a substrate solution of 80 gL^{-1}. An important feature was that the enzyme showed negligible transglycosylation activity; therefore, the risk of reversed reaction or of other unwanted glycosylation is reduced (Wan et al. 2012). In addition, β-galactosidase, a stevioside-specific β-glucosidase purified from Viscozyme L commercial enzyme formulation, also proved effective in the hydrolysis of stevioside to rubusoside. Again, the enzyme displayed minor trans-glycosylation activity as well as negligible hydrolytic activity over the β (1 → 3) glucosyl derivative of the 13-hydroxyl group of rebaudioside A and over the glycosyl linkages at the 19-carboxyl group of steviol glycosides. Accordingly, under optimized conditions, a product yield of 70% from a 220 gL^{-1} substrate solution after 12 h of incubation was reported (Ko et al. 2012).

18.2.2 Difructose Anhydride III

Also, within the scope of the growing interest and dissemination of functional foods, is the production of difructose anhydride III (DFA III). This small cyclic disaccharide, composed of two fructose residues bonded by their reducing carbons, is rarely found in nature. DFA III is gaining relevance as a low-calorie sweetener, thus replacing sugar. Moreover, DFA III has a prebiotic role; this disaccharide is noncario-genic and favors the absorption of minerals and flavonoids and their growth (Hang et al. 2011). DFA III is usually obtained through the enzymatic hydrolysis of inulin, a polyfructan, using inulase II, a depoly-merizing inulin fructotransferase, IFT (EC. 4.2.2.18, formerly EC. 2.4.1.93) (García-Moreno et al. 2008; Haraguchi 2011). This enzyme acts on the linear β–2,1 glycosidic bonds of inulin and related polyfruc-tosides. In the process, kestose (GF2), nystose (GF3), and fructofuranosylnystose (GF4) are also formed as by-products (Hang et al. 2011). An alternative methodology recently presented involves the synthesis of DFA III from sucrose, a much cheaper and more available substrate than inulin, using the combined action of a fructosyltransferase, FT (EC. 2.4.1.9) and IFT (Figure 18.3) (Hang et al. 2013). Actually, FT has been used to produce GF2, GF3, and GF4 from sucrose because it promotes the transfer of a fructo-syl group to a molecule of sucrose (Antosova and Polakovic 2001). In the coupled enzyme reaction, FT promotes this later reaction, and IFT catalyzes the hydrolysis of GF3 and GF4 to DFA III, sucrose, and kestose. Thus, the second reaction displaces the equilibrium of the first reaction toward the product side because these are depleted in the second reaction. Moreover, sucrose and kestose produced in the second reaction can be used as substrates by FT. The overall concept is appealing, but under optimized opera-tional conditions, DFA III yield was only 100 mg g^{-1} sucrose after 36 h of reaction for an initial sucrose concentration of 400 gL^{-1}. The authors consider, however, that if purified and/or genetically engineered enzymes and/or more suitable fructo-oligosaccharides were used as substrates for IFT, a threshold yield of 40% to 50% DFA III could be achieved (Hang et al. 2013).

18.2.3 Sucralose

Sucralose is a high intensity, noncaloric sweetener (600 fold sweeter than sucrose) that is produced on an industrial scale by purely chemical methods through a multistep pathway that involves the synthesis of sucrose-6-acetate, its conversion to sucralose-6-acetate, and finally deacetylation to sucralose (Kerr et al. 2013). Despite continuous technological improvements in the manufacturing process, some concerns still arise mainly due to environmental risks related to chlorinated by-products. Biological methods for the pathway have thus been looked after (Figure 18.4).

A relatively conservative approach involves the synthesis of sucrose-6-acetate from the reaction of glucose-6-acetate, which can be obtained from glucose by fermentation with *Bacillus megaterium*, and fructose, catalyzed by a fructosyltransferase from *A. oryzae*. The enzyme also hydrolyzed sucrose, thus providing for the fructosyl group. Under optimized bioconversion conditions, a 27% molar conversion

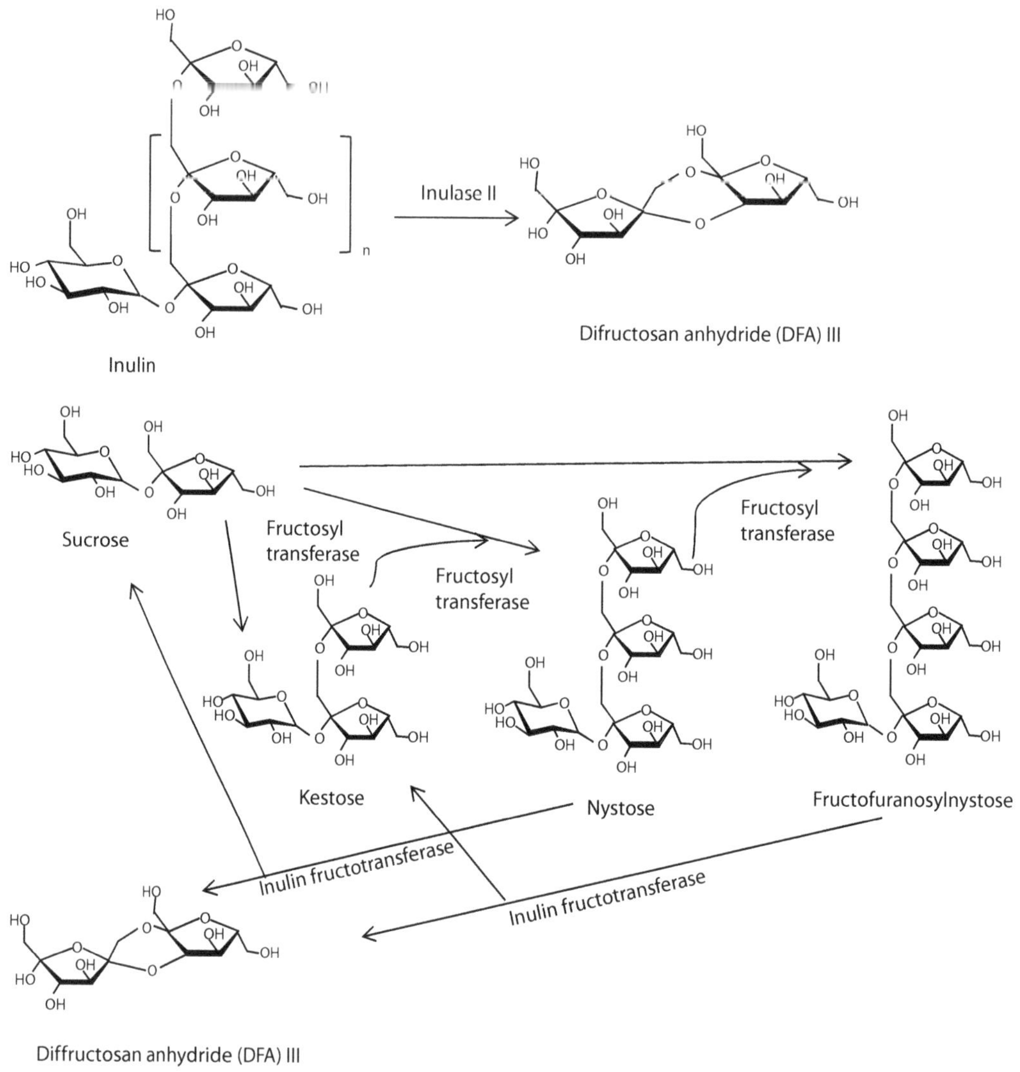

FIGURE 18.3 Examples of enzyme-driven approaches for the production of difructosan anhydride III (DFA III), either involving inulin hydrolysis or synthesis from sucrose.

yield of glucose-6-acetate was observed after 48 h of reaction for an initial sucrose concentration of 600 gL^{-1} and a 3:1 ratio of sucrose to glucose-6-acetate (Han et al. 2011). Still, productivity is relatively low, and further insight into the reaction mechanisms is required. A new proposal for the synthesis of sucrose-6-acetate, a key intermediate in the process, involves the transesterification of sucrose with vinyl acetate in a 2-butanol/Tris-HCl buffer organic–aqueous two-phase system. In a 60:40 volume ratio, a vinyl acetate/sucrose molar ratio of 3.90, and an initial sucrose concentration of 68 gL^{-1}, a 57% yield in sucrose-6-acetate was reported after 4 h of reaction in a relatively low-cost process that can be easily scaled up and that starts from the cheap sucrose substrate. The reaction time requires careful control because hydrolysis and synthesis coexist, and the former gradually overtook control beyond 4 h of reaction. Additionally, the formed sucrose-6-acetate could be transformed to diesters through reaction with the vinyl acetate donor molecules in excess (Zhong et al. 2013). Downstream processing and solvent recovery and reuse should still be considered.

Several approaches are possible for the hydrolysis of sucralose-6-acetate to sucralose. In a recent example, such a reaction was accomplished by lipase, either free or immobilized, or by a protease as

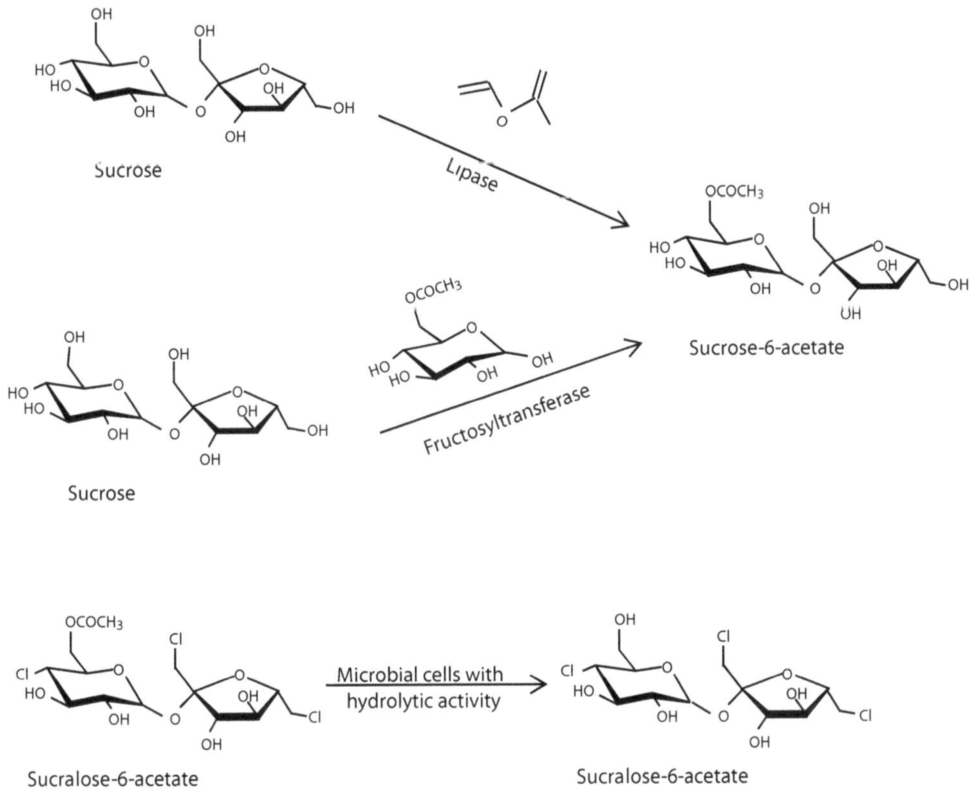

FIGURE 18.4 Representative examples of enzyme-driven steps in the complex conversion of sucrose into sucralose.

summarized in Table 18.1. Hydrolysis of sucralose-6-acetate has also been carried out with whole *B. subtilis* cells with lipase activity. The use of whole cells can cut down costs both regarding biocatalyst preparation and product purification. Operation with free cells allowed for full conversion of the substrate, from 10 to 300 gL^{-1} concentration, albeit at increasingly higher incubation periods from 48 h to 250 h. The overall reaction rate decreased upon whole cell immobilization in calcium alginate beads; still full substrate conversion of a 50 gL^{-1} solution was achieved after 200 h of incubation in a shaken reactor. A recirculating packed-bed reactor was also operated at a flow rate of 10 mLmin^{-1} for the bioconversions to be carried out. Up to three consecutive cycles with complete substrate conversions were performed,

TABLE 18.1

Enzymatic Deacetylation of Sucralose-6-acetate (S-6-ac)

Enzyme Formulation	S-6-ac (gL^{-1})	Temperature (°C)	Comments
Free lipase from A. oryzae	2.5	Room temperature	Operation in stirred reactor. After 12 h, a deacetylation yield of 98.4% was obtained.
Free protease (Alcalase 2.4 L)	3.3	25–30	Operation in stirred reactor. After 36 h, a deacetylation yield of 96.4% was obtained.
Lipase immobilized in Eudragit RL 100	13.3	25–30	Operation in stirred reactor. After 24 h, a deacetylation yield of 98.3% was obtained.
	10.0	Not stated	Operation in recirculating packed-bed reactor with a flow rate of 5 mLmin^{-1}. After 6 h of operation, a deacetylation yield in excess of 98% was observed.

Source: Ratnam, R., Aurora, S., Lali, A. M. and P. Subramaniyam, Enzyme catalyzed de-acylation of chlorinated sugar derivatives, Patent Application WO 2007054973, 2007.

albeit at increasing incubation times of 8, 10, and 12 days (Chaubey et al. 2013). This trend suggests loss of biocatalytic activity, either intrinsic or due to cell leakage, the latter being less likely because no support damage or evidence of cells present in the effluent liquid were reported by the authors.

18.2.4 Human Milk Oligosaccharides

Within the scope of production of functional foods, human milk oligosaccharides (HMO) have received considerable attention. These complex molecules have several beneficial effects on infants, namely by helping in the development of a balanced immune system and by having a prebiotic effect, thus stimulating the growth of given bacteria in the infant's intestine. They are therefore looked after for application in infant food formulation (Bode 2012), particularly because the oligosaccharides used in the currently available formulas differ structurally from HMO, and they are therefore unlikely to have the same structure-specific effects (Bode and Jantscher-Krenn 2012). The commercial production of HMO is not a current reality due to the complexity of the molecules, but efforts are being made to produce galactooligosaccharides that may mimic HMO (Intanon et al. 2014). One of the approaches recently presented involves the production of 3′-sialyllactose from dairy by-products, more specifically casein glycomacropeptide (CGMP), with which sialic acid is a carbohydrate moiety, and lactose. GCMP acts thus as a sialyl donor whereas lactose is the receptor in a reaction catalyzed by a trans-sialidase, displaying transferase activity but roughly devoid of hydrolytic activity, unlike sialidases, which are also used to carry out this reaction (Figure 18.5).

A recent example involves the use of a trans-sialidase derived from *Trypanosoma rangeli* and expressed in *Pichia pastoris* after codon optimization (Holck et al. 2014). The authors used a response surface methodology approach to establish optimal conditions for reaction, such as pH (5.7), temperature (30°C), and sialyl donor-to-acceptor molar ratio (1:4). Under these conditions, the authors established that an increase in substrate concentration from 1.5 mM CGMP and 8 mM lactose to 6.2 mM CGMP and 32 mM lactose increased sialyllactose concentration from 0.15 mM to 0.30 mM but at the cost of

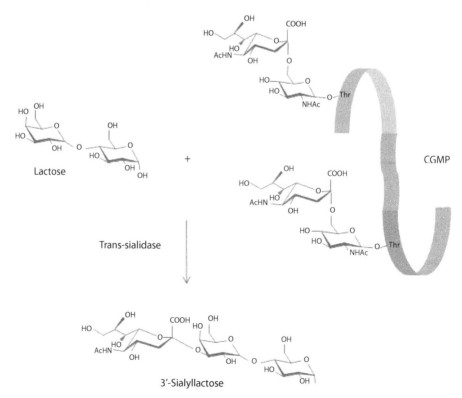

FIGURE 18.5 Schematics for the enzymatic production of 3′-sialyllactose from lactose and casein glycomacropeptide (CGMP) using trans-sialidase.

a decrease in conversion yield, based on CGMP, from 10% to roughly 5%. The authors suggested this trend could be used to establish guidelines for the production according to specific needs. Thus, high-substrate concentrations should be favored when high amounts of product are sought after, whereas lower substrate concentrations should be used in cases in which these are rare or expensive. Moreover, operating at a 6 L scale, the authors reported a product concentration of 2.4 mM (1.54 gL^{-1}), corresponding to a product mass yield of 40 mg g^{-1} CGMP, which exceeded previously obtained results. Long incubation periods (more than 3 h) could be considered because hydrolytic activity was not displayed by the enzyme (Holck et al. 2014).

The enzyme was also able to use lacto-N-tetraose and galactooligosaccharides as acceptors, thus further expanding the range of HMO and HMO-like compounds. As outcomes of the reactions, single sialylated lacto-N-tetraose and doubly sialylated products with a galactooligosaccharide backbone with degrees of polymerization within 3 to 6 were also observed, respectively. Furthermore, the ability of the synthesized compounds to perform the expected role was confirmed experimentally (Holck et al. 2014).

The trans-sialidase from *T. rangeli* and expressed in *P. pastoris* was also engineered through a six-point mutation, aiming to improve the catalytic activity (Michalak et al. 2014). The resulting Tr6 enzyme had pH and temperature optima of 5.5 and 25°C, respectively, for the production of 3'-siallyl-lactose. However, only at a high substrate acceptor:donor molar ratio, namely 25 (and above), corresponding to molar concentrations of 117 mM and 4.6 mM, respectively, could significant amounts of product be obtained without a high release of free sialic acid. This suggests that adequate tuning of reaction conditions is needed for high trans-sialylation while suppressing sialidase activity.

Under the optimized conditions, a transialidase activity of 16.2 µmol min^{-1} mg^{-1} was observed, and the process was scaled up to 5 L, allowing for 5.7 mmol of 3'-sialyllactose after downstream processing, out of the initial 23 mmol sialic acid. Because only half of the sialic acid residues in the CGMP are bound at the 3' position, toward which the enzyme is active, a molar product yield of about 50% was obtained (Michalak et al. 2014). Again, the engineered enzyme displayed some versatility because, in addition to lactose, lacto-N-tetraose and lacto-N-fucopentaoses could also be used as an acceptors although the product yield was lower than that observed when lactose was used as an acceptor molecule (Michalak et al. 2014).

Aiming to enhance the productivity of the synthesis of 3'-sialyllactose based on the catalytic activity of this system, different enzyme immobilization strategies were trialed because this path enables enzyme reutilization (Zeuner et al. 2014). Entrapment of cross-linked trans-sialydase in calcium alginate was discarded due to low recovery (0.1%) on the original enzyme activity, a significant part of which was lost during cross-linking with glutaraldehyde. Yet the reasons underlying this behavior, resulting in blocked active sites and/or mass transfer limitations, were not fully elucidated.

Immobilization onto iminodiacetic acid (IDA) functionalized carbon-coated magnetic nanoparticles loaded with Cu^{2+} through the His-tag residues of the engineered protein that eased enzyme recovery and purification was also assessed. The immobilized enzyme allowed for a higher ratio of rate of sialic acid transfer to CGMP to hydrolysis than that observed for the free enzyme. This made up for the longer optimized reaction time for the immobilized enzyme and, all things considered, resulted in a productivity of 47.6 mg product per mg enzyme, 1.4-fold higher than that of the free enzyme in a 60-min run. The immobilized enzyme was used for seven consecutive runs, resulting in a total productivity of 84.1 mg product per mg enzyme, the latter being illustrative of a progressive decay in enzyme performance. This was partly associated with enzyme leakage and did not compare favorably with similar reuse of the free enzyme, which was implemented through the use of a membrane reactor, equipped with a 10 kDa regenerated cellulose membrane, which allowed for a biocatalytic productivity of 305.6 mg product per mg enzyme (Zeuner et al. 2014). Actually this setup, in which the free enzyme is contained in a membrane reactor, was later implemented in a more complex arrangement, combining membrane filtration steps and enzymatic reaction, with potential for large-scale application (Luo et al. 2014). Thus, CGMP was processed through membrane ultrafiltration and diafiltration steps to remove small peptides and salts and then added to excess lactose and to the genetically engineered sialidase. After incubation (up to 40 min), the reaction medium was processed through centrifugation and nanofiltration steps. The former allowed the recovery and reuse of the enzyme for four runs and simultaneously reduced the risk of product hydrolysis, and the latter allowed for separation of sialyllactose from lactose and concentration of these two compounds. Further improvement of the performance of the bioconversion system for the production of 3' sialylated glycans could be implemented through rational design.

Hence, based on the knowledge of the primary and tertiary structure of Tr6 and of its active site and particularly of the acceptor binding site, mutants derived from Tr6 with point mutations were designed and screened. This allowed for the identification of an enzyme with 13-point mutations as compared to the wild type, displaying impaired trans-sialidase activity and fourfold lower hydrolase activity on 3'-sialyllactose as compared with Tr6. Moreover, it also displayed broader acceptor-substrate specificity, accommodating terminal glucose and monomers of fucose and glucose besides the terminal galactose typical of the wild type (Jers et al. 2014). As mentioned previously, sialic acid is bound to CGMP in α-2,3- and in α-2,6-positions in a ratio of approximately 1:1. Hence, to take full advantage of CGMP as a substrate for the production of HMO, enzymes with trans-sialidase activity on the two positions or a combination of enzymes with trans-sialidase activity on either position are likely to be sought after.

18.3 Sugar Esters

Sugar esters of fatty acids are nonionic surfactants of particular interest in the food industry because they display high emulsifying, stabilizing, and detergency capabilities (Neta et al. 2015). The enzymatic synthesis of these molecules is accomplished either through the direct esterification of sugars with fatty acids or through their transesterification with esters (methyl, vinyl) of the corresponding fatty acid, typically using lipases of different sources (Neta et al. 2015; Yang and Huang 2012). The former type of reaction results in the production of water as a by-product, which is nontoxic and hence may be preferred, provided it is removed from the reaction media in order not to tamper with the equilibrium and thus allow for high yields (Yoo et al. 2007). Methyl and vinyl esters used in transesterification lead to the formation of methanol and of acetaldehyde, both of which are toxic (Adachi and Kobayashi 2005; Gumel et al. 2011).

The enzymatic pathway has some advantages over the chemical approach, such as the milder reaction conditions and a one-step synthesis process, not requiring complex protection/deprotection preparative reactions. In addition, chemical syntheses often require complex separation. Several publications have addressed the lipase-mediated synthesis of fatty acid methyl esters; most have relied on the use of monosaccharides as the acyl acceptor (Gumel et al. 2011; Neta et al. 2015).

Despite the extensive work performed, in this type of bioconversion, it has been acknowledged that several issues are not fully elucidated, namely the reasons underlying the favored synthesis over hydrolysis associated with given organic solvents; the threshold of water activity at which the equilibrium shifts from synthesis to hydrolysis; and the effect of the length of the fatty acid in the reaction (Neta et al. 2015). Accordingly, a significant diversity can be found in the systems described to date. Thus, either the sugar or the lipid are in excess, in ratios up to 3:1; fatty acids with chains from four to 16 carbon atoms have been used in the ratio of lipid to sugar.

A major problem involving the design of a high-yield and high-productivity bioconversion system is the environment required for simultaneously allowing high solubility of both sugars (hydrophilic) and lipids (hydrophobic). To overcome this, operation in nonconventional environments has been evaluated and, alongside synthesis with novel acyl acceptors, has gathered most of the focus in recent years. Moreover, given the large number of variables involved, the use of response surface methodology (RSM) as a tool to optimize operational condition has gained relevance. Some illustrative examples are summarized in Table 18.2.

18.4 Feed Enzymes

Feed enzymes are typically added to animal feed, mostly for swine and poultry, as impure mixtures of active enzymes, and its consistent application is relatively recent although occasional incorporation of enzymes in poultry diet was reported back in the early twentieth century (Choct 2006). Due to lack of uniformity in regulatory affairs in different countries, namely related to safety issues, the use of enzymes as feed additives is not fully widespread (Adrio and Demain 2014; Pariza and Cook 2010). Feed enzymes

TABLE 18.2

Recent Developments in Experimental Methodologies for the Enzymatic Synthesis of Sugar Fatty Ester Acids

Acyl Acceptor/Acyl Donor	Product	Bioconversion Media	Comments	References
Lactose/vinyl laurate (1:3.8, molar)	Lactose monolaurate, LM. An innovative product, from lactose, widely available as a by-product of cheese manufacture	2-Methyl-2-butanol with immobilized lipases. Best results with Lipozyme from *Mucor miehei*	Under RSM optimized conditions, a maximum yield of 99.3% from a lactose titer of 42 mM was reported. LM formation rate and yield bested those obtained for the synthesis of sucrose monolaurate. LM exhibited antilisterial activity when added to milk, yogurt, and cottage cheese.	Chen et al. 2014; Walsh et al. 2009
D-mannose/vinyl myristate (1:10, molar)	Mannosyl mysristate	1-Butyl-1-methylpyrrolidinium trifluoromethanesulfonate with Novozym® 435	Under RSM optimized conditions, a 24 molar yield of 72.2% was obtained out of a total substrate quantity of 0.26 mmol. Identification of key operational variables in the synthesis of sugar fatty acids	Galonde et al. 2013
Fructose/oleic acid (1:4 molar); fructose/mixture of oleic acid-fructose mono- and di-oleate	Fructose oleate	Lipozyme® IM, in solvent-free system. Oleic acid, eventually, together with fructose oleate, is used to resuspend fructose to avoid depletion of the acyl acceptor. In the initial periods of operation, fructose dissolution is favored by concomitant increase in fructose ester.	Operation of closed-loop packed-bed enzymatic reactor with continuous recirculation of a suspension of fructose in a mixture of the acyl donor and its monoester allowed for a productivity of 0.30 mmol ester h^{-1} g^{-1}, for 32 h of operation. Water formed was removed by combining free evaporation with N_2 bubbling and vacuum, so that the water content of the liquid phase was kept around 0.4% (w/w).	Pyo and Hayes 2009; Ye and Hayes 2011
Steric hindered glucose ketal: fatty acid residue acidic residue from the refine process of crude palm oil (equimolar or 1:2)	Glucose esters	Lipozyme IM in continuous flow system, using heptane or tert-butylmethylether (TBME) as solvents of both fatty acid residue and glucose ketal	Solubility of glucose ketal in TBME (40 mM) was twofold higher than in heptane. The acidic residue from the refine process of crude palm oil proved to be the more effective acy donor, among several pure fatty acids. Continuous operation allowed for conversion yields in excess of 90% for residence times of 24 min and a productivity of 76.6 mg product h^{-1} g^{-1} immob. enzyme, the latter exceeding by 20-fold the productivity in batch systems.	Ruela et al. 2013

(Continued)

TABLE 18.2 (CONTINUED)

Recent Developments in Experimental Methodologies for the Enzymatic Synthesis of Sugar Fatty Ester Acids

Acyl Acceptor/Acyl Donor	Product	Bioconversion Media	Comments	References
Maltose:linoleic acid (1:2 molar ratio)	Mono-6 or 6′-O-linoleyl-α-D-maltose as a mixture of two regioisomers in a 1.4:1 ratio	Commercially available lipases, immobilized or as lyophilized powder. Water-miscible and water-immiscible organic solvents, ionic liquids and binary mixtures, combined with molecular sieves	Novozym® 435 and lyophilized lipase from *P. cepacia* proved the most effective among enzymes trialed for maltose acylation. Operation was performed in sealed vessels. After considerations involving technical and economical features, acetone was suggested as the most suitable solvent for industrial application because it can be produced from renewable feedstock and is easily removed from the reaction medium. A maltose conversion of 82% from 0.625 M maltose monohydrate.	Fischer et al. 2013
Lactose:capric acid	Lactose caprate	*C. rugosa* lipase immobilized either by adsorption into acid activated mica and cross-linking with glutaraldehyde or covalently bound into amino-silane activated mica. Reactions were carried out in organic solvents, preferably acetone.	Optimization of operational conditions, namely temperature (55°C) and acceptor donor molar ratio (2:1) were performed. Immobilization enhanced more than twofold the specific activity of the enzyme and allowed for half-lives of at least 10 cycles of 48 h each.	Zaidan et al. 2012

[a] 1.2:5.6-Bis-O-(1-methylethylidene)-α-D-glucofuranose.

are used to enhance the digestibility of nutrients, increasing their bioavailability and therefore improving the efficiency in feed use. Thus, feed enzymes can degrade unwanted compounds in feed, which are otherwise harmful or of little or no value, such as nonstarch polysaccharides (NSPs), and they reduce the viscosity of the digesta. This results in improved absorption of feed nutrients and better digestion (Adrio and Demain 2014; Pariza and Cook 2010). NSPs are processed by several carbohydrases, often termed NSP-degrading enzymes. These are essentially glucanase and xylanase, but other enzymes are also used, such as α-galactosidase, mannanase, and pectinases (which include pectin methyl esterases, pectin transeliminases, and polygalacturonases) (Adrio and Demain 2014; Cowieson and Adeola 2011; Gummadi and Kumar 2005). Also aimed at improving digestibility but through breaking down starch, α-amylase is also used as a feed additive (Kumari et al. 2012). All these feed-relevant carbohydrases belong to the hydrolase/glycosidase family, and they display endo-acting mechanisms so that the products of hydrolysis are low molecular weight oligo- or polysaccharides, but practically no free sugars (Cowieson and Adeola 2011).

The use of enzymes as additives, namely NSP-degrading enzymes, also has a positive impact on the environment. Thus, the fecal nutrient level deployed to the environment is decreased as well as intestinal methane production as diets based on barley, rye, and wheat, rich in NSPs, are processed with NSP-degrading enzymes (Li et al. 2012). Moreover, phosphorus excretion can be minimized with feed enzymes. Thus, the most significant fraction of phosphorus-containing plant-based feedstuff is stored as phytate, indigestible by monogastric animals, unless phytase is added to the feed. The enzyme catalyzes the hydrolysis of phytate in a step-like manner to inositol phosphates, myoinositol, and inorganic phosphate, making phosphorus available for bone growth alongside other minerals, proteins, and amino acids that had become bound to the phytate (Gontia-Mishra and Tiwari 2013). As such, the use of phytase as an additive is suggested not only for pig and poultry but also for fish feed (Afinah et al. 2010).

Proteases are also used as additives, and they have proven effective in the degradation of digestive enzyme inhibitors; they can also reduce the amount of non-nitrogen supplement in the diet, thus allowing for a decrease in urea excretion into the environment (Choct 2006; Li et al. 2012). Yet it has been shown that the same protease from different sources, for example, fungal or bacterial, may have opposite effects. Therefore, for taking advantage of the potential of protease in animals, insight into the interaction with the digestive tract of the host and with other enzymes is advised (Cowieson and Adeola 2011).

Again, as with carbohydrases, phytases and protease have a hydrolytic action over the substrates. Enzymes as feed additives are often delivered as a cocktail composed of different enzymes. Taking full advantage of additive or synergistic effects on digestibility can be a hard task because the matrix is continuously changing, and antagonistic effects of enzyme activity must be avoided, proteolytic action over other enzymes (Choct 2006; Cowieson and Adeola 2011). Improved design of cocktail mixtures is clearly a hot topic, and several recent works have addressed this matter as the associative and synergistic effects resulting from the combined action of different feed enzymes was demonstrated by allowing improved nutrient use and health of the digestive tract and increased gain in body weight in broiler chicks (Narasimha et al. 2013a,b).

18.5 Detoxification

Mycotoxins are secondary metabolites, physically and chemically stable, produced by filamentous fungi that contaminate cereal grains and other feeds. Such contamination has a huge impact on human and animal health worldwide and, accordingly, has raised concerns to the World Health Organization (Adegoke and Letuma 2013). The most disseminated strategy to tackle this problem involves the use of chemisorbents, but although these have proven effective for counteracting aflatoxins, the same has not been reported for other mycotoxins. An alternative method was developed that involves the use of either whole microorganisms or enzyme extracts. In this way, detoxification is achieved through biotransformation (Çelik et al. 2013; EFSA 2014). The strategy is quite adequate, given the specific nature of enzyme action, because the detoxification must not generate toxic products, and the nutritional value of the food as well as the technological properties of the product must be retained (Jard et al. 2011).

Several microorganisms have been shown to present enzymatic pathways that enable biotransformation of different mycotoxins, but particular care has to be given to ensure they comply with these criteria. Hence, thorough identification of the biotransformation products is essential; therefore, the sole fact that the mycotoxin is degraded does not necessarily correspond to a purely effective process (Reddy et al. 2010).

The toxicity of Aflatoxin B1 (AFB1) has been ascribed to a lactone ring; hence, its cleavage is considered to result in detoxification. This biodegradation can be easily monitored because changes in fluorescence take place (Guan et al. 2008; Guan et al. 2010). This pattern was observed when the supernatant of a cell culture of *Myxococcus fulvus* ANSM068 was incubated for 72 h at 30°C in the presence of AFB1, allowing for a bioconversion yield of 77% out of an initial concentration of 100 µgL^{-1} of the mycotoxin (Guan et al. 2010). It was later established that the supernatant degradation ability of a constitutive nature abridged also aflatoxins G1 (AFG1) and M1 (AFM1) with conversion yields of 68% and 64%, respectively, after 48 h of incubation, at which time AFB1 conversion was 72% (Guan et al. 2010; Zhao et al. 2011). Moreover, it was also shown that using the partially purified 32 kDa enzyme, AFG1 and AFM1 conversion yields could be increased to 97% and 96%, respectively (Zhao et al. 2011).

Bacillus subtilis ANSB060 cells also displayed significant constitutive degradation activity over AFB1, AFG1, and AFM1 as reflected by conversion yields of 82%, 60%, and 81%, respectively, were observed for initial mycotoxin concentrations of 100 µgL^{-1} and an incubation period of 72 h. Most of the activity was observed in the supernatant, which clearly suggests an enzymatic mechanism (Gao et al. 2011). A cell-free supernatant of *Bacillus subtilis* UTBSP1 isolated from pistachio nuts was also shown to degrade AFB1. Thus, a conversion yield of 79% out of a 2.5 mgL^{-1} initial concentration of AFB1 after an incubation period of 72 h was reported. Again, the enzymes involved were of a constitutive nature and a significant decline in fluorescence of the enzyme-treated media was observed, suggesting a decrease in toxicity. Further validation of the potential of the approach for detoxification was achieved when the inoculation of pistachio ground kernel with *B. subtilis* UTBSP1 caused a 95% decrease in AFB1 content of kernels after 5 days of incubation (Farzaneh et al. 2012).

An aflatoxin-oxidase, partially purified from the fungus *Armillariella tabescens*, also displayed over AFB1. In this case, given the observed shifts in the polarity of the AFB1 derivative as compared to the AFB1 molecule in addition to fluorescence being retained, the authors suggested that the enzymatic mechanism involved the cleavage of the bis-furan ring, also a key component for AFB1 toxicity, rather than lactone cleavage (Cao et al. 2011). AFB1 degradation was also achieved both through the use of a laccase from *Trametes versicolor* as well as with a recombinant laccase produced by an *Aspergillus niger* strain. Thus, conversion yields of 87% and 55%, respectively, were observed out of an initial concentration of 1.4 mg L^{-1} of AFB1 after 72 h of incubation. Although the products were not clearly identified, the authors suggested that the enzyme preparations altered the double bond of the furofuran ring of the AFB1 molecule, because a significant decrease in the mutagenicity of AFB1 was observed. Moreover, formation of structural analogues during the biotransformation was not detected, further suggesting that the degradation products differed radically from the substrate molecule (Alberts et al. 2009).

More insight on the structure of the products formed as a result of a biotransformation was provided recently (Samuel et al. 2014). By culturing *Pseudomona putida* strains MTCC 1274 and 2445 in mineral salt glucose medium containing AFB1 (0.2 mgL^{-1}), the authors showed that the strains were able to achieve the biotransformation and detoxification of the mycotoxin as well as provide evidence on the structure of the products formed. Thus, biotransformation of AFB1 led to AFD1, AFD2, and AFD3 (Figure 18.1). AFD1 results in the lactone ring opening of AFB1. Further decarbonylation leads to AFD2, corresponding to the removal of both the lactone carbonyl and the cyclopentenone ring of AFB1. These two compounds were reported previously as the outcome of the degradation of AFB1 with citric acid (Mendez-Albores et al. 2008). Moreover, Samuel et al. (2014) also reported the presence of AFD3 as phatalic anhydride but stated that further evidence is required to validate the latter as a product of biotransformation. In any case, the three products of the biotransformation are nonfluorescent and display much lower, if any, toxicity (Samuel et al. 2014).

TABLE 18.3

Whole Cell Biotransformation of Ochratoxin A (OTA)

Biological Catalyst	Comments
Bacteria: *Acinetobacter calcoaceticus, Bacillus* spp., *Lactobacillus* spp., *Phenylobacterium immobile, Streptococcus salivarius* subsp. *thermophilus*	Biodegradation of OTA concentration within 0.1 mgL^{-1} to 10 mgL^{-1} and different conversion yields up to full elimination of OTA. Incubation periods of some hours to 2 days and temperatures within 25°C to 37°C. Data is strain dependent.
Filamentous fungi: *Aspergillus* spp., *Botrytis cinerea, Phanerochaete chrysosporium, Pleurotus ostreatus, Rhizopus* spp	Biodegradation of OTA concentration within 1.0 mgL^{-1} to 7.5 mgL^{-1} and different conversion yields up to full elimination of OTA. Incubation periods of six to 22 days and temperatures within 25°C to 30°C. Data is strain dependent.
Yeasts: *Cryptococcus* spp., *Rhodotorula* spp., *Saccharomyces* spp. and *Trichosporon* spp., *Phaffia rhodozyma*	Biodegradation of OTA concentration within 0.2 mgL^{-1} to 7.5 mgL^{-1} and different conversion yields up to full elimination of OTA. Incubation periods of 5 h to 15 days and temperatures within 20°C to 35°C. Data is strain dependent.

Source: Afinah, S., Yazid, A. M., Anis Shobirin, M. H. and M. Shuhaimi, *Int. Food Res. J.,* 17, 13–21, 2010; McCormick, S. P., *J. Chem. Ecol.,* 39, 907–918, 2013.

Strains from Actinomycetes (viz. *Rhodococcus* spp. and *Mycobacterium* spp.) also evidenced AFB1 and AFG1 detoxification capabilities (McCormick 2013). Recently, the enzymatic mechanism involved in mycobacterial biotransformation was also identified as the reduction of the α,β-unsaturated lactone moiety by F420-dependent reductases, followed by spontaneous hydrolysis to several nontoxic end products (Lapalikar et al. 2012; Taylor et al. 2010).

Evidence for ochratoxin A (OTA) detoxification by microbial/enzymatic sources has also been given (Abrunhosa et al. 2010; McCormick 2013) and relevant information related to whole cell biotransformation is summarized in Table 18.3. Purely enzymatic hydrolysis of OTA was first associated with carboxypeptidase A (CPA), but other enzymes have also evidenced the required catalytic activity, such as pancreatin, lipase, and carboxypetidase Y. CPA is particularly effective with 99.8% of conversion of a 1 mgL^{-1} solution of OTA into OTα in 25 h at 37°C (Abrunhosa et al. 2006). Yet an OTA hydrolytic enzyme has been isolated that displays improved activity as compared to CTA, and conditions for its production through solid-state fermentation have been optimized (Abrunhosa and Venâncio 2007; Abrunhosa et al. 2011).

The use of microorganisms to promote detoxification of other mycotoxins, namely citrinins, fumomisins, patulins, trichothecenes, and zearalenones, have also been described but mostly involving whole cells, and these are described in detail elsewhere (McCormick 2013).

18.6 Conclusions and Future Developments

Enzyme-based methodologies are firmly rooted in the food and feed area as a result of their green nature and specificity. Nonetheless, the endless quest for increased cost-efficiency as well as for novel products and applications keeps the pressure for dedicated research in the field to carry on. Several outcomes of such trends can thus be observed, and they may involve the screening of novel enzymes or activities to promote a set of new reactions, the design of operational conditions toward optimized operation, and the design of enzymes with improved activity. A combination of these approaches may be considered the most promising because a multidisciplinary approach combining complementary knowledge is likely to bring the best results. Moreover, the increasing knowledge of protein structure and on the interactions of the enzyme with the compounds involved in a reaction at the molecular level, combined with high throughput screening methodologies, is foreseen to result in the rational design of enzymes with improved activity and stability so that their application in commercial processes is strengthened and significantly widened.

REFERENCES

Abelyan, V. A., Balayan, A. M., Ghochikyan, V. T. and Markosyan, A. A. 2004. Transglycosylation of stevioside by cyclodextrin glucanotransferases of various groups of microorganisms. *Appl. Biochem. Microbiol.* 40:129–134.

Abrunhosa, L. and Venâncio, A. 2007. Isolation and purification of an enzyme hydrolyzing ochratoxin A from *Aspergillus niger*. *Biotechnol. Lett.* 29:1909–1914.

Abrunhosa, L., Santos, L. and Venâncio, A. 2006. Degradation of ochratoxin A by proteases and by a crude enzyme of *Aspergillus niger*. *Food Biotechnol.* 20:231–242.

Abrunhosa, L., Paterson, R. R. M. and Venâncio, A. 2010. Biodegradation of ochratoxin A for food and feed decontamination. *Toxins* 2:1078–1099.

Abrunhosa, L., Venâncio, A. and Teixeira, J. A. 2011. Optimization of process parameters for the production of an OTA-hydrolyzing enzyme from *Aspergillus niger* under solid-state fermentation. *J. Biosci. Bioeng.* 112:351–355.

Adachi, S. and Kobayashi, T. 2005. Synthesis of esters by immobilized-lipase-catalyzed condensation reaction of sugars and fatty acids in water-miscible organic solvent. *J. Biosci. Bioeng.* 99:87–94.

Adegoke, G. O. and Letuma, P. 2013. Strategies for the prevention and reduction of mycotoxins in developing countries. In *Mycotoxin and Food Safety in Developing Countries*, ed. H. A. Makun, 123–136. Rijeka, Croatia: InTech.

Adrio, J. L. and Demain, A. L. 2014. Microbial enzymes: Tools for biotechnological processes. *Biomolecules* 4:117–139.

Afinah, S., Yazid, A. M., Anis Shobirin, M. H. and Shuhaimi, M. 2010. Phytase: Application in food industry. *Int. Food Res. J.* 17:13–21.

Alberts, J. F., Gelderblom, W. C. A., Botha, A. and van Zyl, W. H. 2009. Degradation of aflatoxin B1 by fungal laccase enzymes. *Int. J. Food Microbiol.* 135:47–52.

Anonymous. 2014. Feed Enzyme Market by Type (Phytase, Protease & NSP), Sub-Type (Xylanase, Cellulase, Pectinase, Beta-Glucanase & Mannase) & Application (Swine, Poultry, Ruminant, Aqua Feed & Others)—Global Trends & Forecasts to 2018. Market Report, Market and Markets. Available at http://www.marketsandmarkets.com/Market-Reports/feed-enzyme-market-1157.html.

Antosova, M. and Polakovic, M. 2001. Fructosyltransferases: The enzymes catalyzing production of fructooligosaccharides. *Chem. Pap.* 55:350–358.

Bode, L. 2012. Human milk oligosaccharides: Every baby needs a sugar mama. *Glycobiology.* 22:1147–1162.

Bode, L. and Jantscher-Krenn, E. 2012. Structure–function relationships of human milk oligosaccharides. *Adv. Nutr.* 3:383S–391S.

Bornscheuer, U. T., Huisman, G. W., Kazlauskas, R. J., Lutz, S. and Moore, J. C. 2012. Engineering the third wave of biocatalysis. *Nature* 485:185–194.

Buchholz, K. and Collins, J. 2013. The roots—A short history of industrial microbiology and biotechnology. *Appl. Microbiol. Biotechnol.* 97:3747–3762.

Cao, H., Liu, D., Mo, X., Xi, C. and Yao, D. 2011. A fungal enzyme with the ability of aflatoxin B1 conversion: Purification and ESI-MS/MS identification. *Microbiol. Res.* 166:475–483.

Çelik, K., Uzatici, A., Coskun, B. and Demir, E. 2013. Current developments in removal of mycotoxins by biological methods and chemical adsorbents. *J. Hyg. Eng. Des.* 3:17–20.

Chatsudthipong, V. and Muanprasat, C. 2009. Stevioside and related compounds: Therapeutic benefits beyond sweetness. *Pharmacol. Ther.* 121:41–54.

Chaubey, A., Raina, C., Parshad, R., Rouf, A., Gupta, P. and Taneja, C. S. C. 2013. Bioconversion of sucralose-6-acetate to sucralose using immobilized microbial cells. *J. Mol. Catal. B: Enzym.* 91:81–86.

Chen, Y., Nummer, B. and Walsh, M. K. 2014. Antilisterial activity of lactose monolaurate in milk, drinkable yogurt and cottage cheese. *Lett. Appl. Microbiol.* 58:156–162.

Choct, M. 2006. Enzymes for the feed industry: Past, present and future. *World's Poultry Sci. J.* 62:5–15.

Cowieson, A. J. and Adeola, O. 2011. Opportunities and challenges in using exogenous enzymes to improve nonruminant animal production. *J. Animal Sci.* 89:3189–3218.

European Food Safety Authority (EFSA). 2009. Review of mycotoxin-detoxifying agents used as feed additives: Mode of action, efficacy and feed/food safety. Available at http://www.efsa.europa.eu/en/scdocs/doc/22e.pdf accessed on July 10, 2014.

Farzaneh, M., Shi, Z.-Q., Ghassempour, A. et al. 2012. Aflatoxin B1 degradation by *Bacillus subtilis* UTBSP1 isolated from pistachio nuts of Iran. *Food Control* 23:100–106.

Fischer, F., Happe, M., Emery, J., Fornage, A. and Schütz, R. 2013. Enzymatic synthesis of 6- and 6′-*O*-linoleyl-α-D-maltose: From solvent-free to binary ionic liquid reaction media. *J. Mol. Catal. B: Enzym.* 90:98–106.

Galonde, N., Brostaux, Y., Richard, G., Notta, K., Jerôme, C. and Fauconnier, M.-L. 2013. Use of response surface methodology for the optimization of the lipase-catalyzed synthesis of mannosyl myristate in pure ionic liquid. *Process Biochem.* 48:1914–1920.

Gao, X., Ma, Q., Zhao, L., Lei, Y., Shan, Y. and Ji, C. 2011. Isolation of *Bacillus subtilis*: Screening for aflatoxins B-1, M-1, and G(1) detoxification. *Eur. Food Res. Technol.* 232:957–962.

García-Moreno, M. I., Benito, J. M., Mellet, C. O. and Fernández, J. M. G. 2008. Chemical and enzymatic approaches to carbohydrate-derived spiroketals: Di-D-Fructose Dianhydrides (DFAs). *Molecules* 13:1640–1670.

Gontia-Mishra, I. and Tiwari, S. 2013. Molecular characterization and comparative phylogenetic analysis of phytases from fungi with their prospective applications. *Food Technol. Biotechnol.* 51:313–326.

Guan, S., Ji, C., Zhou, T., Li, J., Ma, Q. and Niu, T. 2008. Aflatoxin B1 degradation by *Stenotrophomonas Maltophilia* and other microbes selected using coumarin medium. *Int. J. Mol. Sci.* 9:1489–1503.

Guan, S., Zhao, L., Ma, Q. et al. 2010. In vitro efficacy of *Myxococcus fulvus* ANSM068 to biotransform aflatoxin B1. *Int. J. Mol. Sci.* 11:4063–4079.

Gumel, A. M., Annuara, M. S. M., Heidelberg, T. and Chisti, Y. 2011. Lipase mediated synthesis of sugar fatty acid esters. *Process Biochem.* 46:2079–2090.

Gummadi, S. N. and Kumar, S. 2005. Microbial pectic transeliminases. *Biotechnol. Lett.* 27:451–458.

Gurung, N., Ray, S., Bose, S. and Ray, V. 2013. A broader view: Microbial enzymes and their relevance in industries, medicine, and beyond. *BioMed Res. Int.* 2013:329121. doi:10.1155/2013/329121.

Han, Y., Liu, G., Huang, D. et al. 2011. Study on the synthesis of sucrose-6-acetate catalyzed by fructosyltransferase from *Aspergillus oryzae*. *New Biotechnol.* 28:14–18.

Hang, H., Mu, W., Jiang, B. et al. 2011. Recent advances on biological difructose anhydride III production using inulase II from inulin. *Appl. Microbiol. Biotechnol.* 92:457–465.

Hang, H., Miao, M., Li, Y., Jiang, B., Mu, W. and Zhang, T. 2013. Difructosan anhydrides III preparation from sucrose by coupled enzyme reaction. *Carbohyd. Polym.* 92:1608–1611.

Haraguchi, K. 2011. Two types of inulin fructotransferases. *Materials* 4:1543–1547.

Holck, J., Larsen, D. M., Michalak, M. et al. 2014. Enzyme catalysed production of sialylated human milk oligosaccharides and galactooligosaccharides by *Trypanosoma cruzi* trans-sialidase. *New Biotechnol.* 31:156–165.

Hu, A., Lu, J., Zheng, J. et al. 2013. Ultrasonically aided enzymatical effects on the properties and structure of mung bean starch. *Innov. Food Sci. Emerg. Technol.* 20:146–151.

Intanon, M., Arreola, S. L., Pham, N. H., Kneifel, W., Haltrich, D. and Nguyen, T.-H. 2014. Nature and biosynthesis of galacto-oligosaccharides related to oligosaccharides in human breast milk. *FEMS Microbiol. Lett.* 353:89–97.

Jaitak, V., Kaul, V., Bandna Kumar, N. et al. 2009. Simple and efficient enzymatic transglycosylation of stevioside by β-cyclodextrin glucanotransferase from *Bacillus firmus*. *Biotechnol. Lett.* 31:1415–1420.

Jard, G., Liboz, T., Mathieu, F., Guyonvarc'h, A. and Lebrihi, A. 2011. Review of mycotoxin reduction in food and feed: From prevention in the field to detoxification by adsorption or transformation. *Food Addit. Contam. Part A Chem. Anal. Control Expo. Risk Assess.* 28:1590–1609.

Jers, C., Michalak, M., Larsen, D. M. et al. 2014. Rational design of a new *Trypanosoma rangeli* trans-sialidase for efficient sialylation of glycans. *PLoS One* 9:1–11.

Kerr, J., Jansen, R. and Leinhos, D. A. 2013. Method for the production of sucralose. US Patent 8436156.

Ko, J.-A., Kim, Y.-M., Ryu, Y. B. et al. 2012. Mass production of rubusoside using a novel stevioside-specific β-glucosidase from *Aspergillus aculeatus*. *J. Agric. Food Chem.* 60:6210–6216.

Kochikyan, V. T., Markosyan, A. A., Abelyan, L. A., Balayan, A. M. and Abelyan, V. A. 2006. Combined enzymatic modification of stevioside and rebaudioside A. *Appl. Biochem. Microbiol.* 42:31–37.

Kumari, A., Singh, K. and Kayastha, A. M. 2012. α-Amylase: General properties, mechanism and biotechnological applications—A review. *Curr. Biotechnol.* 1:98–107.

Lapalikar, G. V., Taylor, M. C., Warden, A. C., Scott, C., Russell, R. J. and Oakeshott, J. G. 2012. F420H2-dependent degradation of aflatoxin and other furanocoumarins is widespread throughout the Actinomycetales. *PLoS One* 7:1–9.

Lee, Y. M., Kim, S. B., Hong, Y. H., Lee, J. H. and Park, S. W. 2011. Method of manufacturing rebaudioside A in high yield by recycling by products produced from manufacturing process for rebaudioside A. Patent Application US20110256588 A1.

Li, S., Yang, X., Yang, S., Zhu, M. and Wang, X. 2012. Technology prospecting on enzymes: Application, marketing and engineering. *Comput. Struct. Biotechnol. J.* 2:1–11.

Li, S., Li, W., Xiao, Q.-Y. and Xiao, Y. 2013. Transglycosylation of stevioside to improve the edulcorant quality by lower substitution using cornstarch hydrolyzate and CGTase. *Food Chem.* 138:2064–2069.

Liese, A., Seelbach, K., Buchholz, A. and Haberland, J. 2006. Processes. In *Industrial Biotransformations*, 2nd ed., eds. A. Liese, K. Seelbach and C. Wandrey, 147–513. Weinheim: Wiley-VCH Verlag GmbH & Co. KgaA.

Lu, T. and Xia, Y.-M. 2014. Transglycosylation specificity of glycosyl donors in transglycosylation of stevioside catalysed by cyclodextrin glucanotransferase. *Food Chem.* 159:151–156.

Luo, J., Nordvang, R. T., Morthensen, S. T. et al. 2014. An integrated membrane system for the biocatalytic production of 3′-sialyllactose from dairy by-products. *Biores. Technol.* 166:9–16.

McCormick, S. P. 2013. Microbial detoxification of mycotoxins. *J. Chem. Ecol.* 39:907–918.

Mendez-Albores, A., Nicolas-Vazquez, I., Miranda-Ruvalcaba, R. and Moreno-Martínez, E. 2008. Mass spectrometry/mass spectrometry study on the degradation of B-aflatoxins in maize with aqueous citric acid. *Am. J. Agric. Biol. Sci.* 3:482–489.

Michalak, M., Larsen, D. M., Jers, C. et al. 2014. Biocatalytic production of 3′-sialyllactose by use of a modified sialidase with superior trans-sialidase activity. *Process Biochem.* 49:265–270.

Musa, A., Miao, M., Zhang, T. and Jiang, B. 2014. Biotransformation of stevioside by *Leuconostoc citreum* SK24.002 alternansucrase acceptor reaction. *Food Chem.* 146:23–29.

Narasimha, J., Nagalakshmi, D., Ramana Reddy, Y. and Viroji Rao, S. T. 2013a. Synergistic effect of non starch polysaccharide enzymes, synbiotics and phytase on performance, nutrient utilization and gut health in broilers fed with sub-optimal energy diets. *Vet. World* 6:754–760.

Narasimha, J., Nagalakshmi, D., Viroji Rao, S. T., Venkateswerlu, M. and Ramana Reddy, Y. 2013b. Associative effect of non-starch polysaccharide enzymes and probiotics on performance, nutrient utilization and gut health of broilers fed sub-optimal energy diets. *Int. J. Eng. Sci.* 2:28–31.

Neta, N. S., Teixeira, J. A. and Rodrigues, L. R. 2015. Sugar ester surfactants: Enzymatic syntheis and applications in food industry. *Crit. Rev. Food Sci. Nutr.* (e-published ahead of print). doi:10.1080/10408398.2012.667461.

Pariza, M. W. and Cook, M. 2010. Determining the safety of enzymes used in animal feed. *Regul. Toxicol. Pharmacol.* 56:332–342.

Prakash, I., Dubois, G. E., Klucik, J., San Miguel, R. I., Fritsch, R. J. and Chaturvedula, V. S. P. 2011. Sweetness enhancers, compositions thereof, and methods for use. Patent Application WO 2011090709 A1.

Purkayastha, S. 2014. Glucosylated steviol glycosides composition as a taste and flavor enhancer. US Patent Application 2014/0010917 A1.

Pyo, S.-H. and Hayes, D. G. 2009. Designs of bioreactor systems for solvent-free lipase-catalyzed synthesis of fructose–oleic acid esters. *J. Am. Oil Chem. Soc.* 86:521–529.

Ramos-de-la-Peña, A. M., Renard, C. M. G. C., Wicker, L., Montañez, J. C., García-Cerda, L. A. and Contreras-Esquivel, J. C. 2014. Environmental friendly cold-mechanical/sonic enzymatic assisted extraction of genipin from genipap (*Genipa americana*). *Ultrason. Sonochem.* 21:43–49.

Ratnam, R., Aurora, S., Lali, A. M. and Subramaniyam, P. 2007. Enzyme catalyzed de-acylation of chlorinated sugar derivatives. Patent Application WO 2007054973.

Reddy, K. R. N., Farhana, N. I., Salleh, B. and Oliveira, C. A. F. 2010. Microbiological control of mycotoxins: Present status and future concerns. In *Current Research, Technology and Education Topics in Applied Microbiology and Microbial Biotechnology*, vol. 2, ed. A. Mendez-Vilas, 1078–1086. Badajoz: Formatex.

Ruela, H. S., Sutili, F. K., Leal, I. C. R., Carvalho, N. M. F., Miranda, L. S. M. and de Souza, R. O. M. A. 2013. Lipase-catalyzed synthesis of secondary glucose esters under continuous flow conditions. *Eur. J. Lipid Sci. Technol.* 115:464–467.

Samuel, M. S., Sivaramakrishna, A. and Mehta, A. 2014. Degradation and detoxification of aflatoxin B1 by *Pseudomonas putida*. *Int. Biodeterior. Biodegrad.* 86:202–209.

Sun, T., Jiang, B. and Pan, B. 2011. Microwave accelerated transglycosylation of rutin by cyclodextrin glucano-transferase from *Bacillus* sp. SK13.002. *Int. J. Mol. Sci.* 12:3786–3796.

Tanaka, O. 1997. Improvement of taste of natural sweeteners. *Pure Appl. Chem.* 69:675–683.

Taylor, M. C., Jackson, C. J., Tattersall, D. B. et al. 2010. Identification and characterization of two families of F420H2-dependent reductases from *Mycobacteria* that catalyse aflatoxin degradation. *Mol. Microbiol.* 78:561–575.

Walsh, M. K., Bombyka, R. A., Wagha, A., Bingham, A. and Berreau, L. M. 2009. Synthesis of lactose mono-laurate as influenced by various lipases and solvents. *J. Mol. Catal. B: Enzym.* 60:171–177.

Wan, H.-D., Tao, G.-J., Kim, D. and Xia, Y.-M. 2012. Enzymatic preparation of a natural sweetener rubusoside from specific hydrolysis of stevioside with β-galactosidase from *Aspergillus* sp. *J. Mol. Catal. B: Enzym.* 82:12–17.

Wenda, S., Illner, S., Mell, A. and Kragl, U. 2011. Industrial biotechnology—The future of green chemistry? *Green Chem.* 13:3007–3047.

Yang, Z. and Huang Z.-L. 2012. Enzymatic synthesis of sugar fatty acid esters in ionic liquids. *Catal. Sci. Technol.* 2:1767–1775.

Ye, R. and Hayes, D. G. 2011. Optimization of the solvent-free lipase-catalyzed synthesis of fructose-oleic acid ester through programming of water removal. *J. Am. Oil Chem. Soc.* 88:1351–1359.

Yoo, I. S., Park, S. J. and Yoon, H. H. 2007. Enzymatic synthesis of sugar fatty acid esters. *J. Ind. Eng. Chem.* 13:1–6.

Zaidan, U. H., Rahman, M. B. A., Othman, S. S. et al. 2012. Biocatalytic production of lactose ester catalysed by mica-based immobilised lipase. *Food Chem.* 131:199–205.

Zeuner, B., Luo, J., Nyffenegger, C., Aumala, V., Mikkelsen, J. D. and Meyer, A. S. 2014. Optimizing the bio-catalytic productivity of an engineered sialidase from *Trypanosoma rangeli* for 3′-sialyllactose production. *Enzyme Microb. Technol.* 55:85–93.

Zhao, L. H., Guan, S., Gao, X. et al. 2011. Preparation, purification and characteristics of an aflatoxin degradation enzyme from *Myxococcus fulvus* ANSM068. *J. Appl. Microbiol.* 110:147–155.

Zhong, X., Qian, J., Liu, M. and Ma, L. 2013. *Candida rugosa* lipase-catalyzed synthesis of sucrose-6-acetate in a 2-butanol/buffer two-phase system. *Eng. Life Sci.* 13:563–571.

19

Enzymes in Probiotics

T. R. Keerthi, Honey Chandran Chundakattumalayil, and Muthusamy Chandrasekaran

CONTENTS

19.1 Introduction

Probiotics hold immense promise for improved human health benefits on consumption (Table 19.1) and hence have consequent commercial value. Probiotics are live nonpathogenic microorganisms administered to improve microbial balance, particularly in the gastrointestinal tract, which is home to more than 500 different types of bacteria, which help keep the intestines healthy, assist in digesting food, and help the immune system. Researchers believe that some digestive disorders happen when the balance of friendly bacteria in the intestines becomes disturbed after an infection or after taking antibiotics. Intestinal problems can also arise when the lining of the intestines is damaged. Under such circumstances, probiotics can improve intestinal function and maintain the integrity of the lining of the intestines because these friendly organisms help fight bacteria that cause diarrhea. There's also evidence that probiotics help maintain a strong immune system. A sharp increase in autoimmune and allergic diseases is noted in societies with very good hygiene, probably because the immune system is not being properly challenged by pathogenic organisms, whereas introducing friendly bacteria in the form of probiotics is

TABLE 19.1

Health Benefits of Probiotic Bacteria to the Host and Speculated Mechanisms Involved

Health Benefits	Proposed Mechanisms Involved
Resistance to enteric pathogens	Antagonism activity
	Adjuvant effect increasing antibody production
	Systemic immune effect
	Colonization resistance
	Limiting access of enteric pathogens (pH, bacteriocins/defensins, antimicrobial peptides, lactic acid production, and toxic oxygen metabolites)
Aid in lactose digestion	Bacterial lactase acts on lactose in the small intestine
	Small bowel bacterial overgrowth
	Lactobacilli influence the activity of overgrowth flora, decreasing toxic metabolite production
	Normalization of a small bowel microbial community
	Antibacterial characteristics
Immune system modulation	Strengthening of nonspecific and antigen-specific defense against infection and tumors
	Adjuvant effect in antigen-specific immune responses
	Regulating/influencing Th1/Th2 cells, production of anti-inflammatory cytokines
	Decreased release of toxic N-metabolites
Anticolon cancer effect	Antimutagenic activity
	Detoxification of carcinogenic metabolites
	Alteration in procancerous enzymatic activity of colonic microorganisms
	Stimulation of immune function
	Influence on bile salt concentration
Decreased detoxification/ Excretion of toxic microbial metabolites	Increased bifidobacterial cell counts and shift from a preferable protein to carbohydrate-metabolizing microbial community, less toxic and for putrefactive metabolites, improvements of hepatic encephalopathy after the administration of bifidobacteria and lactulose
Allergy	Prevention of antigen translocation into bloodstream
	Prevent excessive immunologic responses to increased amount of antigen stimulation of the gut
Blood lipids, heart disease	Assimilation of cholesterol by bacterial cell
	Alteration in the activity of BSH enzyme
	Antioxidative effect
Antihypertensive effect	Bacterial peptidase action on milk protein results in antihypertensive tripeptides
	Cell wall components act as ACE inhibitors
Urogenital infections	Adhesion to urinary and vaginal tract cells
	Competitive exclusion
	Inhibitor production (H_2O_2, biosurfactants)
Infection caused by *Helicobacter pylori*	Competitive colonization
	Inhibition of growth and adhesion to mucosal cells, decrease in gastric *H. pylori* concentration
Hepatic encephalopathy	Competitive exclusion or inhibition of urease producing gut flora
Neutralization of dietary carcinogens	Production of butyric acid neutralizes the activity of dietary carcinogens
NEC (necrotic inflammation of the distal small intestine)	Decrease in TLRs and signaling molecules and increase in negative regulations
	Reduction in the IL-8 response
Rotaviral gastroenteritis	Increased IgA response to the virus
Inflammatory bowel diseases, type I diabetes	Enhancement of mucosal barrier function
Crohn's disease	Reduction in proinflammatory cytokines, including TNFa, reduction in the number of CD4 cells as well as TNFa expression among intraepithelial lymphocytes
Caries gingivitis	Reduction in gingivitis by *L. reuteri*, affects on streptococcus mutants, colonization of the teeth surface by lactobacilli
	Less caries after the ingestion of living or oral vaccination with heat-killed lactobacilli
Enhanced nutrient value	Vitamin and cofactor production

Source: Nagpal, R., Kumar, A., Kumar, M., Behare, P. V., Jain, S. and Yadav, H., *FEMS Microbiol. Lett.*, 334, 1–15, 2012.

believed to challenge the immune system in healthy ways (http://www.webmd.com/digestive-disorders /features/what-are-probiotics/).

Probiotics exert their beneficial effects through various mechanisms, including lowering intestinal pH, decreasing colonization and invasion by pathogenic organisms, and modifying the host immune response. The molecular mechanisms of probiotics are yet to be understood although "microorganism– host cross talk," such as microorganism-associated molecular patterns (MAMPs) of probiotics, and pattern recognition receptors (PRRs) of the gastrointestinal mucosa are considered to be very important factors (Lebeer et al. 2010).

The first recorded probiotic was fermented milk for human consumption. After that, probiotics became popular with animal nutrition. Since ancient times, probiotic foods, such as fermented products and cultured milk, have been around. But it has been only of late that there has been a great demand for probiotics with the markets flooding with probiotic supplements and foods. For thousands of years, ethnic cuisines across the world have included fermented foods with every meal as condiments, beverages, breads, and protein sources. Beneficial bacteria and yeasts have been an understood necessity in these traditional diets. Fermentation breaks down the fats, proteins, and carbohydrates in food before we eat, and although not all fermented foods are "probiotic" with an ability to colonize the intestine, they do provide enzyme-rich, nutrient-rich foods that are easily digested and healthy to eat. In addition, eating lacto-fermented foods with a meal can boost the nutrient level of all the other foods in that meal.

Scientific studies have been in progress ever since the importance and role of live microbial cultures in health were recognized in the nineteenth century, and researchers focused their investigations on the gut microbiota and the various mechanisms of microflora that facilitate and maintain the health. Recent studies have endorsed the role of probiotics as a part of a healthy diet for humans and animals, providing a safe, cost-effective barrier against microbial infection (Parvez et al. 2006). Moreover, with the recognition and endorsement by the United Nations and World Health Organization, the call that "Efforts should be made to make probiotic products more widely available, especially for relief work and population at high risk of morbidity and mortality," use of probiotics has assumed significant practical importance as a food. Probiotics, prebiotics, synbiotics, and cobiotics are important classes of feed additives that have shown promise in replacing or reducing antibiotics used as prophylactics or growth promoters.

Today, a diverse array of probiotic dietary supplements are available in the market in different forms other than dairy, such as capsules, liquids, powders, and tablets, etc. In the case of dairy other than yogurt, attractive products, such as probiotic ice creams and probiotic-based soft drinks with different flavors have been launched in the market. The global probiotics market is going through a steady phase of growth. In 2013, the market stood at US$58,700.30 million. Supported by the increased demand for probiotics among health-conscious consumers, it is expected to reach a value of US$96,046.80 million by 2020 (http://www.marketsandmarkets.com/market-reports/probiotic-market-advanced-technologies –and–global-market-69.html/). Considering the importance and future prospects of probiotics as one of the major food supplements in the global market, the role and prospects of enzymes in probiotic development and production are discussed in this chapter.

19.2 Concept of Probiotics

The word "probiotic" is derived from the Greek meaning "for life," and the term has had several different meanings over the years. The word probiotic was first used as an anonym of the word antibiotic. Lilly and Stillwell (1965) defined probiotics as microorganisms promoting the growth of other microorganisms. Later, Parker (1974) defined probiotics as "Organisms and substances which contribute to intestinal microbial balance" (Anukam and Reid 2007; Oyetayo and Oyetayo 2005). However, Fuller (1989, 1992) improved the definition by redefining the term probiotic as "A live microbial feed supplement which beneficially affects the host animal by improving its intestinal microbial balance," stressing the importance of live cells as an essential component of the probiotic preparation.

According to the World Health Organization (WHO)/Food and Agriculture Organization (FAO), "probiotics are live microorganisms which when administered in adequate amounts can confer a health benefit on the host," and this definition is presently the most accepted definition.

19.3 Traditional Fermented Foods That Serve the Role of Probiotics

Fermented foods are considered to be so good because the fermentation process increases certain nutrients and the digestibility of food. Further, they are excellent sources of probiotics. Through fermentation, foods are not only preserved, but get transformed into a probiotic supplement in addition to increasing digestibility and vitamin content. Hence, fermented foods are found all over the world in traditional cultures. Some of the well-known traditional fermented foods that serve the purpose of the probiotic role are listed in Table 19.2.

Kombucha is a culture of symbiotic beneficial bacteria and yeasts, which originated in China nearly 2000 years ago. This culture is brewed with tea and sugar and fermented into a sweet and sour, slightly effervescent drink. Kombucha contains many amino acids and B vitamins in addition to its bountiful population of beneficial microorganisms and is believed to be an excellent stimulant to digestion and the immune system.

Kimchi is a traditional Korean lacto-fermented condiment made of cabbage and other vegetables and seasoned with salt, garlic, ginger, and chile peppers. Most Asian diets include a daily portion of some kind of pickled vegetable. Lacto-fermentation occurs when sugars and starches are converted to lactic acid by the lactobacilli that are prevalent in vegetables and fruits. The proliferation of lactobacilli in fermented vegetables enhances their digestibility and increases vitamin levels.

Sauerkraut is a cabbage that has been salted and lacto-fermented over a period of weeks. Latin American cultures make a version of sauerkraut called cortido. The beneficial bacteria so abundant in sauerkraut produce numerous helpful enzymes as well as antibiotic and anticarcinogenic substances. The main by-product, lactic acid, not only keeps vegetables and fruits in a state of preservation, but also promotes the growth of healthy flora throughout the intestine.

Miso is made by adding an enzymatic culture to a base of soybeans and, often, a grain (usually wheat, barley, or rice). Salt and water are the only other ingredients of natural miso. Through aging, the enzymes reduce the proteins, starches, and fats into amino acids, simple sugars, and fatty acids. It also contains lactobacillus bacteria, which aid in digestion. Miso is used as a soup base but is also good in sauces, gravies, dips, spreads, dressings, and marinades. Miso should be unpasteurized and never boiled; high

TABLE 19.2
List of Some Fermented Foods Serving the Purpose of Probiotics

Fermented Food	Probiotic Microbe Involved	Source
Kombucha	A culture of symbiotic beneficial bacteria and yeasts	Brewed tea
Kimchi	Lactobacilli	Cabbage and other vegetables and seasoned with salt, garlic, ginger, and peppers
Sauerkraut	Lactic acid producing bacteria	Cabbage
Miso	Enzymatic culture, lactobacillus bacteria	Soybeans and, often, a grain (usually wheat, barley, or rice)
Umeboshi	–	Pickled plums (ume) from Japan
Pickles	Lactic bacteria	Vegetables (and sometimes fruits, nuts, seeds, animal products, and other ingredients)
Tempeh	Mold	Cooked, split, fermented soybeans bound together
Yogurt and kefir	Bacteria	Milk that has been inoculated with live bacterial cultures
Yogurt	Lactic bacteria	Milk

temperatures will kill the beneficial microorganisms. Miso is a superb source of easily assimilated complete protein.

Umeboshi are salty sour lacto-fermented pickled plums (ume) from Japan. Umeboshi are highly alkaline and used to neutralize fatigue, stimulate the digestive system, and promote the elimination of toxins. They are valued for their natural antibiotic properties and ability to regulate intestinal health.

Pickles include a wide range of vegetables, and sometimes fruits, nuts, seeds, animal products, and other ingredients that can be lacto-fermented using salt, temperature, and a controlled environment for a period of time. Most modern pickles, however, are made using vinegars and/or heat processing, which limits or eliminates the beneficial bacteria and enzymes that result from lacto-fermentation.

Tempeh is an ancient Indonesian staple made from cooked, split, fermented soybeans bound together with a mold that makes soy easier to digest and provides many valuable vitamins. Tempeh is an excellent protein source for calcium and iron, and the mold produces an antibiotic to increase the body's resistance to infections.

19.3.1 Cultured Dairy Products

Yogurt and kefir consist of milk that has been inoculated with live bacterial cultures, which convert the milk's lactose sugar into lactic acid. For people who have difficulty in digesting the lactose in milk, cultured dairy products may be easier to digest because the live, active cultures produce lactase, which predigests the lactose.

Yogurt has been made in cultures around the world for thousands of years. The bacteria that are traditionally used to make yogurt are also responsible for many of yogurt's health benefits, such as improved intestinal health and increased immune function. Yogurts made with soy milk and coconut milk are also available and contain the same active cultures.

Kefir, like yogurt, is a cultured milk product and usually tolerable to those with lactose intolerance. Kefir contains different types of beneficial bacteria than yogurt does as well as beneficial yeasts. Kefir contains more bacterial strains that remain viable in the digestive system, increasing the likelihood of intestinal colonization.

19.4 Microorganisms as Probiotics

Microorganisms considered to be probiotics (Holzapfel et al. 2001) are shown in Table 19.3. Although several microorganisms are known to have potential application as probiotics, the most common

TABLE 19.3

Microorganisms Considered to Be Probiotics

Lactobacillus	Bifidobacterium	Other Lactic Acid Bacteria	Non-Lactic Acid Bacteria
L. acidophilus	*B. adolescentis*	*Enterococcus faecalis*	*Bacillus cereus* var. *toyoi*
L. amylovorus	*B. animalis*	*Enterococcus faecium*	*Bacillus clausii*
L. casei	*B. bifidum*	*Lactococcus lactis*	*Escherichia coli Nissle 1917*
L. crispatus	*B. breve*	*Leuconostoemesenteroldes*	*B. Proplonlbucterlum*
L. delbrueckii subsp.	*B. infantis*	*Pediococcus acidolactici*	*freudenreichii*
Bulgaricus	*B. lactis*	*Streptococcus thermophilus*	*Saccharomyces cerevisiae*
L. gasseri	*B. longum*	*Sporolactobacillus inulinus*	*Saccharomyces boulardii*
L. gallinarum			*Aspergillus oryzae*
L. johnsonii			
L. paracasei			
L. plantarum			
L. reuteri			
L. rhamnosus			

Source: Holzapfel, W. H., Haberer, P., Geisen, R., Björkroth, J., Schillinger, U., *Am. J. Clin. Nutr.*, 73, 365–373, 2001.

organisms of choice as probiotics are lactic acid bacteria (LAB), *Bifidobacterium*. Probiotics consist of *Saccharomyces boulardii* yeast or lactic acid bacteria, such as Lactobacillus and Bifidobacterium species, and are regulated as dietary supplements and foods. LAB exists in humans and in animals. LAB constitutes an integral part of the healthy gastrointestinal microecology and influences the host metabolism (Gibson and Fuller 2000). They are part of the microbiota on mucous membranes, such as the intestine, mouth, skin, and urinary and genital organs of both humans and animals, beneficially influencing the ecosystems (Schrezenmeir and de Vrese 2001). LAB is among the most numerous of the commensal microflora inhabiting the human large intestine and plays an important role in the maintenance of gut homeostasis and is beneficial to health. LABs were first isolated from milk. They can be found in fermented products such as meat, milk products, vegetables, and bakery products and also in decomposing plants and lactic products.

Species of *Lactobacillus* and *Bifidobacterium* are the most common probiotic bacteria used as food adjuvants. These bacteria produce lactic acid as the major metabolic end product of carbohydrate fermentation, which partially enables them to outcompete other bacteria in a natural fermentation as they can withstand the increased acidity from organic acid production (e.g., lactic acid). This characteristic of LAB is attributed to its significant role in food fermentations as acidification inhibits the growth of spoilage agents. Moreover, the proteinaceous bacteriocins produced by several LAB strains do impart an additional hurdle for the proliferation of spoilage and pathogenic microorganisms. Furthermore, lactic acid and other metabolic products contribute to the organoleptic and textural profile of a food item. The industrial importance of the LAB is further evinced by its generally recognized as safe (GRAS) status due to its ubiquitous appearance in food and contribution to the healthy microflora of human mucosal surfaces. Hence, probiotic researchers mainly study LABs as probiotics. LAB-based probiotics used in various products in the world are shown in Table 19.4.

Some of the beneficial effects of LAB consumption include (i) improving intestinal tract health, (ii) enhancing the immune system and synthesizing and enhancing the bioavailability of nutrients, (iii) reducing symptoms of lactose intolerance, and (iv) decreasing the prevalence of allergies in susceptible individuals. Strains isolated from the human intestinal tract are generally recommended as suitable for probiotic use in humans because some health-promoting benefits may be species-specific, and microorganisms may perform optimally in the species from which they were isolated (Stanton et al. 2003; Vinderola et al. 2008). Probiotic efficacy varies from strain to strain and can be enhanced by combinational therapy. Health effects of some specific probiotic strains documented by Sanders (2009) are presented in Table 19.5. The isolation and characterization of new strains are still desirable, in developing countries mainly, for the formulation of probiotic foods because there is still restricted access to probiotic strains to small dairy plants (Vinderola et al. 2008). Other than LAB, the most widely used probiotics are Bacillus and yeast. The genus Bacillus comprises a multiplicity of rod-shaped gram-positive microorganisms naturally found in soil. Some strains of this heterogeneous group have been chosen for the use in animal nutrition because of their beneficial effects (Alexopoulos et al. 2004; Duc et al. 2004; Hoa et al. 2001; Jadamus et al. 2001; Jørgensen and Kürti 2006). *Bacillus* sp.-based probiotics available commercially in the market are shown in Table 19.6.

When Bacillus spores are ingested with feed, they germinate in the digestive tract and grow as vegetative cells but do not proliferate to a larger extent. Bacillus species do not colonize the intestine and are, therefore, by definition, included in the transient flora. As exogenous microorganisms, Bacillus probiotics have a high potential for stimulating local intestinal immunity (Sanders et al. 2003). *Bacilli* was proven to improve health with no visible side effects after a trial conducted in animals (Duc et al. 2004). The inherent resistance of spores to environmental stress is an attractive attribute for commercial probiotic preparations. Moreover, some of the functional foods, such as natto of Japan, which comprises the use of bacilli and varieties of probiotic products that contain spore-formers are available commercially (Duc et al. 2004). Recently, both the vegetative and spore forms of *Bacillus subtilis* isolated from cow milk were observed to have probiotic potency, and it was noted that spores had higher potency (Anu and Keerthi 2012).

Yeast products have been known to be used as health aids for more than 5000 to 8000 years by Babylonians, Egyptians, and Celts, for benefits to the skin and "color" although it was primarily used for alcohol production. Around 370 BC, Hippocrates discovered the diuretic action of yeast and considered

TABLE 19.4

Lactic Acid Bacteria-Based Probiotics Used in Various Products in the World

Probiotic Bacteria	Product	Country
Bifidobacterium bifidum	Infant formula	Turkey
B. breve	Drink	Japan
B. lactis	Infant formula	Israel
	Drink	South Africa, Chile
B. longum SBT-2928	Milk	Japan
B. longum SBS036	Milk	Japan
Bifidobacterium sp	Drink	UK
Lactobacillus acidophilus	Yogurt	Chile, USA
	Drink	UK
	Yogurt, Drink	Austria
Lactobacillus acidophilus 5	Yogurt	UK
L. acidophilus 7	Drink	Austria
L. acidophilus NCFB 1748	Yogurt	Denmark
L. acidophilus SBT-2062	Milk	Japan
L. bulgaricus	Drink	France, Austria
L. casei Shirota	Drink	Argentina, Australia, Belgium, Brazil, Brunet, China, Germany, Hong Kong, Indonesia, Japan, Korea, Luxemberg, Mexico, Netherlands, Philippines, Singapore, Taiwan, Thailand, Uruguay, UK, NewYork/USA
L. casei	Drink	USA
	Yogurt	France, Colorado-Arizona/USA
L. helveticus	Milk drink	Finland
	Drink	Iceland
L. johnsonii Lal	Yogurt	Switzerland, Germany, Japan, Austria
L. lactis LlA	Yogurt	Sweden
L. plantarum 299v	Fruit drink	Sweden
	Ice cream, recovery drink, oat mixture	Sweden
L. reuteri	Infant formula	Israel
	Cheese	Spain, Portugal, Finland
	Milk	Japan, Finland
	Yogurt	USA, Finland
	Yogurt drink	UK
	Ice cream	Finland
	Fruit drink	Finland
L. rhamnose ATCC53103	Yogurt	Australia, Papua New Guinea, Indonesia

it to be a drug. During the middle ages, clergy used yeast against leprosy to prevent contamination. It was also used to cure rubella and scarlet fever. At the start of the twentieth century, the Indochinese used a native Indonesian cure for diarrhea by drinking tea made with tropical fruits (lychee and mango). It has since been discovered that the agent in the tea responsible for stopping diarrhea was live yeast (*Saccharomyces cerevisiae* var. *boulardii*). Probiotic yeasts differ from brewery yeasts by their metabolic activity, the latter being fed in an inactivated form for their nutrient content.

Successful marketing of probiotic products requires a minimal amount of viable probiotic cells guaranteed throughout the shelf life. To obtain the beneficial effects associated with this type of food, the bacteria must remain viable and in a proper concentration when the host consumes the product. This fact could determine the shelf life of the developed product because the survival of the probiotics depends on many factors in the food (Talwalkar and Kailasapathy 2004). There are documented challenges associated with the consumption of probiotics because viability must be maintained in order for them to have

TABLE 19.5

Some Probiotic Strains and Their Associated, Documented Health Effects

Strain/Strain Blend (Alternate Strain Designation)	Health Effects
L. rhamnosus GG (ATCC 53103)	Reduced duration of diarrheal illnesses
	Staying healthy
	Enhanced immune function
B. lactis Bb12	Reduced intestinal infections
	Enhanced immune function
B. lactis HN019 (DR10)	Enhanced immune function
Saccharomyces cerevisiae boulardii	Reduced complications from antibiotics
B. infantis 35264	Reduced symptoms of irritable bowel syndrome
L. casei Shirota	Enhanced immune function
	Extended remission for superficial bladder cancer
	Improved bowel function
L. casei DN-114 001	Enhanced immune function
	Staying healthy
	Reduced complications from antibiotics
B. animalis DN173 010	Improved bowel function
L. reuteri RC-14 + *L. rhamnosus* GR-1	Improved therapeutic outcome for bacterial vaginosis treated with metronidazole
L. reuteri ATCC 55730	Staying healthy
	Reduced crying time in babies with colic
L. fermentum VRI003	Reduced days and severity of respiratory illness
L. johnsonii Lj-1	Enhanced immune function
L. plantarum 299V	Reduced symptoms of irritable bowel syndrome
VSL#3	Extended remission for patients with pouchitis

Source: Sanders, M. E. 2009. Probiotics and health. *The Whitehall-Robins Report.* 18, 1, 2009.

TABLE 19.6

Bacillus sp-Based Probiotics Available Commercially in the Market

Product Name	Manufacturer	Microorganism	Comment
Lactospore	Sabinsa Corp., Piscataway, NJ	*Lactobacillus sporogenes*	Human use
Lacbon, Lacris	Uni-Sankyo	*L. sporogenes*	Human use, approved by the Japanese Ministry of Health and Welfare
Enterogermina	Sanofi-Winthrop	*Bacillus clausii*	Human use
Lactopure	Pharmed Medicare	*L. sporogenes*	Human and animal use
Flora-Balance	Flora-Balance, Montana, USA	*Brevibacillus laterosporus*	Human use
Medilac	Hanmi Pharmaceutical Co., Ltd.	*B. subtilis* R0179 *Enterococcus faecium*	Approved by the Chinese State Drug Authority, also sold OTC in Korea
Biosporin, Subalin, Gynesporin, and others	D. K. Zabolotny Institute of Microbiology and Virology, Ukraine	*B. subtilis* recombinant strains	Human use
Nature's First Food	Nature's First Law, San Diego, CA	42 species listed as pro-human use biotic complex ingredients, including *B. laterosporus*, *B. polymyxa*, *B. subtilis*, *B. pumulis*	Human use

a beneficial effect. For this to occur, they must arrive at the colon at a concentration of 10^7 live cells per gram of intestinal content (Bouhnik 1993). Probiotic viability also decreases during product processing and storage (Mattila-Sandholm et al. 2002).

19.5 Role of Enzymes in Probiotics

Enzymes do have a very important role in the development of probiotics either as a processing aid for preparation of prebiotics or as an additional supplement to a specific probiotic product (Figure 19.1). Some of the enzymes known for their application in probiotic formulations are presented in Figure 19.2. Probiotics and digestive enzymes contribute to the healthy gastrointestinal environment. Nearly 70% of the immune system is located in the digestive tract, supporting intestinal balance, which is important to digestive health. The major digestive enzymes responsible for digestion are proteases that digest proteins, amylases that digest carbohydrates, and lipases that digest fats. These three important enzymes must be present every time food is taken. Prebiotics and digestive enzymes help to boost the growth and functioning of probiotics.

   The action of the probiotics complements the host with vital nutrients that are usually missing from the diet of an individual. However, they can make the food more bioavailable, thereby increasing the digestibility of the subject concerned. Further, appropriate lactobacilli strains in specific amounts were found to provide relief in cases of lactose intolerance (http://probiotics101.probacto.com/effects-of-probiotics -on-nutrient-synthesis/), and *Lactobacillus bulgaricus*, *Streptococcus thermophiles* when used in fermented dairy products provide enough bacterial lactase to the body (Martini et al. 1987). Probiotics even enhance the bioavailability of fats, carbohydrates, and proteins with the help of their different hydrolytic enzymes (Fernandes et al. 1987).

19.6 Prebiotics

Prebiotics are the next class of probiotic nutrients, and they are mostly active in the large intestine where they feed the bifidobacteria. They are just a special type of fiber that promotes bacterial formation. They include soluble, nondigestible fibers made up of long-chain carbohydrates, which are simple sugars (oligosaccharides) found in plants, mainly vegetables. These soluble fiber substances are not digested by the human body's digestive enzymes and make it into the large intestine undigested. They

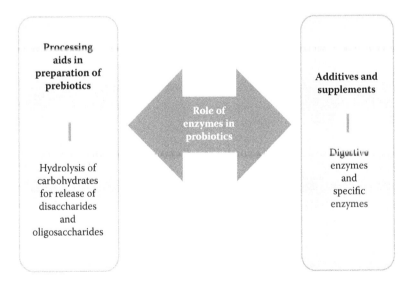

FIGURE 19.1 Role of enzymes in probiotic development.

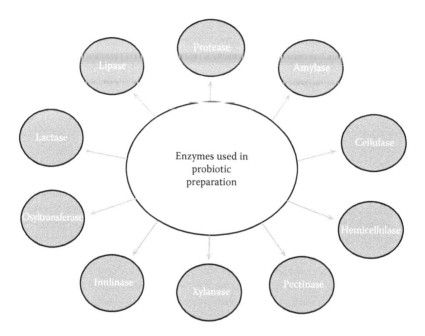

FIGURE 19.2   Enzymes used in production of prebiotics, synbiotics, and cobiotics.

are preferentially utilized by bifidobacteria as carbon and energy sources defined as "bifidogenic factors." They nourish the beneficial bacteria in the gastrointestinal tract and also assist mineral absorption, improve bowel pH, and cleanse the colon of debris. Some of the bifidogenic factors that are of commercial significance include fructooligosaccharides (FOS); lactose derivatives, such as lactulose, lactitol, galactooligosaccharides, isomalto-oligosaccharides, xylooligosaccharides, glucooligosaccharides, and soybean oligosaccharides (O'Sullivan 1996). Among these prebiotics, FOS and inulin have received much attention because they lead to an increase in bifidobacteria numbers and reduce those harmful bacteria in the human colon (Gibson et al. 1995). Butyric acid, a healthy compound that stimulates the growth of colonic epithelial cells, is one of the beneficial end products of bifidobacteria feeding on FOS and inulin. Prebiotics are good when they come in natural packages, such as a banana or an apple, and, particularly, if such foods are consumed with a good quality probiotic supplement rich in bifidobacteria.

For a food ingredient to be classified as a prebiotic, it must meet the following criteria (Gibson and Roberfroid 1995):

1. It cannot be hydrolyzed or absorbed in the upper part of the digestive system.
2. It must cause the growth and/or activation of one or a limited number of beneficial colonic bacteria by being a selective substrate.
3. It must be able to change the microflora of the colon to a healthier composition.
4. It must induce beneficial luminal or systemic effects to the health of the host.

Nondigestible carbohydrates are the most commonly used prebiotics, and they include resistant starch (starch that is not hydrolyzed in the small intestine), nonstarch polysaccharides (hemicellulose, pectins, gums), and oligosaccharides (galactooligosaccharides, fructooligosaccharides [FOS]) (Gibson and Roberfroid 1995; van Dokkum and van den Heuvel 2001).

Oligosaccharides are commonly used in the production of low caloric and diabetic foods. Methods used to produce nondigestible oligosaccharides are (i) hot water extraction of roots (Roberfroid 2000), such as inulin extraction from chicory root; (ii) enzyme hydrolysis of oligosaccharides or polysaccharides, such as xylooligosaccharide production by the action of *xylanase* on pectin; and (iii) the use of *osyl-transferases*, which can be used to synthesize oligosaccharides from a mixture of disaccharides, such as FOS production from sucrose. Fructooligosaccharides or oligofructose is the only oligosaccharide that meets all prebiotic criteria (Gibson and Roberfroid 1995). Inulin is the main polysaccharide for FOS production and is naturally found in chicory and Jerusalem artichoke roots (Roberfroid 2000). FOS can be produced by inulin hydrolysis with acid or *inulase* and from sucrose using *osyl-tranferases*.

Prebiotics have been associated with several health benefits. Some of the more scientifically studied include an increase of beneficial bacteria and a decrease in detrimental bacteria located in the gastrointestinal tract, increased absorption of minerals/cations, reduced plasma triacylglycerols, and constipation relief (Delzenne and Roberfroid 1994; Fiordaliso et al. 1995; Gibson et al. 1995; Tomomatsu 1994). Prebiotics modulate microbial composition and ensure a healthy gastrointestinal tract environment that can prevent colon cancer development (Hijová et al. 2013). The use of prebiotics has also been associated with better absorption of divalent cations in the colon (Gibson and Roberfroid 1995). This was confirmed in rat studies in which a diet that contained FOS or inulin was found to improve intestinal uptake of the minerals Ca^{2+}, Mg^{2+}, and Fe^{2+} by 60%–65% (Delzenne and Roberfroid 1994). High plasma triacylglycerols and cholesterol increase an individual's risk of developing coronary heart disease, a leading cause of death in North America.

By increasing the amount of prebiotics in the diet, it is possible to increase and maintain healthy bacterial gut flora in the host (Gibson et al. 2003; Sanders 1998). Ingredients in certain food products may naturally contain prebiotics, which help to improve the functional efficacy of probiotics. Many other foods, such as dairy and meat products, cereals, beverages, and infant formulas, can be fortified with prebiotics during the manufacturing process to increase probiotic efficacy (Gibson et al. 2004). In addition, a number of other suitable food components, including nonspecific substrates, plants and their extracts, metabolites of microorganisms, and polyunsaturated fatty acids may also be important in probiotic efficacy (Bomba et al. 2006). Some human studies have been conducted to study the benefit of administering prebiotics, predominantly in patients suffering from ulcerative colitis. A couple of studies have reported the benefits of using germinated barley foods as a source of prebiotics to reduce the severity of (Bamba et al. 2002) and prolong the duration of remission in ulcerative colitis (Hanai et al. 2004). Prebiotics may show beneficial effects in human IBD and colitis, and hence, the emerging therapy for such diseases targets the function of intestinal microflora (Looijer-van Langen and Dieleman 2009).

19.6.1 Fructooligosaccharides

Fructooligosaccharides (FOSs) have been shown to have a positive effect in reducing plasma triacylglycerols, phospholipids, and cholesterol (Fiordaliso et al. 1995). These oligosaccharides have the ability to beneficially alter the gut microbiota and also have the potential to reduce the risk of colorectal cancer, stimulate immune response, alleviate symptoms of inflammatory bowel disease, modify serum triglycerides and cholesterol, and enhance mineral absorption in the intestine, thereby reducing the risk of intestinal infectious diseases, cardiovascular disease, non-insulin dependent diabetes, obesity, osteoporosis, and cancer (Roberfroid 2000; Shah et al. 2007; Tuohy et al. 2003; Williams and Jackson 2002). In the human colon, the FOSs are completely fermented mostly to lactate, short-chain fatty acid (SCFA) (acetate, propionate, and butyrate), and gas. The FOSs stimulate bifidobacterial growth and suppress the growth of potentially harmful species in the colon. Other effects of FOSs include a decrease in fecal pH, an increase in fecal or colonic organic acids, a decrease in fecal bacterial enzymatic activities, and a modification in fecal neutral sterols (Bornet et al. 2002). FOSs were also demonstrated to have antagonistic effects of FOSs in humans with ulcerative colitis (Lewis et al. 2005) to enhance magnesium absorption in humans and in animal models and reduction in colon tumor development due to enhancement of both colon butyrate concentrations and local immune system effectors (O'Bryan et al. 2013).

19.6.2 Inulin

Inulin has been shown to stimulate the immune system and can inhibit the growth of harmful bacteria such as *E. coli* and Clostridia in addition to enhancing bifidobacteria in the colon (Zubaidah and Akhadiana 2013). Daily consumption of inulin reduces appetite, making weight loss easier. Inulin's effect on the gut microbiota can help stem obesity. Inulin has a sweet taste, making it very useful as a sugar replacement (Cani et al. 2011), and it increases the regularity and volume of stool and reduces constipation and diarrhea (Roberfroid 2007).

19.6.3 Xylooligosaccharides

Xylooligosaccharides (XOSs) have great prebiotic potential and can be used as ingredients in functional foods, cosmetics, pharmaceuticals, or agricultural products. XOSs can hardly be digested and absorbed but directly enter the large intestine and are preferably utilized by bifidobacterium to proliferate beneficial bacterium for the human body and inhibit growth and proliferation of other harmful bacterium. Xylooligosaccharides have been reported to have immunomodulatory activity, anticancerous activity, antimicrobial activity, growth regulator activity, and other biological activity, such as antioxidant, antiallergic, anti-inflammatory, antihyperlipidemic activity and cosmetics and variety of other properties. It is also used in preparation of micro- or nanoparticles and hydrogels for drug delivery and treatment and prevention of gastrointestinal disorders (Praveen Kumar et al. 2012). XOSs are mildly laxative by stimulation of bacterial growth and fermentation. XOS intake has been reported to be highly effective for reducing severe constipation in pregnant women without any adverse effects (Praveen Kumar et al. 2012).

19.6.4 Pectic Oligosaccharides (POSs)

Pectin is a complex galacturonic acid-rich polysaccharide, which occurs naturally in the cell walls of higher plants and acts as a cement-like material for the cellulosic components of the plant cell wall. Pectic oligosaccharides (POSs) have been proposed as a potential source of prebiotics capable of exerting a number of health-promoting effects (Keawyok et al. 2012). Enzymatic hydrolysis of citrus and apple pectins in membrane reactors produces oligosaccharides of 3–4 kDa molecular weight. A nitric acid hydrolysis of citrus peel produces low molecular weight arabinose-based oligosaccharides. Both of these materials have been evaluated in fecal batch cultures and have both been shown to promote growth of *bifidobacteria* (O'Bryan et al. 2013).

Foods that contain naturally occurring prebiotics are asparagus, beans, rye bread, honey, garlic, onions, pears, apples, most berries, barley, tomatoes, and bananas. But there is a problem; it is hard to get enough FOS or inulin by eating fruits and vegetables. There are several nutritional factors that relate to the bioavailability of naturally occurring prebiotics: Soluble prebiotics, such as FOS, are often chemically bound to or in a matrix of insoluble fiber, such as cellulose, which limits bioavailability. Potential prebiotics, such as fruit pectins, need some help from enzymes before they can perform as effective prebiotics. Other non-FOS oligosaccharides need to be cracked by enzymatic action to be rendered active as prebiotics. Guar gum is an example.

To solve the abovementioned problem, recently an enzymatic consortium named Enzalase was launched to the rescue. Enzalase was developed to improve the bioavailability of naturally occurring prebiotics in whole foods, specifically their bioavailability to bifidobacteria and advertised as containing "bifidogenic enzymes" (http://www.enzylase.net/what-is-enzylase.aspr/). This enzymatic consortium includes the high-potency cellulase (3000 CU/capsule), hemicellulase (6400 HCU/capsule), and pectinase (7500 ADJU/capsule) enzymes. The cellulase in Enzalase digests cellulose that traps prebiotics, such as FOS and inulin, freeing them so they are more available to bifidobacteria. The hemicellulase digests noncellulose polysaccharides such as galactomannoglucans, yielding polysaccharides with lower molecular weight and greater bifidogenic activity, and the pectinase renders fruit pectins more bifidogenic. The overall result of this unique combination of enzymes is a dramatic stimulation of the growth of bifidobacteria.

19.7 Synbiotics

Synbiotics are defined as "a mixture of probiotics and prebiotics (Figure 19.3) that beneficially affect the host by improving the survival and implantation of live microbial dietary supplements in the

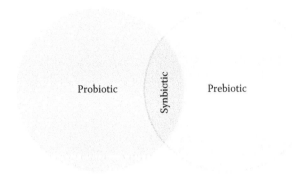

FIGURE 19.3   Concept of synbiotics.

gastrointestinal tract by selectively stimulating their growth and/or by activating the metabolism of one or a limited number of health-promoting bacteria and thus improving welfare" (Gibson and Roberfroid 1995). However, the United Nations Food & Agriculture Organization (FAO) recommends that the term "synbiotic" be used only if the net health benefit is synergistic. The term "synbiotic" is used when a product contains both probiotics and prebiotics. Because the word alludes to synergism, this term should be reserved for products in which the prebiotic compound selectively favors the probiotic component. In this strict sense, a product containing oligofructose and probiotic bifidobacteria would fulfill the definition whereas a product containing oligofructose and a probiotic *Lactobacillus casei* strain would not (Schrezenmeir and de Vrese 2001). The most common synbiotic combinations available include bifidobacteria and fructooligosaccharides (FOSs), Lactobacillus GG and inulins, and bifidobacteria and Lactobacilli with FOSs or inulins. Experiments have provided evidence that synbiotics perform better than either probiotics or prebiotics alone in affecting the blood lipid profile and protecting from colorectal cancer (Gallaher and Khill 1999; Tuohy et al. 2003) but require further investigations. Some of the clinical results showed that preoperative administration of probiotics and synbiotics would upregulate immune function and lead to a decrease in infectious complications (Kanazawa et al. 2005; Rayes et al. 2002).

19.8 Cobiotics

An innovative approach in probiotic research is a new concept: "cobiotics." The cobiotics are more functional than synbiotics, and they are the combination of probiotics, prebiotics, and digestive enzymes (Figure 19.4). A series of research in probiotics and prebiotics led to the development of this new concept of cobiotics. This concept enables boosting of the nutritional value of synbiotics by incorporation of a different type of digestive enzymes and addition of enzymes to liberate prebiotics from their natural sources. This type of product was first made in Belgium (Registered number NUT/PL/AS 1164/22: Federal Public Service, Health Food Chain Safety). The ingredients of this product, other than probiotics, include inulin, dextrin, rice bran, glutamine, amylase, invertase, lactase, xylanase, pectinase, lipase, vitamin A, vitamin B5, vitamin B6, vitamin B9, vitamin B12, vitamin C, vitamin D, vitamin E, and zinc. Suggested use is 5–10 g daily (maximum 15 g/day) with meals or between the meals. The ingredients used in Cobiotics® create a synergy, which reinforces its effectiveness and improves the action of cofactors (Labrador 2013).

Cobiotics are recommended for IBS, leaky gut, or any other intestine dysfunction. It acts on several levels in the small intestine and the colon. It creates optimal conditions for the development of the gut flora and creates necessary conditions for suitable materials to be activated for the renewal of the intestine and colon epithelium.

The presence of the amylolytic and lipolytic enzymes in cobiotics makes it possible to considerably reduce the overload on the digestive system, to improve the absorption of the carbohydrates, the lipids, and proteins on the small intestine, thus helping to control weight and to reduce the viscosity of food not

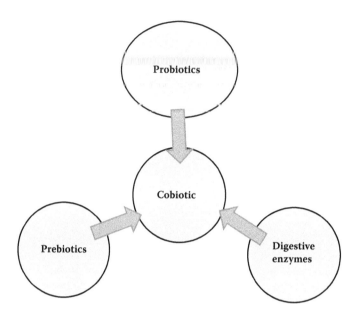

FIGURE 19.4 Concept of cobiotics.

digested in the colon, which allows a more effective action of the gut flora and a reduction in fermentation and production of the toxic metabolites.

Cobiotics contains several types of fibers (prebiotic) necessary for the development, the balance, and the maintenance of the biodiversity of the gut flora. It reinforces synergy with the immune system; helps to reduce the stress of the hepatic, pancreatic, and digestive systems; and thus facilitates an easier and better digestion. It also helps to balance the triglyceride and cholesterol levels by supporting the decomposition and the elimination of fats.

Another cobiotic preparation is Yeasture for animal nutrition formulated by Sretenović et al. (2008). Yeasture directly affects the productive performances of dairy cows as well as udder health. It is composed of live yeast cultures selected from three strains of *Saccharomyces cerevisisiae* in combination with probiotic bacteria and enzymes (*Lactobacillus casei, Streptococcus faecium, Aspergillus oryzae, Lactobacillus acidophilus*, 1,3-b and 1,6 D-Glucan, hemicellulase, protease, cellulase, alpha amylase), which have the ability to modify the fermentation in rumen, stimulating the development of ruminal bacteria and increasing fiber digestion. Effects of this preparation have been investigated on 60 Holstein-Friesian cows and were established. A preparation of Yeasture influences the quantity and composition of the milk (Sretenović et al. 2008).

Synbiotic Plus (an intestinal yeast cleansing formula) contains a professional-grade blend of probiotics, prebiotics, and enzymes that help create and support a highly oxygenated environment; one that can help prevent the growth of large colonies of yeast and promote the growth and maintenance of health-enhancing bacteria. Most of this oxygenating effect is accomplished by the Lactobacillus family of microflora in *Synbiotic Plus*, which produces hydrogen peroxide as a natural by-product. This hydrogen peroxide effectively kills off colonies of excess yeast throughout the intestines. The enzymes (serrapeptase, protease, amylase, glucoamylase, cellulase, and hemicellulase) included in *Synbiotic Plus* can strip away Candida's protective protein shell and cellular infrastructure (https://drbenkim.com/sunshop/index .php?l=product_detail&p=123/). Recently, a biotechnology company, NuMe Health, designed a cobiotic product NM504™ for the prediabetic population (http://www.numehealth.com/). It would help to maintain normal blood glucose levels and a healthy body weight. A case study was conducted and reported by Greenway et al. (2014).

Natural fermentation of LAB produces a range of secondary metabolites, some of which have been associated with the health-promoting properties of which the notable ones are the B vitamins and bioactive peptides. It is now well established that physiologically active peptides are produced from several food proteins during gastrointestinal digestion and fermentation of food by LAB. "Peptides produced in vivo or in vitro by enzymatic hydrolysis of food proteins with biological functions or physiological effects are called bioactive peptides" (Smacchi and Gobbetti 2000). Digestive enzymes, naturally occurring milk enzymes, coagulants, and microbial enzymes, especially from adventitious or starter LAB, generate bioactive peptides during milk fermentation and cheese maturation, thereby enriching the dairy products (Gobbetti et al. 2002; Korhonen and Pihlanto 2006).

Upon oral administration, bioactive peptides may affect the major body systems, namely, cardiovascular, nervous, gastrointestinal, and immune systems, depending on the inherent amino acid composition and sequence. Among these, peptides with blood pressure-lowering effects have received special attention, and considerable significance is being attached to the role of diet in the prevention and treatment of disease (López-Fandiño et al. 2007). Blood pressure regulation is partially dependent on the renin-angiotensin system (Silva and Malcata 2005) in which the angiotensin-I converting enzyme (ACE) regulates the peripheral pressure and hence its inhibition can exert an antihypertensive effect (Gobbetti et al. 2004).

19.9 Commercially Available Probiotics

The number of products in the United States with the descriptor "probiotic" increased from just six in 2004 to 47 in 2008 (Datamonitor's Product Launch Analytics 2009). In 2008, 232 products were introduced worldwide that contained probiotic and/or prebiotic ingredient(s). As of June 30, 2009, 139 products with a probiotic or prebiotic ingredient listed were introduced to the global market. Japan, which has had the Yakult probiotic drink on the market since the 1950s, remains the number one country for probiotic and prebiotic launches, but launches in the United States are gaining significant momentum. The nonalcoholic beverage category is gaining importance in the growth of digestive health products, especially as prebiotic innovation intensifies. The dairy food market is well established as the primary avenue for probiotics and prebiotics, and the maturity of this sector increases the challenge for new products to differentiate when entering this market. Innovation in the area of high-fiber products is strong with manufacturers adding new flavors to products and incorporating fiber and whole grains into new formats (Gray 2009). Apart from yogurt, leading categories of food and beverage introductions containing probiotics and/or prebiotics through the first six months of 2009 included milk, functional drinks, breakfast cereals, cheese, and cookies. According to the Packaged Facts projection, the market exceeded $22 billion in 2013, representing a compound annual growth rate (CAGR) of 12% between 2004 and 2013 (http://www.packagedfacts.com/about/release.aspid1467/).

The presence of Lactobacillus in the oral cavity may correlate to its health status; the more probiotic present, the healthier the mouth. Various strains of *Lactobacillus reuteri* show great potential for the future of oral health (Reddy et al. 2011). Sunstar Americas Inc. offers an over the counter probiotic product—GUM PerioBalance—that contains *L. reuteri Prodentis*, which is specifically engineered to improve oral health. Contained in a mint-flavored lozenge, the goal of PerioBalance is to enhance the health of the oral cavity by supplying "good" bacteria to negatively affect the "bad" disease-causing bacteria in the mouth. GUM PerioBalance is designed to improve the overall health of both the gum tissue and teeth, reduce plaque levels, and fight halitosis. The product is designed for use once daily, immediately following flossing and brushing.

These are some of the variety of probiotic products available in the market with specific application even in a personalized way. Some of the probiotics based on lactic acid bacteria and nonlactic acid bacteria available commercially in the market is presented in Tables 19.4 through 19.7. Examples of some probiotic products are given here.

19.9.1 UAS Life Sciences Up4 Kids Cubes

Kids Cubes Probiotics incorporate the extensively studied probiotic strain *Lactobacillus acidophilus* DDS-1 with 800 IU of vitamin D3 in what tastes like a piece of soft white chocolate. The product is room temperature stable, sugar-free, non-GMO, and gluten free.

TABLE 19.7

Some Commercial Probiotic Strains Used by Various Industries

Strains	Source
L. acidophilus LA 1	Chr. Hansen (Horsholm, Denmark)
L. paracasei CRL 431	
B. lactis Bb-12	Yakult (Tokyo, Japan)
L. casei Shirota	
B. breve strain Yakult	Snow Brand Milk Products Co., Ltd. (Tokyo, Japan)
L. acidophilus SBT-2062	
B. longum SBT-2928	
L. acidophilus R0011	Institut Rosell (Montreal, Canada)
L. rhamnosus R0052	
L. acidophilus NCFM	Rhodia, Inc. (Madison, WI)
L. acidophilus DDS-1	Nebraska Cultures, Inc. (Lincoln, NE)
L. casei DN014001 (Immunitas)	Danone Le Plessis-Robinson (Paris, France)
L. fermentum RC-14	Urex Biotech Inc. (London, Ontario, Canada)
L. rhamnosus GR-1	
L. johnsonii La1 (same as Lj1)	Nestlé (Lausanne, Switzerland)
L. plantarum 299V	Probi AB (Lund, Sweden)
L. rhamnosus 271	
L. reuteri SD2112 (same as MM2)	BioGaia (Raleigh, NC)
L. rhamnosus GG	Valio Dairy (Helsinki, Finland)
L. rhamnosus LB21	Essum AB (Umeå, Sweden)
Lactococcus lactis L1A	
L. salivarius UCC118	University College (Cork, Ireland)
B. longum BB536	Morinaga Milk Industry Co., Ltd. (Zama-City, Japan)
L. delbrueckii subsp. bulgaricus 2038	Meiji Milk Products (Tokyo, Japan)
L. acidophilus LB	Lacteol Laboratory (Houdan, France)
L. paracasei F19	Arla Dairy (Stockholm, Sweden)
L. crispatus	CTV05 Gynelogix, Boulder, CO
L. casei DN 114	Danone, Paris, France
S. boulardii	Biocodex Inc. (Seattle, WA)
B. lactis HN019 (DR10)	New Zealand Dairy Board

Source: Nagpal, R., Kumar, A., Kumar, M., Behare, P. V., Jain, S. and Yadav, H., *FEMS Microbiol. Lett.*, 334, 1–15, 2012.

19.9.2 Jarrow Formulas Jarro-Dophilus EPS

Jarrow Formulas Jarro-Dophilus EPS is a classic among the probiotic set, notable for its stability at room temperature that makes it ideal for travelers (each pill is preserved in individual blister packs). Each capsule is also enteric coated, which protects the fickle friendly bacteria from the acidic stomach environment, allowing them to exert a beneficial effect in the small intestine. The 5 billion CFUs are comprised of eight well-characterized strains.

19.9.3 Enzymatic Therapy Probiotic Pearls

Probiotic Pearls are unique in their tiny round shape that makes them easy to swallow. Its patented triple-layer soft gel protects live probiotics inside from heat, moisture, air, and stomach acid, so vendors and consumers need not refrigerate them. The original acidophilus pearls are marketed as capsules containing "lab 7 bifido" that provide daily relief from occasional gas, bloating, and constipation.

19.9.4 Bigelow Lemon Ginger Herbal Tea Plus Probiotics

Bigelow Lemon Ginger Herbal Tea Plus Probiotics is prepared with the proprietary Ganeden BC30 strain of *Bacillus coagulans*, tastes lovely, and delivers a daily probiotic dose sans supplement pill. Heat usually kills probiotics, but with BC30, about 80% of the microorganisms make it to the intestine alive.

19.9.5 Probium Probiotics Multi Blend 12B

Multi Blend 12B contains 12 billion CFUs of a special blend of *Lactobacillus acidophilus*, *Bifidobacterium lactis*, *B. bifidum*, and *B. longum*. In addition to clinically tested probiotic strains, it is incorporated with two prebiotics, namely fructoollgosaccharides and Sunfiber. The ingredients are packed in a capsule designed to resist stomach acid to maximize live bacteria reaching the intestines and the container Activ-Polymer bottle makes the product shelf stable for 2 years.

19.10 Future Prospects

The current probiotics market has evolved rapidly such that it has moved beyond yogurt to deliver probiotics, including synbiotics and cobiotics in addition to prebiotics in an increasingly wide variety of foods and beverages. In fact, prebiotics are being added to an expanding array of products, from pudding to frozen chicken dinners. In spite of the fact that the probiotic category is more established than prebiotics in the digestive health market, it is the prebiotics sector that has been growing faster recently. In fact, new food products have been formulated with the addition of probiotic cultures. Different types of food matrices have been used, such as various types of cheese, ice creams, milk-based desserts, powdered milk for newborn infants, butter, mayonnaise, powdered products or capsules, and fermented food of vegetable origin (Tamime et al. 2005). Nevertheless, two major trends contribute to the growth of foods and beverages that enhance digestive health: An increase in the numbers of categories of products in which probiotics and prebiotics are included and a growing public awareness of and desire to benefit digestive health and thereby enhance immunity. At the same time, there exist challenges to the probiotic market in terms of consumer confusion and skepticism about digestive health products as well as balancing health benefits with an appetizing product. Consumers find it difficult to get clarity about the numerous strains of probiotics and the health benefits of each, and hence, gaining consumer confidence is a major issue in the long-term profitability of probiotic-/prebiotic-based foods and beverages (Gray 2009). Enzymes have a larger role in probiotic, prebiotic, symbiotic, and cobiotic manufacture in the future because the consumer-oriented food and beverage industry is experiencing a global evolution focusing on safe health. More intensive research and development is needed toward isolation and development of enzymes that can meet the challenges in the development of ideal probiotics that contribute to overall human health.

REFERENCES

Alexopoulos, C., Georgoulakis, I. E., Tzivara, A., Kyriakis, C. S., Govaris, A. and Kyriakis, S. C. 2004. Field evaluation of the effect of a probiotic containing *Bacillus licheniformis* and *Bacillus subtilis* spores on the health status, performance, and carcass quality of grower and finisher pigs. *J. Vet. Med. A Physiol. Pathol. Clin. Med.* 51:306–312.

Anu, P. S. and Keerthi, T. R. 2012. Probiotic effect of wild species of *Bacillus* spore formers and its effect on enteric pathogens. *Int. J. Pharm. Biosci. Bio. Sci.* 3(1):B-327–B-334.

Anukam, K. C. and Reid, G. 2007. Probiotics: 100 years (1907–2007) after Elie Metchnikoff's observation. In *Communicating Current Research and Educational Topics and Trends in Applied Microbiology*, A. Mendez-Vilas (Ed.), pp. 466–474, Formatex, Spain.

Bamba, C., Bobinnec, Y., Fukuda, M. and Nishida, E. 2002. The GTPase Ran regulates chromosome positioning and nuclear envelope assembly in vivo. *Curr. Biol.* 12(6):503–507.

Bomba, A., Jonecová, Z., Koscová, J., Nemcová, R., Gancariková, S. and Mudronfová, D. 2006. The improvement of probiotics efficacy by synergistically acting components of natural origin: A review. *Biologia* 61:729–734.

Bornet, F. R., Brouns, F., Tashiro, Y. and Duvillier, V. 2002. Nutritional aspects of short-chain fructooligosaccharides: Natural occurrence, chemistry, physiology and health implications. *Dig. Liver Dis.* 34(Suppl 2):S111–S120.

Bouhnik, Y. 1993. Survival and effects of bacteria ingested via fermented milks in humans. *Lait* 73:241–247.

Cani, P. D., Delzenne, N. M. and Neyinck, A. M. 2011. Modulation of the gut microbiota by nutrients with prebiotic properties: Consequences for host health in the context of obesity and metabolic syndrome. *Microb. Cell Fact.* 10:1–11.

Datamonitor's Product Launch Analytics 2009. Datamonitor, Stonyfield Farm Case Study Maintaining Growth in the Troubled Organic Food Sector. February 2009. CSCM0228.

Delzenne, N. M. and Roberfroid, M. B. 1994. Physiological effects of non-digestible oligosaccharides. *Lebensm. Wiss. Technol.* 27:1–6.

Duc, L. H., Hong, H. A., Barbosa, T. M., Henriques, A. O. and Cutting, S. M. 2004. Characterisation of *Bacillus* probiotics available for human use. *Appl. Environ. Microbiol.* 70:2161–2171.

Effects of Probiotics on Nutrient Synthesis. Available at http://probiotics101.probacto.com/effects-of -probiotics-on-nutrient-synthesis/ accessed on January 9, 2015.

Fernandes, C. F., Shahani, K. M. and Amer, M. A. 1987. Therapeutic role of dietary lactobacilli and lactobacillic fermented dairy products. *FEMS Microbiol. Rev.* 46:343–356.

Fiordaliso, M. F., Kok, N., Desager, J. P., Goeihals, F., Deboyser, D., Roberfroid, M. and Delzenne, N. 1995. Oligofructose supplemented diet lowers serum and VLDL concentrations of triglycerides, phospholipids and cholesterol in rats. *Lipids* 30:163–167.

Fuller, R. 1989. Probiotics in man and animals. *J. Appl. Bacteriol.* 66:365–378.

Fuller, R. 1992. History and development of probiotics. In *Probiotics: The Scientific Basis*, R. Fuller (Ed.). Chapman and Hall, London, pp. 1–8.

Gallaher, D. D. and Khill, J. 1999. The effect of synbiotics on colon carcinogenesis in rats. *J. Nutr.* 129:1483–1487.

Gibson, G. R. and Roberfroid, M. B. 1995. Dietary modulation of the human colonic microbiota: Introducing the concept of prebiotics. *J. Nutr.* 125(6):1401–1412.

Gibson, G. R. and Fuller, R. 2000. Aspects on *in vitro* and *in vivo* research approaches directed toward identifying probiotics and probiotics for human use. *J. Nutr.* 130:391–395.

Gibson, G. R., Beatty, E. B., Wang, X. and Cummings, J. H. 1995. Selective stimulation of bifidobacteria in the human colon by oligofructose and inulin. *Gastroenterology* 108:975–982.

Gibson, G. R., Rastall, R. A. and Fuller, R. 2003. The health benefits of probiotics and prebiotics. In *Gut Flora, Nutrition, Immunity and Health*, R. Fuller, G. Perdigon (Eds.). Wiley-Blackwell, Oxford, UK, pp. 52–76.

Gibson, G. R., Probert, H. M., Van Loo, J., Rastall, R. A. and Roberfroid, M. B. 2004. Dietary modulation of the human colonic microbiota: updating the concept of prebiotics. *Nutr. Res. Rev.* 17:259–275.

Gobbetti, M., Stepaniak, L., De Angelis, M., Corsetti, A. and Di Cagno, R. 2002. Latent bioactive peptides in milk proteins: Proteolytic activation and significance in dairy processing. *Crit. Rev. Food Sci. Nutr.* 42:223–239.

Gobbetti, M., Minervini, F. and Rizzello, C. G. 2004. Angiotensin I-converting-enzyme-inhibitory and antimicrobial bioactive peptides. *Int. J. Dairy Technol.* 57:173–188.

Gray, L. 2009. Boosting Immunity Through Digestion: The Relation among Probiotics, Prebiotics and Digestive Enzymes. Available at http://www.packagedfacts.com/Boosting-Immunity-Digestion-2286571/ accessed on December 9, 2014.

Greenway, F., Wang, S. and Heiman, M. A. 2014. Novel cobiotic containing a prebiotic and an antioxidant augments the glucose control and gastrointestinal tolerability of metformin: A case report. *Benef. Microbes* 5(1):29–32.

Hanai, N., Nagata, K., Kawajiri, A., Shiromizu, T., Saitoh, N., Hasegawa, Y., Murakami, S. and Inagaki, M. 2004. Biochemical and cell biological characterization of a mammalian septin. *FEBS Lett.* 568:83–88.

Hijová, E., Szabadosova, V., Štofilová, J. and Hrčková, G. 2013. Chemopreventive and metabolic effects of inulin on colon cancer development. *J. Vet. Sci.* 14(4):387–393.

Hoa, T. T., Duc, L. H., Isticato, R., Baccigalupi, L., Ricca, E., Van, P. H. and Cutting, S. M. 2001. Fate and dissemination of *Bacillus subtilis* spores in a murine model. *Appl. Environ. Microbiol.* 67:3819–3823.

Holzapfel, W. H., Haberer, P., Geisen, R., Björkroth, J. and Schillinger U. 2001. Taxonomy and important features of probiotic microorganisms in food and nutrition. *Am. J. Clin. Nutr.* 73:365–373.

Available at https://drbenkim.com/sunshop/index.php?l=product_detail&p=123/ accessed on December 12, 2014.

Available at http://www.enzylase.net/what-is-enzylase.aspr/ accessed on December 31, 2014.

Available at http://www.numehealth.com/ accessed on December 22, 2014.

Available at http://www.packagedfacts.com/about/release.aspid1467/ accessed on November 12, 2014.

Available at http://www.webmd.com/digestive-disorders/features/what-are-probiotics/ accessed on December 5, 2014.

Jadamus, A., Vahjen, W. and Simon, O. 2001. Growth behaviour of a sporeforming probiotic strain in the gastrointestinal tract of broiler chicken and piglets. *Arch. Tierernahr.* 54.17.

Jørgensen, J. N. and Kürti, P. 2006. Novel approach to reduce pre-weaning mortality. *Int. Pig Top.* 19:11–13.

Kanazawa, H., Nagino, M. and Kamiya, S. 2005. Synbiotics reduce postoperative infectious complications: A randomized controlled trial in biliary cancer patients undergoing hepatectomy. *Langenbecks Arch. Surg.* 390:104–113.

Keawyok, K., Youravong, W. and Wichienchot, S. 2012. Feasibility of pectic oligosaccharide production of pectin extracted from dragon fruit peel. pp. 1–7. Available at http://www.ipeyala.ac.th/ipeyala/witjai/pdf/Feasibility-kritsada.pdf.

Korhonen, H., and Pihlanto, A. 2006. Bioactive peptides: Production and functionality. *Int. Dairy J.* 16:945–960.

Labrador, M. 2013. Probiotics and digestive enzymes together. Available at http://www.tacanow.org/blog/probiotics-and-digestive-enzymes-together/ accessed on January 9, 2015.

Lebeer, S., Vanderleyden, J. and De Keersmaecker, S. C. 2010. Host interactions of probiotic bacterial surface molecules: Comparison with commensals and pathogens. *Nat. Rev. Microbiol.* 8:171–184.

Lewis, B. P., Burge, C. B. and Bartel, D. P. 2005. Conserved seed pairing, often flanked by adenosines, indicates that thousands of human genes are microRNA targets. *Cell* 120(1):15–20.

Lilly, D. M. and Stillwell, R. H. 1965. Probiotics: Growth promoting factors produced by microorganisms. *Science* 147:747–748.

Looijer-van Langen, M. A. and Dieleman, L. A. 2009. Prebiotics in chronic intestinal inflammation. *Inflamm. Bowel Dis.* 15:454–462.

López-Fandiño, R., Recio, I. and Ramos, M. 2007. Egg-protein-derived peptides with antihypertensive activity. In *Bioactive Egg Compounds*, R. Huopalahti, R. López-Fandiño, M. Anton, R. Shade (Eds.), pp. 199–211, Springer-Verlag, Heidelberg, Germany.

Martini, M. C., Bollweg, G. L., Levitt, M. D. et al. 1987. Lactose digestion by yogurt β-galactosidase: Influence of pH and microbial cell integrity. *Am. J. Clin. Nutr.* 45:432–436.

Mattila-Sandholm, T., Myllärinen, P., Crittenden, R., Mogensen, G., Fondén, R. and Saarela, M. 2002. Technological challenges for future probiotic foods. *Int. Dairy J.* 12:173–182.

Nagpal, R., Kumar, A., Kumar, M., Behare, P. V., Jain, S. and Yadav, H. 2012. Probiotics, their health benefits and applications for developing healthier foods: A review. *FEMS Microbiol. Lett.* 334:1–15.

O'Bryan, C. A., Pak, D., Crandall, P. G., Lee, S. O. and Ricke, S. C. 2013. The role of prebiotics and probiotics in human health. *J. Prob. Health* 1:108.

O'Sullivan, M. G. 1996. Metabolism of bifidogenic factors by gut flora—An overview. *Bull. Int. Dairy Foundation* 313:23–25.

Oyetayo, V. O. and Oyetayo, F. L. 2005. Potential of probiotics as biotherapeutic agents targeting the innate immune system. *Afr. J. Biotechnol.* 4(2):123–127.

Parker, R. B. 1974. Probiotics, the other half of the antibiotic story. *Anim. Nutr. Health* 29:4–8.

Parvez, S., Malik, K. A., Ah Kang, S. and Kim, H. Y. 2006. Probiotics and their fermented food products are beneficial for health. *J. Appl. Microbiol.* 100:1171–1885.

Praveen Kumar, G., Pushpa, A. and Prabha, H. 2012. A review on xylooligosaccharides. *Int. Res. J. Pharm.* 3:71–74.

Probiotic Market–Advanced Technologies and Global Market (2009–2014). 2009. Markets and markets. com, September. Report code: FB1046. Available at http://www.marketsandmarkets.com/market-reports/probiotic-market-advanced-technologies-and–global-market-69.html/ accessed on October 24, 2014.

Rayes, N., Seehofer, D. and Hansen, S. 2002. Early enteral supply of lactobacillus and fiber versus selective bowel decontamination: A controlled trial in liver transplant recipients. *Transplantation* 74:123–127.

Reddy, N. V., Rao, A. P., Mohan, G. and Kumar, R. R. 2011. Probiotic lacto bacilli and oral health. *Ann. Essences Dent.* 3(2):100–103.

Roberfroid, M. B. 2000. Fructo-oligosaccharide malabsorption; benefit for gastrointestinal functions. *Curr. Opin. Gastroenterol.* 16(2):173–177.

Roberfroid, M. B. 2007. Inulin type fructans: Functional food ingredients 1. *J. Nutr.* 137:2493S–2502S.

Sanders, M. A. 1998. Overview of functional foods: Emphasis on probiotic bacteria. *Int. Dairy J.* 8:341–347.

Sanders, M. E. 2009. Probiotics and health. *The Whitehall-Robins Report* 18(1):1–2

Sanders, M. E., Morelli, L. and Tompkins, T. 2003. A spore formers as human probiotics: *Bacillus, Sporolactobacillus*, and *Brevibacillus*. *Compr. Rev. Food Sci. Food Safety* 2:101–111.

Schrezenmeir, J. and de Vrese, M. 2001. Probiotics, prebiotics, and synbiotics—Approaching a definition. *Am. J. Clin. Nutr.* 73(2 Suppl):361S–364S.

Shah, S. A., Sander, S., White, C. M., Rinaldi, M. and Coleman, C. I. 2007. Evaluation of echinacea for the prevention and treatment of the common cold: A meta-analysis. *Lancet Infect. Dis.* 7(7):473–480.

Silva, S. V., and Malcata, F. X. 2005. Milk caseins as a source of bioactive peptides. *Int. Dairy J.* 15:1–15.

Smacchi, E. and Gobbetti, M. 2000. Bioactive peptides in dairy products: synthesis and interaction with proteolytic enzymes. *Food Microbiol.* 17:129–141.

Sretenovic, L., Petrovic, J., Aleksic, M. P., Pantelic, S., Katic, V., Bogdanovic, V. and Beskorovajni, R. 2008. Influence of yeast, probiotics and enzymes in rations on dairy cows performances during transition. *Biotechnol. Anim. Husbandry* 24(5–6):33–43.

Stanton, C., Desmond, C., Coakley, M., Collins, J. K., Fitzgerald, G. and Ross, R. P. 2003. Challenges facing development of probiotic-containing functional foods. In *Handbook of Fermented Functional Foods*, E. R. Farnworth (Ed.). CRC Press, Boca Raton, FL, pp. 27–58.

Stephen, A. N. 1998. Regulatory aspects of functional products. In *Functional Foods: Biochemical and Processing Aspects*, G. Mazza (Ed.). Lancester, PA: Technomic Publ. Co. Inc. pp. 403–432.

Talwalkar, A. and Kailasapathy, K. 2004. A review of oxygen toxicity in probiotic yoghurts: Influence on the survival of probiotic bacteria and protective techniques. *Compr. Rev. Food Sci. Food Safety* 3:117–124.

Tamime, A. Y., Saarela, M., Korslund Sondergaard, A., Mistry, V. V. and Shah, N. P. 2005. Production and maintenance of viability of probiotic micro-organisms in dairy products. In *Probiotic Dairy Products*, A. Y. Tamime (Ed.). Blackwell Publishing Ltd., London, pp. 39–72.

Tomomatsu, H. 1994. Health effects of oligosaccharides. *Food Technol.* 48:61–65.

Tuohy, K. M., Probert, H. M., Smejkal, C. W. and Gibson, G. R. 2003. Using probiotics and prebiotics to improve gut health. *Drug Discov. Today* 8:692–700.

van Dokkum, W. and van den Heuvel, E. 2001. Nondigestible oligosaccharides and mineral absorption. In *Handbook of Dietary Fiber*, 1st ed., S. S. Cho, M. L. Dreher (Eds.). Marcel Dekker, Inc., New York, pp. 259–268.

Vinderola, G., Capellini, B., Villarreal, F., Suarez, V., Quiberoni, A. and Reinheimer, J. 2008. Usefulness of a set of simple *in vitro* tests for the screening and identification of probiotic candidate strains for dairy use. *Food Sci. Technol.* 41:1678–1688.

Williams, C. M. and Jackson, K. G. 2002. Inulin and oligofructose: Effects on lipid metabolism from human studies. *Br. J. Nutr.* 87:S261–S264.

Zubaidah, E. and Akhadiana, W. 2013. Comparative study of inulin extracts from dahlia, yam, and gembili tubers as prebiotic. *Food Nutr. Sci.* 4:8–12.

20

Enzyme Inhibitors in Regulating Enzyme Processing of Food and Beverages

Manzur Ali Pannippara and Sapna Kesav

CONTENTS

20.1 Introduction

Enzymes are novel biocatalysts employed extensively in various catalytic reactions. The enzyme activities are often inhibited by substances that are present at the site of enzyme action. In fact, enzyme inhibition is the science of enzyme–substrate reaction influenced by the presence of any organic chemical, inorganic metal, or biosynthetic compound due to their covalent or noncovalent interactions with the enzyme active site. The substances that inhibit enzymes are called enzyme inhibitors. Although enzyme inhibitors are used as drugs in therapeutic applications and as biocontrol agents, their role in food processing is less known. Whereas enzymes are extensively used in food and beverage processing, which has been discussed extensively in other chapters in this book, it is important to appreciate the fact that, often, enzyme activities need to be regulated toward optimal performance and desired activities. This regulation can be conveniently carried out employing inhibitor substances, which are available as both natural substances from microbe, plant, and animal resources and as synthetically prepared substances. In this context, the concept of application of enzyme inhibitors for regulating enzyme activities during food and beverage processing and extending their shelf life is discussed in this chapter.

20.2 Enzyme Inhibitors

Enzyme inhibitors are low molecular weight chemical compounds, which can reduce or completely inhibit the enzyme catalytic activity either reversibly or permanently (irreversibly). An inhibitor can modify one amino acid or several side chains required in enzyme catalytic activity.

An enzyme inhibitor is any substance that slows down the rate of an enzyme-catalyzed reaction. Enzyme inhibitors differ from enzyme inactivators, which act by removing cations from metal-dependent proteinases (e.g., chelators), or from protein denaturants, which act by altering the structure of catalytic sites. Most natural inhibitors react reversibly with the enzyme. Conceptually, enzyme inhibitors are classified into two types: nonspecific inhibitors and specific inhibitors. All major classes of enzymes have respective inhibitors that regulate them in natural conditions, and now both natural and synthetic enzyme inhibitors are available commercially for various purposes, such as drugs and biocontrol agents.

A general characteristic of most protein inhibitors is that they are cysteine-rich proteins (oryzacystatin is an exception). The disulfide bridges are important in the stability and active conformation structure of many of these proteins. Many inhibitors are multiheaded, and some of them are bifunctional, having two heterologous domains that inhibit two different enzymes. The most common enzyme inhibitors with a wide range of applications include protease inhibitor, polyphenol oxidase inhibitor, amylase inhibitor, lipase inhibitor, etc. In this chapter, the discussion is restricted to enzyme inhibitors that have been recognized to have a role in the food, vegetable, and seafood processing industries.

The global market for enzyme inhibitors was valued at $104.4 billion in 2010 and reached nearly $104.6 billion in 2011. This market is expected to rise at a compound annual growth rate (CAGR) of 4% and reach nearly $127.4 billion by 2016 (BCC Research 2012, Report code BIO057B). One of the key factors contributing to this market growth is the increasing prevalence of various diseases. The global enzyme inhibitors market has also been witnessing the trend of an increase in mergers and acquisitions among vendors. Some of the major players in the enzyme inhibitor market (especially in the pharmaceutical industry) are AstraZeneca (UK), GlaxoSmithKline-GSK (UK), Bayer (Germany), Novartis (Switzerland), Merck and Co. (USA), Roche (Germany), and Pfizer (USA) (http://www.transparencymarketresearch.com /kinase-inhibitors.html).

20.2.1 Protease Inhibitors

20.2.1.1 Types of Proteases

Protein hydrolysis is achieved by enzymes that come under the major class hydrolases and are collectively called proteases. Proteases bring about the cleavage of long protein chains and form corresponding fragments of amino acids. Proteases are grossly subdivided into two major groups, that is, exopeptidases and endopeptidases, depending on their site of action. Exopeptidases cleave the peptide bond proximal to the amino or carboxy terminal of the substrate whereas endopeptidases cleave peptide bonds distant from the termini of the substrate. Based on the functional group present at the active site, proteases are further classified into four prominent groups, that is, serine proteases (3.4.21), aspartic proteases (3.4.23), cysteine proteases (3.4.22), and metalloproteases (3.4.24) (Hartley 1960). There are a few miscellaneous proteases, which do not precisely fit into the standard classification, for example, ATP-dependent proteases, which require ATP for activity (Menon and Goldberg 1987). Each family of peptidases has been assigned a code letter denoting the type of catalysis, that is, S, C, A, M, or U for serine, cysteine, aspartic, metallo-, or unknown type, respectively.

20.2.1.2 Protease Inhibitors

Protease inhibitors (PIs) are (synthetic/natural) substances that act directly on proteases to lower the catalytic rate. They usually mimic the protein substrate by binding to the active site of the enzyme and are specific for the active site of a given class of proteinases. Protease inhibitors are essential for regulating the activity of their corresponding proteases and play key regulatory roles in many biological processes. The protease inhibitors are of two basic functional classes, the *active-site inhibitors* that bind to the active site of the target protease and inactivate its ability to hydrolyze all substrates and the *α2-macroglobulins*, which operate by the unique process of molecular entrapment. Although most of these protease inhibitors are directed against serine proteases, some target cysteine, aspartyl, or metalloproteases (Bode and Huber 1992).

Protease inhibitors are assigned into families and clans based on the similarity of amino acid sequence or comparison of tertiary structure (Rawlings 2010), and this classification is implemented in the MEROPS database of inhibitors (Table 20.1) (http://www.merops.sanger.ac.uk). The classification of protein peptidase inhibitors is continually revised, and currently the inhibitors are grouped into 71 families based on comparisons of protein sequences. These families can be further grouped into 38 clans based on the highest level of evolutionary divergence such that all the sequences in the same clan are evolutionarily related despite there being limited sequence similarity. It has been shown that structural similarities persist between related proteins despite there being no significant sequence similarity (Chothia and Lesk 1986).

20.2.1.3 Sources of Protease Inhibitors

Protease inhibitors are found abundantly in numerous plants, animals, and microorganisms, owing their significance to their application in the study of enzyme structures, reaction mechanisms, and also their utilization in pharmacology and agriculture. Protease inhibitors, especially serine protease inhibitors, are one of the most abundant classes of proteins in eukaryotes widely distributed in plants, animals, and microorganisms as well as archaea (Silverman et al. 2010).

Microbial protease inhibitors are versatile in their structures and mechanisms of inhibition in ways that differ from those of other sources. The advantage of using microbial and fungal protease inhibitors is that many of them display unique inhibitory profiles and resistance to proteolytic cleavage as well as high thermal and broad pH range stability with the latter being very convenient because harsh conditions may be used for immobilization. They have therefore found countless applications in the fields of medicine, agriculture, and biotechnology (Rawlings 2010; Sabotič and Kos 2012).

The microorganisms of prokaryotic domains of archaeabacteria and eubacteria and the kingdom of fungi, including higher fungi or mushrooms, constitute important sources of protease inhibitors. The number and diversity of proteases found in microorganisms and higher fungi make them an inexhaustible source of novel protease inhibitors with unique features (Sabotič et al. 2007). A majority of the extracellular protein protease inhibitors produced by microorganisms are from the genus *Streptomyces*. The widely distributed and well-characterized proteinaceous inhibitors from *Streptomyces* are the inhibitors of the bacterial serine alkaline protease subtilisin. In addition to the subtilisin inhibitors, there are reports of other related inhibitors of trypsin and other serine proteases from *Streptomyces*. A number of pathogenic gram-negative bacteria, such as *Escherichia coli*, *Klebsiella pneumoniae*, *Serratia marcescens*, or *Erwinia chrysanthemi* seem to be able to protect themselves against their own proteases by producing periplasmic protease inhibitors, such as the protease inhibitor ecotin, which has orthologous sequences widely distributed in the bacterial kingdom (Eggers et al. 2004). Marine microorganisms, with their unique nature, differ very much in many aspects from their terrestrial counterparts and are known to produce a diverse spectra of novel useful substances, including protease inhibitors (Imada 2004; Kanaori et al. 2005; Rawlings 2010). *Chromohalobacter* sp. isolated from sponge *Xestospongia testudinaria* produced a protease inhibitor that was observed to be active against protease produced by *Pseudomonas aeruginosa* (Wahyudi et al. 2010). Several species of fungi are known to secrete protease inhibitors. Some examples include the yeast inhibitors of endogenous proteases A and B, low molecular weight inhibitors of *Pleurotus ostreatus*, serine protease inhibitor from *Lentinus edodes* (Odani et al. 1999), proteinase inhibitor from *Trametes versicolor* (Zuchowski and Grzywnowicz 2006) and *Pleurotus floridanus* (Ali et al. 2014), trypsin specific inhibitors from *Clitocybe nebularis*, CnSPIs (Avanzo et al. 2009), Cospin (PIC1) from *Coprinopsis cinerea* (Sabotič

TABLE 20.1

Families of Proteinaceous Inhibitors of Microbial and Fungal Origin

Family	Common Name	Families of Peptidases Inhibited
I1	Kazal	M10, S1A, S1D, S8A, S9A
I2	Kunitz-BPTI	S1A, S7
I4	Serpin	C1A, C14A, S1A, S7, S8A, S8B
I9	YIB	S8A
I10	Marinostatin	S1A, S8A
I11	Ecotin	S1A
I16	SSI	M4, M7, S1A, S8A, S8B
I31	Thyropin	A1A, C1A, M10A
I32	IAP	C14A
I34	IA3	A1A
I36	SMI	M4
I38	Aprin	M10B
I39		A1A, A2A, C1A, C2A, C11, M4, M10A, M10B, M12A, M12B, S1A, S1B, S8A
I42	Chagasin	C1A
I43	Oprin	M12B
I48	Clitocypin	C1A, C13
I51	IC	S1A, S10
I57	Staphostatin B	C47
I58	Staphostatin A	C47
I63		M43B, S1A
I66	Cnispin	S1A
I69		C10
I75	CIII	M41
I78		S1A, S8A
I79	AVR2	C1A
I85	Macrocypin	C1A, C13, S1A
I87	HflKC	M41

Source: Rawlings, N. D. 2010. Peptidase inhibitors in the MEROPS database. *Biochimie* 92(11), 1463–1483.

et al. 2012), mycocypins, clitocypins, and macrocypins, a group of cysteine protease inhibitors from the mushrooms *Clitocybe nebularis* and *Macrolepiota procera* (Renko et al. 2010).

Plant protease inhibitors are generally small proteins or peptides that occur in storage tissues, such as tubers and seeds, and also in the aerial parts of plants (Macedo et al. 2003; Valueva and Mosolov 2004). There are numerous protease inhibitors isolated and studied from plants, and a few of them are listed in Table 20.2. Of these, the serine PIs are the most studied and have been isolated from various Leguminosae seeds (Macedo et al. 2002; Mello et al. 2002). Plant protease inhibitors have received special attention because of their potential applications in agriculture as bioinsecticide, nematicidal, acaricidal, antifungal, and antibacterial agents. The defensive capabilities of plant protease inhibitors rely on inhibition of proteases present in insect guts or secreted by microorganisms, causing a reduction in the availability of amino acids necessary for their growth and development (Kim et al. 2005). The use of recombinant protease inhibitors to protect plants has emerged as an interesting strategy for insect pest control using genetic engineering (Lawrence and Koundal 2002; Reeck et al. 1997; Whetstone and Hammock 2007).

20.2.1.4 Classification of Protease Inhibitors

Protease inhibitors can be classified according to the source organism (microbial, fungal, plant, animal), according to their structure (primary and three-dimensional), or according to their inhibitory profile

TABLE 20.2

Plant Protease Inhibitors with Some Examples

Common Name	MEROPS Family/ Subfamily	Type Example	Source	References
Kunitz (plant)	13A	Soybean Kunitz trypsin inhibitor Kunitz cysteine peptidase inhibitor 1	*Glycine max* *Solanum tuberosum*	Laskowski and Kato 1980 Gruden et al. 1997
Kunitz (plant)	I3B	Proteinase inhibitor A inhibitor unit Kunitz subtilisin inhibitor	*Sagittaria sagittifolia* *Canavalia lineata*	Laskowski and Kato 1980 Terada et al. 1994
Cereal	I6	Ragi seed trypsin/amylase inhibitor Barley trypsin/factor XIIa inhibitor	*Eleusine coracana* *Hordeum vulgare*	Hojima et al. 1980 Lazaro et al. 1988
Squash	I7	Trypsin inhibitor MCTI-II Macrocyclic squash trypsin inhibitor	*Momordica charantia* *Momordica cochinchinensis*	Huang et al. 1992 Hernandez et al. 2000
Potato type I	I13	Subtilisin/chymotrypsin inhibitor CI-1A Wheat Subtilisin/chymotrypsin inhibitor	*Hordeum vulgare* *Triticum aestivum*	Greagg et al. 1994 Poerio et al. 2003
Mustard	I18	Mustard trypsin inhibitor Mustard trypsin inhibitor-2 Rape trypsin inhibitor	*Sinapis alba* *Brassica hirta* *Brassica napus*	Menengatti et al. 1992 Ceci et al. 1995 Ceciliani et al. 1994
Cystatin	I25B	Onchocystatin Ovocystatin Oryzacystatin II	*Onchocerca volvulus* *Gallus gallus* *Oryza sativa*	Lustigman et al. 1992 Laber et al. 1989 Ohtsubo et al. 2005
Kininogen	I25C	Metalloprotease Sarcocystatin	*Bothrops jararaca* *Sarcophaga peregrina*	Cornwall et al. 2003 Saito et al. 1989
Bowman-Birk	I12	Bowman–Birk plant trypsin inhibitor unit 1 Bowman–Birk trypsin/ chymotrypsin inhibitor Sunflower cyclic trypsin inhibitor	*Glycine max* *Arachis hypogaea* *Helianthus annuus*	Odani and Ikenaka 1976 Suzuki et al. 1987 Mulvenna et al. 2005
Potato type II	I20	Proteinase inhibitor II Tomato peptidase inhibitor II inhibitor unit 2	*Solanum tuberosum* *Solanum lycopersicum*	Greenblatt et al. 1989 Barrette-Ng et al. 2003

Source: Habib, H., and Fazili, K. M., *Biotechnol. Mol. Biol. Rev.*, 2, pp. 68–85, 2007.

(broad range, specific) and reaction mechanism (competitive, noncompetitive, uncompetitive as well as reversible or irreversible). They are commonly classified according to the class of protease they inhibit (cysteine, serine, aspartic, and metalloprotease inhibitors) (Laskowski and Kato 1980).

Protease inhibitors are grouped broadly into two categories: (i) small molecule inhibitors and (ii) proteinaceous inhibitors.

20.2.2 Small Molecule Inhibitors

Small molecule inhibitors (SMIs) include naturally occurring compounds, such as pepstatin, bestatin, and amastatin, as well as synthetic inhibitors generated in a laboratory. SMIs are inhibitors that are not proteins, including peptides and synthetic inhibitors that are generally of microbial origin, and are low molecular weight peptides of unusual structures (Umezawa 1982). Many of them have been synthesized in the laboratory; however, those that occur naturally have been isolated from bacteria and fungi (Rawlings 2010).

20.2.3 Proteinaceous Inhibitors

Protein inhibitors of proteases are ubiquitous and have been isolated from numerous plants, animals, and microorganisms (Birk 1987, Leo et al. 2002). Of these inhibitors, the most extensively studied are the inhibitors of serine proteases. A detailed classification of protein protease inhibitors based on their evolutionary relationship is available in the MEROPS database (http://merops.sanger.ac.uk/inhibitors/), which follows a hierarchy similar to that for proteases. An "inhibitor unit" was defined as the segment of the amino acid sequence containing a single reactive site (or bait region for a trapping inhibitor) after removal of any parts that are known not to be directly involved in the inhibitory activity. A protein that contains only a single inhibitor unit is termed a simple inhibitor, and one that contains multiple inhibitor units is termed a compound inhibitor (Rawlings et al. 2004). Their molecular weight and mechanism of inhibition varies from inhibitor to inhibitor (Rawlings 2010).

20.2.4 Mechanism of Inhibition

The vast majority of protease inhibitors are competitive. Most protease inhibitors bind a critical portion of the enzyme in the active site in a substrate-like manner. Related proteases often show a high degree of homology in the active site; substrate-like binding often leads to inhibitors that can potentially inhibit more than one target protease. Three types of inhibitors can be distinguished based on their mechanism of action: "canonical" (standard mechanism) and "noncanonical" inhibitors and the "serpins." The most thoroughly studied mechanism of protein protease inhibitors is that of the standard (canonical) inhibitors, which are small proteins showing substrate-like binding and blocking the enzyme at the distorted Michaelis complex reaction stage (Otlewski et al. 1999); serpins on the other hand are much larger, typically 350–500 amino acids in size, distributed from viruses to mammals (Gettins 2002; Silverman et al. 2001).

The largest group of protein inhibitors is canonical inhibitors, which act according to the standard mechanism of inhibition. They often accumulate in high quantities, especially in plant seeds, avian eggs, and various body fluids, and comprise proteins from 14 to approximately 200 amino acid residues. The segment responsible for protease inhibition, called the protease-binding loop, surprisingly always has a similar, canonical conformation in all known inhibitor structures (Bode and Huber 1992). This convex, extended, and solvent-exposed loop is highly complementary to the concave active site of the enzyme. The standard mechanism implies that inhibitors are peculiar protein substrates containing the reactive site P1-P1′ peptide bond located in the most exposed region of the protease-binding loop P1, P2, and P1′, P2′ specify inhibitor residues amino and carboxyl terminals to the scissile peptide bond, respectively; S1, S2, and S1′, S2′ denote the corresponding subsites on the protease (Jackson and Russell 2000).

The noncanonical inhibitors are much less abundant than the canonical inhibitors or serpins. Noncanonical inhibitors interact through their N-terminal segment, which binds to the protease active site, forming a short parallel β sheet. These inhibitors also form extensive secondary interactions with the target protease outside the active site, which provides an additional buried area and contributes significantly to the strength, speed, and specificity of recognition (Stubbs et al. 1995; Szyperski et al. 1992). The serpins are a superfamily of proteins with a diverse set of functions, including, but not limited to, inhibition of serine proteinases in the vertebrate blood coagulation cascade (Marshall 1993).

20.3 Applications of Enzyme Inhibitors in Food and Beverage Processing

20.3.1 Protease Inhibitors

Microbial food spoilage is an area of global concern as it has been estimated that as much as 25% of all food produced is lost postharvest owing to microbial activity (Baird-Parker 2003). Food spoilage occurs due to breakdown of proteins, lipids, and carbohydrate content of food. Among the proximate components of food, protein breakdown known as proteolysis occurs rapidly and immediately postharvest in proteinaceous foods, such as meat, seafood, eggs, and milk. Hence, proteolysis assumed significance as a major subject of research because it often results in the loss of structural and functional properties of food, decreased nutritive value, and low market value. In most cases, proteinases are the cause of the loss of many desirable food attributes. In this context, protease inhibition using protease inhibitors is considered to be an ideal option to regulate protease activity and retard or prevent spoilage. Protease inhibitors of microbial origin have already found many different applications (Rawlings 2010; Sabotič and Kos 2012). However, their applications in food processing are very limited.

20.3.1.1 Milk

Milk has many nutritious qualities that make it an important part of the diet consumed worldwide. The highest quality of milk must be obtained without compromising its nutritional benefits. Vulnerability of milk fat and protein to physical and chemical alterations can also lead to deterioration, thus reducing its quality. Bacterial contamination, inadequate packaging systems, and improper temperature control are the major hurdles in preserving the quality of milk. The stability and shelf life of pasteurized milk is highly influenced by the quality of the raw material, the binomial temperature/time pasteurization, resistant microorganisms to pasteurization (particularly psychrotrophic), the presence and activity of postpasteurization contaminants, the packaging system, and the storage temperature postpasteurization (Cromie 1991).

The proteolytic enzymes produced by psychrotrophics in milk are more powerful in their action on milk proteins than those naturally present in milk and those produced by leucocytes even if present in a great amount (Grieve and Kitchen 1985). *Pseudomonas* spp. produces a large number of extracellular toxins, which include phytotoxic factor, pigments, hydrocyanic acid, proteolytic enzymes, phospholipase, and enterotoxins. The high spoilage potential of *Pseudomonas* spp. is not only because of its ability to multiply at refrigeration temperatures but also because of the ability to produce thermostable proteases and lipases (Sorhaug and Stepaniak 1997). One of the spoilage-causing proteases is Protease plasmin (Ismail and Nielsen 2010). Plasmin can survive pasteurization temperatures and can cause degradation of dairy proteins in milk and cause coagulation and gelatinization. Protease inhibitors compatible with milk proteins without altering the texture and taste of the milk can be used for the preservation of milk. However, much research is needed in this respect.

20.3.1.2 Meat Tenderization

Inconsistency and variability in meat texture have been identified as two of the major problems that face the meat industry. For decades, consumers considered tenderness to be the most important quality attribute of meat. The rate and extent of meat tenderization are known to be highly variable. One possible cause of this variability could be the difference in the enzyme content and more likely in the enzyme/inhibitor ratio, a parameter reflecting the efficiency of the proteolytic systems. Usually with proteases such as Papain, when used as a meat tenderizer, meat structure is excessively degraded, giving a spongy texture uncharacteristic of high-quality tender cuts of beef. Decrease of protein degradation by reducing protease activity could be a more efficient way to control the enzyme activity (Goll et al. 1989) and protease inhibitors could be used to solve the problem (Funaki et al. 1991). Studies have indicated that the best predictors of meat tenderness are the concentrations of inhibitors and peptidase inhibitors as in the cases of calpastatin, a calpain inhibitor, and cystatins, a family of cysteine peptidase inhibitors, and serine peptidase inhibitors (Sentandreu et al. 2002) have an essential role in regulating protease activities to optimum levels of choice.

20.3.1.3 Seafood-Surumi Processing

The use of an adequate amount of natural protease inhibitors is an effective way to extend the shelf life of many types of seafood. The protein hydrolysis in fish and shrimp is generally an undesirable process and accounts for the loss of commodity during postharvest treatment. Increased consumer demand for seafood has placed added pressure on traditional fishery stocks. In the seafood industry, endogenous proteases, such as cathepsins, cause severe damage and result in postmortem muscle softening (Haard 1994). Surimi production is the best-known example of food processing with the aid of protease inhibitors. Inhibitors are normally added to surimi before freezing the product. Surimi is a Japanese term for mechanically deboned fish flesh that has been washed with water and mixed with cryoprotectants. Fish muscle softening by protease enzyme is one problem of the surimi industry. The enzyme-catalyzed damage of muscle proteins is irreversible, and the use of inhibitors can prevent this degradation. Pacific whiting, *Merlucciusproductus*, and arrowtooth flounder, *Atheresthesstomias*, which is found on the Pacific coast of the continental United States and Canada, are used in surumi processing. The problem of softening of the flesh of both Pacific whiting and Arrowtooth flounder during cooking have been regarded as low valued species (Greene and Babbitt 1990). This softening is the result of protease activity within the muscle tissue of both species. Although studies have shown that chemical inhibitors are effective, these chemicals do not satisfy the necessary requirements as approved food additives. In view of this, natural sources of protease inhibitors are explored, which have varying degrees of specificity toward protease enzymes. Mammalian blood plasma is an abundant source of inhibitors toward all classes of protease enzymes (Laskowski and Kato 1980). Among food-grade inhibitors used in surimi, beef plasma proteins (BPP) are the most effective in both inhibiting proteolytic activity and enhancing the gel strength of surimi. Serine protease and cathepsins reduce the gel-forming ability of muscle in surimi production. Egg white and whey protein concentrate have also been used as an inhibitor of autolysis of Pacific whiting surimi during processing (Wasson et al. 1992; Weerasinghe et al. 1996).

20.3.1.4 Shrimp Preservation

The presence of protease-producing organisms is responsible for fish and shrimp muscle degradation during preservation (Chandrasekaran 1985). Bijina et al. (2011) reported that the protease inhibitor isolated form *Moringa oleifera* could prevent proteolysis in the commercially valuable shrimp *Paneous monodon* during storage, indicating the scope for its application as a seafood preservative. Similarly, a potential protease inhibitor isolated from *Pseudomonas mendocina* BTMW 301 was assessed for preservation of shrimp, *Peneaus monodon* (Sapna 2013).

20.3.2 Polyphenol Oxidase (PPO) Inhibitors

20.3.2.1 Enzymatic Browning of Fruits, Vegetables, and Seafood

Enzymatic browning remains a major problem in fruits (apples, apricots, pears, bananas, grapes), vegetables (potatoes, mushrooms, lettuce), and seafood (shrimp, lobsters, crabs) with an estimated 50% loss in industry (Martinez and Whitaker 1995). Polyphenol oxidase (PPO; EC. 1.10.3.1) found in fruits, vegetables, fungi, seafood, and mammals has been extensively studied and is primarily responsible for enzymatic browning. PPO is a generic term for the group of enzymes that catalyze the oxidation of phenolic compounds to produce a brown color on the cut surfaces of vegetables and fruits (Whitaker and Lee 1995). This enzyme has also been labeled phenoloxidase, phenolase, monophenol, and diphenol oxidase and tyrosinase. In the presence of oxygen, rapid browning occurs due to enzymatic oxidation of phenols to orthoquinones, which rapidly polymerize to form brown or black pigments, such as melanins.

Browning occurs due to enzymatic and nonenzymatic oxidation. Reactions of amines amino acids, peptides, and proteins with reducing sugars and vitamin C (nonenzymatic browning, often called Maillard reaction), and quinones (enzymatic browning) cause deterioration of food during storage and commercial or domestic processing. Enzymatic browning of fruits, vegetables, beverages, and seafoods takes place in the presence of oxygen when tyrosinase (PPO) and its polyphenolic substrates are mixed after brushing, peeling, and crushing operations, which lead to the rupture of the cell structure (Hurrel and Finot 1984).

However, the browning process is sometimes desirable as it can improve the sensory properties of some products, such as dark raisins, fermented tea leaves, coffee beans, cocoa, etc. Color development in cocoa is facilitated by PPO activity during fermentation and drying. PPOs are also responsible for the development of the characteristic golden brown color in dried fruits, such as raisins, prunes, dates, and figs.

The most important factors that determine the rate of enzymatic browning of vegetables and fruits are the concentrations of active PPO and phenolic compounds present, the pH, the temperature, and the oxygen availability of the tissue. The optimum pH range of PPO is between pH 4 and 7. The adjustment of pH with acids to 4 or below can be used to control browning as long as the acidity can be tolerated tastewise. Heat inactivation of PPO is feasible by applying temperatures of more than 50°C but may produce undesirable colors and/or flavors as well as undesirable changes in texture.

20.3.2.2 Antibrowning of Fruits, Vegetables, and Seafood

To prevent discoloration in fruits, vegetables, and seafood, various inhibitors of enzymatic browning are used, including reducing agents (sulfites, ascorbic acid), chelators (EDTA, organic acids), complexing agents (cyclodextrins), acidulants (citric acid), enzyme inhibitors (peptides, proteins, aromatic carboxylic acid), etc. (Chang 2009; Oms-Oliu et al. 2010). Researchers at the University of Florida have discovered a novel compound in extracts of mussels, identified as hypotaurine, which effectively inhibits browning of food (Schulbach et al. 2013). Honey has great potential to be used as a natural source of antioxidant to reduce the negative effects of PPO browning in fruit and vegetable processing (Chen et al. 2000). Treatment of white grapes and cut fruits with honey has been shown to inhibit PPO activity and browning, and the inhibitory effect is due to a peptide of Mw 600 Da (Oszmianski and Lee 1990). Ascorbic acid is a highly effective inhibitor of enzymatic browning, primarily because of its ability to reduce quinones back to phenolic compounds before they can undergo further reactions to pigments. Citric acid acts as a chelating agent and acidulant, both functionalities inhibiting PPO. 4-Hexylresorcinol (4HR) is a good inhibitor of enzymatic browning for shrimp, apples, potatoes, and iceberg lettuce (Castaner et al. 1996; Monsalve-Gonzalez 1993). EverFresh® is the brand name for 4HR that binds the enzyme (PPO) that causes melanosis or blackspot in shrimp. EverFresh permits delivery of a high-quality product with improved stability against blackspot development without any effect on the flavor or texture of the shrimp. Other applications for EverFresh include melanosis inhibition in lobster, crab, prawns, and crawfish in addition to that in sliced fruit (Peterson 2006). Sporix™, another commercial browning inhibitor containing an acidic polyphosphate mixture (sodium acid pyrophosphate, citric acid, ascorbic acid, and calcium chloride), has been observed to delay the onset of oxidation and enzymatic browning in fruits and vegetables (Gardner et al. 1991).

20.3.3 Lipase Inhibitors

Lipases (EC. 3.1.1.3, triacylglycerol hydrolases) are ubiquitous enzymes performing a crucial role in all aspects of fat and lipid metabolism in a variety of organisms. In humans and other vertebrates, control of the digestion, absorption, and reconstitution of fat as well as lipoprotein metabolism is achieved through a range of lipases (Desnuelle 1986). Pancreatic, endothelial, hepatic, lipoprotein lipases are members of the human lipase super family and possess structural similarity. Other tissues, such as lungs, kidney, skeletal muscles, adipose tissue, and placenta, also secrete lipase enzymes. Pancreatic acinar cells secrete pancreatic lipase (triacylglycerol acyl hydrolase; EC. 3.1.1.3), an important enzyme of pancreatic juice responsible for digestion of dietary triglycerides in the small intestine (Verma et al. 2012). In plants, during postgermination, the metabolism of oil reserves provide energy and a carbon skeleton for embryonic growth and is controlled by the action of lipases (Huang 1987). Bacteria and fungi are also good producers of a wide range of extracellular lipid-degrading enzymes to break down the insoluble lipid into soluble substances to facilitate absorption (Lie et al. 1991).

Lipase inhibitors are substances used to decrease or stop the activity of lipases. Pancreatic lipase inhibition is the most widely studied mechanism for the identification of potential antiobesity agents. Apart from the centrally acting antiobesity drugs, lipase inhibitors act through the pancreatic lipase inhibition. One example is tetrahydrolipstatin (Orlistat), a derivative of the naturally occurring lipase inhibitor produced from *Streptomyces toxytricini* (Ballinger and Peikin 2002). The success of naturally occurring lipase inhibitors for treatment of obesity has influenced the research sector because they lack unpleasant side effects.

20.3.3.1 Source of Lipase Inhibitors

Natural products provide a vast pool of pancreatic lipase inhibitors and various extracts and secondary metabolites derived from plants and microorganisms showed pancreatic lipase inhibitory activity (Birari and Bhutani 2007; Yun 2010).

20.3.3.2 Application of Lipase Inhibitors

Lipase inhibitors are known to be used as antiobesity agents, used in research to have a better understanding of the mechanisms of lipase action and the noncatalytic functions of lipases, and may find applications for the treatment of infectious diseases. However, their applications in food processing industries and feed preparation are yet to gain momentum through systematic research and development. Nevertheless, lipase inhibitors have the potential for application in food processing and feed preparation when lipase activity needs to be regulated.

Lipases have been used directly in the food and/or feed industries, for example, in foods and/or feeds comprising cereals and, in particular, in bread production. Addition of a lipase inhibitor to dough results in an improvement in the antistaling effect so that an improved softness can be obtained without the addition of any additional fat/oil to the dough. It has also been found that under certain conditions the use of lipases in dough may have unfavorable consequences, such as the production of off-flavors, a disadvantageous impact on yeast activity, and/or a negative effect on bread volume. The negative effect on bread volume, known as overdosing, can lead to a decrease in gluten elasticity, which results in too stiff dough with reduced volumes. In addition, such lipases can degrade and result in a shortening of the oil or milk fat added to the dough. The uncontrolled action of lipases can cause problems such as rancidity in edible oils. In order to prevent the uncontrolled action of lipases, lipase inhibitors can be added, which may be advantageous for use in the baking industry. However, with the increased application and use of enzymes in the food and/or feed industries, attention has to be paid to the effect of endogenous or added inhibitors as modifiers of enzyme activity in the food. All food industries in which lipases are used for processing may find the application of lipase inhibitors in order to prevent development of undesired odor, color, taste, and texture due to uncontrolled action of the enzyme.

20.4 Conclusion

Enzyme inhibitors have already earned recognition and significance owing to their effective activities as drugs, particularly in the management of hyperglycemia and hypertension linked to type-2 diabetes, obesity, hyperlipemia, and as biocontrol agents whereas, as has been discussed in this chapter, enzyme inhibitors have immense potential for applications in food and beverage processing as well as in feed preparation for animals. Even though plenty of studies in the field of enzyme inhibitors in pharmaceutical sectors have been conducted, extensive research is necessary for the development of high-performance inhibitors in the agricultural and food fields. The purification and characterization of proteinase inhibitors and the cloning genes will increase our knowledge of the use of these substances in food, vegetable, beverages, and seafood industries. Enzyme inhibitors are plentiful in plants and seeds and also from microorganisms. They need to be isolated and harnessed for varied applications in food and beverage processing, particularly for regulating enzymes that can lead to undesirable effects when used in excess, such as in the case of lipase or protease. We conclude that enzyme inhibitors, both natural and synthetic, have a great future as significant tools in food and beverage industries in the coming years.

REFERENCES

Ali, P. P. M., Sapna, K., Mol, K. R. R., Bhat, S. G., Chandrasekaran, M., and Elyas, K. K. 2014. Trypsin inhibitor from edible mushroom *Pleurotus floridanus* active against proteases of microbial origin. *Appl. Biochem. Biotechnol.* 173, 167–178.

Avanzo, P., Sabotič, J., Anžlovar, S. et al. 2009. Trypsin-specific inhibitors from the basidiomycete *Clitocybe nebularis* with regulatory and defensive functions. *J. Microbiol.* 155, 3971–3981.

Baird-Parker, T. C. 2003. The production of microbiologically safe and stable foods. In B. M. Lund, T. C. Baird-Parker, and G. W. Gould (Eds.), *The Microbiological Safety and Quality of Food*. Aspen Publishers, Inc., Gaithersburg, MD, pp. 3–18.

Ballinger, A., and Peikin, S. 2002. Orlistat: Its current status as an anti-obesity drug. *Eur. J. Pharmacol.* 440, 109–117.

Barrette-Ng, I. H., Ng, K. K. S., Cherney, M. M., and Pearce, G. 2003. Structural basis of inhibition revealed by a 1:2 complex of the two-headed tomato inhibitor-II and subtilisin Carlsberg. *J. Biol. Chem.* 278, 24062–24071.

BCC Research. 2012. Report code BIO057B. Available at http://www.bccresearch.com/market-research/biotechnology/enzyme-inhibitors-global-markets-bio057b.html.

Bijina, B., Chellappan, S., Krishna, J. G., Basheer, S. M., Elyas, K. K., Bakhali, A. H., and Chandrasekaran, M. 2011. Protease inhibitor from *Moringa oleifera* with potential for use as therapeutic drug and as seafood preservative. *Saudi J. Biol. Sci.* 18, 273–281.

Birari, R., and Bhutani, K. 2007. Pancreatic lipase inhibitors from natural sources: Unexplored potential. *Drug Discov. Today* 12, 879–889.

Birk, Y. 1987. Proteinase inhibitors. In A. Neuberger and K. Brocklehurst (Eds.). *Hydrolytic Enzymes* 16, 257–305. Amsterdam: Elsevier Science.

Bode, W., and Huber, R. 1992. Natural protein proteinase inhibitors and their interactions with proteinases. *Eur. J. Biochem.* 204, 433–451.

Castaner, M., Gil, M. I., Artes, F., and Tomas-Barberan, F. A. 1996. Inhibition of browning of harvested head lettuce. *J. Food Sci.* 61, 314–316.

Ceci, L. R., Spoto, N., Virgilio, M. D., and Gallerani, R. 1995. The gene coding for the mustard trypsin inhibitor-2 is discontinuous and wound inducible. *FEBS Lett.* 364, 179–181.

Ceciliani, F., Bortolotti, F., Menengatti, E., Ronchi, S., Ascenzi, P., and Pakmieri, S. A. 1994. Purification, inhibitory properties, amino acid sequence and identification of the reactive site of a new serine proteinase inhibitor from oil-rape (*Brassica napus*) seed. *FEBS Lett.* 342, 221–224.

Chandrasekaran, M. 1985. Studies on microbial spoilage of *Penaeus indicus*. PhD Thesis, Cochin University of Science and Technology, Cochin, India.

Chang, T. S. 2009. An updated review of tyrosinase inhibitors. *Int. J. Mol. Sci.* 10, 2440–2475.

Chen, L., Mehta, A., Berenbaum, M., Zangerl, A. R., and Engeseth, N. J. 2000. Honeys from different floral sources as inhibitors of enzymatic browning in fruit and vegetable homogenates. *J. Agric. Food Chem.* 48, 4997–5000.

Chothia, C., and Lesk, A. M. 1986. The relation between the divergence of sequence and structure in proteins. *EMBO J.* 5, 823–826.

Cornwall, G. A., Cameron, A., Lindberg, I., Hardy, D. M., Cormier, N., and Hsia, N. 2003. The cystatin-related epididymal spermatogenic protein inhibits the serine protease prohormone convertase 2. *Endocrinology* 144, 901–908.

Cromie, S. 1991. Microbiological aspects of extended shelf life products. *Aust. J. Dairy Technol.* 46, 101–104.

Desnuelle, P. 1986. Pancreatic lipase and phospholipase. In P. Desnuelle, H. Sjostrom, and O. Woren (Eds.), *Molecular and Cellular Basis of Digestion*. Elsevier, Amsterdam, Netherlands, pp. 275–296.

Eggers, C. T., Murray, I. A., Delmar, V. A., Day, A. G., and Craik, C. S. 2004. The periplasmic serine protease inhibitor ecotin protects bacteria against neutrophil elastase. *Biochem. J.* 379, 107–118.

Funaki, J., Abe, K., and Hayabuchi, H. 1991. Modulating the conditioning of meat by the use of oryzacystatin, a cysteine proteinase inhibitor of rice seed origin. *J. Food Biochem.* 15, 253–262.

Gardner, J., Manohar, S., and Borisenok, W. S. 1991. Sulfite-free preservative for fresh peeled fruits and vegetables. Patent No. US 4,988,523.

Gettins, P. G. 2002. Serpin structure, mechanism, and function. *Chem. Rev.* 102, 4751–4804.

Goll, D. E., Kleese, W. C., and Szpacenko, A. 1989. Skeletal muscle proteases and protein turn over. In D. R. Campion, G. J. Hausman, and R. J. Martin (Eds.), *Animal Growth Regulation*. Plenun Publishing, New York, pp. 141–181.

Greagg, M. A., Brauer, A. B., and Leatherbarrow, R. J. 1994. Expression and kinetic characterization of barley chymotrypsin inhibitors 1a and 1b. *Biochim. Biophys. Acta* 1222, 179–186.

Greenblatt, H. M., Ryan, C. A., and James, M. N. G. 1989. Structure of the complex of *Streptomyces griseus* proteinase B and polypeptide chymotrypsin inhibitor-1 from Russet Burbank potato tubers at 2.1-A° resolution. *J. Mol. Biol.* 205, 201–228.

Greene, D. H., and Babbitt, J. K. 1990. Control of muscle softening and protease-parasite interaction in arrow-tooth flounder, *Atheresthes stomias*. *J. Food Sci.* 55, 579–580.

Grieve, P., and Kitchen, B. 1985. Proteolysis in milk: The significance of proteinases originating from milk leucocytes and a comparison of the action of leucocyte bacterial and natural milk proteinases on casein. *J. Dairy Res.* 52, 101–112.

Gruden, K., Strukelj, B., Ravnikar, M., and Poljsak-Prijatelj, M. 1997. Potato cysteine proteinase inhibitor gene family: Molecular cloning, characterisation and immunocytochemical localisation studies. *Plant. Mol. Biol.* 34, 317–323.

Haard, N. 1994. Seafoods chemistry, protein hydrolysis in seafoods. In F. Shahidi, and R. Botta (Eds.), *Processing Technology and Quality*. Blackie Academic & Professional, Glasgow, pp. 10–33.

Habib, H., and Fazili, K. M. 2007. Plant protease inhibitors: A defense strategy in plants *Biotechnol. Mol. Biol. Rev.* 2, 68–85.

Hartley, B. S. 1960. Proteolytic enzymes. *Annu. Rev. Biochem.* 29, 45–72.

Hernandez, J. F., Gagnon, J., Chiche, L., and Nguyen, T. M. 2000. Squash trypsin inhibitors from *Momordica cochinchinensis* exhibit an atypical macrocyclic structure. *Biochemistry* 39, 5722–5730.

Hojima, Y., Pirce, J. V., and Pisano, J. J. 1980. Hageman factor fragment inhibitor in corn seeds: Purification and characterization. *Thromb. Res.* 20, 149–162.

Huang, A. H. C. 1987. Lipases. In P. K. Stumpf, and E. E. Cohn. (Eds.), *The Biochemistry of Lipases*, vol. 9. Academic Press Inc., New York, pp. 91–119.

Huang, Q., Liu, S., Tang, Y., Zeng, F., and Qian, R. 1992. Amino acid sequencing of a trypsin inhibitor by refined 1.6 A° X-ray crystal structure of its complex with porcine beta-trypsin. *FEBS Lett.* 297, 143–146.

Hurrel, R. F., and Finot, P. A. 1984. Nutritional consequences of the reactions between proteins and oxidised polyphenolic acids. *Adv. Exp. Med. Biol.* 177, 423–435.

Imada, C. 2004. Enzyme inhibitors of marine microbial origin with pharmaceutical importance. *Mar. Biotechnol.* 6, 193–198.

Ismail, B., and Nielsen, S. S. 2010. Plasmin protease in milk: Current knowledge and relevance to dairy industry—Invited review. *J. Dairy Sci.* 93(11), 4999–5009. doi:10.3168/jds.2010-3122.

Jackson, R. M., and Russell, R. B. 2000. The serine protease inhibitor canonical loop conformation: Examples found in extracellular hydrolases, toxins, cytokines and viral proteins. *J. Mol. Biol.* 296, 325–334.

Kanaori, K., Kamei, K., Taniguchi, M. et al. 2005. Solution structure of marinostatin, a natural ester-linked protein protease inhibitor. *Biochemistry* 44, 2462–2468.

Kim, J. Y., Park, S. C., Kim, M. H., Lim, H. T., Park, Y., and Hahm, K. S. 2005. Antimicrobial activity studies on a trypsin-chymotrypsin protease inhibitor obtained from potato. *Biochem. Biophys. Res. Commun.* 330, 921–927.

Laber, B., Krieglstein, K., Henschen, A., Kos, J., and Turk, V. 1989. The cysteine proteinase inhibitor chicken cystatin is a phosphoprotein. *FEBS Lett.* 248, 162–168.

Laskowski, M. J., and Kato, I. 1980. Protein inhibitors of proteinases. *Annu. Rev. Biochem.* 49, 593–626.

Lawrence, P. K., and Koundal, K. R. 2002. Plant protease inhibitors in control of polyphagous insects. *Electron. J. Biotechnol.* 1, 93–109.

Lazaro, A., Rodriguez-Palenzuela, P., Marana, C., Carbonero, P., and Garcia-Olmedo, F. 1988. Signal peptide homology between the sweet protein thaumatin II and unrelated cereal alpha-amylase/trypsin inhibitors. *FEBS Lett.* 239, 147–150.

Leo, F. D., Volpicella, M., Licciulli, F., Liuni, S., Gallerani, R., and Ceci, L. 2002. PLANT-PIs: A database for plant protease inhibitors and their genes. *Nucleic Acids Res.* 30, 347–348.

Lie, E., Persson, A., and Molin, G. 1991. Screening for lipase producing microorganisms with a continuous cultivation system. *Appl. Microbiol. Biotechnol.* 35, 19–20.

Lustigman, S., Brotman, B., Huima, T., Prince, A. M., and McKerrow, J. H. 1992. Molecular cloning and characterization of onchocystatin, a cysteine proteinase inhibitor of *Onchocerca volvulus*. *J. Biol. Chem.* 267, 17339–17346.

Macedo, M. L. R., Mello, G. C., Freire, M. G. M., Novello, J. C., Marangoni, S., and Matos, D. G. G. 2002. Effect of a trypsin inhibitor from Dimorphandra mollis seeds on the development of *Callosobruchus maculatus*. *Plant Physiol. Biochem.* 40, 891–898.

Macedo, M. L. R., Freire, M. G. M., Cabrini, E. C., Toyama, M. H., Novello, J. C., and Marangoni, S. A. 2003. Trypsin inhibitor from Peltophorum dubium seeds active against pest protease and its affect on the survival of *Anagasta kuehniella*. *Biochim. Biophys. Acta* 1621, 170–182.

Marshall, C. J. 1993. Evolutionary relationships among the serpins. *Philos. Trans. R. Soc. Lond. B Biol. Sci. B* 342, 101–119.

Martinez, M. V., and Whitaker, J. R. 1995. The biochemistry and control of enzymatic browning. *Trends Food Sci. Technol.* 6, 195–200.

Mello, G. C., Oliva, M. L. V., Sumikava, J. T., Machado, O. L. T., Marangoni, S., and Matos, D. G. G. 2002. Purification and characterization of new trypsin inhibitor from *Dimorphandra mollis*. *J. Protein Chem.* 20, 625–632.

Menengatti, E., Tedeschi, G., Ronchi, S. et al. 1992. Purification, inhibitory properties and amino acid sequence of a new serine proteinase inhibitor from white mustard (*Sinapis alba* L.) seed. *FEBS Lett.* 301, 10–14.

Menon, A. S., and Goldberg, A. L. 1987. Protein substrates activate the ATP-dependent protease La by promoting nucleotide binding and release of bound ADP. *J. Biol. Chem.* 262, 14929–14934.

Monsalve-Gonzalez, A., Barbosa-Canovas, G. V., Cavaileri, R. P., McEvily, A. J., and Iyengar, R. 1993. Control of browning during storage of apple slices preserved by combined methods. 4-Hexaylresorcinol as antibrowning agent. *J. Food Sci.* 58, 797–800.

Mulvenna, J. P., Foley, F. M., and Craik, D. J. 2005. Discovery, structural determination, and putative processing of the precursor protein that produces the cyclic trypsin inhibitor sunflower trypsin inhibitor 1. *J. Biol. Chem.* 280, 32245–32253.

Odani, S., and Ikenaka, T. 1976. The amino acid sequences of two soybean double headed proteinase inhibitors and evolutionary consideration on the legume proteinase inhibitors. *J. Biochem.* 80, 641–643.

Odani, S., Tominaga, K., Kondou, S., Hori, H., Koide, T., Hara, S., Isemura, M., and Tsunasawa, S. 1999. The inhibitory properties and primary structure of a novel serine proteinase inhibitor from the fruiting body of the basidiomycete, *Lentinus edodes*. *Eur. J. Biochem.* 262, 915–923.

Ohtsubo, S., Kobayashi, H., Noro, W., Taniguchi, M., and Saitoh, E. 2005. Molecular cloning and characterization of oryzacystatin-III, a novel member of phytocystatin in rice (*Oryza sativa* L. japonica). *J. Agric. Food Chem.* 53, 5218–5224.

Oms-Oliu, G., Rojas-Grati, M. A., Gonzalez, L. A. et al. 2010. Recent approaches using chemical treatments to preserve quality of fresh-cut fruit: A review. *Postharvest Biol. Technol.* 57, 139–148.

Oszmianski, J., and Lee, C. Y. 1990. Inhibition of polyphenol oxidase activity and browning by honey. *J. Agric. Food Chem.* 38, 1892–1895.

Otlewski, J., Krowarsch, D., and Apostoluk, W. 1999. Protein inhibitors of serine proteinases. *Acta Biochim. Pol.* 46, 531–565.

Peterson, C. 2006. The Use of EverFresh® for Preventing Melanosis on Shrimp and Other Crustaceans. Seafood Science & Technology Society of the Americas SST Conference, November. Available at http://fshn.ifas.ufl.edu/seafood/sst/30thAnn/Peterson_SunOpta%20Ever Fresh.pdf.

Poerio, E., Gennaro, S. D., Maro, A. D., and Farisei, F. 2003. Primary structure and reactive site of a novel wheat proteinase inhibitor of subtilisin and chymotrypsin. *Biol. Chem.* 384, 295–304.

Rawlings, N. D. 2010. Peptidase inhibitors in the MEROPS database. *Biochimie* 92 (11), 1463–1483.

Rawlings, D. N., Tolle, P. D., and Barrett, A. J. 2004. Evolutionary families of peptidase inhibitors. *Biochem. J.* 378, 705–716.

Reeck, G. R., Kramer, K. J., Barker, J. E., Kanost, M. R., Fabrick, J. A., and Behnke, C. A. 1997. Proteinase inhibitors and resistance of transgenic plants to insects. In N. Carozzi, and M. Koziel (Eds.), *Advances in Insect Control. The Role of Transgenic Plants*, London: Taylor & Francis, pp. 157–183.

Renko, M., Sabotič, J., Mihelič, M., Brzin, J. E., Kos, J., and Turk, D. A. 2010. Versatile loops in mycocypins inhibit three protease families. *J. Biol. Chem.* 285, 308–316.

Sabotič, J., and Kos, J. 2012. Microbial and fungal protease inhibitors—Current and potential applications. *Appl. Microbiol. Biotechnol.* 93, 1351–1375.

Sabotič, J., Trček, T., Popovič, T., and Brzin, J. 2007. Basidiomycetes harbor a hidden treasure of proteolytic diversity. *J. Biotechnol.* 128, 297–307.

Sabotič, J., Bleuler-Martinez, S., Renko, M. et al. 2012. Structural basis of trypsin inhibition and entomotoxicity of cospin, serine protease inhibitor involved in defense of *Coprinopsis cinerea* fruiting bodies. *J. Biol. Chem.* 287, 3898–3907.

Saito, H., Suzuki, T., Ueno, K., Kubo, T., and Natori, S. 1989. Molecular cloning of cDNA for sarcocystatin A and analysis of the expression of the sarcocystatin A gene during development of *Sarcophaga peregrina*. *Biochemistry* 28, 1749–1755.

Sapna, K. 2013. Isolation, purification, characterization and application of proteinaceous protease inhibitor from marine bacterium pseudomonas mendocina BTMW 301. PhD Thesis, Cochin University of Science and Technology, Cochin, India.

Schulbach, K. F., Johnson, J. V., Simmone, A. H., Kim, J. M., Jeong, Y., Yagiz, Y., and Marshall, M. R. 2013. Polyphenol oxidase inhibitor from Blue Mussel (*Mytilus edulis*) extract. *J. Food Sci.* 78, C425–C431.

Sentandreu, M. A., Coulis, G., and Ouali, A. 2002. Role of muscle endopeptidases and their inhibitors in meat tenderness. *Trends Food Sci. Technol.* 13, 400–421.

Silverman, G. A., Bird, P. I., Carrell, R. W. et al. 2001. The serpins are an expanding superfamily of structurally similar but functionally diverse proteins: Evolution, mechanism of inhibition, novel functions, and a revised nomenclature. *J. Biol. Chem.* 276, 33293–33296.

Silverman, G. A., Whisstock, J. C., Bottomley, S. P. et al. 2010. Serpins flex their muscle: Putting the clamps on proteolysis in diverse biological systems. *J. Biol. Chem.* 285, 24299–24305.

Sorhaug, T., and Stepaniak, J. 1997. Psychrotrophs and their enzymes in milk and dairy products. *Trends Food Sci. Technol.* 8, 35–41.

Stubbs, M. T., Huber, R., and Bode, W. 1995. Crystal structures of factor Xa specific inhibitors in complex with trypsin: Structural grounds for inhibition of factor Xa and selectivity against thrombin. *FEBS Lett.* 375, 103–107.

Suzuki, A., Tsunogae, Y., Tanaka, I., Yamane, T., and Ashida, T. 1987. The structure of Bowman-Birk type protease inhibitor A-II from peanut (*Arachis hypogaea*) at 3.3-A° resolution. *J. Biochem.* 101, 267–274.

Szyperski, T., Guntert, P., Stone, S. R., and Wuthrich, K. 1992. Nuclear magnetic resonance solution structure of hirudin (1-51) and comparison with corresponding three-dimensional structures determined using the complete 65-residue hirudin polypeptide chain. *J. Mol. Biol.* 228, 1193–1205.

Terada, S., Fujimura, S., Katayama, H., Nagasawa, M., and Kimoto, E. 1994. Purification and characterization of two Kunitz family subtilisin inhibitors from seeds of *Canavalia lineata*. *J. Biochem.* 115, 392–396.

Umezawa, H. 1982. Low molecular weight enzyme inhibitors of microbial origin. *Annu. Rev. Microbiol.* 36, 75–99.

Valueva, T. A., and Mosolov, V. V. 2004. Role of inhibitors of proteolytic enzymes in plant defense against phytopathogenic microorganisms. *Biochemistry* 69, 1305–1309.

Verma, N., Thakur, S., and Bhatt, A. K. 2012. Microbial lipases: Industrial applications and properties (a review). *Int. Res. J. Biol. Sci.* 1, 88–92.

Wahyudi, A. T., Qatrunnada, and Mubarik, N. R. 2010. Screening and characterization of protease inhibitors from marine bacteria associated with *Sponge Jaspis* sp. *HAYATI J. Biosci.* 17, 173–178.

Wasson, D. H., Reppond, K. D., Babbitt, J. K., and French, J. S. 1992. Effects of additives on proteolytic and functional properties of arrowtooth flounder surimi. *J. Aquat. Food Prod. Technol.* 1(3/4), 147–165.

Weerasinghe, V., Morrissey, M., Chung, Y., and An, H. 1996. Whey protein concentrate as a proteinase inhibitor in pacific whiting surimi. *J. Food Sci.* 61, 367–371.

Whetstone, P. A., and Hammock, B. D. 2007. Delivery methods for peptide and protein toxins in insect control. *Toxicon* 49, 576–596.

Whitaker, J. R., and Lee, C. Y. 1995. Recent advances in chemistry of enzymatic browning. In C. Y. Lee, and J. R. Whitaker (Eds.), *Enzymatic Browning and Its Prevention*. ACS Symposium Series 600, American Chemical Society, Washington, DC, pp. 2–7.

Yun, J. 2010. Possible anti-obesity therapeutics from nature—A review. *Phytochemistry* 71, 1625–1641.

Zuchowski, J., and Grzywnowicz, K. 2006. Partial purification of proteinase K inhibitors from liquid-cultured mycelia of the white rot basidiomycete *Trametes versicolor*. *Curr. Microbiol.* 53, 259–264.

21

Enzymes in Valorization of Food and Beverage Wastes

Wan Chi Lam, Tsz Him Kwan, and Carol Sze Ki Lin

CONTENTS

21.1 Introduction

Every year, a large quantity of food is wasted globally. It is estimated that approximately one third of food produced for human consumption is lost or wasted worldwide, and it accounts for about 1.3 billion tons per year (Gustavsson et al. 2011). Food wastage can occur during various phases in the food cycle, including food production, processing, distribution, and consumption. Landfill and incineration are the major waste management practices in many countries. However, this also creates a lot of environmental problems, such as air and water pollution and greenhouse gas emissions. Food waste generated from both food and beverage processing industries includes organic materials, which contain significant amount of carbon, nitrogen, and lipids in various forms, including starch, lignocellulose, proteins, and fat or oil. Consequently, hydrolysis of these components from food waste yields significant amounts of glucose, amino acids, and fatty acids that can be subsequently used for fermentative chemicals or biofuel production, and the overall concept is illustrated in Figure 21.1.

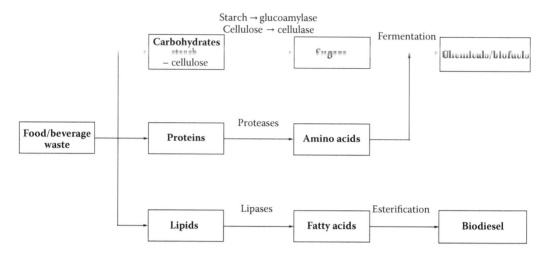

FIGURE 21.1 Valorization of carbohydrates, proteins, and lipids from food and beverage wastes for biodiesel, chemicals, and biofuel production using enzymes.

Enzyme utilization during bioconversion processes is one of the key factors that contributes to the waste-to-energy process in an efficient and environmentally friendly manner. In this chapter, important enzymes, including glucoamylase, protease, lipase, and cellulase, which are commonly used in food waste conversion are first introduced. Then three case studies for food waste valorization using these enzymes are discussed.

21.2 Glucoamylases

Glucoamylase (EC. 3.2.1.3), industrially known as amyloglucosidase, belongs to a family of amylolytic enzymes. It is an exoacting enzyme that catalyzes the hydrolysis of α-(1,4) glycosidic bonds, and α-(1,6) glucosidic bonds of polysaccharides form the nonreducing ends, giving glucose as the end product (Norouzian et al. 2006; Sauer et al. 2000). However, the enzyme cleaves α-(1,6) glucosidic bonds at a much slower rate. It has been revealed that the rate for the enzyme to hydrolyze α-1-6-glycosidic bonds is about 30 times slower than the hydrolysis of α-1-4-bonds (Norouzian et al. 2006; Robertson et al. 2006). Glucoamylase is one of most industrially important enzymes that is used to manufacture crystalline glucose or glucose syrup from starch for wine, biofuel, and chemical production. Therefore, glucoamylase has extensive applications in many fields, including food, chemical, and biotechnological industries. Commercial glucoamylase is produced by filamentous fungi mainly from *Aspergillus* sp., such as *A. niger* (Norouzian et al. 2006) and *A. awamori* (Buchholz and Seibel 2008; James and Lee 1997) due to its ability to produce extracellular thermostable glucoamylases in large quantities.

21.2.1 Kinetic Characteristics

Most fungal glucoamylases from the *Aspergillus* strain are active at a range of acidic pH with optimal pH ranging from 4.5 to 6.5 and at high temperatures at a range of 50°C to 70°C (Flor and Hayashida 1983; Norouzian et al. 2006; Tsujita and Endo 1977). The optimal pH and temperature of glucoamylase from some industrial important fungal producers are summarized in Table 21.1.

21.2.2 Commercial Preparations and Recent Developments in Glucoamylase Preparation

Commercially available glucoamylase is usually produced from species of *Aspergillus* and *Rhizopus* through fermentation. Submerged (SmF) and solid state (SSF) fermentation are two techniques commonly

TABLE 21.1

Kinetic Properties of Glucoamylases from Various *Aspergillus* Strains

Fungal Host	Substrate	Molecular Weight (kDa)	Stable pH	Optimal pH	Stable Temperature	Optimal Temperature	References
A. awamori var. *kawachi* mutant HF-15	Synthetic medium	250	2.0–8.0	N/A	85°C	30°C–65°C	Flor and Hayashida 1983
A. oryzae OSI 1013	Synthetic medium	65	N/A	4.5	up to 46°C (~40 min)	65°C	Tsujita and Endo 1977
	Steamed rice	63–99	N/A	4.0	up to 37°C (40 min)	56°C	Tsujita and Endo 1977
A. niger B-30	Cornstarch	78.3	N/A	4.6	60°C (~120 min)	70°C	Liu et al. 2013
		97.2	N/A	4.0	60°C (~120 min)	70°C	Liu et al. 2013

used for industrial enzyme production (Pandey et al. 1999a). Both SmF and SSF fermentation processes can be used for glucoamylase production with various types of fermenters. In SmF, the fermentation is conducted in free-flowing liquid substrate. Initially, sterile medium with desired pH, carbon (such as soluble starch), and nitrogen concentration is loaded into a fermenter. The medium is then inoculated with *Aspergillus*. Consequently, fermentation is conducted in a controlled pH, aeration, and temperature. Because glucoamylase is secreted to the medium, the spent medium obtained at the end of fermentation is collected for further purification for glucoamylase. In contrast, SSF is conducted on moist solid substrate in the absence of free water. SSF offers numerous advantages for enzyme production, including superior productivity, low capital investment, low energy requirement, less wastewater output, lower contamination risk, and simple substrate treatment (Asgher et al. 2006; Babu and Satyanarayana 1995; Pandey 1995; Regulapati et al. 2007; Viniegra-González et al. 2003). Thus, it is also commonly adopted for enzyme production.

In SSF, it is important to ensure the homogeneity of bakery waste; thus food waste is homogenized and mixed once it has been collected. To facilitate glucoamylase production, the moisture content of the food waste will be adjusted to approximately 60%. Fungal spores from the desired host are then inoculated to the surface of food waste. The mixture is kept in an incubator at 30°C with aeration under static conditions for fermentation. SSF for glucoamylase production usually takes place for 7 days or more. The progress of fermentation for glucoamylase production using pastry wastes at various time intervals is shown in Figure 21.2. In this figure, pretreated bakery waste right after spore inoculation (Figure 21.2a) and a growth of a thick layer of white mycelia on the surface of food waste at 48 h (Figure 21.2b) can be seen. When the fungus continues to grow, sporulation occurs; a layer of black spores covered the entire surface of food waste at 96 hours (Figure 21.2c), and production of glucoamylases from *Aspergillus awamori* on pastry waste over time can be noted (Figure 21.2d). High glucoamylase activity was observed during days 7 to 11.

In general, glucoamylase is produced under glucose-limited condition. Starch has been shown as an inducer for glucoamylase synthesis by some *Aspergillus* producers (Ganzlin and Rinas 2008; Ventura et al. 1995; Zambare 2010), and thus it can be served as a cheap carbon source for fermentative glucoamylase production. The use of sucrose, cassava, sorghum, and wheat flour are known to be used as a carbon source (Beolchini et al. 2006; Fiedurek and Szczodrak 1995; Gomes et al. 2005; Kumar and Satyanarayana 2007; Li et al. 2007; Pacheco-Chávez et al. 2004). Yeast extract, peptone, urea, or ammonium could be used as a nitrogen source to promote the growth of the microorganism. In order to further reduce the production cost of the process, the feasibility of glucoamylase production through fermentation from various municipal food wastes or agriculture residues, such as bakery waste (Lam et al. 2013; Leung et al. 2012; Melikoglu et al. 2013a), restaurant food waste (Wang et al. 2008), wheat by-products (Anto et al. 2006; Du et al. 2008), and cowpea waste (Kareem et al. 2009) have recently been

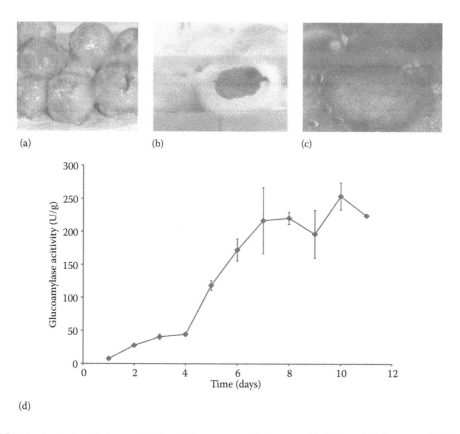

FIGURE 21.2   Production of glucoamylase from bakery waste at (a) time zero, (b) 48 hours, (c) 96 hours, and (d) throughout 11 days fermentation.

TABLE 21.2

Production Yields of Glucoamylase from Various Substrates with *Aspergillus* Strain

Substrate	Concentration (U/ml)	Yield (U/g)	Fungal Host	References
Rice powder	N/A	71.3 ± 2.34[a]	*Aspergillus niger*	Anto et al. 2006
Wheat bran	N/A	110 ± 1.32[a]	*Aspergillus niger*	Anto et al. 2006
Mixed food waste	137	N/A	*Aspergillus niger*	Wang et al. 2008
Cowpea waste	970	N/A	*Aspergillus oryzae*	Kareem et al. 2009
Wheat bran	4.4	48.0	*Aspergillus awamori*	Du et al. 2008
Wheat piece	3.32	81.3	*Aspergillus awamori*	Wang et al. 2009
Waste bread	3.94	78.4	*Aspergillus awamori*	Wang et al. 2009
	N/A	114	*Aspergillus awamori*	Melikoglu et al. 2013a
Pastry waste	76 ± 6.1[a]	253.7 ± 20.4[a]	*Aspergillus awamori*	Lam et al. 2013

[a]   Values indicate means ± standard errors.

investigated. The glucoamylase yields from various wastes presented in Table 21.2 indicate production of a significant amount of glucoamylase.

21.3 Proteases

Proteases are a diverse group of enzymes, which are capable of hydrolyzing the peptide bonds of proteins and giving shorter peptides or amino acids as products. Depending on the site of the hydrolytic actions, proteases can generally be categorized into two main groups: endopetidases and exopetidases. Endopetidases proteolytic enzymes catalyze the cleavage of peptide bonds within the internal region of protein molecules giving shorter peptides as end products, and exopetidases catalyze the cleavage of the peptide bond at or near the terminal of the polypeptides, giving amino acids or dipeptides as end products. Both types of enzymes cooperate for effective digestion of proteins. Moreover, based on the functional groups present on the active site, proteases can also be classified into serine proteases, cysteine proteases, aspartic proteases, and metalloproteases (Hartley 1960). Other protease classifications can be achieved according to their active pH as acidic, neutral, and alkaline proteases.

Proteases are produced from a diversity of sources, including microorganisms, plants, and animals. Microbial proteases are a predominant enzyme source in worldwide sales of industrial enzymes because microorganisms are susceptible to genetic manipulation for overproduction of target enzymes, rapid growth, and require only limited space for cultivation. It is estimated that microbial proteases account for approximately 40% of the total worldwide enzyme sales (Godfrey and West 1996). Proteases from species of *Aspergillus*, *Penicillium*, and *Rhizopus* are commonly used in various food processes for the production of animal feed, fish sauce, or soy sauce due to its being classified as generally regarded as safe (Sandhya et al. 2005).

21.3.1 Kinetic Characteristics

Proteases from *Aspergillus* strains have wide applications in food processing, and they are known to have a high ability to produce an array of proteases (Table 21.3). Production of acid, neutral, and alkaline proteases with *Aspergillus* strains has been reported (Sandhya et al. 2005; Tsujita and Endo 1977; Vishwanatha et al. 2009; Yin et al. 2013). The optimal pH and temperature of proteases from various strains of *Aspergillus* are summarized in Table 21.3.

TABLE 21.3

Kinetic Properties of Acidic, Neutral, and Alkaline Proteases from *Aspergillus* Strains

Fungal Host	Substrate	Molecular Weight (kDa)	Stable pH	Optimal pH	Stable Temperature (°C)	Optimal Temperature (°C)	References
Acidic Proteases							
A. oryzae MTCC 5341	Wheat bran	47	2.5–6.0	3.2	40–57	55	Vishwanatha et al. 2009
A. oryzae BCRC 30118	Malt extract broth	41	–	3.0	<40	60	Yin et al. 2013
A. oryzae (not specified)	3% Malt extract agar	E_1: 60 E_{1a}: 55 E_{1b}: 49 E_2: 42	All: 3.0–6.0	All: 3.0–4.2	E_1: 40 to <60 E_{1a-1b}: 40 to <60 E_2: 40 to <50	–	Tsujita and Endo 1977
Neutral Proteases							
A. oryzae CICIM F0899	Wheat bran and defatted soybean pulp in 4:1 ratio	50	5–9	7.0–9.0	20–40	50	Wang et al. 2013
Alkaline Proteases							
A. oryzae CFR 305	Coffee by-products	35	9.0–12.0	10.0	50–70	60	Murthy and Naidu 2010
A. niger Soil isolate	PDA slant	38	–	10.0	40 for 1 h	45	Kalpana Devi et al. 2008

TABLE 21.4

Production of Proteases from *Aspergillus* Using Various Agroindustrial Wastes

Substrate	Yield (U/g)	Activity (U/ml)	Fungal Host	Protease Type/ Assay pH	References
Wheat bran	31.2[b]	N/A	*A. oryzae* NRRL 1808	Neutral/7	Sandhya et al. 2005
Mixed wheat and rice bran[a]	1256[c]	N/A	*A. oryzae* Ozykat-1	Alkaline/10	Chutmanop et al. 2008
Rice bran	1400[c]	N/A	*A. oryzae* Ozykat-1	Alkaline/10	Chutmanop et al. 2008
Canola cake	371[d]	N/A	*A. oryzae* CCBP 001	Neutral/7	Freitas et al. 2013
Wheat bran	8.3×10^{5e}	N/A	*A. oryzae* MTCC 5341	Acidic/3.2	Vishwanatha et al. 2009
Rice broken	44.7–67.7[c]	N/A	*A. niger* MTCC 281	Neutral/7	Paranthaman et al. 2009
Hulled grain of wheat	N/A	183.1[c]	*A. niger* I1	Acidic/3	Siala et al. 2012
Wine stillage	N/A	200[c]	*A. niger* THA001	Acidic/4	Yang and Lin 1998
Jatropha seed cake	3260.5	N/A	*A. versicolor* CJS-98	Neutral/7.5	Veerabhadrappa et al. 2014

Note: One unit of enzyme activity was defined as the amount of enzyme that liberated incomplete sentence.
[a] Wheat to rice bran ratio of 0.33.
[b] 1 μM tyrosine min^{-1}.
[c] 1 μg tyrosine min min^{-1}.
[d] 0.01 absorbance at 660 nm min^{-1}.
[e] 0.001 absorbance at 280 nm min^{-1} under assay conditions.

21.3.2 Commercial Preparations and Recent Developments in Proteases Preparation

As discussed in the previous section, solid-state fermentation (SSF) and submerged fermentation (SmF) can be applied for fermentative protease production. Moreover, several points should be noted for culture medium used for protease production. Medium with high carbohydrates should be avoided as it is inhibitory to protease production (Kumar and Takagi 1999; Sumantha et al. 2006). Starch and molasses can be used as low-cost carbon sources (Radha et al. 2011). The presence of some amino acids, such as leucine and proline, has been reported to repress proteases production (Çalk et al. 2003). On the other hand, complex nitrogen sources, such as peptides and proteins, are suggested to induce protease production. Corn steep liquor, tryptone, and casein at a certain concentration (approximately 2%) could serve as a cheap nitrogen source for protease production (Kumar and Takagi 1999). Several agroindustrial wastes are recognized as potential low-cost substrates for the production of proteases. The properties and yield of the proteases produced using various agroindustrial wastes with *Aspergillus* strains are summarized in Table 21.4.

21.4 Lipases

Lipases (EC. 3.1.1.3) are a class of hydrolases that catalyzes a number of reactions, including hydrolysis, esterification, interesterification, transesterification, alcoholysis, and acidolysis. The most well-known catalytic reaction is the hydrolysis of triglycerides liberating diglycerides, monoglycerides, free fatty acids, and glycerol. In addition, lipases are active and stable in organic solvents, including supercritical carbon dioxide.

Lipases are monomeric proteins consisting of a central β-sheet surrounded by α-helices. The catalytic mechanism mainly involves interfacial activation. In a homogeneous aqueous medium, they are in the closed and inactive conformation because the active site cleft of the lipases is covered by a lid. In contrast, with the presence of a nonaqueous phase, the lipases are adsorbed on the interface between the aqueous phase and the nonaqueous phase. The lid undergoes conformational displacement and exposes the active site cleft to the nonaqueous phase. As a result, the conformational equilibrium is shifted toward the active form from the closed form of lipases (Guncheva and Zhiryakova 2011).

Lipases can be prepared either by extraction from animal or plant tissue or by cultivation of microorganisms, including bacteria, fungi, and yeast. In view of the rapid growth of microorganisms and absence of seasonal fluctuations, the majority of commercially available lipases are produced by fungi, such as *Candida, Geotrichum, Rhizopus*, and *Thermomyces*, and bacteria, such as *Pseudomonas* and *Bacillus* (Pandey et al. 1999b). *Candida rugosa* is the most frequently used microorganism for lipase production (Pandey et al. 1999b).

21.4.1 Kinetic Characteristics

In general, lipases secreted by bacteria and fungus exhibit the maximal activity at pH 7–9 and 35°C–70°C. Recently, lipases have also successfully been isolated from *Geobacillus stearothermophilus* SB-1, an extremophile, which showed the optimum pH and temperature at 3°C and 100°C, respectively. The optimal pH and temperature for the catalytic activity of lipases produced by different microorganisms are given in Table 21.5.

21.4.2 Commercial Preparation and Recent Development in Lipase Preparation

Solid-state fermentation (SSF) and submerged fermentation (SmF) are the major technologies used in the commercial production of lipases (Pandey et al. 1999b). Lipids, such as natural oils, fatty acids, and fatty esters, are necessary to induce fermentative lipase production. First, the cell-bound lipases hydrolyze the lipids to start cell growth and enzyme induction. As the level of the available lipids decreases along with the fermentation, more lipases are secreted to increase the probability of enzyme–substrate contact for nutrient uptake and cell survival. As a result, the lipase production level reached the maximum in the late stationary growth phase. The feasibility of using agricultural waste residues for the production of lipases has been explored, and several agricultural wastes have appeared as good feedstock for fermentative lipase production (Table 21.6).

Some of the commercially available lipases produced from different microbial origins by different companies are shown in Table 21.7. The commercially available lipases are expensive, and the free enzymes are difficult to be recycled and reused. To overcome this, the majority of current research has been focusing on enzyme immobilization by which the enzymes are immobilized in a support through several methods, such as adsorption, covalent bonding, entrapment, encapsulation, and cross-linking to the support. Ting et al. (2006) found that immobilized lipases exhibited higher thermal stability, storage stability, and enzyme activity than the free lipases. It has been reported that immobilized enzymes retained 80% enzyme activity after six repeated uses.

TABLE 21.5

Kinetic Properties of Lipases from Different Bacterial and Fungal Strains

Host	Optimal pH	Optimal Temperature (°C)	References
Bacillus sp. THL027	7.0	70	Dharmsthiti and Luchai 1999
Bacillus sphaericus 205y	7.0–8.0	55	Sulong et al. 2006
Bacillus pumilus B26	8.5	35	Kim et al. 2002
Pseudomonas fluorescens NS2W	9.0	60	Kanwar and Goswami 2002
Rhizopus oryzae	7.5	35	Hiol et al. 2000
Candida rugosa	7.0	37	Ting et al. 2006
Candida antarctica CalA	7.0	50–70	Pfeffer et al. 2006

TABLE 21.6

Lipases Production from Different Agricultural Wastes

Host	Substrate	Maximum Lipase Activity[a]	References
Aspergillus niger	Wheat bran	630 U/g	Mahadik et al. 2002
Bacillus sp.	Olive mill waste	15 U/ml	Ertuğrul et al. 2007
Bacillus coagulans	Melon waste	78,069 U/g	Alkan et al. 2007
Candida rugosa	Cheese whey	15 U/ml	Tommaso et al. 2011
Penicillium restrictum	Babassu oil waste	27.8 U/g	Palma et al. 2000
Pseudomonas aeruginosa	Jatropha seed cake	1084 U/g	Mahanta et al. 2008
Rhizomucor pusillus	Olive cake and sugarcane bagasse	20.24 U/ml	Cordova et al. 1998

[a] One unit of activity (U) was defined as μmol of *p*-nitrophenol released per minute under assay conditions.

TABLE 21.7

Commercially Available Lipases Produced from Microbial Origins by Different Companies

Host	Trade Name	Supplier	Applications
Alcaligenes sp.	Lipase PL, Lipase PLC, Lipase QLM, Lipase QLC	Meito Sangyo. Co.	Modification of oils and fats
Burkholderia cepacia	Lipase PS	Amano	Chiral synthesis
Aspergillus niger	Lipase DS	Amano	Dietary supplement
Aspergillus oryzae	SPICEIT™ MR	Chr. Hansen A/S	Cheese flavor enhancement
Rhizopus oryzae	Lipopan® F	Novozyme	Dough strengthening
Candida rugosa	Lipase	Sigma-Aldrich	Laboratory chemicals
	Lipase MY	Meito Sangyo. Co.	Flavor production from milk fat
Candida antarctica	Lipase	Sigma-Aldrich	Laboratory chemicals
	Novozym 435	Novozyme	Oil-based specialties
Rhizomucor miehei	Palatase®	Novozyme	Cheese flavor enhancement

21.5 Cellulases

Cellulases are a family of enzymes primarily consisting of endo-(1,4)-β-D-glucanases (EC. 3.2.1.4), exo-(1,4)-β-D-glucanases (EC. 3.2.1.91), and β-glucosidases (EC. 3.2.1.21). They synergistically act to hydrolyze the β-(1,4)-glucan linkages of cellulose and produce glucose subsequently (Kuhad et al. 2011). Cellulases have been regarded as the third largest group of enzymes in the market due to its wide applications in the textile, detergent, paper recycling, and food processing industries. Currently, the actively investigated application of cellulases is sugar recovery from lignocellulosic biomass by enzymatic hydrolysis. The sugars produced can then be used as feedstock in fermentation for chemical production. Nowadays, commercially available cellulases are produced by microorganisms, including fungi and bacteria. Species of *Humicola*, *Thermomonospora*, *Trichoderma*, and *Aspergillus* are the common cellulase producers due to their ability to yield high levels of cellulases (Singhania et al. 2010; Wilson 2009).

21.5.1 Kinetic Characteristics

Generally, cellulase secreted by fungi and bacteria exhibits maximal activity at acidic pH 4–7 with a temperature range of 35°C–80°C. The optimal pH and temperatures for the catalytic activity of cellulase produced by different microorganisms are given in Table 21.8.

TABLE 21.8

Kinetic Properties of Cellulases from Different Bacterial and Fungal Strains

Host	Optimal pH	Optimal Temperature	References
Aspergillus niger	4.0–5.5	35–60	Farinas et al. 2010
Aspergillus terreus M11	2.0–3.0	70	Gao et al. 2008
Bacillus amyoliquefaciens DL-3	7.0	50	Lee et al. 2008
Bacillus pumilus EB3	6.0	60	Ariffin et al. 2006
Chaetomium thermophilum	5.0	50	Rosgaard et al. 2006
Corynascus thermophilus	5.0	50	Rosgaard et al. 2006
Myceliophthora thermophila	5.0	50	Rosgaard et al. 2006
Trichoderma reesei	4.0–5.0	40–60	Ortega et al. 2001
Trichoderma harzianum	5.0	50	Colussi et al. 2012
Thermoascus aurantiacus	3.5–4	75–80	Kalogeris et al. 2003
Streptomyces sp. T3-1	6.5	50	Jang and Chen 2003

21.5.2 Commercial Preparation and Recent Developments in Lipase Preparation

Nowadays, Genencor and Novozyme are the major companies on the market producing cellulases for biomass treatment (Singhania et al. 2010). Some of the cellulases produced by these companies are listed in Table 21.9. *Trichoderma reesei* is the most commonly used and studied commercial cellulase producer due to the high level of extracellular cellulase production.

In fact, cellulases producing microbes primarily metabolize carbohydrates as their sole carbon source for growth. Cellulases are induced in the presence of cellulose and disaccharide during the growth of the microbes. However, production of cellulases is affected by the glucose level in the medium. Low concentrations of glucose cannot activate cell growth and function, and the high concentration of glucose represses cellulase synthesis. Therefore, the low titer production of cellulases and the expensive process control system result in high production costs. Nowadays, it remains as the major challenge for researchers to develop efficient bioprocesses and identify cheap feedstock for high-titer cellulase production.

TABLE 21.9

Information on Commercially Available Cellulase for Biomass Treatment

Trade Name	Supplier	Host	Enzyme Activity (FPAse/ml)[a]	Optimal Condition
Celluclast	Novozymes	*T. longibrachiatum* and *A. niger*	56	pH 5.0, 50°C
Novozymo 188	Novozymes	*A. niger*	<5	pH 5.0, 50°C
Energex L	Novozymes	*T. longibrachiatum/T. reesei*	<5	pH 4.5, 50°C
Ultraflo L	Novozymes	*T. longibrachiatum/T. reesei*	<5	pH 5.0, 50°C
Viscozyme L	Novozymes	*T. longibrachiatum/T. reesei*	<5	pH 5.0, 50°C
Bio-feed beta L	Novozymes	*T. longibrachiatum/T. reesei*	<5	pH 5.0, 50°C
GC 880	Genencor	*T. longibrachiatum/T. reesei*	<5	pH 5.0, 50°C
Spezyme CP	Genencor	*T. longibrachiatum/T. reesei*	49	pH 4.0, 50°C
GC 220	Genencor	*T. longibrachiatum/T. reesei*	116	pH 5.0, 50°C
Multifect CL	Genencor Intl.	*T. reesei*	64	pH 5.0, 50°C
Accellerase® 1500	Genencor	*T. reesei*	2200–2800,[b] 450–775[c]	pH 4.0–5.0, 50°C–60°C

Source: Singhania, R. R., R. K. Sukumaran, A. K. Patel, C. Larroche, and A. Pandey, *Enzyme and Microbial Technology*, 46, 7, 541–549, 2010.

[a] One unit is defined as the amount of enzyme required to release 1 μmol of sugar in 1 min in 1 ml.

[b] One unit is defined as 1 μmol of glucose liberated from carboxymethycellulose in 1 min under specific assay conditions of 50°C and pH 4.8.

[c] One unit is defined as 1 μmol of nitrophenol liberated from para-nitrophenyl-B-D-glucopyranoside per minute at 50°C and pH 4.8.

TABLE 21.10

Substrates and Method Used for Cellulase Production

Host	Substrate	Method	Cellulase Activity	References
Trichoderma reesei RUT C30	Wheat bran	SSF	FPAse 3.8 U/gds[a]	Singhania et al. 2007
Trichoderma reesei ZU-02	Corn cob residue	SSF	158 FPU/g substrates[b]	Xia and Cen 1999
Trichoderma reesei	Xylose/sorbose	SmF (continuous)	0.69 FPU/ml[b]	Schafner and Toledo 1992
Trichoderma reesei RUT C30	Cellulose (Avicell)	SmF	1.8 FPU/ml[b]	Weber and Agblevor 2005
Trichoderma reesei RUT C30	Corrugated cardboard	SmF	2.27 FPU/ml[b]	Szijártó et al. 2004
Trichoderma reesei ZU-02	Corn stover residue	SmF	0.25 FPU/ml[b]	Shen and Xia 2004
Penicillium occitanis	Paper pulp	SmF (fed-batch)	23 FPU/ml[b]	Belghith et al. 2001
Trichoderma reesei	Steam-treated willow	SmF	108 FPU/g cellulose[b]	Reczey et al. 1996
Mixed culture of *Trichoderma reesei* and *Aspergillus niger*	Rice chaff/wheat bran (9:1)	SSF	FPAse 5.64IU/g[c]	Yang et al. 2004

[a] One unit of cellulase activity was defined as the amount of enzyme required for liberating 1 mg of reducing sugar per milliliter per minute and was expressed as U/gds (units per gram dry substrate).

[b] One unit of filter paper activity (IFPU) is the amount of enzyme that forms 1 μmol glucose (reducing sugars as glucose) per min under assay conditions.

[c] One unit of enzyme activity was defined as hydrolyzing filter paper to obtain 1 mg glucose per hour.

Different substrates and methods have been tested to efficiently produce cellulase. Table 21.10 summarizes the substrates and methods used for cellulase production.

21.6 Case Study

21.6.1 Case Study: Valorization of Food Waste with Glucoamylases and Proteases for Fermentative PHB Production

Food waste is a serious global problem in many developed countries. In Hong Kong, more than 3500 tons of food waste are generated every day (Hong Kong SAR Environmental Protection Department). These wastes are commonly delivered for landfilling and incineration in many countries as waste treatment. At the same time, these practices also cause severe environmental pollution and add burden to the economy. Due to environmental awareness, conversion of food waste into value-added products becomes an innovative and alternative approach as food waste treatment (Arancon et al. 2013). The composition of several types of food wastes are shown in Table 21.11.

In general, it is observed that food waste contains significant amounts of carbohydrates and proteins (Arancon et al. 2013; Leung et al. 2012; Pleissner et al. 2013; Zhang et al. 2013). The large amount of carbohydrates and proteins in food waste make it an ideal source of nutrients for fermentative chemical production once the carbohydrates and proteins are hydrolyzed to fermentable sugars and amino acids.

21.6.1.1 Nutrient Recovery from Food Waste

Utilization of commercial glucoamylase for production of glucose using food waste is one of the commonly used approaches (Underkofler 1969; Yan et al. 2012). This method is very rapid; the process can be completed within 30 min with a substrate ratio of 80–140 U/g (Yan et al. 2012). The drawback is the expensive cost of commercial enzymes. Alternatively, crude fungal enzymes (enzyme mixture secreted

TABLE 21.11

Composition of Various Blended Food Wastes per 100 g

Component	Bread[a,b]	Pastry[b,c]	Cake[b,c]	Mixed Bakery Waste[d]	Restaurant Food Waste[d]
Moisture (g)	22.3	34.5	45.0	N/A	N/A
Carbohydrate	46.8	33.5	62.0	62.0	33.3
Starch	59.8	N/A	N/A	N/A	N/A
Lipids	0.9	35.2	19.0	19.0	15.0
Sucrose	3.0	4.5	22.7	N/A	N/A
Fructose	N/A	2.3	11.9	N/A	N/A
Protein (TN × 5.7)	8.9	7.1	17.0	4.3	10.4
Total phosphorus	Trace	1.7	1.5	N/A	N/A
Ash	N/A	2.5	1.6	N/A	N/A

Source: [a]Leung, C. C. J., A. S. Y. Cheung, A. Y. Z. Zhang, K. F. Lam, and C. S. K. Lin, *Biochemical Engineering Journal*, 65, 10–15, 2012; [b]Zhang, A. Y. Z., Z. Sun, C. C. J. Leung et al., *Green Chemistry*, 15, 3, 690–695, 2013; [c]Arancon, R. A. D., C. S. K. Lin, K. M. Chan, T. H. Kwan, and R. Luque, *Energy Science and Engineering*, 1, 2, 53–71, 2013; [d]Pleissner, D., W. C. Lam, Z. Sun, and C. S. K. Lin, *Bioresource Technology*, 137, 139–146, 2013.

by or extracted from fungal culture without purification) could be a low-cost option. *Aspergillus awamori* and *Aspergillus oryzae* are known secretors of glucoamylase and proteases. Therefore, food waste hydrolysis with enzyme extracts from *A. awamori* and *A. oryzae* could offer a cost-efficient method for glucose and amino acid recovery. Figure 21.3 shows the recovery of glucose, free amino acid nitrogen (FAN), and phosphate from food waste when it was hydrolyzed by the mixture of these two fungal enzymes.

In general, two techniques, (i) the addition of the crude enzyme extract form of fungal culture (Lam et al. 2013; Melikoglu et al. 2013a,b), and (ii) the direct addition of fermented solids (fungal mashes) (Dorado et al. 2009; Leung et al. 2012; Murthy and Naidu 2010; Pleissner et al. 2013), can be applied to glucose recovery from food waste. In the first method, hydrolytic enzymes, including glucoamylase and proteases, are extracted from the fermented solids with distilled water at 30°C and used directly to hydrolyze food waste after enzyme activity quantification (Lam et al. 2013). Consequently, this technique

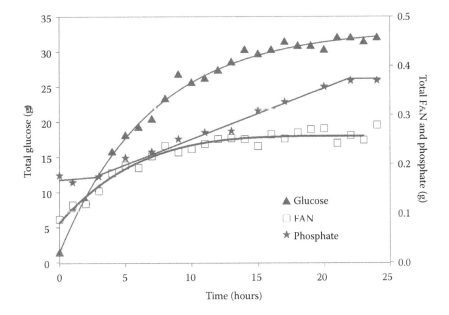

FIGURE 21.3 Recovery of glucose, FAN and phosphate from 100 g food waste in a 1 L stir-tank bioreactor during hydrolysis with *A. awamori* and *A. oryzae*. (From Pleissner, D., Lam, W. C., Sun, Z., and Lin, C. S. K., *Bioresource Technology*, 137, 139–146, 2013.)

TABLE 21.12

Advantages and Disadvantages of Food Waste Hydrolysis Using Commercial and Crude Fungal Enzymes and Commercial Enzymes

Enzyme Source	Typical Conditions	Advantages	Disadvantages
Crude enzyme extract from fungus	• pH 4.5 • 55°C	• Cheap • Easy storage/usage • Adjustable enzyme activities	• Extra steps for enzyme extraction • Long treatment time
Fungal mashes	• pH 4.5 • 55°C	• Cheap • Simple preparation	• Variations in enzymes activity • Long treatment time
Commercial enzyme	• pH 4.5 • 55°C	• Time efficient • Easy storage/usage • Adjustable enzyme activities	• Expensive

allows a consistent hydrolysis method. Nevertheless, crude enzyme solution can be stored for future use. The second method is simpler and only requires the addition of fungal mashes to food waste suspension at a controlled temperature of 55°C in a stir tank bioreactor (Leung et al. 2012; Pleissner et al. 2013, 2014). Therefore, hydrolytic activities of this method depend on the fungal growth and the nature of the substrate. Therefore, variations in the resultant hydrolysate are commonly observed in different batches. The advantages and disadvantages of applying these techniques are summarized in Table 21.12.

High glucose and free amino acid nitrogen recovery yields of 85% and 40% can also be achieved from food waste hydrolysis using crude enzymes from *A. awamori* and *A. oryzae*. Hydrolysate resulting from fungal hydrolysis using the above techniques can lead to a nutrient-complete medium with a high concentration of glucose (more than 100 g/L) and FAN (more than 400 mg/L) (Leung et al. 2012). Consequently, the hydrolysate can be used as the sole feedstock for high value chemical productions, such as succinic acid and polyhydroxybutyrate (PHB) through fermentation.

21.6.1.2 Fermentative PHB Production with Bakery Waste Hydrolysate

Poly(3-hydroxybutyrate) (PHB) belongs to the group of polyhydroxyalkanoates (PHAs), which is microbial polymer that is synthesised by more than 75 different genera of bacteria as intracellular carbon and energy storage compounds (Reddy et al. 2003). The chemical structure of PHB is shown in Figure 21.4.

This biopolymer is 100% biodegradable and nontoxic and has chemical properties compatible with polypropylene (Sudesh et al. 2000). Therefore, its potential applications as renewable and sustainable alternatives to petroleum-based plastic products, such as disposable plastic commodities (bags and containers) and its potential biomedical applications as drug carriers, wound dressings, surgical implants, and scaffolds have been suggested (Chen and Wu 2005; Reddy et al. 2003).

Several bacterial strains, including *Alcaligenes latus* (Yamane et al. 1996), *Azotobacter vinelandii* (Page and Knosp 1989), *Cupriavidus necator* (Koutinas et al. 2007; Tian et al. 2005), and recombinant *Escherichia coli* (Lee and Chang 1994) are regarded as promising hosts for fermentative PHB production on an industrial scale due to their high productivity. PHB accumulation usually occurs under nutrient-limiting conditions with excess carbon. Glucose, fatty acid, or glycerol can be used as a carbon source for fermentative PHB production although glucose is the most frequently reported carbon source used. Although high PHB

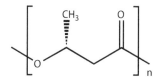

FIGURE 21.4   Chemical structure of poly(3-hydroxybutyrate) PHB.

yield has been reported using glucose as a carbon source, fermentative PHB production on a large scale using glucose is still a cost-inefficient process due to high medium cost (Choi and Lee 1997).

Therefore, food waste hydrolysate could serve as a low-cost nutrient-complete medium for fermentative PHB production and as an alternative method of waste management. Recovered glucose and amino acids from various food wastes are known to be utilized for fermentative PHB production using *Halomonas boliviensis* (Arancon et al. 2013). The overall procedures for fermentative PHB production from food waste are summarized in Figure 21.5.

After the homogenized food waste was subjected to enzymatic hydrolysis using glucoamylase and protease from *A. awamori* and *A. oryzae*, the glucose- and amino acid-rich hydrolysate was recovered. The hydrolysate is further conditioned by adjusting the pH and glucose concentration, sterilized, and serves as a fermentation medium for PHB fermentation. Production of PHB with *H. boliviensis* using conventionally defined medium and food waste hydrolysate are compared in Table 21.13 (Arancon et al. 2013).

In the table, PHB production is observed when bakery waste hydrolysate is used in fed-batch fermentations. Although PHB production is lower than when food waste hydrolysate is used compared to defined medium, it is also observed that PHB production can be greatly enhanced by decreasing the nitrogen concentration in the medium; when the yeast extract concentration decreases from 8 g/L to 5 g/L in defined medium, PHB production increases from 4.3 g to 17.4 g. Therefore, these findings demonstrate the feasibility of large-scale PHB production using various food waste hydrolysate.

At the end of fermentation, the bacterial cells are lysed, and intracellular PHB granules are extracted by using chlorinated solvent, and the extracted PHB is further purified. Figure 21.6 shows a PHB film produced from food waste hydrolysate using PHB granules recovered from *H. boliviensis* culture.

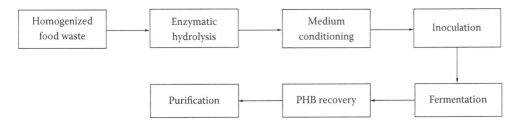

FIGURE 21.5 Process flow of fermentative PHB production from food waste.

TABLE 21.13

Comparison of Overall Performance of Both Defined Medium and Bakery Waste Hydrolysate Fermentation for PHB Production Using *H. boliviensis*

Medium	Mode	Feeding Media	Glucose Consumption (g)	CDW (g)	PHB Production (g)	PHB Content (%)
Defined (40 g/L glucose, 5 g/L yeast extract)	Batch	NIL	24.0	24.9	17.4	17.4
Defined (40 g/L glucose, 8 g/L yeast extract)	Batch	NIL	59.9	9.2	4.3	4.3
Pastry hydrolysate	Batch	NIL	32.8	NIL	NIL	NIL
Pastry hydrolysate	Fed-batch	Glucose	112.3	5.7	2.1	2.1
Pastry hydrolysate	Fed-batch	Pastry hydrolysate	208.8	38.2	3.6	3.6
Pastry hydrolysate	Fed-batch	Pastry hydrolysate	359.9	15.6	0.60	0.59
Cake hydrolysate	Fed-batch	Cake hydrolysate	200.5	11.6	2.9	2.9

Source: Arancon, R. A. D., C. S. K. Lin, K. M. Chan, T. H. Kwan, and R. Luque, *Energy Science & Engineering*, 1, 2, 53–71, 2013.

Note: CDW refers to cell dry weight.

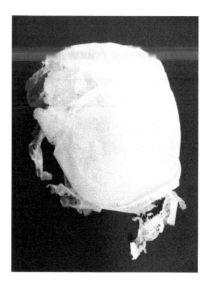

FIGURE 21.6   PHB film produced from *H. boliviensis* fermentation using food waste hydrolysate.

21.6.2 Case Study II: Valorization of Food Waste for Biodiesel Production with Lipases

Biodiesel (fatty acid alkyl ester) is an environmentally friendly liquid fuel made from renewable resources or waste lipid. It is produced by transesterification of triglyceride with an alcohol, which is methanol or ethanol in most cases. A catalyst, which is acid, alkali, or enzyme, is used to improve the reaction rate and yield. The chemical equation of transesterification of triglyceride with an alcohol is shown in Figure 21.7.

Utilization of lipases as a biocatalyst for biodiesel production possesses several distinctive advantages over acid and alkali catalysts, such as milder reaction conditions, a simple purification process, no soap formation, and fewer side reactions. Various vegetable oils have been tested with lipases for biodiesel production. Table 21.14 summarizes the biodiesel production using different substrates and methods. However, due to the long reaction time, high cost of substrates, loss of enzyme activity, and inhibition by more than 1 molar equivalent of alcohol, lipases are not commonly used in the industrial biodiesel production nowadays (Leung et al. 2010). To overcome the disadvantages of a lipase-catalyzed reaction, waste fat and grease as triglyceride sources have been regarded as an economically viable solution (Tan et al. 2010). Various reaction parameters, including reaction temperature, choice of alcohol, and alcohol-to-oil molar ratio, have been optimized to improve biodiesel yield. Meanwhile, stepwise addition of alcohol, immobilization of lipase, ultrasonication, and supercritical CO_2 technology have been found to be promising to enhance the reaction rate (Table 21.14).

Although high conversion rates and short reaction times have been achieved in independent studies, the cost of the lipase-catalyzed process is still higher than that using alkali and acid catalysts. Further research is required to bring lipases with stronger catalytic effects to make the process economically favorable.

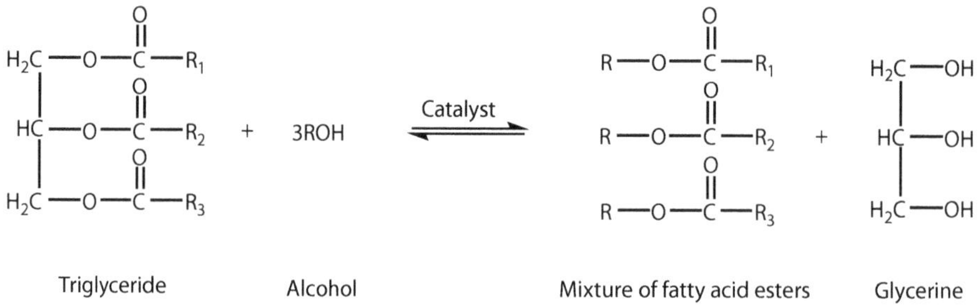

FIGURE 21.7   Chemical equation of transesterification of triglyceride with alcohol.

TABLE 21.14

Summary of the Biodiesel Production Using Different Substrates and Methods

Source of Lipid	Host	Alcohol	Alcohol to Lipid Molar Ratio	Immobilization Medium	Optimum Condition	Other Details	Maximum Yield	References
Soybean oil	*Pseudomonas cepacia*	Methanol	7.5:1	Tetramethoxysilane and isobutyltrimethoxysilane	55°C, 1 h	–	67%	Noureddini et al. 2005
	Candida rugosa	Methanol	1.5:1	Silica	45°C, 3 h	Stepwise addition of methanol, supercritical carbon dioxide process	99.99%	Lee et al. 2011
Jatropha	*Enterobacter aerogenes*	Methanol	4:1	Silica	55°C, 60 h	Ultrasonication (100 W/m^3) applied	94%	Kumar et al. 2011
Tallow and grease	*Thermomyces anuginosa*	Ethanol	2:1	Phyllosilicate sol-gel matrix	40°C, 24 h	–	70%–100%	Hsu et al. 2004
Canola oil	*Thermomyces lanuginosus*	Methanol	6:1	Polyurethane foams	40°C, 24 h	Polyglutaraldehyde used as a cross-linking agent	90%	Dizge and Keskinler 2008
Waste cooking oil	*Bacillus subtilis*	Methanol	3:1	Hydrophobic carrier with magnetic particles (Fe$_3$O$_4$)	40°C, 72 h	Stepwise addition of alcohol	90%	Ying and Chen 2007
Restaurant grease	*Pseudomonas cepacia*	Butanol	4:1	Phyllosilicate sol-gel matrix	40°C, 24 h	–	98%	Hsu et al. 2002

21.6.3 Case Study III: Valorization of Citrus Peel Waste for Ethanol Production with Cellulases

According to the report "Citrus: World Markets and Trade" published by the United States Department of Agriculture, global citrus production was estimated at 51.8 million metric tons for 2013–2014. Approximately 44% of the wet fruit mass results in citrus peel waste (CPW) in juice manufacture (Choi et al. 2013). CPW consisting of peel, cores, segment membranes, juice sacs, and seeds is regarded as a suitable candidate for bioethanol production because of the high level of polysaccharides, such as cellulose and hemicellulose, and fermentable sugars, such as glucose, fructose, and sucrose. Table 21.15 shows the composition profile of different kinds of CPW.

To produce bioethanol using CPW as feedstock, the lignocellulose-rich biomass is first hydrolyzed to produce fermentable C6 and C5 sugars, such as glucose, xylose, arabinose, mannose, and galactose. The sugars obtained are subsequently used as the carbon source for fermentative ethanol production via microbial fermentation. *Saccharomyces cerevisiae* is a common ethanol producer in the fermentation process because of its promising yield of ethanol production.

However, enzymatic hydrolysis of CPW directly with cellulase is not effective due to the highly crystalline structure of cellulose and the presence of a lignin layer on top, which prevents the cellulase from accessing its cellulose. Therefore, pretreatment is required to remove the lignin as well as to decrease the crystallinity of the cellulose in order to improve the accessibility of cellulase to cellulose for efficient hydrolysis. Pretreatment could be categorized into chemical and physical methods. However, chemical pretreatment is economically and environmentally undesirable due to the use of corrosive acid and alkaline. Therefore, physical pretreatments, such as popping and steam explosion have been widely investigated.

The pretreated CPW is further hydrolyzed by commercially available cellulase, β-glucosidase, and pectinases to produce sugars. The two strategies for bioethanol production from CPW based on enzymatic hydrolysis are (i) simultaneous saccharification and fermentation (SSAF), and (ii) separate hydrolysis and fermentation (SHF). In the first strategy, the CPW is hydrolyzed and fermented into ethanol simultaneously in a single vessel in the presence of both cellulolytic enzyme(s) and ethanol producing microorganisms. This method is commonly applied due to the simplicity of operation, improved process economics, and low probability of catabolite repression caused by high glucose concentration. In the second SHF strategy, enzymatic hydrolysis and fermentation are conducted separately in bioreactors. The CPW is first hydrolyzed into sugars, and then the resulting hydrolysate undergoes filtration and evaporation to increase the sugar concentration and remove fermentation inhibitors, such as d-limonene. Finally, the hydrolysate serves as the fermentation medium in ethanol production.

The feasibility of bioethanol production using CPW has been demonstrated (Table 21.16), and further investigation regarding the effect of pretreatment upon ethanol production yield and productivity as well as economic optimization are in progress in order to facilitate the commercialization of the process.

TABLE 21.15

Composition of Mandarin Peel, Orange Peel, and Kinnow Waste

Component	Mandarin Peel Waste (%)	Orange Peel Waste (%)	Kinnow Waste[a] (%)
Cellulose	22.5	37.1	10.10
Hemicellulose	6.0	11.0	4.28
Lignin	8.6	7.5	0.56
Sugars	10.1	9.6	31.58
Protein	7.5	9.1	5.78
Pectin	16.0	23.0	22.6
Ash	5.0	2.6	3.23
Fat	1.6	4.0	–

Source: Oberoi, H. S., P. V. Vadlani, A. Nanjundaswamy et al., *Bioresource Technology*, 102, 2, 1593–1601, 2011; Boluda-Aguilar, M., L. García-Vidal, F. d. P. González-Castañeda, and A. López-Gómez, *Bioresource Technology*, 101, 10, 3506–3513, 2010.

[a] Dry basis.

TABLE 21.16

Fermentative Ethanol Production from Hydrolyzed Citrus Peel Wastes with Different Pretreatment Methods

Substrate	Enzyme Hydrolysis			Fermentation				
	Process	Enzymes (Trade Name/Source)	Host	Ethanol Concentration (g/L)	Yield (g/g Substrate)	Productivity (g/L/h)	Pretreatment	References
Kinnow waste and banana peel waste	SSAF	Cellulase (*Trichoderma reesei* Rut C-30)	*S. cerevisiae* G, *Pachysolen tannophilus* MTCC 1077	26	0.426	0.36	Steam depressurization	Sharma et al. 2007
Kinnow waste	SSF	Cellulase (Celluclast 1.5 L), Pectinase (Sigma–Aldrich), β-glucosidase (Novozym 188)	*S. cerevisiae*	42	—	3.5	Hydrothermal sterilization	Oberoi et al. 2011
Mandarin peel waste	SHF	Pectinase, β-glucosidase, Xylanase	*S. cerevisiae* KCTC 7906	46.2	0.462	3.85	Popping	Choi et al. 2013
Orange peel waste	SHF	Cellulase (Cellic® CTec 2), Pectinase, β-glucosidase	*S. cerevisiae* F15	N/A	0.495	4.85	Acid-catalyzed steam explosion	Santi et al. 2014
Orange peel waste	SSAF	Cellulase (Celluclast 1.5L), Pectinase (Rapidase PTIS), β-glucosidase (Novozym 188)	*S. cerevisiae*	27.0	N/A	0.56	Steam purging	Widmer et al. 2010
Mandarin citrus peel waste	SSF	Cellulase (Celluclast 1.5L), Pectinase (Pectinex Ultra SP), β-glucosidase (Novozym 188)	*S. cerevisiae* CECT 1329	N/A	50 ml/kg substrate	0.417 ml/kg/h	Steam explosion	Boluda-Aguilar et al. 2010

Note: SHF, separate hydrolysis and fermentation; SSAF, simultaneous saccharification and fermentation.

21.7 Conclusions

Conversion of food and beverage wastes to valuable chemicals and biodiesel using different enzymes, including glucoamylase, proteases, lipases, and cellulases, has been discussed. The bioconversion can be achieved efficiently. By recovering glucose and amino acids from the carbohydrate fraction of food and beverage wastes using glucoamylase, cellulase, or proteases, a glucose- and amino acid-rich hydrolysate could be produced, which can be used to produce a range of chemicals or biofuels, including PHB, hydrolytic enzymes, or bioethanol through fermentation. By converting the lipid fraction from food waste to fatty acids using lipases, biodiesel can further be produced, consequently upgrading the value of food waste. Despite waste residues potentially serving as low-cost materials for valuable chemical and biodiesel production, the overall cost of the bioconversion process could still be very high in order to achieve sustainable large-scale production. Therefore, collaborative efforts from the public and government as well as technological development in the future are needed to develop a cost-effective process for waste-to-energy conversion.

ACKNOWLEDGMENTS

The authors acknowledge the Innovation and Technology Funding from the Innovation and Technology Commission (Re: ITS/323/11 & ITS/353/12) in Hong Kong, the donation from the Coffee Concept (Hong Kong) Ltd. for the "Care for Our Planet" campaign as well as a grant from the City University of Hong Kong (Project No. 7200248).

REFERENCES

Alkan, H., Z. Baysal, F. Uyar, and M. Dogru. 2007. Production of lipase by a newly isolated Bacillus coagulans under solid-state fermentation using melon wastes. *Applied Biochemistry and Biotechnology* 136 (2):183–192.

Anto, H., U. B. Trivedi, and K. C. Patel. 2006. Glucoamylase production by solid-state fermentation using rice flake manufacturing waste products as substrate. *Bioresource Technology* 97 (10):1161–1166.

Arancon, R. A. D., C. S. K. Lin, K. M. Chan, T. H. Kwan, and R. Luque. 2013. Advances on waste valorization: New horizons for a more sustainable society. *Energy Science & Engineering* 1 (2):53–71.

Ariffin, H., N. Abdullah, M. S. Umi Kalsom, Y. Shirai, and M. A. Hassan. 2006. Production and characterization of cellulase by Bacillus pumilus EB3. *International Journal of Engineering and Technology* 3 (1):47–53.

Asgher, M., M. J. Asad, and R. L. Legge. 2006. Enhanced lignin peroxidase synthesis by *Phanerochaete Chrysosporium* in solid state bioprocessing of a lignocellulosic substrate. *World Journal of Microbiology and Biotechnology* 22 (5):449–453.

Babu, K. R., and T. Satyanarayana. 1995. α-Amylase production by *Thermophilic Bacillus* coagulans in solid state fermentation. *Process Biochemistry* 30 (4):305–309.

Belghith, H., S. Ellouz-Chaabouni, and A. Gargouri. 2001. Biostoning of denims by Penicillium occitanis (Pol6) cellulases. *Journal of Biotechnology* 89 (2–3):257–262.

Beolchini, F., G. Del Re, G. Di Giacomo, L. Spera, and F. Veglio. 2006. Biological treatment of agro-industrial wastewater for the production of glucoamylase and *Rhizopus* biomass. *Separation Science and Technology* 41:471–483.

Boluda-Aguilar, M., L. García-Vidal, F. d. P. González-Castañeda, and A. López-Gómez. 2010. Mandarin peel wastes pretreatment with steam explosion for bioethanol production. *Bioresource Technology* 101 (10):3506–3513.

Buchholz, K., and J. Seibel. 2008. Industrial carbohydrate biotransformations. *Carbohydrate Research* 343 (12):1966–1979.

Çalk, P., A. Bayram, and T. H. Özdamar. 2003. Regulatory effects of alanine-group amino acids on serine alkaline protease production by recombinant *Bacillus licheniformis*. *Biotechnology and Applied Biochemistry* 37:165–171.

Chen, G. Q., and Q. Wu. 2005. The application of polyhydroxyalkanoates as tissue engineering materials. *Biomaterials* 26 (33):6565–6578.

Choi, J. I., and S. Y. Lee. 1997. Process analysis and economic evaluation for Poly(3-hydroxybutyrate) production by fermentation. *Bioprocess Engineering* 17 (6):335–342.

Choi, I. S., J.-H. Kim, S. G. Wi, K. H. Kim, and H.-J. Bae. 2013. Bioethanol production from mandarin (Citrus unshiu) peel waste using popping pretreatment. *Applied Energy* 102:204–210.

Chutmanop, J., S. Chuichulcherm, Y. Chisti, and P. Srinophakun. 2008. Protease production by *Aspergillus oryzae* in solid-state fermentation using agroindustrial substrates. *Journal of Chemical Technology and Biotechnology* 83:1012–1018.

Colussi, F., W. Garcia, F. Rosseto et al. 2012. Effect of pH and temperature on the global compactness, structure, and activity of cellobiohydrolase Cel7A from Trichoderma harzianum. *European Biophysics Journal* 41 (1):89–98.

Cordova, J., M. Nemmaoui, M. Ismaili-Alaoui et al. 1998. Lipase production by solid state fermentation of olive cake and sugar cane bagasse. *Journal of Molecular Catalysis B: Enzymatic* 5 (1–4):75–78.

Dharmsthiti, S., and S. Luchai. 1999. Production, purification and characterization of thermophilic lipase from Bacillus sp. THL027. *FEMS Microbiology Letters* 179 (2):241–246.

Dizge, N., and B. Keskinler. 2008. Enzymatic production of biodiesel from canola oil using immobilized lipase. *Biomass and Bioenergy* 32 (12):1274–1278.

Dorado, M. P., S. K. C. Lin, A. Koutinas et al. 2009. Cereal-based biorefinery development: Utilisation of wheat milling by-products for the production of succinic acid. *Journal of Biotechnology* 143 (1):51–59.

Du, C., S. K. C. Lin, A. Koutinas et al. 2008. A wheat biorefining strategy based on solid-state fermentation for fermentative production of succinic acid. *Bioresource Technology* 99 (17):8310–8315.

Ertuğrul, S., G. Dönmez, and S. Takaç. 2007. Isolation of lipase producing Bacillus sp. from olive mill wastewater and improving its enzyme activity. *Journal of Hazardous Materials* 149 (3):720–724.

Farinas, C. S., M. M. Loyo, A. Baraldo Jr. et al. 2010. Finding stable cellulase and xylanase: Evaluation of the synergistic effect of pH and temperature. *New Biotechnology* 27 (6):810–815.

Fiedurek, J., and J. Szczodrak. 1995. Immobilization of *Aspergillus niger* mycelium on seeds for glucoamylase production. *Starch—Stärke* 47 (5):196–199.

Flor, P. Q., and S. Hayashida. 1983. Production and characteristics of raw starch-digesting glucoamylase O from a protease-negative, glycosidase-negative *Aspergillus awamori* var. *kawachi* mutant. *Applied and Environmental Microbiology* 45 (3):905–912.

Freitas, A. C., R. J. S. Castro, M. A. Fontenele et al. 2013. Canola cake as a potential substrate for proteolytic enzymes production by a selected strain of *Aspergillus oryzae*: Selection of process conditions and product characterization. *ISRN Microbiology* 2013:8.

Ganzlin, M., and U. Rinas. 2008. In-depth analysis of the *Aspergillus niger* glucoamylase (glaA) promoter performance using high-throughput screening and controlled bioreactor cultivation techniques. *Journal of Biotechnology* 135 (3):266–271.

Gao, J., H. Weng, D. Zhu et al. 2008. Production and characterization of cellulolytic enzymes from the thermo-acidophilic fungal *Aspergillus terreus* M11 under solid-state cultivation of corn stover. *Bioresource Technology* 99 (16):7623–7629.

Godfrey, T., and S. West. 1996. *Industrial Enzymology*, 2nd ed. New York: Macmillan Publishers Inc.

Gomes, E., S. R. D. Souza, R. P. Grandi, and R. D. Silva. 2005. Production of thermostable glucoamylase by newly isolated *Aspergillus flavus* A 1.1 and *Thermomyces lanuginosus* A 13.37. *Brazilian Journal of Microbiology* 36 (1):75–82.

Guncheva, M., and D. Zhiryakova. 2011. Catalytic properties and potential applications of *Bacillus* lipases. *Journal of Molecular Catalysis B: Enzymatic* 68 (1):1–21.

Gustavsson, J., C. Cederberg, U. Sonesson, R. V. Otterdijk, and A. Meybeck. 2011. *Global Food Losses and Food Waste*. Food and Agriculture Organization of the United Nations, Düsseldorf, Germany.

Hartley, B. S. 1960. Proteolytic enzymes. *Annual Review of Biochemistry* 29:45–72.

Hiol, A., M. D. Jonzo, N. Rugani et al. 2000. Purification and characterization of an extracellular lipase from a thermophilic *Rhizopus oryzae* strain isolated from palm fruit. *Enzyme and Microbial Technology* 26 (5–6):421–430.

Hong Kong SAR Environmental Protection Department, Monitoring of solid waste in Hong Kong—Waste statistics for 2011, 2012. Available at https://www.wastereduction.gov.hk/en/materials/info/msw2011.pdf (accessed on April 15, 2014).

Hsu, A. F., K. Jones, T. A. Foglia, and W. N. Marmer. 2002. Immobilized lipase-catalysed production of alkyl esters of restaurant grease as biodiesel. *Biotechnology and Applied Biochemistry* 36 (3):181–186.

Hsu, A. F., K. Jones, T. Foglia, and W. Marmer. 2004. Transesterification activity of lipases immobilized in a phyllosilicate sol-gel matrix. *Biotechnology Letters* 26 (11):917–921.

James, J. A., and B. H. Lee. 1997. Glucoamylases: Microbial sources, industrial applications and molecular biology—A review. *Journal of Food Biochemistry* 21 (6):1–52.

Jang, H. D., and K. S. Chen. 2003. Production and characterization of thermostable cellulases from *Streptomyces transformant* T3-1. *World Journal of Microbiology and Biotechnology* 19 (3):263–268.

Kalogeris, E., P. Christakopoulos, P. Katapodis et al. 2003. Production and characterization of cellulolytic enzymes from the thermophilic fungus *Thermoascus aurantiacus* under solid state cultivation of agricultural wastes. *Process Biochemistry* 38 (7):1099–1104.

Kalpana Devi, M., A. Rasheedha Banu, G. R. Gnanaprabha, B. V. Pradeep, and M. Palaniswamy. 2008. Purification, characterization of alkaline protease enzyme from native isolate *Aspergillus niger* and its compatibility with commercial detergents. *Indian Journal of Science and Technology* 1 (7):1–6.

Kanwar, L., and P. Goswami. 2002. Isolation of a *Pseudomonas* lipase produced in pure hydrocarbon substrate and its application in the synthesis of isoamyl acetate using membrane-immobilised lipase. *Enzyme and Microbial Technology* 31 (6):727–735.

Kareem, S. O., I. Akpan, and S. B. Oduntan. 2009. Cowpea waste: A novel substrate for solid state production of amylase by *Aspergillus oryzae*. *African Journal of Microbiology Research* 3 (12):974–977.

Kim, H. K., H. J. Choi, M. H. Kim, C. B. Sohn, and T. K. Oh. 2002. Expression and characterization of Ca^{2+}-independent lipase from *Bacillus pumilus* B26. *Biochimica et Biophysica Acta (BBA)—Molecular and Cell Biology of Lipids* 1583 (2):205–212.

Koutinas, A. A., Y. Xu, R. Wang, and C. Webb. 2007. Polyhydroxybutyrate production from a novel feedstock derived from a wheat-based biorefinery. *Enzyme and Microbial Technology* 40 (5):1035–1044.

Kuhad, R. C., R. Gupta, and A. Singh. 2011. Microbial cellulases and their industrial applications. *Enzyme Research* 2011:10.

Kumar, C. G., and H. Takagi. 1999. Microbial alkaline proteases: From a bioindustrial viewpoint. *Biotechnology Advances* 17 (7):561–594.

Kumar, P., and T. Satyanarayana. 2007. Production of thermostable and neutral glucoamylase using immobilized *Thermomucor indicae-seudaticae*. *World Journal of Microbiology and Biotechnology* 23 (4):509–517.

Kumar, G., D. Kumar, Poonam, R. Johari, and C. P. Singh. 2011. Enzymatic transesterification of *Jatropha curcas* oil assisted by ultrasonication. *Ultrasonics Sonochemistry* 18 (5):923–927.

Lam, W. C., D. Pleissner, and S. K. C. Lin. 2013. Production of fungal glucoamylase for glucose production from food waste. *Biomolecules* 3 (3):651–661.

Lee, S. Y., and H. N. Chang. 1994. Effect of complex nitrogen source on the synthesis and accumulation of poly(3-hydroxybutyric acid) by recombinant *Escherichia coli* in flask and fed-batch cultures. *Journal of Environmental Polymer Degradation* 2 (3):169–176.

Lee, Y. J., B. K. Kim, B. H. Lee et al. 2008. Purification and characterization of cellulase produced by *Bacillus amyoliquefaciens* DL-3 utilizing rice hull. *Bioresource Technology* 99 (2):378–386.

Lee, J. H., S. B. Kim, S. W. Kang et al. 2011. Biodiesel production by a mixture of *Candida rugosa* and *Rhizopus oryzae* lipases using a supercritical carbon dioxide process. *Bioresource Technology* 102 (2):2105–2108.

Leung, D. Y. C., X. Wu, and M. K. H. Leung. 2010. A review on biodiesel production using catalyzed transesterification. *Applied Energy* 87 (4):1083–1095.

Leung, C. C. J., A. S. Y. Cheung, A. Y. Z. Zhang, K. F. Lam, and C. S. K. Lin. 2012. Utilisation of waste bread for fermentative succinic acid production. *Biochemical Engineering Journal* 65:10–15.

Li, H., Z. Chi, X. Duan et al. 2007. Glucoamylase production by the marine yeast *Aureobasidium pullulans* N13d and hydrolysis of potato starch granules by the enzyme. *Process Biochemistry* 42 (3):462–465.

Liu, Y., Q. S. Li, H. L. Zhu et al. 2013. Purification and characterization of two thermostable glucoamylases produced from *Aspergillus niger* B-30. *Chemical Research in Chinese Universities* 29 (5):917–923.

Mahadik, N. D., U. S. Puntambekar, K. B. Bastawde, J. M. Khire, and D. V. Gokhale. 2002. Production of acidic lipase by Aspergillus niger in solid state fermentation. *Process Biochemistry* 38 (5):715–721.

Mahanta, N., A. Gupta, and S. K. Khare. 2008. Production of protease and lipase by solvent tolerant *Pseudomonas aeruginosa* PseA in solid-state fermentation using *Jatropha curcas* seed cake as substrate. *Bioresource Technology* 99 (6):1729–1735.

Melikoglu, M., C. S. K. Lin, and C. Webb. 2013a. Kinetic studies on the multi-enzyme solution produced via solid state fermentation of waste bread by *Aspergillus awamori*. *Biochemical Engineering Journal* 80:76–82.

Melikoglu, M., C. S. K. Lin, and C. Webb. 2013b. Stepwise optimisation of enzyme production in solid state fermentation of waste bread pieces. *Food and Bioproducts Processing* 91 (4):638–646.

Murthy, P. S., and M. M. Naidu. 2010. Protease production by *Aspergillus oryzae* in solid-state fermentation utilizating coffee by-products. *World Applied Sciences Journal* 8 (2):199–205.

Norouzian, D., A. Akbarzadeh, J. M. Scharer, and M. Moo Young. 2006. Fungal glucoamylases. *Biotechnology Advances* 24 (1):80–85.

Nourcddini, H., X. Gao, and R. S. Philkana. 2005. Immobilized Pseudomonas cepacia lipase for biodiesel fuel production from soybean oil. *Bioresource Technology* 96 (7):769–777.

Oberoi, H. S., P. V. Vadlani, A. Nanjundaswamy et al. 2011. Enhanced ethanol production from Kinnow mandarin (*Citrus reticulata*) waste via a statistically optimized simultaneous saccharification and fermentation process. *Bioresource Technology* 102 (2):1593–1601.

Ortega, N., M. D. Busto, and M. Perez-Mateos. 2001. Kinetics of cellulose saccharification by *Trichoderma reesei* cellulases. *International Biodeterioration & Biodegradation* 47 (1):7–14.

Pacheco-Chávez, R. A., J. C. M. Carvalho, L. C. Tavares et al. 2004. Production of α-amylase and glucoamylase by a new isolate of *Trichoderma* sp. using sorghum starch as a carbon source. *Engineering in Life Sciences* 4 (4):369–372.

Page, W. J., and O. Knosp. 1989. Hyperproduction of poly-β-hydroxybutyrate during exponential growth of *Azotobacter vinelandii* UWD. *Applied and Environmental Microbiology* 55 (6):1334–1339.

Palma, M., A. Pinto, A. Gombert et al. 2000. Lipase production by *Penicillium testrictum* using solid waste of industrial babassu oil production as substrate. In *Twenty-First Symposium on Biotechnology for Fuels and Chemicals*, edited by M. Finkelstein and B. Davison. Humana Press, United States.

Pandey, A. 1995. Glucoamylase research: An overview. *Starch—Stärke* 47 (11):439–445.

Pandey, A., P. Selvakumar, C. R. Soccol, and P. Nigam. 1999a. Solid state fermentation for the production of industrial enzymes. *Current Science* 77 (1):149–162.

Pandey, A., S. Benjamin, C. R. Soccol et al. 1999b. The realm of microbial lipases in biotechnology. *Biotechnology and Applied Biochemistry* 29 (2):119–131.

Paranthaman, R., K. Alagusundaram, and J. Indhumathi. 2009. Production of protease from rice mill wastes by *Aspergillus niger* in solid state fermentation. *World Journal of Agricultural Sciences* 5 (3):308–312.

Pfeffer, J., S. Richter, J. Nieveler et al. 2006. High yield expression of Lipase A from *Candida antarctica* in the methylotrophic yeast *Pichia pastoris* and its purification and characterisation. *Applied Microbiology and Biotechnology* 72 (5):931–938.

Pleissner, D., W. C. Lam, Z. Sun, and C. S. K. Lin. 2013. Food waste as nutrient source in heterotrophic micro-algae cultivation. *Bioresource Technology* 137:139–146.

Pleissner, D., T. H. Kwan, and C. S. K. Lin. 2014. Fungal hydrolysis in submerged fermentation for food waste treatment and fermentation feedstock preparation. *Bioresource Technology* 158:48–54.

Radha, S., V. J. Nithya, R. H. Babu et al. 2011. Production and optimization of acid protease by *Aspergillus* spp. under submerged fermentation. *Archives of Applied Science Research* 3 (2):155–163.

Reczey, K., Z. S. Szengyel, R. Eklund, and G. Zacchi. 1996. Cellulase production by *T. reesei*. *Bioresource Technology* 57 (1):25–30.

Reddy, C. S. K., R. Ghai, Rashmi, and V. C. Kalia. 2003. Polyhydroxyalkanoates: An overview. *Bioresource Technology* 87 (2):137–146.

Regulapati, R., P. N. Malav, and S. N. Gummadi. 2007. Production of thermostable α-amylases by solid state fermentation—A review. *American Journal of Food Technology* 2:1–11.

Robertson, G. H., D. W. S. Wong, C. C. Lee et al. 2006. Native or raw starch digestion: A key step in energy efficient biorefining of grain. *Journal of Agricultural and Food Chemistry* 54:353–365.

Rosgaard, L., S. Pedersen, J. R. Cherry, P. Harris, and A. S. Meyer. 2006. Efficiency of new fungal cellulase systems in boosting enzymatic degradation of barley straw lignocellulose. *Biotechnology Progress* 22 (2): 493–498.

Sandhya, C., A. Sumantha, G. Szakacs, and A. Pandey. 2005. Comparative evaluation of neutral protease production by *Aspergillus oryzae* in submerged and solid-state fermentation. *Process Biochemistry* 40 (8):2689–2694.

Santi, G., S. Crognale, A. D'Annibale et al. 2014. Orange peel pretreatment in a novel lab-scale direct steam-injection apparatus for ethanol production. *Biomass and Bioenergy* 61:146–156.

Sauer, J., B. W. Sigurskjold, U. Christensen et al. 2000. Glucoamylase: Structure/function relationships, and protein engineering. *Biochimica et Biophysica Acta (BBA)—Protein Structure and Molecular Enzymology* 1543 (2):275–293.

Schafner, D. W., and R. T. Toledo. 1992. Cellulase production in continuous culture by *Trichoderma reesei* on xylose-based media. *Biotechnology and Bioengineering* 39 (8):865–869.

Sharma, N., K. L. Kalra, H. Oberoi, and S. Bansal. 2007. Optimization of fermentation parameters for production of ethanol from kinnow waste and banana peels by simultaneous saccharification and fermentation. *Indian Journal of Microbiology* 47 (4):310–316.

Shen, X., and L. Xia. 2004. Production and immobilization of cellobiase from *Aspergillus niger* ZU-07. *Process Biochemistry* 39 (11):1363–1367.

Siala, R., F. Frikha, S. Mhamdi, M. Nasri, and A. Sellami Kamoun. 2012. Optimization of acid protease production by *Aspergillus niger* I1 on shrimp peptone using statistical experimental design. *The Scientific World Journal* 2012:11.

Singhania, R., R. Sukumaran, and A. Pandey. 2007. Improved cellulase production by *Trichoderma reesei* RUT C30 under SSF through process optimization. *Applied Biochemistry and Biotechnology* 142 (1):60–70.

Singhania, R. R., R. K. Sukumaran, A. K. Patel, C. Larroche, and A. Pandey. 2010. Advancement and comparative profiles in the production technologies using solid-state and submerged fermentation for microbial cellulases. *Enzyme and Microbial Technology* 46 (7):541–549.

Sudesh, K., H. Abe, and Y. Doi. 2000. Synthesis, structure and properties of polyhydroxyalkanoates: Biological polyesters. *Progress in Polymer Science* 25 (10):1503–1555.

Sulong, M. R., R. N. Abdul Rahman, A. B. Salleh, and M. Basri. 2006. A novel organic solvent tolerant lipase from *Bacillus sphaericus* 205y: Extracellular expression of a novel OST-lipase gene. *Protein Expression and Purification* 49 (2):190–195.

Sumantha, A., C. Larroche, and A. Pandey. 2006. Microbiology and industrial biotechnology of food-grade proteases: A perspective. *Food Technology and Biotechnology* 44 (2):211–220.

Szijártó, N., Z. Faigl, K. Réczey, M. Mézes, and A. Bersényi. 2004. Cellulase fermentation on a novel substrate (waste cardboard) and subsequent utilization of home-produced cellulase and commercial amylase in a rabbit feeding trial. *Industrial Crops and Products* 20 (1):49–57.

Tan, T., J. Lu, K. Nie, L. Deng, and F. Wang. 2010. Biodiesel production with immobilized lipase: A review. *Biotechnology Advances* 28 (5):628–634.

Tian, J., A. He, A. G. Lawrence et al. 2005. Analysis of transient polyhydroxybutyrate production in *Wautersia eutropha* H16 by quantitative western analysis and transmission electron microscopy. *Journal of Bacteriology* 187 (11):3825–3832.

Ting, W. J., K. Y. Tung, R. Giridhar, and W. T. Wu. 2006. Application of binary immobilized *Candida rugosa* lipase for hydrolysis of soybean oil. *Journal of Molecular Catalysis B: Enzymatic* 42 (1–2):32–38.

Tommaso, G., B. Moraes, G. Macedo, G. Silva, and E. Kamimura. 2011. Production of lipase from *Candida rugosa* using cheese whey through experimental design and surface response methodology. *Food and Bioprocess Technology* 4 (8):1473–1481.

Tsujita, Y., and A. Endo. 1977. Extracellular acid protease of *Aspergillus oryzae* grown on liquid media: Multiple forms due to association with heterogeneous polysaccharides. *Journal of Bacteriology* 130 (1):48–56.

Underkofler, L. A. 1969. Development of a commercial enzyme process: Glucoamylase. In *Cellulases and Their Applications*, edited by George J. Hajny and Elwyn T. Reese. American Chemical Society.

Veerabhadrappa, M. B., S. B. Shivakumar, and S. Devappa. 2014. Solid-state fermentation of Jatropha seed cake for optimization of lipase, protease and detoxification of anti-nutrients in Jatropha seed cake using *Aspergillus versicolor* CJS-98. *Journal of Bioscience and Bioengineering* 117 (2):208–214.

Ventura, L., L. González-Candelas, J. A. Pérez-Gonzáez, and D. Ramón. 1995. Molecular cloning and transcriptional analysis of the *Aspergillus terreus* gla1 gene encoding a glucoamylase. *Applied and Environmental Microbiology* 61 (1):399–402.

Viniegra-González, G., E. Favela-Torres, C. N. Aguilar et al. 2003. Advantages of fungal enzyme production in solid state over liquid fermentation systems. *Biochemical Engineering Journal* 13 (2–3):157–167.

Vishwanatha, K. S., A. G. Appu Rao, and S. A. Singh. 2009. Characterisation of acid protease expressed from *Aspergillus oryzae* MTCC 5341. *Food Chemistry* 114 (2):402–407.

Wang, Q., X. Wang, X. Wang, and H. Ma. 2008. Glucoamylase production from food waste by *Aspergillus niger* under submerged fermentation. *Process Biochemistry* 43 (3):280–286.

Wang, R., L. C. Godoy, S. M. Shaarani et al. 2009. Improving wheat flour hydrolysis by an enzyme mixture from solid state fungal fermentation. *Enzyme and Microbial Technology* 44 (4):223–228.

Wang, D., Z. Y. Zheng, J. Feng et al. 2013. A high salt tolerant neutral protease from *Aspergillus oryzae*: Purification, characterization and kinetic properties. *Applied Biochemistry and Microbiology* 49 (4): 378–385.

Weber, J., and F. A. Agblevor. 2005. Microbubble fermentation of *Trichoderma reesei* for cellulase production. *Process Biochemistry* 40 (2):669–676.

Widmer, W., W. Zhou, and K. Grohmann. 2010. Pretreatment effects on orange processing waste for making ethanol by simultaneous saccharification and fermentation. *Bioresource Technology* 101 (14):5242–5249.

Wilson, D. B. 2009. Cellulases and biofuels. *Current Opinion in Biotechnology* 20 (3):295–299.

Xia, L., and P. Cen. 1999. Cellulase production by solid state fermentation on lignocellulosic waste from the xylose industry. *Process Biochemistry* 34 (9):909–912.

Yamane, T., M. Fukunaga, and Y. W. Lee. 1996. Increased PHB productivity by high-cell-density fed-batch culture of *Alcaligenes latus*, a growth-associated PHB producer. *Biotechnology and Bioengineering* 50 (2):197–202.

Yan, S., J. Yao, L. Yao et al. 2012. Fed batch enzymatic saccharification of food waste improves the sugar concentration in the hydrolysates and eventually the ethanol fermentation by *Saccharomyces cerevisiae* H058. *Brazilian Archives of Biology and Technology* 55:183–192.

Yang, F. C., and I. H. Lin. 1998. Production of acid protease using thin stillage from a rice-spirit distillery by *Aspergillus niger*. *Enzyme and Microbial Technology* 23 (6):397–402.

Yang, Y. H., B. C. Wang, Q. H. Wang, L. J. Xiang, and C. R. Duan. 2004. Research on solid-state fermentation on rice chaff with a microbial consortium. *Colloids and Surfaces B: Biointerfaces* 34 (1):1–6.

Yin, L. J., Y. H. Chou, and S. T. Jiang. 2013. Purification and characterization of acidic protease from *Aspergillus oryzae* BCRC30118. *Journal of Marine Science and Technology* 21:105–110.

Ying, M., and G. Chen. 2007. Study on the production of biodiesel by magnetic cell biocatalyst based on lipase-producing Bacillus subtilis. *Applied Biochemistry and Biotechnology* 137–140 (1–12):793–803.

Zambare, V. 2010. Solid state fermentation of *Aspergillus oryzae* for glucoamylase production on agro residues. *International Journal of Life Sciences* 4:16–25.

Zhang, A. Y. Z., Z. Sun, C. C. J. Leung et al. 2013. Valorisation of bakery waste for succinic acid production. *Green Chemistry* 15 (3):690–695.

Section IV

Future Prospects

22

Emerging Trends and Future Prospectives

Muthusamy Chandrasekaran

CONTENTS

22.1 Introduction

Food research has drawn the attention of the international scientific community in recent years, and the main focus centers around enhanced production of food products, modified foods with enhanced nutritive value, flavor enhancement, extension of shelf life and food protection against food pathogens, quality control, packing technology, and package materials. To match consumer demand in the context of growing interest in convenient foods, food industry researchers are conducting intensive research toward development of modified foods with additives (Chandrasekaran 2012). The World Health Organization (WHO) food safety unit has proposed fermentation potential of food processing involving microorganisms and or their enzymes in order to get nutritive, safe, and secure food, and hence it is time to adapt the same in the food processing industries. Identification, selection, development, and application of suitable fermentative processing technologies in the food industries would be of use not only for preparing and storing food, but also for converting the huge amount of underutilized raw material going as waste causing environmental damage. Application of bioprocesses involving enzymes in food processing may not only solve environmental problems, but may also reduce the toxins and antinutritive factors of the by-products apart from increasing their nutritive value and consumer acceptance. Hence, it is time for the food industries to shift their operations toward green processes. However, extensive research on the consistency of bioprocessed products, risk of food contamination, food-borne illness, fermentation process development, and their control are needed in addition to utilizing the by-products through valorization employing biological processes (Chandrasekaran 2012). Today, industry is looking at enzymatic catalyses as a means to manufacture most of its products not only in environmentally friendly green processes, but also as a tool to generate more added value to the final product and enhance economic production. Consequently, the increased demand for enzymes and enzymatic processes is driving a new wave in enzyme technology, aided by advances in biotechnology and chemical sciences.

In the foregoing chapters, the authors have extensively discussed the role of enzymes in various applications in different segments of the food and beverage processing industries and have also indicated future prospects for enzyme applications in their respective industries. In this chapter, the discussion is restricted to the overall emerging trends in research and development in food enzymes and their applications in food and beverage with the objective of emphasizing the importance in undertaking active research and development activities on food enzymes by food scientists, food technologists, and food biotechnologists in the coming years.

22.2 Emerging Trends in Food and Beverage Processing

As discussed elaborately in this book, enzymes are primarily used as either processing aids or additives in food in addition to their recognition as analytical tools in food quality and safety assessment. Over the years, scientific literature accumulated through intensive research and development on various aspects of enzymes, including their availability, isolation, purification, characteristics, structure, function, role in biosynthesis, metabolism, and clinical importance. Later, the basic information on enzymes was adopted for varied industrial applications, and of very late, their potential applications have drawn the attention of food technologists and food scientists. From the history of enzyme literature, one can infer that there is a continuous evolution of knowledge and its application in enzyme technology through intensive research and development activities. New trends have always emerged when new knowledge is generated as a consequence of advancements in science and technology, particularly in subjects such as molecular biology and chemistry of enzymes and proteins. In food science and technology, it has also happened owing to the advancements in biotechnology and now nanotechnology. In this section, a comprehensive discussion is held on the new trends that have emerged in food enzyme research and technology that have far-reaching impact on the food and beverage processing industries and overall food production.

22.2.1 Enzymes as Analytical Tools: Food Safety Issues and Concerns

Overall quality and safety of food and beverages consumed are of high priority and relevance for both food producers and food consumers who care for the quality and safety of food due to growing concern for healthy food. Food and beverage products have chances of getting contaminated with undesirable pathogens and microbial food poisoning toxins in addition to other toxicants and metals that pose health risks and food safety issues. Consequently, there is a need for accurate, appropriate methods for such food analysis, which is cost-effective, rapid, and easy to perform. The analytical methods must ensure rapid and accurate detection, identification, and monitoring of the presence of chemical (pesticide residues, heavy metals, trace elements), biochemical (aflatoxins), and microbial hazards in foods and beverages. It must be noted that the rapid detection of pathogens and other microbial contaminants in food is critical to assess the safety of food products. Further, consumers are now interested in the credential attributes of the products they consume and the quality linked to geographical origin and the traceability of foods with selected properties. In this context, there is an emerging trend to opt for enzyme-based analytical techniques, such as ELISA and biosensors that meet such needs and requirements of the food and beverage industries. In spite of enormous amounts of scientific literature available on the recognition of many enzymes, their development as suitable enzyme sensors for use in biosensors as well as in ELISA for analyses of varied biomolecules and chemical species are not adequate such that they are commercialized. Only a few enzymes, such as horseradish peroxidase (HRP), have found wide applications in ELISA and glucose oxidase in biosensors. Hence, there is a need for isolation, identification, and development of new and efficient enzymes for such analytical purposes. In fact, there is a promising future for enzymes as analytical tools for instant assessment of food quality and safety, and such a trend may have great implications in the overall growth of the food and beverage industries in addition to contributing to a healthy society in the new millennium.

22.2.2 Emerging Enzyme Technologies, Genetic Engineering, Enzyme Engineering, and Biotransformations

As illustrated by the authors in the various chapters in this book, applications of enzymes in the food and beverage industries not only facilitate process integration, but also allow enhanced performance under operational conditions that minimize the risk of microbial contamination. In fact, profound progress in the application of enzymes in various food production processes has been made due to the outcome of the ever-continuing developments in molecular biology, enzyme engineering, (bio) computational tools, and high-throughput methodologies that assist in the efficient and timely screening/characterization of ideal biocatalysts. Several innovative approaches have been used for the design of new or improved biocatalysts that showed enhanced stability to high and low temperatures and pH levels, reduced dependency on metal ions, and reduced susceptibility to inhibitory agents and to aggressive environmental conditions while maintaining the targeted activity and/or evolving novel activities. This trend is continuing with tremendous impetus toward augmentation of new and novel biocatalysts for newer applications in addition to replacing existing enzymes that need improvement in their activity and stability.

Metagenomics has emerged as a powerful approach for discovering novel enzymes from noncharacterized environmental samples without any need to cultivate and/or isolate the microbial sources. Enzymes suitable for specific applications may now be designed by engineering already known enzymes by rational design or random mutagenesis (Gilbert and Dupont 2011). Several successful cases can be cited, such as the enhancement of the activity and enantioselectivity of *Candida antarctica* A lipase toward a difficult substrate (ibuprofen ester) by combinatorial reshaping of the substrate binding pocket (Sandstrom et al. 2012).

Directed evolution is a powerful tool of protein engineering to design and modify the properties of enzymes (Wang et al. 2012). This technology has become a standard methodology in protein engineering and can be used in combination with rational protein design and other standard techniques to meet the demands for withstanding process conditions, such as high substrate concentrations, high temperatures, and long-term stability as well as presenting desired specificity and/or selectivity (Bottcher and Bornscheuer 2010). Scientists have speculated that this technology has the potential for application in a wide range of proteins, most of which are of interest for biocatalytic processes.

Recombinant DNA technology has been widely employed for quite some time now for cloning the genes of enzymes that are naturally produced at minimal levels by the producing strains available in limited levels in source species in other organisms for large-scale production. Thus, recombinant transglutaminase has found attention in the food industry. Transglutaminase is used as a texturing agent in the processing of, for example, sausages, noodles, and yogurt, with which the cross-linking of proteins provides improved viscoelastic properties of the products (Kuraishi et al. 2001) whereas the current limited availability of the enzyme on an industrial scale limits its wider range of applications. At present, only the transglutaminase from *Streptoverticillium* sp. is commercially available on a reasonable scale, and the availability of the enzyme was increased by recombinant production in *Escherichia coli* (Yokoyama et al. 2000, 2004).

Better understanding of the enzyme structure, function, and know-how on nucleic acid and protein engineering have led to the development of artificial enzymes, hybrid catalysts, and enzyme-based nanoreactors. Advances in chemistry and computational sciences have also facilitated the design of active centers of novel catalysts that do not exist in nature and for modifying existing enzymes to suit the requirements.

A major challenge in enzyme reusability has been solved with the development of ideal immobilization matrices and immobilization techniques that are highly efficient. Immobilization of enzymes has already found enormous applications, including development of biosensors based on enzymes. Further, immobilization of enzymes has not only enabled the improvement of their catalytic properties, but also has facilitated possible industrial applications. Recent advancements in nanotechnology have now led to the recognition of novel nanomaterials, which have immense potential to revolutionize this field of immobilization in which limitations of traditional matrices are being resolved and new possibilities are being explored.

22.2.3 Enzyme Applications in Various Food Industries

The baking industry has been using enzymes extensively for modulating dough performance, improving fresh bread quality, and for extending the shelf life of baked products. This industry is anticipating diversification of products to meet consumer demands for more natural products free of chemical additives and interest in nutrition aspects. Consequently, the search for new enzymes with tailored properties for making baked products using mechanized processing and/or different baked specialties is being pursued. Further, there is a strong tendency toward the replacement of chemicals with enzymes moving to green labels in bakery products (Collar et al. 2000). Intense research and development activities have been initiated toward improving the nutritional and health features of baked products using enzymes. For instance, a recent study reported the combined use of directed evolution and high-throughput screening to improve the performance of a maltogenic α-amylase from *Bacillus* sp. for low pH bread applications (Jones et al. 2008). This has resulted in the development of enzymes with improved properties for established technical applications and in the production of new enzymes tailor-made for entirely new areas of application in which enzymes have not previously been used. Recently, lipases, in particular phospholipases, have found use as a substitute for or supplement to traditional emulsifiers as the enzymes degrade polar wheat lipids to produce emulsifying lipids in situ. Further efforts are being made to understand better bread staling and the mechanisms behind the enzymatic prevention of staling when using α-amylases and xylanases (Medeiros et al. 2014).

Confections and chocolates largely depend on sweeteners, both natural (sucrose) and artificial as an important constituent of confections. Corn (maize) and other useful sweeteners are largely used in soft drinks, candies, baking, jams and jellies, and many other foods. Hence, there is a great demand for sweeteners and production of sugar syrups, sweeteners, and sugars from conventional sources, such as corn, and renewable resources, such as tapioca, form the major activity of the food industry to meet the demands of the confectionery industries. Further enzymes have a larger role in the improvement of flavor in chocolates and production of cocoa butter equivalents.

The fruit juice and beverage industries constitute a major segment of the food industry, and there is demand for developing new products to meet consumer demand. Exoenzymes find a key role in some steps of alcoholic beverage production in which there is the necessity for extraction of important compounds that may present remarkable properties, including functional properties and sensory properties related to the concentration of these compounds. Additionally, the diversity and the high specificity of commercial enzymes pave the way for the development of new processes and new types of both alcoholic and nonalcoholic products, meeting the requirements of these industrial segments that are experiencing strong growth in recent years. Pectinases, cellulases, and tannases have found extensive applications in the fruit juice and beverage industry for clarification purposes, either individually or in combination with other enzymes. Intensive research is being pursued to isolate these enzymes from new sources for large-scale production and application because the organisms that produce such enzymes in high titers are limited. Enzymes are also being experimented with for newer applications, such as extraction of biomolecules that prevent effective processing of the fruit juices and beverages. The use of laccase for clarification of juice (laccases catalyze the cross-linking of polyphenols, resulting in easy removal of polyphenols by filtration) and for flavor enhancement in beer is a recently established application within the beverage industry. It is likely that the functional understanding of different enzyme classes will provide new applications within the food industry in the future.

Several new enzyme-based processes have recently been introduced in the fat and oil industries. Removal of phospholipids in vegetable oils (de-gumming) using a highly selective microbial phospholipase is yet another example in which the introduction of an enzyme-based step has enabled energy and water savings for the benefit of both the industry and the environment (Clausen 2001). However large-scale applications are still limited. Newer applications with different enzymes are being experimented with.

The expanding nutraceutical and functional food market clearly indicates that end users are seeking minimally processed food with extra nutritional benefits and organoleptic value. In this context, enzyme-assisted extraction of biological molecules of nutraceutical importance and functional food value from various plant sources has already marked its beginning, and it has been noted that it can save processing time and energy and potentially provide a more reproducible extraction process at the commercial scale.

Valorization holds the key for successful utilization of food and beverage processing by-products, and recently there has been a growing concern in the food industries. Efforts are being made by scientists and technologists to evolve technological means for diversification of the range of food products or to harness the range of by-products and wastes for further utilization. The scope for using fruit and vegetable peels and pomace, visceral organs of animals and fish, shellfish waste, etc., as potential raw materials for deriving nutraceuticals, functional foods, biomaterial enzymes, micronutrients, etc.; potential biomolecules with commercial value, such as catalytic enzymes, pigments, flavors, functional ingredients, micronutrients, nutraceuticals, active pharmaceutical ingredients, phytochemicals, biofuel, and biomaterials from food processing by-products, which are simply disposed of without assigning any importance are being explored.

22.2.4 Summary of Emerging Trends in Enzyme Research Relevant to Food and Beverage Processing Industries

Food enzyme sources, production, purification, and characteristics:

- Intensive screening and isolation of enzymes from microorganisms for potent industrial applications hitherto unexplored.
- Screening of novel enzymes or activities to promote a set of new reactions, the design of operational conditions toward optimized operation, and the design of enzymes with improved activity.
- Rapid and efficient enzyme purification strategies employing chromatography techniques, such as affinity chromatography combined with other conventional techniques.
- Improvement of various properties of enzymes, such as thermostability specificities, catalytic efficiencies to withstand extreme conditions, such as high temperature, pH, etc., for several industrial processes.
- Harnessing information obtained through environmental genomics, metagenomics, and proteomics for discovering undescribed enzyme activities.
- Functional understanding of different enzyme classes to provide new applications within the food industry in the future.
- Enhancing enzyme production by the source microorganisms. Strategies are needed to enhance production yield and economic production medium for the enzymes currently produced using a limited pool of organisms.
- Development of bioprocesses employing solid-state fermentation (SSF) using cheap solid substrates for economic production of enzymes toward bringing down the cost of enzymes.
- Efficient downstream processing of enzymes produced under SSF.
- Many novel enzyme genes are getting identified, and enzymes with new and exciting or improved properties are likely to be discovered or evolved.

Enzyme engineering:

- Modifying the existing enzyme to suit requirements.
- Rational design of enzymes with improved activity and stability based on knowledge of protein structure and of the interactions of the enzymes with the compounds involved in a reaction at the molecular level combined with high throughput screening methodologies in order to strengthen and significantly widen their application in commercial processes.
- Design and development of tailor-made enzymes, employing novel strategies: directed evolution, rational engineering, recombinant DNA technologies, etc., toward enhancing/improving catalytic activities, efficiency, stability against harsh conditions, extending half-life, specificity, etc.

- Modern biotechnological methods, such as recombinant DNA technologies, are looked upon as a means to improve enzyme yield during production, producing organisms, modifying the enzyme structure to enhance enzyme stability activity resistance to inhibitors withstanding extreme conditions of temperature, pH, metal concentration, substrate concentration, solvent, detergent, etc.
- Development of artificial enzymes, hybrid catalysts, and enzyme-based nano-reactors with the knowledge gained in understanding the enzyme structure function and nucleic acid and protein engineering.
- Design of active centers of novel catalysts that do not exist in nature.

Enzymes as analytical tools:

- Development of enzyme-based biosensors for food analyses, food quality control, food safety, detection of food toxicants, toxins, metals, pesticides, spoilage indices, process-induced toxicants in a range of food and beverage products.
- Application of nanotechnology and use of nanomaterials for developing immobilization matrices.
- Newer applications of enzymes in ELISA technique. Conventionally, HRP is widely used. R&D efforts are moving ahead in experimenting other enzymes for applications in ELISA for rapid detection of target molecules.
- Development of ELISA for enhancing efficiency, performance, and accuracy in detection at trace levels.
- Nanomaterials are poised to revolutionize immobilization processes in which limitations of traditional matrices are being resolved and new possibilities are being explored.
- Compartmentalization on a nanoscale has made it possible to have cofactor regenerating systems, single enzyme nanocatalystic, and even biomimetic multienzyme cascades, which will eventually help in replicating entire biochemical pathways in vitro.

Enzymes as processing aids:

- Search for new enzymes or new sources of enzymes with tailored properties for making baked products using mechanized processing and or different baked specialties.
- Replacement of chemicals with enzymes, moving into green labels in bakery products meeting consumer demands or more natural products free of chemical additives.
- Improving nutritional and health features of baked products.
- Improvements in chocolate flavor with enzyme applications in cocoa bean fermentation.
- Cocoa butter equivalent/substitute production using transesterification with lipase enzymes.
- Sugar/sweetener production from unutilized starch resources (other than corn) employing enzymes.
- Improvements in chocolate flavor with enzyme applications in cocoa bean fermentation.
- Cocoa butter equivalent/substitute production using transesterification with lipase enzymes.
- Enzyme-assisted extraction as an ideal ecofriendly extraction method for the recovery of economically valuable biomolecules that holds potential for applications as pharmaceutical additives, nutraceuticals, and functional foods on a commercial scale.
- Extraction of important compounds that may present remarkable properties, including functional properties and sensory properties related to the concentration of these compounds.
- Use of specific enzymes adapted to fruit processing improves color stability and turbidity increasing the shelf life of juices and concentrates.

- Diversity and specificity of commercial enzymes, develop new processes, new types of both alcoholic and nonalcoholic products.
- New sources of bromelain extraction from pineapple peel for use in the meat industry for meat tenderization.

Enzymes as additives:

- Enzymes have earned significance as additives in foods as digestive aids for human consumption, preparation of cobiotic–probiotics and in feed formulations. As a result, newer enzymes for these kinds of applications have gained increased attention.

22.3 Strategies to Be Adopted toward Successful Harnessing of Enzymes in Food and Beverage Processing in the Future

It is imperative that appropriate strategies be adopted for successful implementation of enzyme technology in the food and beverage processing industries. In spite of the fact that several strategies have been devised and adopted by the food industries, the following strategies merit due consideration in respect to their significance.

- Database development on the potentials of food-grade enzymes, their characteristics, sources, and production technologies.
- Screening and isolation of novel enzymes with unique characteristics for applications in food and beverage processing from unexplored sources, such as extreme environments, archaebacterial, thermophilic, and psychrophilic fungi, etc.
- Harnessing the latest developments in metagenomics and proteomics for design and development of novel enzyme catalysts for deriving new products with potential applications in food and beverages.
- Design and development of novel enzyme catalysts with specificity and stability for activity under challenging processing conditions.
- Exploiting the latest developments in nanomaterials and nanobiotechnology for their applications in enzyme engineering and modifications toward efficient use of enzymes under immobilized conditions.
- Exploration and exploitation of potential enzymes for enzyme-mediated biotransformations in deriving a new range of biomolecules for applications in food and beverages.
- Discovery of new enzymes from microbial and plant resources that have the potential for application as additives in food and beverage processing in addition to use as digestive aids for humans.
- Development of appropriate enzyme technologies for cost-effective production of sweeteners and sugar syrups from underutilized and unutilized starch-containing plant resources.
- Recognition of potential enzymes hitherto underutilized for use in fruit and vegetable processing toward improving nutritive value, extending shelf life, and deriving valuable biomolecules of varied industrial applications, such as in pharmaceuticals.
- Development of ideal enzyme-based extraction strategies and technologies for isolation and recovery of novel and potential bioactive molecules that have applications as functional food components and as nutraceuticals.
- Development of novel enzymes for applications in feed processing and use in animal husbandry, the dairy sector, and seafood processing.
- Knowledge of quality of food processing by-products with particular reference to their chemical, physicochemical, and biological characteristics that decide their potential as prospective raw materials for deriving value-added products through valorization.

- Life cycle analyses on materials and by-products of food processing industries.
- Development of enzyme-based alternate technologies and methods for utilization of by-products from agroproduce that goes as waste.
- Scope for utilization of by-products toward diversification of food- and feed-related products employing enzyme technologies.
- Good manufacturing practices (GMP) in food and beverage industries involving enzymes.

22.4 Future Prospects

Nature is a rich source of several enzymes, which may have hidden potential for varied applications in human endeavors whereas only a few of them have been utilized in the past 100 years as discussed in this book, and the rest remain unutilized. There are several reasons for this scenario, which may include factors such as that the source could be unreliable or undesired, such as pathogenic microorganisms, or from inaccessible environments, such as the deep sea; difficulties in large-scale isolation and purification, commercial scale production, and downstream processing; requirements for activity under harsh conditions, etc. Hence, scientists and technologists are continuing their efforts in discovering new enzymes from new sources and experimenting with their possible utilization, meeting all challenges through research efforts on a laboratory scale. However, only a few of them have tasted success and reached commercial scale production and application. As a result, extensive application of enzymes in food processing industries happened in the latter half of the twentieth century, and this emerging trend opened up renewed interest in researchers to discover a new range of enzymes from unconventional resources and to describe enzymes for possible application. Consequently, now, enormous activities are going on in the search for novel enzymes and their novel applications.

As discussed earlier, advancements in molecular biology, instrumentation, biotechnology, bioinformatics, and nanotechnology have opened up new horizons in the design and development of artificial enzymes, hybrid catalysts, and tailor-made properties in enzymes for varied applications in food processing industries. Further, it is possible through biotechnology to isolate the gene coding for novel enzymes from organisms that cannot be cultivated on a large-scale for deriving copious amounts of enzymes and cloning easily cultivable microorganism(s) for large-scale/commercial production. Hence, it is anticipated that future initiatives hold immense promise for valuable new enzymes, which are to be safe and commercially available at affordable prices. New enzymes through modern biotechnology will lead to enzyme products with improved effects at diverse physiological conditions, such as low or high temperatures, which may allow various industrial processes to operate at low temperatures; less harm to the environment; greater efficiency; lower costs; lower energy consumption; and the enhancement of product properties. Further, environmental genomics and proteomics are anticipated to transform the enzyme industry with rich novel genes and novel enzymes from microbial resources of extreme environments and underexplored biological resources that remain hitherto unutilized for applications. Modern tools of biotechnology hold the key for efficient harnessing of enzymes from nature that are sufficiently robust to be useful under harsh processing conditions, such as extremes of pH and temperature, and thus hold great promise for replacing certain chemical processes in the future with much cleaner protein-catalyzed processes. Already there is a major drive for adoption of greener technologies that are ecofriendly, and the need for conservation of the environment and sustainable utilization of natural resources have necessitated future initiatives that hold the key for solutions in the efficient management and valorization of enormous wastes and by-products disposed into the environment by the food processing industries.

It is envisaged that the future could witness some major prospective initiatives listed below, which would transform food and beverage processing, significantly contributing to quality and healthy food, meeting consumers' ever-growing needs and demands, in addition to taking care of environmental, health, and socioeconomic considerations. Key market drivers for such initiatives would be consumer demands, large-scale availability of source materials and biocatalysts, optimal technologies, economic downstream processing, and overall production costs among others.

- Discovery of genes coding for novel enzymes with competent properties that will meet the demands of the food and beverage processing industries.

- Understanding of structure, functions, and interactions of enzymes with other biomolecules and chemical species, which may pave ways and means to design and develop suitable biocatalysts for varied applications with high specificity, stability, improved catalytic activities, mode of action, optimal operational conditions, and for developing ideal enzyme combinations with which multienzyme complex(es) need to be used.

- Design and development of ideal commercial production process(es) for large-scale manufacture of enzymes, which are cost-effective and affordable. For example, solid-state fermentation using cheap agro-residues and wastes as substrates for production of fungal exoenzymes that are extensively used in the food and beverage industries.

- Development of cost-effective downstream processing of enzymes from production media that determine the yield and consequent cost of enzymes and properties of enzymes.

- Development of ideal immobilized biocatalyst(s) with suitable chemical, physical, and geometric characteristics, which can be produced under mild conditions, used in different reactor configurations, and comply with the economic requirements for large-scale application. Nanomaterials hold the key in this respect, and considering the recent developments in this field, this trend is foreseen to be further implemented.

- Newer applications of enzymes in production of modified starch, oligosaccharides, sugar, and sugar syrup, utilizing underutilized starch sources for use in the manufacture of a diverse range of products in bakeries, confectioneries, and prebiotics.

- New enzymes and improved enzymes for removing undesired chemical compounds and biomolecules from fruit juices, beverages, and fats and oils derived from vegetables.

- Development of enzymes for enhancing the final quality of meat and its products, which are much consumed worldwide. For example, the tenderness is the main characteristic considered by consumers; strategies to improve meat tenderness could increase the value of non-prime cuts, making possible the use of by-products from the meat industry and reducing the waste and environmental impact generated in this supply chain.

- In the seafood industry, enzyme technology is still in its early stages. Enzymatic approaches will become an alternative solution to circumvent the environmental problems caused by the mechanical and chemical processing of seafood.

- The enzyme-assisted extraction of biomolecules of interest and micronutrients from fruits, vegetables, and other plant materials of interest for use as nutraceuticals, functional foods, and pharmaceuticals.

- Evaluation of the effect of enzyme processing on the biological availability and effectiveness of nutraceuticals.

- Enzymic bioconversions or biotransformations of food processing by products and waste into functional compounds.

- Discovery and development of new enzymes that have the potential for use as additives in probiotics, prebiotics, symbiotics, and cobiotics manufactured in the future because the consumer-oriented food and beverage industry is experiencing a global evolution focusing on safe health.

- Discovery and development of novel enzyme inhibitors from plants, seeds, and microorganisms for varied applications, such as regulating enzymes that lead to undesirable effects when used in excess as in the case of lipase or protease in food and beverage processing.

- Valorization of fruit and vegetable peels and pomace; visceral organs of animals and fishes, shellfish waste, etc., for deriving catalytic enzymes, pigments, flavors, functional ingredients, micronutrients, nutraceuticals, active pharmaceutical ingredients, phytochemicals, biofuel, and biomaterials employing enzyme processes.

- Many of the investigations and studies are conducted on a laboratory scale whereas successful commercialization of any bioprocess solely relies on appropriate pilot-scale studies that ensure success at industrial-scale production levels. Thus, the future could witness the development of ideal bioprocesses based on enzymes based on pilot-scale studies and subsequent technology transfer to industries.

- It is envisaged that there will be an increase in trained human resources capable of adopting and implementing enzyme-based food and beverage processing who will account for the success of these industries in their endeavors.

22.5 Conclusion

The intensity of current research and developmental activities are rather limited in the case of utilization of enzymes in food and beverage processing. This is evidenced by the limited availability of research literature on applications of enzymes in food and beverage processing industries. Of course, there are intense research and development activities being pursued in the food industries currently. It is apparent that the growth in development of enzyme catalysts will be enormous in the coming years, especially in the areas of novel enzyme development including de novo development of artificial and hybrid enzymes and in development of nanobiocatalysts. In the immediate future, it is expected that better and greener synthetic processes based on enzyme catalysis will be available to the industry. Efforts will be made toward the development of newer immobilized biocatalysts with suitable chemical, physical, and geometric characteristics that can be used in different reactor configurations and that respond to the economic requirements for large-scale application. The development of newer chemical and genetic strategies aiming to modify the protein structures in order to create semisynthetic and artificial enzymes, strengthening the enzyme properties (activity, selectivity, and stability), minimizing, at the same time, their flaws will undoubtedly be a feasible alternative. It could be foreseen that the application in food and feed processing of all the abovementioned strategies either isolated or, preferably, suitably integrated, will undoubtedly be the main road to improve existing processes and implement new ones.

REFERENCES

Bottcher, D., and Bornscheuer, U. T. 2010. Protein engineering of microbial enzymes. *Curr Opin Microbiol* 13: 274–282.

Chandrasekaran, M. 2012. Future prospects and need for research. In *Valorization of Food Processing By-products*. Ed. M. Chandrasekaran. CRC Press, Taylor & Francis Group, Boca Raton, FL, pp. 753–767.

Clausen, K. 2001. Enzymatic oil degumming by a novel microbial phospholipase. *Eur J Lipid Sci Technol* 3103: 333–340.

Collar, C., Martinez, J. C., Andreu, P., and Armero, E. 2000. Effect of enzyme associations on bread dough performance. A response surface study. *Food Sci Technol Int* 6: 217–226.

Gilbert, J. A., and Dupont, C. L. 2011. Microbial metagenomics: Beyond the genome. *Annu Rev Mar Sci* 3: 347–371.

Jones, A., Lamsa, M., Frandsen, T. P., Spendler, T., Harris, P., Sloma, A., Xu, F., Nielsen, J. B., and Cherry, J. R. 2008. Directed evolution of a maltogenic α-amylase from *Bacillus* sp. TS-25. *J Biotechnol* 134: 325–333.

Kuraishi, C., Yamazaki, K., and Susa, Y. 2001. Transglutaminase: its utilization in the food industry. *Foods Rev Int* 17: 221–246.

Medeiros, A. B. P., Rossi, S. C., Bier, M. C. J., Vandenberghe, L. P. S., and Soccol, C. R. 2014. Enzymes for flavor, dairy, and baking industries. In *Food Composition and Analysis: Methods and Strategies*. Eds. A. K. Haghi, and E. Carvajal-Millan. CRC Press, Boca Raton, FL, pp. 37–48.

Sandstrom, A. G., Wikmark, Y., Engstrom, K., Nyhlen, J., and Backvall, J. E. 2012. Combinatorial reshaping of the *Candida antarctica* lipase A substrate pocket for enantioselectivity using an extremely condensed library. *Proc Natl Acad Sci U S A* 109: 78–83.

Wang, M., Si, T., and Zhao, H. 2012. Biocatalyst development by directed evolution. *Biores Technol* 115: 117–125.

Yokoyama, K., Nakamura, N., Seguro, K., and Kubota, K. 2000. Overproduction of microbial transglutamin ase in *Escherichia coli*, *in vitro* refolding, and characterization of the refolded form. *Biosci Biotechnol Biochem* 64: 1263–1270.

Yokoyama, K., Nio, N., and Kikuchi, Y. 2004. Properties and applications of microbial transglutaminase. *Appl Microbiol Biotechnol* 64(4): 447–454.

Index

Page numbers followed by f and t indicate figures and tables, respectively.

Milton Keynes UK
Ingram Content Group UK Ltd.
UKHW051537141024
449569UK00028B/1516